Late-Quaternary Environments of the United States
Volume 2
The Holocene

Late-Quaternary Environments of the United States

H. E. Wright, Jr., Editor

Volume 2

The Holocene

H. E. Wright, Jr., Editor

University of Minnesota Press, Minneapolis

Library of Congress Cataloging in Publication Data

Main entry under title:
Late-Quaternary environments of the United States.

  Includes bibliographies and indexes.
    Contents: v. 1. The late Pleistocene / Stephen C.
Porter, editor — v. 2. The Holocene / H. E.
Wright, Jr., editor.
    1. Geology, Stratigraphic — Quaternary — Congresses.
2. Geology — United States — Congresses. I. Wright,
H. E. II. Porter, Stephen C.
QE696.L29   1983   551.7'9'0973   83-5804

ISBN 0-8166-1252-8 (set)
ISBN 0-8166-1169-6 (v. 1)
ISBN 0-8166-1171-8 (v. 2)

Contents

Environmental Archaeology

Climatology

Contributors to This Volume

Thomas A. Ager, *U.S. Geological Survey, Reston, Va. 22092*

C. Melvin Aikens, *Department of Anthropology, University of Oregon, Eugene, Oreg. 97403*

David A. Baerreis, *Department of Anthropology, University of Wisconsin, Madison, Wis. 53706*

Richard G. Baker, *Department of Geology, University of Iowa, Iowa City, Iowa 52242*

Subir K. Banerjee, *Department of Geology and Geophysics, University of Minnesota, Minneapolis, Minn. 55455*

P. W. Birkeland, *Department of Geological Sciences, University of Colorado, Boulder, Colo. 80309*

Arthur L. Bloom, *Department of Geological Sciences, Cornell University, Ithaca, N.Y. 14853*

Linda B. Brubaker, *School of Forestry, University of Washington, Seattle, Wash. 98195*

Richard B. Brugam, *Department of Biological Sciences, Southern Illinois University, Edwardsville, Ill. 62026*

R. M. Burke, *Department of Geology, Humboldt State College, Arcata, Calif. 95521*

Duane E. Champion, *U.S. Geological Survey, Menlo Park, Calif. 94025*

Edward R. Cook, *Tree-Ring Laboratory, Lamont Doherty Geological Observatory, Palisades, N.Y. 10964*

E. J. Cushing, *Department of Ecology and Behavioral Biology, University of Minnesota, Minneapolis, Minn. 55455*

Jonathan O. Davis, *Desert Research Institute, University of Nevada, Reno, Nev. 89506*

Margaret Bryan Davis, *Department of Ecology and Behavioral Biology, University of Minnesota, Minneapolis, Minn. 55455*

Pieter M. Grootes, *Quaternary Isotope Laboratory, University of Washington, Seattle, Wash. 98195*

J. C. Knox, *Department of Geography, University of Wisconsin, Madison, Wis. 53706*

John E. Kutzbach, *Center for Climatic Research, University of Wisconsin, Madison, Wis. 53706*

Robert V. Ruhe, *Water Resources Research Center, Indiana University, Bloomington, Ind. 47405*

Andrei M. Sarna-Wojcicki, *U.S. Geological Survey, Menlo Park, Calif. 94025*

Holmes A. Semken, Jr., *Department of Geology, University of Iowa, Iowa City, Iowa 52242*

James B. Stoltman, *Department of Anthropology, University of Wisconsin, Madison, Wis. 53706*

T. Webb, III, *Department of Geological Sciences, Brown University, Providence, R.I. 02912*

H. E. Wright, Jr., *Limnological Research Center, University of Minnesota, Minneapolis, Minn. 55455*

Preface

The Holocene is that portion of the Quaternary period of geologic time that follows the last major phase of continental glaciation; it extends from about 10,000 or 12,000 years ago up to the present day. It is of scientific interest especially because of the challenge it affords to Quaternary scientists reconstructing from the stratigraphic and geomorphic record the changing environments brought about by the basic shift from a glacial to a nonglacial climatic mode. The Holocene record is abundantly accessible to study, relatively easy to date by isotopic methods, and sufficiently complex to reveal environmental conditions that in some cases differ from those apparent today. The Holocene is also the time of rapid development of human cultures under changing environmental conditions.

The Quaternary period has seen a great many interglacial intervals during its 2-million-year history, and virtually all theories of climatic change presume that the Holocene is a representative interglaciation that will be followed in due time by another episode of continental glaciation. Although the long ocean-sediment records indicate that interglacial episodes altogether constituted a relatively minor proportion of Quaternary time, the temperate climatic conditions that prevailed over large areas of the world during these intervals led to the formation of soils, fossiliferous sediments, landforms, and other elements of the geologic record that can best be understood by detailed study of the Holocene.

In 1972, the United States and the Soviet Union entered into the Bilateral Agreement on Cooperation in the Field of Environmental Protection. Under the auspices of this agreement a program in paleoclimatology has sponsored three conferences on Quaternary environments, in Moscow and Baku (1976), in New York and Salt Lake City (1978), and in Irkutsk and Yakutsk (1980) in order to exchange information on Quaternary research in the two countries, which together occupy a large portion of the temperate latitudes of the Northern Hemisphere. Because the conferences involved only a few scientists, a plan was devised to prepare for a broader audience several volumes reviewing the current knowledge of Quaternary environments in the two countries, with simultaneous publication in English and Russian. The two American volumes deal separately with the late Pleistocene and the Holocene. The Soviet volume is concerned with late-Quaternary glacial and periglacial features and with the nature of the landscapes in different parts of the country.

Acknowledgments should be expressed to Academician I. P. Gerasimov and Dr. A. A. Velichko, leaders of the Soviet group, and Dr. John Imbrie and Dr. Alan Hecht, leaders of the American group, for facilitating the organization of the conferences and to John Mirabito for assisting in the arrangements.

Introduction
H. E. Wright, Jr.

The last retreat of the Pleistocene continental ice sheets and the climatic change that intiated this event ushered in the Holocene epoch of the Quaternary. In most natural systems controlled primarily by climatic change, the Holocene probably has not differed substantially from earlier interglacial intervals, at least so far, and there is little doubt that another major glaciation will ensue. Because the Holocene is so close to us and because so much is known about its environmental history, it deserves its separate designation—and a volume devoted to the interpretation of its geologic record.

The most recent of the Pleistocene episodes of continental glaciation is called the Wisconsin, at least for North America east of the Rocky Mountains. After a long buildup during the Early and Middle Wisconsin, with some fluctuations in its extent, the Laurentide ice sheet reached its maximum in the Late Wisconsin—before 20,000 years ago in the case of parts of its southern margin but as late as 14,000 years ago for ice lobes west of the Mississippi River, at a time when more easterly lobes were already in retreat. This diverse behavior in itself says something about the complex responses of a continental ice sheet to climatic change and to other factors that control glacier regime and glacier dynamics.

Soon after 14,000 years ago, the Laurentide ice front was in full retreat throughout its southern perimeter, and great efforts have been made to correlate ice-front position across the continent (Mickelson et al., 1983). By about 11,000 years ago, the Gulf of St. Lawrence had been opened, leaving a residual ice mass in northern New England, and the proglacial Great Lakes had already had a complex history of different outlets and lake levels. By 10,000 years ago, the ice front had withdrawn north of the Great Lakes, and Lake Agassiz and other proglacial lakes had formed in Manitoba and northwestern Ontario, also with a complex history related to ice-margin fluctuations, crustal tilting, and shifting of outlets. The last of the fluctuations produced the Cochrane moraine south of Hudson Bay about 7500 years ago. Rapid ice retreat by iceberg calving at the Hudson Straits opened Hudson Bay, thus splitting the thinning ice sheet into two remnants, of which the eastern survived on the Quebec/Labrador plateau until about 6500 years ago.

The retreat of the Laurentide ice sheet from the northern United States—and of the Cordilleran ice sheet as well as alpine glaciers in the western mountains—left vast areas open for colonization by plants and animals, development of soils, and adjustment of streams and slopes to the new geomorphic setting. Adjustments of different natural systems to the local scene involve complex interactions that leave an equally complex statigraphic and geomorphic record, and the complexity of this record provides the focus for many early-Holocene studies. The regional climate during the early Holocene apparently involved the response of the general circulation of the atmosphere to rapidly changing patterns of solar radiation, but it also may have involved the persistent periglacial influence of the waning ice sheet. By the middle Holocene, the latter factor was no longer important, and many of the local adjustments of the early-Holocene transitional phase had been completed: the deglaciated landscape was fully covered by relatively stable vegetation, and in many areas the various geomorphic processes had established a dynamic equilibrium. During the late Holocene, on the other hand, new climatic settings—this time not involving any periglacial components—caused certain adjustments in natural systems, and in the case of some mountain areas the long-diminished alpine glaciers advanced in the culminating Little Ice Age.

Most studies of Holocene history involve interdisciplinary approaches not only because of the implicit interaction of the various natural systems involved but because these systems have different sensitivities and different response times to climatic change. Although most of the chapters in the present volume deal with a particular system, many interactions and complications are revealed in the interpretation of the geologic history. In this introduction, some of the complexities are highlighted. The approach is chronologic because of the natural tripartite subdivision of the Holocene over much of the country: the early Holocene may still have reflected the lingering effects of the waning ice sheets, the middle Holocene seems to have represented maximum expression of the interglacial climatic mode, and the late Holocene has displayed the climatic reversal leading to the Little Ice Age.

The focus of discussion for the early Holocene is, first, on the transitional conditions related to the receding ice sheet and its record in marine systems and, second, on the record of vegetational change. Vegetation and climate are perhaps the dominant factors in determining the distribution of faunas and human populations, the development of soils, and the erosional processes on the landscape and thus the sedimentation in rivers and lakes. All these facets are treated in some detail in the present volume, and references to relevant chapters are made where appropriate.

It must be acknowledged that the essence of history is chronology, and some of the most interesting aspects of the Holocene are the time-transgressive events that have affected the landscape on the broad scales of time and space. Just as today the climate and vegetation and dominant geomorphic processes are different from one place to another, so also were they in the past, and it is a challenge to make the most appropriate reconstructions from stratigraphic or geomorphic records. Because of the importance of chronology in these efforts, a separate chapter is devoted to a review of isotope dating as applied to the Holocene (Grootes, 1983), and another to the special field of tree-ring analysis, which also has strong climatological as well as chronologic applications (Brubaker and Cook, 1983). Although the focus of most of this volume concerns the climate and its effects on landscape processes, two additional subjects are included because of their importance in correlation and chronology: Holocene variations in the paleomagnetic record in lake sediments (Banerjee, 1983) and volcanic-ash stratigraphy in the western United States (Sarna-Wojcicki et al., 1983).

Early Holocene

THE GLACIER SYSTEM

The transformation from glacial climatic conditions to postglacial conditions was not instantaneous, because of the gradual nature of the presumed driving mechanisms, that is, variations in the distribution of solar radiation according to the Milankovitch hypothesis. Elsewhere in this volume, Kutzbach (1983) makes the case that about 10,000 years ago the radiation curves for tilt and precession reinforced each other, resulting in the global average solar radiation for July being 7% greater than it is today. His model experiments for 9000 years ago (the ice sheets were not included in the model) show that Northern Hemispheric land areas were 0.7°C warmer in summer than they are today and that precipitation was 7% greater. The effects were particularly great in the Old World, where the July temperature was 5°C higher in parts of central Asia than it is today, and the enhanced monsoonal circulation brought an increase in precipitation of as much as 60%. The great expansion of pluvial lakes in North Africa during the early Holocene can be explained by this increase.

It is not certain how the model results would differ if the great ice sheets were included in the stipulation of the boundary conditions, and future experiments must be undertaken to evaluate the effects of the waning ice sheet on the periglacial climate in North America in relation to changing radiation conditions. However, the tendency should certainly be for higher summer temperatures for inland areas compared to those of the preceding interval of lesser solar radiation, thus accounting for the basic climatic change that brought about the recession of the ice sheets at the end of the Pleistocene.

In North America, the ice sheets that stretched across the continent must have had a significant effect on the climate far to the south. Certainly in the summer, their high albedo engendered the development of a cool air mass that must have spread southward and kept away for most of the time the maritime tropical air that today dominates the country east of the Rocky Mountains. For the winter, the case has been made by Bryson (1966) that during the Late Wisconsin glacial maximum the massive ice sheets trapped the frigid arctic air in the polar regions and prevented the outbreaks that bring blizzards to the western Plains today. Furthermore, such northern air as did reach the Plains was warmed adiabatically as it descended from the ice-sheet crest. The corollary to this hypothesis is that the arctic air thus blocked by the North American ice sheets spread instead to the other side of the world and brought the excessive cooling and expansion of permafrost documented for Siberia (Velicho, 1983) and the rest of Eurasia. It may also have escaped through the broad corridor of the Norwegian Sea, and in this way it may have contributed as much to the well-documented cooling of the North Atlantic as did the proximity of the ice sheet itself in eastern Canada. With the retreat and thinning of the Laurentide ice sheet along the Rocky Mountain front, both the entrapment of cold air and the factor of adiabatic warming were eliminated, so that early-Holocene average winter temperatures may already have reached today's levels. This change, combined with the reduction of summer cooling, must have resulted in increased seasonality during the early Holocene.

Early-Holocene glacial events in the midcontinent were largely confined to Canada, as the southern margin of the Laurentide ice sheet fluctuated in its general retreat. At least four distinct major moraines can be identified in northwestern Ontario north of Minnesota, covering the time from about 10,000 to 8500 years ago. These moraines presumably record climatically controlled changes in the balance between winter accumulation and summer ablation. Some lag in the response of a glacier terminus to climatic change may be involved if the controlling factor is increased snowfall rather than decreased melting, because of the time necessary for the ice to flow from the accumulation area to a distant margin. The several distinct moraines, however, imply that the ice margin was sensitive to nearly contemporaneous controlling forces, i.e. that little lag was involved. For moraines formed later, glacial surging has been suggested as a mechanism for an advance of particular ice lobes, for by 7500 yr B.P. not only was the ice front bordered by big proglacial lakes but Hudson Bay to the north had opened enough and sea level had risen enough to assure that the thinning ice was no longer frozen to its bed. But a surging mechanism cannot explain the sequence of major moraines in northwestern Ontario, which have the usual complex of meltwater features implying long-lasting ice, and in lieu of other hypotheses a climatic explanation seems most reasonable.

In the western mountains, alpine glaciers had retreated from their Late Wisconsin maxima relatively rapidly, as did the Cordilleran ice sheet itself. Recent evidence indicates that local glaciers readvanced in the early Holocene (about 8400 yr B.P. in the North Cascade Range of Washington (Beget, 1981), so here also the climatic trend to warmer conditions was not completely steady.

Despite the glacial evidence for early-Holocene climatic fluctuations in both midcontinental and western North America, other lines of evidence do not yet support the climatic interpretations. The curve for rising sea level (Bloom, 1983) records the addition of glacial meltwaters to the sea rather than any direct climatic trend. The fluctuations of major ice lobes such as those in the Great Lakes area or northwestern Ontario were not sufficiently great to be reflected even in the best-documented sea-level curves, which have a liberal scatter of controlling radiocarbon dates.

The stratigraphic records of ocean cores also cannot be used to reconstruct early-Holocene climatic events directly, although they do provide insight about the secondary effects of ice-sheet retreat on the oceans. The local influx of glacial meltwater to the Gulf of Mexico via the Mississippi River during the retreat of the Laurentide ice sheet is recorded by anomaly in the oxygen–isotope stratigraphy. By early-Holocene time, the major drainage had shifted to the Gulf of St. Lawrence and the adjacent northwestern Atlantic, where the sedimentary regime was too irregular to permit detailed stratigraphic analysis and correlation. Ruddiman and McIntyre (1981) postulate that the meltwater influx to both the Gulf of Mexico and the North Atlantic before 13,000 years ago was caused primarily by thinning of the ice sheet rather than by frontal retreat. The late-glacial readvance of the Scandinavian ice sheet between 11,000 and 10,000 yr B.P. is well matched in the North Atlantic record by the recurrence of a polar faunal assemblage. Subsequent events in the early Holocene, however, imply only a steady retreat of the polar front.

THE VEGETATIONAL AND RELATED NATURAL SYSTEMS

The early-Holocene glacial retreat discussed above involved fluctuations probably related to minor climatic perturbations, but the effect of glacial wastage on sea level and on ocean-bottom sediments reflects the gross trends, not the details that might correlate with the record of recessional moraines. Does the terrestrial record of other climatic sensors provide more information on early-Holocene conditions? What aspects of environmental history are found in the areas exposed by retreat of the glaciers and in the former periglacial areas no longer affected by the proximity of ice sheets?

For any consideration of these questions, a review of vegetational history must be the starting point because of its direct or indirect control on other natural systems, such as the faunas, soils, and numerous geomorphic processes. In the paragraphs that follow, the discussion for the early Holocene will be confined to the Middle West, for this is the area most likely to show the residual climatic effects of the waning ice sheet. The early vegetational histories for other parts of the country are referred to in the section on the Middle Holocene.

The modern vegetational zonation of North America east of the western cordillera clearly reflects the climatic zonation. The latitudinal arrangement that prevails in northern and eastern North America—tundra, boreal forest, mixed conifer/hardwood forest, temperate deciduous forest, and southern pine forest—reflects the basic southward increase in atmospheric temperature. At the same time, the westward change to grasslands and steppe is a response to decreasing moisture in that direction, reflecting greater distance from the Caribbean moisture source. Expressed in other terms, the tundra/forest border reflects the mean summer position of the arctic air mass; the boreal forest/mixed forest, the winter position; and the Prairie Peninsula, the frequency of summer drought caused by the prevalence of Pacific air made dry by its transit over the western mountains (Borchert, 1950; Bryson 1966).

Early-Holocene vegetational history might be expected to reflect the northward movement of the major climatic zones as the climate warmed, but several factors may complicate the picture: the existence of fresh glacial sediments that provide special habitats for pioneer plants (at least in the glaciated area), the periglacial influence of the waning ice sheet, and the differential rates of migration (range extension) for many important trees from their ice-age distributions.

Colonization of plants on newly exposed land is controlled by several ecologic factors, including availability of suitable habitats and the dispersal mechanisms of propagules from seed sources. Plant succession on deglaciated terrain has been studied in a few areas of modern glaciers, for example, in the St. Elias Mountains of southwestern Yukon (Birks, 1980), where the pollen stratigraphy implies a succession from herbs to shrubs to *Picea* forest matching the recent plant succession observed in a series of progressively older moraines. In

this case, the area became forested within less than a century, in part because seed sources for spruce were located on nearby valley slopes.

In the case of the retreating Laurentide ice sheet, the distances to seed sources were presumably much greater, so the stages of primary succession may have lasted for a longer time before *Picea* forest became established. In this case, opportunity existed for pioneer species of plants and animal to dominate the settings where today they would seem to be far out of place. Thus, in reconstructions from fossil pollen assemblages, a distinction has been made between a *tundralike vegetation* consisting of herbs and shrubs that today are found in forested as well as in tundra regions and a *true tundra* in which some plants occur that are found today *only* in tundra areas. Evidence from the fossil beetle fauna has been used to support the hypothesis that climatic conditions might have been suitable for trees well before the trees actually arrived on the scene (Ashworth et al., 1981).

Actually, many lake sites in the area uncovered by the retreating Laurentide ice sheet originated as small ponds on forest-coverd stagnant ice covered by forest, and such ponds were quite ephemeral until all the ice melted away. In this research, the analogy with the Yukon example described above is more valid (Wright, 1980). The pond enlarges and deepens as the ice melts, but by this time the pioneer stage has long passed, and the first pollen record is of spruce forest.

One might expect the vegetation on newly deglaciated terrain to be different from that beyond the glacial limit, where vegetation and soils were well established and where delayed plant immigration was not necessarily a factor. The periglacial area in much of the Middle West, however, was covered with glacial loess, and in much of the East by frost disturbed soils, both of which may also have provided nutrient-rich substrates similar to the fresh materials of direct glacial origin. The paucity of pollen sites peripheral to the Late Wisconsin ice limits makes it difficult to compare the early postglacial vegetational history on newly deglaciated areas with that outside the glacial boundary, although for a pair of sites in eastern Pennsylvania Watts (1979) suggests that the treeless vegetation on the deglaciated area was dominated by sedges, whereas the contemporaneous vegetation outside the glacial border was grasses.

Following the development of the first forest on deglaciated terrain, continued climatic warming led to many changes in vegetation. In the western Plains, the *Picea* forest shifted to prairie as early as 11,000 years ago. Farther east, a brief interval of deciduous forest—probably in a parkland structure—intruded before prairie began its dominance about 9500 years ago. Still farther east, the interval of deciduous forest was longer, extending to about 7000 years ago. It was, in fact, preceded by an interval of *Pinus* forest, which interceded between the time of spruce forest and the time of deciduous forest. The succession can be represented on isopoll maps, which show the changing pollen percentages throughout the Great Lakes region (Webb et al., 1983).

The sequence of early-Holocene vegetational change at the prairie/forest border in the Middle West provides strong evidence for contemporaneous climatic change—more specifically for progressive decrease in moisture, presumably reflecting the increasing frequency of dry Pacific air masses. This shift coincided in time with the retreat of the Laurentide ice sheet across central and eastern Canada and thereby, presumably, with the decreasing frequency of Arctic air, thus making way for dry Pacific air. The deciduous forest that bordered the prairie at that time was not the same as that of today, for it was dominated by *Ulmus, Fraxinus, Ostrya, Acer,* and other mesic trees rather than by *Quercus,* perhaps because of cooler summers related to the persistent influence of the waning ice sheet. An alternative hypothesis attributes the more mesic forest to a higher level of soil nutrients than exists today. The greater availability of nutrients in the early Holocene is also a possible explanation for the higher level of lake productivity, as inferred from the stratigraphy of fossil pigments in lake sediments. This subject is discussed in the chapter on paleolimnology in this volume (Brugam, 1983).

Vegetation is the principal element in controlling the distribution of animals, for it provides not only food but cover. The most striking phenomenon concerning late-Quaternary faunas was the extinction of dozens of species of large mammals, as well as of many birds and certain other vertebrates. The chronology of megafaunal extinction is discussed in this volume by Semken (1983), who points out that most extinctions occurred before 10,000 yr B.P. and that only a few occurred later in the Holocene. Thus, the major extinctions occurred precisely at the time of maximum climatic and vegetational change, as well as at the time when the Rocky Mountains corridor, opened by retreat of the Laurentide ice sheet, allowed massive colonization of areas south of the ice sheets by human hunters. The characteristics of these early hunting cultures are described elsewhere in this volume by Aikens (1983) for the West and especially by Stoltman and Baerreis (1983) for the Plains and the East.

The problems of faunal extinction per se are considered at greater length in volume 1 by West (1983) with respect to the hypothesis of overkill, and by Lundelius and associates (1983) in relation to the significance of environmental changes at the end of the Pleistocene. The latter group point out that many of the Late Wisconsin local faunas were "disharmonious," that is, they contain elements that today have nonoverlapping ranges. The suggestion is made that the seasonality factor in climate may have been the cause. The same reasoning can be used in an explanation for the anomalous Late Wisconsin pollen assemblages that have puzzled paleoecologists for decades. The climatic mechanism for increased seasonality at the end of the Pleistocene has been presented above—retreat of the Laurentide ice sheet from the Rocky Mountain front allowed the escape of frigid winter air previously trapped north of the ice sheet, and the thinning of the ice sheet removed the factor of adiabatic warming of air descending from the ice sheet. Warmer and perhaps shorter winters during the glacial period had permitted temperate deciduous trees like ash and oak to mix in the boreal spruce forest, and they thus provided a greater range of habitats at that time for a more diversified fauna, including species that are now temperate. At the same time, the summer temperatures must have been cooler than they are today because of the cold air mass emanating from over the ice sheet. The contrast in atmospheric pressure between the cold air mass and the warm air of non-glacial origin engendered strong winds that kept the boreal forest somewhat open, thus providing additional faunal habitats.

The early-Holocene adjustments of vegetation and fauna to the changed climatic setting reveal some of the details of environmental change. The warming climate set in motion a series of progressive changes not only in the biologic sphere but also in soil development, landscape erosion, and other geomorphic processes. By middle-Holocene time, many of these trends came to a kind of dynamic equilibrium, and they can best be mentioned in the context of the middle-Holocene setting.

Middle Holocene
MIDDLE WEST

The increasingly dry conditions that are substantiated in the Middle West by the shift from spruce forest to deciduous forest to prairie continued to a culmination in the middle Holocene, when the prairie/forest border was more than 100 km east of its present position. The border changed its orientation somewhat during its general shift to the east, in the area of Minnesota to Illinois and Iowa, and the Prairie Peninsula was sharper or blunter at different times during its maximum extent about 8000 to 5000 years ago (Webb et al., 1983). This shift in orientation may have been related to the waning effects of the Laurentide ice sheet, which lingered on the Quebec/Labrador plateau until 6500 years ago.

The general trend of the prairie/forest border in the Middle West today is meridional; the exception is the Prairie Peninsula, which is a mosaic of prairie and forest. It is clearly controlled by precipitation derived primarily from Caribbean air masses. Its shift to the east in the middle Holocene thus signals a decrease in the moisture available to plants rather than an increase in temperature, although the two trends usually go together in the midcontinent (Borchert, 1950). The greater incidence of summer drought reflects the more frequent occurrence of warm, dry, westerly air flow.

Confirmation of this climatic interpretation comes from the evidence for low lake levels in the prairie/border area in the middle Holocene (Brugam, 1983; Winter and Wright, 1977). It also comes from the distribution of soil types, which shows that certain nonforested areas in this transitional prairie/forest border region have prairie soils, as a relict of the middle-Holocene time of prairie expansion (Ruhe, 1983).

The middle-Holocene history of river systems also provides some insights into climatic history, although the relations are highly complex and variable geographically because of the interactions of climate and vegetation and because of the different response of tributary streams and trunk streams to storm events (Knox, 1983). For example, wooded areas are somewhat protected from hillslope erosion because much of the rain is intercepted by the canopy and infiltrates the forest floor, whereas grassland areas show more surface runoff and greater sediment yields despite having less rainfall. The direct result of these relations is seen in the lithostratigraphy of lake sediments in the prairie/forest border area of Iowa, where mineral sediment records the middle-Holocene and late-Holocene phases of landscape stability under forest cover (Ruhe, 1983).

Convectional storms of high intensity and small area, commonly associated in the Middle West with the dominance of Pacific air masses, cause floods on tributary streams, which develop the competence to entrench the valley floors and to carry coarse sediment (Knox, 1983). What may be eroded from tributary streams may be deposited by trunk streams. On the other hand, more protracted frontal rainfall of greater extent, brought by meridional air flow during domance of moist Caribbean air masses, causes floods on trunk streams, which may incise their channels or cut them laterally.

Knox (1983) interprets the alluvial history of Middle-Western streams in the context of Holocene climatic trends: episodes of strong erosion and deposition in small watersheds during the middle-Holocene dry period, when forest cover was reduced and sediment yield was high, are attributed to an increase in convectional storms associated with westerly air flow. During subsequent time, with more extensive forest cover and with stronger meridional air flow and frontal storms, the larger floods and smaller sediment concentrations caused incision and lateral migration along trunk streams. For other parts of the country, the response of river systems to climatic changes was different; in the continuously forested eastern areas, for example, rivers probably responded more directly through changes in flood frequency than indirectly through interaction with vegetative cover and sediment yields from the drainage basins.

The faunal record for the middle Holocene in the Middle West is consistent with the vegetational record: prairie species of mammals invaded areas in Missouri and Illinois that are now forested, and the inferrred climatic gradient was steeper than it is today (Semken, 1983). Consistent also is the archaeological record, which shows changes in dependence on bison as a principal food resource (Stoltman and Baerreis, 1983). A case has been made by some archaelogists that the relative aridity of the Great Plains during the middle Holocene caused a major reduction in human populations.

EAST

The eastern United States, except for parts of Florida, was forested throughout the Holocene, and ecotones of climatic significance are less distinct than they are in the Middle West. Such ecotones or tension zones that do exist, however, are more likely to reflect parameters of temperature than parameters of precipitation. Thus, the general northward movement of the southern limit of northern conifers like *Picea*, *Abies*, and northern *Pinus* species in the early Holocene implies a trend toward warmer climatic conditions, and the middle-Holocene upward expansion of *Pinus strobus* and *Tsuga* in the White Mountains of New Hampshire to elevations above their present range suggest a long interval of higher temperature (Davis, 1983). Otherwise in the East, the paleoclimatic interpretation of Holocene pollen diagrams is complicated by the differential rates of migration of major trees from Pleistocene distributions in the southern Appalachian Mountains, so that the progressive appearance of differrent deciduous trees (e.g., *Fagus*, *Carya*, *Castanea*) may not reflect contemporaneous climatic changes as much as different modes of seed dispersal or different results of competition among vegetational dominants. This subject is under discussion, however, because, when competition and other ecologic factors are important, it is difficult to relate the ranges of particular trees solely to climatic parameters.

In the Southeast, the Holocene vegetational history seems to feature a bipartite pattern rather than a tripartite pattern: a general expansion of southern *Pinus* species at the expense of *Quercus* about 5000 years ago is apparent at most sites, with uncertain paleoclimatic meaning (Watts, 1980). At the same time in Florida an open oak vegetation with low shrubs and herbs gave way to the *Pinus/Quercus* assemblage that prevails today (Watts, 1975). The Late Wisconsin and early-Holocene climate there apparently was quite dry, and only the deepest lake basins contained water. Lowered sea level may have been a factor in controlling the ground-water level and thus the lake levels, but the pollen sequence indicates that upland vegetation was at least partially treeless in the early Holocene.

The history of soil development and river alluviation in the East is less distinct than in the Middle West, again because of the continuous forest cover. The cultural history, however, at least in the Northeast, shows trends consistent with the northward retreat of coniferous forest, which contains relatively few food resources compared to the nuts, berries, and edible herbs of the decidous forest (Stoltman, and Baerreis, 1982). Human populations in the coniferous forests

of the early Holocene had to depend largely on coastal or riverine habitats, with the exception of those groups that utilized woodland caribou. Some secondary effects of middle-Holocene geomorphic changes, such as faunal diversification on coastal streams and the coasts themselves, have been adduced to explain cultural shifts, as stream gradients were reduced during the rise of sea level and as coastal marshes became established with the stabilization of sea level and the warming of seawater. The transition of the lower Mississippi River from a braided to a meandering pattern with backwater swamps also diversified the food resources for human populations.

WEST

The Holocene vegetational history of the western United States (Baker, 1983) is not easy to summarize because of the very large and diverse area involved, the existence of different climatic zones (e.g., winter rainfall maximum in the west-coast area, summer maximum in the Rocky Mountains), the lack of suitable pollen sites in ecotonal areas that are most sensitive to climatic change, and the ease with which pollen is blown from one vegetation zone to the next. As in the mid-continent, where the prairie border is controlled basically by precipitation and the boreal-forest borders by temperature, it should be possible to summarize the vegetational history by tracing the history of the moisture-controlled lower treeline (against the steppe) and of the temperature-controlled upper treeline (against the alpine zone). Unfortunately, a transect of suitable pollen sites up an individual major mountain from steppe to alpine zone has not been found, so it has been necessary to piece together results from scattered localities.

Balanced against these difficulties, the very productive studies of fossil packrat middens in the desert and semidesert areas, where pollen sites are extremely sparse, provide a great deal of additional paleoecologic information. All in all, certain consistent paleoclimatic trends are suggested in the West that at least can serve as the basis for interpreting cultural history and for comparing with the paleoclimatic record elsewhere in the country.

Most sites in the western mountains indicate that the upper treeline was lowered substantially during the time of glaciation and that alpine vegetation prevailed over larger areas. Lower-elevation forest zones were accordingly depressed by the cooler climate and, in some cases, were forced southward, but the present associations apparently did not necessarily exist intact, and species changed their ranges as individuals rather than in groups (Spaulding et al., 1983). With the change in climatic conditions at the end of the Pleistocene, the areas of alpine vegetation succeeded to subalpine forest in the early Holocene, and this to lower-elevation associations in the middle Holocene, with a slight reversal to subalpine forest in the late Holocene (Baker, 1983). Thus, a middle-Holocene warm interval is inferred. Other Rocky Mountain sites, however, do not exhibit this sequence.

In the Pacific Northwest, the inferred climatic sequence is distinctly different. In the Puget lowland of Washington State, an early-Holocene open woodland of *Pseudotsuga* prevailed from about 10,000 to 5500 yr B.P.; this implies relatively warm and dry conditions compared to the following interval, when the forest became closed with *Tsuga* and *Thuja* (Barnosky, 1981). The bipartite interpretation of Holocene climatic history resembles that of the southeastern United States (especially Florida) more than that of the Middle West, which may have still been affected climatically in the early Holocene by the retreating Laurentide ice sheet.

Alaska is so far removed from the rest of the western United States that it has its own Holocene history. The Late Wisconsin vegetation of interior Alaska was marked by herbaceous tundra that many investigators believe has few modern analogues in the Arctic today. It is described as steppe tundra to emphasize the dry conditions that apparently prevailed (Ager, 1983). After a transitional interval of birch shrub tundra that started in some areas as early as 14,000 years ago, *Picea* reached the interior by 9500 years ago, followed by *Alnus*. But in western Alaska, *Alnus* preceded *Picea*, which did not arrive until 5500 years ago. A middle-Holocene interval of warm, dry climate can be recognized in a few areas.

In the southwestern United States, Holocene pollen sites are rare, and much more paleoecologic information has come from plant-macrofossil analysis of cemented packrat middens. Results show that the Late Wisconsin vegetation of most desert areas was grassland or woodland and that desert scrub developed only in the early Holocene, after a transitional interval as the climate became drier (Baker, 1983). The middle Holocene was a time of maximum warming at many sites, especially in the mountains.

The archaeological record for the Southwest can be interpreted in paleoecologic terms (Aikens, 1983). During the terminal Wisconsin and early-Holocene transitional interval, the Clovis paleo-Indians hunted large mammals. As moisture decreased and as the large mammals disappeared, the human population was reduced, until finally the big game was restricted to bison on the Great Plains. Meanwhile in the Southwest, the cultural economy turned to mixed hunting and foraging as conditions became drier. Many sites were abandoned during the middle Holocene, apparently because of dry conditions (as indicated by the faunal remains in caves), but other sites continued to be occupied. Total abandonment of the region did not take place, as previously postulated.

During the late-Holocene reversal of climatic conditions, populations expanded, and local maize cultivation began. Many cultural changes during the last 2000 years are attributed to environmental fluctuations, with maxima in human populations centering around A.D. 650 and A.D. 1050, in correlation with moist intervals documented by tree-ring studies.

Late Holocene

The late Holocene in the western mountains is marked in many areas by regrowth of alpine glaciers (Burke and Birkeland, 1983). Dates for these glacial advances are not consistent from one mountain district to another, so a detailed chronology of late-Holocene climatic fluctuations is not possible. A possible exception are the advances of the last few centuries, collectively grouped into the Little Ice Age. Great efforts have been made in recent years to quantify measurements of soil formation on glacial deposits, weathering rates on exposed boulders, and erosional modification of depositional landforms, so that these criteria can be used to differentiate glacial features. In some regions, volcanic ash deposits assist in correlation, for these can now be identified to source by their mineralogy or chemistry (Sarna-Wojcicki et al., 1983). Otherwise, radiocarbon analysis is the standard dating method, although lichenometry and tree-ring counts can be used to date Little Ice Age features.

The vegetational and associated archaeological records for the late Holocene have already been referred to in part in the section describing middle-Holocene trends. In general, the clarity of the pollen evidence for late-Holocene climatic changes depends on the sensitivity of the

local vegetation, and many areas show little change during the last few thousand years. Ecotones provide the best setting. Thus, the climatic reversal terminating the middle-Holocene dry interval started as early as 7000 years ago at the prairie/forest border of the day, and it then shows up progressively younger farther west as the ecotone moved to the west.

The southward shift of the boreal forest indicators (e.g., *Picea*), as a manifestation of cooler conditions, comes somewhat later (Webb et al., 1983). A combination of the trends of these two climatic parameters is perhaps best seen in the gradual westward expansion of the big peatlands on the plain of Glacial Lake Agasiz in northern Minnesota. This is near the area where the eastern deciduous forest between the prairie and the mixed coniferous/hardwood forest pinches out, so that the prairie comes up virtually against the boreal forest. Peat accumulation in this area involves both wetter and cooler climate. The lake plain was apparently a wet meadow bordered by uplands of prairie or oak woodland during the middle Holocene, but then, as the climate changed and coniferous forests developed on the uplands, the wet meadows were converted to peatlands, for the wetter conditions prevented the vegetation from drying out and the cooler conditions inhibited decomposition. Basal radiocarbon dates on the peat grade from about 5000 yr B.P. on the east to 2000 yr B.P. in the western part of the vast Red Lake peatland, which consists of a complex pattern of fens and raised bogs (Glaser et al., 1981).

Conclusions

The most eventful portion of the Holocene was the very beginning, when the glacial climatic mode was rapidly changing to the interglacial. The retreating North American ice sheets still had an effect on areas to the south, and they still manifested temporary stillstands or advances. On the marine scene, sea level was rising, and meltwater from the wasting ice sheet in the North Atlantic affected organic productivity and other marine functions. Vegetation on the deglaciated terrain went through rapid pioneer succession, but vegetational associations remained somewhat anomalous in the Middle West through the early Holocene in comparison with modern associations, perhaps because of the continuing influence of the waning Laurentide ice sheet. Dry conditions reached there maximum in the Middle West about 7000 years ago, when the prairie/forest border was more than 100 km east of its present position. At the same time, the coniferous forest in the eastern states shifted north in response to warmer conditions and rose higher on the New England mountains; temperate trees were still expanding their ranges at different rates from ice-age distributions in the southern Appalachian Mountains. Reversal of the climatic trends in the late Holocene led to the modern conditions.

Holocene vertebrate faunas start with a flood of extinctions as Pleistocene conditions drew to a close and as human hunters expanded over the country. Middle-Holocene faunal changes are best documented near the prairie/forest border in the Middle West, but the rise of sea level stabilized coastal areas and coastal streams and created more food resources for human populations.

The Middle-Western prairie/forest border region is also the area where soils show the shift from prairie to forest in recent millennia. The record of alluviation and dissection in Middle-Western streams has also been correlated with the extent and nature of vegetation and with the intensity of storms.

In the West, the interval of dry climate is generally in the early part of the Holocene, particularly in the Pacific Northwest, perhaps because the area was far removed from the possible effects of the retreating ice sheet. Faunal and archaeological changes are consistent with the climatic sequence inferred from the vegetational history.

References

Ager, T.A. (1983). Holocene vegetational history of Alaska. *In* ''Late-Quaternary Environments of the United States,'' vol. 2, ''The Holocene'' (H. E. Wright, Jr., ed.), pp. 128-41. Minneapolis, University of Minnesota Press.

Aikens, C.M. (1983). Environmental archaeology of the western United States. *In* ''Late-Quaternary Environments of the United States,'' vols. 2, ''The Holocene'' (H. E. Wright, Jr., ed.), pp. 239-51. Minneapolis, University of Minnesota Press.

Ashworth, A. C., Schwert, D. P., Watts, W. A., and Wright, H. E., Jr. (1981). Plant and insect fossils at Norwood in south-central Minnesota: A record of late-glacial succession. *Quaternary Research* 16, 66-79.

Baker, R. G. (1983). Holocene vegetational history of the western United States. *In* ''Late-Quaternary Environments of the United States,'' vol. 2, ''The Holocene'' (H. E. Wright, Jr., ed.), pp. 109-27. Minneapolis, University of Minnesota Press.

Banerjee, S. K. (1983). The Holocene paleomagnetic record in the United States. *In* ''Late-Quaternary Environments of the United States,'' vol. 2, ''The Holocene'' (H. E. Wright, Jr., ed.), pp. 78-85. Minneapolis, University of Minnesota Press.

Barnosky, C. W. (1981). A record of late Quaternary vegetation from Davis Lake, southern Puget lowland, Washington. *Quaternary Research* 16, 221-339.

Baulin, V. V., and Danilova, N. S. (1983). Dynamics of the permafrost in the Late Pleistocene: Asian part of the USSR. *In* ''Late-Quaternary Environments of the Soviet Union'' (A. A. Velichko, ed.). Minneapolis, University of Minnesota Press.

Beget, J. E. (1981). Early Holocene glacier advance in the North Cascade Range, Washington. *Geology* 9, 409-13.

Birks, H. J. B. (1980). The present flora and vegetation of the moraines of the Klutlan Glacier, Yukon Territory, Canada: A study in plant succession. *Quaternary Research* 14, 60-86.

Birks, H. J. B. (1981). Late Wisconsin vegetational and climatic history at Kylen Lake, northeastern Minnesota. *Quaternary Research* 16, 322-55.

Bloom, A. L. (1983). Sea level and coastal changes. *In* ''Late-Quaternary Environments of the United States,'' vol. 2, ''The Holocene'' (H. E. Wright, Jr., ed.), pp. 42-51. Minneapolis, University of Minnesota Press.

Borchert, J. R. (1950). Climate of the central North American grassland. *Association of American Geographers Annals* 40, 1-39.

Brubaker, L. B., and Cook, E. R. (1983). Tree-ring studies of Holocene environments. *In* ''Late-Quaternary Environments of the United States,'' vol. 2, ''The Holocene'' (H. E. Wright, Jr., ed.), pp. 222-35. Minneapolis, University of Minnesota Press.

Brugam, R. B. (1983). Holocene paleolimnology. *In* ''Late-Quaternary Environments of the United States, vol. 2, ''The Holocene'' (H. E. Wright, Jr., ed.), pp. 208-21. Minneapolis, University of Minnesota Press.

Bryson, R. A. (1966). Air masses, streamlines, and the boreal forest. *Geographical Bulletin* 8, 228-69.

Burke, R. M., and Birkeland, P. W. (1983). Holocene glaciation in the mountain ranges of the western United States. *In* ''Late-Quaternary Environments of the United States, vol. 2, ''The Holocene'' (H. E. Wright, Jr., ed.), pp. 3-11. Minneapolis, University of Minnesota Press.

Davis, M. B. (1983). Holocene vegetational history of the eastern United States. *In* ''Late-Quaternary Environments of the United States, vol. 2, ''The Holocene'' (H. E. Wright, Jr., ed.), pp. 166-81. Minneapolis, University of Minnesota Press.

Glaser, P. H., Wheeler, G. A., Gorham, E., and Wright, H. E., Jr. (1981). The patterned mires of the Red Lake peatland, northern Minnesota: Vegetation, water chemistry and landforms. *Journal of Ecology* 69, 575-99.

Grootes, P. M. (1983). Radioactive isotopes in the Holocene. *In* ''Late-Quaternary Environments of the United States, vol. 2, ''The Holocene'' (H. E. Wright, Jr., ed.), pp. 86-105. Minneapolis, University of Minnesota Press.

Knox, J. C. (1983). Responses of river systems to Holocene climates. *In* ''Late-Quaternary Environments of the United States, vol. 2, ''The Holocene'' (H. E. Wright, Jr., ed.), pp. 26-41. Minneapolis, University of Minnesota Press.

Kutzbach, J. E. (1983). Modeling of Holocene climates. *In* ''Late-Quaternary Environments of the United States, vol. 2, ''The Holocene'' (H. E. Wright, Jr., ed.), pp. 271-77. Minneapolis, University of Minnesota Press.

Mickelson, D. M., Clayton, L., Fullerton, D. S., and Borns, H. W., Jr. (1983). The Late Wisconsin glacial record of the Laurentide ice sheet in the United States. *In* ''Late-Quaternary Environments of the United States, vol. 1, ''The Late Pleistocene'' (S. C. Porter, ed.), pp. 3-37. Minneapolis, University of Minnesota Press.

Ruddiman, W. F., and McIntrye, A. (1981). The mode and mechanism of the last deglaciation: Oceanic evidence. *Quaternary Research* 16, 125-34.

Ruhe, R. V. (1983). Aspects of Holocene pedology in the United States. *In* ''Late-Quaternary Environments of the United States, vol. 2, ''The Holocene'' (H. E. Wright, Jr., ed.), pp. 12-25. Minneapolis, University of Minnesota Press.

Sarna-Wojcicki, A. M., Champion, D. E., and Davis, J. O. (1983). Holocene volcanism in the conterminous United States and the role of silicic volcanic ash layers in correlation of latest-Pleistocene and Holocene deposits. *In* ''Late-Quaternary Environments of the United States'', vol. 2, ''The Holocene'' (H. E. Wright, Jr., ed.), pp. 52-77. Minneapolis, University of Minnesota Press.

Semken, H. A., Jr. (1983). Holocene mammalian biogeography and climatic change in the eastern and central United States. *In* ''Late-Quaternary Environments of the United States, vol. 2, ''The Holocene'' (H. E. Wright, Jr., ed.), pp. 182-207. Minneapolis, University of Minnesota Press.

Spaulding, W. G., Leopold, E. B., and Van Devender, T. R. (1983). Late Wisconsin paleoecology of the American Southwest. *In* ''Late-Quaternary Environments of the United States'', vol. 1, ''The Late Pleistocene'' (S. C. Porter, ed.), pp. 259-93. Minneapolis, University of Minnesota Press.

Stoltman, J. B., and Baerreis, D. A. (1983). The evolution of human ecosystems in the eastern United States. *In* ''Late-Quaternary Environments of the United States, vol. 2, ''The Holocene'' (H. E. Wright, Jr., ed.), pp. 252-68. Minneapolis, University of Minnesota Press.

Velichko, A. A. (1983). Spatial paleoclimatic reconstructions. *In* ''Late-Quaternary Environments of the Soviet Union'' (A. A. Velichko, ed.). Minneapolis, University of Minnesota Press.

Watts, W. A. (1975). A late Quaternary record of vegetation from Lake Annie, south-central Florida. *Geology* 3, 344-46.

Watts, W. A. (1979). Late Quaternary vegetation of central Appalachia and the New Jersey Coastal Plan. *Ecological Monographs* 49, 427-69.

Watts, W. A. (1980). The late Quaternary vegetation history of the southeastern United States. *Annual Review of Ecology and Systematics* 11, 387-409.

Webb, T., III, Cushing, E. J., and Wright, H. E., Jr. (1983). Holocene changes in the vegetation of the Midwest. *In* ''Late-Quaternary Environments of the United States, vol. 2, ''The Holocene'' (H. E. Wright, Jr., ed.), pp. 142-65. Minneapolis, University of Minnesota Press.

West, F. H. (1983). The antiquity of man in America. *In* ''Late-Quaternary Environments of the United States'', vol. 1, ''The Late-Pleistocene'' (S. C. Porter, ed.), pp. 364-82. Minneapolis, University of Minnesota Press.

Winter, T. C., and Wright, H. D., Jr. (1977). Paleohydrologic phenomena recorded by lake sediments. *American Geophysical Union EOS* 58, 188-96.

Wright, H. E., Jr. (1980). Surge moraines of the Klutlan Glacier, Yukon Territory, Canada: Origin, wastage, vegetation succession, lake development, and application to the late-glacial of Minnesota. *Quaternary Research* 14, 2-18.

Physical Geology

CHAPTER 1

Holocene Glaciation in the Mountain Ranges of the Western United States

R. M. Burke and P. W. Birkeland

Introduction

Virtually all of the major mountain ranges of the western United States possess Holocene glacial or rock-glacial deposits nested up-valley from the latest Wisconsin moraines. Very few of these deposits have been studied in detail, so it is impossible to construct a comprehensive regional picture of Holocene glacial activity in the western United States. However, the few detailed studies that have been made during the past three decades present a coherent sample of Holocene glacial activity from widely spaced localities in the western United States.

Several recent reviews are concerned with the timing of Holocene glaciation in North America and the possible synchroneity of Holocene glaciation both within North America and throughout the world (Benedict, 1973; Denton and Karlén, 1973; Denton and Porter, 1970; Grove, 1979; Porter and Denton, 1967). We do not dwell on the synchroneity/nonsynchroneity arguments in this chapter, but, instead, we present an up-to-date review of research in individual mountain areas. In addition to a review of the literature, we discuss many of our own ideas about relative-dating techniques and offer our own conclusions (based to a great extent on unpublished data) about the stratigraphy of individual drainage basins.

Relative-dating methods (Birkeland, Colman, et al., 1979) measure the properties of a deposit that change through time and, therefore, can be used to estimate the ages of deposits when absolute dating is not available. These methods include measuring the amount of lichen growth, the degree of soil development, and the amount of stone weathering. In some areas of Holocene deposition, the basic stratigraphic framework is provided by radiocarbon dating. However, in most areas, there is no datable material available, and the more common relative-dating parameters provide the basic stratigraphic framework. Even where radiocarbon dates exist, they may provide only minimum or maximum limits on the timing of a depositional event.

The nomenclature in this chapter is based on the assumption that the Holocene extends through the last 10,000 years, that the Altithermal was a time of glacial retreat during the middle Holocene, and that the Neoglacial followed the Altithermal time transgressive-ly, encompassing the last 5000 or 6000 years (Denton and Karlén, 1973; Porter and Denton, 1967). We further informally subdivide the Neoglacial into early, intermediate, and late phases in order to accommodate numerous local stratigraphies.

For this chapter, we chose to review areas for which there were detailed records and/or with which we are familiar. These areas exemplify the deposits within the Rocky Mountains of Colorado, Wyoming, and Utah (Figure 1-1); several areas within the Sierra Nevada of California; the Wallowa Mountains of Oregon; and the Cascade Range of Washington. The Rocky Mountain sequences have been the most extensively studied, and yet they present some of the greatest problems in nomenclature. The Sierra Nevada contains many valleys with deposits of different ages, but only a few stratigraphic sequences have been worked out in detail. The Wallowa Mountains and the Cascade Range both have multiple tephra layers, which are very helpful in the dating of the glacial deposits.

Rocky Mountains

The work of Moss (1951) in the Wind River Mountains of Wyoming was a landmark because he used weathering and soil data to differentiate the deposits. For at least the next two decades, part of the stratigraphic nomenclature of Moss (1951) was widely used for work throughout the Rocky Mountains (Figure 1-2). The ages he assigned were based partially on correlations with the midcontinental region of the United States.

Richmond (1965) adequately summarized the Holocene glacial history for a large part of the Rocky Mountains, mostly on the basis of his own work. Radiocarbon dates were few, so he made extensive use of soils in subdividing tills locally and in correlating them over a large region. He placed the end of the Pinedale Glaciation (Late Wisconsin equivalent) at about 8000 yr. B.P. and recognized several tills of Neoglacial age. Unlike Moss (1951), Richmond (1960, 1962, 1965) considered the Temple Lake deposits to be Neoglacial (post-

We are grateful to J. C. Yount for reviewing the manuscript. Part of the work was supported by U. S. Geological Survey Grant number 14-08-0001-G-202, and part was supported by the Soil Correlation and Dating Project of the U.S. Geological Survey.

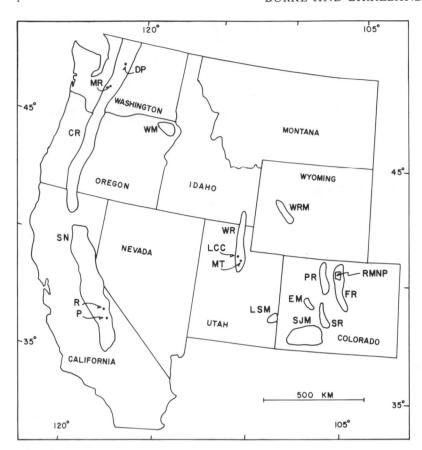

Figure 1-1. Location map of study areas in the western United States. Abbreviations for localities mentioned in text are as follows: WRM, Wind River Mountains; RMNP, Rocky Mountain National Park; FR, Front Range; PR, Park Range; EM, Elk Mountains; SR, Sawatch Range; SJM, San Juan Mountains; LSM, La Sal Mountains; WR, Wasatch Range; LCC, Little Cottonwood Canyon; MT, Mount Timpanogos; SN, Sierra Nevada; R, Recesses area; P, Palisades area; CR, Cascade Range; WM, Wallowa Mountains, MR, Mount Rainier; and DP, Dome Peak.

Altithermal) in age (Figure 1-2) and to be represented by two tills deposited between about 3100 and 1000 year B.P. and followed by the Gannett Peak stade during the last several centuries (Birkeland et al., 1971). A considerable amount of the work carried out after Richmond's 1965 review is summarized in the remainder of this section.

COLORADO

Holocene glacial deposits are found above the treeline in many cirques throughout the state of Colorado. Nelson (1954) did the first detailed work in the state, and, on the basis of estimated weathering rates and correlation with the midcontinent, he suggested a middle Holocene age for the deposits he named Chapman Gulch in the Sawatch Range (Figure 1-2). Richmond (1960) mapped deposits in Rocky Mountain National Park (Figure 1-1), where he applied the nomenclature from the Wind River Mountains of Wyoming (Figure 1-1) and used radiocarbon dates from Colorado and Utah to estimate ages for Holocene events. More recently, in the Front Range (Figure 1-1), Benedict (1967, 1968, 1973, 1981; Benedict and Olson, 1978) combined lichenometric data, rock-weathering and soil data, and numerous radiocarbon dates to derive the most detailed history of Holocene glacial advances so far constructed. Benedict proposed local names for the several Holocene units because of the uncertainties of long-distance correlations; they are the Satanta Peak, Ptarmigan, Triple Lakes, Audubon, and Arapaho Peak glacial advances within about the last 10,000 years (Figure 1-2). The middle Altithermal age of the Ptarmigan advance in one Front Range valley (Benedict, 1981) supports Benedict's previous findings (Benedict, 1973) that the Front Range Holocene glacial chronology does not appear to fit the proposed worldwide glacial fluctuations, and further research is in pro-

gress by Benedict and other workers (e.g., Davis and Waterman, 1979; Davis et al., 1979). In the San Juan Mountains of Colorado (Figure 1-1) (Carrara and Andrews, 1976), glacial deposits younger than an estimated 8000 years are lacking, and some cirques have been ice free for at least the last 14,000 years (Carrara and Mode, 1979). This suggests that within Colorado the Front and Park Ranges have the only morainal record of ice advances during the Neoglaciation, although some Colorado ranges have not yet been investigated. In the San Juan Mountains, high basins that have not been glaciated during the Holocene contain interbedded organic and clastic sediments that may correlate with Holocene climatic fluctuations recorded elsewhere by tills deposited during minor ice advances (Andrews et al., 1975).

There have been various interpretations of how closely a radiocarbon assay dates an associated Holocene deposit, and work within the Front Range exemplifies the variety of interpretations made from radiocarbon and relative-dating results. For example, Benedict (1973) interpreted some minimum dates for the Triple Lakes advance as being close to the age of the underlying till. However, near-basal dates for a lake core inside an inner Triple Lakes moraine are 1400 years older than the previously supposed age of the enclosing Triple Lakes moraines (Davis and Waterman, 1979). Estimates based on relative-dating methods also suggest an older age, perhaps one as old as the type Satanta Peak (Birkeland and Shroba, 1974).

Parameters of lichenometry, soil investigations, and rock-weathering studies work reasonably well for dating and correlating the main groups of Holocene glacial deposits (Benedict, 1968, 1973; Birkeland and Shroba, 1974; Carroll, 1974; Mahaney, 1973, 1974). The youngest deposits (Gannett Peak and Arapaho Peak [Figure 1-2])

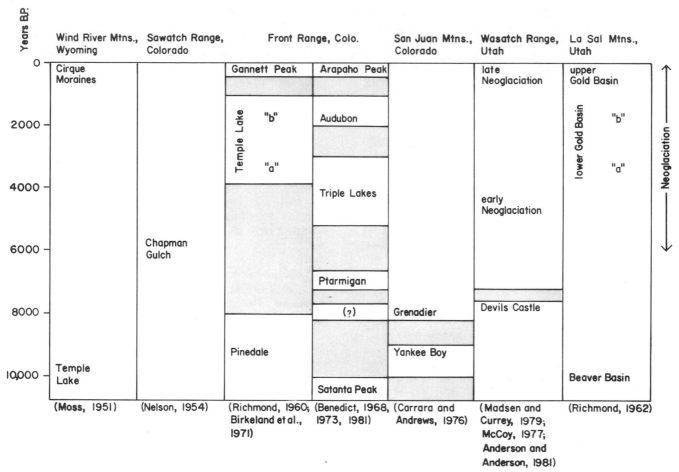

Figure 1-2. Principal stratigraphic names applied to Holocene glacial deposits in the Rocky Mountains. Shading indicates the approximate time span of the nonglacial intervals in areas with some radiocarbon control.

generally have few or no lichens and unweathered rocks and little or no soil development. These parameters change rapidly enough with time so that Audubon and early Neoglacial deposits (Figure 1-2) can be readily distinguished. Of the relative-dating methods, only rock weathering and soil development are useful in differentiating the early-Neoglacial deposits from the early-Holocene Satanta Peak deposits, but there is sufficient overlap in the data and/or in the various opinions of workers so that the differentiation is not easily made. Reconnaissance suggests that Satanta Peak and late-Pleistocene Pinedale deposits (Figure 1-2) are difficult to separate with relative-dating parameters.

Rock glaciers are the most common Holocene glacial deposits in the cirques of the Colorado Rocky Mountains (e.g., P. G. White, 1979; S. E. White, 1971, 1976). Equilibrium-line depression was apparently sufficient to form both glaciers and rock glaciers in the Front Range and the Park Range (Madole, 1978), but only rock glaciers were formed in the more southerly high mountains of Colorado. The situation probably is analogous to that in Alaska (Wahrhaftig and Cox, 1959), where the equilibrium-line altitude for ice glaciers is higher than that for rock glaciers.

The rock-glacial deposits have been difficult to date, but a few radiocarbon dates and relative-dating results have been used to group them into broad age classes (Benedict, 1968, 1973; Birkeland, 1973; Carrara and Andrews, 1973; Miller, 1973). In general, rock glaciers

are approximately the same age as the glacial advances, although the response time of an ice glacier to climatic change is much less than that of a rock glacier. Not all suspected cold periods are represented by rock-glacial deposits in all valleys. Late-Pleistocene or early-Holocene rock-glacial deposits are common throughout the high mountains of Colorado, and in many cirques they provide the only age of rock glacial deposits recognized. Rock-glacial deposits of Neoglacial age are less common, but they have been recognized in some of the cirques in most major mountain ranges.

Rock glaciers can be classified as either active or inactive (Wahrhaftig and Cox, 1959). Few movement data are available, but maximum rates are near 10 cm per year in the Front Range (White, 1971), 12.5 cm per year in the Sawatch Range (Miller, 1973), and 60 cm per year in the Elk Mountains (Bryant, 1971). Present activity or inactivity, however, has nothing to do with the age of debris on the rock-glacier surface, and some active rock glaciers have surface debris as old as Pinedale just upvalley from their fronts.

WIND RIVER MOUNTAINS, WYOMING

Moss (1951) estimated that moraines of the Temple Lake stade (Figure 1-2) in the southern part of the Wind River Mountains are about 10,000 years old and that younger cirque moraines were supposedly formed during the last several centuries. Richmond (1965), working farther north in the range, considered moraines equivalent to the Temple Lake moraines to be of Neoglacial age because of their soil development. He defined the Gannett Peak stade (Figure 1-2) for moraines formed during the last several centuries (Richmond, 1957).

More recent work in the Wind River Mountains has altered the Holocene chronology. An important radiocarbon date of 6500 ± 230 yr B.P. (GX-3166D) was obtained from the base of a bog on the type Temple Lake Till (Currey, 1974). Hence, the designation Temple Lake should be reserved for pre-Altithermal early-Holocene tills in the Wind River Mountains. Currey (1974) and Miller and Birkeland (1974) used relative-dating methods to show that the cirque moraines in the southern Wind River Mountains were much more complex than had been envisioned by Moss (1951). The youngest till in the cirque where Moss worked borders the present-day ice, and unpublished work by Miller and Birkeland indicates that its relative weathering parameters are identical to those for the type Gannett Peak Till of Richmond (1957) in the northern mountains. Fronting the Gannett Peak moraines in this and many other nearby cirques are moraines that could be composed of till equivalent to the Audubon Till of Colorado (Figure 1-2) (Miller and Birkeland, 1974); in places, the Gannett Peak moraines seem to have buried this older deposit. Finally, Currey (1974) and Miller and Birkeland (1974) agree that the oldest cirque moraines described by Moss (1951) are probably of early-Neoglacial age.

Miller and Birkeland (1974; unpublished data) recognized Temple Lake (pre-Altithermal) and three Neoglacial deposits over a large portion of the southern Wind River Mountains. Several limiting radiocarbon dates are available, but closer limiting dating is necessary to support the local correlations and absolute ages of the deposits. Mahaney (1978) presents a persuasive case for the existence in the northern Wind River Mountains of the same four units recognized by Miller and Birkeland in the South.

UTAH

The work on Holocene glacial activity in Utah has been focused on two Mountain Ranges: the La Sal Mountains and the Wasatch Mountains (Figure 1-1). Richmond's (1962) contribution to the work in the La Sal Mountains was notable because of the numerous depositional facies mapped and because of the extensive use made of soils in both local and regional correlations. He recognized the early-Holocene upper Beaver Basin Till high in the valleys as well as two ages of Neoglacial till and rock-glacial debris, the lower and upper Gold Basin deposits (Figure 1-2). He considered the lower Gold Basin unit to have undergone two episodes of deposition, one he considered to have occurred between 3100 and 2800 yr. B.P. and the other between 1800 and 1000 yr B.P., whereas he considered the upper Gold Basin unit to date from the last several centuries. A brief reconnaissance made by Birkeland and a group of students suggests that some of the ages assigned by Richmond may be too young; that is, Neoglacial deposits may be more restricted than indicated by the Gold Basin deposits mapped by Richmond (1962).

For the Little Cottonwood Canyon area of the Wasatch Range (Figure 1-1), Ives (1950) was the first to assign an age to the moraines formed within 2.4 km of the cirque headwalls; he considered them to have been laid down during the "Little Ice Age" (Neoglacial in present terminology) as defined by Matthes (1939, 1942). In the same area, Richmond (1964) distinguished Holocene tills of two ages and, following his regional scheme, assigned the older one to the Temple

Lake stade (post-Altithermal) and the younger one to the historical stade of Neoglaciation.

McCoy (1977) and Madsen and Currey (1979) provide the relative-dating criteria and radiocarbon dates that suggest that some of the high-altitude tills are older than Richmond (1964) thought. For example, a radiocarbon basal bog date near altitude 2500 m (the highest peak in the Cottonwood Canyon area is 3502 m) of 12,300 ± 330 yr B.P. (GX-3481) suggests upper-valley deglaciation by at least that time. Still higher in the valley, near 3000 m, Madsen and Currey (1979) report a minimum limiting radiocarbon date of 7515 ± 180 yr B.P. (GX-4644) for the newly named Devils Castle Till (Figure 1-2). In contrast, Richmond had assigned the till to the Temple Lake stade of Neoglaciation. The Devils Castle Till may indeed be equivalent to the Temple Lake stade as the term is currently applied to pre-Altithermal deposits. Finally, McCoy (1977) has been able to distinguish early- and late-Neoglacial moraines and rock glaciers, but their extent is more restricted than Richmond (1964) initially showed.

The latest work in the Wasatch Mountains was done by Anderson and Anderson (1981), who worked in the Mount Timpanogos area (Figure 1-1), the only other area in the mountains that experienced Holocene glacial activity. Relative dating involved mainly making thickness measurements of weathering rinds developed on quartzarenite clasts in till and rock glaciers, but no absolute dates exist to construct a growth-rate curve for the rinds. The oldest deposit, a moraine 1.5 to 2.0 km from the cirque, may correlate with the type Devils Castle deposit. Upvalley from the moraine is a rock glacier made up of at least two ages of Neoglacial debris.

In summary, the study of Holocene glacial activity in the Rocky Mountains has progressed from the regional studies of Richmond during the 1950s and 1960s to the more detailed studies undertaken by other investigators during the late 1960s and 1970s. Richmond's work was very important in setting up a regional scheme. Subsequent mapping, radiocarbon dating, and detailed relative dating have necessitated changes in the stratigraphic nomenclature. Benedict (1981) cautions against making correlations between the glacial chronology for the Front Range in Colorado and those elsewhere within in the region or the world (e.g., Denton and Karlén, 1973; Grove, 1979). His persuasive arguments are backed by many data and radiocarbon dates. The Holocene glacial history in the Wasatch Mountains resembles that of Colorado, with the youngest tills dating from close to the Pleistocene/Holocene boundary and with the Holocene record being marked mainly by rock-glacial activity. In contrast, the Wind River Mountains of Wyoming provide a record of Neoglacial glacial activity, with the youngest tills dating from about the last century. Only detailed work in more widely scattered mountains, combined with adequate dating, will determine the degree of Holocene glacial synchroneity over the Rocky Mountain region.

Sierra Nevada Mountains, California

Except for a few deposits that have been located in the mountains of the northern part of California (Sharp, 1960), the Holocene glacial record in the state is essentially restricted to the cirques of the central and northern Sierra Nevada (Figure 1-1). Even the highest peaks (up to 4200 m) in the southern parts of this predominantly granitic mountain range appear to have escaped significant glacial or rock-glacial activity during the last few millennia. The Holocene deposits have received far less attention than those in the Rocky Mountains, and much of the present interpretation originates from recent work by Yount, Birkeland, and Burke (1979; unpublished data). Few radiocarbon dates have come from this work, and most of the estimated ages and correlations come from relative-dating criteria.

The Sierra Nevada underwent extensive glacial and rock-glacial activity during the Holocene. The depositional record of this activity lies at and above the present treeline in or just below cirques at elevations commonly above 3300 m. The original stratigraphic nomenclature (Figure 1-3) is based on the work of Birman (1964), who subdivided deposits primarily according to their positions in the valley and the degree of weathering on surface stones. Because Birman considered all deposits to be within what Matthes (1939, 1942) originally defined as the "Little Ice Age" and because Matthes had considered the "Little Ice Age" as having begun about 4000 years ago, Birman initially placed the Hilgard, Recess Peak, and Matthes in the post-Altithermal (Figure 1-3).

Curry (1969, 1971) also recognized these three advances, but he tentatively placed the age of the Hilgard at about 9000 to 10,000 years (Figure 1-3) on the basis of a comparison of soil development between an undated till and a radiocarbon-dated landslide deposit (7030 ± 130 yr B.P. [I-2287]) as well as on a worldwide climatic curve suggesting a cold period near the beginning of the Holocene. His Neoglacial age assignments and associated dating (Figure 1-3) are based on lichenometry. He describes two advances within the Recess Peak (2600 and 2000 yr B.P.), an unnamed advance at about 1100 yr

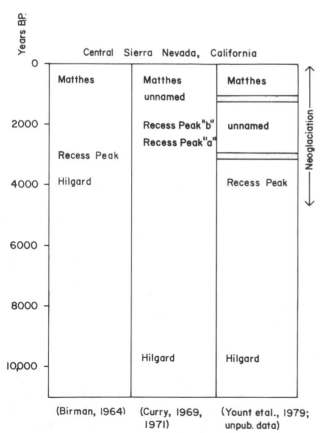

Figure 1-3. Principal stratigraphic names applied to Holocene glacial deposits in the Sierra Nevada, California. Shading indicates the preliminary radiocarbon-dating control from intraglacial deposits.

B.P., and several Matthes advances dating from 700 yr B.P. to the present.

Recent work over the last decade by the authors and their colleagues (Birkeland et al., 1976; Birkeland, Burke, and Walker, 1979; Birkeland, Walker, Benedict, and Fox, 1979; Burke, 1978; and Yount et al., 1979, and unpublished data) using a multiparameter relative-dating approach has refined the stratigraphy. Future radiocarbon dates from bog deposits should help to provide limiting absolute dates on the deposits. In general, we feel that the subdivision of deposits within valleys and the correlation of deposits from valley to valley are fairly straightforward. We suggest retaining the nomenclature used by Birman (1964), but we also agree with Curry (1971) that the Hilgard is latest Pleistocene or earliest Holocene. We recognize three post-Altithermal deposits: Recess Peak, the oldest, is thought to be early Neoglacial in age; an unnamed advance is intermediate Neoglacial in age; and Matthes, the youngest, includes those deposits that were laid down in the last several hundred years (Figure 1-3).

In the Palisades area (Figure 1-1), the deposits in cirques are not common, lie at the highest elevations, and are dominated by rock-glacial debris rather than till (Yount et al., 1979; unpublished data). We recognize the Recess Peak and Matthes, but other Holocene deposits are generally absent. To the north (Recesses area [Figure 1-1]), however, we recognize the full suite of deposits, and till is common. Apparently, the unnamed intermediate Neoglacial ice advance was limited to the northern part of the central Sierra Nevada, or in the south its deposits were overridden during the Matthes advance.

Although lichenometry has worked well in subdividing the Neoglacial deposits, it is not useful in distinguishing pre-Altithermal deposits from early-Neoglacial deposits. Pre-Altithermal deposits, however, are typified by soils with about a meter of oxidation and with well-defined color-B-horizons; early-Neoglacial deposits are much less oxidized. Intermediate-Neoglacial and Matthes deposits have very little or no soil development. Preliminary radiocarbon dates of bog deposits on the Recess Peak (about 3000 yr B.P.) and the unnamed intermediate advance (about 1100 yr B.P.) should be regarded as limiting dates only, because the stratigraphic relationships between the tills and the bogs are somewhat complex. Where we have applied multiparameter relative-dating techniques to deposits previously studied by Birman (1964) or Curry (1969, 1971), we commonly do not agree with the original age assignments (Yount et al., 1979; unpublished data). Thus, although we retain the stratigraphic order, we believe the individual deposits may be older or younger than originally estimated.

Although the glacial record of the Sierra Nevada has received a great deal of study, work on deciphering the Holocene record has only just begun. Radiocarbon dates are few, but preliminary dating of bog deposits supports the initial age assignments of recent investigations. Data on soil development and rock weathering support a threefold subdivision for the Neoglacial and an older pre-Altithermal advance. Although valley-to-valley correlation is best accomplished though the use of weathering characteristics, the timing of the glacial advances and possible correlations with Holocene glaciations in other mountain ranges should not depend solely on relative dating; rather, it should be postponed until radiocarbon dates become available.

Oregon

Holocene glacial and rock-glacial deposits exist in both the Cascade Range and the Wallowa Mountains of Oregon (Figure 1-1). Although the Cascade Range was undoubtedly the more heavily glaciated during the Holocene, little work has been done to decipher its stratigraphic history. Carver (1972), Scott (1977), and Dethier (1980) report on Holocene deposits in the central part of the Cascade Range (Figure 1-4). The presence or absence of widespread tephra provides time control to help date glacial and rock-glacial deposits. Each worker recognizes deposits older than the 6600-year-old Mount Mazama ash (Table 1-1), and in each area the deposits date from near the Pleistocene/Holocene boundary. Two or three Neoglacial events are recognized, but the timing of each event and, therefore, the correlation of deposits must wait for better dating control to be achieved.

The most detailed Holocene record in Oregon comes from the Wallowa Mountains, where cirque deposits have long been recognized (Lowell, 1939; Stovall, 1929). Recently, workers have independently arrived at similar Holocene stratigraphic interpretations. Williams (1974) defines pre-Altithermal, early-Neoglacial, and latest-Neoglacial glacial advances. Kiver (1974), whose nomenclature has been adopted in this chapter (Figure 1-5), also recognizes three Holocene advances but finds that the Neoglacial is represented by an intermediate and a younger glacial episode, with no early-Neoglacial deposits. Burke (1978; unpublished data) supplies abundant stone-weathering and soil data to support Kiver's general subdivisions. However, the data do not support a hiatus between the Eagle Cap and Prospect Lake deposits (Figure 1-5) as large as that in other ranges between the intermediate and youngest Neoglacial deposits, and cor-

Table 1-1. Holocene Tephra Units Discussed in Text, Their Source Volcanoes, and Approximate Ages

Tephra Unit	Source Volcano	Approximate Age (yr B.P.)
W	Mount St. Helens	450
C	Mount Rainier	2200
Y	Mount St. Helens	3000-4000
O	Mount Mazama (Crater Lake)	6600
R	Mount Rainier	> 8750

Sources: Crandell and Miller, 1974: Table 1; Mullineaux, 1974: Table 4.

relation with other ranges would be premature.

The Cascade Range has supplied the northwestern United States with several widespread tephrostratigraphic markers over the past few millennia (Crandell et al., 1962; Mullineaux, 1974), and at least two of these have been identified within northeastern Oregon. The Mazama ash, dated at 6600 yr B.P., and the Mount St. Helens Y (?) layer, dated at 4000 to 3000 yr B.P. (Table 1-1), are both present on the Glacier Lake deposits and absent from Prospect Lake and Eagle Cap deposits. If Williams's (1974) age assignment of early Neoglacial is correct, the apparent lack of the Mount St. Helens Y tephra layer on those deposits either is anomalous or provides a maximum limiting date for them.

Cascade Range, Washington

The Cascade Range in Washington includes several composite volcanoes veneered by alpine glaciers, and the mountains show ample evidence of having undergone more extensive glaciation earlier in the Holocene. The local maritime climate and the steep valley profiles combine to make this one of the most heavily glaciated areas in the western United States during the Holocene. The most detailed stratigraphic work has been carried out on the flanks of Mount Rainier and on Dome Peak (Figure 1-1), near the central portion of the range.

Dendrochronology, lichenometry, and tephrochronology have been used in correlating and dating the tills. The terminal features of many Holocene glaciers lie below the treeline, making dendrochronology a useful dating tool, and the youthful volcanic centers located along the Cascade crest produce abundant tephra layers for dating control. The tephrostratigraphy (Table 1-1) has been well established, because the individual layers are mineralogically unique. Generally, the age of the tephra layers has been determined through the radiocarbon dating of interbedded organic debris both near the volcanic sources and from distant low-lying areas.

Crandell and Miller (1964, 1974) and Crandell (1969) applied a local nomenclature (Figure 1-6) to the deposits of Mount Rainier, and it has remained essentially unchanged. High in some cirques are limited deposits overlain by tephra layer R (8750 ± 280 yr B.P. [W-950]) (Table 1-1) and thought to date from latest Pleistocene time (Crandell and Miller, 1974). All the Holocene glacial deposits on Mount Rainier are Neoglacial in age, and all are considered part of the Winthrop Creek glaciation (Figure 1-6). The Winthrop Creek is divided into the older Burroughs Mountain and the younger Garda stades, which are best distinguished and dated by the presence or absence of tephra layers. The Burroughs Mountain deposits are older than tephra layer C and yet younger than tephra layer Y (Table 1-1), which lies on the valley floor outside the Burroughs Mountain

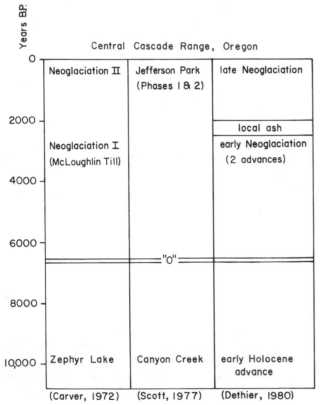

Figure 1-4. A summary of stratigraphic names applied to Holocene glacial deposits in the central Cascade Range of Oregon. The band at 6600 yr B.P. represents the Mazama tephra layer O, which is present in all three study areas. Dethier (1980) uses a slightly older age for the Mazama tephra and also uses a local tephra layer as a stratigraphic marker.

moraines. Thus, the Burroughs Mountain stade deposits are between 3500 and 2000 years old. The oldest Garda stade deposits are overlain by tephra layer W (450 yr B.P.) (Table 1-1), but most are ash free. The oldest tree found growing on a Garda deposit dates from the early 13th century (Crandell and Miller, 1964).

Dome Peak also has yielded a detailed Neoglacial record, but no early-Holocene (pre-Altithermal) deposits have been found (Miller, 1969). Miller studied the Holocene deposits within four valleys, and in two of these he found tephra layer O (6600 yr B.P.) (Table 1-1) on the valley floors beyond deposits dated from the 16th century by dendrochronology. Therefore, these glaciers have not been more extensive than their 16th-century position for 6600 years. If early-Holocene advances took place, their deposits were overrun by late-Neoglacial advances. The oldest Holocene advance located in the

Dome Peak area is also younger than 6600 years and is thought to be early Neoglacial (Miller, 1969). In one of the valleys, trees sheared off by advancing ice yield radiocarbon dates of 4700 ± 300 (W-1030) and 4960 ± 90 yr B.P. (VW99), thus providing a maximum age for the Dome Peak deposits. Either the other three valleys did not experience this early-Neoglacial advance, or evidence for it was destroyed by later glacial advances. Evidence for Neoglacial deposits of intermediate age does not exist on Dome Peak, but ample evidence is present for a twofold Neoglacial record when the older event (about 5000 yr B.P.) is considered to be Neoglacial and not Altithermal. Dated primarily by dendrochronology, the youngest Neoglacial glacial episode can be considered a twofold glaciation; a major advance occurred at about A.D. 1600, and the last major Holocene advance occurred at about A.D. 1850 (Miller, 1969).

Two recent studies provide a valuable addition to the Cascade Holocene chronology, because they demonstrate the existence of early-Holocene glacial deposits. On Glacier Peak, approximately 25 km south of the Dome Peak area (Figure 1-1), Beget (1980) applied radiocarbon dating and tephrochronology to date an early-Holocene till between about 8400 (W-4227) and 6700 yr B.P. Waitt, Yount, and Davis (1980) recognized an early-Holocene advance (about 8500 to 7500 yr B.P.) as well as a late-Neoglacial advance (about 450 yr B.P.) in a valley on the eastern side of the Cascade crest.

In the heavily glaciated Cascade Mountains of Washington, the Holocene record is generally restricted to the Neoglacial. The pre-Altithermal advance so common in other western ranges was either limited to a few valleys or restricted in extent so that evidence for it was destroyed by the more extensive Neoglacial advances. The suggestion that early-Holocene deposits were overrun by Neoglacial ice has been discussed by Miller (1969) and Kiver (1974), and the case for an early-Holocene advance is upheld by recent work by Beget (1980)

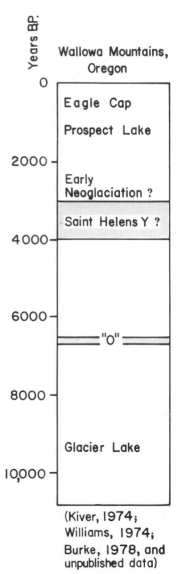

Figure 1-5. Stratigraphic names applied to Holocene glacial deposits in the Wallowa Mountains, Oregon. The early Neoglacial advance was reported only by Williams (1974), and its deposits are not overlain by tephra. Tephra layers O and Y are shown by dark bands, but the Y layer has been only tentatively identified (Kiver, 1974). The nomenclature is taken from Kiver (1974) and differs from that applied by Williams (1974).

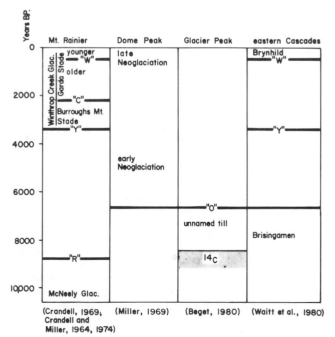

Figure 1-6. Principal stratigraphic names and correlation of Holocene glacial deposits in the Cascade Range, Washington. Shading (Glacier Peak) indicates limiting radiocarbon control; dark bands are labeled to indicate some of the tephra layers identified.

and Waitt, Yount, and Davis (1980). Thus, in Washington a more complex glacial record may exist for the Cascade Range than can be documented by the deposits that remain.

Conclusion

During the last four decades, investigations of Holocene glacial chronologies in the mountain ranges of the western United States have become more detailed and less regional in scope. The ability to date deposits by both relative and absolute methods has been refined during this time, and the prospects for the future advancement of our understanding of Holocene glacial activities are good. The deposits can be subdivided through the use of a multiparameter relative-dating approach. Where absolute dating control is available, there is usually depositional evidence for glacial activity during the early Holocene and multiple advances during the Neoglacial. A comparison of relative-dating data between chronosequences with absolute dating and those without absolute dating allows for an extension of age control into areas where radiocarbon dates are not available. In at least one area, the Front Range of Colorado, a middle-Altithermal glacial advance has been documented (Benedict, 1981), and we are cautioned against assuming a synchroneity of Holocene glacial activity. Many local glacial sequences throughout the western United States have been subdivided, but an indisputable correlation among these sequences cannot yet be demonstrated.

References

Anderson, L. W., and Anderson, D. S. (1981). Weathering rinds on quartzarenite clasts as a relative-age indicator and the glacial chronology of Mount Timpanogos, Wasatch Range, Utah. *Arctic and Alpine Research* 13, 25-31.

Andrews, J. T., Carrara, P. E., King, F. B., and Stuckenrath, R. (1975). Holocene environmental changes in the alpine zone, northern San Juan Mountains, Colorado: Evidence from bog stratigraphy and palynology. *Quaternary Research* 5, 173-97.

Beget, J. E. (1980). Tephrochronology of deglaciation and latest Pleistocene or early Holocene moraines at Glacier Park, Washington. *Geological Society of America, Abstracts with Programs* 12, 96.

Benedict, J. B. (1967). Recent glacial history of an alpine area in the Colorado Front Range, U.S.A.: I. Establishing a lichen-growth curve. *Journal of Glaciology* 6, 817-32.

Benedict, J. B. (1968). Recent glacial history of an alpine area in the Colorado Front Range, U.S.A.: II. Dating the glacial deposits. *Journal of Glaciology* 7, 77-87.

Benedict, J. B. (1973). Chronology of cirque glaciation, Colorado Front Range. *Quaternary Research* 3, 584-99.

Benedict, J. B. (1981). ''The Fourth of July Valley: Glacial Geology and Archaeology of the Timberline Ecotone.'' Center for Mountain Archaeology Research Report 2.

Benedict, J. B., and Olson, B. L. (1978). ''The Mount Albion Complex: A Study of Prehistoric Man and the Altithermal.'' Center for Mountain Archaeology Research Report 1.

Birkeland, P. W. (1973). Use of relative age dating methods in a stratigraphic study of rock glacier deposits, Mt. Sopris, Colorado. *Arctic and Alpine Research* 5, 401-16.

Birkeland, P. W., Burke, R. M., and Walker, A. L. (1979). Variation in chemical parameters of Quaternary soils with time and altitude, Sierra Nevada, California. *Geological Society of America, Abstracts with Programs* 11, 388.

Birkeland, P. W., Burke, R. M., and Yount, J. C. (1976). Preliminary comments on late Cenozoic glaciations in the Sierra Nevada. *In* ''Stratigraphy of North America: Proceedings of a Symposium'' (W. C. Mahaney, ed.), pp. 283-95. Dowden, Hutchinson, and Ross, Stroudsburg, Pa.

Birkeland, P. W., Colman, S. M., Burke, R. M., Shroba, R. R., and Meierding, T. C. (1979). Nomenclature of alpine glacial deposits, or, what's in a name? *Geology* 7, 532-36.

Birkeland, P. W., Crandell, D. R., and Richmond, G. M. (1971). Status of correlation of Quaternary stratigraphic units in the western conterminous United States. *Quaternary Research* 1, 208-27.

Birkeland, P. W., and Shroba, R. R. (1974). The status of the concept of Quaternary soil-forming intervals in the western United States. *In* ''Quaternary Environments: Proceedings of a Symposium'' (W. C. Mahaney, ed.), pp. 241-76. Geographical Monographs York University-Atkinson College, Toronto.

Birkeland, P. W., Walker, A. L., Benedict, J. B., and Fox, F. B. (1979). Morphological and chemical trends in soil chronosequences: Alpine and arctic environments. Agronomy Abstract, p. 188, 71st Annual Meeting of the American Society of Agronomy, Madison, Wis.

Birman, J. H., (1964). ''Glacial Geology across the Crest of the Sierra Nevada, California.'' Geological Society of America Special Paper 75.

Bryant, B., (1971). ''Movement Measurements on Two Rock Glaciers in the Eastern Elk Mountains, Colorado.'' U. S. Geological Survey, Professional Paper 750-B, pp. B108-B116.

Burke, R. M. (1978). Comparison of relative age dating (RAD) data from eastern Sierra Nevada cirque deposits with those from the tephrochronologically age controlled deposits of the Wallowa Mountains, Oregon. *Geological Society of America, Abstracts with Programs* 10, 211.

Carrara, P. E., and Andrews, J. T. (1973). Problems and applications of lichenometry to geomorphic studies, San Juan Mountains, Colorado. *Arctic and Alpine Research* 5, 373-84.

Carrara, P. E., and Andrews, J. T. (1976). Holocene glacial/periglacial record: Northern San Juan Mountains, southwestern Colorado. *Zeitschrift für Gletscherkunde und Glazialgeologie* 11, 155-74.

Carrara, P. E., and Mode, W. N. (1979). Extensive deglaciation in the San Juan Mountains, Colorado, prior to 14,000 yr B.P. *Geological Society of America, Abstracts with Programs* 11, 399.

Carroll, T. (1974). Relative age dating techniques and a late Quaternary chronology, Arikaree cirque, Colorado. *Geology* 2, 321-25.

Carver, G. A. (1972). Glacial geology of the Mountain Lakes Wilderness and adjacent parts of the Cascade Range, Oregon. Ph.D. dissertation, University of Washington, Seattle.

Crandell, D. R. (1969). ''Surficial Geology of Mount Rainier National Park, Washington.'' U.S. Geological Survey Bulletin 1288.

Crandell, D. R., and Miller, R. D. (1964). ''Post-hypsithermal Glacier Advances at Mount Rainier, Washington.'' U.S. Geological Survey Professional Paper 501-D, pp. D110-D114.

Crandell, D. R., and Miller, R. D. (1974). ''Quaternary Stratigraphy and Extent of Glaciation in the Mount Rainier Region, Washington.'' U.S. Geological Survey Professional Paper 847.

Crandell, D. R., Mullineaux, D. R., Miller, R. D., and Rubin, M. (1962). ''Pyroclastic Deposits of Recent Age at Mount Rainier, Washington.'' U.S. Geological Survey Professional Paper 450-D, pp. D64-D68.

Currey, D. R. (1974). Probably pre-Neoglacial age of the type Temple Lake moraine, Wyoming. *Arctic and Alpine Research* 6, 293-300.

Curry, R. R. (1969). Holocene climatic and glacial history of the central Sierra Nevada, California. *In* ''United States Contributions to Quaternary Research'' (S. A. Schuman and W. C. Bradley, eds.), pp. 1-47. Geological Society of America Special Paper 123.

Curry, R. R. (1971). Glacial and Pleistocene History of the Mammoth Lakes Sierra, California: A Geologic Guidebook.'' Montana Department of Geology, Geological Serial Publication 11.

Davis, P. T., Upson, S., and Waterman, S. E. (1979). Lacustrine sediment variation as an indicator of late Holocene climatic fluctuations, Arapaho cirque, Colorado Front Range. *Geological Society of America, Abstracts with Programs* 11, 410.

Davis, P. T., and Waterman, S. E. (1979). New radiocarbon ages for type Triple Lakes moraines, Arapaho cirque, Colorado Front Range. *Geological Society of America, Abstracts with Programs* 11, 270.

Denton, G. H., and Karlén, W. (1973). Holocene climatic variations: Their pattern and possible cause. *Quaternary Research* 3, 155-205.

Denton, G. H., and Porter, S. C. (1970). Neoglaciation. *Scientific American* 222, 101-10.

Dethier, D. P. (1980). Reconnaissance study of Holocene glacier fluctuations in the Broken Top area, Oregon. *Geological Society of America, Abstracts with Programs* 12, 104.

Grove, J. M. (1979). The glacial history of the Holocene. *Progress in Physical Geography* 3, 1-54.

Ives, R. L. (1950). Glaciations in Little Cottonwood Canyon, Utah. *Science Monthly* 71, 105-17.

Kiver, E. P. (1974). Holocene glaciation in the Wallowa Mountains, Oregon. *In* "Quaternary Environments: Proceedings of a Symposium" (W. C. Mahaney, ed.), pp. 169-96. Geographical Monographs 5, York University-Atkinson College, Toronto.

Lowell, W. R. (1939). Glaciation in the Wallowa Mountains. M.S. thesis, University of Chicago, Chicago.

McCoy, W. D. (1977). A reinterpretation of certain aspects of the late Quaternary glacial history of Little Cottonwood Canyon, Wasatch Mountains, Utah. M.A. thesis, University of Utah, Salt Lake City.

Madole, R. F. (1978). Chronology of Pinedale deglaciation in the southern Park Range, Colorado. *Geological Society of America, Abstracts with Programs* 10, 220.

Madsen, D. B., and Currey, D. R. (1979). Late Quaternary glacial and vegetation changes, Little Cottonwood Canyon area, Wasatch Mountains, Utah. *Quaternary Research* 12, 254-70.

Mahaney, W. C. (1973). Neoglacial chronology in the Fourth of July cirque, central Colorado Front Range. *Geological Society of America Bulletin* 84, 161-70.

Mahaney, W. C. (1974). Soil stratigraphy and genesis of Neoglacial deposits in the Arapaho and Henderson cirques, central Colorado Front Range. *In* "Quaternary Environments: Proceedings of a Symposium" (W. C. Mahaney, ed.), pp. 197-240. Geographical Monographs York University-Atkinson College, Toronto.

Mahaney, W. C. (1978). Late-Quaternary stratigraphy and soils in the Wind River Mountains, western Wyoming. "Quaternary Soil" (W. C. Mahaney, ed), pp. 223-64. Geological Abstracts, Ltd, Norwich, England.

Matthes, F. E. (1939). Report of the committee on glaciers. *In* "American Geophysical Union Transaction," part III, 518-23.

Matthes, F. E. (1942). Glaciers. *In* "Physics of the Earth: Hydrology" (O. E. Meinzer, ed.), pp. 142-219. McGraw-Hill, New York.

Miller, C. D. (1969). Chronology of Neoglacial moraines in the Dome Peak area, North Cascade Range, Washington. *Arctic and Alpine Research* 1, 49-66.

Miller, C. D. (1973). Chronology of Neoglacial deposits in the northern Sawatch Range, Colorado. *Arctic and Alpine Research* 5, 385-400.

Miller, C. D., and Birkeland, P. W. (1974). Probably pre-Neoglacial age of the Type Temple Lake moraine, Wyoming: Discussion and additional relative-age data. *Arctic and Alpine Research* 6, 301-16.

Moss, J. H. (1951). Late glacial advances in the southern Wind River Mountains, Wyoming. *American Journal of Science* 249, 865-83.

Mullineaux, D. R. (1974). "Pumice and other pyroclastic deposits in Mount Rainier National Park, Washington." U.S. Geological Survey Bulletin 1326.

Nelson, R. L. (1954). Glacial geology of the Frying Pan River drainage, Colorado. *Journal of Geology* 62, 325-43.

Porter, S. C., and Denton, G. H. (1967). Chronology of Neoglaciation in the North American cordillera. *American Journal of Science* 265, 177-210.

Richmond, G. M. (1957). Correlation of Quaternary deposits in the Rocky Mountain region, U.S.A. *In* "Résumés des Communications," p. 157. Fifth Congress of the International Association for Quaternary Research, Madrid-Barcelona.

Richmond, G. M. (1960). Glaciation of the east slope of Rocky Mountain National Park, Colorado. *Geological Society of America Bulletin* 71, 1371-82.

Richmond, G. M. (1962). "Quaternary Stratigraphy of the La Sal Mountains, Utah." U.S. Geological Survey Professional Paper 324.

Richmond, G. M. (1964). "Glaciation of Little Cottonwood and Bells Canyons, Wasatch Mountains, Utah." U.S. Geological Survey Professional Paper 4454-D.

Richmond, G. M. (1965). Glaciation of the Rocky Mountains. *In* "The Quaternary of the United States" (H. E. Wright, Jr., and D. G. Frey, eds.), pp. 217-30. Princeton University Press, Princeton, N.J.

Scott, W. E. (1977). Quaternary glaciation and volcanism, Metolius River area, Oregon. *Geological Society of America Bulletin* 88, 113-24.

Sharp, R. P. (1960). Pleistocene glaciation in the Trinity Alps of northern California. *American Journal of Science* 258, 305-40.

Stovall, J. C. (1929). Pleistocene geology and physiography of the Wallowa Mountains with special reference to the Wallowa and Hurricane canyons. M.S. thesis, University of Oregon, Eugene.

Wahrhaftig, C., and Cox, A. (1959). Rock glaciers in the Alaska Range. *Geological Society of America Bulletin* 70, 383-436.

Waitt, R. B., Yount, J. C., and Davis, P. T. (1980). Early and late Holocene glacier advances in Upper Enchantment Basin, eastern North Cascades, Washington. *Geological Society of America, Abstracts with Programs* 12, 158.

White, P. G. (1979). Rock glacier morphometry, San Juan Mountains, Colorado. *Geological Society of America Bulletin* 90, 515-18, 924-52.

White, S. E. (1971). Rock glacier studies in the Colorado Front Range, 1961 to 1968. *Arctic and Alpine Research* 3, 43-64.

White, S. E. (1976). Rock glaciers and block fields, review and new data. *Quaternary Research* 6, 77-97.

Williams, L. D. (1974). Neoglacial landforms and Neoglacial chronology of the Wallowa Mountains, northeastern Oregon. M.S. thesis, University of Massachusetts, Amherst.

Yount, J. C., Birkeland, P. W., and Burke, R. M. (1979). Glacial and periglacial deposits of the Mono Creek recesses, west-central Sierra Nevada, California: Measurement of age-dependent properties of the deposits. *In* "Field Guide to Relative Dating Methods Applied to Glacial Deposits in the Third and Fourth Recesses and along the Eastern Sierra Nevada, California, with Supplementary Notes on Other Sierra Nevada Localities" (R. M. Burke and P. W. Birkeland, eds.) Field Trip Guidebook for the Friends of the Pleistocene, Pacific Cell.

CHAPTER 2

Aspects of Holocene Pedology in the United States

Robert V. Ruhe

Introduction

Any reasonable exposé of Holocene pedology in the United States would require a bulky monograph because of a simple fact: there are more than 10,000 soil series, or soil-taxonomic units, recognized in the country; and, if soil-mapping units were considered, the number of series would be multiplied by a factor of N. Obviously, an in-depth treatment of such a large subject is not feasible in a brief chapter, so certain aspects of the subject that may have broad application in principle have been chosen for discussion.

First, the relations between soils and Holocene climates are treated from the viewpoint of soil classification. Next, the relations between soils and Holocene vegetation are examined with respect to the organic matter systems in mineral soils and the environmental records contained in organic soils of wetlands. Then, soil development during the Holocene is examined from a broad perspective of soil-geographic relations. Finally, soil development during the Holocene is discussed in the restricted context of soil-geomorphic relations. By following this approach, I hope to demonstrate the principles of Holocene pedology.

Alaska and Hawaii are not being used as examples in this chapter, but readers can refer to a recent and excellent discussion of polar soils and particularly to the soils of Alaska (Tedrow, 1977: 282-348). The soils of Hawaii are described in a soil survey report (Cline, 1955).

Soil Classification and Historical Climate

For better or for worse, formal soil classification is directly involved in Holocene pedology in the United States, and the impact of the Holocene on soils is defined climatically and applied to every soil in the nation. In the soil taxonomy for the United States (Soil Survey Staff, 1975), all soils are grouped in categories from highest to lowest rank: 10 orders, 47 suborders, approximately 206 great groups, numerous subgroups and families, and approximately 10,000 series. The climatic impact of soil temperature and moisture regimes is introduced at the level of the suborder.

Five main soil-moisture regimes (Table 2-1) are based on the depth to groundwater or on the moisture held at a tension of less than 15 bars for a specified time. Six major soil-temperature regimes (Table 2-2) are based on mean annual air temperature because it is recognized as closely relating to mean annual air temperature. Soil-temperature regimes are further defined by specific temperatures at specific depths in the soil.

The climate's impact on soils can be illustrated by a comparison of the geographic distribution of the soil orders (Figure 2-1) with average annual temperatures (Figure 2-2) and rates of precipitation (Figure 2-3). As an example, the main body of Mollisols (grassland

Table 2-1. Soil-Moisture Regimes

Regime	Moisture State and Duration
Aquic (wet)	Saturated by groundwater; lack of dissolved oxygen, soil temperature > 5°C
Aridic (hot and dry)	Dry in all parts 50% cumulative time; soil temperature at depth of 50 cm > 5°C
	Never moist in part or whole 90 consecutive days; soil temperature at 50 cm > 8°C
Udic (humid)	Not dry in any part 90 cumulative days; not dry in all parts 45 consecutive days
	Mean annual soil temperature <22°C; mean annual winter and summer soil temperatures at 50 cm vary > 5°C
Ustic (dry)	Dry in part or whole 90 cumulative days; mean annual soil temperature > 22°C; mean annual winter and summer temperatures at 50 cm vary <5°C
	Moist in some part 180 cumulative days or 90 consecutive days
Xeric (dry)	Moist in all parts > 45 consecutive days; winters moist and cold
	Dry in all parts > 45 consecutive days; summers dry and warm

Source: Soil Survey Staff, 1975.
Note: Moisture regimes are based on groundwater level or moisture held at tension of less than 15 bars through specified time.

Table 2-2. Soil-Temperature Regimes

Regime	Mean Annual Temperature
Pergelic (permafrost)	< 0 °C
Cryic (very cold)	0 °C-8 °C
Frigid (cold)	< 8 °C
Mesic (moderate)	8 °C-15 °C
Thermic (hot)	15 °C-22 °C
Hyperthermic (very hot)	> 22 °C

Source: Soil Survey Staff, 1975.

Note: Mean annual soil temperature relates closely to mean annual air temperature. Mean winter and summer temperatures at depth of 50 cm vary more than 5 °C for frigid, mesic, thermic, and hyperthermic regimes. If they vary less than 5 °C, then the regime is isofrigid or isohyperthermic.

soils of steppes and prairies) extends from the Central Lowlands (Figure 2-1, longitude 88 °W, latitude 41 °N) westward across the Great Plains to the Rocky Mountain front (at about longitude 105 °W). Along meridian 100 °W, the latitudinal extent is about 20 °, from 29 °N to 49 °N. Considerable variation in climate is to be expected in this large region. Mean annual temperatures are latitudinally and uniformly banded from 21 °C in the South to 2 °C in the North (Figure 2-2). Mean annual rates of precipitation are generally longitudinally banded from 90 to 100 cm in the East to 40 cm in the West. Precipitation gradients are steeper along parallels 35 °N and 45 °N. Along parallel 40 °N, the gradient is less steep and is related to the eastward extension of the Prairie Peninsula in Illinois and Indiana (Figure 2-2).

The combination of soil temperature and moisture regimes is recognized in the suborders of the Mollisols (Table 2-3). Borolls (cool and moist soils) generally occur north of latitude 45 °N (Figure 2-1). Udolls (warm and moist soils) extend from latitude 35 °N to 45 °N and east of about longitude 97 °W. The remainder of the main Mollisol region has Ustolls (warm and dry soils).

A similar analysis can be made for the Alfisols (high-base-status forest soils) in the central and eastern states (Figures 2-1, 2-2, 2-3). The temperature range is about the same as that for Mollisols, but the moisture range differs and is 60 to 165 cm.

The pedologic-climatic system in modern times appears to be neatly bound, but was it during the Holocene? The climatic parameters built in to the soil-classification system are data from the last 40 to 80 years and represent less than 1% of the Holocene when the round number of 10,000 years is used for its duration. An admonition is in order for the unwary: do not invert the system and read the climatic parameters from the soils.

Soils and Vegetation

If the modern episode of Holocene climate is acknowledged in the recognition of soil properties, an intimate associate of climate (namely, vegetation) must be accorded the same status. The vegetation in the United States can be broken down into four major categories: western forest, eastern forest, prairie or grassland, and shrubland (Figure 2-4). In the eastern forest, spruce, fir, pine, and northern hardwoods (Figure 2-4: 1, 2, 10 [Latin names are listed in the figure caption]) dominate to the north; and pine and hardwoods grow in the

I Inceptisols
E Entisols
H Histosols
A Alfisols
S Spodosols
M Mollisols
D Aridisols
V Vertisols
U Ultisols

Figure 2-1. Soil orders in the United States. (Generalized from Soil Conservation Service, 1970).

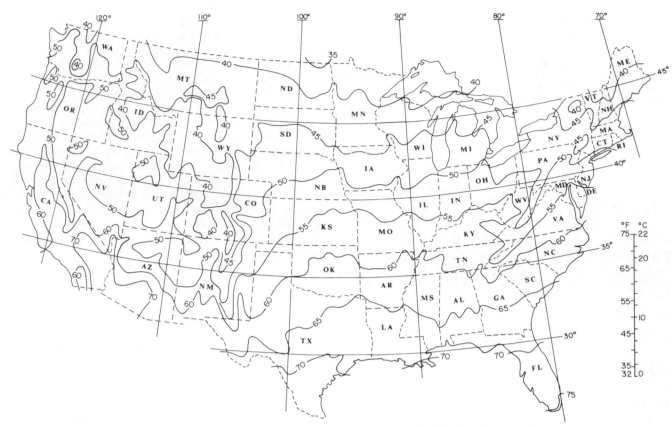

Figure 2-2. Average annual air temperatures in °F in the United States. (Conversion scale given for °F to °C.) (Generalized from U.S. Department of Commerce, 1968.)

extreme southern areas (Figure 2-4: 8, 9). Between these two zones are the broadleaf forests (without conifers). In the central grasslands and along latitude 40 °N, bluestem prairie with oak-hickory forest on slopes characterize eastern vegetation (Figure 2-4: 11); farther to the west are wheatgrass-bluestem-needlegrass prairie, bluestem prairie, and grama-buffalo grass prairie (Figure 2-4: 8, 9, 12). An obvious correlation exists between climate (Figures 2-2, 2-3) and vegetation (Figure 2-4).

An obvious correlation also exists between the broad geographic distributions of soils (Figure 2-1) and vegetation (Figure 2-4). Worthy of note are the relations of Mollisols to the central grasslands, of Spodosols (forest soils with subsoil accumulations of sesquioxides and humus) and Alfisols to the eastern forest in the north-central and northeastern states, and of Alfisols and Ultisols (low-base-status forest soils) to the eastern forest in the south-central and southeastern states. In the southwestern states, the Aridisols (soils of arid regions) and shrubland are compatible as could be expected.

The role of vegetation in soil formation is the production of organic matter, whether in horizon differentiation as additions, removals, transfers, and transformations (Simonson, 1959) or as part of an energy model in soil development (Runge, 1973). One should expect that the organic matter in most soils is Holocene in age.

ORGANIC MATTER SYSTEMS

The addition of organic matter and its transformation to humus in the A-horizon of a soil is a continuing phenomenon, whereas the removal and transfer are subordinate processes. In systematic studies

of hillslope and soil formation in Iowa (Kleiss, 1970; Ruhe, 1974; Vreeken, 1973; Walker, 1966; Walker et al., 1968a, 1968b), the hillslopes can be traced down into alluvial fills radiocarbon dated as 6600 to 2000 years old. In Mollisols on the hillslopes, the dark-colored, humus-rich surface horizons can be as thick as 50 cm and may contain 4% to 5% organic carbon.

In Aridisols on alluvial fans along mountain fronts in the desert of southern New Mexico, the organic carbon in the surface horizons is related to an orographic-climatic system (Gile, 1975; Ruhe, 1967). Average annual precipitation increases from about 20 to 40 cm in an elevation range of 1190 to 1830 m. Above an elevation of 1585 m, the surface horizon of soils is darker and thicker than at lower elevations. The higher soils contain 1% to 1.5% organic carbon, and the lower soils rarely exceed 0.5% organic carbon in this arid region. The alluvial-fan sediments, radiocarbon dated from the charcoal of buried hearths, are 6400 to 1130 years old.

The additions of organic matter can be much younger. In the same area bordering the Rio Grande Valley, where a dated alluvial-fan sediment is less than 2600 years old, soils on the campus of New Mexico State University were flood irrigated from 1919 to 1952 and sprinkler irrigated since that time and have been under Bermuda grass (*Cynodon dactylon*). In 1967, the soils had 1.4% to 2% organic carbon in the top 15 cm and as much as 0.5% organic carbon at a depth of 85 cm. In contrast, adjacent unirrigated soils of the same type had only 0.1% to 0.15% organic carbon in the top 15 cm.

The radiocarbon ages of soil-organic carbon also attest to the Holocene additions and transformations. The ages of organic carbon

Figure 2-3. Average annual precipitation in inches in the United States. (Conversion scale given for in. to cm.) (Generalized from U.S. Department of Commerce, 1968.)

from the A1-horizons of a well-drained Mollisol (Hapludoll) and a more poorly drained Mollisol (Haplaquoll) in Iowa are 440 ± 120 and 270 ± 120 years. The organic carbon from the A1-horizon of an Argiudoll (a Mollisol with textural B-horizon) is less than 100 and 210 ± 130 years old (Ruhe, 1969). These dates are the mean residence times for unfractionated organic carbon. They can be compared with dates for various fractions of organic carbon after chemical extractions. The mean residence time of carbon from the A-horizon of a Boroll in Saskatchewan, Canada (Campbell et al., 1967a, 1967b), was 870 ± 50 years. The radiocarbon dates of 14 chemical extractants ranged from 1410 to less than 100 years.

The mean residence time of unfractionated soil-organic carbon is an approximate average value that represents the continued additions of younger carbon to older carbon within the soil system. The radiocarbon date does not represent the age of the soil. It represents only a value of the organic carbon system. The organic carbon, however, is sometimes associated with another pedogenic component such as silicate clay, and so, indirectly, other pedogenic phenomena can be dated. In most soils, the radiocarbon dates for organic carbon are Holocene in age. (For a review of the problems involved, see Scharpenseel [1977] and Scharpenseel and Schiffman [1977a, 1977b].)

The transfer of organic carbon downward from the surface horizon in soils is also a continuing Holocene phenomenon. In an Argialboll in south-central Iowa, a Mollisol with a bleached horizon beneath a dark-surface horizon and above a textural B-horizon (formerly Planosol), the soil peds in the B-horizon have black coatings of humus

and clay. Such coatings (clay skins) are evidence for translocated clay (Soil Survey Staff, 1975: 21-24). The maximum clay content in the B-horizon is at depths of 56 to 66 cm. The radiocarbon dates (mean residence times) of organic carbon in this soil are 410 ± 110, 840 ± 200, and 1545 ± 110 years at mean sampling depths of 15, 38, and 61 cm, respectively. The increase in age with depth in this soil is a feature common to many kinds of soils. Organic matter

Table 2-3. Climatic Criteria for Selected Soil Orders

Suborder	Great Groups	Moisture	Temperature
Alfisols:			
Aqualfs	9	Aquic	Frigid to hyperthermic
Boralfs	5	Udic	Cryic to frigid
Udalfs	9	Udic	Mesic to thermic
Ustalfs	6	Ustic	Mesic to hyperthermic
Xeralfs	6	Xeric	Mesic to thermic
Mollisols:			
Albolls	2	Udic	Frigid to thermic
Aquolls	6	Aquic	Pergelic to hyperthermic
Borolls	7	Udic	Cryic to frigid
Rendolls	1	Udic	Cryic to thermic
Udolls	4	Udic	Mesic to thermic
Ustolls	7	Ustic	Mesic to thermic
Xerolls	6	Xeric	Mesic to thermic

Source: Soil Survey Staff, 1975.

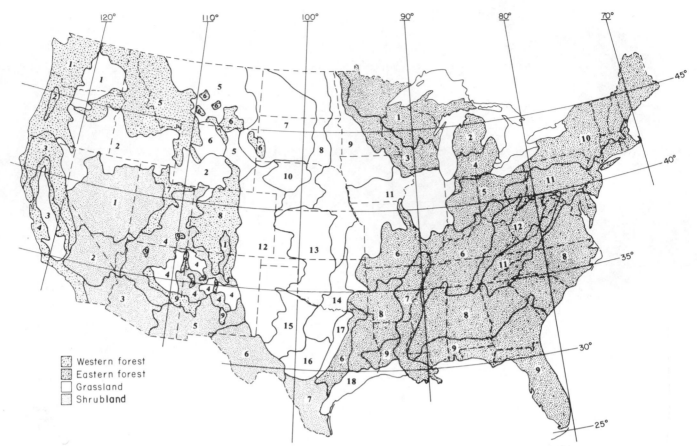

Figure 2-4. Distribution of vegetation in the United States. (Generalized from Küchler, 1970.) Latin names are added parenthetically after first listing of common names.

Western forest

1. Spruce (*Picea*), cedar (*Thuja*), hemlock (*Tsuga*), Douglas-fir (*Pseudotsuga*), fir (*Abies*), pine (*Pinus*)
2. Oak (*Quercus*), cedar, hemlock, Douglas-fir
3. Cedar, hemlock, Douglas-fir, fir, pine, redwood (*Sequoia*)
4. Oak-chapparel
5. Pine, Douglas-fir, cedar, hemlock, fir, spruce
6. Pine
7. Spruce, fir
8. Spruce, fir, pine, Douglas-fir
9. Pine, Douglas-fir

Eastern forest

1. Spruce, fir, pine
2. Pine, maple (*Acer*), birch (*Betula*), beech (*Fagus*), hemlock
3. Maple, basswood (*Tilia*), bluestem (*Andropogon*)
4. Oak, hickory (*Carya*), beech, maple
5. Beech, maple

6. Oak, hickory
7. Oak, tupelo (*Nyssa*), cypress (*Taxodium*)
8. Oak, hickory, pine
9. Beech, sweetgum (*Liquidambar*), magnolia (*Magnolia*), pine oak
10. Maple, birch, beech, spruce, fir
11. Oak
12. Maple, walnut (*Juglans*), beech, tulip (*Liriodendron*), poplar (*Populus*), oak, basswood

Grassland

1. Wheatgrass (*Agropyron*), bluegrass (*Poa*), sagebrush (*Artemisia*)
2. Sagebrush, wheatgrass
3. Needlegrass (*Stipa*)
4. Grama (*Bouteloua*), galleta (*Hilaria*)
5. Grama, needlegrass, wheatgrass
6. Wheatgrass, needlegrass, sagebrush
7. Wheatgrass, needlegrass
8. Wheatgrass, bluestem, needlegrass
9. Bluestem
10. Bluestem, sandreed (*Calamovilfa*)

11. Bluestem, oak, hickory
12. Grama, buffalo grass (*Buchloë*)
13. Bluestem, grama
14. Bluestem, oak
15. Buffalo grass, mesquite (*Prosopis*)
16. Bluestem, juniper (*Juniperus*), oak
17. Bluestem, needlegrass
18. Bluestem, cordgrass (*Spartina*)

Shrubland

1. Sagebrush, saltbush (*Altriplex*), greasewood (*Sarcobatus*)
2. Creosotebush (*Larrea*)
3. Creosote bush, bur sage (*Franseria*), Palo verde (*Cercidium*), cactus (*Opuntia*)
4. Blackbrush (*Coleogyne*), saltbush, greasewood, juniper, pinyon (*Pinus*)
5. Creosotebush, grama, tobosa (*Hilaria*)
6. Creosotebush, tarbush (*Flourensia*)
7. Mesquite, acacia (*Acadia*), millet (*Setaria*)

nearer the surface is a younger addition not yet subject to thorough decomposition and subsequent transfer to depth. In the Argialboll, the humus in the B-horizon is complexed with clay in the ped coatings; this indicates that clay and humus transfers occurred during the Holocene even though the soil was formed in late-Pleistocene loess. At this site, the loess overlies a paleosol whose Ab-horizon is dated at 16,500 ± 500 years (Ruhe, 1969: 227).

The Holocene translocation of organic carbon in Argiudolls is demonstrated in Illinois (Ballagh and Runge, 1970). The organic carbon complexed with clay (less than 0.2 μm and 2 μm fractions) has been traced downward below the textural B-horizons in three soils. In two of the soils, the radiocarbon dates of organic carbon were inverted: (1) 9330 ± 190 years at 104 to 105 cm deep and 4270 ± 95 years at 116 to 117 cm and (2) 2935 ± 105 years at 89 to 103 cm and 2500 ± 105 years at 103 to 111 cm. In the third soil, the dates were 3070 ± 105 years at 86 to 97 cm and 3090 ± 105 years at 97 to 104 cm. Even though these soils were formed in late-Pleistocene sediments (Ballagh and Runge, 1970), the organic carbon *and silicate clay* components of the pedogenic systems are early to late Holocene in age.

HISTOSOLS

Histosols are organic soils commonly called peats and mucks. Pollen contained in the profiles reflects the kinds and amounts of vegetation that grew in the region at the time that the materials accumulated in the bog (Davis, 1963). Generalized climatic inferences can be drawn from the pollen record, such as cool, warm, moist, dry, et cetera (Wright, 1972). Through sophisticated statistical techniques, interpretations are made of air-mass durations, precipitation, and even mean July temperatures (Webb and Bryson, 1972; Webb and Clark, 1977). The organic nature of the soils permits their radiocarbon dating. The focus here is on the record contained in Histosols and its use in explaining some of the more extensive associated mineral soils.

All summary studies of pollen profiles from the central grasslands (McComb and Loomis, 1944; Wells, 1970; Wright, 1968a, 1971) agree that the prairie was established during the Holocene. Boreal spruce forest existed during the latest-Pleistocene glacial episode as far south as Kansas and Missouri (Wells, 1970) and extended eastward from Nebraska into Iowa, Illinois, and Indiana (Wright, 1968a). Prairie vegetation succeeded almost immediately in Kansas, Nebraska, and South Dakota and extended eastward to Illinois by 8000 years ago (Wright, 1968a). Prairie was thoroughly established in Iowa at that time (Ruhe, 1969). It expanded about 120 km northeast of its present border in Minnesota about 7000 years ago and then receded to its present position by about 2000 years ago (Bernabo and Webb, 1977; Wright, 1968a). The expansion of the prairie is generally attributed to warmer and drier conditions during the middle Holocene, and the recession is considered to represent cooler and moister conditions at a later time.

The central prairies (Figure 2-4) constitute a response to climate. The northeastern boundary conforms generally to an arcuate front of Arctic air, and the southeastern boundary conforms generally to the arcuate front of maritime tropical air from the Gulf of Mexico. Between these air masses is a wedge of Pacific air that is dried and warmed in crossing and descending the eastern front of the Rocky Mountains (Bryson, 1966; Bryson and Wendland, 1967). Thornthwaite (1931) demonstrated the relation of the prairie to precipitation and evaporation (P.E. index). The grassland's index is 23 to 63,

and the index of the western steppe is 16 to 31. The eastern-forest region's index is 64 to 128 or more. Borchert (1950) added the residence time of dry westerly air and precipitation patterns during drought years to the explanation of the prairie. Conditions in the 1930s were used as the model. The 6-months isochrone of mean transport of dry westerly air and the isopleths of 50%, 60%, and 70% of normal July rainfall during drought years correspond closely with the distribution of prairie (Figure 2-5).

A paradox enters when drought years are introduced into the explanation of the prairie. During the 1930-to-1939 drought in western Iowa, numerous trees on suitable or even protected sites died or were severely damaged, but the oak-hickory forests were not noticeably affected (McComb and Loomis, 1944). Perennial grasses were destroyed and surface-soil conditions became favorable for invasion by tree seedlings, but the invasion did not occur. McComb and Loomis concluded that factors other than climate are involved in the persistence of prairie. One factor not considered in their study was the persistence of a particular kind of climate. The relatively "droughtlike" conditions inferred from the pollen record may have persisted for 5000 years, not just for a decade (Ruhe, 1969).

Regardless of the paradox, Histosols do contain a record of the history of vegetation, and climatic inferences can be drawn from the record. A difficult problem involves the application of the record to the development of mineral soils that are associated with the Histosols. This problem is discussed later in the chapter.

REFORESTATION AND SOILS

As noted previously, the pollen record indicates that about 7000 years ago the prairie reached its maximum expansion about 120 km northeast of its present border in Minnesota. The maximum limit falls within a forest-prairie zone today (Figure 2-4: eastern forest, 3) that is characterized by maple, basswood, and bluestem. In northern Minnesota, the reforestation included an expansion of white pine (*Pinus strobus*) about 2700 years ago that was followed by jack pine (*P. banksiana*) and red pine (*P. resinosa*) about 1000 years ago (Wright, 1968b).

These shifts of vegetation and their effects on soils during the late Holocene are well known in Iowa. There, a biosequence of soils formed in loess includes the Tama, Downs, and Fayette silt loams. The Tama soil is considered as the typical "prairie soil" (Smith et al., 1950) or today as the typical Mollisol (a Typic Argiudoll). The Fayette silt loam is a forest soil, or Alfisol (a Typic Hapludalf). The Downs is a transitional soil, or a Mollic Hapludalf. Photographs of and data on profiles of the two end members are reproduced elsewhere (Ruhe, 1969: 170-73). The biosequence is positioned topographically, with the Fayette soil lower and the Downs soil higher on the valley slopes and the Tama soil on upland summits.

In 1849, when the land was originally surveyed in northeastern Iowa, forest covered 70% of a township (93 km²) in Clayton County (Loomis and McComb, 1944). In a later soil survey, Tama soil occupied 80% of the township, Downs soil occupied 10%, and Fayette soil occupied 10%. Consequently, about 56% of the total expanse of Tama (the modal Mollisol) was under forest 130 years ago, but the occupation apparently was not long enough to seriously modify the grassland properties of the soil. Reforestation, however, continues today.

The state of Iowa is in the forest-prairie transitional zone, and the main encroachment of forest is from the southeast toward the north-

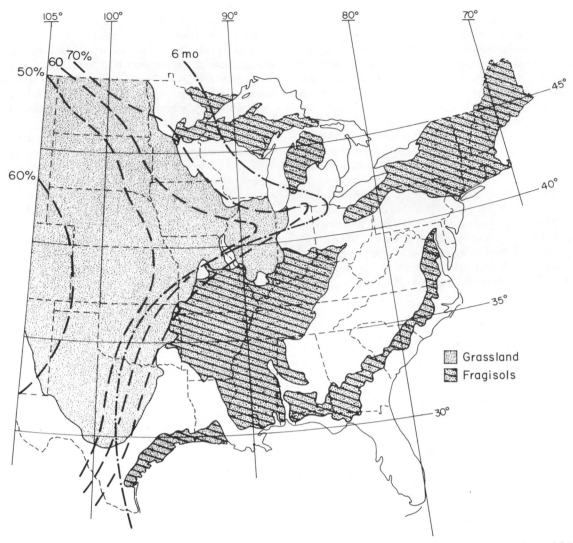

Figure 2-5. Distributions of grassland, soils with fragipans (fragisols), and climatic parameters in central and eastern United States: percentages of normal July rainfall in drought years (50% to 70%) and months per year of mean transport of westerly air (6 months). (Grassland from Figure 2-4; soils from Grossman and Carlisle, 1969; and climatic data from Borchert, 1950.)

west. The southeastern region is occupied by bluestem, oak, and hickory, but the northwestern region is under bluestem grass (Figure 2-4: grassland zones 11 and 9). The climatic balance appears to be in favor of the warmer and moister southerly air mass. Biosequences of soils are recognized in most of the soil-association areas in the state (McComb and Riecken, 1961; Oschwald et al., 1965).

Soil Geography and Soil Development

Soil development during the Holocene in the United States can be examined geographically for a broad perspective and geomorphically for a more restricted viewpoint. It is discussed in the soil-geographic context first. The geomorphic discussion begins on p. 20.

MOLLISOLS OF THE CENTRAL PRAIRIES

Throughout the central prairies, Mollisols occur on the hillslopes and summits that bound the depressions that sometimes contain Histosols. The question is: how much of the simple tripartite (cool-moist, warm-dry, cool-moist) record of the Holocene (Wright, 1976)

contained in the Histosols applies to the Mollisols? The distributions of the Mollisols (Figure 2-1) and the prairie (Figure 2-4) overlap geographically. As previously noted, the soil-organic carbon systems are compatible. This conclusion is enhanced when the broad grouping of the Mollisols is compared to regional groupings of the soils of the prairie in an older system of soil classification (U.S. Department of Agriculture, 1938). North of about latitude 37°N and from east to west within the Mollisols, soils were banded longitudinally in the order: Prairie, Chernozem, Chestnut, and Brown Soils. Annual precipitation (Figure 2-3) progressively decreases in the same direction from about 90 cm to 40 cm. A-horizons in the soil progressively decrease in thickness and in organic carbon content. In addition, textural B-horizons are weaker, soils have greater base saturation, and carbonates (ca-horizons) are at shallower depths. All of these soils seem to have developed under the prairie environment, but there is still room for doubt and study.

Along latitude 40°N these soils can be restricted to a common parent material, late-Pleistocene loess. If pollen records are accepted, the loess seems to have weathered originally under woodland and then under prairie. What properties of the soils relate to the former environment? How much eluviation and illuviation are affected by it? Additions and transfers of organic matter and recycling of bases under grass may mask former eluvial and illuvial horizons. The carbonate system also may be a mask. Although the parent materials may be calcareous leading to inference regarding depths of leaching, carbonate may be added to the soils from the atmosphere. High concentrations of calcium as aerosols are in the atmosphere above the Great Plains today (Junge and Werby, 1958). Additions of carbonate from the atmosphere are known in Mollisols of the southern Great Plains (Figure 2-1) (Brown, 1956) and in Aridisols to the west (Gile et al., 1970; Ruhe, 1967). Perhaps these questions may be resolved in a study along the 40°N parallel of latitude from Indiana to Colorado.

The Mollisols (Figure 2-1) also can be subdivided from north to south on the Great Plains. South of latitude 37°N in Kansas, Oklahoma, Texas, and New Mexico, the former great soil groups were called Reddish Prairie Soils in the east and Reddish Chestnut Soils in the west (U.S. Department of Agriculture, 1938). These soils are in part paleosols in that they emerge on the land surface from beneath Wisconsin loess in southern Kansas and may be the emergent Sangamon Soil (Thorp et al., 1951). Some of them are formed in Illinoian "cover sands" and may have weathered from Sangamon to the present. A study covering the area around a north-south traverse along the 100°W meridian in southern Kansas might be useful. When any postburial alteration is excluded, what is the nature of the red soils where they are buried under loess? Their characteristics should relate to their preburial development. What is the nature of the red soils where they are not buried? The differences between the two sets of soils should be attributable to late-Pleistocene/Holocene modification.

SOILS WITH FRAGIPANS OF THE EASTERN FORESTS

A fragipan (from Latin *fragilis*, brittle) is a soil subsurface horizon that appears to be cemented when dry and has a hard to very hard consistency. When moist, the fragipan has a brittleness that is exhibited by peds that tend to shatter when pressure is applied rather than to deform slowly. Dry peds slake when placed in water (Soil Survey Staff, 1975: 42-45). The origin of fragipans is obscure and controversial. Their nature and problems associated with their nature have been reviewed extensively (Grossman and Carlisle, 1969). Liberal use is made here of Grossman and Carlisle's review, but additional commentary is also included.

The fragipan horizon, designated by the suffix X, may occur in A2-, B-, or C-horizons, so there are many possible horizon sequences. In one case, a sequence in a solum may be A2, B, BX. In some soils, the profile may be bisequal, as A2, B, A′2, B′X.

Soils with fragipans, informally called fragisols (Ruhe et al., 1975), include Spodosols, Inceptisols, Alfisols, and Ultisols, and they occur mainly in the eastern states (Figure 2-1). They are unknown in Mollisols in the central states and Great Plains and occur infrequently in the western states. Their geographic distribution indicates a genetic association with forest vegetation (Figure 2-5). The bounding of the Prairie Peninsula by fragisols is particularly pertinent. Holocene boundaries of climate and vegetation regionally separate soils with or without fragipans (Figure 2-5). This relation suggests that prairie climate and vegetation were not conducive to the formation of fragipans but that forest climate and vegetation were (Ruhe et al., 1975).

Soil-stratigraphic relations have a bearing on the argument. The consistent depth of fragipans indicates that they were controlled by something related to the land surface (Grossman and Carlisle, 1969). Across southern Indiana and Ohio (Figure 2-5), fragipans occur in Alfisols formed in thin Wisconsin loess that generally overlies a textural Bb-horizon of a paleosol formed in glacial till, material derived from bedrock, or weathered bedrock. The fragipan consistently occurs near the lithologic contact. The Bb-horizon of the paleosol is a hydrologic impediment in the vertical profile, and soil water perches above it during wetter times of the year. Seep lines are common on hillslope shoulders at the paleosol contact.

In bisequal soils, the lower eluvial A′2-horizon always occurs between a textural B-horizon of the upper part of the sequence and a textural B-horizon, B′X (fragipan), of the lower part of the sequence. The textural B-horizons are degraded by the intervening eluvial A′2-horizon, which must be younger. The upper solum formed in Wisconsin loess, whose deposition ceased 14,000 to 12,000 years ago. The upper solum is younger, and the A′2-horizon must be still younger, or Holocene in age. In any event, the fragipan confuses the morphology and contacts of sediments and the horizons of ground and buried soils and causes "soil welding" (Ruhe and Olson, 1980).

In profiles of Indiana soils, the maximum of free silica coincides with the fragipan horizon; this suggests that silica is a bonding agent between grains that imparts brittleness to the material (Harlan et al., 1977; Norton and Franzmeier, 1978). The association of the fragipan with a paleosol contact, the wetness of the zone caused by the hydrologic impediment of the paleosol, and the high silica content all suggest a mechanism for fragipan formation. Silica released during weathering accumulates in the wet zone within reach of tree roots. With seasonal drying and extraction of moisture by roots, silica concentrates and precipitates, binding soil particles and forming the fragipan (Harlan et al., 1977). This explanation seems reasonable, and the process must be operating today. Here, then, is an example of more severe weathering and soil development during the Holocene.

Holocene weathering also may be a dominant factor in the formation of fragipans of the northern states (Figure 2-5), where the fragipans in Alfisols, Inceptisols, and Spodosols are formed in Late Wisconsin glacial sediment and so must be younger. The same reasoning can be applied to fragisols in Late Wisconsin loess in the south-central states along the 90°W meridian (Figure 2-5). A different view is given, however, in Tennessee (Buntley et al., 1977), where the conclusion was reached that fragipans were formed during two episodes of weathering. An earlier weathering occurred between two thin loesses of Wisconsin age, and a later weathering was superimposed on the earlier one following deposition of the upper thin loess. This explanation was rejected for formation of fragipans in similar loess systems in Indiana (Norton and Franzmeier, 1978).

Like the development of every other soil, the development of fragipans cannot be ascribed to any restricted period of time. In the broad, arcuate belt of fragisols in the southeastern states from Mississippi to Maryland (Figure 2-5), the soils are Ultisols (Figure 2-1). In North Carolina, the fragisols occur on the Coastal Plain terraces (Daniels et al., 1966), and in Wilson County about 95% of the Sunderland Terrace has fragisols. That terrace is Pliocene or early Pleistocene in age (Daniels et al., 1978).

Such an ancient age for the fragipans may not be real. The Hofmann Pocosin, an upland bog in Jones and Onslow Counties, North Carolina, covers about 25,000 ha on the Wicomico and Talbot Terraces, which are Pleistocene in age (Daniels et al., 1977). Soils buried beneath the bog have little or no horizonation and are poorly developed, but where trees are sparse they have brittle horizons that are similar to the fragipan horizons in the ground soils of the region. There is a radiocarbon date of 7930 years from the base of the bog sediments (Daniels et al., 1977). The Histosols certainly are Holocene in age, and the buried, poorly developed soil with brittle horizons, in part, may also be Holocene in age.

Another line of evidence indicates younger fragipans. Eolian sands also bury the Sunderland, Wicomico, and Talbot surfaces and their soils (Daniels et al., 1969). Where there is a radiocarbon date of 29,300 years from the base of the eolian sand, the buried soil is weakly developed. Where there is a radiocarbon date of 10,790 years, the buried soil is strongly horizonated and is comparable to the ground soils on the Sunderland surface. Although it is suspected that the radiocarbon samples were contaminated, the younger date is considered reasonable (Daniels et al., 1969). A Holocene age is approached again.

ENTISOLS AND INCEPTISOLS

Entisols are soils that show little or no evidence of having developed pedogenic horizons (Soil Survey Staff, 1975). With one rare exception, these soils are Holocene in age. The exception occurs where parent materials consist of mineral material, such as quartz sand, that does not weather readily to form soil horizons. On a broad geographic basis, Entisols are more abundant in the West than elsewhere in the United States (Figure 2-1). They relate mainly to alluvial and eolian deposits, where intermittent sedimentation obscures weathering, or they relate to sloping topography, where erosion is active.

Inceptisols have an unavoidably complicated definition (Soil Survey Staff, 1975) but can be considered as embryonic soils with few diagnostic features (Buol et al., 1973). They are a bit farther along the genetic pathway than Entisols in that bases, iron, and aluminum may have been removed, transferred, and accumulated. They do not have illuvial horizons of silicate clay (Soil Survey Staff, 1975). They are generally restricted to steep slopes or depressions, or they occur on young geomorphic surfaces with a corresponding limitation of soil development (Buol et al., 1973). On a broad geographic basis, they occur in humid, mountainous regions and in alluvial valleys (Figure 2-1).

In general, Entisols and Inceptisols are Holocene soils, and from the genetic viewpoint they are interesting as the initial and early stages of soil formation. Viewing soils from the soil-geographic approach only leads to broad generalizations about soils development. Greater insight can be gained from the soils-geomorphic approach.

Soil Geomorphology and Soil Development

The simple subdivision of the Holocene on the basis of pollen data and inferred climatic regimes comes into serious question when the results of the work in soil geomorphology are considered. Numerous studies have been made in Iowa during the past three decades (they are listed in Ruhe, 1974). A common thread among all of the studies has been an analysis of the geomorphic evolution of hillslopes as related to valley or wetland sedimentation, radiocarbon dating of the system, and formation of soils, whether on the surface or buried.

HILLSLOPES AND SOIL FORMATION

Erosion on hillslopes produces sediment that accumulates lower on the slope and on adjacent valley bottoms or closed depressions. The record may be only partial where sediment is transported downvalley, or it may be complete in depressions when eolian blowout or subsurface loss (as in karst topographies) can be precluded. Organic matter contained in the sediments can be radiocarbon dated to provide a time framework not only for the sediments but also for the hillslopes from which the sediments were derived. Soils, in turn, can be fitted to the entire landscape and then evaluated within the soil-geomorphic system.

In the Thompson Creek watershed in southwestern Iowa (Figure 2-6, A), alluvial fills occur in a valley (Daniels et al., 1963). The lowest bed contains spruce logs dated at 14,300 years near the base and 11,120 years at the top. The top of the fourth bed has a red elm (*Ulmus rubra*) log dated at 2020 years. The third bed terminates with a willow log dated at 1800 years. The top of the second bed has a box-elder (*Acer negundo*) stump rooted in place and dated at 250 years. The uppermost bed is postsettlement alluvium. The sedimentation record shows three Holocene presettlement episodes, with the oldest extending from 11,120 to 2020 years ago.

In the Treynor watershed to the south (Figure 2-6, C), the oldest episode was interrupted by accumulation of mixed organic and mineral sediment 8740 years ago. In a gully fill between the two watersheds (Figure 2.6, B), the base is 6800 years old. Five episodes of hillslope erosion and sedimentation are recorded at these three sites in southwestern Iowa.

In the Four-Mile Creek watershed in east-central Iowa (Figure 2-6, J), four alluvial fills occur in tributary valleys; the two older fills are 7710 and 6200 years old (Vreeken, 1968). In a nearby terrace deposit along Wolf Creek (Figure 2-6, K), an American elm (*Ulmus americana*) log buried at a depth of 2.7m is 2080 years old (Ruhe, 1974). To the northeast (Figure 2-6, L), hillslopes that truncate a high-level peat are less than 2930 years old (Kleiss, 1970). Within the peat and above and below a thin silt bed, the dates are 5520 and 9270 years (Van Zant and Hallberg, 1976). At a depth of 3 m in alluvium along the Chariton River in south-central Iowa (Figure 2-6, I), a red elm log was dated as 1830 years old.

Near Cherokee, in northwestern Iowa (Figure 2-6, D), the hillslope record is more complex. An alluvial fan descends from the valley wall to the floodplain of the Little Sioux River (Hallberg et al., 1974). The base of the fan is 10,030 years old. A soil (soil 1) and five buried soils (soils 2 through 6) are on and in the fan. Soils 4, 5, and 6 have been radiocarbon dated at 6100, 7400, and 8500 years, respectively. Soils 1, 2, and 3 are above soil 4 and must be less than 6100 years old and are more strongly developed than the other soils. The somewhat confused report that mixes sediments, soils, and archaeological remains describes three more soils (7, 8, and 9) below soil 6 but still above the base of the fan. In any event, numerous hillslope erosion-sedimentation and associated soil-formation episodes are recorded, and all of them occurred in the Holocene from 10,000 to 6100 years ago or less.

The numerous episodes of erosion have resulted in the formation of a widespread erosion surface that covers 18,500 km^2 in northeastern Iowa (Figure 2-6). Wisconsin loess has been almost completely stripped from that region. It occurs only as thick caps on isolated hills, or paha (Scholtes, 1955). The paha are loess-mantled erosion remnants that stand above the erosion surface at interstream divides (Ruhe et al., 1968). The erosion surface is generally marked by thin loam sediments above a stone line on glacial till, and the surface ranges in age from 12,700 years (Figure 2-6, M) to less than 2900 years (Figure 2-6, L).

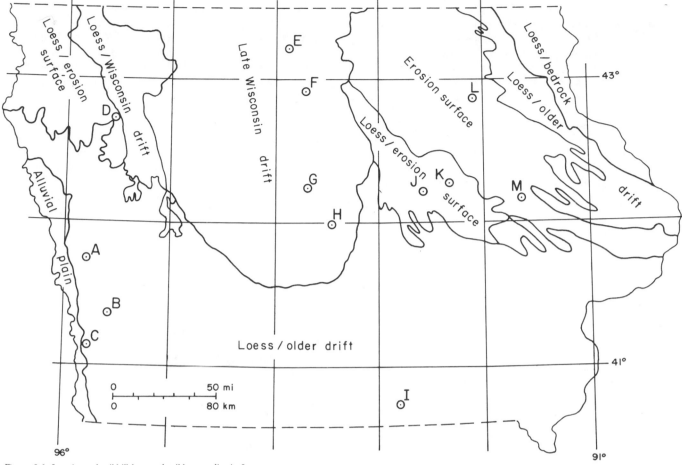

Figure 2-6. Locations of soil-hillslope and soil-bog studies in Iowa.

The episodes of hillslope erosion and sedimentation and the cutting of the widespread erosion surface in Iowa indicate that the Holocene was and is a time of numerous episodes of instability in the landscape. Intermittent soil formation like that at Cherokee also indicates numerous episodes of stability in the landscape. The results of the detailed soil-geomorphic studies do not agree with the simple tripartite history of gradual changes shown by the pollen records of the region.

This paradox can be examined with examples from Iowa. Sediments in depressions on the Late Wisconsin drift (Figure 2-6, E through H) have a double stratigraphic sequence of muck, silt, muck, and silt (Walker, 1966). The relations of the sediments, vegetation generalized from pollen, hillslope erosion, and radiocarbon chronology are generalized in Table 2-4. Two paired episodes of hillslope instability and stability are recorded. The silts indicate that mineral sediment dominated organic sediment from 13,000 to 10,500 and from 8000 to 3000 years ago. The first episode occurred under conifer cover, and the second episode occurred under prairie cover. Organic sediments dominated mineral sediments from 10,500 to 8000 years ago and from 3000 years ago to the present. Hillslopes were more stable, and rates of erosion on hillslopes were 5 to 70 times less than they were during times when mineral sediments were accumulating. The presence of pollen in the lower peat and muck indicates an upward transition from conifers to hardwoods, and pollen in the upper silt shows the dominance of prairie. Pollen in the upper muck shows the invasion of prairie by trees. There is little evidence in these

sediments of the numerous episodes of hillslope erosion and erosion-surface evolution that were recorded in other parts of Iowa during the same span of Holocene time.

Soils on the various Holocene hillslopes in Iowa range from those with little evidence of pedogenic horizons to those with strong horizonation. The common soil great groups (Soil Survey Staff, 1975) are Udorthents, Hapludolls, Haploquolls, Hapludalfs, and Argiudolls. The less-developed soils show only an accumulation of organic car-

Table 2-4. Holocene History of Double Bogs in Iowa

Zone	Environment	Sedimentation Rate Bog Center (cm/100 years)	Erosion Rate Hillslope (cm/100 years)	Date (years ago)
Peat/muck	Oak invading prairie	2.9	0.2	Present
				3000
Silt	Prairie	4.8	1.0	
				8000
Peat/muck	Hardwood forest, conifers	2.7	0.02	
				10,500
Silt	Conifers	3.3	1.4	
				13,000
Glacial till	—	—	—	—

Source: Modified from Walker, 1966.

bon and a loss of carbonates. The moderately developed soils are characterized by accumulations of organic carbon; clayey B-horizons with structural peds; and losses of carbonates, bases, and sesquioxides. In addition, in the more strongly developed soils, eluvial and illuvial horizons are present in the soil profiles. Along a hillslope transect from summit to shoulder, back-slope, foot-slope, and toe-slope soil development can range from weak to strong. A good example of the fit of soils to hillslope is described by Vreeken (1973). Soils with little development and those with illuvial B-horizons can be on the same hillslope component that is less than 2000 years old (Ruhe, 1969).

PEDIMENTS, ALLUVIAL FANS, AND SOIL FORMATION

During the Holocene, unstable landscapes have also characterized the arid and semiarid regions, and an example of the soil geomorphology of such an area is the classic study performed by the Desert Project in southern New Mexico (Gile and Grossman, 1979). The study area extends along the Rio Grande Valley and includes 1062 km². Mountains with Holocene slopes occupy 13% of the area. Holocene alluvial plains, alluvial fans, basin fills, pediments, and terraces cover 40% of the area, and the remaining 47% is occupied by multileveled alluvial fans and pediments that date back to the middle Pleistocene (Ruhe, 1967).

The Holocene sediments and surfaces (Figure 2-7) include the Rio Grande alluvial plain, which covers 132 km² and which includes historical abandoned channels across its extent (Ruhe, 1967). Adjacent to the alluvial plain are the multileveled valley-border terraces, alluvial fans, and pediments that are inset below Pleistocene sediments and surfaces. The cuts and fills are episodic. Charcoal from buried hearths has been radiocarbon dated at 2620, 2850, 3960, 4910, and 7340 years (Gile, 1975; Ruhe, 1967).

Analogues to the valley-border sediments and surfaces occur along the mountain fronts, where the heads of alluvial fans are inset below older Pleistocene fans. Downslope, the younger alluvium buries older fans and the soils on them. Charcoal horizons in the younger sediments have been dated at 1130, 2120, 2220, 4035, 4200, 4570, 4640, 4700, 4960, and 6400 years.

The valley-border group occupies 115 km², and the mountain-front group occupies 138 km². In addition, a closed basin of 11 km² has been filled, and low-gradient pediments occupying 24 km² have extended headward from the basin into the Holocene alluvial fans (Figure 2-7).

The magnitude of Holocene events is striking in that 57% of the area has been affected. More pertinent is the fact that the geomorphic surfaces and associated sediments can be readily traced and correlated along the Rio Grande trench from El Paso, Texas, to Albuquerque, New Mexico (Hawley, 1965; Kottlowski, 1958; Ruhe, 1967), a distance of about 435 km. Complexity within the Holocene is regional.

In the Las Cruces area (Figure 2-7), soil development on Holocene surfaces and deposits ranges from very weak (Entisols) to reasonably strong (Aridisols: Haplargids). The developmental sequence is based on the soil's age and the nature of the parent material, whether it is high (more than 15%) or low (less than 2% in carbonate content and high (more than 50%) or low (less than 20%) in gravel content (Gile, 1975). Soils less than 100 years old have only thin, vesicular A-horizons with a slight accumulation of organic carbon. Soils 100 to 1100 years old have a slight accumulation of carbonate as filaments in

the soil matrix or as coatings on pebbles. Soils 1100 to 2100 years old have carbonate filaments and pebble coatings in high-carbonate, low-gravel material. In high-gravel material, the profile has A', thin Bca, and Cca-horizons. In low-carbonate, low-gravel material, soils have noncalcareous brown or reddish brown B-horizons and weak carbonate filaments and coatings in their C-horizons. In high-gravel material, reddish brown B-horizons overlie Cca-horizons, and weak clay skins are present in the textural B-horizon, indicating a weak stage of illuviation. Soils 2200 to 4600 years old have B-horizons with prismatic and subangular blocky structures in high-carbonate, low-gravel material. In high-gravel material, the profile have A-, Bca-, and Cca-horizons. In low-carbonate material, regardless of its gravel content, brown or reddish brown textural B-horizons have stronger clay skins and overlie Cca-horizons, indicating a more pronounced illuviation. Soils 7000 years old have pronounced textural B-horizons with illuvial clay. They are noncalcareous and reddish brown, and they overlie pronounced Cca-horizons. These soils are Haplargids. This range of soil development through 7000 years is in an area that today is arid, where the annual rainfall ranges from 20 cm in the Rio Grande Valley to 40 cm in the mountains.

A sequence of Holocene soil development under greater rainfall is well expressed in the Willamette Valley in Oregon (Balster and Parsons, 1968; Parsons and Herriman, 1970; Parsons et al., 1970). Mollisols occur in the valley, and Ultisols are on the bounding Coast Range to the west and the foothills of the Cascades Mountains to the east (Figure 2-1). A channel belt along the Willamette River is as much as 3.2 km wide, and two higher terraces occupy valleys as wide as 16 km. These surfaces include an area of 7800 km². Current annual precipitation in the valley is about 100 cm.

The sediments of the higher terrace apparently are a late-Quaternary valley fill that dates from 34,410 years near the base of the fill to 10,850 and 5250 years beneath the higher terrace. The low terrace is 3290 to 555 years old (Balster and Parsons, 1968).

Soils on the lower terrace have as much as 4% organic matter in the A-horizon and have color and structural B-horizons. Clay (less than 2 μm) ratios of the B- to A-horizons are 1:1, so textural B-horizons have not formed (Parsons and Herriman, 1970; Parsons et al., 1970). There have been only slight losses of bases in the sola.

Soils on the higher terraces have textural B-horizons. Clay ratios of the B- to A-horizons are 1.3:1 and 1.4:1, and illuviation is indicated by clay skins in the B-horizons. The loss of bases through leaching is great, and base saturation in the upper parts of the soils can be 15% to 30% less than in the lower parts.

A comparable development characterizes soils on pediments in the adjacent Coast Range (Balster and Parsons, 1966). Under current precipitation of 150 to 180 cm, the soils have strong textural and structural B-horizons and are the most strongly horizonated soils on the landscape. The pediments are less than 9570 years old according to radiocarbon dating.

Summary and Conclusions

Holocene soils in the United States range in development from very weak to strong. Within this range, soils in 9 of the 10 soils orders (Figure 2-1) have formed wholly or partially during the Holocene. These 9 soil orders are the Inceptisols, Entisols, Histosols, Alfisols, Spodosols, Mollisols, Aridisols, Vertisols, and Ultisols. Examples have been given throughout the chapter for all these orders except

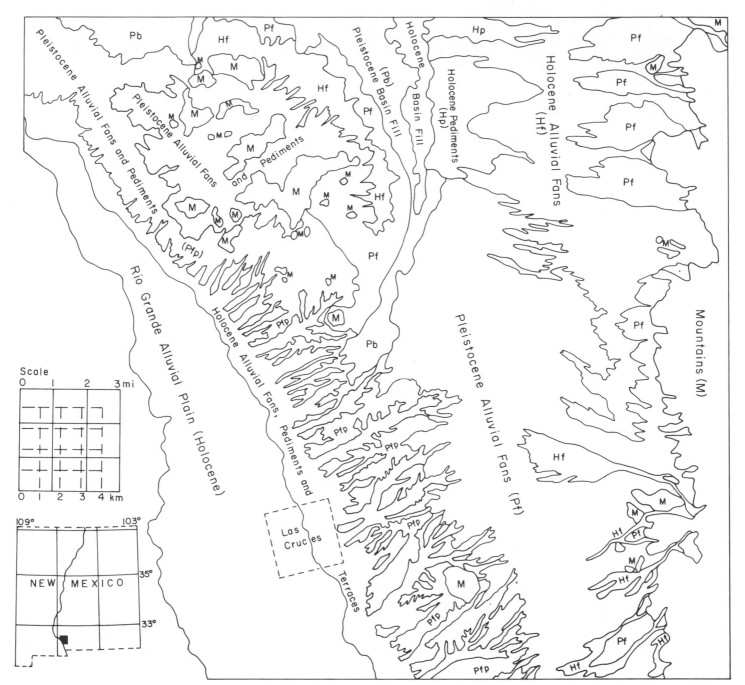

Figure 2·7. Holocene and Pleistocene sediments and surfaces in Las Cruces area, New Mexico. (Generalized from Ruhe, 1967.)

the Vertisols (self-mulching soils), which generally have expandable clay minerals so that upon wetting and drying they heave and turn over and produce a gilgai topography.

Overall, the first three orders are relatively weakly developed soils. The Histosols are accumulations of organic matter with little or no mineral alteration. The Entisols show little or no evidence or pedogenic horizons but may have small accumulations of organic carbon and slight losses of carbonates where parent materials are calcareous. The Inceptisols are a bit farther along the genetic path, with an accumulation of organic carbon and a loss of bases, carbonates, iron, and aluminum. They do not have illuvial horizons of

silicate clay or textural B-horizons, but they may have horizons of silica, iron, or base accumulation. The accumulation of silica can result in the development of a fragipan.

The Alfisols and Spodosols, occurring under trees, are still farther along the genetic path. Where soil-geomorphic control demonstrates Holocene age, these soils have eluvial horizons from which sesquioxides have been removed and then have accumulated in a subjacent illuvial horizon. The Alfisols, in addition, have an accumulation of silicate clay as clay skins in the illuvial horizon. Fragipans are common in these soil orders and are associated with the eluvial-illuvial horizons. In some cases, bisequal soils form with a sequence of A1-,

A2-, Bt-, A′2-, B′X-horizons. This association indicates a more advanced weathering than occurs in the simple accumulation of silica (as in fragipans in Inceptisols). The loss of carbonates and bases also is more severe in Alfisols than in Inceptisols.

The Mollisols, occurring under grass, also may be farther along the genetic path. They are dark-colored, base-rich soils, but their status results from the grass's recycling. They sometimes have Bt-horizons with clay skins and horizons of salt or carbonate accumulation. Where soil-geomorphic control demonstrates Holocene age, the combination of properties and horizonation of the soils can range from weak to strong.

The Aridisols (dry-land soils) on demonstrable Holocene sediments and surfaces also display a range of development that often is related to their age. They can have eluvial horizons, textural B-horizons with clay skins, and reasonably strong carbonate horizons. The carbonate horizon, however, is not as well developed as the K-horizon, which has been suggested as a master soil horizon of carbonate accumulation (Gile et al., 1965).

If the climatic parameters of temperature and moisture are acceptable as evidence of the Holocene's impact on soils, then that impact affected the Ultisols (Figure 2-1) as well as the other soil orders. Excluding this historical effect, little is known about Holocene weathering and the Ultisols. The landscape of these soils emerges from beneath Wisconsin loess in Mississippi (Simonson, 1954) and from beneath Wisconsin and Illinoian (Loveland) loesses in Tennessee (Thorp et al., 1951; Wascher et al., 1948). The Ultisols occur on Pleistocene and older surfaces of the Coastal Plain in North Carolina (Daniels et al., 1978) and are mainly relict soils of pre-Holocene age.

References

Ballagh, T. M., and Runge, E. C. A. (1970). Clay-rich horizons over limestone—Illuvial or residual? *Soil Science Society of America Proceedings* 34, 534-36.

Balster, C. A., and Parsons, R. B. (1966). "A Soil-Geomorphic Study in the Oregon Coast Range." Oregon Agricultural Experiment Station Technical Bulletin 89.

Balster, C. A., and Parsons, R. B. (1968). "Geomorphology and Soils, Willamette Valley, Oregon." Oregon Agricultural Experiment Station Special Report 265.

Bernabo, J. C., and Webb, T. (1977). Changing patterns in the Holocene pollen record of northeastern North America: A mapped summary. *Quaternary Research* 8, 94-96.

Borchert, J. R. (1950). The climate of the central North American grassland. *Annals of the Association of American Geographers* 40, 1-39.

Brown, C. N. (1956). The origin of caliche on the northeastern Llano Estacado, Texas. *Journal of Geology* 64, 1-15.

Bryson, R. A. (1966). Air masses, streamlines, and the boreal forest. *Geographical Bulletin* 8, 228-69.

Bryson, R. A., and Wendland, W. M. (1967). Tentative climatic patterns for some late glacial and postglacial episodes in central North America. *In* "Life, Land, and Water" Studies of Glacial Lake Agassiz. University of Manitoba Press, Winnipeg.

Buntley, G. J., Daniels, R. B., Gamble, E. E., and Brown, W. T. (1977). Fragipan horizons in soils of the Memphis-Loring-Grenada sequence in West Tennessee. *Soil Science Society of America Journal* 41, 400-407.

Buol, S. W., Hole, F. D., and McCracken, R. J. (1973). "Soil Genesis and Classification." Iowa State University Press, Ames.

Campbell, C. A., Paul, E. A., Rennie, D. A., and McCullem, K. J. (1967a). Applicability of the carbon-dating method of analysis to soil humus studies. *Soil Science* 104, 217-24.

Campbell, C. A., Paul, E. A., Rennie, D. A., and McCullem, K. J. (1967b). Factors affecting the accuracy of the carbon-dating method in soil humus studies. *Soil Science* 104, 81-85.

Cline, M. G. (1955). "Soil Survey of the Territory of Hawaii. Islands of Hawaii, Kauai, Lanai, Maui, Molokai, and Oahu." U.S. Department of Agriculture, Soil Conservation Service, Series 1939, 25.

Daniels, R. B., Gamble, E. E., and Buol, S. W. (1969). Eolian sands associated with Coastal Plain river valleys: Some problems in their age and source. *Southeastern Geology* 11, 97-110.

Daniels, R. B., Gamble, E. E., and Wheeler, W. H. (1978). Age of soil landscapes in the Coastal Plain of North Carolina. *Soil Science Society of America Journal* 42, 98-105.

Daniels, R. B., Gamble, E. E., Wheeler, W. H., and Holzhey, C. S. (1977). The stratigraphy and geomorphology of the Hofmann Forest Pocosin. *Soil Science Society of America Journal* 41, 1175-80.

Daniels, R. B., Nettleton, W. D., McCracken, R. J., and Gamble, E. E. (1966). Morphology of soils with fragipans in parts of Wilson County, North Carolina. *Soil Science Society of America Proceedings* 30, 376-80.

Daniels, R. B., Rubin, M., and Simonson, G. H. (1963). Alluvial chronology of the Thompson Creek watershed, Harrison County, Iowa. *American Journal of Science* 261, 473-87.

Davis, M. B. (1963). On the theory of pollen analysis. *American Journal of Science* 261, 897-912.

Gile, L. H. (1975). Holocene soils and soil-geomorphic relations in an arid region of southern New Mexico. *Quaternary Research* 5, 321-60.

Gile, L. H., and Grossman, R. B. (1979). "The Desert Project Soil Monograph." U.S. Department of Agriculture, Soil Conservation Service, Washington, D. C.

Gile, L. H., Hawley, J. W., and Grossman, R. B. (1970). "Distribution and Genesis of Soils and Geomorphic Surfaces in a Desert Region of Southern New Mexico." Soil Science Society of America Guidebook, Soil-Geomorphology Field Conference, August 21-22 and 29-30, 1970.

Gile, L. H., Peterson, F. F., and Grossman, R. B. (1965). The K horizon: A master soil horizon of carbonate accumulation. *Soil Science* 99, 74-82.

Grossman, R. B., and Carlisle, F. J. (1969). Fragipan soils of the eastern United States. *In* "Advances in Agronomy," vol. 21, pp. 237-79. Academic Press, New York.

Hallberg, G. R., Hoyer, B. E., and Miller, G. A. (1974). The geology and paleopedology of the Cherokee sewer site. *Journal of the Iowa Archaeological Society* 21, 17-49.

Harlan, P. W., Franzmeier, D. P., and Roth, C. B. (1977). Soil formation on loess in southwestern Indiana: II. Distribution of clay, free oxides, and fragipan formation. *Soil Science Society of America Journal* 41, 17-49.

Hawley, J. W. (1965). Geomorphic surfaces along the Rio Grande Valley from El Paso, Texas, to Caballo Reservoir, New Mexico. *In* "Guidebook of Southwestern New Mexico II" (J. P. Fitzsimmons and C. Lochman-Balk, eds.), pp. 188-98. New Mexico Geological Society, 16th Field Conference, October 15-17, 1965.

Junge, C. E., and Werby, R. T. (1958). The concentration of chloride, sodium, potassium, calcium, and sulfate in rain water over the United States. *Journal of Meteorology* 15, 417-25.

Kleiss, H. J. (1970). Hillslope sedimentation and soil formation in northeastern Iowa. *Soil Science Society of America Proceedings* 34, 287-90.

Kottlowski, F. E. (1958). Geologic history of the Rio Grande near El Paso. *In* "West Texas Geological Society Guidebook, 1958 Field Trip, Franklin and Hueco Mountains, Texas." pp. 46-54.

Küchler, A. W. (1970). Potential natural vegetation. *In* "National Atlas of the United States of America," sheets 89-91. U.S. Geological Survey, Washington, D.C.

Loomis, W. E., and McComb, A. L. (1944). Recent advances of forest in Iowa. *Iowa Academy of Science Proceedings* 51, 217-24.

McComb, A. L., and Loomis, W. E. (1944). Subclimax prairie. *Torrey Botanical Club Bulletin* 71, 46-76.

McComb, A. L., and Riecken, F. F. (1961). Effect of vegetation on soils in the forest-prairie region. *In* "Recent Advances in Botany," pp. 1627-31. University of Toronto Press, Toronto.

Norton, L. D., and Franzmeier, D. P. (1978). Toposequences of loess-derived soils in southwestern Indiana. *Soil Science Society of America Journal* 42, 622-27.

Oschwald, W. B., Riecken, F. F., Dideriksen, R. I., Scholtes, W. H., and Schaller, F. W. (1965). "Principal Soils of Iowa." Iowa State University Cooperative Extension Service Special Report 42.

Parsons, R. B., and Herriman, R. C. (1970). Haploxerolls and Argixerolls developed in recent alluvium, southern Willamette Valley, Oregon. *Soil Science* 109, 299-309.

Parsons, R. B., Balster, C. A., and Ness, A. O. (1970). Soil development and geomorphic surfaces, Willamette Valley, Oregon. *Soil Science Society of America Proceedings* 34, 485-91.

Ruhe, R. V. (1967). "Geomorphic Surfaces and Surficial Deposits in Southern New Mexico." New Mexico Bureau of Mines and Mineral Resources Memoir 18.

Ruhe, R. V. (1969). "Quaternary Landscapes in Iowa." Iowa State University Press, Ames.

Ruhe, R. V. (1974). Holocene environments and soil geomorphology in midwestern United States. *Quaternary Research* 4, 487-95.

Ruhe, R. V., Brunson, K. L., and Hall, L. E. (1975). Fragisols and Holocene environments in midwestern U.S.A. *Anais da Academia Brasiliera de Ciências* 47 (Suplemento), 119-26.

Ruhe, R. V., Dietz, W. P., Fenton, T. E., and Hall, G. F. (1968). "Iowan Drift Problem, Northeastern Iowa." Iowa Geological Survey Report of Investigations 7.

Ruhe, R. V., and Olson, C. G. (1980). Soil Welding. *Soil Science* 130, 132-39.

Runge, E. C. A. (1973). Soil development sequences and energy models. *Soils Science* 115, 183-93.

Scharpenseel, H. W. (1977). The search for biologically inert and lithogenic carbon in recent soil-organic matter. *In* "Soil Organic Matter Studies," vol. 2, pp. 193-200. International Atomic Energy Agency, Vienna.

Scharpenseel, H. W., and Schiffman, H. (1977a). Radiocarbon dating of soils: A review. *Zeitschrift Pflanzanernaehr Bodenkunde* 140, 159-74.

Scharpenseel, H. W., and Schiffman, H. (1977b). Soil radiocarbon analysis and soil dating. *Geophysical Surveys* 3, 143-56.

Scholtes, W. H. (1955). Properties and classification of the paha loess-derived soils in northeastern Iowa. *Iowa State College Journal of Science* 30, 163-209.

Simonson, R. W. (1955). Identification and interpretation of buried soils. *American Journal of Science* 252, 705-32.

Simonson, R. W. (1959). Outline of a generalized theory of soil genesis. *Soil Science Society of America Proceedings* 23, 152-56.

Smith, G. D., Allaway, W. H., and Riecken, F. F. (1950). Prairie soils of the upper Mississippi Valley. *In* "Advances in Agronomy," vol. 2, pp. 157-205. Academic Press, New York.

Soil Conservation Service (1970). Distribution of principal kinds of soils: Orders, suborders and great groups. *In* "National Atlas of the United States of America," sheets 85-88. U.S. Geological Survey, Washington, D.C.

Soil Survey Staff (1975). "Soil Taxonomy: A Basic System of Soil Classification for Making and Interpreting Soil Surveys." U.S. Department of Agriculture Handbook 436, Washington, D.C.

Tedrow, J. C. F. (1977). "Soils of the Polar Landscapes." Rutgers University Press, New Brunswick, N.J.

Thornthwaite, C. W. (1931). The climates of North America according to a new classification. *Geographical Review* 21, 633-55.

Thorp, J., Johnson, W. M., and Reed, E. C. (1951). Some post-Pliocene buried soils of central United States. *Journal of Soil Science* 2, 1-19.

United States Department of Agriculture (1938). Soil Associations of the United States. *In* "Soils and Men." U.S. Department of Agriculture Yearbook of Agriculture, Washington, D.C.

United States Department of Commerce (1968). "Climatic Atlas of the United States." U.S. Department of Commerce, Environmental Data Base, Washington, D.C.

Van Zant, K. L., and Hallberg, G. R. (1976). "A Late-Glacial Pollen Sequence from Northeastern Iowa: Sumner Bog Revisited." Iowa Geological Survey Technical Information Series 3.

Vreeken, W. J. (1968). Stratigraphy, sedimentology, and moisture contents in a small loess watershed in Tama County, Iowa. *Iowa Academy of Science Proceedings* 75, 225-33.

Vreeken, W. J. (1973). Soil variability in small loess watersheds: Clay and organic carbon content. *Catena*, 181-96.

Walker, P. H. (1966). "Postglacial Environments in Relation to Landscape and Soils on the Cary Drift, Iowa." Iowa Agricultural Experiment Station Research Bulletin 549, pp. 838-75.

Walker, P. H., Hall, G. F., and Protz, R. (1968a). Relation between landform parameters and soil properties. *Soil Science Society of America Proceedings* 32, 101-4.

Walker, P. H., Hall, G. F., and Protz, R. (1968b). Soil trends and variability across selected landscapes in Iowa. *Soil Science Society of America Proceedings* 32, 97-101.

Wascher, H. L., Humbert, R. P., and Cady, J. G. (1948). Loess in the southern Mississippi Valley: Identification and distribution of the loess sheets. *Soil Science Society of America Proceedings* 12, 389-99.

Webb, T., and Bryson, R. A. (1972). Late- and postglacial climatic change in the northern Midwest, U.S.A.: Quantitative estimates derived from fossil pollen spectra by multivariate statistical analysis. *Quaternary Research* 2, 70-115.

Webb, T., and Clark, D. R. (1977). Calibrating micropaleontological data in climatic terms: A critical review. *Annals of the New York Academy of Science* 288, 93-118.

Wells, P. V. (1970). Postglacial vegetational history of the Great Plains. *Science* 167, 1574-82.

Wright, H. E. (1968a). History of the Prairie Peninsula. *In* "The Quaternary of Illinois" (R. E. Bergstrom, ed.), pp. 78-88. University of Illinois College of Agriculture Special Publication 14.

Wright, H. E. (1968b). The roles of pine and spruce in the forest history of Minnesota and adjacent areas. *Ecology* 49, 937-55.

Wright, H. E. (1971). Late Quaternary vegetational history of North America. *In* "The Late Cenozoic Glacial Ages" (K. K. Turekian, ed.), pp. 425-64. Yale University Press, New Haven, Conn.

Wright, H. E. (1972). Interglacial and postglacial climates: The pollen record. *Quaternary Research* 2, 274-82.

Wright, H. E. (1976). The dynamic nature of Holocene vegetation: A problem in paleoclimatology, biogeography, and stratigraphic nomenclature. *Quaternary Research* 6, 581-96.

CHAPTER 3

Responses of River Systems to Holocene Climates

J. C. Knox

Introduction

The characteristics of runoff and sediment yield are the principal determinants of the physical properties of alluvial channels and floodplains. The frequency and magnitude of water and sediment yields are adjusted to climate, vegetative cover, and physiography. A vast literature pertains to relations between alluvial histories and Holocene climates. Not surprisingly, many of the interpretations and conclusions contradict each other. In this chapter, I first identify a few hypotheses that characterize the responses of river systems to Holocene climates. I then examine the relations among climate, vegetation, and fluvial activity in cases where causal connections can be sought. Finally, I describe alluvial chronologies from several areas of the conterminous United States and discuss their relations to Holocene climates.

Early Hypotheses

Before about 1890, nearly all alluvial terraces were attributed to movements of the Earth's crust. Soon afterward, alluvial terraces of nonglaciated as well as glaciated regions were ascribed to climatic variations (Davis, 1902; Gilbert, 1900; Johnson, 1901). Davis (1902: 97), for example, postulated that the gradient of a river's longitudinal profile represents a balance between the erosion and the transportation of sediments, and he indicated that the volume and character of the sediment load are strongly adjusted to climate. He suggested that a change of climate from humid to arid would lead to a steepening of river profiles and to valley aggradation and that a change of climate from arid to humid would lead to a reduction in gradient and to valley trenching.

The humid versus arid contrast of Davis's model is typical of early perceptions of riverine responses to climatic changes. The tendency to dichotomize climate probably resulted from the early emphasis on the semiarid-to-arid Southwest and on other arid regions. There, it was readily apparent that changes in the amount of available moisture could produce significant changes in vegetative cover and thus influence the magnitude and frequency of floods and sediment yields.

Although there was consensus among the early investigators that the morphology and sedimentology of alluvial channels and floodplains were strongly related to the characteristics of floods and sediment yields, disagreement emerged regarding the relative importance of each factor. To characterize this disagreement, I select the hypotheses of Ellsworth Huntington and Kirk Bryan, both prominent researchers whose ideas were expanded and modified by many later investigators.

Huntington (1914: 32) hypothesized that valley alluviation in the Southwest occurred during dry episodes, when vegetation was minimal and sediment yields were very high. He suggested that a shift to a wetter climate would improve the vegetative cover, reduce fluvial sediment load, and cause channel entrenchment. Bryan (1928), on the other hand, associated channel entrenchment in the Southwest with prolonged periods of drought. He suggested that minimal vegetative cover during dry episodes resulted in large floods that initiated channel entrenchment, which expanded upstream.

The hypotheses of Huntington and Bryan also illustrate the different perceptions of time scales for fluvial responses to climatic change. Huntington's hypothesis indicates that episodes of aggradation or degradation result principally from long-term changes in sediment concentration in stream flow. Bryan's model, on the other hand, implies relatively rapid rates of degradation during and briefly after large floods but more gradual aggradation during intervening episodes, when large floods are rare or nonexistent. Excellent summaries of subsequent refinements of these hypotheses, although biased in orientation toward the semiarid southwestern United States, have been presented by Antevs (1952), Tuan (1966), and Cooke and Reeves (1976). Both original hypotheses are supported by field studies at specific localities, but generalizations are difficult to formulate. The prevalence of one mode of hydraulic response over another apparently depends upon the prevailing environmental characteristics of the drainage system and upon the relations among existing and antecedent climatic events. It also depends upon geographic location within a given drainage hierarchy, for tributary and trunk channel systems sometimes respond differently to the same climatic conditions.

Episodic versus Gradual Holocene Environmental Change

Most Quaternary scientists probably believe that Holocene climates changed gradually from being relatively cool and moist during the early Holocene to being warm and dry during the middle Holocene and then to being cool and moist during the late Holocene. Others suggest, however, that Holocene climates can be divided into many distinct episodes with synchronous boundaries over wide regions. Although the tripartite concept of Holocene climates apparently originated in Europe (Wright, 1976: 593), in North America it usually is associated with Antevs (1955), who characterizes the period between around 7500 and 4000 yr B.P. in the western United States as the Altithermal Long Drought. Wright (1976: 593) concludes that a tripartite division in many parts of the United States was the most realistic concept because pollen evidence indicates that the boundaries between the three climatic divisions are gradational and time transgressive geographically.

Those who have argued for subdividing the Holocene into many distinct and synchronous climatic episodes usually base their arguments on models of atmospheric circulation. For example, Bryson and others (1970: 55-56) suggest that even a smooth change in an external climatic control (e.g. radiation input) would force atmospheric circulation to shift through a series of quasi-stable states, with rapid transitions between states. Bryson and Wendland (1967) and Wendland and Bryson (1974) suggest a stepwise model of Holocene climatic change in which stable climatic episodes are separated from each other by abrupt transitions. They argue that each episode of Holocene climate can be characterized by prominent patterns in the atmospheric circulation systems and that abrupt discontinuities between quasi-stable episodes would be recorded as globally synchronous events in the environmental record. They suggest that the Blytt-Sernander environmental terminology of northern Europe be adopted to represent Holocene climatic episodes, for an analysis of North American and European Holocene radiocarbon dates reveals that the boundaries of the Blytt-Sernander intervals agree closely with widespread discontinuities in geologic and botanical conditions.

Although climate was the driving force behind Holocene environments, vegetational adjustments to changing climates produced equally important hydrogeomorphic effects. Bryson and others (1970: 70-71) estimate that Holocene vegetational changes in Minnesota and Wisconsin lagged about 50 to 200 years behind climatic changes. Their estimates are based upon time intervals in which percentages of Holocene pollen types changed halfway from an earlier level to a later level. Wright (1976: 591-92) cautions that recognizing the abruptness in pollen zones at a few sites can be deceptive because a change of pollen percentages in a profile may only reflect the passage of an ecotone or the range limit of a particular species as it passes a locality.

The more significant Holocene changes of vegetation occurred before 7000 yr B.P., and particularly before about 9000 yr B.P. (Bernabo and Webb, 1977; Davis et al., 1980; Van Devender and Spaulding, 1979; Wells, 1979; Wright, 1970, 1976) in association with rapid disintegration of the Laurentide and Cordilleran ice masses. Subsequent vegetational changes have been modest in comparison.

The evolution of Holocene river systems can be evaluated in relation to both the direct effects of climatic events (e.g., storms and floods) and the indirect effects of vegetation as it controls runoff and erosion. The rapid changes in climate and vegetation before about 7000 yr B.P. imply important adjustments in water and sediment yields. By comparison, the gradual shift of ecotones since about 7000 yr B.P. indicates that any widespread and synchronous adjustments in Holocene river systems may more strongly reflect direct responses to individual climatic events than indirect responses to large-scale changes in average climate and vegetation.

Fluvial Processes

CHANNEL AND FLOODPLAIN PROCESSES

Cross-sectional areas of alluvial channels tend to be most strongly related to the water volume during frequently recurring flood flows (Gilbert, 1877: 113; Rubey, 1952: 130; Wolman and Miller, 1960: 66). Cross-sectional shapes, on the other hand, are mainly related to the volume and textural characteristics of the flood-related sediment loads (Lane, 1937: 135-38; Leopold and Maddock, 1953: 19-21; Schumm, 1960). The channel and floodplain system, therefore, tends to develop a quasi-equilibrium morphology adjusted to the relative proportions of water and sediment discharges in the drainage network (Leopold and Maddock, 1953: 49-51). Climatic change—through its effects on the magnitude, frequency, seasonal occurrence, and duration of short-term variations of storms and surface runoff—can cause major changes in sediment yields followed by the destabilization of the channel and floodplain system. During a period of disequilibrium, sediment delivery (the ratio of transported to eroded sediment) can vary widely as sediments are either temporarily stored or flushed from the channels or floodplains of the drainage systems.

GEOGRAPHIC VARIATIONS OF HIGH FREQUENCY FLUVIAL PROCESSES

Responses of river systems to Holocene climatic and vegetational changes can be evaluated by comparing them to contemporary relationships among the regional characteristics of climate and vegetative cover and the regional yields of water and sediment. Langbein and others (1949) demonstrate that annual variations of runoff predominantly reflect variations of precipitation but that temperature, through its influence on evapotranspiration, is also a major influence. They use temperature-weighted precipitation/runoff curves to illustrate how runoff for a given annual precipitation decreases as temperature rises and how for a given temperature the runoff increases as precipitation increases.

More central to the present discussion, however, is the role of climatic and vegetational change in affecting the portion of runoff derived from overland flow (surface runoff), which is the principal contributor to flood flow and is the main factor responsible for erosion and sedimentation in river systems. Subsequent work by Langbein and Schumm (1958) has determined a relationship between watershed mean annual sediment yield and mean annual "effective" precipitation, based on a reference temperature of 10°C. Effective precipitation was defined as the amount of precipitation required to produce a known amount of runoff at 10°C as defined by the 10°C precipitation/runoff curve of Langbein and others (1949). Langbein and Schumm (1958) have found that with a mean annual temperature of 10°C regional sediment yields reach a maximum of about 280 tonnes per square kilometer where average annual precipitation is about 300 mm and that sediment yields decrease with lesser or greater amounts of precipitation at that temperature (Figure 3-1). They suggest that, as precipitation increases above zero in desert regions, sediment yields initially increase rapidly because limited vegetative cover offers little protection against erosion from raindrop impact and surface runoff. However, according to Langbein and Schumm, as annual precipitation continues to increase above about 300 mm in a region of transition between desert shrub and grassland, a corresponding increase in the density of vegetative cover provides more protection so that sediment yields decrease systematically. As mean annual precipitation increases above about 1000 mm into forested regions, further increases in precipitation have little effect on sediment yields.

The Langbein and Schumm curve of Figure 3-1 was derived from sediment loads in a small sample of rivers that have basin areas that average about 2400 km². Dendy and Bolton (1976) have evaluated regional variations of sediment yields based on estimates of sediment stored in approximately 800 reservoirs in American watersheds that range in size from about 2.5 to 80,000 km². They report that the mean annual sediment yield per unit area increases quite rapidly to about 650 tonnes per square kilometer as mean annual runoff increases from 0 to about 50 mm; subsequently, it decreases as runoff increases from 50 mm to about 1300 mm. The conversion of units indicates that, at mean annual temperature of 10 °C, the peak sediment yield occurs at mean annual precipitation of about 470 mm rather than at the 300 mm calculated by Langbein and Schumm (1958). Other differences probably result from Dendy and Bolton's use of sediment data from reservoirs rather than sediment data from rivers as used by Langbein and Schumm.

The regional yields of sediment shown on the Langbein and Schumm curve in Figure 3-1 emphasize the strong influence of vegetative cover on surface runoff. Because excess surface runoff is a principal cause of floods, flood magnitudes of moderate to high recurrence frequencies can be expected to display similar geographic associations with vegetation. Such a relationship is illustrated by the increasing magnitude of 50-year floods in watersheds along a humid to semiarid transect in the north-central United States (Figure 3-2).

As in Figure 3-1, the curve of Figure 3-2 displays a sharp rise as the magnitude of mean annual precipitation drops below about 500 mm.

The concentration of sediment in rivers helps to determine whether a channel system will aggrade, degrade, or remain stable. A curve depicting the relation between mean annual sediment concentration and mean annual precipitation would be similar in many ways to Figures 3-1 and 3-2. Figure 3-3, for example, shows that, in regions of the United States receiving more than about 800 mm mean annual precipitation, moderate changes in mean annual precipitation have little effect on mean annual sediment concentration. As with floods and sediment yields, mean annual sediment concentration becomes increasingly responsive to changes in mean annual precipitation as precipitation diminishes, especially below about 400 to 500 mm. However, unlike mean annual sediment yields, which eventually decline rapidly as the climate becomes very arid, mean annual sediment concentration tends to increase with increasing aridity, probably because river channels become increasingly responsive to climatic events as the climate becomes increasingly arid.

The regional differences in the magnitudes of floods, sediment yields, and sediment concentrations largely represent adjustments to average conditions of climate and vegetation. Vegetation probably is the dominant control, because it influences the infiltration capacity and velocity of surface runoff. The regional curves represent less well the hydrologic activity associated with floods resulting from snowmelt or from extreme amounts or durations of rainfall, since vegetation has a limited influence on such events. Nevertheless, the regional relationships provide a useful framework for estimating how river systems may have responded to large-scale Holocene environmental changes.

Particularly noteworthy in Figures 3-1, 3-2, and 3-3 are the regions where mean annual precipitation is about 300 to 500 mm and about 800 to 900 mm. These precipitation zones represent thresholds that delimit fluvial responsiveness to climatic change. Below a rate of about 300 to 500 mm, vegetation is sparse. Noble (1965: 118) shows in a subalpine-herbaceous region of Utah that the quantity of eroded soil in storm runoff increases very gradually as the percentage of ground covered by vegetation is reduced from about 90% to 70% but then increases very rapidly as ground cover is further reduced. Above about 800 to 900 mm precipitation, grassland grades to forest and undisturbed forest lands yield relatively little overland flow, even from relatively large storms (Bates and Zeasman, 1930). Although

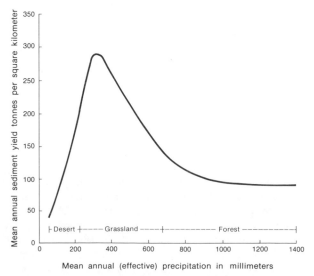

Figure 3-1. The relation between mean annual sediment yield and mean annual effective precipitation where mean annual temperature is 10 °C. (After Langbein and Schumm, 1958.)

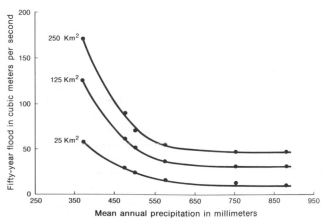

Figure 3-2. The relation between flood size and mean annual precipitation in the north-central United States. (After Knox, 1972.)

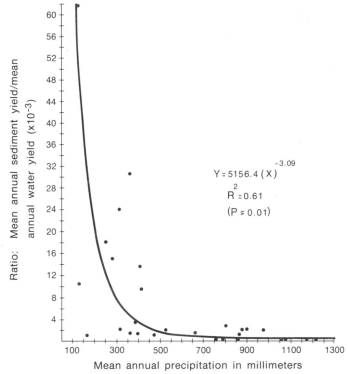

Figure 3-3. The relation between mean annual sediment concentration and mean annual precipitation (yields of water and sediment both in tonnes per square kilometer). The data are for rivers including the Scantic, Conococheague, Monocacy, Ruppahannock, South Yadkin, Edisto, Michigamme, Portage, Killbuck, Tradewater, Big Raccoon, White Breast, Penbina, Willow, Heart, South Fork of the White, Floyd, Vermillion, Kiowa, Walnut, Ute, North Fork of the Red, Nueces, Hondo, Santa Clara, San Pedro, Tucannon, and San Antonio. (Data from U.S. Geological Survey.)

grasslands in good condition are nearly as effective as forest cover in reducing overland flow (Sartz, 1970), grasslands lose much of their effectiveness during droughts. Brice (1966: 263), in reviewing the effects of drought on short-grass prairie in the Great Plains of Kansas, reports that plant cover was between 80% and 95% during the early 1930s but that after a sequence of drought years the cover was commonly reduced to 20% or less.

Thus, on the basis of contemporary environmental associations, it can be concluded that changes of Holocene climates resulting in displacements of major vegetative formations caused significant changes in the magnitude and frequency of floods and in the yield and concentration of sediment. Such changes probably resulted in the metamorphosis of river systems. Furthermore, it is apparent that Holocene environmental changes involving changes of forest composition probably produced relatively modest adjustments in fluvial behavior, but, where grasslands or desert shrubs replaced woodlands, the adjustments were more substantial.

TEMPORAL VARIATIONS OF LOW-FREQUENCY FLUVIAL PROCESSES

So far the emphasis has been on strong fluvial responses to average characteristics of climate and vegetation. Less frequent but larger floods, however, may be more critical to the evolution of alluvial deposits and alluvial landforms. Schumm and Lichty (1963), for example, report that frequent moderate to extensive flooding after a

very large flood on the Cimarron River in southwestern Kansas caused the wholesale destruction of the Cimarron floodplain and that channel narrowing and floodplain construction occurred only after a period of low-magnitude floods. Similar fluvial responses are documented for the Gila River in Safford Valley, Arizona (Burkham, 1972), and along the Missouri River adjacent to western Iowa (Hallberg et al., 1979). Stevens and others (1975) show that river systems with a wide range of peak-flood discharges are susceptible to frequent changes in form.

Contrary to a common assumption that the probability of a flood of a given magnitude remains constant over time, many long historical flood records reveal clustering, especially of large floods (Knox et al., 1975). This tendency reflects the fact that floods are associated with large-scale upper-atmosphere regimes with westerly circulations. During periods when westerly circulation is weak and meridional regimes prevail, polar air masses often extend to low latitudes and tropical air masses often extend to high latitudes. Such extreme latitudinal ranges frequently result in the development of intense cyclones and storms in the middle latitudes that are followed by large floods. During periods when zonal regimes prevail, a strong westerly flow predominates and intense cyclones and storms are less likely. Hence, most areas of the United States that are relatively far removed from sources of moisture tend to become drier and experience fewer large floods during periods of zonal dominance. An example is the historical record of Mississippi River floods at St. Paul, Minnesota (Figure 3-4). The record of all floods exceeding a discharge of 368 m^3 per second is partitioned into four categories defined by changes in the dominant patterns of large-scale atmospheric circulation (Kalnicky, 1974: 106; Knox et al., 1975: 27-30; Kutzbach, 1970: 716; Lamb, 1966: 183; 1974: 27-30; 1977: 540). In general, meridional pattern are more dominant during the flood-episodes represented by the pre-1896 and post-1949 categories. Zonal patterns are shown to have been more dominant during the intervening years. Figure 3-4 shows that high-frequency, low-magnitude floods differ little among the four climatically defined episodes but that low-frequency, high-magnitude floods are considerably different. Thus, over longer time spans, such as the Holocene, the recurrence intervals of floods of a given magnitude probably have changed dramatically in response to changes in prevailing patterns of atmospheric circulation over various regions of the United States. As noted above, frequent recurrence of moderate to extensive flooding may result in the destabilization of river systems and the removal of sediments that accumulated gradually over a long time. Thus, major morphological responses in a fluvial system may result without any change in vegetation, for vegetative cover reflects average rather than extreme climatic conditions.

Holocene Environments

HOLOCENE ATMOSPHERIC CIRCULATION

The Holocene environment of the United States is often viewed as one changing from being cool and moist during the early Holocene to being warm and dry during the middle Holocene and then back to being cool and moist in the late Holocene. The tripartite division inadequately represents the changing environmental conditions that are most important in the evolution of Holocene alluvial river systems. A specific example is the apparently greater importance of zonal atmospheric circulation patterns during the early Holocene versus the increased importance of meridional atmospheric regimes during the

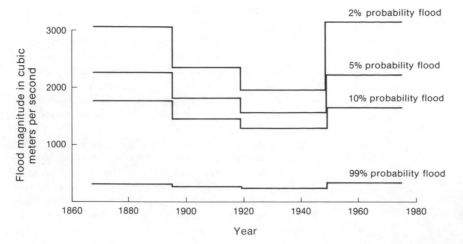

Figure 3-4. Climatic influence on Mississippi River floods. The estimates were determined with the Log-Pearson Type-III method applied to partial duration series floods from the U.S. Geological Survey gauging station at St. Paul, Minnesota. The series boundaries are based on dates of change in prevailing atmospheric circulation patterns. Meridional circulation prevailed frequently before about 1895 and since about 1950, but zonal circulation prevailed frequently between 1895 and 1950.

late Holocene. The early Holocene patterns were associated with the disintegration of the Laurentide ice sheet.

About 10,000 yr B.P., the main part of the Cordilleran ice sheet had disintegrated, but the Laurentide ice sheet continued to occupy most of eastern and north-central Canada (Figure 3-5). As the elevation of the Laurentide ice sheet was lowered during wastage, the corresponding atmospheric circulation patterns were analogous to those that occur frequently today in late winter (Hare, 1976), when an extensive, cold, snow-covered surface in north-central North America intensifies the steepness of the latitudinal temperature gradient across the United States and favors strong westerly surface winds. Thus, present-day March patterns of circulation approximate the zonal westerly wind patterns that probably were very common during the early Holocene before about 7000 yr B.P. (Figure 3-6). Such patterns probably were favored throughout the year during the early Holocene

because of the continuous effects of the Laurentide ice sheet. Hence, summer climates during the early Holocene were undoubtedly quite different from those during the late Holocene. At the beginning of the Holocene, much of the northern Midwest and Northeast probably was dominated in summer by cool/dry air masses from Canada, but, as the elevation and southern extent of the Laurentide ice sheet diminished during the middle Holocene, air masses from the Pacific presumably became increasingly dominant during summer in the Midwest and West. The wedge of warm/dry Pacific air would have displaced northeastward the southern extent of Canadian air masses and restricted the warm/moist Gulf of Mexico air masses to the South and Southeast (Figure 3-6).

After the dissipation of the Laurentide ice sheet about 7000 to 6000 yr B.P. (Bryson et al., 1969; Prest, 1969), the large difference in summer temperatures between central Canada and the southern

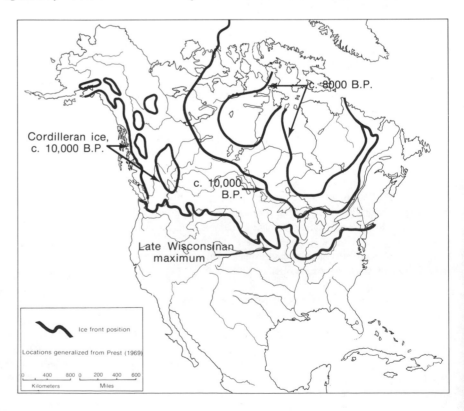

Figure 3-5. Positions of ice fronts during the waning of the Laurentide and Cordilleran ice sheets.

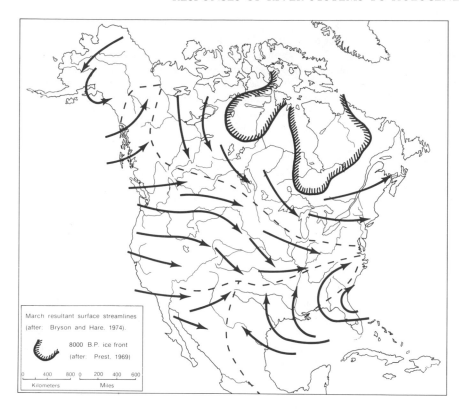

March resultant surface streamlines
(after: Bryson and Hare, 1974).

8000 B.P. ice front
(after: Prest, 1969)

0   400   800   0   200   400   600
Kilometers   Miles

Figure 3-6. Late-winter resultant surface streamlines of the present-day approximating a variety of zonal westerly flows that probably were dominant throughout the year during the early Holocene. The persistent westerly flow allows Pacific air masses to dominate in much of the United States, and it minimizes the collision of Canadian and Gulf air masses over the interior.

United States no longer existed. The strength of the summer's westerly circulation was greatly weakened, and both polar and tropical air masses penetrated deeply into the United States. For example, the warm, moist tropical air masses that reach the latitude of Lake Superior frequently during present-day summers (Figure 3-7) probably would have occurred only rarely before about 7000 to 6000 yr B.P. Because the magnitude, frequency, and geographic distribution of storms and floods are directly linked to characteristics of the large-scale atmospheric circulation patterns (Knox, 1976: 182-86), the indirect climatic effects of the Laurentide ice sheet contributed significantly to the evolution of Holocene river systems. The increased frequency of meridional patterns of atmospheric circulation after about 6000 yr B.P. is significant because the associated slow-moving storm systems often generate persistent heavy rains followed by large floods. Stratigraphic evidence indicates increased occurrences of large floods after about 6000 yr B.P. in the northern Midwest (Knox et al., 1981).

Although the present-day March resultant surface streamlines are presented as an approximation of wind-flow patterns at about 8000 yr B.P. (Figure 3-6), the streamlines should be viewed as a generalization of various zonal patterns of westerly winds that probably were prevalent throughout the early Holocene. The intensity and orientation of the westerly winds undoubtedly changed systematically with the disintegration of the ice. The rapid northeastward migration of the prairie/forest ecotone in the northern Midwest between 12,000 and 9000 yr B.P. and its continued but slower northeastward movement between about 9000 and 7000 yr B.P. was associated with the disintegration of the Laurentide ice sheet. The major movements of vegetative communities during the early Holocene compared to their relative stability during the late Holocene indicate that many fluvial responses to early-Holocene environments probably displayed a relatively strong time-transgressive characteristic in comparison to fluvial responses after about 6000 yr B.P.

BIOTIC RESPONSE

The change of atmospheric circulation from frequent zonal dominance in the early Holocene to mixed zonal and meridional dominance in the late Holocene influenced the regional distribution of vegetative communities and thus the activity of river systems. The contributions of vegetational change to fluvial activity can be inferred from the fluvial information in Figures 3-1, 3-2, and 3-3 combined with the inferred history of vegetation as reconstructed from the fossil record. For convenience of discussion, I have divided the United States into six regions on the basis of the characteristics of contemporary natural vegetation (Figure 3-8). These regions include areas where early-Holocene vegetational changes probably produced broadly similar effects on fluvial processes. In two large regions, the High-Elevation Western Woodlands and the Eastern Woodlands, changes in composition probably have had rather modest fluvial significance, because the Holocene vegetation probably has always been dominated by woodland. However, in the other four regions, the biotic response to early-Holocene climatic change involved shifts from woodland to grassland or from grassland to steppe. These changes favored widespread valley alluviation in many areas, because vegetational change probably was accompanied by increases both in mean annual sediment yield and sediment concentration in rivers. As mean annual temperatures continued to increase toward a middle-Holocene maximum, mean annual sediment yield and sediment concentration responded differently to vegetational change in some regions, for mean annual sediment yield depends on flood frequency whereas mean annual sediment concentration depends on the effectiveness of vegetation in preventing erosion. It is apparent from the following discussion that the greatest fluvial responses to biotic change occurred during the early Holocene, especially before about 8000 yr B.P. Although changes in the type and density of vegetation apparently closely followed changes in the climate, the changes in vegetation appear time transgressive when viewed at time scales shorter than about 500 years.

Holocene Alluvial Chronologies

Because vegetation reflects prevailing climatic conditions, vegetational regions provide a convenient framework for describing the responses of river systems to Holocene climates. The following discussion of alluvial chronologies is organized according to the six broad-scale natural vegetational regions presented in Figure 3-8. I attempt to show that each of the six regions experienced unique Holocene changes of climate and vegetation resulting in specific widespread fluvial responses.

EASTERN WOODLANDS

Most of the region east of the Mississippi River has been dominated by forest vegetation throughout the Holocene (Davis et al., 1980; Wright, 1976). Even though important changes in forest composition have occurred, the effects of these changes on fluvial processes probably were conservative compared to those characteristic of many other regions, because the magnitudes of floods, sediment yields, and sediment concentrations show only modest responses to changes in forest type (Figures 3-1, 3-2, and 3-3). In the New England region, the interval of maximum warmth during the Holocene started 9000 yr B.P. and lasted at least until 5000 yr B.P., and mean annual temperatures may have been 2°C warmer and mean annual precipitation 400 mm less than present (Davis et al., 1980). Modern forest communities in the New England White Mountains are markedly dif-

ferent in species composition and relative abundance from those present earlier in the Holocene.

In the southeastern United States, the differential migration of various tree types prevents a clear interpretation of paleoclimates (Wright, 1976), for the major expansion of southern pines occurred thoughout the region about 6000 yr B.P. Pollen diagrams from several Florida sites, however, imply that upland vegetation was locally treeless before about 6000 years ago and that the dry conditions of the late Pleistocene extended through to the middle Holocene in the extreme Southeast, a condition that differs sharply from that in the northern regions described above (Wright, 1976).

Some of the apparent conflicts in interpretations of paleoclimates can be resolved through considerations of the regional effects of large-scale atmospheric circulation patterns. The present-day north-to-south climatic gradient across the region varies from very gentle in summer to very steep in winter, but during the early Holocene the waning Laurentide ice sheet favored strong zonal westerly circulation over much of the United States (Figure 3-6). Such zonal flow throughout the year probably restricted the northward penetration of moist tropical air masses to the southern and southwestern parts of the region; this same circulatory pattern occurs during present-day late winters (Figure 3-6). The zonal flow also concentrated frontal precipitation along a convergence zone between Pacific and tropical air masses; this pattern frequently affects present-day winter conditions. Borchert (1950: 16) shows that July rainfall in the southeast tends to be above average when a strong zonal westerly flow predominates. During the early 1930s, zonal westerly atmospheric circulation was dominant in summers, and drought characterized much of the Midwest and the Great Plains (Borchert, 1950). At the same time, annual runoff magnitudes in rivers throughout much of the southern region of the Eastern Woodlands were typically among the top 25% of flows for a 30-year period from 1931 to 1960 (Busby,

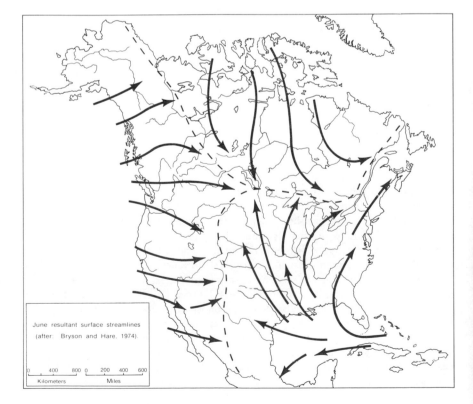

Figure 3-7. Resultant surface streamlines of present-day June illustrating a strong meridional component of the large-scale atmospheric circulation. Meridional flows result in extensive latitudinal ranges of Canadian and Gulf air masses and often are associated with intense, slow-moving cyclones followed by large floods in the middle latitudes. Meridional patterns in summer probably became frequent only after about 6000 yr B.P., after the Laurentide ice sheet had disintegrated.

June resultant surface streamlines
(after: Bryson and Hare, 1974).

0   400   800 0   200   400   600
     Kilometers   Miles

Figure 3-8. Natural vegetational regions where Holocene vegetational changes contributed similar hydrologic effects.

1960), but the extreme Southeast, especially Florida, usually had runoff well below normal. Apparently, the extreme Southeast was south of the mean frontal zone and was frequently dominated by dry weather associated with a semipermanent subtropical high-pressure system. If the zonal circulation patterns of the 1930s are reasonable analogues for the time before 6000 yr B.P., then the persistence of dryness in the extreme Southeast during the early Holocene can be accounted for.

Because most of the Eastern Woodlands region has been woodland throughout the Holocene, the responses of rivers to Holocene environmental changes may be closely related directly to climatic controls such as storms and floods rather than indirectly to broad-scale changes in vegetative cover. The early-Holocene environmental changes probably produced modest valley alluviation in response to modest increases in yields and concentrations of sediment in rivers. There have been few detailed studies of Holocene alluvial chronologies within the region. One of the more detailed investigations is that by Grissinger and others (1981), who have studied late-Quaternary valley-fill deposits in north-central Mississippi. They identify four distinct stratigraphic units: bog-type sediments composed of fine-grained organic materials deposited in low-energy fluvial systems, channel-lag sediments composed of coarse-grained and cross-bedded materials deposited in high-energy fluvial systems, massive silty-sediments grading upward from slightly bedded silty sand or sandy silt to massive silt, and meander-belt alluvium with occasional oxbow deposits.

Radiocarbon dates indicate that coarse-grained channel-lag deposits and the bog-type sediments accumulated between about 12,000 and 8500 yr B.P., when northern Mississippi probably experienced greater precipitation and runoff than it does today. The massive silt accumulated between about 10,000 and 6100 yr B.P., when climate was warming and drying and when large floods were uncommon. Some

investigators suggest that the massive silts relate to ponding in the tributaries resulting from a rising baselevel associated with the aggradation of the lower Mississippi River between 12,000 and 8500 yr B.P. (S. A. Schumm, personal communication, 1981). Sand deposits dated at 6100 to 4000 yr B.P. represent relict channels set into the massive silts. The greatest entrenchment into the massive silts, however, occurred from 3000 yr B.P. until the time of agricultural settlement, which was associated with lateral channel migration and the accumulation of meander-belt alluvium composed of both lateral and vertical accretion deposits. The incision of the massive silts apparently began about 6000 yr B.P., coincident with the increased importance of meridional circulation in summer, a condition that probably favors more frequent large floods. Thus, climatic change, not baselevel, was the dominant control of depositional processes (Grissinger et al., 1981).

Along the lower Ohio River in southern Illinois, floodplain alluviation has occurred at an increasing rate during the last 4000 to 8000 years probably as a result of either larger floods or a gradual decline in the flow of excess Ohio River floodwaters through the adjacent Cache Valley (Alexander and Prior, 1971). For the Whitewater Basin of southern Indiana, Gooding (1971) concludes, from a few radiocarbon-dated alluvial deposits, that with the decrease in meltwater volume and outwash debris the Ohio River adjusted its channel about 9000 yr B.P. and that these adjustments were extended into the river's southeastern Indiana tributaries by 7000 yr B.P. Most Holocene alluvium in the tributaries accumulated between about 7000 and 1000 yr B.P., when valley-bottom elevations were relatively stable (Gooding, 1971).

Farther northeast in the Eastern Woodlands, in the upper Susquehanna River basin of New York State, Scully and Arnold (1981) have identified an episode of incision that occurred before about 9700 yr B.P.; the episode corresponds with a change in pollen stratigraphy from predominantly spruce to pine. A low terrace of lateral accretion deposits was formed when the channel elevation was stable 8000 to 4400 yr B.P. Deposition by vertical accretion predominated after about 4700 yr B.P. but then apparently slowed after about 3000 yr B.P., when channel incision and renewed lateral migration were initiated as a possible response to cooler and wetter conditons. The present floodplain level was established about 1200 yr B.P.

It is difficult to generalize the Holocene alluvial chronologies described above for the Eastern Woodlands. The studies vary in representation from local to regional scale and represent different sizes of drainage basins. In general, valley alluviation characterized the early Holocene, when the climate was apparently on a trend toward warmer and drier conditions and channel incision seems usually to have followed shifts of climate toward cooler and wetter conditions. The removal of older deposits during a change to apparently wetter climates was associated with intensified meandering and accumulation of lateral accretion deposits at a new lower elevation.

TALL-GRASS PRAIRIE AND MIDWEST

In the Tall-Grass Prairie and Midwest, the late-glacial boreal forest that had extended southward into the southern Great Plains rapidly deteriorated beginning about 12,000 yr B.P. in the southern Plains and ending about 9500 yr B.P. in the northern Plains (Wright, 1970: 163). Spruce forest gave way to a brief interval of deciduous woodlands and then to grasslands. In the central part of the region, in northwestern Iowa, deciduous forest predominated until about 9000 yr B.P. (Van Zant, 1979). The timing of the change from forest to prairie ranged from about 10,000 yr B.P. in the southern Plains to about 8000 yr B.P. in the northern Plains, and in the central Plains grasslands succeeded spruce forest almost directly (Wright, 1970: 170, 1976: 589).

The rapid northward migration of the prairie-forest ecotone represented a response to the retreat of the ice sheet. The strong westerly atmospheric circulation associated with the waning ice sheet brought Pacific air masses warmed and dried by their passage over the western mountains. The dry Pacific air acted as a wedge to prevent the intrusion and collision of polar and tropical air masses over the region. Borchert (1950: 16-17) shows that average July temperatures in the Great Plains during years with strong zonal westerly wind flow were about 2°C warmer than the long-term average and that July rainfall was about 50% or 60% or less of the long-term average. Under these conditions, localized convectional thunderstorms are the dominant precipitation process. Floods on large rivers result mainly from the contributions of frontal rains and/or snowmelt, whereas floods on small streams result mainly from high-intensity convectional storms. Thus, flood hydrology probably changed significantly after about 6000 yr B.P., when the incidence of zonal westerly circulation was greatly reduced in summers.

Alluvial chronologies of the eastern Great Plains are best documented for sites along its eastern border in southwestern Wisconsin and in Missouri. Climatic change is the dominant cause of a number of fluvial episodes in the driftless area of southwestern Wisconsin (Knox et al., 1981). After the area's fluvial adjustments to the environmental changes associated with the Pleistocene/Holocene transition, the climatic trend toward warmer and drier conditions was

associated with valley alluviation until about 7500 or 7000 yr B.P. Early-Holocene sandy gravels are overlain by 2 to 3 m of massive silts similar to the early-Holocene deposits in north-central Mississippi (Grissinger et al., 1981). Large floods were very rare during the early Holocene, probably because moist air masses were seldom able to penetrate the region before 7000 or 6000 yr B.P. Maximum warmth and dryness occurred about 7200 B.P., when the prairie expanded into the woodlands of the area. Between about 7500 and 6000 yr B.P., convectional thunderstorms probably were a dominant cause of floods, because the erosion and deposition of alluvial sediments were concentrated in small watersheds. Since about 6000 yr B.P., the climate has generally followed a cooling trend, and sedimentary sequences indicate that floods larger than those of the early Holocene occasionally occurred. These floods apparently contributed to intense lateral channel migration and slight channel incision, especially during the periods 6000 to 4400, 3100 to 1800, and 1200 to 800 yr B.P. Although channel conditions at other times were apparently more stable, some lateral migration has occurred continuously throughout the Holocene. The relatively gradual changes in Holocene vegetation implied in pollen diagrams for the region correspond poorly with the relatively abrupt changes in the alluvial record. The abruptness of change in fluvial activity appears to be better related to long-term variations in the recurrence intervals of large floods (Knox et al., 1981).

The Holocene alluvial chronology of the Pomme de Terre River in the Ozark uplands of western Missouri as reported by Ahler (1973a, 1973b), Haynes (1976), and Brakenridge (1981) is nearly identical to the chronology for southwestern Wisconsin. An erosional episode during the Pleistocene/Holocene transition was followed by the alluviation of a brown, clayey silt 10,000 to 8100 yr B.P., a brief episode of erosion 8100 to 7500 yr B.P., and then the deposition of alluvium to the former floodplain level until about 5000 yr B.P. (Brakenridge, 1981). Episodes of floodplain erosion occurred about 5000, 2900, 1500, and 350 yr B.P. Brakenridge concludes that no obvious parallels exist between the local alluvial history and the record of baselevel changes induced by changes in sea level or glacial outwash. He supports the hypothesis (Knox, 1976) that some episodes of Holocene stream erosion or of intensified lateral channel migration may have been caused by increased occurrences of large floods resulting from intensified meridional upper-atmospheric circulation.

Holocene alluvial chronologies for other areas in the Tall-Grass Prairie and Midwest are less well understood, because erosion has removed much of the earlier record. Entrenched streams in Harrison County, western Iowa, reveal that much of the Holocene alluvial fill is less than about 2000 years old (Daniels and Jordan, 1966). Ruhe (1969: 161-65) shows that floodplain alluvium adjacent to many Iowa streams is younger than about 2000 years. A small alluvial fan along the Little Sioux River in northwestern Iowa was formed during intermittent episodes of rapid deposition followed by relatively long periods of stability and soil formation (Hallberg et al., 1974; Hoyer, 1980). Boundaries of major sedimentary units within the fan are dated at about 10,050, 8400, 7200, 6350, and 4600 yr B.P. The coarsest sediments in the fan, deposited between about 7400 and 6100 yr B.P., are correlated with the period of maximum dryness during the Holocene in western Iowa (Hallberg et al., 1974). Ruhe (1969: 147-48) documents accelerated hillslope erosion throughout much of Iowa during the period of maximum warmth and dryness in the middle Holocene. Butzer (1977, 1980) also identifies very active hillslope erosion that occurred about 10,000 to 7700 yr B.P. along a small tributary to the Illinois River in western Illinois. He suggests that the lower Illinois River was aggraded at least 15 m about 8000 to 5500 yr B.P. and that it probably stabilized near its present level about 5500 to 5000 yr B.P. He further suggests that the Illinois River changed from a braided channel with a wide, sandy floodplain about 7700 to 5500 yr B.P. to a meandering system with some large floods about 5500 to 2100 yr B.P.

On the western margin of the Tall-Grass Prairie and Midwest, Holocene alluvial fills show evidence of major aggradation and degradation, but the sequences are poorly defined by radiocarbon dating. Here, the very rapid transition from late-Pleistocene forest to grasslands at the beginning of the Holocene was associated with major valley alluviation (Schultz and Martin, 1970). In the Medicine Creek drainage basin of southwestern Nebraska, early-Holocene deposits range from about 10 m thick in the small tributaries to about 30 m thick downstream in the main valley (Brice, 1966). Early-Holocene terraces 10 to 15 m above the present river channels in Nebraska were exposed by channel entrenchment starting at about 5000 yr B.P. (Brice, 1964; 1966). Low depositional terraces dated at 3500 to 2200 yr B.P. indicate that widespread entrenchment had ended by then (Brice, 1966; Schultz and Martin, 1970). Similar dates for low terraces are documented in Kansas and Oklahoma. Johnson and others (1980) report an episode of soil formation on floodplains of Kansas, Oklahoma, and northwestern Missouri streams that ended about 2000 yr B.P. and was followed by alluviation until about 1000 to 800 yr B.P., when stream channels were entrenched to their present levels. Hall (1980) reports that rapid valley filling was in progress about 2600 yr B.P. in Oklahoma and that it was followed by soil formation 2050 yr B.P. and modest alluviation since then, with a brief period of soil development between about 600 and 400 yr B.P. Unlike the more humid regions described earlier, where erosional episodes appear to be correlated with shifts toward moister climates and where alluviation is correlated with shifts toward drier climates, for this region the reverse correlations are made (Hall, 1980).

In summary, the river systems of the Tall-Grass Prairie and Midwest experienced major responses to both the direct and indirect effects of Holocene climates. Vegetational change from forest to grassland or from forest to mixed forest and grassland probably was the most important cause of active valley alluviation from about 10,000 to 7500 yr B.P. Early-Holocene alluviation appears to have slowed between about 7500 and 6000 yr B.P. and ended when climatic conditions became cooler and moister after about 6000 yr B.P. Although climatically induced erosion of the early-Holocene alluvial fill may have been initiated as early as 8000 to 7500 yr B.P., the first widespread episode of incision and removal occurred between about 6000 and 5000 yr B.P. Subsequent removal has been accomplished largely by lateral channel migration and modest vertical incision, especially 6000 to 4500, 3000 to 2000, and 1200 to 800 yr B.P. These generalized episodes of active erosion and sedimentation do not rigidly correspond with the fossil record of late-Holocene vegetational change. Erosion tends to be associated with vegetational changes that imply shifts to cooler and moister conditions, and alluviation tends to be associated with vegetational changes that imply shifts to warmer and drier conditions. Intensified lateral channel migration probably results from increased frequencies of large floods associated with the anomalous meridional patterns of the large-scale atmospheric circulation.

SHORT-GRASS PRAIRIE AND WESTERN GREAT PLAINS

The Short-Grass Prairie and Western Great Plains region extends from north-central Texas to the Canadian border (Figure 3-8). Its western sector includes the present zone of maximum annual sediment yields in the United States (Figure 3-1). It is not known whether the late-Pleistocene boreal forest that characterized the Great Plains of Kansas until about 12,000 yr B.P. also covered this region (Wright, 1970). Because of the rain-shadow effect of the adjacent Rocky Mountains on the west, it is possible that much of the western sector of the region was never forested during the Holocene. However, Baker and Penteado-Orellana (1977), in a summary of research on fossil fauna and pollen, report the existence of mixed forest and grassland in north-central Texas before 9000 yr B.P. and a rapid demise of forest between about 9000 and 7000 yr B.P. This change is consistent with the idea that the central portion of the region was characterized by very active sand dunes during the early to middle Holocene (Madole et al., 1981). Predominating strong zonal westerly atmospheric circulation, as hypothesized for the early Holocene prior to 7000 to 6000 yr B.P. (Figure 3-6), probably resulted in extreme drought in much of the region. The changes of climate and vegetation associated with a shift to warmer and drier conditions during the late-Pleistocene/Holocene transition greatly accelerated hillslope erosion and apparently caused major early-Holocene alluviation. During the period of peak warmth and dryness about 8000 to 7000 yr B.P., hillslope sediment yields and valley alluviation may have declined during extended droughts. The early-Holocene alluvial fills were deeply entrenched and eroded throughout the region, probably after about 6000 to 5000 yr B.P., when the climate became cooler and wetter.

The early-Holocene alluviation common throughout the region of the Short-Grass Prairie and Western Great Plains occurred mainly because the plant cover diminishing as the climate became warmer and drier was less effective in protecting hillslopes from erosion. Baker and Penteado-Orellana (1977) have found that the middle reach of the Colorado River in Texas was incised near the end of the cooler and moister late-Pleistocene climatic period but then aggraded with coarse gravels between about 10,000 and 7000 yr B.P. as the climate became warmer and drier. In the northwestern part of the region in Wyoming, there is evidence that local slope wash continued to be an important source of sediments for valley aggradation until at least about 8300 yr B.P. (Albanese and Wilson, 1974).

The early-Holocene alluvial fills have been extensively incised. Leopold and Miller (1954), without the benefit of radiocarbon dates, postulate soil formation and erosion on high terrace deposits in Wyoming during the warm/dry middle Holocene. Albanese (1974) uses radiocarbon dates to show that an episode of soil formation occurred between about 5800 and 4400 yr B.P., and he suggests that widespread valley cutting was initiated about 4500 yr B.P. in the northwestern Great Plains. Scott (1963) shows that at the edge of the Plains in northeastern Colorado, incision of early-Holocene alluvium had occurred at least by 5800 yr B.P. because alluvial deposits resting on the eroded surface are dated 5800 to 5500 yr B.P.

Radiocarbon-dated sites throughout the region demonstrate that alluvium accumulated between about 3000 and 2000 yr B.P. Scott (1963) suggests that alluviation began about 2800 yr B.P. in northeastern Colorado, and Madole (1976) indicates that gravels filled the valley floor of St. Vrain Creek in eastern Boulder County, Colorado, between about 3400 and 2000 yr B.P. Baker and Penteado-Orellana (1977) refer to a brief more-mesic period in northwestern Texas between about 3000 and 2000 yr B.P., when the Colorado River became narrow and sinuous and transported relatively fine grained sediments (in comparison to early-Holocene conditions). They attribute slack-water deposition by the Pecos River in eastern New Mexico and western Texas to more uniform floods with lower magnitudes during this same time interval. The fluvial activity of this mesic period was followed by entrenchment and then by accumulation of coarse gravels as the climate became more arid during the last 2000 years. Although Baker and Penteado-Orellana recognize two minor episodes of entrenchment during the last 1000 years, radiocarbon control is not sufficient to place these events accurately in time. In the northwestern Great Plains, valley alluviation has been dominant for most of the last 700 years and may have ended with the entrenchment of channels during the late 19th century (Albanese and Wilson, 1974).

It can be concluded that valley alluviation dominated the early Holocene until at least about 8000 yr B.P. Although some incision may have started shortly after about 8000 yr B.P., the primary erosion of the early-Holocene alluvium occurred during the periods 6000 to 4500 and 3400 to 2000 yr B.P. and since about 700 yr B.P. When the alluvial chronologies are compared with the Holocene glacial chronology of the Front Range in the Rocky Mountains (Benedict, 1973), it is apparent that, since about 6000 yr B.P., ice buildup has tended to be correlated with river entrenchment on the Plains; and glacial retreat, with alluviation. The late-Holocene alluvial chronologies of the southern Plains rivers are sometimes out of phase with the alluvial chronologies of the northern Plains, but radiocarbon dates that bracket alluvial episodes indicate a widespread, general synchroneity of change in fluvial activity, which suggests a climatic cause.

SOUTHWEST DESERT SHRUBLAND

The Southwest Desert Shrubland region extends in a broad zone from southeastern California to southwestern Texas except in local areas of high elevation (Figure 3-8). Creosote bush, mesquite, and desert grasses are typical vegetation. Today, the area is very arid and includes portions of the Mohave, Sonoran, and Chihuahuan Deserts. In contrast to many regions of the United States where maximum dryness occurred during the period 7500 to 6000 yr B.P., maximum dryness in the Southwest was delayed until the late Holocene. Radiocarbon-dated macrofossils (Van Devender and Spaulding, 1979; Wells, 1979) document the persistence of woodlands until about 8000 yr B.P. in what are now deserts of middle and low elevation. Van Devender and Spaulding suggest that grasslands dominated during the middle Holocene in association with a relatively warm and moist climate, which is consistent with an atmospheric circulation pattern that is strongly zonal (Figure 3-5). This conclusion is supported by the observation that during the early 1930s, when zonal atmospheric circulation patterns were common in summer (Borchert, 1950), rivers throughout much of the Southwest experienced flows that were above the long-term average (Busby, 1963). The present desert vegetational communities have, therefore, developed only during the late Holocene. Apparently, the development of a stronger meridional component in the summer atmospheric circulation after about 6000 yr B.P. permitted the Southwest to come under the summer dominance of subtropical high-pressure systems and dry weather.

As the climate became warmer and drier during the early Holocene and the vegetation became less extensive, surface runoff and erosion resulted in major valley alluviation before about 8000 yr B.P. (Haynes, 1968). Hence, the early-Holocene fluvial response to climate was similar to that of most other regions of the United States previously discussed. However, because the development of maximum aridity in the Southwest was out of phase with many other regions of the United States, the later Holocene alluvial chronologies can be expected to display relationships similarly out of phase.

When Haynes (1968) conducted his pioneering correlation of alluvial chronologies in the western and southwestern United States, he assumed that Antevs's model of maximum warmth and dryness in the middle Holocene was also applicable to the Southwest. Haynes identified groupings of depositional units for a region extending from Wyoming to Arizona. But, because it now appears that the middle- and late-Holocene climates of the Southwest were out of phase with those farther north in the West and Great Plains, caution should accompany direct correlations.

Because Haynes observed no alluvial units representing the period between about 7500 and 6000 yr B.P., he concludes that entrenchment began then. Hall (1977), however, demonstrates that two episodes of early-Holocene erosion prior to 7000 yr B.P. were followed at Chaco Canyon in northwestern New Mexico by alluviation 6700 to 2400 yr B.P. as aridity increased and pinyon woodlands and ponderosa forest rapidly diminished in abundance and range on the uplands. This arid episode probably correlates with the late-Holocene aridity that produced the present-day desert vegetation of the Southwest (Van Devender and Spaulding, 1979). Meanwhile, at the Gardner Spring arroyo in southern New Mexico, alluviation began by at least 6400 yr B.P. but apparently stopped or was greatly reduced some time after about 4600 yr B.P., when soil formed on the deposit (Gile, 1975). Alluviation was renewed about 2200 yr B.P. and was followed by entrenchment after 1100 yr B.P. At Chaco Canyon, the alluviation that ended at about 2400 yr B.P. was followed by erosion until about 2200 yr B.P. (Hall, 1977). Alluviation prevailed again 2000 to 850 yr B.P. under climatic conditions slightly more arid than those of the present. Hall suggests that increased rainfall and runoff beginning about 850 yr B.P. resulted in channel entrenchment more than 3 m deep at Chaco Canyon but that by about 600 yr B.P. the entrenched channel began to fill with sand. Alluviation apparently continued until the time of late-19th-century trenching. Similarly, the entrenched channel that is younger than 1100 yr B.P. at the Gardner Spring arroyo is also filled with alluvium.

The alluvial chronologies from New Mexico are very similar to the chronology of the upper San Pedro Valley in southern Arizona (American Quaternary Association, 1976). There, alluviation occurred from about 10,000 to 8000 yr B.P. and erosion is suggested between 8000 to 6000 yr B.P. Renewed alluviation was especially active from about 6000 to 4000 yr B.P. and cut-and-fill sequences, with occasional brief episodes of soil formation, dominate the late Holocene.

In summary, the available alluvial chronologies for the Southwest Desert Shrubland indicate that vegetational responses to early-Holocene warming produced an episode of alluviation similar to that which occurred in many regions of the United States. However, it seems that by about 8000 yr B.P. responses of southwestern rivers were beginning to be somewhat different and sometimes out of phase with rivers in other parts of the United States. The relatively widespread evidence for the erosion of alluvial fills between about

8000 and 7000 yr B.P. apparently is associated with the replacement of woodlands by grasslands as the climate changed from being relatively cool and moist to being warm and moist. The persistence of strong zonal westerly atmospheric circulation in summers often results in increased summer rainfall over the Southwest (Bryson et al., 1970). Perhaps the incision of alluvial fills between 8000 and 7000 yr B.P. resulted from increased runoff and occasional large floods during summers as rainfall increased in response to middle-Holocene zonal westerly atmospheric circulation. Hall (1977), for example, concludes that the channel entrenchment at Chaco Canyon about 850 yr B.P. resulted from an increase in rainfall and runoff. Environmental evidence from the Great Plains indicates that zonal westerly atmospheric circulation became dominant in summers about 850 yr B.P. and persisted for a few centuries (Bryson et al., 1970). The onset of major alluviation after about 6700 to 6000 yr B.P. apparently is related to the development of the present-day arid regime and the introduction of desert vegetation to many parts of the landscape. The extensive alluviation from that time until about 4000 yr B.P. is out of phase with the tendency for channel entrenchment and erosion of alluvial fills in most other regions of the United States. Since about 4000 yr B.P., the alluvial chronologies of the Southwest have been sometimes in phase and sometimes out of phase with the alluvial chronologies of other regions. Presumably, the more complex relationships relate to the greater prevalence of meridional circulation patterns during the late Holocene, which often result in rather complex interregional climatic relationships. Nevertheless, radiocarbon dates that bracket late-Holocene fluvial episodes in the Southwest are in general agreement with those for other regions of the United States. For example, the dates of about 2200, 1100, and 800 yr B.P. are common to many alluvial chronologies. These similarities suggest that instability in river systems is very sensitive to climatic change, although the type of fluvial response may vary from region to region.

GREAT BASIN AND NORTHERN WEST SAGEBRUSH STEPPE

The region of the Northern West Sagebrush Steppe is represented by the northern Great Basin and includes most of Nevada, southeastern and western Utah, central and southwestern Wyoming, eastern Oregon, southern Idaho, and southeastern Washington (Figure 3-8). The characteristic vegetation of the lower-elevation landscape is sagebrush and grass. Butler (1976) concludes that the modern sagebrush-grass biome probably did not emerge on the eastern Snake River Plain of southern Idaho until approximately 7000 yr B.P. He also suggests that pine forests extended to the lowland margins of the eastern Snake River Plain until about 10,800 yr B.P. and were gradually replaced by sagebrush and grass during the Holocene period of warming. Madsen (1976) indicates that late-Pleistocene vegetational zones remained depressed until about 8000 yr B.P. in the southern Great Basin but that by 7000 yr B.P. a shift to markedly warmer conditions introduced the present vegetational communities. A mosaic of grasslands and shrubs probably dominated much of the present sagebrush landscape between about 10,000 and 8000 yr B.P. In the southern part of the region, woodlands may have been absent from the lowlands even during the late-Pleistocene because of rain-shadow effects from high mountains and because the area is so remote from its main moisture source, the Gulf of Mexico (Wells, 1979). Wells (1979) hypothesizes that the aridity of the region intensified during the middle and late Holocene in response to a northward shift in the average summer position of the subtropical high-pressure

cell. In contrast to the Southwest, where maximum aridity occurred during the late Holocene, the Great Basin region became less arid after about 5000 yr B.P. Benson (1978) reports that pluvial Lake Lahontan and its modern successors in Nevada experienced extremely low stands from 9000 to 5000 yr B.P. and that lake levels have increased during the last 5000 years.

In comparison to other regions of the United States little is known about responses of rivers to climatic changes in the Great Basin region. Valley alluviation probably occurred during the early Holocene as the climate became more arid and the landscape became less protected by vegetation. There is not enough information to specify whether aggradation continued through the maximum warm/dry period of the middle Holocene. Although Bryan (1928) and Antevs (1955) argue that increased aridity results in the increased storm runoff and major floods that initiate channel entrenchment, most recent research suggests that entrenchment more commonly follows a climatic shift to more humid conditions. Hence, it seems likely that major entrenchment of Holocene alluvium occurred after about 5000 yr B.P.

A late-Holocene alluvial chronology is available for the Meadow Valley wash in southern Nevada (Madsen, 1976), where very brief episodes of degradation occurred about 3000, 1700, 1250, 900, and 250 yr B.P. and separated longer intervals of aggradation. Madsen suggests that changes in the periodicity of rainfall, caused by shifts in the relative position of weather systems based in the Gulf of Mexico and the Pacific, may account for the episodes of cutting and filling.

HIGH-ELEVATION WESTERN WOODLANDS

Alluvial chronologies have not been studied in the High-Elevation Western Woodlands. Although the upper and lower treelines experienced modest fluctuations, most of the region has been dominated by forest cover during the Holocene. It seems unlikely that Holocene changes in vegetation caused major adjustments in fluvial activity. In contrast, the direct effects of changes in temperature and precipitation probably influenced the physical behavior of many rivers that drain high-elevation landscapes. Madsen and Currey (1979) conclude that Holocene warming is indicated at many sites in the Rocky Mountains until about 8000 yr B.P., when minor glaciation was renewed. They report that most sites illustrate a period of maximum Holocene warmth between about 7500 and 5000 yr B.P. Synchronous minor expansions and contractions of Holocene glaciers (Benedict, 1973; Denton and Karlén, 1973) imply long-term episodic changes in water and sediment discharges in rivers draining the western mountains. Some large rivers that originate in mountain areas but pass through extensive lowland regions, such as the Great Valley of California, probably also have alluvial chronologies that are dominated by climatic events in the mountains.

In summary, the paleoclimatic evidence implies that the beginning and ending dates of alluvial episodes on river systems in the High-Elevation Western Woodlands probably differ little from those in large-scale lowland areas in the same regions. However, because the elevations and orientations of mountains can greatly affect local climatic conditions responding to a particular atmospheric circulation pattern, the alluvial chronologies may sometimes be out of phase with each other and with the lowland regional chronologies.

Discussion and Conclusions

This chapter examines the alluvial chronologies of rivers systems in order to determine the responses of rivers to Holocene climates. The alluvial chronologies indicate that climatic changes and associated vegetational changes have contributed significantly to widespread synchronous episodes of aggradation or degradation in the river systems of the conterminous United States.

The general interest in understanding the possible contributions of climate as a cause of aggradation or degradation in river systems began during the late 19th century, but there still is a lack of consensus on the cause-and-effect relationships involved (Leopold, 1976; Patton and Schumm, 1981). Most of the alluvial chronologies reviewed here support the concept that channel and/or valley alluviation occurs after a shift to drier conditions and that channel and/or valley erosion or entrenchment occurs after a shift to wetter conditions. This relationship supports Huntington's (1914) hypothesis that valley alluviation occurs during dry episodes when vegetation is minimal and sediment yields are high. The observed relationship is consistent with the general rule that persistent increased sediment concentration in river flow usually results in channel aggradation.

Huntington's hypothesis is only partly supported, however, because his concept implies long-term gradual change in aggradation or degradation. The alluvial chronologies reviewed here support the concept that aggradation may often be a long-term, slow process, but they also indicate that episodes of erosion often begin abruptly throughout wide areas and may be of shorter duration that episodes of alluviation. The widespread triggering of erosional episodes probably is the result of the onset of conditions that favor more frequent recurrences of moderate to extensive flooding capable of destabilizing a channel system. Bryan (1928) hypothesizes that minimal vegetation during dry periods resulted in increased surface runoff and large floods, which initiated the channel entrenchment that ultimately expanded upstream through a drainage system. The importance of large floods in channel entrenchment is well documented by historical examples. Unfortunately, most alluvial chronologies have not been radiocarbon-dated in sufficient detail to determine whether significant erosion and entrenchment occurred near the end of a dry episode, when increased rainfall fell on a landscape with vegetation still adjusted to dry conditions, or later within the wet-climate episode. Recent surveys in small southwestern headwater channels support Bryan's hypothesis, but the record is too short to specify long-term implications, especially for larger drainage systems (Leopold, 1976).

The controversy over the hypotheses of Huntington and Bryan might be resolved if alluvial sequences were carefully radiocarbon dated throughout drainage systems. Fluvial responses to climate and vegetation often differ between tributary and trunk channels (Antevs, 1955: 318; Schumm, 1965: 790-92). The lack of conformance between tributary and trunk channels probably is most important in arid and semiarid regions (Schumm, 1965: 792), but even the Holocene alluvial chronology of the humid driftless area in Wisconsin shows that tributary and trunk channels were responding differently to middle-Holocene dry climates (Knox et al., 1981). Because of these uncertainties, I have emphasized dates of discontinuities in fluvial sequences rather than tendencies for aggradation or degradation.

Schumm (1977: 74-81) shows that cutting and filling may depend on geomorphic thresholds such as slope stability and that one initial episode of cutting can result in multiple terrace levels. Patton and Schumm (1981) suggest that, where sediment yields are high and the resulting ratio of sediment discharge to water discharge is high, channel cutting and filling are a natural sequence of events by which sedi-

ment is episodically transported out of a drainage system. They conclude that, though major components of an alluvial chronology reflect major climatic changes, details of an alluvial chronology may only reflect episodic cut-and-fill transport processes. I agree with these interpretations. In many parts of the humid United States, episodes of intensified lateral stream migration after about 6000 yr B.P., which were associated with active cut-bank erosion and point-bar deposition, produced similar detailed chronologies that probably are no more environmentally significant than deposits produced by the episodic cut-and-fill transport processes described by Patton and Schumm (1981). What is most significant, of course, is that during certain periods of the Holocene cutting and filling and/or the relative intensity of lateral channel migration were very active, whereas during other periods of the Holocene they were relatively inactive. These periods define the major components of alluvial chronologies.

The major alluvial episodes did not begin and end everywhere at precisely the same time, but they represent long-term intervals when similar types of fluvial activity were dominant within large regions. At the same time, the type and intensity of dominant fluvial activity in one region may be out of phase with the type and intensity of dominant fluvial activity in another region because of their relationships to the large-scale atmospheric circulation.

The responses of rivers to Holocene climates can be summarized in the following way (Figure 3-9). Between about 10,000 and 8000 yr B.P., most regions were rapidly becoming warmer and drier and valley alluviation was dominant. The magnitude of valley alluviation generally increased westward in parallel with increased drying and increased vegetational change. Between about 8000 and 7500 yr B.P., valley alluviation apparently was interrupted by erosion that was relatively minor in the East and the humid Midwest but very significant in the Southwest. Between about 8000 and 6000 yr B.P., warm/dry conditions dominated in the northern West and in the Midwest, but the southern Southwest and parts of the East and Southeast were warm/wet. During the period 8000 to 6000 yr B.P., valley alluviation apparently slowed in the northern West, the Midwest, and the East, but in the warm/wet Southwest major erosion of valley fills occurred.

Vegetational change in response to warmer temperatures probably was the predominant factor responsible for alluviation before 8000 yr B.P. The geographic locations of large-scale vegetational communities became relatively stabilized after about 8000 yr B.P., and the direct effects of climate became more dominant in alluvial chronologies. Hence, warm/wet conditions produced in the South by the persistent zonal atmospheric circulation of the early Holocene contributed to valley trenching in the Southwest, where grasslands had emerged, but had little effect in the Southeast, where forest continued to dominate.

Between about 6000 and 4500 yr B.P., significant erosion of early-Holocene alluvial fills was occurring in most regions, except in the Southwest, where active alluviation prevailed. There are two major causes of the widespread erosion after 6000 yr B.P. First, the long-term late-Holocene cooling trend had begun and probably improved vegetative cover in some areas so that sediment yields were reduced and entrenchment was favored. Second, and probably more important, meridional patterns of large-scale atmospheric circulation became much more frequent in summers after about 6000 yr B.P. They replaced the strong zonal westerly summer circulation of the early Holocene. Whereas the zonal regimes of the early-Holocene summers dominated the northern West and the Midwest with dry

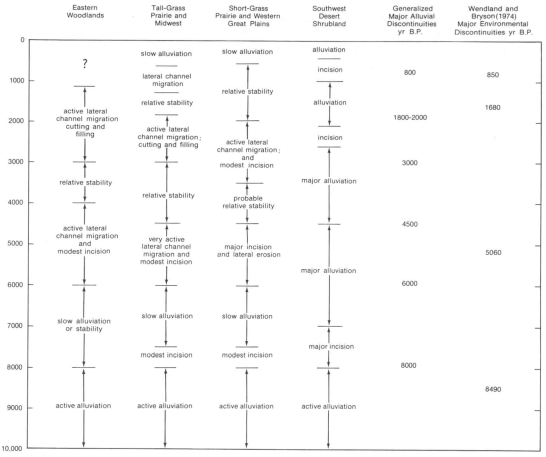

Figure 3-9. Regional alluvial chronologies. This illustration is a very generalized representation of Holocene fluvial activity derived from many site-specific alluvial sequences. There was not enough information to generalize the Great Basin and Northern West Sagebrush Steppe and the High-Elevation Western Woodlands. Vegetational change and rapid warming caused widespread alluviation before 8000 yr B.P. Since about 8000 yr B.P., the Southwest's climate trends and fluvial activity apparently have differed from those common to other large regions of the United States. Alluviation dominated during much of the early Holocene, but degradation, apparently associated with occasional episodes of intense lateral channel migration, probably dominated during the late Holocene, except in the Southwest Desert Shrubland, where maximum aridity developed during the late Holocene.

Pacific air masses, the meridional regimes produced deep penetrations of polar and tropical air masses into these same regions. Precipitation and large floods undoubtedly increased significantly because of frequent frontal activity in the collision zone between unlike air masses. More frequent large floods may have been the most critical factor in initiating erosional activity on valley floors, which probably were oversteepened by early-Holocene alluviation.

Meanwhile in the Southwest, where active alluviation was occurring between about 6000 and 4500 yr B.P., the climate was changing to increased aridity as a result of a northward displacement of the summer position of the subtropical high-pressure cell after the early-Holocene zonal circulation was replaced by more frequent occurrences of meridional circulation.

The intensity of erosional and depositional processes subsided somewhat in most regions between about 4500 and 3000 yr B.P., but between about 3000 and 1800 yr B.P. many regions experienced the processes with renewed intensities. The type of renewed fluvial activity varied from region to region. For example, in the northern Midwest, it involved very active lateral channel migration with erosion and deposition of sediments; on the western edge of the Great Plains, many sites experienced alluviation; and, in the southern Great

Plains of Texas, erosion and entrenchment occurred. The intensity of fluvial activity again appears to have slowed after about 1800 yr B.P. in many regions and remained at a modest level until 1200 to 800 yr B.P., when fluvial activity represented by cutting and filling, active lateral channel migration, et cetera, again became more evident. Since about 800 yr B.P., modest alluviation seems to have dominated most regions until it was ended by late-19th-century trenching in some regions.

In summary, the significant generalized dates in radiocarbon years that bracket alluvial episodes appear to be 8000, 6000, 4500, 3000, 2000 to 1800, and 800 (Figure 3-9). These dates are not widely different from the dates of major discontinuities identified by Wendland and Bryson (1974) in an analysis of North American and European radiocarbon dates associated with changes in geologic and botanical conditions. These similarities lend support to those investigators who favor subdividing the Holocene into more than three categories. The tripartite system also gives the false impression that the cool/moist early Holocene was similar to the cool/moist late Holocene. It has been shown here that the large-scale dominant atmospheric circulation patterns of the two periods were very different and that they produced rather different responses in river systems. The relative responsiveness of rivers to climatic changes showed major regional differences closely related to regional vegetative cover. The responsiveness of rivers to climatic change increased as vegetative cover became less effective as a control on surface runoff and sediment yield. These observations support the concept of using contemporary regional relationships of floods, sediment yields, and sediment concentrations to approximate the relative effectiveness of vegetational change over time in influencing river activity. Although vegetation seems to have been the most important factor in determining the relative degree of responsiveness to climatic change, the widespread synchroneity of fluvial episodes in different vegetational regions suggests that the magnitudes and frequencies of floods are a critical direct climatic control.

References

Ahler, S. A. (1973a). Chemical analysis of deposits at Rodgers Shelter, Missouri. *Plains Anthropologist* 18, 116-31.

Ahler, S. A. (1973b). Post-Pleistocene depositional change at Rodgers Shelter, Missouri. *Plains Anthropologist* 18, 1-26.

Albanese, J. P. (1974). Holocene alluvial chronology and climate change on the northwestern Plains. *In* "American Quaternary Association, Third Biennial Meeting, Abstracts,"p. 88. University of Wisconsin, Madison.

Albanese, J. P., and Wilson, M. (1974). Preliminary description of the terraces of the North Platte River at Casper, Wyoming. *In* "Applied Geology and Archaeology: The Holocene History of Wyoming" (M. Wilson ed.), pp. 8-18. Geological Survey of Wyoming Report of Investigations 10.

Alexander, C. S., and Prior, J. C. (1971). Holocene sedimentation rates in overbank deposits in the Black Bottom of the lower Ohio River, southern Illinois. *American Journal of Science* 270, 361-72.

American Quaternary Association (1976). "Guidebook, San Pedro Valley Field Trip, October 11, 1976." American Quaternary Association, Fourth Biennial Meeting, Arizona State University, Tempe.

Antevs, E. (1952). Arroyo-cutting and filling. *Journal of Geology* 60, 375-85.

Antevs, E. (1955). Geologic-climatic dating in the West. *American Antiquity* 20, 317-35.

Baker, V. R., and Penteado-Orellana, M. M. (1977). Adjustment to Quaternary climatic change by the Colorado River in central Texas. *Journal of Geology* 85, 395-422.

Bates, C. G., and Zeasman, O. R. (1930). "Soil Erosion: A Local and National Problem." University of Wisconsin Agricultural Experiment Station Bulletin 99.

Benedict, J. B. (1973). Chronology of cirque glaciation, Colorado Front Range. *Quaternary Research* 3, 584-99.

Benson, L. V. (1978). Fluctuations in the level of pluvial Lake Lahontan during the last 40,000 years. *Quaternary Research* 9, 300-318.

Bernabo, J. C., and Webb, T. W. (1977). Changing patterns in the Holocene pollen record or northeastern North America: A mapped summary. *Quaternary Research* 8, 64-96.

Borchert, J. R. (1950). The climate of the central North American grassland. *Annals of the Association of American Geographers* 40, 1-39.

Brakenridge, G. R. (1981). Late Quaternary floodplain sedimentation along the Pomme de Terre River, southern Missouri. *Quaternary Research* 15, 62-76.

Brice, J. C. (1964). "Channel Patterns and Terraces of the Loup Rivers in Nebraska." U.S. Geological Survey Professional Paper 422-D.

Brice, J. C. (1966). "Erosion and Deposition in the Loess-Mantled Great Plains, Medicine Creek Drainage Basin, Nebraska." U.S. Geological Survey Professional Paper 352-H.

Bryan, K. (1928). Historic evidence on changes in the channel of the Rio Puerco, a tributary of the Rio Grande in New Mexico. *Journal of Geology* 36, 265-82.

Bryson, R. A., Baerreis, D. A., and Wendland, W. M. (1970). The character of late-glacial and post-glacial climatic changes. *In* "Pleistocene and Recent Environments of the Central Great Plains" (W. Dort, Jr., and J. K. Jones, Jr., eds.), pp. 53-74. University of Kansas Press, Lawrence.

Bryson, R. A., and Hare, F. K. (1974). "Climates of North America." Elsevier, New York.

Bryson, R. A., and Wendland, W. M. (1967). Tentative climatic patterns for some late-glacial and post-glacial climatic episodes in central North America. *In* "Life, Land, and Water, Lake Agassiz Region" (W. J. Mayer-Oakes, ed.) pp. 271-98. University of Manitoba Press, Winnipeg.

Bryson, R. A., Wendland, W. M., Ives, J. D., and Andrews, J. T. (1969). Radiocarbon isochrones on the disintegration of the Laurentide ice sheet. *Arctic and Alpine Research* 1, 1-14.

Burkham, D. E. (1972). "Channel Changes of the Gila River in Safford Valley, Arizona 1846-1970." U.S. Geological Survey Professional Paper 655-G.

Busby, M. W. (1963). "Yearly Variations in Runoff for the Conterminous United States." U.S. Geological Survey Water-Supply Paper 1669-S.

Butler, B. R. (1976). The evolution of the modern sagebrush-grass steppe biome on the eastern Snake River Plain. *In* "Holocene Environmental Change in the Great Basin" (R. Elston, ed.), pp. 5-39. Nevada Archaeological Survey Research Paper 6.

Butzer, K. W. (1977). "Geomorphology of the Lower Illinois Valley as a Spatial-Temporal Context for the Koster Archaic Site." Illinois State Museum Report of Investigations 34.

Butzer, K. W. (1980). Holocene alluvial sequences: Problems of dating and correlation. *In* "Timescales in Geomorphology" (R. A. Cullingford, D. A. Davidson, and J. Lewin, eds.), pp. 131-42. John Wiley and Sons, London.

Cooke, R. V., and Reeves, R. W. (1976). "Arroyos and Environmental Change in the American South-West." Oxford University Press, London.

Daniels, R. B., and Jordan, R. H. (1966). "Physiographic History and the Soils, Entrenched Stream Systems, and Gullies, Harrison County, Iowa." U.S. Department of Agriculture Technical Bulletin 1348.

Davis, M. B., Spear, R. W., and Shane, L. C. K. (1980). Holocene climate of New England. *Quaternary Research* 14, 240-50.

Davis, W. M. (1902). Base level, grade, and peneplain. *Journal of Geology* 10, 77-111.

Dendy, F. E., and Bolton, G. C. (1976). Sediment yield-runoff-drainage area relationships in the United States. *Journal of Soil and Water Conservation* 31, 264-66.

Denton, G. H., and Karlén, W. (1973). Holocene climatic variations: Their pattern and possible cause. *Quaternary Research* 3, 155-205.

Gilbert, G. K. (1877). "Report on the Geology of the Henry Mountains." U.S. Geographical and Geological Survey of the Rocky Mountain Region, Washington, D.C.

Gilbert, G. K. (1900). Rhythms in geologic time. *Proceedings of the American Association for Advancement of Science* 49, 1-19.

Gile, L. H. (1975). Holocene soils and soil-geomorphic relations in an arid region of southern New Mexico. *Quaternary Research* 5, 321-60.

Gooding, A. (1971). Postglacial alluvial history in the upper Whitewater Basin, southeastern Indiana, and possible regional relationships. *American Journal of Science* 271, 389-401.

Grissinger, E. H., Murphy, J. B., and Little, W. C. (1981). Late-Quaternary valley-fill deposits in north-central Mississippi. Unpublished Manuscript, U.S. Department of Agriculture, Agricultural Research Service, Oxford, Miss.

Hall, S. A. (1977). Late Quaternary sedimentation and paleoecologic history of Chaco Canyon, New Mexico. *Geological Society of America Bulletin* 88, 1593-1618.

Hall, S. A. (1980). Corresponding geomorphic, archaeologic, and climatic change in the southern Plains: New evidence from Oklahoma. *In* "American Quaternary Association, Sixth Biennial Meeting, Abstracts and Program, 18-20 August 1980," p. 89. Institute for Quaternary Studies, University of Maine, Orono.

Hallberg, G. R., Harbaugh, J. M., and Witinok, P. M. (1979). "Changes in the Channel Area of the Missouri River in Iowa 1879-1976." Iowa Geological Survey Special Report Series 1.

Hallberg, G. R., Hoyer, B. E., and Miller, G. A. (1974). The geology and paleopedology of the Cherokee sewer site. *Journal of the Iowa Archaeological Society* 21, 17-50.

Hare, F. K. (1976). Late Pleistocene and Holocene climates: Some persistent problems. *Quaternary Research* 6, 507-17.

Haynes, C. V. (1968). Geochronology of late-Quaternary alluvium. *In* "Means of Correlation of Quaternary Successions" (R. B. Morrison and H. E. Wright, Jr., eds.), pp. 591-631. University of Utah Press, Salt Lake City.

Haynes, C. V. (1976). Late Quaternary geology of the lower Pomme de Terre Valley. *In* "Prehistoric Man and His Environments" (W. R. Wood and R. B. McMillan, eds), pp. 47-61. Academic Press, New York.

Hoyer, B. E. (1980). The geology of the Cherokee sewer site. *In* "The Cherokee Excavations: Holocene Ecology and Human Adaptations in Northwestern Iowa" (D. C. Anderson and H. A. Semken, Jr., eds.), pp. 21-66. Academic Press, New York.

Huntington, E. (1914). "The Climatic Factor as Illustrated in Arid America." Carnegie Institution Publication 192.

Johnson, W. C., Dort, W., and Sorenson, C. J. (1980). An episode of late-Holocene soil development on floodplains of the central Plains. *In* "American Quaternary Association, Sixth Biennial Meeting, Abstracts and Program," p. 116. Institute for Quaternary Studies, University of Maine, Orono.

Johnson, W. D. (1901). "The High Plains and Their Utilization." U.S. Geological Survey, 11th Annual Report, part IV, pp. 601-741.

Kalnicky, R. A. (1974). Climatic change since 1950. *Annals of the Association of American Geographers* 64, 100-112.

Knox, J. C. (1972). Valley alluviation in southwestern Wisconsin. *Annals of the Association of American Geographers* 62, 401-10.

Knox, J. C. (1976). Concept of the graded stream. *In* "Theories of Landform Development" (W. N. Melhorn and R. C. Flemal, eds.), pp. 169-98. State University of New York at Binghamton Publications in Geomorphology.

Knox, J. C., Bartlein, P. J., Hirschboeck, K. K., and Muckenhirn, R. J. (1975). "The Response of Floods and Sediment Yields to Climate Variation and Land Use in the Upper Mississippi Valley." University of Wisconsin Institute for Environmental Studies Report 52.

Knox, J. C., McDowell, P. F., and Johnson, W. C. (1981). Holocene fluvial stratigraphy and climatic change in the driftless area, Wisconsin. *In* "Quaternary Paleoclimate" (W. C. Mahaney, ed.), pp. 107-27. Geo Abstracts Limited, Norwich, England.

Kutzbach, J. E. (1970). Large-scale features of monthly mean Northern Hemisphere anomaly maps of sea-level pressure. *Monthly Weather Review* 98, 708-16.

Lamb, H. H. (1966). Climate in the 1960s. *The Geographical Journal* 132, 183-212.

Lamb, H. H. (1974). "The Current Trend of World Climate: A Report on the Early 1970s and a Perspective." University of East Anglia Climate Research Unit Research Publication.

Lamb, H. H. (1977). "Climate: Present, Past and Future: Climatic History and the Future." Vol. 2. Methuen, London.

Lane, E. W. (1937). Stable channels in erodible materials. *American Society of Civil Engineers, Transactions* 102, 123-94.

Langbein, W. B., and others (1949). "Annual Runoff in the United States." U.S. Geological Survey Circular 52.

Langbein, W. B., and Schumm, S. A. (1958). Yield of sediment in relation to mean annual precipitation. *Transactions of the American Geophysical Union* 39, 1076-84.

Leopold, L. B. (1976). Reversal of erosion cycle and climatic change. *Quaternary Research* 6, 557-62.

Leopold, L. B., and Maddock, T. (1953). "The Hydraulic Geometry of Stream Channels and Some Physiographic Implications." U.S. Geological Survey Professional Paper 252.

Leopold, L. B., and Miller, J. P. (1954). "A Postglacial Chronology for Some Alluvial Valleys in Wyoming." U.S. Geological Survey Water Supply Paper 1261.

Madole, R. F. (1976). Differentiation of upper Pleistocene and Holocene gravels along St. Vrain Creek, eastern Boulder County, Colorado. *In* "American Quaternary Association, fourth Biennial Meeting, Abstracts," p. 146. Arizona State University, Tempe.

Madole, R. F., Swinehart, J. B., and Muhs, D. R. (1981). Correlations of Holocene dune sands on the central Great Plains and their paleoclimatic implications. *Geological Society of America, Programs with Abstracts* 13, 287.

Madsen, D. B. (1976). Pluvial-post pluvial vegetation changes in the southeastern Great Basin. *In* "Holocene Environmental Change in the Great Basin" (R. Elston, ed.), pp. 104-19. Nevada Archaeological Survey Research Paper 6.

Madsen, D. B., and Currey, D. R. (1979). Late Quaternary glacial and vegetation changes, Little Cottonwood Canyon area, Wasatch Mountains, Utah. *Quaternary Research* 12, 254-70.

Noble, E. L. (1965). "Sediment Reduction through Watershed Rehabilitation." U.S. Department of Agriculture, Miscellaneous Publication 970, pp. 114-23.

Patton, P. C., and Schumm, S. A. (1981). Ephemeral-stream processes: Implications for studies of Quaternary valley fills. *Quaternary Research* 15, 24-43.

Prest, V. (1969). Retreat of Wisconsin and recent ice in North America. Geological Survey of Canada Map 1257A.

Rubey, W. W. (1952). "Geology and Mineral Resources of the Hardin and Brussels Quadrangles (in Illinois)." U.S. Geological Survey Professional Paper 218.

Ruhe, R. V. (1969). "Quaternary Landscapes in Iowa." Iowa State University Press, Ames.

Sartz, R. S. (1970). Effect of land use on hydrology of small watersheds in southwestern Wisconsin. *In* "Symposium on the Result of Research on Representative and Experimental Basins," pp. 286-95. International Association of Scientific Hydrology Publication 96.

Schultz, C. B., and Martin, L. D. (1970). Quaternary mammalian sequence in the central Great Plains. *In* "Pleistocene and Recent Environments of the Central Great Plains" (W. Dort, Jr., and J. K. Jones, Jr., eds.), pp. 341-53. University of Kansas Press, Lawrence.

Schumm, S. A. (1960). "The Shape of Alluvial Channels in Relation to Sediment Type." U.S. Geological Survey Professional Paper 352-B.

Schumm, S. A. (1965). Quaternary paleohydrology. *In* "The Quaternary of the United States" (H. E. Wright, Jr., and D. G. Frey, eds.), pp. 783-94. Princeton University Press, Princeton, N.J.

Schumm, S. A. (1977). "The Fluvial System." John Wiley and Sons, New York.

Schumm, S. A., and Lichty, R. W. (1963). "Channel Widening and Flood-Plain Construction along the Cimarron River in Southwestern Kansas." U.S. Geological Survey Professional Paper 352-D.

Scott, G. R. (1963). "Quaternary Geology and Geomorphic History of the Kassler Quadrangle, Colorado." U.S. Geological Survey Professional Paper 421-A.

Scully, R. W., and Arnold, R. W. (1981). Holocene alluvial stratigraphy in the upper Susquehanna River basin, New York. *Quaternary Research* 15, 327-44.

Stevens, M. A., Simons, D. B., and Richardson, E. V. (1975). Non-equilibrium river form. *Proceedings of the American Society of Civil Engineers, Journal Hydraulics Division* HY 5, 101, 551-66.

Tuan, Y-F. (1966). New Mexican gullies: A critical review and some recent observations. *Annals of the Association of American Geographers* 56, 573-97.

Van Devender, T. R., and Spaulding, W. G. (1979). Development of vegetation in the southwestern United States. *Science* 204, 701-10.

Van Zant, K. (1979). Late-glacial and postglacial pollen and plant macrofossils from Lake West Okoboji, northwestern Iowa. *Quaternary Research* 12, 358-80.

Wells, P. V. (1979). An equable glaciopluvial in the West: Pleniglacial evidence of increased precipitation on a gradient from the Great Basin to the Sonoran and Chihuahuan Deserts. *Quaternary Research* 12, 311-25.

Wendland, W. M., and Bryson, R. A. (1974). Dating climatic episodes of the Holocene. *Quaternary Research* 4, 9-24.

Wolman, M. G., and Miller, J. P. (1960). Magnitude and frequency of forces in geomorphic processes. *Journal of Geology* 68, 54-74.

Wright, H. E., Jr. (1970). Vegetational history of the central Plains. *In* "Pleistocene and Recent Environments of the Central Great Plains" (W. Dort, Jr., and J. K. Jones, Jr., eds.), pp. 157-72. University of Kansas Press, Lawrence.

Wright, H. E., Jr. (1976). The dynamic nature of Holocene vegetation: A problem in paleoclimatology, biogeography, and stratigraphic nomenclature. *Quaternary Research* 6, 581-96.

CHAPTER 4

Sea Level and Coastal Changes

Arthur L. Bloom

Introduction

The first day of the Holocene Epoch 10,000 years ago was not noticeably different from the days that preceded or followed it. The ice sheets covering Canada and Scandinavia were rapidly disintegrating. Immature soils, still only a few thousand years old, were forming on the loess and drift of recently deglaciated regions. Permafrost was disappearing from the surficial soil layers of a rapidly shrinking periglacial region. Plants and animals were shifting into new habitats as the climate warmed. And sea level was rising.

From a glacially lowered level of some 120 ± 60 m below present sea level (Bloom, 1983), the sea surface rapidly rose in reciprocal proportion to the volume of the shrinking late-glacial ice sheets and mountain glaciers. By 10,000 yr B.P., the area of former ice cover had been reduced to about 50% of its maximum extent during the Wisconsin Glaciation. The volume of remaining ice is much more difficult to estimate, but it probably had been reduced to 30% to 50% of the ice volume of the glacial maximum (Bloom, 1971: Figure 4). The remaining one-half of the deglaciated area was to be uncovered and the remaining one-half to one-third of the ice volume was to return to the sea within the next 3000 to 4000 years.

At the first dawn of the Holocene, sea level was about 40 ± 10 m below its present level on most coasts. It was rising rapidly, about 1 cm per year, or almost 1 mm per month. The decision of many modern Quaternary scientists to define the beginning of the Holocene Epoch at 10,000 yr B.P. would have aroused little interest among the inhabitants of the late-glacial world (had they known): they were involved with survival, migration, extinction, and a chaotic, rapidly evolving biogeographic environment. Especially from the viewpoint of modern coastal research, the placement of the latest epochal boundary at the time of the maximum rate of late-glacial sea-level change has little meaning.

Late Wisconsin sea-level positions and coastal paleogeography are based on a minimum of facts learned with great difficulty from evidence that is mostly submerged. (See Bloom, 1983.) It has been agreed that almost no radiocarbon dates reliably document sea-level positions deeper than − 60 m and older than 10,000 yr B.P. However, for times since 10,000 yr B.P. and certainly since about 7000 yr B.P., many more data are available. Most Holocene sea-level curves extend back to about 7000 yr B.P., when they were in steep descent (Bloom, 1977). Fewer than 40% of the published sea-level curves even suggest a position or trend as early as 10,000 yr B.P.; these show regional consistency attributable to either converging evidence or pervasive regional or national bias. The data are too few for statistical analyses, but the position of sea level on the east coast of the United States at 10,000 yr B.P. is usually estimated at − 25 to − 32 m, on the Gulf Coast at − 30 to − 40 m, in the Caribbean Sea at − 32 to − 36 m, on the California coast at − 40 to − 60 m, and on the Alaskan coast at − 20 to − 30 m. No averaging technique is valid for such generalizations.

At 10,000 yr B.P., recently deglaciated areas of the northeastern United States were experiencing strong postglacial isostatic uplift and related gravitational and geoidal adjustments (Clark et al., 1978). The coast of New England north of Boston (lat. 42.5° N) had been submerged by the late-glacial transgression over isostatically depressed terrain, but the reemergence seems to have been complete by 12,500 yr B.P. (Stuiver and Borns, 1975). The extent of the reemergence seaward of the present shoreline has not been determined. Late-glacial marine sediments are oxidized to 11 m below present sea level in coastal Maine, suggesting at least that much reemergence (Bloom, 1963: 869). By 4000 to 3000 yr B.P., slow marine transgression was again in progress on the coast of Maine, and it has continued into the present time.

Along almost all other coasts of the conterminous states and Hawaii (but excluding Alaska and a few California localities), the Holocene has been a time of marine transgression, although the rates and amounts have varied from place to place and as time has passed. Most of this chapter will be a review of the effects of the Holocene transgression on various coastal terrains in the United States.

This is a contribution to the U.S. National IGCP Project 61, the Sea-Level Project and Cornell University Department of Geological Sciences contribution number 734.

Causes of Variation in the Rates
of Holocene Submergence

The primary cause of the approximately 40 m of Holocene submergence on most coasts of the United States and elsewhere was the ablation of the remaining ice of the vanished Wisconsin glaciers. If that had been the only phenomenon in operation, Holocene sea-level change would have been uniform in all nonglaciated regions. The demonstration that each coastal segment has had a unique Holocene sea-level history has led to several decades of scientific debate about the relative tectonic stability of various coastal regions. However, it has become generally understood as a result of many national and international conferences and field excursions that tectonic movement is only a minor factor in the Holocene history of most coasts, exclusive of those at tectonically active margins and in areas of postglacial isostatic uplift. More important are factors related to isostatic adjustments.

A superficial and not entirely accurate insight into the causes of different Holocene submergence on various coasts can be gained by noting that the density of the lithosphere is about three times as great as the density of water. As melting ice sheets returned an amount of water to the ocean, the suboceanic lithosphere should have responded to the water load by subsiding an amount approximately one-third of the thickness of the new water layer. At the edge of the oceans, this hydro-isostatic response of the sea floor must have been complex, depending on the planimetric shape of the coast and the offshore depth profile (Bloom, 1967; Higgins, 1965). Postglacial isostatic uplift of deglaciated areas demonstrates the viscous time dependence involved in lithospheric response to changing superincumbent loads, and a similar time lag should be expected in the response of coastal lithosphere to adjacent and superjacent water loads.

More sophisticated models of postglacial sea-level history include factors such as the loss of gravitational attraction of seawater toward former ice sheets (Clark, 1976, 1977). Another complex factor includes the changes in the geoidal figure of the Earth when large ice sheets, asymetrically arranged in high northern latitudes relative to the Earth's axis of rotation, melted and distributed water mass over a largely tropical oceanic surface. The mathematical analysis of glacial unloading and ocean-floor loading by meltwater and the related geoidal and gravitational responses of a layered viscoelastic Earth has been greatly advanced in the past decade (Cathles, 1975; Chappell, 1974; Clark, 1976, 1977, 1980; Clark et al, 1978; Clark and Lingle, 1979; Farrell and Clark, 1976; Walcott, 1972). It is now much easier to understand how even a hypothetical, linear eustatic rise of sea level, ending at 5000 yr B.P., could have produced global differences in late-Holocene sea-level history that range from 10 m of submergence to 4 m of emergence, even excluding the regions of obvious postglacial isostatic uplift (Figure 4-1). We no longer seek a single eustatic ("worldwide") curve of sea-level change but recognize instead that every coast has its own unique Holocene sea-level history. However, regional trends can be recognized, and at least six characteristic types of Holocene sea-level curves have been defined for a hypothetical (but reasonable) eustatic model (Figure 4-1).

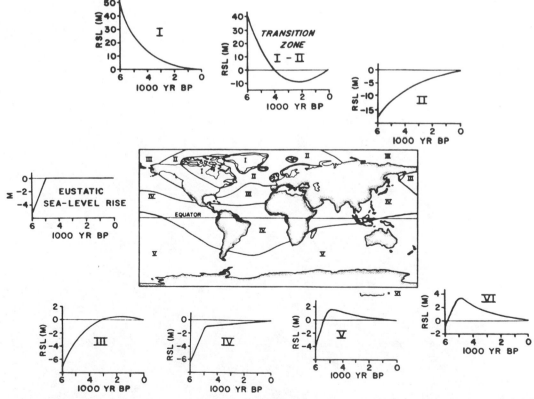

Figure 4-1. Typical Holocene relative sea-level (RSL) curves predicted for each of six global zones. Predictions are based on Northern Hemisphere deglaciation and a resulting eustatic sea-level rise of 75.6 m between 17,000 and 5000 years ago. No eustatic change is assumed during the last 5000 yeras. Significant Holocene deglaciation in Antarctica would modify the boundaries of the six zones. (From Clark and Lingle, 1979: Figure 1.)

Paleogeography of Holocene Coasts

Depending upon their geologic terrane, sediment supply, and energy regimes, various coasts evolved differently as the Holocene submergence approached present sea level. Details differ, but examples chosen from muddy, sandy, and coral-reef coasts will illustrate the Holocene evolution of the coasts of the United States. The periglacial coast of Alaska had a Holocene history different enough to deserve a fourth category of description.

MUDDY COASTS

The coast of New England south and west from Maine to New York City (but excluding the sandy shores of Long Island) is the drowned margin of a glaciated metamorphic terrane of low relief. Structurally controlled fluvial and glacial erosion have etched the northeastward-striking metamorphic grain oblique to the trend of the coast, so that a ria coast has evolved (Figure 4-2). South of Boston, the bedrock relief along the coast is rarely more than 10 m. Northeastward from Boston, the coastal relief increases to 50 m. Mount Desert Island, Maine, has an anomalous height of 467 m and the only fiords on the Atlantic coast of the United States.

The New England coast is characterized by shallow bays and estuaries eroded along weak-rock zones. The coastal sediments are generally muddy, with the exception of Cape Cod and other smaller areas where sand and gravel derived from glacial drift have built barriers. The "New England-type" tidal salt marsh has been made well known by the early writings of Douglas Johnson (1919, 1925) as well as by plant geographers (Chapman, 1960; Davis, 1910). A New England tidal marsh has a broad expanse of salt-tolerant grasses such as *Spartina patens*, *S. alterniflora*, and *Distichlis spicata*, which form a high-tide meadow drained by meandering tidal creeks and artifical ditches.

Davis (1910) was the first to show that the New England type of tidal marsh is a late-Holocene phenomenon. By coring the marshes around Boston, he demonstrated that the peat of the high-tide grass assemblage is only about 3 m thick and overlies either peat of less heavily vegetated midtide marshes or mud of an open estuary. At the base of a New England tidal marsh, usually at a depth of no more than 10 to 15 m, a layer of fresh- or brackish-water sedge peat can be found, which records the Holocene transgression into a former swampy lowland. The basal sedge-peat layer can sometimes be traced landward to the inner edge of a tidal marsh continuous with its living counterpart. The continuity of a basal sedge-peat layer on the sloping floor of an estuarine marsh testifies to the continuity of the Holocene transgression. Radiocarbon dates on the basal transgressive peat beds have abundantly documented the Holocene submergence history of many New England tidal marshes (Bloom, 1964; Bloom and Stuiver, 1963; McIntire and Morgan, 1962; Newman et al., 1971; Redfield and Rubin, 1962). Although the total amount of submergence, and therefore the rate, varies from place to place for reasons discussed in the previous section, the submergence curves for New England tidal marshes typically become less rapid in the most recent millennia. In

Figure 4-2. Structurally controlled coast of northern New England. Glacier flow was toward the south or southeast, oblique to the strike of the metamorphic rocks. (NASA photograph ERTS E-1472-14530.)

Connecticut, for instance, the average submergence rate from 7000 to 3000 yr B.P. (calendar years corrected for radiocarbon errors) was about 1.2 m per 1000 years. For the last 3000 years, the average rate was only 0.85 m per 1000 years (Bloom, 1964; Bloom and Stuiver, 1963). The decrease in submergence rate coincided with a change from open muddy estuaries to New England-type tidal marshes.

One can hypothesize that the nearshore open-water sedimentation rate was slightly less than the earlier and more rapid submergence rate, so that bays and estuaries were not filled to intertidal mud flats but continued to be open water. After 3000 yr B.P., however, the sedimentation rate exceeded the rate of submergence, and intertidal mud flats became widespread. The mud flats were quickly colonized by salt-marsh grasses, and within 1000 years the muddy estuaries had become high-tide grassy meadows. For the last 2000 years, the high-tide meadows have persisted in spite of an additional submergence of 2 m. Once established, a tidal marsh can trap sediment so efficiently that it can grow upward at 2 to 6 mm per year (Harrison and Bloom, 1977), equal even to short-term historical pulses of extremely rapid submergence. Support for the idea of a dramatic change in coastal scenery because of a slight decrease in submergence rate is provided by measurements of the modern sedimentation rate in Long Island Sound seaward of the Connecticut tidal marshes. The modern rates range from 0.33 to 0.7 mm per year (Gordon, 1979: 636) and should be higher nearer shore. Before human intervention, there were about 116 km^2 of tidal salt marsh along the 145 km straight-line length of the Connecticut coast. Extensive straightening of the shoreline had occurred through marsh accumulation, even though bay-mouth barriers of sand are rare. The seaward edges of the marshes erode landward even as their peaty sediments accrete upward at a rate that balances the sea-level rise plus compaction (Bloom, 1965). On the landward edge of the tidal salt marshes, the continuing progress of the leading edge of the marine transgression can be observed in shallow excavations.

The subtle decrease in the Holocene submergence rate until it balanced or was exceeded by the nearshore sedimentation rate might have been responsible for significant changes in coastal prehistoric human occupancy in southern New England. Archaeological sites of the Archaic people who were in the coastal region before 3500 yr B.P. primarily show an upland food supply based on hunting and gathering, although there is evidence that oysters were sometimes collected (Brennan, 1977: 426; Ritchie, 1969: 142-44; Wyatt, 1977). It is possible that the Archaic sites, even though they are near the coast today, might actually have been seasonal inland camps and that shellfish were seasonally collected at sites that are now submerged. Nevertheless, the much greater utilization of shellfish by the later Transitional and Woodland people after 3300 yr B.P. is notable (Ritchie, 1969: 166). The enormous productivity of the newly evolved tidal salt marshes may have been the basis for the dietary change. By about 2500 yr B.P., some of the Woodland sites on Long Island, New York, show a decline in emphasis on shellfish eating, perhaps attributable to the completed infilling of local estuaries by marshes (Wyatt, 1977: 407).

Within certain muddy bays and estuaries southward from Long Island, New York, even though the sediments on that part of the Atlantic coast are predominantly sandy, the Holocene stratigraphy is essentially identical to the sequence in Connecticut just described. Radiocarbon dates from basal peat beds in bays and in the nearshore zone on the south coast of Long Island show that the Holocene submergence rate there was about 2.5 m per 1000 years from 7000 to 3000 yr B.P. but reduced abruptly to only 1.0 m per 1000 years dur-

ing the last 3000 years (Rampino, 1979). Even though the reported submergence rates are higher than those of the Connecticut coast 45 km farther to the north, the inferred change from open muddy lagoons to *Spartina patens* salt marshes was at 3000 yr B.P. Presumably, a higher sedimentation rate on the south shore of Long Island happened to coincide with the higher submergence rate, so that the changes from lagoon to marsh occurred at the same time in southern Long Island and in Connecticut. Most of the data points for the southern Long Island study are derived from published reports of other localities, so the rates cannot be absolutely confirmed.

An even more instructive example of the evolution of the New England type of tidal salt marsh can be found in Delaware (latitude 39°N). The decreasing rate of Holocene submergence there has been approximated by three straight-line segments, with sea level rising at 2.96 m per 1000 years prior to 5000 years ago, about 2.07 m per 1000 years from about 5000 to 2000 years ago, and 1.25 m per 1000 years for the last 2000 years (calendar years corrected for radiocarbon errors; see Belknap and Kraft, 1977: 621). An estuary with a high rate of mud deposition within the larger Delaware Bay responded to the decreasing rate of submergence by recording a marine transgression during the initial rapid submergence, then a regression as the submergence rate decreased and rapid sedimentation dominated, and finally a renewed transgression during the last 2000 years possibly related to tidal changes. By contrast, a bay with a low supply of muddy sediment built a stratigraphic record of rapid, then slow, then again rapid transgression during the same tripartite rate of submergence (Belknap and Kraft 1977: 627). The important principle is that either a marine transgression or regression can occur or they can alternate during continuing submergence. The critical factors are the relative rates of submergence and sediment accretion. A transgressive-regressive-transgressive sedimentary record should not be accepted as proof of an oscillation in a time-depth plot of a sea-level curve.

The most-studied muddy coast of the United States is the Mississippi Delta plain. (See Bloom, 1983: Figure 11-6.) At least six named Holocene distributary systems have developed, been abandoned, and subsided during the last 5000 years before the establishment of the modern digitate distributary system, which is less than 550 years old (Morgan, 1970). Although the Mississippi Delta plain is entirely of Holocene age, as are most of the world's deltas, it was described in the first volume of this series because of its intimate dependence on late-Pleistocene events along the Gulf Coast.

San Francisco Bay is the only large muddy estuary on the California coast. Its threshold near the Golden Gate is 60 to 70 m below present sea level and was probably entered by the sea about 11,000 to 10,000 years ago. By 8000 yr B.P., most of the present Bay area was flooded. Before 8000 yr B.P., the submergence rate was about 2 cm per year, double the inferred rate for the Atlantic coast of the United States (Figure 4-3). Since 6000 yr B.P., the rate has been only about one-tenth as fast, about 1 to 2 mm per year (Atwater et al., 1977). The earlier rate greatly exceeded the rate of sediment accretion, but, as the rate of sea-level rise declined, mudflats and tidal salt marshes prograded into southern San Francisco Bay. Most of the progradation has been in the past few thousand years. Tectonic subsidence in southern San Francisco Bay may have been a factor contributing to the unusually rapid Holocene submergence.

SANDY COASTS

From the south shore of Long Island, New York, southward to beyond Cape Canaveral, Florida, the Atlantic coast of the United

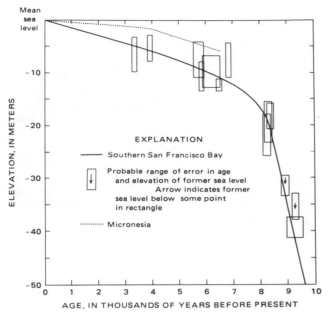

Figure 4-3. Holocene sea-level history for southern San Francisco Bay. Tectonic subsidence of the Bay area is suggested by the downward displacement of the curve relative to a similar curve for Micronesia and also by other geomorphic and stratigraphic evidence. (From Atwater et al., 1977: Figure 6.)

States is a complex of sandy barriers, including barrier spits that trail downdrift from eroding headlands and freely migrating barrier islands backed by lagoons or tidal marshes. Long segments of the Gulf Coast and significant segments of the Pacific coast also have sandy beaches and barriers.

Because of the extreme mobility of sand in the coastal environment, sandy coasts are particularly dynamic environments. The Holocene evolution of the barrier coasts of the United States, particularly the middle-Atlantic coast and the Gulf Coast, has been a topic of persistent interest because of the residential and recreational use of the beaches. A recent position paper prepared by a group of concerned coastal geologists (Kerr, 1981; Pilkey and Howard, 1981) noted that current methods of coastal management are ineffective and human interference with coastal processes is creating dangerous living conditions on inhabited barriers and other sandy coasts of the United States.

Until the Holocene marine transgression slowed about 7000 to 4000 years ago, barriers and other constructional nearshore landforms on the inner continental shelf were probably small and transient. Either the rapid rise of sea level would have driven shorezone sand toward land, up the gentle shelf gradient, or else it would have overtopped incipient protobarriers before they could build to a size that could be subsequently preserved. As noted by Swift (1975: 19-24), the emerged middle-Atlantic continental shelf of late-glacial time must have developed numerous ephemeral barriers as sea level rose, but the only record of these hypothetical transgressive barriers is a discontinuous sheet of clean sand on the inner continental shelf. The sand sheet has neither the internal structures nor the littoral or intertidal fauna of a barrier system; instead, shelf sand with a nearshore bottom fauna directly overlies back-barrier deposits without an intervening barrier-sand unit. Littoral sand obviously was in motion during the early and rapid period of Holocene

marine transgression, but it was completely reworked after submergence.

As the late-Holocene submergence rate decreased, incipient barriers were not driven so rapidly landward. Thus, more time was available to erode an equilibrium shore-face profile on their frontal slopes. On some, littoral transport even caused net seaward progradation. The tendency of the shore face to equilibrate with available wave and current energy meant that barriers grew in place instead of migrating (Swift, 1975: 14-16). Erosional deepening of the lower shore-face profile, combined with littoral drift along the upper part of the surface, provided abundant sand for intertidal and eolian processes to build large subaerial landforms that essentially became the modern barriers. The debate continues over the relative importance of longshore drifting and spit migration as opposed to onshore sand transport from deeper water as mechanisms for barrier construction, but certainly both were effective (Fisher, 1968; Hoyt, 1967, 1968, 1970; Otvos, 1970a, 1970b).

There are numerous indications in the oxygen-isotope record (Shackleton and Opdyke, 1973, 1976) and in coral-reef stratigraphy (Bloom et al, 1974) that Holocene sea level has returned to a level within a few meters of that repeatedly occupied in earlier interglacial times. It is highly likely, therefore, that Holocene barriers would be superimposed on older coastal landforms. Pierce and Colquhoun (1970) demonstrated that about 39% of the barrier coast of North Carolina rests on weathered sediments of older barrier, lagoon, and nearshore environments. They postulated an ancestral coastal landscape similar to the Holocene one on which the Holocene landforms have been built.

With near-stabilization of sea level and the construction of the modern barriers during the last few thousand years, the sandy portions of the Atlantic coast and the Gulf Coast of the United States assumed their present form. Changes have continued, however. After major storms have eroded the shore face of coastal barriers, sediments of back-barrier or lagoon origin usually can be observed on the lower foreshore. This is strong evidence that landward retreat of sandy coasts is the dominant Holocene process. Along the 630 km of the middle-Atlantic coast between New Jersey and North Carolina (lat. 39.5°N to 34.5°N), net erosion for the last 15 to 30 years has averaged 1.5 m per year (Dolan et al., 1979). The average rate is deceptive in that many barrier islands have different rates at opposite ends or may even show net erosion on one end and net deposition at the other. During a comparable 32-year time interval between A.D. 1940 and 1972, the submergence trend on nine tide gauges within the middle-Atlantic region ranged from 2.4 to 3.9 mm per year (Hicks and Crosby, 1974). These historical and regionally consistent submergence rates recorded by tide gauges are two or three times more rapid than the longer-term averages for the last few thousand years that were discussed in the previous section of this chapter, so the modern shoreline recession rate is probably also faster than the average late-Holocene rate. Nevertheless, net landward retreat of the middle-Atlantic sand barriers has probably been the norm since shortly after they were formed.

The large barrier islands of the Texas coast have had more complex late-Holocene histories. Matagorda Island, which is 54 km long and 1.6 to 6.4 km wide, began to build as a sand shoal when sea level reached the level of −8 m. It migrated shoreward in response to the remaining rise of sea level, and it reached the present position of the landward part of Matagorda Island as sea level stabilized about 4000 years ago (Wilkinson, 1975). Thereafter, for about 2000 years, it expanded to its present width by seaward progradation of about 1.6 km, combined with the growth of large flood-tide deltas on the lagoon side of tidal inlets. At present, it is nearly in equilibrium with wind, wave, and current energy.

Galveston Island, northeast of Matagorda Island, is regarded as the "classic regressive barrier of the world" (Kraft, 1978: 371). Regression in this case means that the barrier has enlarged seaward, so that the sea has retreated from the mainland by the frontal progradation of the barrier. In their original demonstration of the regressive or progradational late-Holocene history of Galveston Island, Bernard and LeBlanc (1965: 158) sketched time lines through many radiocarbon dates on a diagramatic cross section of Galveston Island and inferred that it had built to its present width of 4.5 km during the last 3500 years. They discounted the numerous older radiocarbon dates of mollusk shells collected from many boreholes through Galveston Island, because they regarded most of the shells as reworked. Kraft (1978: 374) reinterpreted the age of the cross section of Galveston Island and suggested that it had been built during approximately the last 5300 years (Figure 4-4).

Sand movement on the California coast is dominantly southward. Because the narrow continental shelf is broken by many tectonic lineaments and by submarine canyons that head almost within the modern surf zone, sand transport is discontinuous (Inman and Brush, 1973). Several sediment-circulation cells have been defined, each originating at a sediment source at the updrift or northern end of the cell and ending at a sediment trap at the downdrift end. The sediment traps on the California coast are usually the submarine canyons that permanently remove sand from coastal circulation cells and direct it to abyssal basins within the chaotic block-faulted submarine terrain. Strong correlation has been proved between the direction of incoming wave energy and the rate of longshore sand transport (Ingle, 1966).

Multiple emerged abrasion platforms on the California coast demonstrate continuing tectonic uplift, but the most recent controlling event of coastal evolution was the postglacial rise of sea level. This is a submerged coast in spite of numerous emerged terraces of Wisconsin interstadial (about 80,000 to 45,000 years ago), last-interglacial (about 125,000 years ago), and older ages (Lajoie et al., 1979). During the period of low sea level, the generally small, high-gradient rivers that drain the semiarid coastal mountain ranges eroded valleys that extend to at least 30 m below present sea level (Upson, 1949). In the early Holocene, the rivers became small estuaries as longshore sand transport sealed their mouths with bay-mouth barriers. Most of the estuaries have subsequently been filled with sediment by the seasonal floods that characterize the semiarid or Mediterranean climate. Unfortunately, almost no work has been done on the Holocene sea-level history of southern California. Only one sea-level curve has been published for the entire coast (Figure 4-3). From the

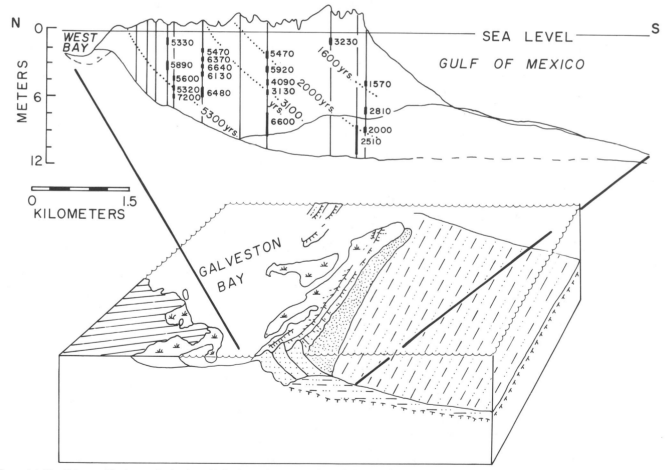

Figure 4-4. The Galveston Island regressive barrier, with isochrons that suggest about 5300 years of progradation. Numerous radiocarbon ages are from mollusk shells, and many of them may be too old because of redeposition. (Modified from Kraft, 1978: Figure 9 and Bernard and LeBlanc, 1965: Figure 20.)

differential altitudes of the older emerged terraces, one can predict that the Holocene depositional histories of the various small bays and lagoons of California will vary greatly in relation to the local tectonic history of each coastal segment.

CORAL-REEF COASTS

Many sandy beaches are composed of a high percentage of molluscan shell fragments. In some, the calcareous fragments readily cement themselves together to form beach rock or coquina. But, for having completely limestone coastal deposition, the tropical coral-reef coasts are unsurpassed. Whereas muddy and sandy coasts require a supply of the appropriate sediment grain sizes, either from the land behind the beach or the sea floor in front of it, coral reefs are typically constructed of biogenic limestone precipitated directly from seawater. The reef-building organisms have pelagic larval stages during which they can be widely distributed by water currents. If the larva settle on a suitable substrate in clear, warm, shallow water, reef growth can begin.

Reef growth can be extremely rapid under suitable conditions. A maximum vertical accretion rate of 3 to 5 mm per year was calculated by Smith and Kinsey (1976); from this they inferred that coral-reef communities would be unable to persist as three-dimensional structures if sea level were rising more rapidly than about 3 to 5 mm per year. Adey (1978), however, cited several examples of radiocarbon-dated Holocene reefs in which the upward growth had been at rates of 9 to 15 mm per year (equal to 9 to 15 m per 1000 years) in Caribbean reefs and 8 to 10 mm per year in Indo-Pacific reefs. Even if sea level rose at a typical rate of 10 m per 1000 years during early-Holocene time, reef growth might have been able to keep up, maintaining steep-sided, perhaps ribbonlike structures at or near sea level during the entire final 30 to 40 m of postglacial sea-level rise that characterized the Holocene Epoch. Lighty, Macintyre, and Stuckenrath

(1978, 1979) have emphasized that, under conditions of rapidly rising sea level before about 7000 years ago, reef growth was predominantly vertical rather than horizontal. The result was a series of narrow, submerged ridges on the outer parts of the continental shelf of Florida and the insular shelves around many Caribbean islands. Instead of being driven landward during rising sea level like sand barriers, coral reefs tended to remain fixed and to grow upward.

On both the Florida shelf and the windward insular shelf of St. Croix, Virgin Islands, the early-Holocene shelf-edge reefs began to grow before 9000 years ago as sea level rose over the shelf margins that are now submerged 30 m or more. However, by about 9000 to 7000 years ago, the shelf-edge reefs had died or ceased vigorous growth (Adey, 1978; Lighty et al., 1978). On the coast of Florida, reef growth has never recovered in the later Holocene. The reefs of St. Croix resumed growth about 5000 to 4000 years ago but in new positions, either as fringing reefs of branching corals on shallow benches closer to the island or as bulky detrital rubble mounds entrapped on the inner shelf by communities of head corals (Figure 4-5).

It is illogical to suppose that the early-Holocene shelf-edge reefs died or became moribund because of rising sea level, for they grew vigorously during the most rapid part of the sea-level rise. After 7000 years ago, they should have been easily able to maintain their upper surfaces close to the less rapidly rising sea surface. Adey (1978) and Lighty, Macintyre, and Stuckenrath (1978) attributed the abrupt reduction of reef growth in the middle Holocene to excessive turbidity induced by the flooding of the inner shelves. The early and most rapid part of the Holocene sea-level rise went up the steep front of the outer shelves. As the rising sea spread over the shelves, the former soil cover was eroded and suppressed reef growth. By the time the inner shelf was swept clean, the most favorable place for reef growth was in much shallower water. On the Florida coast, the interval of middle-Holocene turbidity occurred during the Hypsithermal interval. By the time reef growth could resume, cold winter storms and

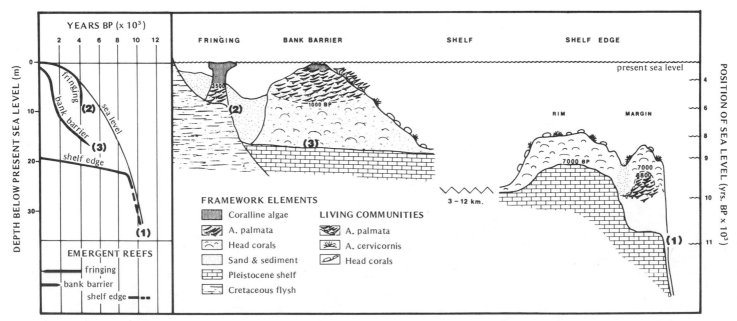

Figure 4-5. Idealized section across the St. Croix windward platform showing Holocene reef development relative to sea-level rise. (1) Sea level reached bench on shelf edge prior to 10,000 years ago. (2) Fringing reefs began growing on nearshore benches about 4000 years ago. (3) Bank-barrier reefs began growing on inner-shelf detrital mounds about 5000 years ago. (From Adey, 1978: Figure 1. Copyright 1978 by the American Association for the Advancement of Science.)

upwelling cold water had assumed the role of coral-growth inhibitors. Lighty, Macintyre, and Stuckenrath (1978) offered the intriguing note that, if the relict shelf-edge reefs of the tropical western Atlantic had survived the middle-Holocene flooding of the shelves behind them, barrier reefs would now separate deep ocean waters from shelf seas and lagoons along much of the Florida and Caribbean island coasts. The modern coastal geography would have been very different.

The Holocene history of Hawaiian coral reefs can be inferred from a detailed study of the Hanauma Reef on Oahu (Easton and Olson, 1976). Ten boreholes were drilled across the reef flat, and numerous coral samples were radiocarbon dated (Figure 4-6). Reef growth began about 7000 years ago, when a breached volcanic crater was invaded by the sea. Most of the vertical reef growth was during the interval from 5800 to 3500 years ago, when the upward growth rate of the reef was about 3.3 m per 1000 years. During the last 3000 years, the reef has built seaward and less rapidly upward. Some minor reversals in the ages of successive borehole samples can be logically attributed

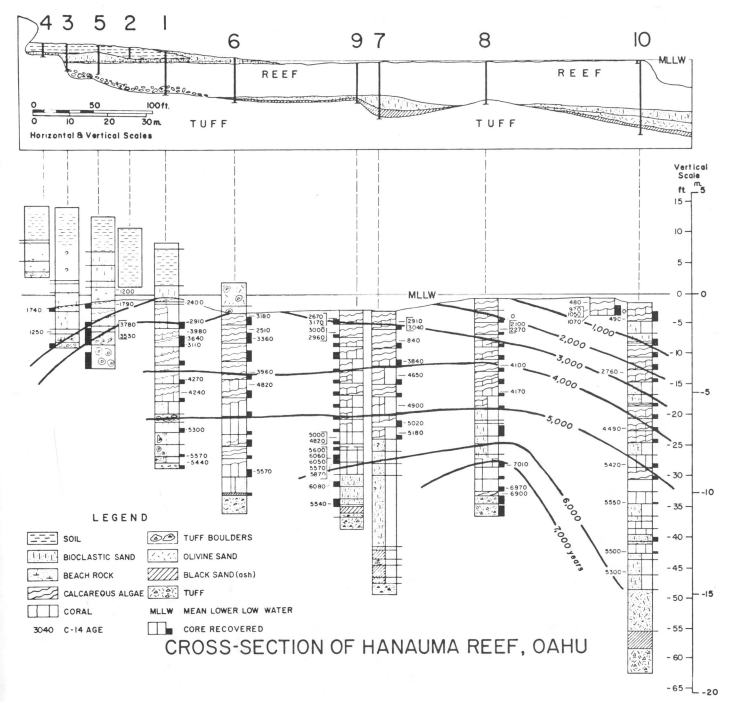

LEGEND

SOIL		TUFF BOULDERS	
BIOCLASTIC SAND		OLIVINE SAND	
BEACH ROCK		BLACK SAND (ash)	
CALCAREOUS ALGAE		TUFF	
CORAL		MLLW	MEAN LOWER LOW WATER
3040	C-14 AGE		CORE RECOVERED

CROSS-SECTION OF HANAUMA REEF, OAHU

Figure 4-6. Cross section of Hanauma Reef, Oahu. Upper section shows position of boreholes; vertical and horizontal scales are equal. Stratigraphy, radiocarbon ages, and growth isochrons are shown on lower figure, with ×5 vertical exaggeration. (From Easton and Olson, 1976: Figure 3.)

to younger corals that grew on older reef rock beneath overhanging ledges or within pools. No reversals or oscillations in the generally smoothly rising trend of Holocene sea level can be inferred, nor was sea level ever any higher during the Holocene than it is today (Easton and Olson, 1976: 717).

THE PERIGLACIAL COAST OF ALASKA

The late-Pleistocene and Holocene history of sea level along the coast of Alaska is reviewed in another chapter in this series (Bloom, 1983). Freshwater peat was still accumulating on the Bering Sea shelf at a depth of − 20 m as recently as 10,000 years ago, but present sea level had been reached by 5000 years ago. Relict permafrost is still preserved beneath sea level on the Arctic coast of Alaska; the seawater temperature is cold enough to maintain the frozen fresh ground water. On both the Bering Sea and the Arctic Ocean coasts, shore ice and wind-driven ice floes are significant geomorphic agents. Coarse, poorly sorted gravel is a common component of barrier spits and islands. Large deltas have prograded seaward during the Holocene, with much of the sediment being supplied during the early summer breakout of the frozen rivers.

Umnak Island in the eastern Aleutians had a Holocene history of sea level 2 to 3 m higher than present from about 8250 years ago until about 3000 years ago (Black, 1974). It provides an interesting case history of cultural adaptation by the ancient Aleut people, who must have settled the region by boat, for all the major passes in the eastern Aleutians were flooded and did not freeze over in the winter. Furthermore, before 3000 years ago, the major food resources of the present strandflat would have been much less available because of the higher sea level. The area had been settled by Aleuts about 8400 years ago. The causes for the Holocene sea-level changes could be diverse, including glacioeustatic, isostatic, volcanic, and tectonic movements.

Conclusions

The sea-level and coastal changes of the Holocene are minor compared to the much greater impact of rising sea level during Late Wisconsin time. However, modern coastal landscapes assumed their present forms during the Holocene time of decelerating sea-level rise. Secondary effects such as isostatic and tectonic movements of only a few meters became noticeable in the Holocene as the great glacioeustatic transgression ended. Modern deltas prograded; muddy sediments filled shallow estuaries and simplified the coastal outlines; and sandy barriers stabilized, grew, and then generally resumed their slow landward migration. Coral reefs experienced rapid vertical growth and then adjusted to nearly stable sea level or died back under the impact of the post-Hypsithermal cooling. During all this time, a steadily increasing human population made progressively greater use of coastal resources. With continuing sea-level rise and increasing population pressures, coastal zone occupancy is becoming more and more hazardous.

References

Adey, W. H. (1978). Coral reef morphogenesis: A multidimensional model. *Science* 202, 831-37.

Atwater, B. F., Hedel, C. W., and Helley, E. J. (1977). "Late Quaternary Depositional History, Holocene Sea-Level Changes, and Vertical Crustal Movement, Southern San Francisco Bay, California." U.S. Geological Survey Professional Paper 1014.

Belknap, D. F., and Kraft, J. C. (1977). Holocene relative sea-level changes and coastal stratigraphic units on the northwest flank of the Baltimore Canyon trough geosyncline. *Journal of Sedimentary Petrology* 47, 610-29.

Bernard, H. A., and LeBlanc, R. J. (1965). Résumé of the Quaternary geology of the northwestern Gulf of Mexico Province. *In* "The Quaternary of the United States" (H. E. Wright, Jr., and D. G. Frey, eds.), pp. 137-85. Princeton University Press, Princeton, N.J.

Black, R. F. (1974). Late-Quaternary sea level changes, Umnak Island, Aleutians: Their effects on ancient Aleuts and their causes. *Quaternary Research* 4, 264-81.

Bloom, A. L. (1963). Late-Pleistocene fluctuations of sea level and postglacial crustal rebound in coastal Maine. *American Journal of Science* 261, 862-79.

Bloom, A. L. (1964). Peat accumulation and compaction in a Connecticut coastal marsh. *Journal of Sedimentary Petrology* 34, 599-603.

Bloom, A. L. (1965). The explanatory description of coasts. *Zeitschrift für Geomorphologie* 9, 422-36.

Bloom, A. L. (1967). Pleistocene shorelines: A new test of isostasy. *Geological Society of America Bulletin* 78, 1477-94.

Bloom, A. L. (1971). Glacial-eustatic and isostatic controls of sea level since the last glaciation. *In* "The Late Cenozoic Glacial Ages" (K. K. Turekian, ed.), pp. 355-79. Yale University Press, New Haven, Conn.

Bloom, A. L. (comp.) (1977). "Atlas of Sea-Level Curves." International Geological Correlation Program Project 61. Department of Geological Sciences, Cornell University, Ithaca, N.Y.

Bloom, A. L. (1983). Sea level and coastal morphology through the Late Wisconsin glacial maximum. *In* "Late-Quaternary Environments of the United States," vol. 1, "The Late Pleistocene" (S. C. Porter, ed.), pp. 215-29. University of Minnesota Press, Minneapolis.

Bloom, A. L., Broecker, W. S., Chappell, J. M. A., Matthews, R. K., and Mesolella, K. J. (1974). Quaternary sea level fluctuations on a tectonic coast: New ^{230}Th/^{234}U dates from the Huon Peninsula, New Guinea. *Quaternary Research* 4, 185-205.

Bloom, A. L., and Stuiver, M. (1963). Submergence of the Connecticut coast. *Science* 139, 332-34.

Brennan, L. A. (1977). The Lower Hudson: The Archaic. *In* "Amerinds and Their Paleoenvironments in Northeastern North America" (W. S. Newman and B. Salwen, eds.), pp. 411-30. Annals of the New York Academy of Sciences 288.

Cathles, L. M., III (1975). "The Viscosity of the Earth's Mantle." Princeton University Press, Princeton, N.J.

Chapman, V. J. (1960). "Salt Marshes and Salt Deserts of the World." Wiley Interscience, New York.

Chappell, J. (1974). Late Quaternary glacio- and hydro-isostasy, on a layered Earth. *Quaternary Research* 4, 405-28.

Clark, J. A. (1976). Greenland's rapid postglacial emergence: A result of ice-water gravitational attraction. *Geology* 4, 310-12.

Clark, J. A. (1977). An inverse problem in glacial geology: The reconstruction of glacier thinning in Glacier Bay, Alaska between A.D. 1910 and 1960 from relative sea-level data. *Journal of Glaciology* 18, 481-503.

Clark, J. A. (1980). Reconstruction of the Laurentide ice sheet of North America from sea level data: Method and preliminary results. *Journal of Geophysical Research* 85, 4307-23.

Clark, J. A., Farrell, W. E., and Peltier, W. R. (1978). Global changes in postglacial sea level: A numerical calculation. *Quaternary Research* 9, 265-87.

Clark, J. A., and Lingle, C. S. (1979). Predicted relative sea-level changes (18,000 years B.P. to present) caused by late-glacial retreat of the Antarctic ice sheet. *Quaternary Research* 11, 279-98.

Davis, C. A. (1910). Salt marsh formation near Boston and its geological significance. *Economic Geology* 5, 623-39.

Dolan, R., Hayden, B. P., Rea, C., and Heywood, J. E. (1979). Shoreline erosion rates along the middle Atlantic coast of the United States. *Geology* 7, 602-6.

Easton, W. H., and Olson, E. A. (1976). Radiocarbon profile of Hanauma Reef, Oahu, Hawaii. *Geological Society of America Bulletin* 87, 711-19.

Farrell, W. E., and Clark, J. A. (1976). On postglacial sea level. *Geophysical Journal of the Royal Astronomical Society* 46, 647-67.

Fisher, J. J. (1968). Barrier island formation: Discussion. *Geological Society of America Bulletin* 79, 1421-26.

Gordon, R. B. (1979). Denudation rate of central New England determined from estuarine sedimentation. *American Journal of Science* 279, 632-42.

Harrison, E. Z., and Bloom, A. L. (1977). Sedimentation rates on tidal salt marshes in Connecticut. *Journal of Sedimentary Petrology* 47, 1484-90.

Hicks, S. D., and Crosby, J. E. (1974). Trends and variability of yearly mean sea level 1893-1972. U.S. National Oceanic and Atmospheric Administration Technical Memorandum NOS 13.

Higgins, C. G. (1965). Causes of relative sea-level changes. *American Scientist* 53, 464-76.

Hoyt, J. H. (1967). Barrier island formation. *Geological Society of America Bulletin* 78, 1125-36.

Hoyt, J. H. (1968). Barrier island formation: Reply. *Geological Society of America Bulletin* 79, 1427-31.

Hoyt, J. H. (1970). Development and migration of barrier islands, northern Gulf of Mexico: Discussion. *Geological Society of America Bulletin* 81, 3779-82.

Ingle, J. C. (1966). Movement of Beach Sand, "Advances in Sedimentology," vol. 5, Elsevier, New York.

Inman, D. L., and Brush, B. M. (1973). The coastal challenge. *Science* 181, 20-32.

Johnson, D. W. (1919). "Shore Processes and Shoreline Development." John Wiley and Sons, New York.

Johnson, D. W. (1925). "New England-Acadian Shoreline." John Wiley and Sons, New York.

Kerr, R. A. (1981). Whither the shoreline? *Science* 214, 428.

Kraft, J. C. (1978). Coastal stratigraphic sequences. *In* "Coastal Sedimentary Environments" (R. A. Davis, Jr., ed.), pp. 361-83. Springer-Verlag, New York.

Lajoie, K. R., Kern, J. P., Wehmiller, J. F., Kennedy, G. L., Mathieson, S. A., Sarna-Wojcicki, A. M., Yerkes, R. F., and McCrory, P. F. (1979). Quaternary marine shorelines and crustal deformation, San Diego to Santa Barbara, California. *In* "Geological Excursions in the Southern California Area" P. L. Abbott, ed.), pp. 1-15. Department of Geological Sciences, San Diego State University, San Diego.

Lighty, R. G., Macintyre, I. G., and Stuckenrath, R. (1978). Submerged early Holocene barrier reef south-east Florida shelf. *Nature* 275, 59-60.

Lighty, R. G., Macintyre, I. G., and Stuckenrath, R. (1979). Holocene reef growth on the edge of the Florida shelf: Reply. *Nature* 278, 281-82.

McIntire, W. G., and Morgan, J. P. (1962). "Recent Geomorphic History of Plum Island, Massachusetts and Adjacent Coasts." Atlantic Coastal Studies Technical Report 19, Part A. Coastal Studies Institute, Louisiana State University, Baton Rouge.

Morgan, J. P. (1970). Deltas: a résumé. *Journal of Geological Education* 18, 107-17.

Newman, W. S., Fairbridge, R. W., and March, S. (1971). Marginal subsidence of glaciated areas: United States, Baltic and North Seas. *In* "Etudes sur le Quater-naire dans le Monde" (M. Ters, ed.), pp. 795-801. Eighth Congress of the International Association for Quaternary Research, 1969, Paris.

Otvos, E. G. (1970a). Development and migration of barrier islands, northern Gulf of Mexico. *Geological Society of America Bulletin* 81, 241-46.

Otvos, E. G. (1970b). Development and migration of barrier islands, northern Gulf of Mexico: Reply. *Geological Society of America Bulletin* 81, 3783-88.

Pierce, J. W., and Colquhoun, D. J. (1970). Holocene evolution of a portion of the North Carolina coast. *Geological Society of America Bulletin* 81, 3697-3714.

Pilkey, O. H., Jr., and Howard, J. D. (1981). Old solutions fail to solve beach problem. *Geotimes* 26, 18-22.

Rampino, M. R. (1979). Holocene submergence of southern Long Island, New York. *Nature* 280, 132-34.

Redfield, A. C., and Rubin, M. (1962). Age of salt marsh peat and its relation to recent changes in sea level at Barnstable, Massachusetts. *Proceedings of the National Academy of Sciences* 48, 1728-35.

Ritchie, W. A. (1969). "Archaeology of New York State." (Rev. Ed.) Natural History Press, Garden City, N.Y.

Shackleton, N. J., and Opdyke, N. D. (1973). Oxygen isotope and paleomagnetic stratigraphy of equatorial Pacific core V28-238: Oxygen istope temperatures and ice volumes on a 10^5 year and 10^6 year scale. *Quaternary Research* 3, 39-55.

Shackleton, N. J., and Opdyke, N. D. (1976). Oxygen isotope and paleomagnetic stratigraphy of equatorial Pacific core V28-239, late Pliocene to latest Pleistocene *In* "Investigation of Late Quaternary Paleoceanography and Paleoclimatology" (R. M. Cline and J. D. Hays, eds.), pp. 449-64. Geological Society of America Memoir 145.

Smith, S. V., and Kinsey, D. W. (1976). Calcium carbonate production, coral reef growth, and sea level change. *Science* 194, 937-39.

Stuiver, M., and Borns, H. W., Jr. (1975). Late Quaternary marine invasion in Maine: Its chronology and associated crustal movement. *Geological Society of America Bulletin* 86, 99-103.

Swift, D. J. P. (1975). Barrier-island genesis: Evidence from the central Atlantic shelf, eastern U.S.A. *Sedimentary Geology* 14, 1-43.

Upson, J. E. (1949). Late Pleistocene and recent changes of sea level along the coast of Santa Barbara County, California. *American Journal of Science* 247, 94-115.

Walcott, R. I. (1972). Past sea levels, eustasy and deformation of the Earth. *Quaternary Research* 2, 1-14.

Wilkinson, B. H. (1975). Matagorda Island, Texas: The evolution of a Gulf Coast barrier complex. *Geological Society of America Bulletin* 86, 959-67.

Wyatt, R. J. (1977). The Archaic on Long Island. *In* "Amerinds and Their Paleoenvironments in Northeastern North America" (W. S. Newman and B. Salwen, eds.), pp. 400-410. Annals of the New York Academy of Sciences 288.

CHAPTER 5

Holocene Volcanism in the Conterminous United States and the Role of Silicic Volcanic Ash Layers in Correlation of Latest-Pleistocene and Holocene Deposits

Andrei M. Sarna-Wojcicki, Duane E. Champion, and Jonathan O. Davis

Introduction

In the first part of this chapter, we summarize information on the areal distribution and positions of Holocene volcanic fields; the composition of the materials erupted; the available age controls; and the relations of tectonic setting, heat flow, and hydrothermal activity to the spatial distribution of volcanism. In the second part, we summarize information on the widespread silicic ash layers of latest-Pleistocene and Holocene age; their inferred areal distributions, volumes, and frequencies; some of the petrographic and chemical characteristics by which they can be identified; and their use in late-Quaternary correlation and age dating.

We exclude Alaskan volcanism from our discussion because much work remains to be done to document the age and areal distribution of volcanism in this region. Rates of crustal subduction beneath the Aleutian Islands and southern Alaska are greater than in the Pacific Northwest. However, not enough information is available at present to assess accurately the frequency, areal extent, and volume of volcanic materials erupted in Alaska during latest-Quaternary time. We also exclude Hawaii from our discussion because volcanism in Hawaii during this same period has been frequent; furthermore, the topic is well documented (e.g. Porter, 1979; Holcomb, 1980), and it deserves separate treatment beyond the scope of this report. We refer the reader to Simkin and others (1981) for information on active volcanoes in areas outside the conterminous United States. These authors have tabulated data on the active volcanoes of the world, and they provide a regional directory, gazetteer, and chronology of volcanism during the last 10,000 years.

The study of volcanism and volcanic materials is relevant to an understanding of latest-Quaternary history in at least two ways. First, volcanic activity may be an important if not a major agent of climatic change. Increased amounts of carbon dioxide and other volcanogenic gases in the atmosphere as a result of volcanic activity can warm world climate, whereas an increase in such aerosols as sulfuric acid droplets and fine glass shards in the upper atmosphere may have an initial minor warming effect that is followed by net long-term cooling effect (Davitaya, 1969; Lamb, 1970, 1977; Pollack et al., 1976). Some workers have suggested that increased volcanic activity may trigger glaciation (Bray, 1974, 1977; Kennett and Thunell, 1975; Porter, 1981), and others that increased volcanic activity may be an indirect result of climatic change caused by the loading and unloading of the crust by ice sheets or seawater (Lamb, 1977). In any case, volcanism and climatic change may be causally related.

Frequent, prolonged, or large-magnitude eruptions may appreciably affect climate (Bray, 1977; Rose et al., 1981); evidence for such effects can be found by determining the areal distributions, thicknesses, and ages of major volcanic-flow rocks and ash layers in the uppermost Quaternary stratigraphic record. Through comparisons of this record of eruptive activity with the record of climatic fluctuations over the same period, the influence of volcanism on climate, (or conversely, the potential indirect effects of climate on volcanism through such mechanisms as crustal loading) can be assessed. Thus, the study of volcanic activity during latest-Quaternary time may contribute to a worldwide data base from which the relation of volcanism to climatic change can be evaluated.

An understanding of volcanism and volcanic materials is important to studies of latest-Quaternary history in a second way: volcanic strata, particularly volcanic-ash layers, are highly useful for making correlations and age determinations of uppermost Quaternary deposits. Although the extent of volcanic-flow rocks of latest-Quaternary age is limited in the United States, the widespread occurrence of silicic volcanic-ash layers provides a method by which a wide variety of uppermost Quaternary deposits can be correlated and dated. Widespread silicic volcanic-ash layers, commonly produeced by explosive eruptions of large magnitude, are fairly numerous in the latest-Quaternary stratigraphic record, particularly in the West, where the volcanic source areas are situated. Such layers can be dated by isotopic

D. P. Adam, C. R. Bacon, J. M. Donnelly-Nolan, D. R. Harden, K. R. Lajoie, N. S. MacLeod, D. R. Mullineaux, R. E. Wilcox, and H. E. Wright, Jr., offered numerous helpful suggestions and generally improved the manuscript for this chapter. N. S. MacLeod, C. D. Miller, E. M. Shoemaker, and G. H. Ulrich generously provided unpublished material that made our data more nearly complete. A. H. Lachenbruch, J. H. Sass, and C. R. Bacon provided useful ideas and information.

or other methods and correlated over long distances by petrographic and chemical methods (Wilcox, 1965; Jack and Carmichael, 1967; Izett et al., 1970; Sarna-Wojcicki, 1976; Sarna-Wojcicki et al., 1980). Through positive identification of a widespread ash layer, deposits of diverse depositional environments can be correlated over long distances. Once such an ash layer has been accurately dated, it becomes an important time horizon at all localities in which it occurs. The use of such time markers and stratigraphic markers may make it possible to test whether such late-Quaternary events as alpine-glacial advances and retreats are synchronous over broad areas and, in turn, provide information regarding the relationship between climate and volcanism. We list the upper-most Quaternary tephra units (those erupted during the past 13,000 years) that have been identified as widespread stratigraphic markers (Table 5-1), as well as some of their petrographic and chemical characteristics (Table 5-2), and we describe some of the diverse age and correlation problems that can be resolved with the aid of volcanic-ash chronology.

Regional Distribution of Latest-Quaternary Volcanism

Latest-Quaternary volcanism in the conterminous United States is restricted to the western states. All major and minor volcanic fields and isolated individual volcanic flows and vents are restricted to the region west of the 104th meridian (Figure 5-1). Although several widespread layers of early- and middle-Pleistocene ash erupted from volcanic centers in the western states extend to the central Great Plains (Nebraska, Kansas, and Texas) (Izett et al., 1970, 1972), no widespread latest-Quaternary ash beds have been found east of the 108th meridian (Figure 5-2).

Major silicic volcanism during latest-Quaternary time is restricted to two main areas: the Cascade volcanic range in the Pacific Northwest (Washington, Oregon, and northern California) and the Mono Craters-Long Valley area of east-central California (Figure 5-2). There is a third, minor field of latest-Quaternary silicic volcanic activity in

Table 5-1. Volcanic-Ash Layers of Moderate to Large Volume Erupted during Latest Pleistocene and Holocene Time from Source Areas in the Conterminous United States

Ash Layer	Age	Name or other Designation of Ash Layer or Set	Volcanic Source Area	Area of Distribution	References(s)
1	1 (1980 eruption sequence)	Set D	Mount St. Helens, Washington	Washington, Idaho, Montana, Oregon; Alberta, Canada	Sarna-Wojcicki, Meyer, et al., 1981
2	~180 (A.D.~1800)	T	Mount St. Helens, Washington	Washington, Idaho, Montana	Crandell and Mullineaux, 1978
3	~500 (A.D.~1500)	We, Wn	Mount St. Helens, Washington	Washington, Idaho	Crandell and Mullineaux, 1978
4	640 ± 40 (A.D.~1270)	—	Panum Crater, Mono Craters, California	East-central California	Wood and Brooks, 1979
5	725 ± 60; 760 ± 60 (A.D.~1250)	Tephra 1	Inyo Craters, California	East-central California	Wood, 1977a
6	914-918 (A.D.~1063-1067)	—	Sunset Crater, San Francisco Peaks, Arizona	Arizona	Pilles, 1979
7	1190 ± 80 (A.D.~760)	Tephra 2	Mono Craters, California	East-central California	Wood, 1977b
8	~1150-1350 (A.D.~430-630)	—	Newberry Volcano, Oregon	Eastern Oregon, Idaho (?)	Sherrod and McLeod (1979)
9	1990 ± 200 (~50 B.C.)	—	Mono Craters, California	East-central California, Nevada (?)	Wood, 1977a
10	2930 ± 250-2580 ± 250 (~600-1000 B.C.)	Set P	Mount St. Helens, Washington	Washington	Mullineaux et al., 1975
11	3510 ± 230-3350 ± 250 (~1400-1600 B.C.)	Ye, Yn	Mount St. Helens, Washington	Washington, Oregon; British Columbia, Alberta, Canada	Mullineaux et al., 1975
12	~7000-6700 (~5000 B.C.)	Set O, or Mazama ash	Crater Lake, Oregon	Oregon, Washington, Idaho, Montana, Wyoming, Utah, Nevada, California; British Columbia, Alberta, Saskatchewan, Canada	Williams, 1942; Wilcox, 1965; Williams and Goles, 1967; Mullineaux et al., 1975, 1980
13	11,700 ± 400-8300 ± 300	Set J	Mount St. Helens, Washington	Washington	Mullineaux et al., 1975
14	11,200 ± 100; 11,300 ± 230	B	Glacier Peak, Washington	Washington, Idaho, Wyoming	Porter, 1978
15	≤ 12,000 (?)	M	Glacier Peak, Washington	Washington	Porter, 1978
16	12,750 ± 350	G	Glacier Peak, Washington	Washington, Montana; British Columbia, Alberta, Canada	Porter, 1978
17	> 11,900 ± 300; < 13,130 ± 350	So, Sg	Mount St. Helens, Washington	Washington	Mullineaux et al., 1978

Note: Ages are given in years before present or in radiocarbon years before 1950 with analytical errors. Ages in calendar years are given in parentheses.

Table 5-2. Petrographic and Chemical Characteristics of Ash Layers of Moderate to
Large Volume Erupted during Latest Pleistocene and Holocene Time
from Sources in the Conterminous United States

Ash Layer	Eruptive Source and Name or Other Designation of Ash Layer or Sequence	Characteristic[a] Mineralogy (in order of Abundance)	Major-element Chemistry of Glass (Electron Microprobe Analysis Unless Otherwise Noted)								Reference(s) (Mineralogy/Glass Chemistry)
			SiO_2	Al_2O_3	Fe_2O_3	MgO	CaO	TiO_2	Na_2O	K_2O	
1	Mount St. Helens (May 18, 1980)	pl, hy, hb, ag	71.5	15.0	2.48	0.52	2.32	0.37	4.68	2.03	Sarna-Wojcicki, Meyer, et al., 1981/ Sarna-Wojcicki, Shipley, et al., 1981
2	Mount St. Helens T	pl, hy, hb, ag	70.4	15.7	2.97	0.72	2.82	0.42	4.48	2.50	Crandell and Mullineaux, 1978/ Sarna-Wojcicki, Meyer, et al., 1981
3	Mount St. Helens We	pl, hy, hb	74.2	13.7	1.66	0.24	1.45	0.24	4.40	2.41	Crandell and Mullineaux, 1978/
	Mount St. Helens Wn	pl, hy, hb	72.8	14.1	1.73	0.30	1.56	0.20	4.32	2.28	Sarna-Wojcicki, Meyer, et al., 1981
5	Inyo Craters tephra 1	pl, hb, bi, hy, zr	72.6	14.3	1.35	0.05	0.62	0.08	4.00	4.65[b]	Wood, 1977/this report
6	Sunset Crater	—	46.2	16.2	15.58	7.80	9.60	1.80	3.60	0.90[c]	G. H. Ulrich and E. M. Shoemaker, personal communication, 1981
7	Mono Craters tephra 2	pl, bi, hb, hy	74.0	12.5	1.12	0.03	0.52	0.07	3.90	4.48[d]	Davis, 1978/this report
8	Newberry Volcano	aphyric	72.8	14.5	2.34	0.20	0.89	0.22	5.10	3.98[e]	N. S. MacLeod, personal communication, 1981
10	Mount St. Helens Pm	pl, hy, hb	74.9	13.2	1.02	0.15	0.97	0.23	4.58	2.10	Crandell and Mullineaux, 1978/ Sarna-Wojcicki, Meyer, et al., 1981
11	Mount St. Helens Ye	pl, cm, hb	73.1	13.8	1.34	0.34	1.62	0.15	4.16	2.03	Crandell and Mullineaux, 1978/
	Mount St. Helens Yn	pl, cm, hb	72.9	14.1	1.38	0.35	1.75	0.15	3.65	1.87	Sarna-Wojcicki, Meyer, et al., 1981
12	Crater Lake set O (Mazama ash)	pl, hy, hb, ag, ap	70.9	13.9	2.12	0.36	1.50	0.42	4.53	2.76	Davis, 1978/this report
13	Mount St. Helens Jy	pl, hy, hb	74.4	13.1	1.31	0.26	1.36	0.17	4.33	2.26	Mullineaux et al., 1975/ Sarna-Wojcicki, Meyer, et al., 1981
14	Glacier Peak B	hy, hb, pl	72.3	12.2	1.16	0.24	1.24	0.25	2.79	2.86	Porter, 1978/this report
15	Glacier Peak M	hy, hb, pl	73.8	12.4	1.02	0.17	0.99	0.20	3.54	3.40	Porter, 1978/this report
16	Glacier Peak G	hy, hb, pl	72.9	12.1	0.98	0.19	1.05	0.20	3.59	3.30	Porter, 1978/this report
17	Mount St. Helens So	pl, cm, hb, hy	72.8	12.3	1.27	0.28	1.33	0.17	4.21	2.20	Mullineaux et al., 1978/
	Mount St. Helens Sg	pl, cm, hb, hy	73.0	13.8	1.29	0.31	1.48	0.17	4.06	2.09	Sarna-Wojcicki, Meyer, et al., 1981

[a] *Abbreviations:* pl, plagioclase; hy, hypersthene; hb, hornblende; ag, augite; zr, zircon; bi, biotite; cm, cummingtonite; ap, apatite.
[b] Analysis on surface or near-surface pumice pebbles in Reds Meadow, Devils Postpile quadrangle, several kilometers south of Inyo Craters; inferred to correlate with tephra layer 1.
[c] Conventional wet-chemical analysis on basaltic cinders collected from the Sunset Crater area.
[d] Average of five downwind tephra samples inferred to correlate with tephra layer 2.
[e] Wavelength-dispersive X-ray fluorescence analysis.

the Salton Buttes area of southern California (Figure 5-1). Although no latest-Quaternary activity has been recorded for the Yellowstone volcanic field of northwestern Wyoming and adjacent Idaho or for the Jemez Mountains area of northern New Mexico (Figure 5-4), these areas have also been active centers of widespread and explosive volcanism throughout much of Quaternary time (Doell et al., 1968; Christiansen and Blank, 1972). There have been six great catastrophic eruptions in the western United States within the past 2 my (million years), each of which produced about 300 km³ or more of silicic tephra (both ash fall and ash flow) (Izett, 1981); five of these tephra eruptions have been from the Yellowstone and Jemez Mountains areas (Table 5-3). The average interval between such gigantic eruptions in the conterminous United States is about 350,000 years. The Yellowstone and Jemez Mountains volcanic areas can be considered dormant and potentially active, because significant high heatflow and hydrothermal anomalies are associated with them at the present time (Figure 5-3 and 5-4). Holocene time (the past 10,000 years) represents less than 1% of Quaternary time (the past 2 my), and so the period

is too short for assessments to be made of long-term sporadic volcanic activity, such as that from the Jemez Mountains and the Yellowstone area.

Basaltic volcanism is much more widespread in the western states than silicic volcanism. In addition to the major basaltic volcanic fields (such as the Snake River Plain of Idaho, the basaltic shields of central and eastern Oregon, the Cascade Range, and the San Francisco Mountains volcanic field of Arizona) (Figures 5-1 and 5-2), numerous small, isolated basaltic flows and individual cinder cones are scattered throughout several of the western states. Notable examples are Ubehebe Craters in Death Valley (Crowe and Fisher, 1973), Amboy Crater in the Mojave Desert of southeastern California (Parker, 1963), the Dotsero flow in central Colorado (Giegengack, 1962), and the Black Rock Desert area in Utah (Hoover, 1974). Basaltic flows have also been erupted at or near major centers of silicic volcanic activity in the Cascade Range and east-central California. In addition, some of the major Cascade Range volcanoes (for instance, Mount Adams, Mount Rainier, and Mount Jefferson) are mostly formed of

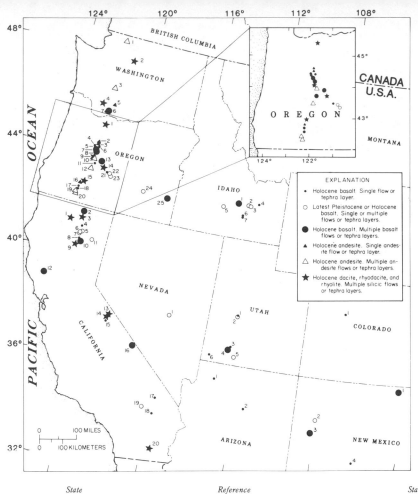

Figure 5·1. Areal distribution of Holocene volcanic vents and fields in the conterminous United States. Names of Holocene volcanic fields in the conterminous United States and principal references are given below. Numbers for volcanic fields listed below for each state correspond to numbers within the same state shown on the map. Also see Smith and Shaw (1975).

State	Reference
Arizona	
1. Unikaret	Hamblin (1970)
2. Sunset Crater	Pilles (1979); Moore and Wolfe (1976)
California	
1. Mount Shasta	Miller (1980); Christiansen and Miller (1976)
2. Modoc	Powers (1932); A. E. Jones (unpub. data)
3. Medicine Lake Highland	Anderson (1941); Heiken (1978b); Donnely-Nolan et al. (1980)
4. Brushy Butte	Peterson and Martin (1980)
5. Big Cave	C. D. Miller (unpub. data)
6. Sixmile Hill	C. D. Miller (unpub. data)
7. Cinder Butte	Anderson (1940); C. D. Miller (unpub. data)
8. Hat Creek	Anderson (1940)
9. Mount Lassen	Williams (1932); Heiken and Eichelberger (1980)
10. Cinder Cone	Heiken (1978a); Finch and Anderson (1930)
11. Brockman Flat	Gester (1962); C. D. Miller (unpub. data)
12. Clear Lake	Sims and Rymer (1975)
13. Mono Lake Islands	Lajoie (1968); Wood (1977b); Kilbourne et al. (1980)
14. Mono and Inyo Creaters	Rinehart and Huber (1965); Lajoie (1968); Dalrymple (1968); Wood (1977b); Kilbourne et al. (1980); Friedman (1968)
15. Red Cones	Huber and Rinehart (1967)
16. Ubehebe Craters	Crowe and Fisher (1973)
17. Cima	Barca (1966); Katz and Boettcher (1980)
18. Amboy	Parker (1963)
19. Pisgah	Dibblee (1966); Wise (1969)
20. Salton Buttes	Robinson et al. (1976); Friedman and Obradovich (1981)
Colorado	
1. Dotsero	Giengengack (1962)
Idaho	
1. Craters of the Moon	Kuntz et al. (1980)
2. North and South Robbers	Kuntz (1978)
3. Cerro Grande	Kuntz (1978)
4. Hell's Half Acre	Karlo (1977); Kuntz (1977)
5. Shoshone	LaPoint (1977)
6. King's Bowl	Greeley et al. (1977)
7. Wapi	Champion and Greeley (1977); Champion (1973)
Nevada	
1. Lunar Crater	Scott and Trask (1971)
New Mexico	
1. Capulin	Collins (1949); Baldwin and Muehlberger (1959)
2. El Tintero	Thaden and Ostling (1967)
3. Bandera	Hathcway (1971); Laughlin ct al. (1972)
4. Carrizozo	Allen (1951); Smith (1964)

State	Reference
Oregon	
1. Mount Hood	White (1980); Crandell (1980)
2. Forked Butte	Scott (1977); Greene (1968)
3. South Cinder Peak	Scott (1977); Greene (1968)
4. Nash Crater	Taylor (1968)
5. Sand Mountain	Taylor (1968); Nieland (1970)
6. Blue Lake Crater	Taylor (1968)
7. Belknap Crater	Taylor (1968)
8. Ahalapam cinder field	Williams (1944); Taylor (1968)
9. South Sister area	Williams (1944); Taylor (1978)
10. Bachelor Butte	N. S. MacLeod (unpub. data); Williams (1957)
11. Cascade Lakes Highway	N. S. MacLeod (unpub. data); Williams (1957)
12. Black Rock Butte area	N. S. MacLeod (unpub. data); Williams (1957)
13. Newberry NW fissure	MacLeod et al. (1981); Peterson and Groh (1969)
14. Newberry Crater	MacLeod et al. (1981)
15. Crater Lake	Williams (1942); Kittleman (1973); Bacon (1981)
16. Goosenest	J. G. Smith (unpub. data)
17. Imagination Peak	J. G. Smith (unpub. data)
18. Big Bunch Grass	J. G. Smith (unpub. data)
19. Mount McLoughlin	J. G. Smith (unpub. data); Maynard (1974)
20. Brown Mountain	J. G. Smith (unpub. data); Maynard (1974)
21. Devils Garden	Peterson and Groh (1963; 1965)
22. Squaw Ridge	Peterson and Groh (1963); Walker et al. (1967)
23. Four Craters	Peterson and Groh (1963); Walker et al. (1967)
24. Diamond Crater	Peterson and Groh (1964)
25. Jordan Craters	Otto and Hutchison (1977); Russell (1903)
Utah	
1. Ice Springs	Condie and Barsky (1972); Hoover (1974); Valastro et al. (1972)
2. Tabarnacle Hill	Condie and Barsky (1972); Hoover (1974)
3. Miller Knoll	Gregory (1949)
4. Markagunt Plateau	Gregory (1949)
5. Bald Knoll	Goode (1973); Gregory (1951)
6. Santa Clara	Hamblin (1963); Cook (1960)
Washington	
1. Mount Baker	Coombs (1939); Easterbrook (1975); Hyde and Crandell (1978)
2. Glacier Peak	Crowder et al. (1966); Beget (1980)
3. Mount Rainier	Fiske et al. (1964); Crandell (1971)
4. Mount St. Helens	Crandell and Mullineaux (1978); C. A. Hopson (unpub. data)
5. Mount Adams	Hopkins (1976); Hammond 1973)
6. Indian Heaven fissures	Hammond (1977); Wise (1970)
7. West Crater	Hammond (1977); Wise (1970)

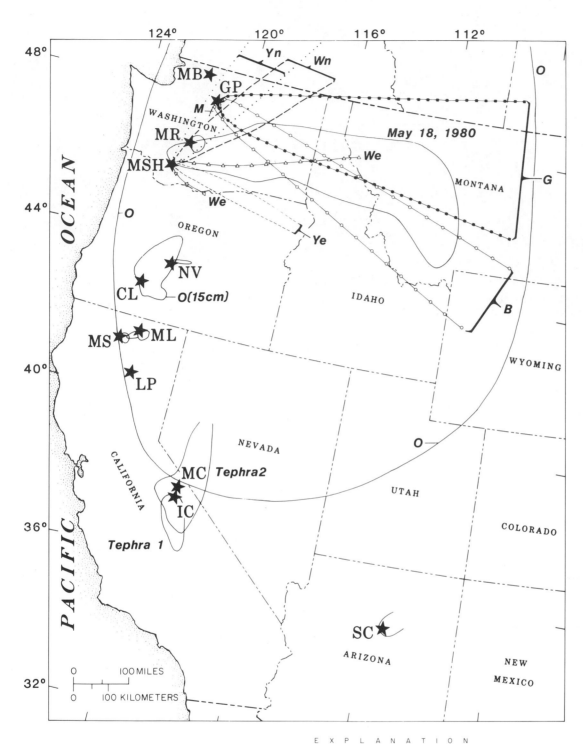

124° 120° 116° 112°

48°

Yn *Wn*

MB ★
GP ★
M
WASHINGTON
MR ★
MSH ★

May 18, 1980

OCEAN

We

MONTANA

O

G

44°

We
O

OREGON

Ye

NV ★
CL ★
O(15cm)

IDAHO

B

MS ★ ○ ○ ML

40°

LP ★

WYOMING

NEVADA

Tephra2

MC ★
★
IC

UTAH

COLORADO

36°

CALIFORNIA

Tephra 1

O

PACIFIC

SC ★

ARIZONA

NEW
MEXICO

32°

0   100 MILES
0   100 KILOMETERS

Figure 5-2. Areal distribution of latest Quaternary tephra units in the conterminous United States. Distributions shown for tephra units are probably minimal. Distribution shown for Mazama ash (layer O) is a composite of several lobes; although this distribution probably is also minimal, some small areas between adjacent lobes within the fallout region shown on the map may not have been covered by ash.

E X P L A N A T I O N

SOURCE AREAS OF TEPHRA UNITS

		NAME OR SYMBOL FOR TEPHRA UNIT
MB	Mount Baker, Washington	None shown
GP	Glacier Peak, Washington	***G, M, B***
MR	Mount Rainier, Washington	Total mapped area, nine units
MSH	Mount St. Helens, Washington	***Yn, Ye, Wn, We, May 18, 1980***
NV	Newberry Volcano, Oregon	Total mapped area, one unit
CL	Crater Lake, Oregon	***O*** (Mazama ash)
ML	Medicine Lake Highland, California	Total mapped area, two units
MS	Mount Shasta, California	Total mapped area, one unit
LP	Lassen Peak, California	None shown

CASCADE RANGE

SOURCE AREAS OF TEPHRA UNITS (cont.)

		NAME OR SYMBOL FOR TEPHRA UNIT (cont.)
MC	Mono Craters, California	***Tephra 2***
IC	Inyo Craters, California	***Tephra 1***
SC	Sunset Crater, Arizona	Total mapped area, multiple units

MONO AND INYO CRATERS

Table 5-3. Six Great Catastrophic Eruptions (Eruptive Volumes of about 300 km³ or Greater) That Have Occurred in the Conterminous United States during Quaternary Time

Name	Vent Locality	Age (my)	Volume (km³)	Reference
Lava Creek tephra	Yellowstone area, Wyoming	0.6	1000	Christiansen, 1979
Bishop tephra	Long Valley caldera, California	0.7	500	Bailey et al., 1976
Upper Bandelier	Valles caldera, New Mexico	1.15	300	Smith, 1979
Mesa Falls tephra	Yellowstone area, Wyoming	1.27	280	Christiansen, 1979
Lower Bandelier	Valles caldera, New Mexico	1.47	300	Smith, 1979
Huckleberry Ridge	Yellowstone area, Wyoming	2.01	2000	Christiansen, 1979

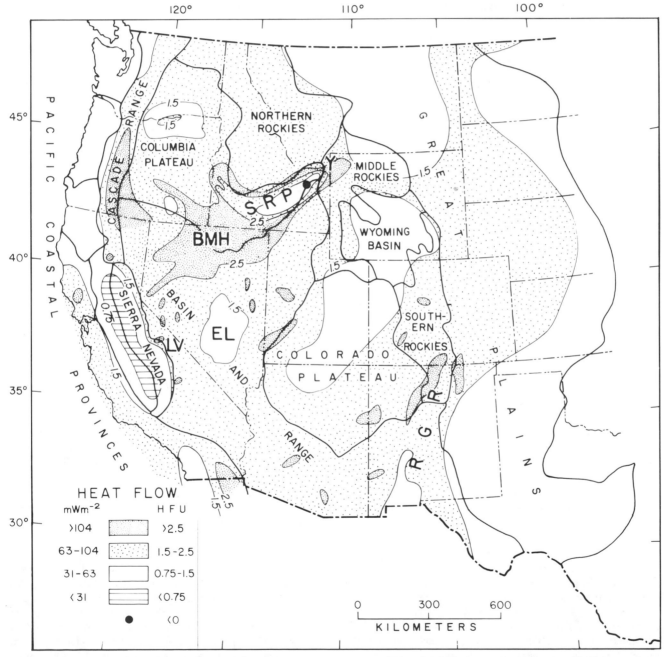

Figure 5-3. Map of the western United States showing heat-flow contours (in heat-flow units; 1 HFU = 41.8 milliwatts per square meter, or mWm⁻²), heat-flow provinces, and major physiographic divisions. (SRP, Snake River Plain; BMH, Battle Mountain high; EL, Eureka low; RGR, Rio Grande rift; Y, Yellowstone; LV, Long Valley.) (After Sass et al., 1980.)

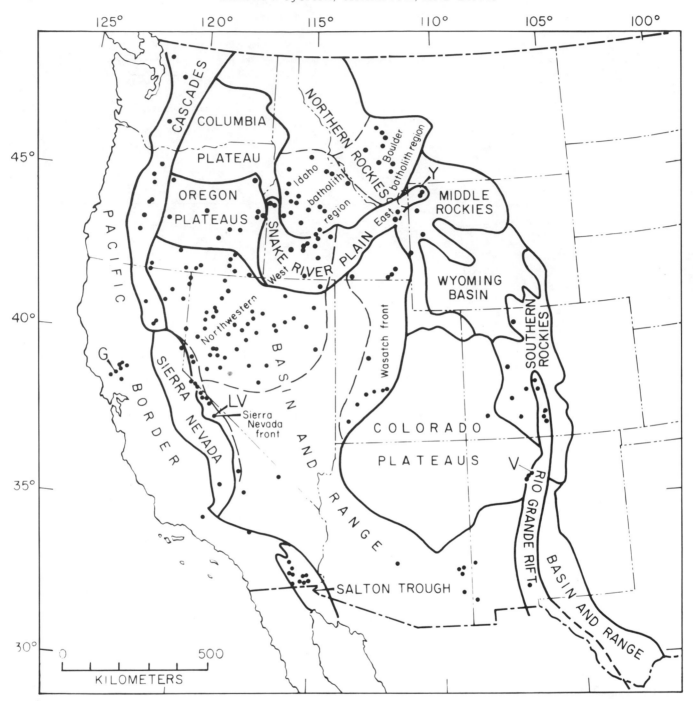

Figure 5-4. Localities of identified hydrothermal convection systems (dots) with reservoir temperatures greater than 90 °C. (G, the Geysers; V, Valles caldera; LV, Long Valley caldera; Y, Yellowstone area). (After Brook et al., 1979.)

flows and minor local pyroclastic deposits of basalt and andesite. The occurrence of basaltic or andesitic pyroclastic material is generally restricted to the vicinity of volcanic centers. Notable exceptions are the eruptions of basaltic cinders and ash from Sunset Crater in the San Francisco Mountains volcanic center of Arizona (Pilles, 1979). The known area (1500 km₂) covered by volcanic ash from this center is, nevertheless, relatively small in comparison with the widespread layers of silicic ash erupted from the major silicic volcanic fields.

Some uncertainty exists in the age assignments of latest-Quaternary volcanic rocks presented here. Accurate isotopic and other age determinations on young volcanic rocks are few, and many investigators have made age assignments on the basis of youthful morphology, stratigraphic relations, or other criteria that are not strictly age diagnostic. Consequently, some vents and fields of Holocene age may have been omitted form our compilation (Figure 5-1), and, conversely, some volcanic vents or fields included in our compilation may be of pre-Holocene age.

Areal Distribution of Holocene Volcanic Rocks and Its Relationship to Heat Flow

tive to compare the areal distribution of Holocene volcanism with that of measured heat flow. The distribution of Holocene volcanic rocks in the western United States generally coincides with the areal distribution of heat flow (Figure 5-3), especially with the distribution of reduced heat flow (total measured heat flow minus the estimated contribution from radioactive heat generated in crustal rocks) (Lachenbruch and Sass, 1978). The Cascade Range, Salton Trough, and Rio Grande rift all show a close coincidence of high heat flow (more than 2.5 heat-flow units [HFU]) with the occurrence of young volcanism. Volcanism in the Basin and Range province seems to be confined to the periphery of a region of moderate heat flow (1.5 to 2.5 HFU) (Figure 5-3) but commonly is situated close to isolated smaller areas of high heat flow. The dominant heat-flow feature of western North America, the Battle Mountain high, however, is not obviously related to surface Holocene volcanism. Heat flow in this region may be associated with underplating (accretion of solidifying basalt at the base of a stretching lithosphere) (Lachenbruch, 1978) rather than with extrusion of basaltic magma. Also, no Holocene volcanic activity has taken place in the Great Plains, an area of moderate heat flow east of the Rocky Mountains (Figure 5-3).

Three areas of low heat flow, especially the elongate low (0.75 to 1.5 HFU) in the eastern Snake River Plain, are hydrologically determined. Lows associated with the Snake and Columbia Rivers are the effects of active subsurface-aquifer systems that mask otherwise high heat flows in these areas (Brott et al., 1976). The regional heat flow of the eastern Snake River Plain, for instance, is estimated at between 2.5 and 3.5 HFU when the hydrologic effect is subtracted (Brott et al., 1978). The Eureka low in southern Nevada, surrounded by an area of moderate heat flow (Lachenbruch and Sass, 1977), is also thought to result from hydrologic effects.

Areal Distribution of Holocene Volcanic Rocks and Its Relationship to Hydrothermal-Convection Systems

The areal distribution of Holocene volcanic rocks generally coincides with that of hot ($\geqslant 90\,°C$; 3-km depth) hydrothermal-convection systems (Figure 5-4) (Brook et al., 1979). The major physiographic provinces in which volcanic rocks occur are also those areas where hot hydrothermal-convection systems exist (e.g., the Cascade Range, Oregon Plateaus, Snake River Plain, Sierra Nevada front, Wasatch front, and Salton Trough [Figure 5-4]), although considerable differences exist in detail. Thus, the hot hydrothermal systems shown on the map (Figure 5-4) are much more common in the western United States than Holocene volcanic rocks largely because, by definition, we have identified only the surface expression of Holocene igneous activity. Many hydrothermal systems that extend to some depth are associated with intrusive rocks that have not reached the surface. Some hydrothermal systems coincide with batholiths or other regions of high heat flow, such as the Battle Mountain high, where hydrothermal convection may extend to considerable depth but where no Holocene volcanism has occurred. In some areas, hydrothermal-convection systems and Holocene volcanic rocks do

not appear to be related, particularly in the southern Basin and Range and the Rio Grande rift areas, where the limited availability of ground water or unfavorable rock structure may account for the absence of surface or near-surface hydrothermal-convection systems near Holocene volcanic fields.

Volcanism in the western United States appears to be much more rare and sporadic than heat flow or hydrothermal convection; particularly when viewed within such a short time frame as the Holocene. Nevertheless, these three phenomena—heat flow, hydrothermal convection, and volcanism—are surface manifestations of the same driving mechanism: the outward transfer of heat from within the Earth. And they are spatially related, particularly when viewed over a long interval.

Relationship of Tectonic Setting to Areal Distribution of Holocene Volcanic Rocks

Holocene volcanism in the western United States is principally restricted to four physiographic and tectonic provinces: (1) the Cascade Range of western Washington, western Oregon, and northern California; (2) the periphery of the Great Basin (Basin and Range province), including eastern and northeastern California, the eastern Oregon Plateaus, Nevada, the Snake River Plain of southern Idaho, western Utah, and northern Arizona; (3) the periphery of the Rio Grande rift area of New Mexico, possibly extending northward into west-central Colorado; and (4) the Salton Trough of southernmost California and northern Baja California (Figures 5-1 and 5-4).

According to several investigators, volcanism in the Cascade Range of the Pacific Northwest is caused by the subduction of the Juan de Fuca plate beneath the North American plate (Figure 5-5) (Lipman et al., 1972; Snyder et al., 1976; Cross and Pilger, 1978). The north-south alignment of Cascade Range volcanoes, their position east of the inferred subduction zone, and the recency of volcanic activity in this province support the notion of northeast-directed subduction of the Juan de Fuca plate. According to White and McBirney (1978), however, compositional variations observed in late-Cenozoic volcanic rocks of the Cascade Range cannot be interpreted in terms of the steady subduction and partial fusion of an oceanic plate.

Common rock types erupted in the Cascade Range during late-Cenozoic time were basaltic andesite and basalt, which produced widespread flows, low volcanic-shield volcanoes, and cinder cores. According to White and McBirney (1978), basalt is the most voluminous rock type in the Cascade Range of northern California and central Oregon and andesite is the next most common; dacite and rhyolite are of minor significance. In central Washington, andesite predominates over basalt and dacite and rhyolite again make up a small percentage of the total. Several volcanic centers within the Cascade chain, however, have evolved chemically to produce more-silicic magmas (Mount Lassen, Mount Shasta, Medicine Lake Highland, Crater Lake, Newberry Volcano, the Three Sisters, Mount St. Helens, and Glacier Peak). These silicic magmas were extruded to form dacitic to rhylitic domes or flows, and some vents have produced silicic air-fall tephra deposits from explosive eruptions. The density of Holocene volcanic vents and associated deposits in the Cascade Range is high, with the exception of Washington State, which has fewer but larger volcanic centers.

Volcanism in the Great Basin of the western United States, according to current plate-tectonic theory (Armstrong et al., 1969; Chris-

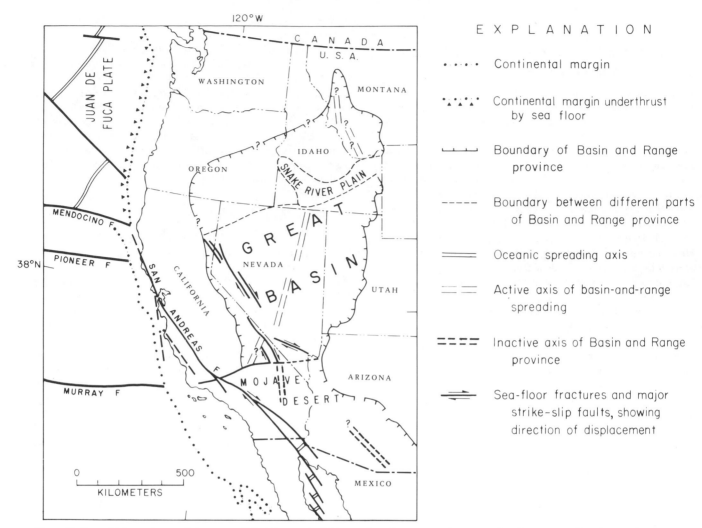

EXPLANATION

· · · · ·   Continental margin

ᵛᵛᵛ   Continental margin underthrust by sea floor

└──┴──┘   Boundary of Basin and Range province

- - - - -   Boundary between different parts of Basin and Range province

═══════   Oceanic spreading axis

══ ══   Active axis of basin-and-range spreading

══ ══ ══   Inactive axis of Basin and Range province

⇄   Sea-floor fractures and major strike-slip faults, showing direction of displacement

Figure 5-5. Generalized tectonic map of the western United States and adjoining sea floor showing approximate axes of crude symmetry of broad zones of spreading. (Modified from Proffett, 1977, and Kilbourne et al., 1980.)

tiansen and Lipman, 1972; Proffett, 1977; Cross and Pilger, 1978) is due to crustal spreading and attenuation of continental crust within this region. According to Lachenbruch and Sass (1978), volcanism results from the upward intrusion of basaltic magma into the continental crust to compensate for mass lost as a result of lateral spreading. Basalt intruded at high temperatures may react with the continental crust in several ways. (1) Basaltic magma intruded to high levels in the crust may cause the partial melting of silicic crustal rock and result in the formation of a more-silicic magma. (2) Basaltic magma intruded upward into the crust may differentiate and become zoned vertically, with a silicic upper part of the chamber and a more basaltic lower part. Or (3) basalt may be intruded directly through the crust and erupted on to the surface.

Volcanism in the Great Basin during Holocene time is largely restricted to the periphery of the basin. This volcanism is predominantly basaltic, except at Craters of the Moon, Idaho, and Miller Knoll, Utah, where some lava flows of andesitic composition have been erupted, and at Mono and Inyo Craters, California, where more highly differentiated rhyolitic and dacitic rocks have been erupted. Mono and Inyo Craters are spatially and genetically related to the Long Valley caldera, a large volcanotectonic depression formed after the

eruption of the Bishop Tuff 0.73 my ago (Gilbert, 1938; Dalrymple et al., 1965). This area, which has been a locus of volcanic activity during at least the past 2 my, has generated large volumes of silicic tephra and flows ranging widely in composition (Bailey, 1980; Bailey et al., 1976).

Eruptions of basalt around the periphery of the Basin and Range province have been generally small in volume and short in duration. The Craters of the Moon lava field in the eastern Snake River Plain is a notable exception; as much as 31 km³ of basaltic lava has been erupted during eight short eruptive episodes in the past 15,000 years (Kuntz et al., 1980). The duration of each of these eruptive episodes was less than 100 years, and the volume of lava that was erupted varied (between about 1.5 and 6.3 km³). Plots of cumulative lava volume versus time show that lava was produced at two different but constant rates in this area: 1.6 km³ per 1000 years before about 7000 yr B.P. and 2.8 km³ per 1000 years after 7000 yr B.P. This increase in the rate of eruption coincides with an increase in the silica content of the lavas erupted after about 7000 yr B.P.

Very young volcanism has also occurred in the late-Cenozoic volcanic field of the San Francisco Mountains of northern Arizona. Activity at Sunset Crater and subsidiary vents in this area began about A.D. 1065 (Smiley, 1958; Breternitz, 1967), as dated by tree rings. the eruption at Sunset Crater was probably similar in style to the historical eruption of Paricutin Volcano, Mexico, which continued for a period of 8 years. Sporadic activity at Sunset Crater, however, probably continued for a period of about 200 years, as determined from stratigraphic and paleomagnetic studies (E. M. Shoemaker and D. E. Champion, personal communication, in Pilles, 1979). Evidence for the timing of this activity is based on the correlation of isolated lava flows by means of superposed tephra layers and the comparison of the magnetization directions of individual flows by means of the archaeomagnetic secular-variation curve of DuBois (1974). Sternberg and McGuire (1981) suggest that this curve gives age estimates that are about 100 years too young. Thus, the duration of the eruptive episode from Sunset Crater would be about 100 years. Holocene volcanism in the Great Basin, in addition to being confined to the periphery of the province, also closely follows the trends of present-day seismicity and recently active faulting.

Volcanism in the Rio Grande rift of New Mexico and Colorado is predominately basaltic (Christiansen and Lipman, 1972), and this trend has continued to Holocene time with the formation of a few small basaltic fields. These fields, however, are not always within the rift itself, but are usually peripheral to it. Holocene volcanism in the Salton Trough of southern California (Robinson et al., 1976) is exclusively rhyolitic, producing small domes along segments of an active subsurface spreading ridge.

Role of Volcanic Ash Layers in the Correlation of Uppermost Quaternary Deposits in the Western United States

Volcanic-ash chronology (Wilcox, 1965), also referred to as tephrochronology (S. Thorarinsson, in Westgate and Gold, 1974), is a stratigraphic and geochronologic technique that permits the temporal correlation and age dating of sedimentary and volcanic deposits by means of volcanic ejecta, primarily ash and tuff ("tephra," as used here, is a collective term introduced by S. Thorarinsson to refer to all ejecta, solid or liquid, erupted from a volcanic vent and transported by air.) Tephrochronology, in a broad sense, can be subdivided into two types of studies: tephrostratigraphy, the temporal correlation of tephra layers by whatever appropriate means (physical, petrographic, or chemical), and tephrochronology in the restricted sense, the actual dating of tephra layers, either directly by such methods as potassium-argon or fission-track dating or indirectly by stratigraphic bracketing of tephra layers by isotopic or other types of age control (e.g., radiocarbon dating of overlying and underlying beds).

These two techniques, tephrostratigraphy and tephrochronology, constitute a powerful geochronologic tool when used in combination, because the age of a given well-dated tephra layer at an individual site can be extended to any other site where this same layer occurs. Thus, a tephra layer that has been identified, dated, and correlated in multiple localities over a broad area becomes a virtual time horizon. Because widespread tephra units are found in various depositional environments, it is often possible to select those sites where age determinations can be most successful (in other words, where the most precise technique can be employed and where the age analysis can be performed on the best available material with the least possibility of contamination or analytical error). Once dated, a tephra layer then becomes a critical link by which deposits of diverse depositional en-

vironment can be correlated and dated, a feat that is difficult to accomplish by the usual stratigraphic methods, such as faunal correlation.

An important additional use of tephrochronologic correlation is as a check on isotopic and other types of age analysis. The temporal correlation of a volcanic ash layer can be documented by the use of several different methods and various parameters that can be measured with a high degree of precision. Furthermore, the problem of contamination of tephra by other materials can be minimized or completely eliminated by the careful preparation of samples and the selection of the analytical methods to be used in the correlation. Consequently, discordant ages obtained for an ash layer that has been carefully correlated by tephrostratigraphic methods indicate a problem in the age analysis rather than in the tephra correlation.

Several assumptions, implicit or explicit, are made in the use of tephra layers as a tool for correlation purposes. The first and most important assumption is that tephra layers possess diagnostic characteristics by which they can be distinguished from each other. This assumption implies that the specific characteristics of a particular tephra unit (e.g., its mineralogy, mineral chemistry, and glass chemistry) are not exactly duplicated in another unit of demonstrably different age. Although it is difficult to prove this assumption conclusively, presently available data on multiple superposed tephra units (Sarna-Wojcicki et al., 1980; Sarna-Wojcicki, Meyer, et al., 1981) indicate that this is indeed the case. Tephra units derived from the same volcanic province, particularly those erupted from the same volcanic vent over short intervals, can be petrographically and chemically similar. Available data on analyzed tephra layers where independent age control is available, however, indicate that tephra units are not identically replicated within the limits of presently available analyses, except for multiple units (or tephra "sets") erupted from the same vent within short eruptive episodes as long as perhaps several tens of years. Corollary assumptions made in temporal correlation of tephra layers are (1) that the parameters used in the correlation of a tephra layer are essentially constant or vary little from site to site, regardless of its distance from the volcanic source (or, if variations exist, they are significantly smaller within a given tephra layer than between tephra layers), (2) that the downwind dispersal and deposition of tephra are rapid, and (3) that the reworking and redeposition of ash, when they do occur, take place shortly after initial deposition.

All these assumptions have been tested to some degree, and, although they are generally valid (Sarna-Wojcicki et al., 1979, 1980), exceptions have also been reported, particularly with respect to reworked and mixed tephras (Cerling et al., 1978). The eruption of Mount St. Helens on May 18, 1980, however, presents a modern case for which some of these assumptions can be tested.

The Eruption of Mount St. Helens and the Areal Distribution and Characteristics of the Downwind Ash

The downwind dispersal and deposition of volcanic ash from the May 18, 1980, eruption of Mount St. Helens in Washington State were well documented by reports from eyewitnesses, satellite surveillance, and mapping and observations (Lipman and Mullineaux, 1981; Sarna-Wojcicki, Shipley, et al., 1981). Within 13 minutes after the start of the eruption, the tephra plume had grown vertically to an altitude of about 23 km above the vent; eventually, it reached an altitude of

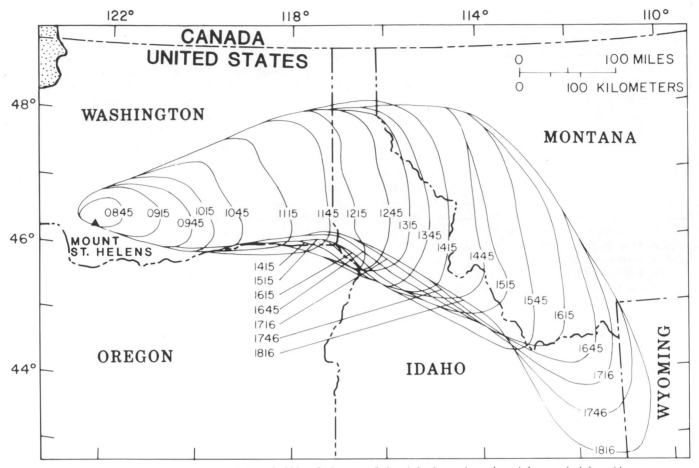

Figure 5-6. Ash fall from May 18, 1980, eruption of Mount St. Helens. Isochron map (in hours) showing maximum downwind extent of ash from airbourne-ash plume carried by fastest-moving wind layer, as observed on satellite photographs. (After Sarna-Wojcicki, Shipley, et al., 1981.)

about 27 km. It attained an average vertical velocity of 160 km per hour during the first 10 minutes of its growth (Sarna-Wojcicki, Shipley, et al., 1981). From an initial lateral velocity of more than 350 to 400 km per hour (Rice, 1981), which was an effect of the initial gas-propelled lateral blast, the tephra plume decelerated to an average lateral velocity of about 100 km per hour. The progress of the plume was tracked by satellite imagery as airborne ash was carried downwind toward the east-northeast; the plume covered the first 1000 km of its journey in about 10 hours (Figure 5-6). The main body of ash was carried east-northeastward and then eastward across central Washington and into Idaho and western Montana.

The ash plume was transported by the wind first east-northeastward, then southeastward across the midwestern states, then eastward and northeastward again to the eastern seaboard, and then over the Atlantic Ocean (McCormick, 1981). Although the main body of ash was transported within the high-velocity wind layer between about 10 and 16 km in altitude (Figure 5-7), ash erupted to altitudes of 16 to 19 km was carried to the north and northeast into Canada and looped to the east and southeast toward the Great Lakes region. Ash erupted to altitudes of about 20 to 23 km moved southeastward over Oregon and then looped sharply to the northwest and out over the Pacific Ocean (Figure 5-7) (Danielsen, 1980). Ash that fell or was erupted below about 2200 m was carried northward and northeastward. Light detection and ranging (LIDAR) measurements by atmospheric backscattering of a laser beam indicate that the ash travelled

completely around the Earth in about 2 weeks (Danielsen, 1980; McCormick, 1981).

The areal distribution of the ash on the ground and the mass per area of the ash deposited were determined by observations and measurements made shortly after the eruption, during the period between May 19th and 22nd (Figure 5-8) (Sarna-Wojcicki, Shipley, et al., 1981). Samples of volcanic ash collected at various distances from the volcano were analyzed for bulk chemistry as well as for composition of the volcanic glass. Although the bulk-chemical analyses of downwind ash showed progressive compositional variations as a function of distance from the volcano (owing to preferential fallout closer to the volcano of denser crystals and lithic fragments relative to volcanic glass), the glass composition was essentially identical for all the samples analyzed, no matter where they were collected (Table 5-4).

A comparison of the chemical composition of the May 18, 1980, ash with that of other Holocene and latest-Pleistocene ashes erupted from Mount St. Helens indicates that the composition of each ash set is distinctive (Table 5-4) (e.g., compare the glass composition of the May 18, 1980, ash with that of Mount St. Helens T, erupted about A.D. 1800). In many instances, individual layers within the ash sets, such as Mount St. Helens We and Wn, can also be distinguished on the basis of stratigraphic or petrographic criteria (Mullineaux et al., 1975, 1978; Crandell and Mullineaux, 1978) or on the basis of chemical properties (Table 5-2) (Sarna-Wojcicki, Meyer, et al., 1981).

Distinctions are more difficult within than among sets, however, because the parent magmas of tephra erupted from the same vent within short periods tend to show less chemical variance. In distal areas, making distinctions among such layers also is difficult, because stratigraphic control from other ash layers is generally sparse and characteristic minerals are very fine or absent. In such instances, correlations must be made on the basis of glass chemistry.

Observations of the May 18, 1980, eruption and comparisons of the ash with that from previous eruptions of Mount St. Helens confirm some of the assumptions used in tephrochronologic correlations. Stratigraphic characteristics documented for this tephra layer provide criteria that can be applied to the study of prehistoric tephra layers. Of particular importance are stratigraphic criteria by which the thicknesses of initial air-fall ash can be distinguished from secondary depositional thickening due to reworking. From studies such as these, as well as from mapping of the minimum areal extents of ancient tephra layers, we may be able to estimate volumes of prehistoric eruptions, which, together with estimates of initial volatile content (particularly SO_2, CO_2, and Cl_2), are basic to an assessment of the effect of volcanism on climate.

Characteristics of Tephra Layers

Several characteristics are useful in identifying tephra layers. Such field criteria as superposition, sequence, bedding characteristics, color, and megascopic pyroclastic composition, combined with radiometric-age control, are useful in areas close to eruptive sources. For tephra units at both proximal and distal localities, these methods are supplemented by petrography, which includes glass-shard morphology, refractive indexes of glass, and mineralogy (the mineral species present, their relative abundances, and refractive indexes in the principal optical directions). Field and petrographic methods are

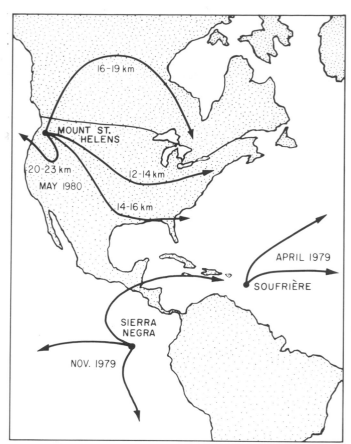

Figure 5-7. Movement of material injected into the stratosphere at different altitudes by eruptions of Soufrière, Sierra Negra, and Mount St. Helens. (After McCormick, 1981.)

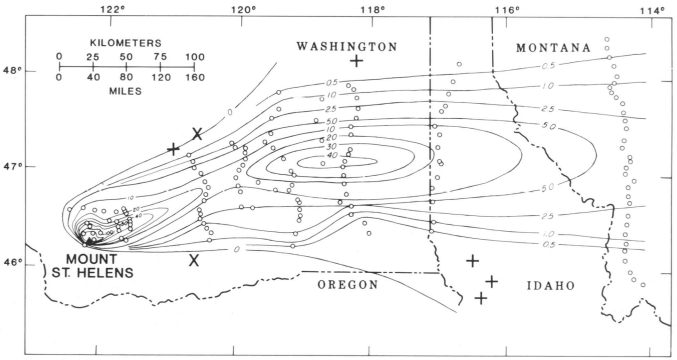

Figure 5-8. Ash fall from May 18, 1980, eruption of Mount St. Helens. Isopach map of air-fall ejecta. Lines represent uncompacted thicknesses (in millimeters). Plus signs indicate light dustings of ash; capital letter exes mean that no ash observed; lowercase letter ohs mark observation sites. (After Sarna-Wojcicki, Shipley, et al., 1981.)

Table 5-4. Elemental Spectral Intensities of May 18 Volcanic Glass Samples Showing Compositional Homogeneity
of Glass Downwind (Intensities of May 18 Bulk Ash, Older Glass Samples from
Mount St. Helens, and Standard Granite Shown for Comparison)

Sample	Description	Distance from Volcano (km)	Normalized Peak Intensities										
			$K_{k\alpha}$	$Ca_{k\alpha}$	$Ca_{k\beta}$	$Ti_{k\alpha}$	$Mn_{k\alpha}$	$Fe_{k\alpha}$	$Fe_{k\beta}$	$Rb_{k\alpha}$	$Sr_{k\alpha}$	$Y_{k\alpha}$	$Zr_{k\alpha}$
						May 18 Glass Samples							
1	Glass from ash layer	18	234	437	77	173	113	6195	1011	600	2831	410	2834
2	Glass from ash layer	248	239	444	80	177	115	6188	1000	590	2694	406	2701
3	Glass from ash layer	305	236	439	80	169	112	6115	981	622	2702	413	2722
4	Glass from ash layer	308	233	441	83	180	114	6198	994	573	2816	399	2697
5	Glass from ash layer	320	233	448	78	177	114	6182	998	602	2711	419	2625
6	Glass from ash layer	400	234	438	77	173	111	6211	999	605	2826	411	2850
7	Glass from ash layer	420	234	438	79	173	114	6191	1003	597	2785	421	2730
8	Glass from ash layer	632	239	441	77	171	111	6116	991	601	2652	427	2707
	Average intensity		235	441	79	174	113	6175	997	599	2752	413	2733
	Standard deviation		±2	±4	±2	±4	±2	±37	±9	±14	±70	±9	±74
						Comparative Intensities							
9	Layer T glass (~180 yr B.P.)[a]	8	174	427	76	170	121	6320	1016	541	2813	412	2422
10	Layer Pm glass (~3000-2400 yr B.P.)[a]	8	391	523	92	168	140	5859	954	682	2875	400	2562
11	Bulk ash from dark basal layer (sample 7)	420	132	542	95	176	95	6286	1023	393	3789	355	2342
12	Bulk ash from light upper layer (sample 7)	420	175	494	86	172	103	6265	1017	489	3464	358	2583
13	Granite; USGS standard rock G-1	—	1040	440	73	211	86	5264	836	1527	2043	513	2429

Source: Modified from Sarna-Wojcicki, Meyer, et al., 1981.
Note: Analyses were by energy-dispersive X-ray fluorescence; analyst was Marta J. Woodward. Values given are integrated peak intensities divided by total counts in each of two spectral regions (windows).
[a] From pumice-lapilli-lithic ash collected by D. R. Mullineaux, A. M. Sarna-Wojcicki, and R. B. Waitt, Jr. (1977).

further supplemented by chemical analyses of glass or specific pyrogenic minerals.

Chemical methods of identification are particularly useful for tephra layers far from their eruptive sources. In those places, such field criteria as superposition may be absent, other criteria such as bedding characteristics or refractive indexes of glass may not be diagnostic, and pyrogenic minerals may be too small, too sparse, or absent. Chemical data can be readily adapted to quantitative and statistical methods of comparison to document the similarities and differences among tephra layers in probabilistic terms.

Tephra can be characterized chemically in several ways. Electron microprobe analysis of volcanic glass is the easiest and quickest method. Components of tephra layers such as glass, minerals, and lithic fragments need not be separated, and individual glass shards can be analyzed so that the homogeneity of the glass in individual layers or the mixing of glass in several tephra layers can be tested (Cerling et al., 1978). Only the major elements and a few minor elements can be analyzed precisely by this method, however, owing to its low sensitivity, and tephra units of similar composition usually cannot be distinguished by electron probe analysis. Other chemical methods of characterization rely on the analysis of separated glass; among these methods are wavelength- and energy-dispersive X-ray flourescence spectrometry, atomic-absorption and atomic emission spectroscopy,

and instrumental neutron-activation analysis. Though more expensive and time-consuming, these methods provide greater resolution in distinguishing among tephra layers of similar composition—a problem encountered in attempts to distinguish among tephra layers erupted from a single source or derived from a single volcanic province.

Major Source Areas and Dispersal Patterns of Silicic Ash Layers

Widespread layers of latest-Quaternary silicic ash have been erupted from two volcanic provinces in the western United States, the Cascade Range of the Pacific Northwest and the Mono and Inyo Craters area of east-central California (Figure 5-2). Basaltic or andesitic tephra layers of more limited areal extent have also been erupted from these two provinces, as well as from Sunset Crater, Arizona, and from numerous smaller basaltic and andesitic volcanic fields and vents (Figures 5-1 and 5-2).

Tephra layers erupted in the western United States have typically been dispersed by prevailing westerly winds (Figure 5-2). The greatest number of superposed tephra layers of latest-Quaternary age occur in the Cascade Ranges and to the east, in central and eastern Washington, eastern Oregon, northern Idaho, and western Montana (Figure 5-2); and so the best available age control from tephra is in these areas. Some ash plumes were also carried northward, westward, southward, and southwestward from eruptive centers. Individual tephra plumes from the 1980 eruption sequence at Mount St. Helens, for instance, were dispersed in every direction of the compass, and even the dispersal patterns from individual eruptions were complex and multidirectional (Figure 5-7) (Sarna-Wojcicki, Shipley, et al., 1981). Similar conditions must have prevailed during many late-Quaternary eruptions.

Tephra erupted from the Mono and Inyo Craters volcanic province in east-central California is potentially useful for dating Holocene deposits throughout east-central California and northwestern Nevada (Figure 5-2). Most major plumes from these silicic eruptions were transported eastward, although the southward or southeastward transport of some lobes suggests that some eruptions may have occurred during seasonal southward Santa Ana winds.

With further investigation, fallout areas for the major tephra layers will undoubtedly be widened, and new layers will probably be recognized. We show minimum areal distributions on the map (Figure 5-2), and several layers that have been recognized only since about 1960 or 1970 are not shown even though they are probably widely distributed, also. Thin tephra layers and disseminated fine ash of latest-Quaternary age are undoubtedly present in areas beyond those shown on the map (Figure 5-2). For instance, ash from the May 18, 1980, eruption of Mount St. Helens was deposited as a thin layer, or dusting, across the entire North American continent, and the volume of this eruption was small in comparison with the volumes of several widespread uppermost Quaternary tephra layers (Sarna-Wojcicki, Shipley, et al., 1981). Most of this fine material will not be preserved in the geologic record as a discrete layer; and, where it is preserved, it may not be recognizable through the methods currently available. A determined effort to find ash beds in new areas, combined with the development of new techniques for the separation and analysis of very fine particles, may extend these time horizons considerably beyond their presently known area.

Tephra layers of basaltic and andesitic composition are generally not as widespread as those of more silicic composition. Locally, as in the vicinity of Mount Rainier, Washington, or Sunset Crater, Arizona, such layers are useful time and stratigraphic markers. In addition, basaltic vents are much more numerous and widely distributed (Figure 5-1) and thus can provide age control in areas where widespread silicic tephra layers are absent.

Widespread Silicic Ashes from Cascade Range Vents

During latest-Quaternary time (approximately the past 13,000 years), widespread layers of silicic ash have been erupted from three main volcanoes of the Cascade Range: Mount St. Helens, Washington, Glacier Peak, Washington, and Crater Lake, Oregon (Figure 5-2). Of these three volcanoes, Mount St. Helens has had the longest and most frequent record of explosive activity, which began about 37,500 yr B.P. (Mullineaux et al., 1975) and has continued sporadically to the present.

Tephra Layers Derived from Mount St. Helens, Washington

Nine major widespread tephra sets were erupted from Mount St. Helens before 1980 (Crandell and Mullineaux, 1973, 1978;

Mullineaux et al., 1975, 1978), six of which were erupted during latest-Quaternary time (from oldest to youngest: sets S, J, Y, P, W, and T) (Table 5-1). The tephra set erupted in 1980, here referred to as set D, represents a seventh.

Several criteria have been used to identify and to distinguish the layers of tephra erupted from Mount St. Helens. In the vicinity of the volcano and for distances as far as several tens of kilometers away, stratigraphic superposition and other field criteria, such as color and megascope pyroclastic composition, have been used (Crandell et al., 1962; Crandell and Mullineaux, 1973; Mullineaux et al., 1975, 1978). Because the layers of tephra erupted from Mount St. Helens radiate outward in generally rather narrow lobes, stratigraphic control becomes sparse beyond several tens of kilometers from the volcano, except where these layers overlap those erupted from other vents (Figure 5-2). Petrographic and mineralogical criteria are useful in identifying tephra layers for intermediate distances as far as several hundreds of kilometers from the volcano (Okazaki et al., 1972; Mullineaux et al., 1975, 1978). These criteria are supplemented by chemical methods of identification for layers at intermediate to great distances (Westgate et al., 1970; Randle et al., 1971; Smith et al., 1977a, 1977b; Sarna-Wojcicki, Meyer, et al., 1981).

Widespread layers of tephra erupted from Mount St. Helens contain glass shards that are dominantly pumiceous and generally highly vesiculated. Common phenocrysts are plagioclase, hypersthene, hornblende, augite, and minor magnetite, ilmenite, and apatite (Table 5-2) (Mullineaux et al., 1975; Crandell and Mullineaux, 1978; Sarna-Wojcicki, Meyer, et al., 1981). Some layers (for instance, those of sets S and Y) contain the amphibole cummingtonite, which has not been reported in tephra of other Holocene volcanoes in the conterminous United States and is consequently diagnostic of Mount St. Helens tephra.

SET S ASH LAYERS

The oldest of the tephra sets discussed here, set S, was erupted between about 13,600 and 12,000 yr B.P. (Table 5-1) (Mullineaux et al., 1975, 1978). The two most widespread tephra layers, So and Sg, were erupted between about 13,000 and 12,000 yr B.P. Set S has been mapped in all directions away from the volcano, although the thickest layers are situated to the east-northeast (Mullineaux et al., 1975). The identification of an ash bed, tentatively correlated with set S on the basis of the electron microprobe analysis of glass and mineralogy, at Double Hot Spring in northwestern Nevada suggests that one lobe of set S may have been carried southeastward. Small crystals of cummingtonite, characteristic of set S (Table 5-2) and several other layers of tephra erupted from Mount St. Helens, have been found in ash from this locality.

Significance of Set S Ash Layers as Time and Stratigraphic Markers

Ash is interbedded with late-Pleistocene flood deposits of the channeled scablands of central and eastern Washington at several localities; these deposits were formed by repeated catastrophic emptying (jökulhlaups) of glacial Lake Missoula in northwestern Montana (Bretz, 1923, 1969). Mullineaux and others (1978) correlate these ash layers, situated stratigraphically near the top of the flood deposits, with Mount St. Helens set S on the basis of mineralogy and glass chemistry. Therefore, the last major scabland floods are dated at about 13,000 yr B.P., considerably later than previously suggested from indirect evidence that generally correlates the floods with late-

Table 5-5. Comparison of Neutron-Activation Analysis Data For Glass in
Mount St. Helens Tephra Sets S and M

Sample	Sc	Fe	La	Ce	Sm	Tb	Yb	Lu	Th	U
1	2.45	0.88	15.4	33.8	2.44	0.28	1.02	0.16	4.51	1.66
2	2.59	0.91	15.7	34.7	2.55	0.28	1.06	0.16	4.49	1.74
3	2.16	0.96	15.0	32.7	2.34	0.24	0.92	0.14	4.25	1.64
4	2.26	0.96	15.1	32.9	2.38	0.27	0.94	0.14	4.25	1.62
5	1.76	0.92	16.8	34.1	1.96	0.18	0.76	0.11	4.79	1.89
6	2.40	1.07	17.1	33.4	1.95	0.20	0.75	0.11	4.58	1.79
7	± 0.02	± 0.02	± 0.04	± 0.05	± 0.01	± 0.01	± 0.02	± 0.01	± 0.04	± 0.03

Note: Element concentrations are in parts per milion except for iron, which is in weight percentages. Analysts were H. R. Bowman and Frank Asaro, University of California, Lawrence Berkeley Laboratory.
Samples: (1) Tephra layer So (upper) near Mount St. Helens. (2) Tephra layer (upper) in scabland flood deposits, southeastern Washington. (3) Tephra layer Sg (lower) near Mount St. Helens. (4) Tephra layer (lower) in scabland flood deposits, southeastern Washington. (5) Tephra layer Mm (upper) near Mount St. Helens. (6) Tephra layer Mp (lower) near Mount St. Helens. (7) Average analytical error.

glacial events in the northern Rocky Mountains. The neutron-activation analysis of glass (Waitt, 1980) supports this correlation. Ash beds in the channeled-scabland deposits more nearly resemble Mount St. Helens set S than the older Mount St. Helens set M, erupted between about 20,350 and 18,500 yr B.P. (Table 5-5). The comparison of results from the electron microprobe analysis and neutron-activation analysis of the volcanic glass in set S ash beds indicates that correlations based on neutron-activation analysis are more definitive because of the greater precision of this method and the larger number of elements that can be detected. This conclusion has also been documented for many older Quaternary tephra units analzed from southern California (Sarna-Wojcicki et al., 1980).

SET J ASH LAYERS

Set J, which was erupted from Mount St. Helens between about 11,700 ± 400 and 8300 ± 300 yr B.P., has been mapped in a continuous sector from the north through the east to the south of the volcano, although it is thickest in areas between the northeast and the southeast (Mullineaux et al., 1975). The areal distribution of this set is not well defined at present. The most distal ash locality reported to date is in eastern Washington, at Lind Coulee (Smith et al., 1977b). However, because of the thickness of set J and the thickness and coarseness of some of the layers in this set near its source, these ash layers are probably widespread.

According to Mullineaux and others (1975), pumice from set J is mineralogically similar to that erupted from Glacier Peak, Washington, but differs from it in glass chemistry (R. E. Wilcox and J. A. Westgate, personal communication, 1973, *in* Mullineaux et al., 1975). These differences are further confirmed in this report: in chemical composition the layer Jy resembles that of the three most widespread layers of tephra erupted from Glacier Peak and is most similar to that of layer B from that volcano, but it can be distinguished from that of layer B by its iron, calcium, titanium, and alkali concentrations (Table 5-2).

SET Y ASH LAYERS

Tephra set Y was erupted from Mount St. Helens between about 4000 and 3000 yr B.P. (Mullineaux et al., 1975); layers Ye and Yn of this set were erupted between about 3510 ± 230 and 3350 ± 250 yr B.P., whereas layer Yb, less widespread than layers Ye and Yn, was erupted about 3900 ± 250 yr B.P. (Table 5-1) (Mullineaux et al.,

1975; Crandell and Mullineaux, 1978). Two major lobes showing the minimum areal extent of this ash (Figure 5-2) were mapped by Mullineaux and others (1975). Layer Ye, the youngest voluminous tephra layer of the set, was carried in a narrow lobe to the east-southeast across central Washington and into northeastern Oregon, where it was identified by Norgren and others (1970) and tentatively by Borchardt and others (1973) on the basis of radiocarbon ages and neutron-activation analysis of glass.

The older layer, Yn, the most extensive layer of the set, was carried in a rather narrow lobe to the north-northeast, where it covered parts of central and eastern Washington, British Columbia, and Alberta to a distance of about 900 km northeast of Mount St. Helens (Westgate et al., 1970). This layer, the thickest and coarsest of all the layers of Holocene tephra erupted from Mount St. Helens (Mullineaux et al., 1975), probably represents the most voluminous Holocene eruption from this volcano.

Set Y ash layers, which contain the characteristic ferromagnesian minerals cummingtonite and hornblende (Table 5-2), can be distinguished from those of set S by the presence of hypersthene in the latter and from other layers of cummingtonite-bearing tephra erupted from Mount St. Helens by a combination of mineralogical criteria (Mullineaux et al., 1975, 1978) as well as by glass chemistry. Layers Ye and Yn are difficult to distinguish on the basis of glass chemistry, even by neutron-activation analysis; small and unconvincing differences in calcium content from electron microprobe analysis (Table 5-2) may be significant, but more replicate analyses are needed to test these differences.

Set Y layers, together with Mount Mazama layer O and Glacier Peak ashes B, M, and G, are particularly useful in studies of late-glacial and neoglacial deposits, as well as in archaeological studies of the northwestern United States. Layer Yn is a particularly useful time and stratigraphic marker in northern Washington and southwestern Canada, where it has been used to help date postglacial and neoglacial deposits (Westgate et al., 1970; Porter, 1978).

SET P ASH LAYERS

Set P, erupted from Mount St. Helens between about 2930 ± 250 and 2580 ± 250 yr B.P. (Mullineaux et al., 1975), is a compound unit that near the volcano consists of two subsets of coarse to fine lithic and pumiceous tephra layers separated by an interval of sand and silt (Crandell and Mullineaux, 1973). Four thin ash beds of this set were identified in Mount Rainier National Park, about 80 km north-northeast of Mount St. Helens (Mullineaux et al., 1974). Set P tephra in this area is interbedded with several less-silicic tephra layers of limited areal extent erupted from Mount Rainier (Mullineaux et al., 1974). Two other Mount St. Helens layers are found in stratigraphic relation to set P in this area: layer Wn, which overlies set P, and layers of set Y, which underlie set P. Set Y, in turn, is underlain in this area by layer O, the widespread Mazama ash described later in this section.

Although the mineralogy of set P layers resembles that of sets J and W, the sets can be distinguished on the basis of the refractive indexes of hypersthene (Mullineaux et al., 1975), as well as the major- and minor-element chemistry of the glass (Table 5-2).

SET W ASH LAYERS

Widespread tephra of set W was erupted from Mount St. Helens about A.D. 1500, as determined by tree-ring counts (Mullineaux et al., 1975). Layers We and Wn, the two major tephra layers of set W, are widespread and were probably both erupted about 450 years ago (Mullineaux et al., 1975). Layer Wn, the coarsest, thickest, and most widespread of the set, extends northeastward from the volcano into central and northern Washington, British Columbia, and Alberta (Figure 5-2) (Mullineaux et al., 1975; Smith et al., 1977a; Crandell and Mullineaux, 1978). The youngest layer, We, extends due east from the volcano across south-central Washington (Mullineaux et al., 1975) and into Idaho (Figure 5-2) (Smith et al., 1977a).

In proximal areas, set W can be distinguished from both the older and younger tephra layers of Mount St. Helens by field criteria. Pumice from set W, however, contains the same mineral suite as sets P and J (Mullineaux et al., 1975) and Glacier Peak layers B, M, and G (Table 5-4) (Porter, 1978). In distal areas, where stratigraphic control is absent, set W can be distinguished from layers containing similar mineral suites by the refractive indexes of hypersthene (Mullineaux et al., 1975), as well as by electron microprobe analysis of its glass (Table 5-2) (Smith et al., 1977a; Sarna-Wojcicki, Meyer, et al., 1981). The glass compositions of sets Wn and We, as determined by electron microprobe, overlap, although Smith and others (1977a) suggest that these two layers can be distinguished by their calcium-iron-potassium ratios; the two layers can also be readily distinguished on the basis of the more sensitive neutron-activation analysis of glass (Table 5-6).

T ASH LAYER

Ash layer T was erupted from Mount St. Helens about A.D. 1800, as determined by tree-ring counts (Lawrence, 1954; see also Crandell and Mullineaux, 1978, and Majors, 1980). Layer T tephra was carried to the northeast of the volcano (Crandell and Mullineaux, 1978), across central and northeastern Washington, and into the northernmost part of Idaho and the northwestern corner of Montana (Okazaki et al., 1972). The ash is readily distinguishable from other widespread tephra units by its stratigraphic position (above set W and below the May 18, 1980, tephra) as well as by its mineralogy and chemistry (Table 5-2). Layer T is relatively mafic among the widespread layers of ash erupted from Mount St. Helens and contains high concentrations of aluminum, iron, magnesium, and calcium and the lowest concentration of silica (Table 5-2).

Tephra Layers Derived from Glacier Peak, Washington

Glacier Peak volcano in northern Washington (Figure 5-2) was active during a relatively short interval in latest-Pleistocene time, from about 12,750 to 11,250 yr B.P. (Fryxell, 1965; Mehringer et al., 1977; Porter, 1978). Plinian eruptions from this source produced nine tephra layers, three of which (G, M, and B) were spread over large areas east and south of the volcano (Porter, 1978).

Although the nine Glacier Peak tephra layers can be identified by a combination of field characteristics in proximal areas, all these layers contain the same phenocrysts: hypersthene, hornblende, and plagioclase. The layers' mineralogical characteristics, therefore, are not useful for differentiating the layers. The major- and minor-element chemistry of the volcanic glass permits the characterization of the three most widespread layers (Smith et al., 1977b).

G ASH LAYER

The oldest layer of the sequence, Glacier Peak G, was carried nearly due east (S 84°E) in a sector about 20° wide across eastern Washington, northern Idaho, and Montana, as well as into southern British Columbia and Alberta (Figure 5-2) (Porter, 1978). Mollusk shells interbedded with Glacier Peak tephra inferred to be layer G at Diversion Lake in western Montana gave a radiocarbon age of 12,750 ± 350 yr B.P. (Porter, 1978).

M ASH LAYER

Glacier Peak M ash layer, which overlies G in the sequence, was erupted about 12,000 yr B.P. or later but before about 11,250 yr B.P., the age of overlying layer B (Porter, 1978). The tephra plume from the eruption trended south-southeastward into central Wash-

Table 5-6. Comparison of Neutron-Activation-Analysis Data for Glass
in Mount St. Helens Tephra Set W

Sample	Sc	Fe	La	Ce	Sm	Tb	Yb	Lu	Th	U
1	3.44	1.23	16.8	35.2	3.05	0.37	1.48	0.22	4.39	1.70
2	3.47	1.23	17.2	36.0	3.10	0.42	1.52	0.21	4.58	1.71
3	3.43	1.21	17.0	36.1	3.09	0.40	1.48	0.21	4.60	1.72
4	3.45	1.27	16.6	36.7	3.01	0.37	1.51	0.21	4.60	1.72
5	4.57	1.26	17.6	38.8	3.37	0.46	1.61	0.24	4.74	1.77
6	± 0.02	± 0.02	± 0.5	± 0.5	± 0.03	± 0.02	± 0.02	± 0.01	± 0.05	± 0.03

Note: Element concentrations are in parts per million except for iron, which is in weight percentages. Analysts were H. R. Bowman and Frank Asaro, University of California, Lawrence Berkeley Laboratory.
Samples: (1-4) Tephra layer Wn, near Mount St. Helens. Four samples from two localities. (5) Tephra layer We, near Mount St. Helens. (6) Average analytical error.

ington (Figure 5-2). The areal distribution of this unit has not been well defined.

B ASH LAYER

The youngest layer of the Glacier Peak sequence, B, was carried to the southeast (S 65°E) in a rather narrow lobe covering a sector of about 10° across central and southeastern Washington, northern Idaho, and southwestern Montana and into northwestern Wyoming (Figure 5-2) (Porter, 1978). The age of layer B is constrained by two radiocarbon ages, one below (11,200 ± 100 yr B.P.) and one above (11,300 ± 230 yr B.P.) the ash at a downwind locality in western Montana (Mehringer et al., 1977; Porter, 1978).

SIGNIFICANCE OF GLACIER PEAK ASH LAYERS AS TIME AND STRATIGRAPHIC MARKERS

The areal mapping of Glacier Peak tephra layers and their stratigraphic relationships to late-glacial moraines and outwash deposits in the North Cascade Range has made it possible to date the late-Wisconsin retreat and minor readvance of alpine glaciers and the Cordilleran ice sheet (Porter, 1978). The presence of layer G in some alpine valleys of northern Washington indicates that ice had retreated upvalley before the eruption about 12,750 yr B.P. Because ash layers deposited on glaciers were not preserved after the glaciers retreated, the absence of layer G in parts of several valleys in this region within the known fallout areas suggests that ice was present and retreated only some time after the eruption of layer G (Porter, 1978). Stratigraphic relations between glacial moraines and volcanic-ash beds have also enabled Porter to bracket the timing of a Latest-Wisconsin glacial readvance, represented by the Rat Creek moraine. This moraine, near the northern Cascade Range crest, is overlain by the Mazama ash (layer O) and Mount St. Helens layer Yn but not by Glacier Peak ash beds; this relationship suggests an age for the re-advance intermediate between that of layer G and layer O (Porter, 1978).

Mazama Ash (Layer O), Crater Lake, Oregon

The most widespread latest-Quaternary tephra layer in the United States is the ash of Mount Mazama, also known as the Mazama ash, Mount Mazama ash, pumice, and layer O (Williams, 1942; Crandell et al., 1962; Powers and Wilcox, 1964; Wilcox, 1965; Randle et al., 1971; Mullineaux, 1974; Mullineaux and Wilcox, 1980; Bacon, 1981). The tephra was erupted from a composite stratovolcano (named Mount Mazama by Williams [1942]) that stood at the present site of Crater Lake. After the eruption of voluminous air-fall tephra and pumice-ash flows, the summit of the volcano foundered and formed a circular caldera that eventually filled and formed Crater Lake. The volume of air-fall ash generated by the eruption is estimated at between 30 and 38 km³; however, the total volume of material erupted may have been as much as 70 km³ (Williams and Goles, 1968).

The age of the Mazama ash is between about 7000 and 6700 years. An age of about 6600 years was previously assigned to this tephra layer on the basis of several radiocarbon ages (Rubin and Alexander, 1960; Powers and Wilcox, 1964; Wilcox, 1965). More recently, two radiocarbon ages of 6720 ± 120 (Mehringer et al., 1977) and 6990 ± 300 yr B.P. (Adam, 1967) were obtained on peat directly beneath the ash, and three ages of 6700 ± 100 (Mehringer et al.,

1977), 6710 ± 110 (Davis, 1978), and 6730 ± 250 yr B.P. (Mullineaux et al., 1974) were obtained on peat and organic material immediately above or between layers of the ash. In addition, two ages, 6940 ± 120 and 7000 ± 120 yr B.P. were obtained on a sample of charcoal from proximal pumice-ash flows from the climatic eruption of Mount Mazama (Randle et al., 1971; Kittleman, 1973). Most recently, a radiocarbon age of 7015 ± 45 yr B.P. has been obtained by Bacon (1981) for twigs in soil underlying a proximal southeast-trending tephra layer, which he suggests represents an early Plinian eruption preceding the climatic eruption. The weighted-mean age of four analyses of charcoal from within or beneath ash-flow and air-fall deposits from the climactic eruption (D. R. Crandell, in Bacon, 1981) is 6840 ± 50 yr B.P.

The minimum area covered by the Mazama ash (Figure 5-9) is nearly 1.7 million km² (estimate based on the planimetry of a distribution map in Mullineaux, 1974: Figure 12). The ash covered most of Oregon and Washington, all of Idaho, northeastern California, northern Nevada, northwestern Utah, western Wyoming and Montana, southern British Columbia and Alberta, and southwestern Saskatchewan.

The Mazama ash actually consists of several tephra layers presumably erupted within a relatively short period (Williams, 1942) estimated to have been as short as 3 years (Mehringer et al., 1977) or as long as 100 or 200 years (Bacon, 1981). Near Crater Lake, as many as six separate layers of air-fall ash have been recognized (Mullineaux and Wilcox, 1980), and in distal areas four major lobes have been mapped (R. E. Wilcox, personal communication, 1980).

Samples of the Mazama ash contain plagioclase, hypersthene, hornblende, augite, and apatite in varying proportions. Crystal habit and shard types also vary; for example, some samples are highly microporphyritic, and others contain more glassy shards. Davis (1978) describes two Mazama-like ash beds closely superposed at several locations in northwestern Nevada. The upper bed, which Davis identifies as the Mazama bed, contains a mineral assemblage of 6% plagioclase, 10% hypersthene, 6% hornblende, 2% augite, and 1% apatite. The underlying bed, named the "Tsoyawata bed" by Davis, contains 47% hornblende, 30% plagioclase, 3% hypersthene, and rare augite and apatite; the hornblende occurs in small, abundant green laths. The major- and minor-element chemistry of the glass, however, is essentially identical in the two beds (Table 5-7).

The chemical composition of many samples identified as the Mazama ash varies (Table 5-7), probably owing to repose intervals between eruptions, which allowed the parent magma to differentiate, as well as to zoning in the magma chamber during the climactic eruption. Some differences in the chemical composition of Mazama ash samples may also reflect the eolian sorting of glass shards during transport.

The variations in the Mazama ash are not wide enough to cause confusion with most other widespread late-Quaternary tephra layers. Glass of the Mazama ash can be distinguished from that of other tephra units by its relatively low silica and high iron and titanium content in electron probe analysis, as well as by a combination of trace elements in neutron-activation analysis (Table 5-7). Rare earth elements and several other elements, such as scandium, iron, thorium, and uranium, are particularly useful in distinguishing the Mazama ash from other tephra layers.

The chemical composition of glass of the Mazama ash resembles that of the Wono bed, an ash found at several localities in northwestern Nevada (Davis, 1978). The Wono bed can be distinguished

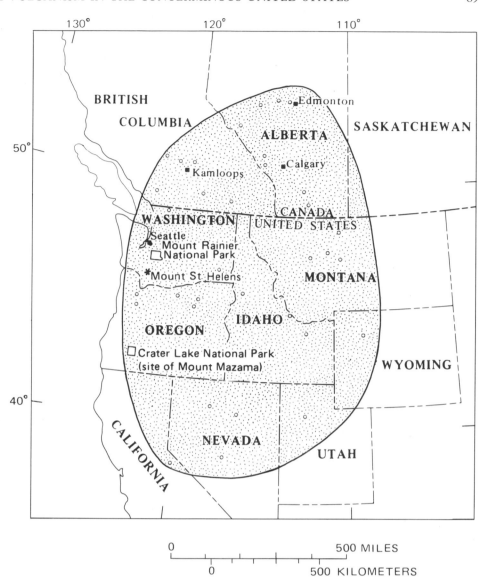

Figure 5-9. Minimum extent of Mazama ash (patterned) in the western United States and Canada. Circles mark sites where ash layer has been identified. (After Mullineaux, 1974: Figure 12.)

from the Mazama ash, however, on the basis of glass chemistry (aluminum, magnesium, titanium, and chlorine) and mineralogy (plagioclase, hypersthene, minor apatite but, unlike the Mazama ash, no hornblende or augite). The Wono bed is interbedded with late-Pleistocene pluvial lakebeds in the lower part of the Sehoo Formation, in the pluvial Lake Lahontan basin of northwestern Nevada (Davis, 1978). The Wono bed is older than the Mazama ash, because it stratigraphically underlies deposits dated at 24,480 ± 430 yr B.P.; it also overlies several ash beds in the basal part of the Sehoo Formation, the lowermost of which overlies deposits with a radiocarbon age of 33,650 ± 1720 yr B.P. (Davis, 1978). The ash of the Wono bed was probably erupted from the vicinity of Crater Lake since it is chemically very similar to the Mazama ash.

SIGNIFICANCE OF THE MAZAMA ASH AS A TIME AND STRATIGRAPHIC MARKER

The Mazama ash is the most useful time and stratigraphic marker in the western United States for Holocene deposits because it is widespread and readily identifiable. The timing of Mazama ash eruptions allows it to be used to distinguish generally between latest-

Pleistocene and neoglacial deposits. Mullineaux (1974) points out that the Mazama ash and Mount St. Helens set Y together are useful in the study of neoglacial and other Holocene deposits in the northwestern United States because they divide Holocene time into three roughly equal intervals. The Mazama ash has also been found in sediment on the continental slope off Washington, where it was most likely transported by streams and ocean-bottom currents (Royse, 1967), and thus it enables the correlation of marine with terrestrial deposits. Davis (1978) summarizes the areal distribution of the Mazama ash in Nevada and shows that alluvial fills at least 10 m thick accumulated in uplands and mountain ranges after the eruption of the Mazama ash.

Another important use of the Mazama ash is as a chronologic datum in archaeological and faunal studies in the northwestern United States. The Mazama ash overlies and underlies numerous habitation sites and individual artifact localities in this area and thus, together with Glacier Peak and Mount St. Helens tephra sets, provides a framework for studies of population influx and dispersal and cultural variation over time. Critical sites where the Mazama ash has been reported include Marmes Rockshelter in Washington (Fryxell and

Table 5-7. Chemical Composition of Volcanic Glass in Tephra Layers
and Pumice Identified as the Mazama Ash

Electron Microprobe Analyses

Sample	SiO$_2$	Al$_2$O$_3$	Fe$_2$O$_3$	MgO	MnO	CaO	TiO$_2$	Na$_2$O	K$_2$O
1	72.1	14.7	2.18	0.46	0.08	1.59	0.43	5.01	2.74
2	70.0	14.0	2.29	0.30	—	1.65	0.50	4.04	2.59
3	71.8	14.1	1.93	0.33	0.06	1.25	0.34	4.86	2.91
4	70.2	13.9	2.26	0.27	—	1.51	0.45	3.90	2.60
5	71.4	14.1	2.03	0.35	0.06	1.39	0.38	4.81	2.81
6	70.7	13.8	2.13	0.42	0.06	1.51	0.42	4.80	2.75

Neutron-Activation Analyses

Sample	Sc	Fe	La	Ce	Sm	Tb	Yb	Lu	Th	U
5	5.68	1.28	20.1	44.1	4.21	0.58	2.72	0.37	5.58	2.36
6	6.25	1.57	21.4	45.7	4.26	0.60	2.63	0.40	5.78	2.27
7	6.16	1.47	20.2	46.3	4.25	0.58	2.66	0.36	5.64	2.30
8	± 0.03	± 0.03	± 0.5	± 0.7	± 0.1	± 0.02	± 0.04	± 0.02	± 0.05	± 0.03

Note: Electron microprobe analyses are in weight percentages; analyst was C. E. Meyer, U.S. Geological Survey. Neutron-activation analyses are in parts per million except for iron, which is in weight percentages; analysts were H. R. Bowman and Frank Asaro, University of California, Lawrence Berkeley Laboratory.
Samples: (1, 2) Mazama ash, northwestern Nevada. (3, 4) Tsoyowata Bed of Davis (1978), northwestern Nevada. (5) Mazama ash, east-central California. (6) Coarse proximal pumice from climactic eruption, Crater Lake, Oregon. (7) Mazama ash, east-central Washington. (8) Average analytical error for neutron-activation analysis.

Daugherty, 1962), Fort Rock Cave in Oregon (Bedwell and Cressman, 1971), the Wadsen site in Idaho (Butler, 1972), and Hidden Cave and Gatecliff Shelter in Nevada (Davis, 1978). The Mazama ash has also been found in hundreds of less well known archaeological localities.

The Mazama ash has been used in archaeology primarily as a chronologic aid. The Mazama ash is most useful, however, in relating paleoenvironmental data to archaeological chronology. A layer of the Mazama ash occurs in almost every palynologic sequence studied in the northwestern United States within its fallout area, including those at Osgood Swamp, California (Adam, 1967; Davis, 1978), Lost Trail Bog, Idaho (Mehringer et al., 1977), Curelom Cirque, Utah (Mehringer et al., 1971), and Wildcat Lake, Washington (Davis et al., 1977), and in several peat bogs described by Hansen (1947). In these pollen records, the Mazama ash marks conditions similar to historical conditions but perhaps slightly cooler and wetter.

Climatic conditions before the Mazama ash fall differed significantly from those of the present, and pre-Mazama archaeological assemblages differ strikingly from later Holocene assemblages in both style and composition. Presumably, these variations reflect different adaptive patterns during pre-Mazama time, although the number of described pre-Mazama assemblages is still too small to allow ecologic analysis.

Broad changes in archaeofaunas in eastern Oregon and adjacent areas of the northern Basin and Range province during the period between about 7000 and 5000 yr B.P. are attributable to gradual climatic change, possibly a decrease in effective precipitation, rather than to the direct impact of the Mount Mazama eruption (Grayson, 1979). Such climatic changes, however, may actually represent a worldwide (or, at least, hemispheric) climatic response to the Mount Mazama eruption and other broadly contemporaneous volcanic activity (Benedict, 1981). For instance, Porter (1981) shows that short-term climatic variations and neoglacial advances coincide with periods of increased historical volcanic activity. Such variations, according to Porter (1981), may affect only the limited latitudinal belt within which the volcanism takes place or a specific hemisphere and need not necessarily be worldwide. One important problem that needs to be examined is whether the Mount Mazama eruption and broadly synchronous volcanic activity could effectively decelerate, perhaps even reverse, the warming trend that followed the last major glaciation. Antev's (1948) discussion of the Altithermal geologic-climate "unit" is based in part on the stratigraphic position of the Mazama ash relative to the pollen record at Summer Lake, Oregon. His discussion, however, is confused by the misidentification of a Pleistocene tephra layer at this locality as the Mazama ash (Allison, 1945, 1966).

Late-Holocene Tephra Layer from Newberry Volcano

Silicic pumice and ash erupted from Newberry Volcano about 1600 yr B.P. (1550 ± 120 and 1720 ± 250 yr B.P.) form a straight and narrow east-trending lobe (Sherrod and MacLeod, 1979). Although the mapped area covered by this tephra layer is small (Figure 5-2), the mapped outer boundary represents the 25-cm isopach, and so the ash must actually be much more extensive. The measured thickness along the lobe axis of as much as 25 cm at a downwind distance of 60 km suggests that this eruption was probably larger than the May 18, 1980, eruption of Mount St. Helens and perhaps as large or larger than most other Holocene layers of tephra erupted from Mount St. Helens except layer Yn. Ash from this eruption of the Newberry Volcano will probably be found in eastern Oregon and Idaho and perhaps still farther east.

Tephra Layers from Other Cascade Range Volcanoes

Tephra layers covering areas of small to moderate size (100 to 5000 km²) have been erupted during Holocene time from other Cascade Range volcanoes. Tephra layers of moderate volume have been erupted from Mount Baker (Hyde and Crandell, 1978), Mount Rainier (Mullineaux, 1974), South Sister (Williams, 1944; Taylor, 1968), Medicine Lake Highland (Heiken, 1978b), Mount Shasta (Miller, 1980), and Lassen Peak (Williams, 1932; Crandell, 1972; Crandell et al., 1974). Among the more extensive of these layers is andesitic ash erupted from Mount Rainier (Figure 5-2). Mullineaux (1974) maps 11 layers of tephra erupted from this volcano, 8 of which are dated between 6500 and 4000 yr B.P. These tephra layers, which lie primarily within a quadrant northeast to southeast of the volcano, together with the more widespread Mount St. Helens and Mazama ash layers, have been used to decipher the late-glacial and neoglacial history of Mount Rainier (Crandell and Miller, 1964; Crandell, 1969; Mullineaux, 1974).

Two Holocene tephra layers of moderate extent were erupted about 1100 years ago from the Medicine Lake Volcano (Heiken, 1978b, 1981). One pumiceous tephra erupted from Little Glass Mountain was carried to the southwest and northeast, and the other tephra erupted from Glass Mountain was carried to the east-northeast (Figure 5-2). Ash from each eruption was carried downwind for several tens of kilometers.

The remaining Cascade Range volcanoes—Mount Adams, Mount Hood, Mount Jefferson, Mount Thielson, and Mount McLaughlin—are not known to have dispersed tephra beyond their flanks during Holocene time.

Widespread Silicic Ash Layers from the Mono and Inyo Craters, East-Central California

Several silicic tephras of moderate to large areal extent have been erupted from the Mono and Inyo Craters of east-central California during latest-Pleistocene and Holocene time (Rinehart and Huber, 1965; Wood, 1977a, 1977b; Kilbourne et al., 1980). Mono and Inyo Craters are a chain of small volcanic vents, explosion pits, and associated flows, domes, and tephra extending north-south for a distance of about 30 km south of Mono Lake, just east of the central Sierra Nevada (Figure 5-2 and 5-10).

The Mono Craters volcanic chain extends from Panum Crater just south of Mono Lake southward to the vicinity of Wilson Butte (Figure 5-10). The Inyo Craters chain extends from Wilson Butte southward to the vicinity of Mammoth Lakes. The southern half of the Inyo Craters chain is situated in the western part of the Long Valley caldera, a large volcanotectonic depression formed about 0.7 my ago after the eruption of the Bishop Tuff (Gilbert, 1938; Dalrymple et al., 1965; Bailey et al., 1976). Volcanism within and around the periphery of the caldera has continued intermittently since that time (Dalrymple et al., 1968; Bailey et al., 1976; Kilbourne et al., 1980; Kilbourne and Anderson, 1981), although the Inyo and Mono Craters themselves are comparatively young features formed during the past 10 or more millennia. The most recent eruptions from this source occurred several hundred years ago (Kilbourne and Anderson, 1981).

The Mono and Inyo Craters lie in a rather linear, north-trending chain. The southern end of the chain, the Inyo Craters, is expressed as several young phreatic explosion pits along a linear system of open fissures extending north from the flanks of Mammoth Mountain and across the western part of the Long Valley caldera. Holocene domes and pyroclastic deposits were erupted from the northern part of this fissure system, both in the western part of the caldera and also north of the caldera's rim.

The Mono Craters and domes extend north of the fissure system that defines the Inyo Craters and have a distinctly arcuate shape. The presence of a concentric system of faults and fissures about 15 km in diameter that includes the Mono Craters chain suggests that these features define a ring-fracture system that outlines a large magma chamber at depth (Bailey, 1980). Silicic tephra erupted from the Mono Craters is homogeneous and depleted in many trace and minor elements relative to that of the Inyo Craters.

Paoha and Negit Islands (about 1000 yr B.P.) and Black Point (13,800 yr B.P.) north of Mono Craters are young volcanic vents and associated flows and tephra of more heterogeneous composition and more limited areal extent. Black Point, a broad, flat cinder cone, is of particular interest because it erupted entirely beneath an expanded

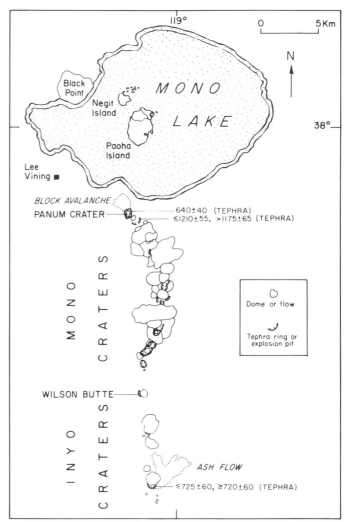

Figure 5-10. The Mono and Inyo Craters, east-central California. (Radiocarbon dates in yr B.P.) (Modified from Lajoie, 1968, and Wood, 1977b.)

Mono Lake and is terraced by recessional strandlines on its flanks (Christianson and Gilbert, 1964; Lajoie, 1968; Custer, 1971).

The volcanic domes, flows, and tephra of Mono Craters range in age from about 680 to about 42,000 yr B.P., as determined by numerous potassium-argon, ionium, hydration-rind, and radiocarbon dates (Dalrymple, 1968; Friedman, 1968; Lajoie, 1968; Wood, 1977b; Wood and Brooks, 1979). Most of the ages on Mono Craters extrusive rocks cluster toward the younger end of this range, about 12,000 to 1000 yr B.P. As Lajoie (1968) points out, however, more than 20 ash beds within the age range of 36,000 to 12,000 radiocarbon years are interbedded with late-Pleistocene lacustrine deposits throughout the Mono Lake basin. These ash beds are correlated with the various Mono Craters on the basis of minor- and trace-element chemistry; therefore, explosive volcanism from this chain must have been more or less continuous over the entire eruptive period of about 40,000 years even though this range is not reflected in ages of the proximal units, probably because older units are covered by younger (K. R. Lajoie, personal communication, 1981). Ages on extrusive units of Inyo Craters encompass a range of about 12,000 to 550 years (Lajoie, 1968; Wood, 1977a).

In contrast to such prolific Cascade Range sources as Mount St. Helens, few known widespread tephras were erupted from Mono and Inyo Craters. However, latest-Quaternary deposits of the Great Basin east and south of the Mono and Inyo Craters, in the direction of the prevailing winds, have not been studied as extensively or in such detail as those of the Cascade Range. At least three Mono and Inyo Craters tephra units are interlayered with cultural deposits in the Borealis mine area near Hawthorne, Nevada (Pippen, 1980), and Mono and Inyo Craters ash layers lie above the Mazama ash at Hidden Cave, Mount Jefferson, and Eastgate, Nevada. More widespread tephra units from Mono and Inyo Craters sources may be found in the future, and work is continuing to define more closely the areal distribution and correlation of late-Pleistocene and Holocene tephra units erupted from this area (K. R. Lajoie and A. M. Sarna-Wojcicki; C. D. Miller; J. O. Davis; and S. H. Wood, unpublished data, 1981).

TEPHRA LAYER 2, MONO CRATERS

Wood (1975) has mapped the areal distribution of two widespread Holocene ash beds in mountain meadows, marshes, and lakes of the central and southern Sierra Nevada. The older bed, which he labels tephra 2, was erupted about 1200 yr B.P. (Wood, 1977a); this ash is widespread in the central Sierra Nevada of east-central California (Figure 5-2). We have recently identified samples of this ash from a bed in the Yosemite Valley (mapped and collected by R. Milestone, U.S. National Park Service), which overlies a recessional moraine of Tioga age and younger lakebeds of the valley. This ash has also been identified by Davis (1978) from several localities in northwestern Nevada, including beach dunes of late-Holocene Fallon Lake. Wood (1977a) initially suggested that this ash was erupted from Panum Crater, the northern most vent of the Mono Craters. A subsequent radiocarbon age (640 ± 40 yr B.P.), however, indicates that the Panum eruption was much later and that tephra layer 2 must have been erupted from another Mono Craters vent, probably a large explosion pit immediately southwest of Panum Crater (Wood and Brooks, 1979). Tephra erupted from Mono Craters, presumably tephra layer 2, is found in archaeological contexts in Hidden Cave, Nevada, in sand dunes near Fallon, Nevada, and in lacustrine deposits of Deep Springs Valley, California (Lajoie, 1968). Tephra at some of these sites northeast of Mono Craters, however, may have been

erupted from Panum Crater about 640 ± 40 B.P. (Wood and Brooks, 1979; S. H. Wood, personal communication, 1981). Further work is needed to distinguish between the compositionally similar young tephra layers of Mono Craters.

TEPHRA LAYER 1, INYO CRATERS

The second ash bed identified by Wood (1977a, 1977b), Inyo Craters tephra 1, overlies Mono Craters tephra layer 2 at several sites in the east-central and southern Sierra Nevada. Wood (1977a, 1977b) suggests, on the basis of the chemical similarity between the glass in this ash layer and that from the Inyo Craters, that the source of this ash is one of the Inyo Craters. The age of tephra 1 at the inferred proximal site as well as at distal sites is about 750 yr B.P. (Wood, 1977a, 1977b; Kilbourne et al., 1980). Because the pumice of tephra layer 1 is widely distributed to the south and southeast of Long Valley in the central and southern Sierra Nevada (Figure 5-2), the eruption probably occurred during late summer or fall, when the prevailing southward Santa Ana winds were blowing. Pumice clasts as large as several centimeters in diameter attributed to this eruption occur in the soil or at the surface for several tens of kilometers south of Long Valley.

We have recently identified samples of tephra 1 from a new locality, found and sampled by J. C. Tinsley III, in the Redwood Creek basin of Kings Canyon National Park in the southwestern Sierra Nevada. Ash there is interbedded with 1 or 2 m of alluvium in a partly filled sinkhole developed on calcareous metamorphic rocks. This correlation is based on energy-dispersive X-ray fluorescence of volcanic glass from the ash in the sinkhole and surface pumice found several kilometers south of the Inyo Craters.

Pumice Erupted from Salton Buttes, California

Besides Cascade Range volcanoes and Mono and Inyo Craters, the only other source area known to have erupted silicic extrusive rocks during Holocene time is Salton Buttes, situated at the southern shore of the Salton Sea in southern California (Figure 5-1) (Robinson et al., 1976; Friedman and Obradovich, 1981). Although pumice has been erupted and pumice clasts are distributed around the shores of the Salton Sea, no tephra layers from these vents have been described or mapped.

Basaltic Tephra Erupted from Sunset Crater, Arizona

Cinders and ash of basaltic composition were erupted from Sunset Crater, Arizona, about A.D. 1065 and sporadically thereafter for perhaps as long as about 100 years (Pilles, 1979). This tephra was carried downwind to the northeast for at least several tens of kilometers (Figure 5-2) and for an undetermined distance beyond.

Notable changes in the Sinagua culture of north-central Arizona—specifically, increases in population related to increases in agricultural productivity due to the mulching effect and increased soil nutrients of the volcanic ash—were attributed to the eruptions from Sunset Crater (Colton, 1932, 1968). Recent interpretations (Pilles, 1979), however, suggest that the impact of these eruptions was minor and local and that major cultural changes were a response to broader climatic changes.

Conclusions

Holocene volcanism in the conterminous United States is restricted to the western states. Most Holocene volcanic vents and fields are situated along the Cascade Range, the periphery of the Basin and Range province, the periphery of the Rio Grande rift, and the Salton Trough. Volcanism in the Cascade Range tectonic province is probably a result of crustal-plate convergence and subduction of the oceanic Juan de Fuca plate beneath the North American continent; volcanism in the other three provinces is probably a consequence of crustal spreading. The spatial distribution of Holocene volcanism roughly coincides with that of heat flow and hydrothermal activity; all three of these phenomena reflect the outward transfer of heat from within the Earth.

Major silicic volcanism during Holocene time has taken place in two major areas: the Cascade Range of the Pacific Northwest and the Mono and Inyo Craters of east-central California. Explosive eruptions of silicic tephra from these sources have produced numerous widespread tephra layers that provide age calibration for various geologic studies. Among these studies are the timing of Late Wisconsin glacial retreats; neoglacial fluctuations; and archaeofaunal migration, dispersion, and differentiation, as well as temporal correlation of uppermost Quaternary deposits of diverse environments.

Additional work is needed to better define the areal extent, mass, and volume of major late-Quaternary eruptions. Such data, together with estimates of volcanigenic gasses released during these eruptions, may permit an assessment of the impact of volcanic activity on climatic variation, as well as provide information on the frequency of major volcanic eruptions. Such information is also important in assessing volcanic hazards in the western United States.

References

Adam, D. P. (1967). Late-Pleistocene and recent palynology in the central Sierra Nevada, California. *In* "Quaternary Paleoecology" (E. J. Cushing and H. E. Wright, Jr., eds.), pp. 275-301, Yale University Press, New Haven, Conn.

Allen, J. E. (1951). The Carrizozo Malpais. *In* "Roswell Geological Society Guidebook, Fifth Field Conference, Carrizozo-Capitan-Chupadera Mesa Region, Lincoln and Socorro Countries, New Mexico," pp. 9-11.

Allison, I. A. (1945). Pumice beds at Summer Lake, Oregon. *Geological Society of America Bulletin* 56, 789-807.

Allison, I. A. (1966). Pumice at Summer Lake, Oregon: A correction. *Geological Society of America Bulletin* 77, 329.

Anderson, C. A. (1940). Hot Creek lava flow. *American Journal of Science* 238, 477-92.

Anderson, C. A. (1941). "Volcanoes of the Medicine Lake Highland, California." University of California Publications, Department of Geological Sciences Bulletin 25 (7), pp. 347-422.

Antevs, E. (1948). Climatic changes and pre-white man. *In* "The Great Basin, with Emphasis on Glacial and Postglacial Times," pp. 68-191. University of Utah Bulletin 38 (20).

Armstrong, R. L., Ekren, E. B., McKee, E. H., and Noble, D. C. (1969). Space-time relations of Cenozoic silicic volcanism in the Great Basin of the western United States. *American Journal of Science* 267, 478-90.

Bacon, C. R. (1981). Eruptive history of Mount Mazama (abstract). *In* "Symposium on Arc Volcanism, International Association of Volcanology and Chemistry of the Earth's Interior, Tokyo, 1981," pp. 22-23.

Bailey, R. A. (1980). Structural and petrologic evolution of the Long Valley, Mono Craters, and Mono Lake volcanic complexes, eastern California (abstract). *EOS (American Geophysical Union Transactions)* 61, 1149.

Bailey, R. A., Dalrymple, G. B., and Lanphere, M. A. (1976). Volcanism, structure and geochronology of Long Valley caldera, Mono Country, California. *Journal of Geophysical Research* 81, 725-44.

Baldwin, B., and Meuhlberger, W. R. (1959). "Geologic Studies of Union County, New Mexico." New Mexico Bureau of Mines and Mineral Resources Bulletin 63.

Barca, R. A. (1966). Geology of the northern part of the Old Dad Mountain quadrangle, San Bernardino County, California. California Division of Mines and Geology Map Sheet 7.

Bedwell, S. F., and Cressman, E. R. (1971). Fort Rock report: Prehistory and environment of the pluvial Fort Rock Lake area of south-central Oregon. *In* "Great Basin Anthropological Conference, 1970, Selected papers" (C. M. Aikens, ed.), pp. 1-25. Oregon University Anthropological Papers 1.

Beget, J. E. (1980). Eruptive history of Glacier Peak volcano, Washington State. *Geological Society of America, Abstracts with Programs* 12, 384.

Benedict, J. B. (1981). Prehistoric man, volcanism, and climatic change: 7500-5000 ¹⁴C yr. B.P. *Geological Society of America, Abstracts with Programs* 13, 407.

Borchardt, G. A., Norgren, J. A., and Harward, M. E. (1973). Correlation of ash layers in peat bogs of eastern Oregon. *Geological Society of America Bulletin* 84, 3101-08.

Bray, J. R. (1974). Volcanism and glaciation during the past 40 millenia. *Nature* 252, 679-80.

Bray, J. R. (1977). Pleistocene volcanism and glacial initiation. *Science* 197, 251-54.

Breternitz, D. A. (1967). The eruptions of Sunset Crater: Dating and effects. *Plateau* 40, 72-75.

Bretz, J H. (1923). The channeled scablands of the Columbia Plateau. *Journal of Geology* 31, 617-49.

Bretz, J H. (1969). The Lake Missoula floods and the Channeled Scabland. *Journal of Geology* 77, 505-43.

Brook, C. A., Mariner, R. H., Mabey, D. R., Swanson, J. R., Guffanti, M. and Muffler, L. J. P. (1979). Hydrothermal convection systems with reservoir temperatures ≥ 90° C. *In* "Assessment of Geothermal Resources of the United States-1978" (L. J. P. Muffler, ed.), pp. 18-43. U.S. Geological Survey Circular 790.

Brott, C. A., Blackwell, D. D., and Mitchell, J. C. (1967). Heat flow study of the Snake River Plain region, Idaho. "Geothermal Investigations in Idaho." Idaho Department of Water Resources Water Information Bulletin 30, part 8.

Brott, C. A., Blackwell, D. D., and Mitchell, J. C. (1978). Tectonic implications of the heat flow of the western Snake River Plain. *Geological Society of America Bulletin* 89, 1697-1707.

Butler, B. R. (1972). The Holocene in the desert West and its cultural significance. *In* "Great Basin Cultural Ecology: A symposium" (D. D. Fowler, ed.), pp. 5-12. University of Nevada, Desert Research Institute Publications in the Social Sciences 8.

Cerling, T. E., Cerling, B. W., Curtis, G. H., Drake, R. E., and Brown, F. H. (1978). Correlation of reworked ash deposits: The K B S tuff, northern Kenya. *In* "Short Papers of the Fourth International Conference, Geochronology, Cosmochronology, Isotope Geology, 1978" (R. E. Zartman, ed.), pp. 61-63. U.S. Geological Survey Open-File Report 78-701.

Champion, D. E. (1973). The relationship of large scale surface morphology to lava flow direction, Wapi lava field, south-eastern Idaho. M.A. thesis, State University of New York at Buffalo.

Champion, D. E. (1980). "Holocene Geomagnetic Secular Variation in the Western United States: Implications for the Global Geomagnetic Field." U.S. Geological Survey Open-File Report 80-824.

Champion, D. E., and Greeley, R. (1977). Geology of the Wapi lava field, Snake River Plain, Idaho. *In* "Volcanism of the Eastern Snake River Plain, Idaho: A Comparative Planetary Geology Guidebook," pp. 133-52. National Aeronautics and Space Administration Publication CR-154621.

Christiansen, R. L. (1979). Cooling units and composite sheets in relation to caldera structure. *In* "Ash-Flow Tuffs" (C. E. Chapin and W. E. Elston, eds.), pp. 29-42. Geological Society of America Special Paper 180.

Christiansen, R. L., and Blank, R. H., Jr. (1972). "Volcanic Stratigraphy of the Quaternary Rhyolite Plateau in Yellowstone National Park." U.S. Geological Survey Professional Paper 729-B.

Christiansen, R. L., and Lipman, P. W. (1972). Cenozoic volcanism and plate tectonic evolution of the western United States: II. Late Cenozoic. *Royal Society of London Philosophical Transactions*, series A, 271, 249-84.

Christiansen, R. L., and Miller, C. D. (1976). Volcanic evolution of Mt. Shasta, California. *Geological Society of America, Abstracts with Programs* 8, 360-61.

Christiansen, M. N., and Gilbert, C. M. (1964). Basaltic cone suggests constructional origin of some guyots. *Science* 143, 240-42.

Collins, R. I. (1949). Volcanic rocks of northeastern New Mexico. *Geological Society of America Bulletin* 60, 1017-40.

Colton, H. S. (1932). The effect of a volcanic eruption on an ancient Pueblo people. *Geographical Review* 22, 582-90.

Colton, H. S. (1962). "Archaeology of the Flagstaff Area." Guidebook of the Mogollon Rim Region, New Mexico Geological Society, 13th Field Conference.

Colton, H. S. (1968). Frontiers of the Sinagua. *In* "Collected Papers in Honor of Lyndon Laud Hargrave" (A. H. Schroeder, ed.) Archeological Society of New Mexico, Santa Fe.

Condie, K. C., and Barsky, C. K. (1972). Origin of Quaternary basalts from the Black Rock Desert region, Utah. *Geological Society of America Bulletin* 83, 333-52.

Cook, E. F. (1960). "Geologic atlas of Utah, Washington County." Utah Geologic and Mineralogic Survey Bulletin 70.

Coombes, H. A. (1939). Mt. Baker, a Cascade volcano. *Geological Society of America Bulletin* 50, 1493-1510.

Crandell, D. R. (1969). "The Geologic Story of Mount Rainier." U.S. Geological Survey Bulletin 1292.

Crandell, D. R. (1971). "Postglacial Lahars from Mount Rainier Volcano, Washington." U.S. Geological Survey Professional Paper 677.

Crandell, D. R. (1972). "Glaciation near Lassen Peak, Northern California." U.S. Geological Survey Professional Paper 800-C, pp. C179-C188.

Crandell, D. R. (1980). "Recent Eruptive History of Mount Hood, Oregon, and Potential Hazards from Future Eruptions." U.S. Geological Survey Bulletin 1492.

Crandell, D. R., and Miller, R. D. (1964). Post-Hypsithermal glacier advance of Mount Rainier, Washington. *In* "Geological Survey Research 1964," pp. D110-D114. U.S. Geological Survey Professional Paper 501-D.

Crandell, D. R., and Mullineaux, D. R. (1973). "Pine Creek Volcanic Assemblage of Mount St. Helens, Washington." U.S. Geological Survey Bulletin 1383-A.

Crandell, D. R., and Mulineaux, D. R. (1978). "Potential Hazards from Future Eruptions of Mount St. Helens Volcano, Washington." U.S. Geological Survey Bulletin 1383-C.

Crandell, D. R., Mullineaux, D. R., Miller, R. D., and Rubin, M. (1962). Pyroclastic deposits of recent age at Mount Rainier, Washington. *In* "Short Papers in Geology, Hydrology, and Topography," pp. D64-D68. U.S. Geological Survey Professional Paper 450-D.

Crandell, D. R., Mullineaux, D. R., Sigafoos, R. S., and Rubin, M. (1974). Crags eruptions and rockfall-avalanches, Lassen Volcanic National Park, California. *Journal of Research of the U.S. Geological Survey* 2, 49-59.

Cross, T. A., and Pilger, R. H. (1978). Contraints on absolute motion and plate interaction inferred from Cenozoic igneous activity in the western United States. *American Journal of Science* 278, 865-902.

Crowder, D. F., Tabor, R. W., and Ford, A. B. (1966). Geologic map of the Glacier Peak quadrangle, Snohomish and Chelan Counties, Washington. U.S. Geological Survey Geologic Quadrangle Map GQ-473.

Crowe, B. M., and Fisher, R. V. (1973). Sedimentary structures in base-surge deposits with special reference to cross-bedding, Ubehebe Craters, Death Valley, California. *Geological Society of America Bulletin* 84, 663-82.

Custer, S. G. (1971). Stratigraphy and sedimentation of Black Point volcano, Mono Basin, California. M.S. thesis, University of California, Berkeley.

Dalrymple, G. B. (1968). Potassium-argon ages of recent rhyolites of the Mono and Inyo Craters, California. *Earth and Planetary Sciences Letters* 3, 289-98.

Dalrymple, G. B., Cox, A., and Doell, R. R. (1965). Potassium-argon age and paleomagnetism of the Bishop Tuff, California. *Geological Society of America Bulletin* 76, 665-74.

Danielsen, E. (1980). Mount St. Helens plume dispersion based on trajectory analysis. *In* "Mount St. Helens Eruption: Its Atmospheric Effects and Potential Climatic Impact," p. 21. Institute for Atmospheric Optics and Remote Sensing, a NASA-sponsored workshop, November, 1980, Washington, D.C.

Davis, J. O. (1978). "Quaternary Tephrochronology of the Lake Lahontan Area, Nevada and California." Nevada Archeological Survey Research Paper 7.

Davis, O. K., Kolva, D. A., and Mehringer, P. F., Jr., (1977). Pollen analysis of Wildcat Lake, Whitman County, Washington: The last 1000 years. *Northwest Science* 51, 13-30.

Davitaya, F. F. (1969). Atmospheric dust content as a factor affecting glaciation and climatic change. *Association of American Geographers Annals* 59, 552-60.

Dibblee, T. W., Jr. (1966). Geologic map of the Lavic quadrangle, San Bernadino County, California. U.S. Geological Survey Miscellaneous Geologic Investigations Map I-472.

Doell, R. R., Dalrymple, G. B., Smith, R. L., and Bailey, R. A. (1968). Paleomagnetism, potassium-argon ages and geology of rhyolites and associated rocks of the Valles caldera, New Mexico. *In* "Studies in Volcanology: A Memoir in Honor of Howel Williams (R. R. Coats, R. L. Hay, and C. A. Anderson, eds.), pp. 211-48. Geological Society of America Memoir 116.

Donnelly-Nolan, J. M., Ciancanelli, E. V., Eichelberger, J. C., Fink, J. C., and Heiken, G. (1980). Roadlog for field-trip to Medicine Lake Highland. *In* "Guides to Some Volcanic Terranes in Washington, Idaho, Oregon, and Northern California" (D. A. Johnston, and D. M. Donnelly-Nolan, eds.), pp. 141-49. U.S. Geological Survey Circular 838.

Drummond, R. J. (chair.) (1981). Plate tectonic map of the circum-Pacific region, northeast quadrant, scale 1:10,000,000. Circum-Pacific Council for Energy and Mineral Resources, American Association of Petroleum Geologists.

DuBois, R. L. (1974). Secular variation in southwestern United States as suggested by archeomagnetic studies. *In* "Proceedings of the Takesi Nagata Conference" (R. M. Fisher, M. D. Fuller, V. A. Schmidt, and P. I. Wasilewski, eds.), pp. 133-34. University of Pittsburgh.

Easterbrook, D. J. (1975). Mt. Baker eruptions. *Geology* 3, 679-82.

Finch, R. H., and Anderson, C. A. (1930). "The Quarts Basalt Eruptions of Cinder Cone, Lassen Volcanic National Park, California." University of California Publications, Department of Geological Sciences Bulletin 19 (10), pp. 243-73.

Fiske, R. S., Hopson, C. A., and Waters, A. C. (1964). Geologic map and section of Mount Rainier National Park, Washington. U.S. Geological Survey Miscellaneous Geologic Investigations Map I-432.

Friedman, I. (1968). Hydration rind dates rhyolite flows. *Science* 159, 878-81.

Friedman, I. and Obradovich, J. D. (1981). Obsidian hydration dating of volcanic events. *Quaternary Research* 16, 37-47.

Fryxell, R. (1965). Mazama and Glacier Peak volcanic ash layers: Relative ages. *Science* 147, 1288-90.

Fryxell, R. and Daugherty, R. D. (1962). "Interim Report: Archaeological Salvage in the Lower Monumental Reservoir, Washington." Washington State University, Laboratory of Anthropology and Geochronology Reports of Investigations 21.

Gester, G. C. (1962). "The Geological History of Eagle Lake, Lassen County, California." California Academy of Sciences Occasional Papers 34.

Giegengack, R. F., Jr. (1962). Recent volcanism near Dotsero, Colorado. M.S. thesis, University of Colorado, Boulder.

Gilbert, C. M. (1938). Welded tuff in eastern California. *Geological Society of America Bulletin* 49, 1829-62.

Goode, H. D. (1973). Preliminary geologic map of the Bald Knoll quadrangle, Utah. U.S. Geological Survey Miscellaneous Field Studies Map MF-520.

Grayson, D. K. (1979). Mount Mazama, climatic change, and Fort Rock Basin archaeofaunas. *In* "Volcanic Activity and Human Ecology" (P. D. Sheets and D. K. Grayson, eds.), pp. 427-57. Academic Press, New York.

Greeley, R., Theilig, E., and King, J. S. (1977). Guide to the geology of King's Bowl lava field. *In* "Volcanism of the Eastern Snake River Plain, Idaho: A Comparative Planetary Geology Guidebook," pp. 171-88. National Aeronautics and Space Administration Publication CR-154621.

Greene, R. C. (1968). "Petrography and Petrology of Volcanic Rocks in the Mount Jefferson Area, High Cascade Range, Oregon." U.S. Geological Survey Bulletin 1251-G.

Gregory, H. E. (1949). Geologic and geographic reconnaissance of eastern Markagunt Plateau, Utah. *Geological Society of America Bulletin* 60, 969-98.

Gregory, H. E. (1951). "The Geology and Geography of the Pawnsaugunt Region, Utah." U.S. Geological Survey Professional Paper 226.

Hamblin, W. K. (1963). Late Cenozoic basalts of the St. George Basin, Utah, *In* "Intermountain Association of Petroleum Geologists Guidebook, 12th Annual Field

Conference, Southwestern Utah,'' pp. 84-89. Utah Geological and Mineralogical Survey.

Hamblin, W. K. (1970). Late Cenozoic basalt flows of the western Grand Canyon. *In* ''The Western Grand Canyon District'' (W. K. Hamblin and M. G. Best, eds.), pp. 21-38. Utah Geological Society Guidebook to the Geology of Utah 23.

Hammond, P. E. (1973). Quaternary basaltic volcanism in the southern Cascade Range, Washington. *Geological Society of America, Abstracts with Programs* 5, 49.

Hammond, P. E. (1977). Reconnaissance geologic map and cross-sections of southern Cascade Range, Washington. *Geological Society of America, Abstracts with Programs* 9, 1003-4.

Hansen, H. P. (1947). Postglacial forest succession, climate, and chronology in the Pacific Northwest. *American Philosophical Society Transactions* 37, 1-130.

Hatheway, A. W. (1971). Lava tubes and collapse depressions. Ph.D. dissertation, University of Arizona, Tucson.

Heiken, G. (1978a). Characteristics of tephra from Cinder Cone, Lassen Volcanic National Park, California. *Bulletin Volcanologique* 41-2, 119-30.

Heiken, G. (1978b). Plinian-type eruptions in the Medicine Lake Highlands, California, and the nature of the underlying magma. *Journal of Volcanology and Geothermal Research* 4, 375-402.

Heiken, G. (1981). Holocene Plinian tephra deposits of the Medicine Lake Highland, California. *In* ''Guides to Some Volcanic Terranes in Washington, Idaho, Oregon, and Northern California'' (D. A. Johnston and J. M. Donnelly-Nolan, eds.), pp. 177-81. U.S. Geological Survey Circular 838.

Heiken, G., and Eichelberger, J. C. (1980). Eruptions at Chaos Crags, Lassen Volcanic National Park, California. *Journal of Volcanology and Geothermal Research* 7, 443-81.

Holcomb, R. T. (1980). ''Kilauea Volcano, Hawaii, Chronology and Morphology of the Surficial Lava Flows.'' U.S. Geological Survey Open-File Report 80-796.

Hoover, J. D. (1974). Periodic Quaternary volcanism in the Black Rock Desert, Utah. *Brigham Young University Geology Studies* 21 (part 1), 3-72.

Hopkins, K. D. (1976). Geology of the south and east slopes of Mount Adams volcano, Cascade Range, Washington. Ph.D. dissertation, University of Washington, Seattle.

Huber, N. K., and Rinehart, C. D. (1967). ''Cenozoic Volcanic Rocks of the Devil's Postpile Quadrangle, Eastern Sierra Nevada, California.'' U.S. Geological Survey Professional Paper 554-D.

Hyde, J. H., and Crandell, D. R. (1978). ''Postglacial Volcanic Deposits at Mt. Baker, Washington, and Potential Hazards from Future Eruptions.'' U.S. Geological Survey Professional Paper 1022-C.

Izett, G. A. (1981). Volcanic ash beds: Recorders of upper Cenozoic silicic pyroclastic volcanism in the western United States. *Journal of Geophysical Research* 86, 10200-222.

Izett, G. A., Wilcox, R. E., and Borchardt, G. A. (1972). Correlation of a volcanic ash bed in Pleistocene deposits near Mount Blanco, Texas, with the Guaje pumice bed of the Jemez Mountains, New Mexico. *Quaternary Research* 2, 554-78.

Izett, G. A., Wilcox, R. E., Powers, H. A., and Desborough, G. A. (1970). The Bishop ash bed, a Pleistocene marker bed in the western United States. *Quaternary Research* 1, 121-32.

Jack, R. N., and Carmichael, I. S. E. (1968). ''The Chemical 'Fingerprinting' of Acid Volcanic Rocks.'' California Division of Mines and Geology Special Report 100, pp. 17-31.

Jones, A. E. (1939). Progress report on the geology of the Lava Beds National Monument. Unpublished manuscript.

Karlo, J. F. (1977). Geology of the Hell's Half Acre lava field. *In* ''Volcanism of the Eastern Snake River Plain, Idaho: A Comparative Planetary Geology Guidebook,'' pp. 121-31. National Aeronautics and Space Administration Publication CR-154621.

Katz, M., and Boettcher, A. L. (1980). The Cima volcanic field. *In* ''Geology and Mineral Wealth of the California Desert'' (D. L. Fife and A. R. Brown, eds.), pp. 236-41. South Coast Geological Society, Santa Ana, Calif.

Kennett, J. P., and Thunell, R. C. (1975). Global increase in Quaternary explosive volcanism. *Science* 187, 497-503.

Kilbourne, R. T., and Anderson, C. L. (1981). Volcanic history and ''active''

volcanism in California. *California Geology* 34, 159-68.

Kilbourne, R. T., Chesterman, C. W., and Wood, S. H. (1980). Recent volcanism in the Mono Basin-Long Valley region of Mono County, California. *In* ''Mammoth Lakes, California, Earthquakes of May 1980'' (R. W. Sherburne, ed.), pp. 7-22. California Division of Mines and Geology Special Report 150.

Kittleman, L. R. (1973). Mineralogy, correlation, and grain-size distributions of Mazama tephra and other postglacial pyroclastic layers, Pacific Northwest. *Geological Society of America Bulletin* 84, 2957-80.

Kuntz, M. A. (1977). Extensional faulting and volcanism along the Arco rift zone, eastern Snake River Plain, Idaho. *Geological Society of America, Abstracts with Programs* 9, 740-41.

Kuntz, M. A. (1978). ''Geologic Map of the Arco-Big Southern Butte Area, Butte, Blaine, and Bingham Counties, Idaho.'' U.S. Geological Survey Open-File Report 78-302.

Kuntz, M. A., Lefebvre, R. H., Champion, D. E., McBroome, L. A., Mabey, D. P., Stanley, W. D., Covington, H. R., Ridenour, J., and Stotelmeyer, R. B. (1980). ''Geological and Geophysical Investigations and Mineral Resources Potential of the Proposed Great Rift Wilderness Area, Idaho.'' U.S. Geological Survey Open-File Report 80-475.

Lachenbruch, A. H. (1978). Heat flow in the Basin and Range province and thermal effects of tectonic extension. *Pure and Applied Geophysics* 117, 34-50.

Lachenbruch, A. H., and Sas, J. H. (1977). Heat flow in the United States and the thermal region of the crust. *In* ''The Earth's Crust: Its Nature and Physical Properties'' (J. G. Heacock, G. V., Keller, and J. E. Oliver, eds.), pp. 626-75. American Geophysical Union Geophysical Monograph 20.

Lachenbruch, A. H., and Sass, J. H. (1978). Models of an extending lithosphere and heat flow in the Basin and Range province. *In* ''Cenozoic Tectonics and Regional Geophysics of the Western Cordillera'' (R. B. Smith and J. P. Eaton, eds.), pp. 209-50. Geological Society of American Bulletin Memoir 152.

Lajoie, K. R. (1968). Quaternary stratigraphy and geologic history of Mono basin, eastern California. Ph.D. dissertation, University of California, Berkeley.

Lamb, H. H. (1970). Volcanic dust in the atmosphere; with a chronology and assessment of its meteorological significance. *Royal Society of London Philosophical Transactions* series A, 266, 425-533.

Lamb, H. H. (1977). ''Climate: Present, Past and Future.'' Vol. 2. Methuen, Barnes, and Noble, New York.

LaPoint, P. J. I. (1977). Preliminary photogeologic map of the eastern Snake River Plain, Idaho. U.S. Geological Survey Miscellaneous Field Studies Map MF-850.

Laughlin, A. W., Brookins, D. G., and Causey, J. D. (1972). Late Cenozoic basalts from the Bardea lava field, Valencia County, New Mexico. *Geological Society of America Bulletin* 83, 1543-52.

Lawrence, D. B. (1954). Diagrammatic history of the northeast slope of Mount St. Helens. *Mazama* 36, 49-54.

Lipman, P. W., and Mullineaux, D. R. (eds.) (1981). ''The 1980 Eruptions of Mount St. Helens, Washington.'' U.S. Geological Survey Professional Paper 1250.

Lipman, P. W., Prostka, H. J., and Christiansen, R. L. (1972). Cenozoic volcanism and plate-tectonic evolution of the western United States: I. Early and Middle Cenozoic. *Royal Society of London Philosophical Transactions* series A, 271, 217-48.

McCormick, P. M. (1981). Lidar measurements of Mount St. Helens effluents. *Proceedings of the International Society for Optical Engineering* 278, 19-22.

MacLoed, N. S., Sherrod, D. R., Chitwood, L. ., and McKee, E. H. (1981). Newberry Volcano, Oregon. *In* ''Guides to Some Volcanic Terranes in Washington, Oregon, and Northern California'' (D. A. Johnston and J. M. Donnelly-Nolan, eds.), pp. 85-91. U.S. Geological Survey Circular 838.

Majors, H. M. (1980). Ash deposit from the *circa* 1802 eruption, in Mount St. Helens series. *Northwest Discovery* 1, 25-31.

Maynard, L. C., (1974). Geology of Mt. McLoughlin. M.S. thesis, University of Oregon, Eugene.

Mehringer, P. J., Jr., Blinman, E., and Peterson, K. L. (1977). Pollen influx and volcanic ash. *Science* 198, 257-61.

Mehringer, P. J., Jr., Nash, W. P., and Fuller, R. H. (1971). A Holocene volcanic ash from Northwestern Utah. *Utah Academy of Sciences, Arts, and Letters Proceedings* 48, 46-51.

Miller, C. D. (1980). "Potential Hazards from Future Eruptions in the Vicinity of Mount Shasta Volcano, Northern California." U.S. Geological Survey Bulletin 1503.

Moore, R. B., and Wolfe, E. W. (1976). Geologic map of the eastern San Francisco volcanic field, Arizona. U.S. Geological Survey Miscellaneous Investigations Map I-953.

Mullineaux, D. R. (1974). "Pumice and Other Pyroclastic Deposits in Mount Rainier National Park, Washington." U.S. Geological Survey Bulletin 1326.

Mullineaux, D. R., Hyde, J. H., and Rubin, M. (1975). Widespread late glacial and postglacial tephra deposits from Mount St. Helens volcano, Washington. U.S. Geological Survey Journal of Research 3, 329-35.

Mullineaux, D. R., and Wilcox, R. E. (1980). Stratigraphic subdivision of Holocene air-fall tephra from the climatic series of Mount Mazama, Oregon (abstract). EOS (American Geophysical Union Transactions) 61, 66.

Mullineaux, D. R., Wilcox, R. E., Ebaugh, W. F., Fryxell, R., and Rubin, M. (1978). Age of the last major scabland flood of the Columbia Plateau in eastern Washington. Quaternary Research 10, 171-80.

Nieland, J. (1970). Spatter cone pits, San Mountain lava field, Oregon Cascades. Ore Bin 32, 231-36.

Norgren, J. A., Borchardt, G. A., and Harward, M.E. (1970). Mount St. Helens ash in northeastern Oregon and south central Washington. Northwest Science 44, 66.

Okazaki, R. Smith, H. W., and Gilkeson, R. A. (1972). Correlation of West Blacktail ash with pyroclastic layer T from the 1800 A.D. eruption of Mount St. Helens. Northwest Science 46, 77-89.

Otto, B. R., and Hutchinson, D. A. (1977). The geology of Jordan Craters, Malheur County, Oregon. Ore Bin 39, 125-40.

Parker, R. B. (1963). Recent volcanism at Amboy Crater, San Bernadino County, California. California Division of Mines and Geology Special Report 76.

Peterson, J. A., and Martin, L. M. (1980). Geologic map of the Baker-Cypress BLM Roadless Area and Timbered Crater RARE II areas, Modoc, Shasta, and Siskiyou Counties, California. U.S. Geological Survey Miscellaneous Field Investigations Map MF-1214-A.

Peterson, N. V., and Groh, E. A. (1963). Recent volcanic landforms in central Oregon. Ore Bin 25, 33-45.

Peterson, N. V., and Groh, E. A. (1964). Diamond Craters, Oregon. Ore Bin 26, 17-34.

Peterson, N. V., and Groh, E. A. (1965). Hole in the Ground, Fort Rock, Devils Garden field trip. In "Lunar Geological Field Conference Guide Book" (N. V. Peterson and E. A. Groh, eds.), pp. 19-28. Oregon Department of Geology and Mineral Industries Bulletin 57.

Peterson, N. V., and Groh, E. A. (1969). The ages of some Holocene volcanic eruptions in the Newberry Volcano area, Oregon. Ore Bin 31, 73-87.

Pilles, P. J., Jr. (1979). Sunset Crater and the Sinagua: A new interpretation. In "Volcanic Activity and Human Ecology" (P. O. Sheets and D. K. Grayson, eds.), pp. 459-81. Academic Press, New York.

Pippin, L. C. (1980). "Prehistoric and Historic Patterns of Lower Pinyon-Juniper Woodland Ecotone Exploitation at Borealis, Mineral County, Nevada." University of Nevada, Desert Research Institute Social Sciences Center Technical Report 17.

Pollack, J. B., Toon, O. B., Sagan, C., Summers, A., Baldwin, B., and Van Camp, W. (1976). Volcanic explosions and climatic change: A theoretical assessment. Journal of Geophysical Research 81, 1071-83.

Porter, S. C. (1978). Glacier Peak tephra in the north Cascade Range, Washington: Stratigraphy, distribution, and relationship to late-glacial events. Quaternary Research 10, 30-41.

Porter, S. C. (1979). Quaternary stratigraphy and chronology of Mauna Kea, Hawaii: A 380,000-yr record of mid-Pacific volcanism and ice-cap glaciation. Geological Society of America Bulletin 90, 609-16.

Porter, S. C. (1981). Recent glacier variations and volcanic eruptions. Nature 291, 139-42.

Powers, H. A. (1932). The lavas of the Modoc Lava Bed quadrangle, California. American Mineralogist 17, 253-94.

Powers, H. A., and Wilcox, R. E. (1964). Volcanic ash from Mount Mazama (Crater Lake) and from Glacier Peak. Science 144, 1334-36.

Proffett, J. M., Jr. (1977). Cenozoic geology of the Yerrington district, Nevada, and implications for the nature and origin of Basin and Range faulting. Geological Society of America Bulletin 88, 247-66.

Randle, K., Goles, G. G., and Kittleman, L. R. (1971). Geochemical and petrological characterization of ash samples from Cascade Range volcanoes. Quaternary Research 1, 261-82.

Rice, C. J. (1981). "Satellite Observations of the Mount St. Helens Eruption of 18 May 1980." Aerospace Corporation, Space Sciences Laboratory Report SSL-81 (6640).

Rinehart, C. D., and Huber, N. K. (1965). "The Inyo Crater Lakes: A Blast in the Past." California Division of Mines and Geology Mineral Information Service 18 (9).

Robinson, P. T., Elders, W. A., and Muffler, L. J. P. (1976). Quaternary volcanism in the Salton Sea geothermal field, Imperial Valley, California. Geological Society of America Bulletin 87, 347-60.

Rose, W. I., Harris, D., Heiken, G., Sarna-Wojcicki, A. M., and Self, S. (1981). Volcanological description of the May 18, 1980, eruption of Mount St. Helens. In "Mount St. Helens Eruptions of 1980: Atmospheric Effects and Potential Climatic Impact," pp. 1-35. Report of National Aeronautic and Space Administration Workshop, Washington, D.C., November 1980.

Royse, C. F., Jr. (1967). Mazama ash from the continental slope off Washington. Northwest Science 41, 103-9.

Rubin, M., and Alexander, C. (1960). U.S. Geological Survey radiocarbon dates, part V. Radiocarbon (Supplement) 2, 129-85.

Russell, I. C. (1903). "Geology of Southwestern Idaho and Southeastern Oregon." U.S. Geological Survey Bulletin 217.

Sarna-Wojcicki, A. M. (1976). "Correlation of Late Cenozoic Tuffs in the Central Coast Ranges of California by Means of Trace and Minor Element Chemistry." U.S. Geological Survey Professional Paper 972.

Sarna-Wojcicki, A. M., Bowman, H. R., Meyer, C. E., Russell, P. C., Asaro, F., Michael, H., Rowe, J. J., Jr., Baedecker, P. A., and McCoy, G. (1980). "Chemical Analyses, Correlations, and Ages of Late Cenozoic Tephra Units of East-Central and Southern California." U.S. Geological Survey Open-File Report 80-231.

Sarna-Wojcicki, A. M., Bowman, H. R., and Russell, P. C. (1979). "Chemical Correlation of Some Late Cenozoic Tuffs of Northern and Central California by Neutron Activation Analysis of Glass and Comparison with X-ray Fluorescence Analysis." U.S. Geological Survey Professional Paper 1147.

Sarna-Wojcicki, A. M., Meyer, C. E., Woodward, M. J., and Lamothe, P. L. (1981). Composition of air-fall ash erupted on May 18, May 25, June 12, July 22, and August 7. In "The 1980 Eruptions of Mount St. Helens" (P. W. Lipman and D. R. Mullineaux, eds.), pp. 667-81. U.S. Geological Survey Professional Paper 1250.

Sarna-Wojcicki, A. M., Shipley, S., Waitt, R. B., Jr., Dzurisin, D., and Wood, S. H. (1981). Areal distribution, thickness, mass, volume, and grain size of air-fall ash from the six major eruptions of 1980. In "The 1980 Eruptions of Mount St. Helens" (P. W. Lipman and D. R. Mullineaux, eds.), pp. 577-600. U.S. Geological Survey Professional Paper 1250.

Sass, J. H., Blackwell, D. D., Chapman, D. S., Costain, J. K., Decker, E. R., Lawver, L. A., and Swanberg, C. A. (1980). Heat flow from the crust of the United States. In "Physical Properties of Rocks and Minerals" (Y. S. Touloukian, W. R. Judd, and R. F. Roy, eds.), pp. 503-45. McGraw-Hill, New York.

Scott, D. H., and Trask, N. J. (1971). "Geology of the Lunar Crater Volcanic Field, Nye County, Nevada." U.S. Geological Survey Professional Paper 599-I.

Scott, W. E. (1977). Quaternary glaciation and volcanism, Metolius River area, Oregon. Geological Society of America Bulletin 88, 113-24.

Sherrod, D. R., and MacLeod, N. S. (1979). The last eruption at Newberry volcano, central Oregon. Geological Society of America Abstracts with Programs 11, 127.

Simkin, T., Siebert, L., McLelland, L., Bridge, D., Newhall, C., and Latter, J. H. (1981). "Volcanoes of the World: A Regional Directory, Gazeteer, and Chronology of Volcanism during the Last 10,000 Years." Dowden, Hutchinson, and Ross, Stroudsburg, Pa.

Sims, J. D., and Rymer, N. J. (1975). "Preliminary Description and Interpretations of Cores and Radiographs from Clear Lake, California, Core 7" U.S. Geological Survey Open-File Report 75-144.

Smiley, T. L. (1958). The geology and dating of Sunset Crater, Flagstaff, Arizona. In "Guidebook of the Black Mesa Basin, Northeastern Arizona" (R. Y. Anderson and J. W. Harshbarger, eds.), pp. 186-90. New Mexico Geological Society, Ninth Field Conference.

Smith, C. T. (1964). Geology of the Little Black Peak quadrangle, Socorro and Lincoln Counties, New Mexico. *In* "New Mexico Geological Society Guidebook, 15th Field Conference, Ruidoso County," pp. 92-99.

Smith, H. W., Okazaki, R., and Knowles, C. R. (1977a). Electron microprobe analysis of glass shards from tephra assigned to set W, Mount St. Helens, Washington. *Quaternary Research* 7, 207-17.

Smith, H. W., Okazaki, R., and Knowles, C. R. (1977b). Electron microprobe data for tephra attributed to Glacier Peak, Washington. *Quaternary Research* 7, 197-206.

Smith, R. L. (1979). Ash-flow magmatism. *In* "Ash-Flow Tuffs" (C. E. Chapin and W. E. Elston, eds.), pp. 1-27. Geological Society of America Special Paper 180.

Smith, R. L., and Shaw, H. R. (1975). Igneous-related geothermal systems. *In* "Assessment of Geothermal Resources of the United States—1975" (D. E. White and D. L. Williams, eds.), pp. 58-83. U.S. Geological Survey Circular 726.

Snyder, W. S., Dickinson, W. R., and Silberman, M. L. (1976). Tectonic implications of space-time patterns of Cenozoic magmatism in the western United States. *Earth and Planetary Science Letters* 32, 91-106.

Sternberg, R. S., and McGuire, R. H. (1981). Archeomagnetic secular variation in the American Southwest (abstract), *EOS (American Geophysical Union Transactions)* 62, 852.

Taylor, E. M. (1968). Roadside geology, Santiam and McKenzie Pass highways, Oregon. *Oregon Department of Geology and Mineral Industries Bulletin* 62, 3-33.

Taylor, E. M. (1978). "Field Geology of Southwest Broken Top Quadrangle, Oregon." Oregon Department of Geology and Mineral Industries Special Paper 2.

Thaden, R. E., and Ostling, E. J. (1967). Geologic map of the Bluewater quadrangle, Valencia and McKinley Counties, New Mexico. U.S. Geological Survey Geologic Quadrangle Map GQ-679.

Valastro, S., Davis, E. M., and Varela, A. G. (1972). Craters of the Moon lava flows, Idaho. *In* University of Texas at Austin radiocarbon dates IX. *Radiocarbon* 14, 468-70.

Waitt, R. B., Jr. (1980). About forty last-glacial Lake Missoula jokulhaups through southern Washington. *Journal of Geology* 88, 653-79.

Walker, G. W., Peterson, N. V., and Greene, R. C. (1967). Reconnaissance geologic map of the east half of the Crescent quadrangle, Lake Deschutes and Crook Counties, Oregon. U.S. Geological Survey Miscellaneous Geologic Investigations Map I-493.

Waters, A. C. (1962). Basalt magma types and their tectonic associations: Pacific Northwest of the United States. *In* "The Crust of the Pacific Basin" (G. A. MacDonald and H. Kuno, eds.), pp. 158-70. American Geophysical Union Geophysical Monograph 6.

Westgate, J. A., and Gold, C. M. (1974). "World Bibliography and Index of Quaternary Tephrochronology." University of Alberta, Edmonton.

Westgate, J. A., Smith, D. G. W., and Nichols, H. (1970). Late Quaternary pyroclastic layers in the Edmonton area, Alberta. *In* "Proceedings of the Pedology and Quaternary Research Symposium, May 13 and 14, 1969, Alberta, Canada," (S. Pawluk, ed.), pp. 179-86.

White, C. M. (1980). "Geology and Geochemistry of Mt. Hood Volcano." Oregon Department of Geology and Mineral Industries Special Paper 8.

White, C. M., and McBirney, A. R. (1978). Some quantitative aspects of orogenic volcanism in the Oregon Cascades. *In* "Cenozoic Tectonics and Regional Geophysics of the Western Cordillera" (R. B. Smith and G. P. Eaton, eds.), pp. 369-88. Geological Society of America Memoir 152.

Wilcox, R. E. (1965). Volcanic ash chronology. *In* "The Quaternary of the United States" (H. E. Wright, Jr., and D. G. Frey, eds.), pp. 807-16. Princeton University Press.

Williams, H. (1932). "Geology of the Lassen Volcanic National Park, California." University of California Publications, Department of Geological Sciences Bulletin 21, pp. 195-385.

Williams, H. (1942). "The Geology of Crater Lake National Park, Oregon, with a Reconnaissance of the Cascade Range Southward to Mount Shasta." Carnegie Institution of Washington Publication 540.

Williams, H. (1944). "Volcanoes of the Three Sisters Region, Oregon Cascades." University of California Publications, Department of Geological Sciences Bulletin 27, pp. 37-63.

Williams, H. (1957). A geologic map of the Bend quadrangle, Oregon, and a reconnaissance geologic map of the central portion of the High Cascade mountains. Oregon Department of Geology and Mineral Industries.

Williams, H., and Goles, G. (1968). Volume of the Mazama ash-fall and the origin of Crater Lake caldera. *In* "Andesite Conference Guidebook" (H. M. Dole, ed.), pp. 37-41. Oregon Department of Geology and Mineral Industries Bulletin 62.

Wise, W. S. (1969). Origin of basaltic magmas in the Mojave Desert area, California. *Contributions to Mineral and Petrology* 23, 53-64.

Wise, W. S. (1970). "Cenozoic Volcanism in the Cascade Mountains of Southern Washington." Washington Division of Mines and Geology Bulletin 60.

Wood, S. H. (1975). Holocene stratigraphy and chronology of mountain meadows, Sierra Nevada, California. Ph.D. dissertation, California Institute of Technology, Pasadena.

Wood, S. H. (1977a). "Chronology of Late Pleistocene and Holocene Volcanics, Long Valley and Mono Basin Geothermal Areas, Eastern California." Final Technical Report, Contract 14-08-0001-15166, to U.S. Geological Survey Geothermal Research Program.

Wood, S. H. (1977b). Distribution, correlation, and radiocarbon dating of late Holocene tephra, Mono and Inyo Craters, eastern California. *Geological Society of America Bulletin* 88, 89-95.

Wood, S. H., and Brooks, R. (1979). Panum Crater dated 640 ± 40 radiocarbon yrs. B.P., Mono Craters, California. *Geological Society of America, Abstracts with Programs* 11, 543.

CHAPTER 6

The Holocene Paleomagnetic Record in the United States

Subir K. Banerjee

Introduction

Research on the paleomagnetism of rocks and sediments began during the 1950s and has been responsible for dramatic discoveries in the areas of Phanerozoic continental drift and plate tectonics. It is difficult, however, to point to an equal maturity in paleomagnetic research for the Holocene period. The reasons for this are threefold. First, rocks and sediments with stable natural remanent magnetizations (NRM) and suitably detailed temporal records have been hard to find. Second, it was difficult to measure weak NRMs until the cryogenic magnetometers were developed. Third, accurate and high-resolution age dating of such rocks and sediments is difficult. The last 10 years, however, have seen a marked change in this state of affairs. Since the first successful study (Mackereth, 1971) of the record of the horizontal component of the Earth's magnetic field in the soft sediments of Lake Windermere in northwestern England, many new paleomagnetic studies on lake sediments have been reported from different countries and continents (Banerjee et al., 1979; Barton and Polach, 1980; Creer et al., 1972; 1979; Dodson et al., 1977; Nakajima and Kawai, 1973; Noltimier and Colinvaux, 1976; Turner and Thompson, 1981). In spite of the many sources of error inherent in paleomagnetic studies of soft sediments from present-day lakes, it has been possible to confirm the reliability of a large number of these records by various magnetic and nonmagnetic techniques.

In addition to lake sediments, there are two other sources for the Holocene paleomagnetic record: archaeological samples such as bricks and pottery and lava flows that can be dated by determining the radiocarbon age of organic material overridden by them. Paleomagnetic work on archaeological samples, or archaeomagnetism, is not new, but until recently there had been few archaeomagnetic studies with high temporal resolution (such as the work of Walton [1979] on the Agoran potsherds from Athens). On the other hand, the paleomagnetic investigation of Holocene radiocarbon-dated lava flows is very new, and only one detailed study has been made to date (Champion, 1980). At the present time, however, it is possible to make a preliminary assessment of the Holocene paleomagnetic records presently available in the United States from all the three sources mentioned above (lake sediments, archaeological samples, and radiocarbon-dated lava flows).

A new potential stratigraphic tool is based not on paleomagnetic information but on rock-magnetic information from lake sediments (Thompson et al., 1980). The distinction arises between the two types of magnetic information, because rock magnetism provides information about depth-dependent changes in the basic nature of the magnetic mineral influx at a site whereas paleomagnetism under the proper circumstances provides a direct, time-dependent record of the secular variation in direction and the relative intensity of the geomagnetic field. it is easier and less time-consuming to measure the "as-is" rock-magnetic stratigraphy than to prove the fidelity of the paleomagnetic record by use of intralake and interlake comparisons. New areas where rock-magnetic stratigraphy can be applied successfully may exist.

This chapter reviews three main areas of study: the paleomagnetic record, the rock-magnetic record, and the application of these two techniques to paleoclimatic and paleoecologic problems.

The Paleomagnetic Record

Paleomagnetic and rock-magnetic techniques and the necessary instrumentation have recently been reviewed (Banerjee, 1981), and an earlier article by Collinson (1975) provides a more detailed discussion of the instruments. A few words about the paleomagnetic method and its limitations are in order, however.

Three assumptions are made in any paleomagnetic study of rocks, sediments, or archaeological bricks or pottery. First, it is assumed

This is contribution number 1041 from the School of Earth Sciences and number 245 from the Limnological Research Center from the University of Minnesota. I am indebted to the members of the geomagnetism and paleomagnetism group at Minnesota, especially J. King, S. Lund, and J. Marvin, for allowing me to use their published and unpublished data for preparing this chapter. My colleagues H. E. Wright, Jr., and Duane Champion reviewed the manuscript and made valuable suggestions for its improvement. Our research is supported by U.S. National Science Foundation grants EAR-7919985 and EAR-8019466.

that the sample in question, roughly a 2-cm cube, has a large number of magnetic grains of small size (submicron to hundreds of microns, μm), for example, magnetite grains, whose intrinsic remanent magnetizations form a cone with a small angle (10° to 20°) about a mean direction (the direction of the NRM of the whole sample). (The direction coincides approximately with that of the geomagnetic field when the sample acquired its "primary" or "characteristic" NRM.) The second assumption is that the instantaneous magnetic field at the site can be approximated by a geocentric axial dipole, whose intersections at points on the surface of the Earth are called virtual geomagnetic poles (VGP). The third assumption is that, even though the sample may acquire secondary magnetizations at a date or dates later than the time when the primary component was acquired, it is possible to employ appropriate physicochemical methods of selective demagnetization, or "cleaning," to isolate the primary component. A major problem in work on Holocene paleomagnetism is the overconfidence of those investigators who assume that the last condition can be easily fulfilled. The literature is filled with claims of the geomagnetic field's very unusual behavior (for example, a reversal of the geomagnetic field lasting only 5 years), even though the stratigraphic level of occurrence of such unusual behavior, such as the boundary between glacial till and postglacial mud, suggests that the primary component of NRM is hard to separate in such cases of deposition in a relatively high-energy environment.

In rock-magnetic studies, the chief problem seems to be our inability to separate from a bulk sample those grains that are truly responsible for the observed remanence (NRM, ARM [anhysteretic remanent magnetization], IRM [isothermal remanent magnetization], and VRM [viscous remanent magnetization]; for definitions and detailed descriptions of these and other rock-magnetic parameters, see Stacey and Banerjee, 1974).

Geomagnetic Field Behavior

Given the preconditions of the accuracy and reliability of paleomagnetic records, what type of information can we expect to find from them? The answer lies in the short-term behavior (or secular variation) of the Earth's magnetic field (over hundreds to thousands of years). At any given time, 80% or more of the Earth's total magnetic field over all of its surface can be approximated by a geocentric axial dipole, while the rest of the field is better modeled by up to eight dipoles of smaller intensity located at eight different sites near the core/mantle boundary and directed radially (some "up" and some "down") toward the Earth's surface, therefore, includes a major contribution from the geocentric dipole and a minor contribution from one or more of these nearby smaller-intensity radial dipoles,called "nondipoles" in order to be consistent with conventional geomagnetic terminology. A comprehensive discussion of the sources of the geomagnetic field can be found in the work of Jacobs (1975), and a briefer description is available in the first few chapters of a book by McElhinny (1973).

When the time variation of the geomagnetic field is considered, globally or at a particular site, we observe that, averaged over 10,000 to 100,000 years, the field has truly shown geocentric axial dipole behavior, at least for the last 5 million years (McElhinny, 1973). During the 10,000 years of the Holocene, however, we find that the virtual geomagnetic pole (VGP) at a particular site at a given time does not agree with that at another site on another continent for the same

time. This is so not only because the dipole field is suspected of having periodic fluctuations in intensity with a time constant of about 8000 to 9000 years but because superposed on it are a range of periodicities in nondipole field intensity and direction, from roughly 2500 years down to 500 years or less (Burlatskaya, 1978; Cain, 1979; Cox, 1968). (The origins of such periodic variations are not discussed here.) Two reasons for these events have been suggested: a purely latitudinal westward drift of the local nondipole field centers (or foci) or a combination of partially westward drift and a partially pure intensity fluctuation of some of the nondipole sources in a "standing" mode (Bullard et al., 1950; Yukutake and Tachinaka, 1969). Because there are presently only eight nondipole foci and because they are widely distributed in space, each nondipole focus affects large, continentwide areas. For paleoclimatic and paleoecologic studies, therefore, it is possible to use characteristic time-dependent geomagnetic field features (secular variation) as relative-age-dating tools. It is unlikely that there has been a true reversal of the geomagnetic field in the last 10,000 years or even a so-called "excursion" during which near-equatorial positions of the VGP can be observed. In their analysis of the many recent claims of paleomagnetically observed excursions, Verosub and Banerjee (1977) conclude that most were simply the results of poor paleomagnetic recording processes; the most recent candidate for an excursion is at least 15,000 years old, that is, older than the Holocene.

Paleomagnetic Measurements and Data Analyses

The two basic types of instruments employ either an active or a passive method of sensing the weak magnetic flux emanating from the sample. The NRM of some organic sediments may be of the order of 10^{-7} cgs emu per cubic centimeter (10^{-4} A per meter in SI units), whereas a typical value for the Earth's magnetic field at latitude 45°N is about 5×10^{-1} cgs units. The NRM measurements are, therefore, carried out in a field-free space usually obtained by enclosing the sensor and the sample with multiple high-permeability (μ-metal) shields and sometimes with large sets of Helmholtz coils with automatic gain control. In an active-type instrument, such as a spinner magnetometer, the sample is spun at a low frequency about an axis while a fluxgate sensor combined with a phase-sensitive detector measures two components of NRM. The sample is placed in different orientations to obtain a total of four independent determinations for each of the three components of NRM.

In the modern versions of these instruments, the data are fed into an on-line minicomputer and given the latitude and longitude of the site and the field orientation of the samples. Not only the ancient declination (D) and inclination (I) but also the ancient latitude and longitude of the VGP can be obtained from the data collected from a suite of samples of a given age. In the case of a passive sensing mechanism, the insertion of the sample into the field-free sensing region of an astatic or a cryogenic magnetometer produces the necessary readings corresponding to the components of NRM. The cryogenic magnetometer with sensors based on the Josephson effect has a noise level of only 2×10^{-8} cgs emu per cubic centimeter. For weakly magnetic samples and samples that are so fragile that they cannot be spun even at 5 Hz, the cryogenic magnetometer is the preferred instrument.

Figure 6-1 shows the orientation of the total geomagnetic vector F at a site in the Northern Hemisphere (vector pointing down when the field is of normal polarity). Z and H are the vertical and horizontal components, respectively. The total horizontal component H can be further decomposed into X and Y, the northward and eastward components of H. Thus, declination (D) from true north is given by arc tan Y/X, and inclination (I) is arc tan Z/H. These two geomagnetic elements, D and I, when combined with an independent field-intensity determination at the site (either H or Z or F), provide a full description of the local geomagnetic field. Frequently, these are the parameters quoted by paleomagnetists interested in the secular variation of the geomagnetic field for the last 10,000 years. Another way to present the data is to calculate the latitudes and longitudes of the VGP that would produce the observed D and I at a local site. The advantage of such an approach can be appreciated when data from samples of the same age but from different sites (say 1000 km apart) are being compared. Identical or very close locations of the VGPs indicate that the two locations are influenced by the same total field (i.e., dipole with or without additional local nondipole components of the total field).

A detailed discussion of the errors involved in measuring the past values of D and I and in determining the latitude and longitude of the ancient VGP is beyond the scope of this chapter. (See McElhinny, 1973.) Here, we indicate the mathematical relationships between the observed D and I values from a site (latitude λ and longitude ϕ) and the ancient VGP latitude (λ') and longitude (ϕ'). The primary equation in paleomagnetism is:

$$\tan 1 = 2 \tan \lambda = 2 \cot \phi,$$

where I = observed past value of inclination, λ = deduced past value of latitude of the site, and $\phi = (90° - \lambda)$, or past colatitude.

Because past values of longitudes are not independently measurable, the equation for λ' and ϕ' of the VGP utilizes directly the past values of D. Thus, from McElhinny (1973: 25):

$$\text{Sin } \lambda' = \text{Sin } \lambda \, \cos \phi + \text{Cos } \lambda \, \text{Sin } \phi \, \text{Cos } D$$
$$\text{and } \phi' = \phi + \beta \text{ when Cos } \phi \geqslant \text{Sin } \lambda \, \text{Sin } \lambda'$$
$$\text{or } = \phi + 180 - \beta \text{ when Cos } \phi < \text{Sin } \lambda \, \text{Sin } \lambda',$$

where Sin β = Sin ϕ Sin D/Cos λ' and $-90° \leqslant \beta \leqslant +90°$.

Most of the literature on Holocene paleomagnetism presents direct D and I values at a site as a function of time, although the ancient intensity (or paleointensity) when determinable preferably should be converted to a virtual axial dipole moment (VADM), which would produce the observed paleointensity (F) at the site. Thus:

$$\text{VADM} = Fa^3/(1 + 3 \text{ Sin}^2\lambda)^{1/2},$$

where a = the mean radius of the Earth.

Results of Paleomagnetic Studies in North America

Figure 6-2 shows the distribution of the sites from which Holocene paleomagnetic records have been studied. There are three types of material studied from these 12 sites: lake sediments, archaeological material, and radiocarbon-dated lava flows (only one site). The lake sediments, in turn, can be divided into two groups: sediments collected from large lakes such as Superior, Michigan, Erie, and Ontario and those collected from smaller postglacial lakes.

To take the lake sediments first, there are advantages and disadvantages in studying sediments from both the Great Lakes and the.

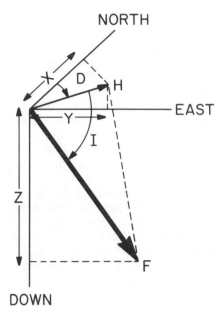

Figure 6-1. Orientation of the total geomagnetic field F at a site in the Northern Hemisphere. H and Z are the horizontal and vertical components of F. Declination D and inclination I are also shown. X and Y are the northern and eastern components of the total horizontal field H.

smaller postglacial lakes. The sediments of the Great Lakes are rich in detrital magnetite, the mineral responsible for a strong and stable NRM, but they lack the organic carbon useful for obtaining accurate radiocarbon ages. The smaller lakes have the opposite characteristics: they contain sediments rich enough in carbon to provide accurate radiocarbon ages (error of ± 100 years or less), but their NRM values may be weak (10^{-6}cgs emu/cm³). For especially organic sediments, the mechanism by which the NRM is acquired is not as well understood as the mechanism for the mineral-rich sediments, for which the process is a postdepositional remanent magnetization, or PDRM (Kent, 1973), which is known to occur within the first 10 to 20 cm below the sediment/water interface. The sediments from the smaller lakes are collected in segments by piston corers, and the problems associated with piston coring are now understood well enough to allow correction for possible errors owing to twisting and nonvertical penetration. The sediment cores from the Great Lakes were collected with a variety of techniques (e.g., Creer, Anderson, and Lewis, 1976), and the results can have complex bearings on the paleomagnetic data.

Lake-sediment studies in North America have not progressed to the point where we can provide composite master curves of past D and I variations and associated reliable ages for the whole of the Holocene. In general, the Great Lakes data suffer from the lack of radiocarbon dates. For one study of Lake Michigan sediments (Creer, Gross, and Lineback, 1976), no radiocarbon dates at all were available. Three independent studies of Lake Michigan sediments in principle permit a test of the validity of the paleomagnetic method. Vitorello and Van der Voo (1977) have compared their data with those of Creer, Gross, and Lineback (1976) by using idealized composites from each study. However, even though the two relative declination series seem to agree with each other, the inclination series generally do not, and an apparently large excursion ($\Delta D = 100°$, $\Delta I = 70°$) recorded in the Winnetka Member (estimated age 7000 radiocarbon years) in two out of three cores by Vitorello and Van der

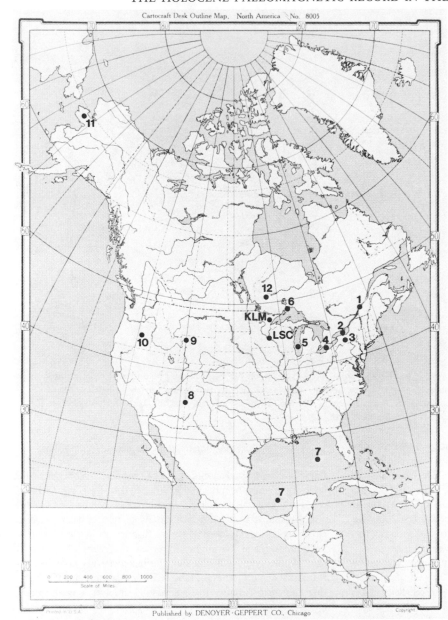

Figure 6-2. Distribution of North American sites from which Holocene paleomagnetic records have been studied to date. (From [1] Mott and Foster, 1977; [2] Creer et al., 1976b; [3] Woodrow and Russell, 1979; [4] Anderson et al., 1976; [5] Creer et al., 1976a; Dodson et al., 1977; Vitorello and Van der Voo, 1977; [6] Mothersill, 1979; [7] Clark and Kennett, 1973; Freed and Healy, 1974; [8] Bucha et al., 1970; Dubois, 1974; Dubois and Watanabe, 1965; Nagata et al., 1965; Sternberg and Butler, 1978; Watanabe and Dubois, 1965; Weaver, 1967; [9] Shuey et al., 1977; [10] Champion, 1980; [11] Noltimier and Colinvaux, 1976; [12] Schwarz and Christie, 1967; Lake St. Croix [LSC] and Kylen Lake [KLM] both from discussion in this chapter.)

Voo does not appear in the other data set. Only the study by Dodson and others (1977) shows a detailed Holocene record although it also lacks direct dating. Approximate pollen dates and dates obtained by other methods assigned to certain levels of the Michigan record by Evenson (1973) allow a comparison to be made of the Dodson and others (1977) study with one made from Holocene sediments from two small postglacial lakes in Minnesota, Lake St. Croix, and Kylen Lake (Lund, 1981; Lund and Banerjee, 1981). Figure 6-3 shows the comparison of declination (D) and inclination (I) on a time scale of radiocarbon years.

In a previous publication (Lund and Banerjee, 1979), the criteria for the acceptability of paleomagnetic data from lake sediments are specified. The most important of these are an agreement between at least two independent data sets from each lake and objective age information from multiple levels of at least one of the sediment cores. Radiocarbon ages by themselves do not necessarily provide the most reliable ages because of the influence of graphite (mineral carbon).

Thus, whenever possible, radiocarbon dates should be supplemented by pollen dates, mineral and biostratigraphic markers, geomorphic or paleogeographic information, et cetera.

The measurements on the St. Croix and Kylen samples were made by subsampling at every 5 to 10 cm, and the discrete data sets were then analyzed by the Clark-Thompson procedure (1978); the double lines for the smoothed data sets (Figure 6-3) include a 95% confidence region. Levels of radiocarbon dates are indicated with solid triangles on the left. The Lake Michigan data were obtained from two parallel cores pushed through a cryogenic magnetometer for continuous measurements of D and I.

The first conclusion to be drawn is that, although the detailed data for these two postglacial lakes and Lake Michigan do not show an exact correspondence, they are indeed comparable, giving credence to both approaches. The lack of independent and reliable age information for the Lake Michigan sediments is still a problem, however.

The second conclusion is that the geomagnetic elements, D and I,

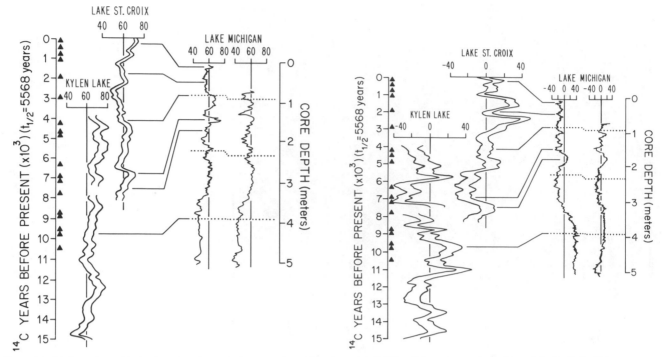

Figure 6-3. (A) Comparison of inclination values (shown in degrees at the top) for the period 15,000 yr B.P. to present obtained from Lake St. Croix, Kylen Lake, and Lake Michigan. The St. Croix and Kylen sediments have been dated by the radiocarbon method, and the dated horizons are shown by solid triangles on the left. The Michigan sediments can only be dated approximately by the use of pollen analysis and other methods. (B) Comparison of declination values (shown in degrees at the top) obtained for the same period from the same three lakes.

although periodic, do not show a single periodicity. This is in complete disagreement with the conclusion drawn by Creer and his colleagues (1976) for Lake Michigan and Lake Erie. The Lake Erie data are not considered here because of the large noise inherent in the data and because of a lack of objective age information. The 2100-year periodicity in D claimed by Creer, Gross, and Lineback (1976) in their Lake Michigan data was derived by assigning ages to the top and bottom of the sediment and *assuming that the declination is periodic*. It is a circular argument, then, to claim that the magnetic data show a periodicity of 2100 years.

The third conclusion is that the lack of a uniform periodicity makes it possible to use characteristic D and I fluctuations at some levels for stratigraphic correlation. Further work on postglacial lake sediments from New York and Pennsylvania is in progress to study whether these magnetic features are a continentwide phenomenon. Poor age control on sediments and/or a lack of replicate cores from the eastern Canadian lakes (Mott and Foster, 1977) and Lake Ontario (Anderson et al., 1976: 203-6) do not allow for such a comparison at this time. High resolution D and I series are available from Thunder Bay in Lake Superior, however (Mothersill, 1979).

It does not appear that there were any sharp geomagnetic excursions during the Holocene or even as far back as 16,000 yr B.P. as is shown by combining the data from Figure 6-3 with the results of Banerjee and others (1979). Such claims for North America had been made by Mörner and Lanser (1975) for about 12,000 yr B.P. and by Creer, Anderson, and Lewis (1976) for the period 14,000 to 7000 yr B.P. Opdyke (1976) discusses the mechanical sources of error that could have led to the observation made by Mörner and Lanser, and I argue elsewhere (Banerjee et al., 1979) that the Lake Erie excursion is also the result of poor paleomagnetic recording.

Paleointensity Studies

Either a relative or an absolute value of the ancient geomagnetic field intensity (F) must be measured in order to describe comprehensively the behavior of the geomagnetic vector. It can then be used simply as a stratigraphic tool or in a search for any correlative behavior, for example, a possible relationship between fluctuations in the geomagnetic field intensity and a stratigraphic record of climatic changes in the past (Wollin et al., 1977). Using the paleointensity compilation of Burlatskaya and Nachasova (1977), Barton and others (1979) show that for North America the data base is simply too small to determine with any reliability what the overall past paleointensity has been for the Holocene. The North American paleointensities are presently most reliable for the last few millennia, for which an archaeomagnetic record is available for comparison.

The total magnetic field intensity (F) at the equator is about 0.3 Oe (23.9 A per meter in SI units) and about 0.6 Oe (47.8 A per meter) at the poles. These magnitudes are consistent with a virtual axial dipole moment (VADM) of 8.0×10^{25} cgs emu per cubic centimeter (8.0 $\times 10^{21}$ A per meter). In a determination methods for absolute paleointensity such as that of Thellier and Thellier (1959), a sample (e.g., an archaeological brick or potsherd or a lava sample) is heated in a magnetic-field-free space to demagnetize it thermally; it is then reheated to the same temperature and cooled in the presence of a known field (H) to acquire a new thermoremanence (TRM). Because the original magnetization (NRM) of a brick or a lava flow was acquired by cooling in the past field of paleointensity (H_p), the magnitude of H_p can be determined from the relationship:

$$H_p = \frac{NRM}{TRM} \, H,$$

all other variables remaining constant. It is often the case, however, that reheating in the laboratory causes subtle psysicochemical changes, and detailed tests have to be carried out in order to be sure that the requisite variables have indeed remained constant.

Pottery and bricks from the southwestern United States have been studied for their paleointensities by various workers, and the latest comparison of such studies has been published by Sternberg and Butler (1978), who cover the period from 2000 years ago to the present. Figure 6-4 shows a compilation of the paleointensity results of Sternberg and Butler (1978), Bucha and others (1970), and Schwarz and Christie (1967) on archaeological samples from the southwestern United States and Champion (1980) on radiocarbon-dated lava flows from the western United States. (See Figure 6-2 for locations.) Figure 6-4 also shows relative paleointensity results for Lake St. Croix sediments in Minnesota. The method of relative paleointensity determination for sediments is discussed by Levi and Banerjee (1976); the results shown here should be considered preliminary.

The absolute intensity axis shows estimated paleointensities at Latitude 45°N obtained by converting the data for other latitudes by use of a virtual axial dipole approximation. Furthermore, at every site the determined paleointensity is the sum of dipole and nondipole components. And, because the nondipole intensity component is not necessarily the same at each site, disagreements can be expected in detailed paleointensity behavior. Thus, even though some disagreements among different data sets are seen, especially for ages older than 2000 yr B.P., it is not certain whether they result from poor data or from variable nondipole contributions. Further studies should show whether the high-frequency behavior of the data obtained from lake sediments reflects true geomagnetic field fluctuations. Walton (1979) has observed similar high-frequency behavior in an absolute paleointensity study of Agoran potsherds from Athens.

Rock-Magnetic Studies

The intrinsic rock-magnetic properties of sediments and soil samples lead to a magnetic stratigraphy that does not depend on past geomagnetic field behavior. Thompson and others (1980) point to a variety of approaches utilizing rock-magnetic stratigraphy.

Unoriented samples of a given age contain magnetic minerals whose chemical composition and grain size can serve as important markers. Because rock-magnetic parameters such as low field susceptibility (χ) or various kinds of laboratory-imparted remanent magnetizations (ARM, IRM, VRM) reflect changes in chemical composition and grain size, a rock-magnetic study may replace more time-consuming conventional sedimentologic, X-ray, or optical-microscopy studies. Thompson and others (1975) show that low field susceptibility (χ) can act as a quick correlational tool, and Blomendal and others (1979) apply susceptibility stratigraphy very successfully in correlating 20 cores for rapidly assessing sediment influx in a Welsh lake covering 6.2 ha.

Figure 6-5 shows another example of rock-magnetic stratigraphy for Long Lake, Minnesota (Banerjee et al., 1981). The rock-magnetic parameters (ARM and χ), which are proportional to the relative fractions of fine and coarse grains, respectively, are plotted against each other. It was known from previous studies that the magnetic mineral present was magnetite alone. Thus, a plot of ARM versus χ shows the relative proportions of the coarse and fine grains in the sediment column (Figure 6-5). Two paleocologic disturbances occurred during the time period represented by the length of the Long Lake sediments shown here. One is the Little Ice Age (600 to 400 yr B.P.) that resulted in the colder and wetter conditions that led to the expansion of the "big woods"; the other is the onset of large-scale land clearance associated with European immigration (100 yr B.P.) in the area, which led to the increase in the pollen of *Ambrosia* (ragweed). The two events are identified in the sediments through changes in pollen assemblages, and Figure 6-5 shows characteristic changes in the ARM/χ slope that coincide with these changes. A detailed description of the theoretical model that explains the magnetic behavior is published elsewhere (King et al., 1981).

Rock-magnetic stratigraphic studies are in their infancy, but it appears very likely that, when combined with other paleoecologic, paleogeographic, and paleohydrologic data, they will prove to be a sensitive as well as rapid method for assessing past geologic changes in local regions.

Conclusions

The construction of an accurate and independently dated master curve of geomagnetic secular variation for North America will provide a reliable and sensitive tool for dating. Dickson and others (1978)

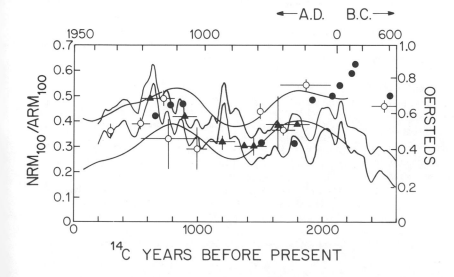

Figure 6-4. Comparison of relative paleointensity data obtained from Lake St. Croix sediments (double lines show high-frequency behavior) with absolute paleointensity data obtained from radiocarbon-dated lava flows (solid circles) (Champion, 1980) and potsherds from Ontario (open circles) (Schwarz and Christie, 1967) and the southwestern United States (triangles) (Bucha et al., 1970; Sternberg and Butler, 1978). (Bucha and others' data are shown as double lines with low-frequency behavior.) The double lines in this and the preceding figure denote regions of 95% confidence calculated according to the method of Clark and Thompson (1978).

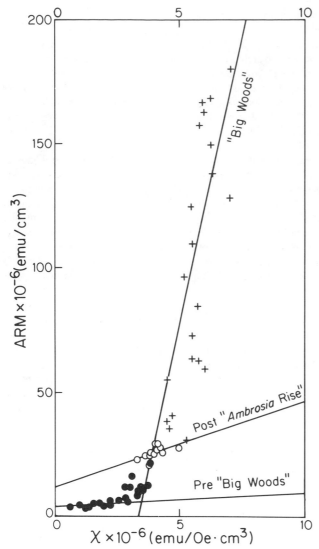

Figure 6-5. Anhysteretic Remanent Magnetization (ARM) versus low-field susceptibility (χ) for sediments from Long Lake, Minnesota (Banerjee et al., 1981). The distinctly different slopes are associated with different ratios of coarse- and fine-grained magnetite, which are themselves indicators of paleohydrologic and paleoecologic changes.

and Barton and Polach (1980) have already shown the value of such magnetic age-dating methods in correcting errors in radiometric dating of sediments in Scotland and Australia. If future work can result in the confirmation of the very high frequency (that is, a time resolution of the order of 100 years) records of geomagnetic field behavior (in D, I, and F), the magnetic method may well become superior to other dating methods because of its small standard error and its applicability to a variety of sediment types, whether they are rich or poor in organic content. Rock-magnetic stratigraphy is still an untapped resource. The method is swift and does not require the lengthy tests so necessary to prove the integrity of paleomagnetic records.

References

Anderson, T. W., Richardson, R. J., and Foster, J. H. (1976). "Late Quaternary Paleomagnetic Stratigraphy from East-Central Lake Ontario." Geological Survey of Canada Paper 76-1C, pp. 203-6.

Banerjee, S. K. (1981). Experimental techniques in rock magnetism and paleomagnetism. *Advances in Geophysics* 23, 25-99.

Banerjee, S. K., King, J., and Marvin, J. (1981). A rapid method for magnetic granulometry with applications to environmental studies. *Geophysical Research Letters* 8, 333-36.

Banerjee, S. K., Levi, S., and Lund, S. P. (1979). Geomagnetic record in Minnesota lake sediments: Absence of the Gothenberg and Erieau excursions. *Geology* 7, 588-91.

Barton, C. E., Merrill, R. T., and Barbetti, M. (1979). Intensity of the Earth's magnetic field over the last 10,000 years. *Physics of the Earth and Planetary Interiors* 20, 96-110.

Barton, C. E., and Polach, H. A. (1980). ¹⁴C ages and magnetic stratigraphy in three Australian maars. *Radiocarbon* 22, 728-39.

Blomendal, J., Oldfield, F., and Thompson, R. (1979). Magnetic measurements used to assess sediment influx at Llyn Goddionduon. *Nature* 280, 50-53.

Bucha, V., Taylor, R. E., Berger, R., and Haury, E. W. (1970). Geomagnetic intensity: Changes during the past 3,000 years in the Western Hemisphere. *Science* 168, 111.

Bullard, E. C., Freedman, C., Gellman, H., and Nixon, J. (1950). The westward drift of the Earth's magnetic field. *Philosophical Transactions of the Royal Society of London* series A, 243, 67.

Burlatskaya, S. P. (1978). Characteristics of the spectrum of secular geomagnetic field variations in the past 8,500 years. *Geomagnetism and Aeronomy* 18, 621-23.

Burlatskaya, S. P., and Nachasova, I. ye. (compilers) (1977). "Archaeomagnetic Determinations of Geomagnetic Field Elements." Mater. Mirov. Tsentra Dannykh B., Sov. Geophys. Comm. Acad. Sci. U.S.S.R., Moscow.

Cain, J. C. (1979). Main field and secular variation. *Reviews of Geophysics and Space Physics* 17, 273-77.

Champion, D. E. (1980). Holocene geomagnetic secular variation in the western United States: Implications for the global geomagnetic field. Ph.D. dissertation, California Institute of Technology, Pasadena.

Clark, H. C., and Kennett, J. P. (1973). Paleomagnetic excursion recorded in latest Pleistocene deep-sea sediments, Gulf of Mexico. *Earth and Planetary Science Letters* 19, 267-74.

Clark, R. M., and Thompson, R. (1978). An objective method for smoothing paleomagnetic data. *Geophysical Journal* 52, 205-13.

Collinson, D. W. (1975). Instruments and techniques in paleomagnetism and rock magnetism. *Reviews of Geophysics and Space Physics* 13, 659-86.

Cox, A. (1968). Length of geomagnetic polarity intervals. *Journal of Geophysical Research* 73, 3247-60.

Creer, K. M., Anderson, T. W., and Lewis, C. F. M. (1976). Late Quaternary geomagnetic stratigraphy recorded in the Lake Erie sediments. *Earth and Planetary Science Letters* 31, 37-47.

Creer, K. M., Gross, D. L., and Lineback, J. A. (1976). Origin of regional geomagnetic variations recorded by Wisconsin and Holocene sediments from Lake Michigan and Lake Windermere. *Geological Society of America Bulletin* 87, 531-40.

Creer, K. M., Hogg, E., Malkowski, Z., Mojski, J. E., Niedziolka-Krol, E., Readman, P. W., and Tucholka, P. (1979). Paleomagnetism of Holocene lake sediments from north Poland. *Geophysical Journal* 59, 287-313.

Creer, K. M., Thompson, R., Molyneaux, L., and Mackereth, F. J. H. (1972). Geomagnetic secular variation recorded in the stable magnetic remanence of recent sediments. *Earth and Planetary Science Letters* 14, 115-27.

Dickson, J. H., Stewart, D. A., Thompson, R., Turner, G., Baxter, M. S., Dondarsky, N. D., and Rose, J. (1978). Palynology, paleomagnetism and radiometric dating of Flandrian marine and freshwater sediments of Lock Lomond. *Nature* 274, 548-53.

Dodson, R. E., Fuller, M. D., and Kean, W. F. (1977). Paleomagnetic records of secular variations from Lake Michigan sediment cores. *Earth and Planetary Science Letters* 34, 387-95.

Dubois, R. L. (1974). Secular variation in southwestern United States as suggested by archaeomagnetic studies. In "Proceedings of the Takesi Nagata Conference;" University of Pittsburgh. R. M. Fisher, M. Fuller, V. A. Schmidt, and P. J. Wasilewski, (eds.), pp. 133-44.

Dubois, R. L., and Watanabe, N. (1965). Preliminary results of investigations made to study the use of Indian pottery to determine the paleointensity of the geomagnetic field for the United States 600-1400 A.D. *Journal of Geomagnetism and Geoelectricity* 17, 417-23.

Evenson, E. B. (1973). Late Pleistocene shorelines and stratigraphic relations in the Lake Michigan basin. *Geological Society of America Bulletin* 84, 2281-98.

Freed, W. K., and Healy, N. (1974). Excursions of the Pleistocene geomagnetic field recorded in Gulf of Mexico sediments. *Earth and Planetary Science Letters* 24, 99.

Jacobs, J. A. (1975). ''The Earth's Core.'' Academic Press, New York.

Kent, D. V. (1973). Post-depositional remanent magnetization in a deep-sea sediment. *Nature* 246, 32-34.

King, J., Banerjee, S. K., Marvin, J., and Özdemir, Ö. (1982). A comparison of different magnetic methods for determining the relative grain size of magnetite in natural materials: Some results from lake sediments. *Earth and Planetary Science Letters* 59, 404-19.

Levi, S., and Banerjee, S. K. (1976). On the possibility of obtaining relative paleointensities from lake sediments. *Earth and Planetary Science Letters*, 29, 219-26.

Lund, S. P. (1981). High resolution secular variation studies from lake sediments in Minnesota and Tennessee. Ph.D. dissertation, University of Minnesota, Minneapolis.

Lund, S. P., and Banerjee, S. K. (1979). Paleosecular variation from lake sediments. *Reviews of Geophysics and Space Physics* 17, 244-49.

Lund, S. P., and Banerjee, S. K. (1981). Time-series analysis of late Quaternary paleomagnetic records from two North American lakes (abstract). *In* ''Proceedings of the Fourth IAGA Scientific Assembly,'' p. 175, Edinburgh.

McElhinny, M. W. (1973). ''Paleomagnetism and Plate Tectonics.'' Cambridge University Press, London.

Mackereth, F. J. H. (1971). On the variation in direction of the horizontal component of remanent magnetism in lake sediments. *Earth and Planetary Science Letters* 12, 332-38.

Mörner, N. A., and Lanser, J. (1975). Paleomagnetism in deep-sea core A179-15. *Earth and Planetary Science Letters* 26, 121-24.

Mothersill, J. S. (1979). The paleomagnetic record of the late Quaternary sediments of Thunder Bay. *Canadian Journal of Earth Sciences* 16, 1016-23.

Mott, R. J., and Foster, J. H. (1977). Preliminary paleomagnetic studies of lake sediments in eastern Canada. *Géographie physique et Quaternaire* 31, 379-87.

Nagata, T., Kobayashi, K., and Schwarz, E. J. (1965). Archaeomagnetic intensity studies of South and Central America. *Journal of Geomagnetism and Geoelectricity* 17, 399-405.

Nakajima, T., and Kawai, N. (1973). Secular geomagnetic variation in the recent 60,000 years found from the Lake Biwa sediments. *Rock Magnetism and Paleogeophysics* 17, 34-38.

Noltimier, H. C., and Colinvaux, P. A. (1976). Geomagnetic excursion from Imuruk Lake, Alaska. *Nature* 259, 197-200.

Oldfield, F., Rummery, T. A., Thompson, R., and Walling, D. E. (1979). Identification of suspended sediment sources by means of magnetic measurements: Some preliminary results. *Water Resources Research* 15, 211-18.

Opdyke, N. D. (1976). Discussion of paper by Mörner and Lanser concerning the paleomagnetism of deep-sea core A179-15. *Earth and Planetary Science Letters* 29, 238-39.

Schwarz, E. J., and Christie, K. W. (1967). Original remanent magnetization of Ontario potsherds. *Journal of Geophysical Research* 72, 3263-69.

Shuey, R. T., Ugland, R. O., and Schmitt, C. R. (1977). Magnetic properties and secular variation in core from Yellowstone and Jackson Lakes, Wyoming. *Journal of Geophysical Research* 82, 3739-46.

Stacey, F. D., and Banerjee, S. K. (1974). ''The Physical Principles of Rock Magnetism.'' Elsevier, Amsterdam.

Sternberg, R., and Butler, R. F. (1978). An archaeomagnetic paleointensity study of some Hohokam potsherds from Snaketown, Arizona. *Geophysical Research Letters* 5, 101-4.

Thellier, E., and Thellier, O. (1959). Sur l'Intensité du champ magnétique terrestre dans la passé historique et géologique. *Annales des Géophysique* 15, 285-376.

Thompson, R., Battarbee, R. W., O'Sullivan, P. E., and Oldfield, F. (1975). Magnetic susceptibility of lake sediments. *Limnology and Oceanography* 20, 687-98.

Thompson, R., Blomendal, J., Dearing, J. A., Oldfield, R., Rummery, T. A., Stober, J. C., and Turner, G. M. (1980). Environmental applications of magnetic measurements. *Science* 207, 481-86.

Turner, G. M., and Thompson, R. (1979). Behavior of the Earth's magnetic field as recorded in the sediment of Loch Lomond. *Earth and Planetary Science Letters* 42, 412-26.

Turner, G. M., and Thompson, R. (1981). Lake sediment record of the geomagnetic secular variation in Britain during Holocene times. *Geophysical Journal* 65, 703-25.

Verosub, K. L., and Banerjee, S. K. (1977). Geomagnetic excursions and their paleomagnetic record. *Reviews of Geophysics and Space Physics* 15, 145-55.

Vitorello, I., and Van der Voo, R. (1977). Magnetic stratigraphy of Lake Michigan sediments obtained from cores of lacustrine clay. *Quaternary Research* 7, 398-412.

Walton, D. (1979). Geomagnetic intensity of Athens between 2000 BC and AD 400. *Nature* 277, 643-44.

Watanabe, N., and Dubois, R. L. (1965). Some results of an archaeomagnetic study on the secular variation in the southwest of North America. *Journal of Geomagnetism and Geoelectricity* 17, 395-97.

Weaver, G. H. (1967). Measurements of the past intensity of the Earth's magnetic field. *Archaeometry* 9, 174-86.

Wollin, G., Ryan, W. B. F., and Ericson, D. B. (1977). Paleoclimate, paleomagnetism and the eccentricity of the Earth's orbit. *Geophysical Research Letters* 4, 267-70.

Woodrow, D. L., and Russell, A. M. (1979). Paleomagnetic stratigraphic framework for eastern North America to 12,000 B.P.: II. Six lakes in New York State. *EOS (Transactions of the American Geophysical Union)* 60, 237-38.

Yukutake, T., and Tachinaka, H. (1969). Separation of the Earth's magnetic field into drifting and standing parts. *Bulletin of the Earthquake Research Institute* 47, 66-97.

CHAPTER 7

Radioactive Isotopes in the Holocene

Pieter M. Grootes

Introduction

Since the 1965 publication of Broecker's paper in "The Quaternary of the United States," the use of isotopes in geologic research has increased vastly. Large amounts of radioactive isotopes produced in the atmosphere by nuclear weapons tests resulted in the rapid improvement of detection techniques for studying and monitoring fallout. These techniques have been utilized in studies of the isotopic record in sediments as it pertains to paleoclimatic fluctuations, as well as in investigations of such problems as mineralization history.

Isotopic studies can be divided into three classes: stable isotopes, natural radioactive isotopes, and radioactive isotopes produced by nuclear explosions. Stable isotopes are generally studied by mass spectrometry. The abundance ratios of stable isotopes provide information on the formation and the geochemistry of the sample. Frequently used are the isotope ratios ^{2}H (D)/^{1}H, ^{13}C/^{12}C, ^{15}N/^{14}N, ^{18}O/^{16}O, and ^{34}S/^{32}S. Their main applications are in the study of climatic fluctuations and of vegetation and mineralization history.

Natural radioactive isotopes are studied by detecting their decay with proportional counters, Geiger counters, scintillation-crystal detectors, liquid-scintillation detectors, and solid-state detectors. The isotopes are long-lived (e.g., ^{40}K, ^{238}U, ^{235}U, ^{87}Rb, ^{187}Re, and ^{176}Lu), continuously produced by cosmic radiation in the upper atmosphere (^{3}H [T], ^{10}Be, ^{14}C, ^{26}Al, ^{32}Si, ^{36}Cl, and ^{39}Ar), or continuously produced by the decay of parent nuclei (uranium decay series, ^{230}Th, ^{231}Pa, ^{226}Ra, ^{210}Pb).

The detection of radioactive isotopes produced during weapons testing follows the same procedure as the detection of natural isotopes. Most of the "unnatural" isotopes were produced during a short period of time, less than a decade. They entered the atmosphere and from there the oceans, biosphere, and sediments as a sharp spike. They are, therefore, suitable for studying accumulation, mixing rates, and residence times in various reservoirs (the atmosphere, the oceans, the sediments, and the groundwater).

This chapter focuses on the short-lived natural and the bomb-produced radioactive isotopes. The still relatively fresh and complete geologic record for the Holocene allows us to study geologic processes in detail and with a high time resolution by using the shorter-lived natural radioactive isotopes. The use of bomb-produced isotopes as tracers in the study of active processes such as sedimentation may lead to a more accurate interpretation of the older sedimentary record. The isotopes discussed in this chapter are T, ^{210}Pb, ^{137}Cs, ^{32}Si, and ^{14}C. The newly developed technique of accelerator mass spectrometry is explored in the last section.

Most of the longer-lived natural isotopes discussed by Broecker (1965) are not included here because their half-lives are too long to make them useful tools for Holocene studies. Stable isotopes, which are powerful tools in paleoclimatic and geochemical studies, are discussed by Friedman (1983).

Also outside the scope of this chapter are some dating techniques, all in the exploratory stage, that are still being developed. Examples are the use of ^{39}Ar (with a half-life [$T_{1/2}$] of 269 years) and ^{85}Kr ($T_{1/2}$ of 10.76 years) for groundwater dating (Loosli and Oeschger, 1979, 1980; Rózański and Florkowski, 1979). Unexpectedly large discrepancies between ^{14}C and ^{39}Ar dates stress the need for the further study of the geochemistry of isotopes used in groundwater dating. In addition, several nonisotopic dating techniques have been developed that are based on processes such as amino acid racemization, thermoluminescence, the formation of fission tracks, and obsidian hydration. These methods have become increasingly useful for older Pleistocene samples but are little used for Holocene material.

Distribution of the Isotopes

The distribution of the cosmogenic and fallout isotopes over the different reservoirs is determined by their source, by the processes that determine their transport between the reservoirs, and by their residence time (concentration) in the different reservoirs. A knowledge of the distribution patterns of the different isotopes is important for their use as a tracer.

Cosmogenic isotopes such as ^{3}H, ^{10}Be, ^{14}C, ^{26}Al, ^{32}Si, ^{36}Cl, and ^{39}Ar are produced mostly in the lower stratosphere. Radioactive isotopes

This work was supported by the National Science Foundation (grant EAR-7904523, geochemistry program).

produced by nuclear-weapons testing such as ^{3}H, ^{14}C, ^{55}Fe, ^{90}Sr, ^{137}Cs, and ^{239}Pu have been injected into the stratosphere. Since most of the tests occurred in the Northern Hemisphere, the highest concentrations of the fallout isotopes were observed there. Concentrations in the Southern Hemisphere increased later and never were as high because the residence time (about 5 years) in the stratosphere for interhemispheric exchange (Schell et al., 1974) is of the same order as that for the injection from the stratosphere into the troposphere. The injection of stratospheric air with cosmogenic and fallout isotopes into the troposphere takes place during the spring at the middle latitudes (30° to 50°). In the troposphere, the residence time of most isotopes is only a week to a month because of the scavenging by particles and precipitation. This period is short compared with the tropospheric mixing time, and so the localized input during the spring results in seasonal and latitudinal variations in the concentrations of the isotopes. Seasonal variations of a factor of 5 have been observed for ^{32}Si (Dansgaard et al., 1966) and a factor of 3 for ^{10}Be (Raisbeck, Yiou, Fruneau, Loiseaux, Lieuvin, and Ravel, 1979). Bomb-produced tritium increased tropospheric concentrations by a factor of 1000 in the Northern Hemisphere and produced seasonal concentration fluctuations of a factor of about 10 (Schell et al., 1974). An exception is ^{14}C, which, in the form of carbon dioxide, has a much longer tropospheric residence; the atmosphere acts as a reservoir attenuating the relative concentration variations resulting from the seasonal injection. The largest tropospheric seasonal ^{14}C concentration fluctuation caused by the injection of bomb ^{14}C was only 20% of normal, prebomb activity, and the maximum concentration was about 2 times normal. Under normal conditions, little or no seasonal effects would be detectable for ^{14}C.

Lead-210 differs from the other isotopes in that it is not produced in the stratosphere but near ground level by the decay of ^{222}Rn. This noble gas produced by the decay of ^{226}Ra in the soil escapes to the atmosphere from the land surface at an average rate of about 42 atoms per minute per square centimeter.

Tritium (^{3}H or T)

PRODUCTION AND DETECTION

Tritium is produced naturally by cosmic radiation, mainly by the spallation reaction ^{14}N (n, ^{3}H) ^{12}C and in "stars." Through the emission of a beta particle and a neutrino, it decays to ^{3}He (^{3}H → ^{3}He + β^- + $\bar{\nu}$). Its short half-life (12.26 years) limits its usefulness as a dating tool. (A recent determination of the tritium half-life gave 12.43 ± 0.05 years [Unterweger et al., 1980]. The use of this new half-life has been recommended by the International Atomic Energy Agency's group of consultants [1981].) Tritium is generally measured through the detection of its beta decay in a proportional counter or in a liquid-scintillation counter. The detection limit is a few tritium units (TU) for proportional and liquid-scintillation counters. One TU is equivalent to a concentration of 1 tritium atom in 10^{18} hydrogen atoms. Liquid scintillation is somewhat less sensitive and requires a larger sample size than does gas counting. The sensitivity of detection can be increased routinely by a factor of at least 10 through preenrichment of the tritium in the sample. The enrichment method most commonly used is electrolysis (Östlund and Werner, 1962; Taylor, 1981).

In a recently developed technique (Clarke et al., 1976), the tritium concentration of a sample is determined mass spectrometrically by measuring the amount of ^{3}He produced by its decay in a previously degassed water sample. Several months of waiting time are required to allow a sufficiently large fraction of the ^{3}H atoms to decay and to produce a measurable amount of ^{3}He. The measurement, however, is much faster than the counting of beta decays. The sensitivity of the ^{3}He method can be an order of magnitude greater than that of the radiometric method because large sample volumes (liters) and long waiting times can be used. This eliminates the need for electrolytic enrichment and long counting times.

APPLICATIONS

Tritium in the form of HTO is an ideal tracer for water. The early work was directed toward establishing T levels in nature (Faltings and Harteck, 1950; Grosse et al., 1951; Kaufman and Libby, 1954; Von Buttlar and Libby, 1955). Natural T levels in water were very low (3 to 8 TU). Test explosions of thermonuclear bombs, however, produced large amounts of T. The first effects of the Castle series of tests in 1954 were evident in Chicago rain by March 1954, resulting in a 10-to 100-fold increase in T concentrations (Von Buttlar and Libby, 1955). Bomb-produced T was subsequently found throughout the Northern Hemisphere (Begemann and Libby, 1957; Brown, 1961; Brown and Grummitt, 1956). T concentrations in air moisture increased by three orders of magnitude (over 2000 TU average, with peaks up to 6000 TU in precipitation in 1963) in the Northern Hemisphere and by two orders in the tropics (200-250 TU in 1963) and the Southern Hemisphere (about 100 TU in 1964 and later years). (See Schell et al., 1974.) Much higher T values occurred in the Northern Hemisphere because all the thermonuclear detonations before the Partial Test Ban Treaty (1963) took place in the Northern Hemisphere. The large tropospheric tritium spike of 1958, 1959, and especially 1963 makes groundwater chronologic studies impossible, but it produced a body of water that could be traced through the hydrologic cycle. The spike was used to study the movement of water in the soil through the vadose zone into the groundwater in different soil types and areas (e.g., Andersen and Sevel, 1974; Atakan et al., 1974; Münnich et al., 1967; Smith et al., 1970). These studies showed that groundwater replenishment occurred almost exclusively through precipitation during the autumn and winter because high evapotranspiration in the summer prevents summer rain from contributing to the groundwater. The downward movement of the infiltrating water occurs as a displacement flow (piston action) with dispersion.

The presence of bomb T in spring and well waters indicates that the water comes at least partly from local rapid infiltration of surface waters. Large underground reservoirs or groundwater flows with slow or distant recharge did not show the bomb spike immediately (International Atomic Energy Agency, 1967, 1970, 1974, 1979). Studies of groundwater reservoirs and flow and replenishment rates have been the most common application of tritium decay determinations. Tritium has also been used to identify the origin of spring waters in a karst region, where water sources with a distinct difference in T concentrations were involved (e.g., Guizerix et al., 1967; Kirkov et al., 1974). In the United States, tritium was used to study mixing in deep lakes (Crater Lake and Lake Tahoe) and water-vapor exchange between a lake and the atmosphere (Imboden et al., 1977; Leventhal and Libby, 1970; Simpson, 1970). Because almost 20 years have passed since the injection of the large spike of T into the stratosphere, tritium has become more evenly distributed in nature and its activity levels have dropped. This makes the tritium-spike study in terrestrial environments increasingly difficult. In the oceans, however, large-scale mixing processes are slow, so the penetration of the bomb-tritium spike has provided and will continue to provide valuable information on the oceanic water circulation (Michel and Suess, 1975; Östlund et al., 1974, 1976; Roether, 1974; Roether et al., 1980; Rooth and Östlund, 1972).

The development of the ^{3}He-detection technique has opened up interesting, new hydrologic applications. The decay of T in a water body such as a lake or the groundwater produces ^{3}He in excess of the concentration in which it normally occurs as a dissolved ^{3}He-^{4}He mixture in equilibrium with atmospheric helium. The excess radiogenic ^{3}He will be small as long as the water can exchange with the atmosphere. Excess ^{3}He, therefore, provides a measure for the effective isolation of the water from the atmosphere. The comparison of the concentration of T with that of its daughter ^{3}He thus can be used to calculate how long the water parcel has been closed off from the atmosphere or, when the time of last mixing is known, the degree to which equilibration with the atmosphere took place. Torgersen and others (1977) have applied the technique to Lake Erie, Lake Huron, and Lake Ontario; average tritium values were 129 ± 4; 118 ± 5, and 137 ± 2 TU, respectively. The ages calculated from the T-^{3}He concentrations ranged from a few days near the surface to about 100 days in the hypolimnion (the deep, cold layer in a thermally stratified lake). A study of two chemically stratified lakes (Lake 120 near Kenora, Ontario, Canada, and Green Lake near Syracuse, New York) by Torgersen and others (1979) shows a clear stratification, with T-^{3}He ages of less than 170 days in the upper 15 m of Lake 120 and 3 ± 1.5 years below and ages of less than about 100 days in the upper 13 m of Green Lake and between 2 and 9 years below. Apart from its usefulness in lake studies, the T-^{3}He pair may also be applicable to the dating of groundwater, subpermafrost groundwater, and permafrost.

Lead-210 (^{210}Pb)

PRODUCTION AND DETECTION

This isotope is a member of the ^{238}U decay series (endpoint ^{206}Pb). It has a half-life of 22.26 ± 0.22 years (Robbins, 1978) and decays mainly via beta emission to ^{210}Bi. The noble gas parent of ^{210}Pb, ^{222}Rn (T$_{1/2}$ of 3.8 days), produced by decay of ^{226}Ra (T$_{1/2}$ of 1620 years), escapes to the atmosphere mainly from the land surface. Via a series of short-lived daughters, ^{222}Rn decays to ^{210}Pb, which is effectively removed from the atmosphere by precipitation (atmospheric residence time, a few weeks). When precipitated, it becomes separated from its parent ^{222}Rn, which starts the ^{210}Pb clock. From the surface water ^{210}Pb is rapidly removed to the sediments by scavenging particulate matter. The sediments, however, also receive ^{226}Ra and ^{210}Pb from soil erosion (surface runoff and dust particles) and from ^{226}Ra leached from rock and soil. For sediments, therefore, one must consider the amount of ^{210}Pb not in equilibrium with a radioactive parent (the ''unsupported'' fraction). Lead-210 is extracted from sediments by leaching with hydrochloric acid or by total dissolution of the sample using hydrofluoric acid. The resulting solution is radiochemically purified, and sources for decay counting are prepared from it.

Because the beta energy of the ^{210}Pb decay is very low, the ^{210}Pb concentration is often measured via beta counting in a proportional flow counter of the decay of its daughter ^{210}Bi (T$_{1/2}$ of 5.0 days, max-

imum beta energy, 1.17 MeV). Alternatively, alpha-spectrometry of the 5.3-MeV alpha line of its granddaughter ^{210}Po (T$_{1/2}$ of 138.4 days) is used.

MODELS

To derive a sample age from a measured concentration of a radioactive isotope, one has to know the ratio of the measured concentration to that originally present in the sample material. Furthermore, one has to assume that the only process changing this original concentration has been radioactive decay (closed-system assumption). When these two conditions are not met, ages can only be derived under special conditions and with additional assumptions. For the ^{210}Pb dating of ice or sediment, two models for deriving sample ages are commonly used. (For a detailed discussion, see Robbins, 1978).

The simplest model assumes that constancy of the rate of sediment or snow supply and of the flux of unsupported ^{210}Pb results in a constant initial concentration of unsupported ^{210}Pb (the c.i.c. model). When the profile does not contain reworked material and is undisturbed, an age or a sedimentation rate can be calculated from a normal decay equation.

$$A = A_{obs} - A_f = A_0 e^{-\lambda t} = A_0 e^{-\lambda m/r}, \qquad (1)$$

Where A_{obs} is the measured (total) activity of ^{210}Pb at a given depth, A_f is the activity of the ^{210}Pb supported by the decay of ^{226}Ra, A_0 is the original unsupported activity, and λ is the radioactive decay constant for ^{210}Pb (ln2/T$_{1/2}$ = 1/32.11 year $^{-1}$). The cumulative weight of dry sediment—m (grams per square centimeter)—and the mass sedimentation rate—r (grams per square centimeter per year)—are used since compaction in the upper layers of the sediment leads to a variable sediment thickness for constant mass of dry sediment. The supported ^{210}Pb activity A_f is usually taken equal to the (constant) value of A_{obs} at greater depth; thereby it is assumed that the ^{226}Ra concentration is constant throughout the profile. A_0 can be measured directly at the top of the sediment or determined (more accurately) from a series of measurements at different depths in the sediment by calculation or extrapolation to depth zero. When the sedimentation rate and the ^{210}Pb flux vary in the same way so that the specific activity of unsupported ^{210}Pb in the sediment is constant, the situation is equivalent to the c.i.c. model.

A less restrictive model allows a variable sedimentation rate but assumes a constant rate of supply of unsupported ^{210}Pb (the c.r.s. model) (Appleby and Oldfield, 1978). The age of a layer at depth x is calculated in the c.r.s. model from:

$$t = \frac{1}{\lambda} \ell n(A(O)/A(x)), \qquad (2)$$

where $A(O)$ is the total unsupported ^{210}Pb in the sediment column and $A(x)$ is its residual below depth x, both calculated from a measured unsupported ^{210}Pb-depth profile. In practice, matters can be complicated for both the c.i.c. model and the c.r.s. model by the incorporation of older reworked material, by the mobility of isotopes, and by the mixing through bioturbation of the upper few centimeters of sediment.

APPLICATIONS

The ^{210}Pb dating method was developed by Goldberg (1963). Initially, it was used to study snow deposition rates in Greenland and Antarctica (Crozaz and Langway, 1966; Crozaz et al., 1964; Picciotto et al., 1968) and in alpine glaciers (Picciotto et al., 1967). The first successful application to lacustrine sediments was reported by Krishnaswami and others in 1971. The study of coastal marine sediments followed shortly (Koide et al., 1972, 1973). Varved deposits from the center of the Santa Barbara Basin (lat. 34°14.0′N, long. 120°01.5′W) from a 575-m water depth yield ^{210}Pb concentrations that decrease more or less uniformly with increasing depth and varve age over a period of about 80 varve years (Koide et al., 1972). Subsequent measurements on a nonvarved core from Baja California (lat. 25°13.8′N, long. 112°40.6′W, 520 m depth) have given comparable results. Irregular variation of ^{210}Pb with depth was attributed to slump (Koide et al., 1973). The ^{210}Pb sedimentation rate for the Santa Barbara Basin of 0.39 cm per year is in good agreement with an estimated rate of 0.4 cm per year based on the varves. In lake (and ocean) studies, problems can be experienced with sediment loss at the top of the core. (See Koide et al., 1973.) Average sedimentation rates for two low-altitude Wisconsin lakes of 0.63 cm per year (Trout Lake) and 0.60 cm per year (Lake Mendota) are much higher than those of two high-altitude lakes (about 0.1 cm per year). This finding is interpreted as indicating the importance of river-borne solids to the sedimentary deposits. Similar low rates obtained for the Great Lakes show, however, that, in addition to altitude and river influx, the topography of the lake basin must be considered.

Sedimentation in the Great Lakes has been studied with ^{210}Pb, often in combination with ^{137}Cs and/or pollen studies. Robbins and Edgington (1975) report sedimentation rates between 0.066 and 0.28 cm per year for a number of cores from Lake Michigan and a satisfactory agreement between the ^{210}Pb and ^{137}Cs results. At the water-

sediment interface, a well-defined zone of rapid, steady-state mixing of sediments has to be assumed to explain the observed profiles. The same process applies to Lake Huron (Robbins et al., 1977). Lake Superior yields average sedimentation rates of 0.017 and 0.013 cm per year, but the curve for unsupported ^{210}Pb concentration versus depth seems to show a kink (Bruland et al., 1975). From Lake Ontario and Lake Erie a number of cores have been obtained in which no surficial mixing of sediments is evident (Robbins et al., 1978). Because the percentage of solids in these cores increases with depth by as much as a factor of four, mass sedimentation rates have had to be used (Table 7-1). For Lake Ontario mass sedimentation rates derived from ^{210}Pb measurements are in good agreement with those resulting from ^{137}Cs measurements. The sedimentation rate based on the increase in ragweed (*Ambrosia*) pollen as a mark of agricultural land clearance agrees for one core but is almost three times lower in a second core. For Lake Erie, again, agreement with ^{137}Cs and pollen-derived rates is evident except where a discontinuity in the ^{210}Pb profile suggests sediment loss. Durham and Joshi (1980) have found in Lake Huron evidence of varying lead concentrations and discrepancies between sedimentation rates based on ^{210}Pb and on ^{137}Cs (^{137}Cs rates are generally larger). Additional ^{210}Pb work on coastal sediments by several groups is described in Table 7-2.

The increasing amount of data on both lacustrine and marine sediments has made it obvious that the simple assumption of constant initial concentration (c.i.c. model) is frequently incorrect because changes in sedimentation rate occur. A model with constant ^{210}Pb flux independent of sedimentation rate (c.r.s. model) (Appleby and

Table 7-1. ^{210}Pb Studies in North America

Lake Sediments	Author(s)/Year	Sedimentation Rate	
		Centimeters per Year[a]	Grams per Square Centimeter per Year
Lake Tahoe, Calif.-Nev.	Koide et al., 1973	0.1	—
Lake Mendota, Wis.	Koide et al., 1973	0.60	—
Trout Lake, Wis.	Koide et al., 1973	0.63	—
Lake Michigan	Robbins and Edgington, 1975	0.066-0.28	0.0121-0.0938
Lake Superior	Bruland et al., 1975	0.017; 0.013	—
Lake Huron	Robbins et al., 1977	0.097; 0.11	0.021; 0.051
Lake Huron	Durham and Joshi, 1980	0.02-0.213	0.0048-0.0444
Lake Ontario	Robbins et al., 1978	≃ 0.28 and 0.34[b]	0.057; 0.078
Lake Ontario	Farmer, 1978	0.031-0.143	0.031-0.056
Lake Erie	Robbins et al., 1978	1.4; 0.46 and 0.20[b]	0.440; 0.083 and 0.096
Lake Union, Wash.	Barnes et al., 1979	0.16-0.8	—
Lake Washington, Wash.	Barnes et al., 1979	0.063-0.83	0.015-0.180
Lake Sammamish, Wash.	Barnes et al., 1979	0.11-0.56	0.023-0.123
Chester Morse Reservoir, Wash.	Barnes et al., 1979	0.073-0.23	0.015-0.036
Findley Lake, Wash.	Barnes et al., 1979	≃ 0.037	≃ 0.005
Lake Meridian, Wash.	Barnes et al., 1979	0.16-0.39	0.008-0.026
Linsley Pond, Conn.	Brugam, 1978	0.33-0.94[c]	—
Lake Whitney, Conn.	Krishnaswami et al., 1980	0.73	0.16

[a]The linear sedimentation rates are unreliable when compaction occurs; then, the use of the mass sedimentation rate is necessary.
[b]Values for upper 10 cm only.
[c]Discontinuous curve; values below and above breakpoint, respectively.

Table 7-2. ^{210}Pb Studies in North America

Coastal and Estuarine Environments	Author(s)/Year	Sedimentation Rate (cm/year)
Santa Barbara Basin, Calif.	Koide et al., 1972	0.39
Baja California	Koide et al., 1973	0.27
Coast, Southern California	Bruland et al., 1974	0.07; 0.09
Mississippi River Delta	Shokes, 1976	—
Salt marsh, Long Island	Armentano and Woodwell, 1975	0.47; 0.63
Salt marsh, Conn.	McCaffrey, 1977	—
Long Island Sound	Benninger, 1976, 1978	—
Long Island Sound	Krishnaswami et al., 1980	—
Strait Juan de Fuca	Schell, 1977	—
Saguenay Fjord	Smith and Walton, 1980	0.1-7
Narragansett Bay	Santschi et al., 1979	—

Oldfield, 1978) has been used successfully by Oldfield and others (1978), Appleby and others (1979), and Barnes and others (1979).

Barnes and others (1979) have studied the sedimentation history of six lakes in the west-central part of Washington State. Their detailed sampling provided profiles of unsupported ^{210}Pb as a function of overlying sediment mass. In all but one case (Findley Lake), they report two or more regions of different sedimentation rates. In general, the change from one sedimentation regime to another coincides with human activity in the area. Supporting evidence has been obtained from stable lead profiles. Brugam (1978) reports similar results for Linsley Pond in southern Connecticut.

Postdepositional redistribution of sediments by detritus-feeding organisms occasionally appears to be important. Measurements of these mixing effects by ^{137}Cs analysis have been published by Robbins and others (1977) for Lake Huron and by Kipphut (1978) for a number of Canadian lakes. Benninger and others (1979) have found in Long Island Sound a roughly homogeneous mixing zone in the top 2 to 4 cm, below which unsupported ^{210}Pb decreases quasi-exponentially to zero at or above 15 cm depth. Below this, ^{210}Pb excess is observed in identifiable burrow fillings intersected at 70 and 115 cm depth. This shows that large burrows are filled by surficial sedimentation rather than by collapse. Krishnaswami and others (1980) have studied sediment mixing in Long Island Sound and Lake Whitney (New Haven, Connecticut) by using ^{7}Be,^{239,240}Pu, and ^{210}Pb, and they report general agreement among the different isotopic techniques. These results imply that short-lived and abrupt variations are smoothed out and may be lost when their duration is much shorter than the time required to accumulate a sediment layer with the thickness of the mixing zone. In addition, failure to identify the filling of burrows with surficial sediments may lead to erroneous interpretation of the climate in a core intersecting them.

Cesium-137 (^{137}Cs)

PRODUCTION AND DETECTION

Cesium-137 was produced by nuclear-weapons testing in the atmosphere. Most of the production occurred during the extensive nuclear-weapons test series preceding the signing of the Partial Test Ban Treaty of 1963. Fallout was greatest during the years 1962 to 1966 and has decreased since then.

The half-life of ^{137}Cs is 30 years, and it decays via beta emission to ^{137}Ba, with a maximum beta energy of 1.18 MeV. Most (92%) of the decay occurs to an excited state ^{137}Ba^m, which returns to the ground state ($T_{1/2}$ of 2.6 minutes) under emission of a gamma quantum of 0.661 MeV. Cesium-137 is generally detected by this gamma line with a scintillation-crystal (sodium iodide) or a solid-state germanium lithium detector connected to a multichannel analyzer. Because of the penetrating gamma emission of ^{137}Cs, its concentration can be measured directly in the dried sediment.

APPLICATIONS

Ritchie and others (1974) have studied the ^{137}Cs balance of three small northern Mississippi State watersheds, and they report that most (97%, 88%, 85%) of the ^{137}Cs has been retained in the soil. The amount of ^{137}Cs transported as well as the amount recovered in the sediment of the three reservoirs is directly proportional to the amount of erosion. This suggests that ^{137}Cs is adsorbed on soil particulate matter (clay particles) once it enters the soil, and after erosion it moves into lakes with the particulate fraction, rather than in solution. Cesium-137 precipitated directly into lakes is also adsorbed as ions onto clay particles and then is readily transported into the sediments (residence time in the upper Great Lakes, 1 to 2 years [Wahlgren and Nelson, 1974]). As a consequence, one expects in the sediment a ^{137}Cs concentration that increases with depth to a peak value corresponding to the year 1964 and abrupt reduction below this level to zero. Cesium-137 thus provides a marker horizon in the sediment that can be used to study sedimentation over the past 15 to 20 years.

Pennington and others (1973) have studied ^{137}Cs in five lakes in the English Lake District, and they report concentration profiles in good agreement with those expected from direct wet precipitation of ^{137}Cs and the virtual absence of mixing processes. The sedimentation rates, however, are significantly higher than long-term averages. This could be due to increasing human influence.

Robbins and Edgington (1975) have used ^{137}Cs in combination with ^{210}Pb to study recent sedimentation rates in Lake Michigan. They report a well-mixed zone in the upper few centimeters of sediment. Generally, sedimentation rates based on ^{137}Cs are higher than those based on ^{210}Pb because of sediment mixing and some diffusion in the sediment (^{137}Cs is more mobile than ^{210}Pb). Mixing and diffusion both displace the lower boundary of the ^{137}Cs spike downward. Thus, a sedimentation rate based on the position of this boundary is strongly affected. For ^{210}Pb, supplied by nature at a constant rate, sedimentation rates are calculated from a concentration-depth profile of a much longer sediment column and, therefore, are much less sensitive to the effects of mixing and diffusion. Robbins and others (1977) have studied ^{210}Pb and ^{137}Cs profiles from southern Lake Huron in conjunction with the distribution of benthic macroinvertebrates (mainly Tubificidae). They report finding a well-mixed zone of 3 cm at one location and 6 cm at another. The 3-cm zone contained 90% of the total number of invertebrates in the profile; the 6-cm zone, 95%. This clearly indicates that the depth of the mixed zone is determined by biologic mixing. The profiles further show a gradual decrease of ^{137}Cs concentration below the well-mixed zone. Although molecular diffusion cannot be ruled out, the decrease of ^{137}Cs more likely results from a decreasing probability of mixing with increasing depth.

Durham and Joshi (1980) have found in Lake Huron ^{137}Cs below the 1958-through-1964 depth as determined by ^{210}Pb measurements.

Studies by Kipphut (1978) in the Experimental Lakes Area of Canada show a well-mixed layer 2 to 6 cm thick followed by a zone of less intense mixing (mixing coefficients 10^{-8} to 10^{-9} cm^2 per second) to considerable depth (more than 10 cm). Because of the observed biologic mixing of ^{137}Cs in lake sediments, ^{137}Cs dating is often unsuitable for the derivation of sedimentation rates in lake deposits. When sedimentation rates are obtained by other means (e.g., from ^{210}Pb), the ^{137}Cs concentration profile identifies the zones of rapid and slow mixing and determines their extent. Since the thickness of these zones seems to be fairly constant at a given locality (Robbins et al., 1977), a knowledge of their extent aids the interpretation of the older sediment record. When no well-mixed zone is apparent in the ^{210}Pb or ^{137}Cs record, sedimentation rates derived by both methods agree (Robbins et al., 1978). Laboratory experiments with tubificid worms and ^{137}Cs by Robbins and others (1979) show that a well-localized amount of ^{137}Cs tracer adsorbed on clay particles (illite) is distributed more or less evenly to a depth of 6 cm, with a decrease to zero at 9 cm within half a year. This is accompanied by an initial downward movement of the concentration peak and, after about 45 days, the generation of a second, smaller concentration peak at the surface when the first has reached a depth of about 3 cm.

Cesium-137 in coastal marine sediments has been studied by Livingston and Bowen (1979), who also report finding indications of bioturbation, and by Smith and Walton (1980) in conjunction with their ^{210}Pb work on Saguenay Fjord.

Silicon-32 (^{32}Si)

PRODUCTION AND DETECTION

Silicon-32 is produced by spallation of argon by the primary and secondary particles of cosmic radiation. It decays with beta emission to ^{32}P, which is also a beta emitter, and decays with a 14.3-day half-life to ^{32}S. The half-life of ^{32}Si has not yet been well established. Values range from around 60 years (Turkevich and Samuels, 1954) to (Lindner, 1953) 710 years. More recent determinations have yielded values close to 300 years (Clausen, 1973; DeMaster, 1980; Jantsch, 1967; Lal et al., 1970). In conflict with these estimates are two recent measurements using the new technique of accelerator mass spectrometry (108 ± 18 years [Elmore et al., 1980] and 101 ± 18 years [Kutschera, Henning, Paul, Smither, et al., 1980]). A careful, critical reevaluation of the assumptions and measurements used in the determination of the different half-lives is needed. Until a reliable half-life has been established, absolute ages derived from ^{32}Si activities will be largely uncertain.

A major problem in the use of ^{32}Si as a dating tool is its very low concentration in nature (from about 1 dpm [decay per minute] per ton in rainwater in Denmark [Dansgaard et al., 1966] to 0.02-0.04 dpm per 10^3 ℓ in Indian subsurface waters [Lal et al., 1970; Nijampurkar et al., 1966]). In beta-decay-counting silica from water volumes of 10^3 to 10^4 ℓ needs to be collected for a datable sample. Where the water can be easily collected, silica is scavenged out by ferric hydroxide. Otherwise (for example, in the deep-ocean water samples), special filters filled with chemically pretreated natural sponges (Lal et al., 1964) or Acrylan fiber converted into a cation-exchange resin (Krishnaswami et al., 1972) are used to collect silica in situ from a large volume of water. Radiochemically pure silica is then extracted from the ferric hydroxide slime or the filters. The silicon daughter ^{32}P is allowed to grow to near equilibrium by storage for 2

to 3 months. Then, the ^{32}P is "milked" from the silicon extract and deposited as a radiochemically pure sample on a backing. The ^{32}P beta activity (E_{max} equaling 1.71 MeV) is measured in a special low-background four-pi beta counter for 30 to 45 days (two to three ^{32}P half-lives). The elaborate sample preparation and radiometric counting procedure required have seriously limited the use of ^{32}Si in geologic studies.

APPLICATIONS

The application of ^{32}Si dating to groundwater studies has been carried out by Nijampurkar and others (1966) and Lal and others (1970), who surveyed ^{32}Si activities in the Ganges and Godavari Rivers, Lake Tansa, and a number of wells in India. Mean annual ^{32}Si concentrations in rainwater and snow are fairly constant over the Indian subcontinent at 0.3 dpm per 10^3 ℓ (variations with time and place are less than 0.1 dpm per 10^3 ℓ). These same values are found in runoff in the absence of biologic activity. In rivers ^{32}Si concentrations range between 0.1 and 0.2 dpm per 10^3ℓ, and in lakes they are around 0.2 dpm per 10^3ℓ. The reduction in ^{32}Si content may be due to biologic activity. Thus, groundwater initially has a ^{32}Si concentration of 0.3 dpm per 10^3 ℓ when recharged by rainwater and about 0.15 dpm per 10^3 ℓ when fed from lakes and streams. An additional complication in ^{32}Si groundwater dating is the fact that the dissolved silicon concentration in equilibrium with soils varies with soil type between 5 and 20 ppm. When water passes from a low- to a high-concentration soil, dissolution of silicon occurs, without exchange, so that the ^{32}Si concentration is unaffected. However, when moving from high to low, precipitation takes place and ^{32}Si is lost. Despite these problems, the ^{32}Si "ages" obtained for 13 groundwater samples provide useful information supplementing the T and ^{14}C measurements (Lal et al., 1970). In two cases in which bomb ^{14}C levels and a measurable T concentration have been found, the ^{32}Si activities, which are not affected by bomb tests, showed that the water body was several hundreds of years old (based on a ^{32}Si half-life of 500 years). This proves that limited recent recharge into an older body of groundwater had occurred. The T and ^{14}C (produced in nuclear-weapons tests) that entered the groundwater with this recharge then led to erroneously young ^{14}C and T ages.

Silicon-32 measurements on marine sediments have been performed by Kharkar and others (1963, 1969). Sedimentation rates derived from ^{32}Si in three Antarctic sediment cores based on a half-life of 500 years (Kharkar et al., 1969) are 40 to 80 times higher than those based on ^{230}Th. This can be interpreted as an indication that the long-term sedimentation rate (^{230}Th, 10^5-year time scale) for the three cores is low because of occasional sediment loss through slumping and/or dissolution. The use of a more recently determined, lower ^{32}Si half-life does not remove the discrepancy and thus leaves this interpretation intact. DeMaster (1980) has used varved sediments in a core from the Gulf of California to estimate the ^{32}Si half-life at 276 ± 32 years. Krishnaswami and others (1971) have measured ^{32}Si levels in two lake sediments (Pavin, France, and Tansa, India). Measuring ^{32}Si has not been further developed into a dating technique, probably because in lake sediments the radiocarbon dating method is much easier and more reliable.

The accuracy of ^{32}Si applied to ice dating (Clausen, 1973; Dansgaard et al., 1966) is lower than that of most stratigraphic methods in its time range. This has led Clausen (1973) to a determination of the ^{32}Si half-life based on observed ^{32}Si concentrations and ice age determined by counting summer maxima in a complete, 741-year-long $^{18}O/^{16}O$ record. The estimated half-life so derived is 330 ± 40 years.

Oceanographic studies by Lal and others (1960, 1964, 1970, 1976), Schink (1962), Kharkar and others (1963), and Somayajulu and others (1973) have shown that ^{32}Si may be a useful tool in the study of ocean-mixing processes. The ^{32}Si supply to surface ocean waters seems to be more or less constant and independent of the supply of dissolved silicon. This leads to ^{32}Si-specific activities in the surface ocean that are generally inversely proportional to the dissolved silica concentrations and differ from ocean to ocean. A difference in specific ^{32}Si activity between surface waters (25 to 99 dpm per kilogram SiO_2) and deep water (5 to 20 dpm per kilogram SiO_2) (Lal et al., 1976) may be used to study vertical mixing.

The applicability of the ^{32}Si dating method will probably be increased greatly when ^{32}Si can be measured directly with accelerator mass spectrometry, because the much smaller sample size will facilitate sampling, the requirements for sample purity are much less stringent, and the measuring time is much shorter. In addition, it may be possible to determine a more accurate value for its half-life.

Carbon-14 (^{14}C)

The radiocarbon dating technique and its applications are discussed by Broecker (1965). Since 1965, a large number of dates have been produced through essentially the same technique, with continuing improvements and refinements. A radically new approach to radiocarbon dating by direct detection of ^{14}C using an accelerator as an ultrasensitive mass spectrometer is discussed in a separate section of this chapter. This section compares ^{14}C dating by liquid-scintillation counting, which is increasingly being used, with the use of proportional carbon dioxide counters. The precision, accuracy, and range of the radiocarbon dating method are discussed in some detail. These are limited not only by technical factors such as the recent and background counting rate of the counter but also by the degree to which basic assumptions of the ^{14}C dating method are correct. The three assumptions discussed are these. (1) *The sample has been in a closed system with respect to carbon exchange with its environment.* Often this is not true. Added or exchanged carbon results in the problem known as "contamination." (2) *The atmospheric ^{14}C concentration has been constant through time.* Significant variations have occurred that can be measured on dendrochronologically dated samples. (3) *The initial ^{14}C concentration is independent of place and environment.* Local effects such as the upwelling of "old" ocean water or the dissolution of "dead" carbonate in groundwater can affect the initial ^{14}C concentration.

These factors do not limit the precision of radiocarbon assays, but they do affect the accuracy of their interpretations. However, when properly considered, valid results can be obtained.

LIQUID SCINTILLATION

For routine measurements of ^{14}C activities, liquid-scintillation counting is increasingly being used. The sample carbon is first converted to benzene, and then, mixed with a scintillator, it is counted. The advantages of this method are these. (1) The complete liquid-scintillation-counting setup, including the sample-preparation line, can be purchased commercially. (2) Although the efficiency of detection depends on the purity of the sample, the purity of the benzene prepared routinely is usually adequate and measurements of ^{14}C concentration are

easily reproducible in routine operations. (3) A large amount of carbon can be contained in a relatively small volume of benzene. This increases the signal-to-noise ratio since the activity of a sample is proportional to the amount of carbon and the background counting rate is proportional to both the volume and the surface area of the counter. (See the next section of this chapter.) (4) The alternate counting of sample, background, and standards can be automated and gives a "quasi-simultaneous" signal-to-noise reading.

Disadvantages are these. (1) The number of steps needed to convert carbonaceous sample material to benzene (carbon dioxide→carbide→acetylene→benzene) are a problem. Each step has to be quantitative or else it may give rise to isotopic fractionation. Overall yields are usually in the range of 90% to 95% for careful operation. (2) Commercial liquid-scintillation counters lack the heavy lead shielding and anticoincidence shielding of proportional counter setups and, therefore, have relatively high background counting rates. (3) The hydrogen used in the benzene synthesis has to be free of tritium; otherwise, part of the T decay may be counted as ^{14}C decay. (4) The detection efficiency of ^{14}C decay in the benzene-scintillator mixture is less than 100% (typically about 70% to 90%) and depends on the purity of the benzene preparation (quenching).

Altogether, liquid-scintillation counting is easier because less specialized technical knowledge is required, and its average precision is comparable to that of the average proportional gas counting. For high-precision work (less than 0.5%), proportional gas counting has generally been used. Careful study of the factors affecting the precision of liquid-scintillation counting enabled Pearson (Pearson, 1979, 1980; Pearson et al., 1977) to obtain a precision of 0.25% for a modern sample containing about 17 g of carbon. This approaches the precision of the best proportional counting setups (0.15% to 0.2% [Stuiver, 1978a; Tans and Mook, 1979]). Continued work on large-volume liquid-scintillation counters (Eichinger et al., 1980) and their improved shielding from cosmic and environmental radiation (Schotterer and Oeschger, 1980) make it likely that the precision and the dating range for liquid-scintillation counting will equal that for proportional gas counting.

PRECISION AND RANGE OF ^{14}C DATING

The age of a radiocarbon sample is calculated as:

$$T = -\frac{1}{\lambda} \ell n \frac{A}{A_0} = -8033 \, \ell n \frac{(N/t) \cdot B}{(N_0/t_0) \cdot B}, \qquad (3)$$

Where A and A_0 are the decay rate (in counts per minute [cpm]) of the sample and the recent standard, respectively; and λ is the decay constant equal to $\ell n \, 2/T_{1/2}$ equaling 1/8033, where $T_{1/2}$ is the conventional ^{14}C half-life of 5568 years. A and A_0 are obtained by filling the counter with sample or standard and observing N and N_0 decays in t and t_0 minutes, respectively, and correcting the count rates thus obtained for the counter background B. Because no standard sample of the undisturbed atmospheric ^{14}C concentration exists, it has to be defined in relation to a generally available standard. On the basis of results of 19th century tree-ring measurements, 95% of the ^{14}C activity of a carbon dioxide sample prepared from a special batch of oxalic acid from the U.S. National Bureau of Standards and corrected for radioactive decay to the year A.D. 1950, has been adopted as representative of the standard atmospheric ^{14}C activity. The carbon dioxide should have a fixed stable carbon isotopic composition $\delta_{PDB}^{13}C$ equals -19.0‰, where $\delta_{PDB}^{13}C$ is the relative difference in carbon isotopic composition between the carbon dioxide from the oxalic acid and the carbon dioxide from the international PDB carbonate standard. The standard ^{14}C activity defined above fixes the zero point of the radiocarbon time scale at A.D. 1950 and ^{14}C ages in yr B.P. (years before present, that is, before A.D. 1950). Because A_0 is nonexistent and A_{ox}, the oxalic acid activity, is measured instead, one has to replace in equation 3 A_0 by 0.95 A_{ox} and N_0 by 0.95 N_{ox}.

The *precision* of a radiocarbon age or, as generally quoted, its standard deviation, is calculated as:

$$\sigma(T) = -\frac{1}{\lambda} \ell n \left[1 \mp \left\{ \left(\frac{\sigma(N/t)}{A} \right)^2 + \left(\frac{\sigma(N_0/t_0)}{A_0} \right)^2 + \left(\frac{(A_0 - A)\sigma(B)}{A_0 A} \right)^2 \right\}^{1/2} \right] \qquad (4)$$

Where $\sigma(N/t)$ equals $(N/t^2)^{1/2}$, $\sigma(N_0/t_0)$ equals $(N_0/t^2_0)^{1/2}$, and $\sigma(B)$ equals $(B/t_B)^{1/2}$ and Bt_B is the number of counts accumulated as the background is counted for t_B minutes. The $\sigma(B)$ term of equation 4 shows that the background uncertainty is unimportant for recent samples when A_0 and A are approximately equal. Since both A_0 and B are measured regularly, the accumulated times t_0 and t_B will be very large and, therefore, the uncertainty in A_0 and B (when the results were fully reproducible within expected statistics) will be negligible compared to that in A. Measurements of the NBS-oxalic-acid-standard carbon dioxide and the background counting rate at the University of Washington over a 3-year period show that this can indeed be the case (Stuiver, 1978b). Equation 4 can then be simplified to:

$$\sigma(T) = -1/\lambda \ell n \left[1 \mp (N/t^2)^{1/2}/A \right] = -1/\lambda \ell n \left[1 \mp ((A + B)/t)^{1/2}/A \right]. \qquad (5)$$

Obviously, high-precision ^{14}C dating [small $\sigma(T)$] requires a large A (large samples and counters) and a long counting time t while B is much smaller than A. For example, a Quaternary Isotope Laboratory counter having a recent activity of 85 cpm and a background B of 1.6 cpm (Stuiver et al., 1979) will give in 4000 minutes a relative precision of 0.17% for a recent sample.

The *age range* of radiocarbon dating (T_{max}) is determined by the ratio of the ^{14}C activity of a recent sample (A_0) and the minimum ^{14}C activity that can reliably be measured in a particular counting setup over a given period of time. For a finite age, this minimum activity should be larger than twice the standard deviation of the background counting rate B [$\sigma(B) = (B/t)^{1/2}$; 2σ criterion giving 98% confidence].

$$T_{max} = + 1/\lambda \ell n A_0/2(B/t)^{1/2} \qquad (6)$$

For old samples, obviously, it is important to minimize B. In construction of the counters and the building in which they are housed, materials should be used that contain very few radioactive impurities. Materials usable for counters are high-purity copper and quartz; for buildings, especially selected low gamma-activity aggregates. In addition, heavy shielding can lower B by reducing the level of cosmic radiation. Most practical in this respect is the use of a deep-underground counting room, as in the Quaternary Isotope Laboratory

in Seattle, the U.S. Geological Survey laboratory in Menlo Park, California, the University of Bern Radiocarbon Laboratory in Bern, Switzerland, and the Natural Isotopes Division of the Council for Scientific and Industrial Research in Pretoria, South Africa. For the very best counting setup, the range of the ^{14}C dating method is about 60,000 years (Stuiver et al., 1979), but most setups are limited to 45,000 years.

ENRICHMENT OF ^{14}C

An extension to the dating range has been obtained by the enrichment of ^{14}C through thermal-diffusion isotope separation. With an enrichment system involving hot-wire columns and carbon monoxide (Haring et al., 1958), a range of about 75,000 years was reached in Groningen (Grootes, 1977; Grootes et al., 1975) and Seattle (Grootes and Stuiver, 1979; Stuiver et al., 1978). The Quaternary Isotope Laboratory enrichment system is illustrated in Figure 7-1. A small difference in isotope concentration between the carbon monoxide gas at the heated central wire and at the water-cooled wall is generated by the temperature gradient, and this causes the convective flow of gases in each tube to concentrate the heavy isotopic species at the bottom of the column and the light at the top. Depending on the ratio of total volume to bottom volume, ^{14}C enrichments by a factor of 8 (Seattle) to a factor of 14 (Groningen) are obtained in about 5 weeks. The long enrichment time and the large sample size required (50 to 100 g of pure, pretreated carbon) limit the use of this technique.

So far, 12 enrichment dates have been published for North America (Grootes, 1978; Stuiver et al., 1978; Vogel and Waterbolk, 1972). Because the residual ^{14}C activity of these old samples is only 10^{-3} to 10^{-4} times the modern activity, the measured age can easily be too young because of contamination of the sample with younger carbon. Great care needs to be taken when the very old dates are being interpreted.

CONTAMINATION

Between the moment of formation and the moment of sampling, ^{14}C samples may receive an admixture of carbonaceous material having a specific ^{14}C activity (age) different from that of the original sample material. This admixture can include humic substances or carbonates carried in groundwater, old detrital carbon (e.g., graphite), and plant rootlets, and it can be caused by the action of soil organisms. Apart from contamination in situ, the admixture can also be introduced during sampling, transport, and storage and during handling in the dating laboratory. All but in situ contaminations of samples can be avoided when proper procedures are instituted. To minimize the likelihood of contamination in situ, it is important to make a careful selection of the sample and a thorough study of the environment in which it was found. Signs of extensive leaching, disturbance of the soil profile near the sample, or recent plant root growth all mark potential contamination problems, as do powdery surfaces on materials such as shells and other carbonates.

Figure 7-2 shows the influence of an admixture of a certain percentage of "dead" (0% activity) or "recent" (100% activity) contaminant on the measured sample age. The effect of a contaminant with a specific ^{14}C activity intermediate between 0% and 100% can also be estimated from Figure 7-2. One can arrive at the specific ^{14}C activity of the contaminant by mixing sample material with the right amount of 0% activity carbon for older contaminants or 100% activity carbon for younger contaminants. Thus, the effect of the

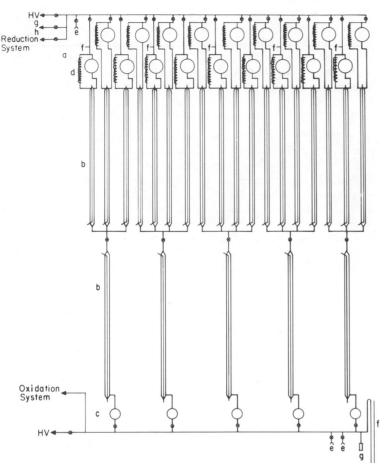

Figure 7-1. Thermal-diffusion radiocarbon-enrichment system: (a) storage volume, (b) thermal-diffusion columns, (c) enriched sample volume, (d) thermal circulation, and (e) sampling point. (From Grootes and Stuiver, 1979.)

intermediate-activity contaminant becomes equal to that of its 0% or 100% part, and the amount of the intermediate contaminants that can be tolerated is greater. From the figure it is obvious that, whereas a "dead" contaminant gives the same age error irrespective of the sample's age, recent contaminants especially influence the measured ^{14}C age of an old sample. For measurements close to the limit of ^{14}C age determination—75,000 years, which has been achieved through the use of ^{14}C enrichment—recent contamination must be well below 0.01%, preferably not more than 0.001%. This means that no more than 10 mg can contaminate a sample containing 1000 g of carbon!

The relative errors in the measured age for both active and dead contaminants are shown in Figure 7-3. In the same graph, the relative age uncertainties based on counting statistics alone are given for two proportional counters (one in use at the Quaternary Isotope Laboratory, one hypothetical). For samples less than a thousand years old, the relative age uncertainty of ^{14}C dates caused by counting statistics is about 1% or more. It is large because the statistical uncertainty in the measured sample activity becomes large in relation to the decrease in ^{14}C activity caused by decay over such a short period. The smallest relative age uncertainties are about 0.25% to 0.5% and are obtained for samples between about 4000 and 30,000 years old. Because contamination with dead carbon produces a constant age er-

ror, the relative age error is inversely proportional to the sample age. Relative age errors caused by recent contaminants depend little on sample age for ages below 10,000 years but increase rapidly for older samples. The level of contamination that is tolerable is determined by the counting equipment that is used and by the precision required.

Considering the very low contamination levels that can be tolerated when full use is to be made of the measurement's precision, the likelihood of obtaining sufficiently clean samples just by careful selection in the field is small. Every organic sample should, therefore, be subjected to a check for rootlets and then a chemical extraction. Frequently, the extraction is made with 1% to 4% hydrochloric acid at 80°C, followed by a 1% to 3% sodium hydroxide solution and again 1% to 4% hydrochloric acid, plus washing with distilled water until the extract is nearly neutral after each extraction. This procedure eliminates carbonates and soluble contaminants introduced into the sample by groundwater. The resulting loss of sample material (often 20% to 80%) is an acceptable price for obtaining a purer and more reliable sample. When the contaminant was introduced in particulate form by groundwater or soil organisms, however, it may be only partially soluble and so would not be eliminated completely from the sample. The intrusive contaminant in this case may be younger material from overlying layers or reworked material from older layers. The latter may have been added at the time the sample material was deposited in, for example, peat beds and soils. Carbonate samples (such as shell, coral, limestone, and concretions) also need to be examined for contamination involving geochemical processes such as recrystallization.

Although experience has shown that reliable ^{14}C dates can be produced and that contamination can be reduced to a level below the detection limit, it is obvious that for reliable ^{14}C dates a careful sample selection has to be combined with an examination of the sample for extraneous material and a thorough physical and chemical purification. In addition to the purified sample, the material extracted from the sample toward the end of the purification process should be dated. This can provide an indication of the sample's purity. When the extracted material's specific activity is the same as that of the sample, the extraction of contaminants is probably complete and the measured age, therefore, may be meaningful.

FLUCTUATIONS OF THE ATMOSPHERIC ^{14}C CONCENTRATION

Dendrochronology

The development of dendrochronologies extending some 8000 years into the past (Ferguson, 1970, 1973) has provided us with ^{14}C samples of precisely known ages. Precise measurements yield the atmospheric ^{14}C level during the year the tree ring was formed and thus supply us with a record of the atmosphere's ^{14}C fluctuations in the past.

After early measurements performed by de Vries (1958) showed that atmospheric ^{14}C fluctuations do exist, extensive measurements on bristlecone pine and sequoia tree rings were carried out by the ^{14}C laboratories at La Jolla, California, (Suess, 1967, 1970, 1978), Tucson, Arizona (Damon et al., 1970, 1973), and Philadelphia, Pennsylvania (Michael and Ralph, 1970, 1973; Ralph and Michael, 1970). The atmospheric ^{14}C level turned out to have been above the chosen standard of 95% of the activity of NBS oxalic acid (making ages appear too young) for the last 400 years, below the standard (making samples appear too old) between 500 and 2000 yr B.P., and then in-

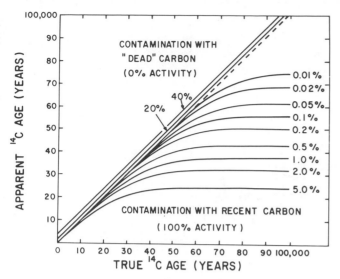

Figure 7-2. Apparent ^{14}C age measured as a function of true ^{14}C age and percentage of contamination. The broken line represents no contamination. Above it, the apparent (too-old) ages result from contamination with "dead" (0% activity) carbon; below it, the too-young ages are caused by different amounts of a "recent" (100% activity) contaminant.

creasingly above the standard (making samples appear too young) before that time (Figure 7-4). The measured deviations from the adopted standard ^{14}C activity have been used to construct graphs and tables to convert radiocarbon ages to calendar ages (Damon et al., 1973; Michael and Ralph, 1973; Ralph et al., 1973; Suess, 1970). This increased the usefulness of ^{14}C dates for accurate studies of absolute time. On the other hand, the existence of several slightly different correction curves and tables has caused confusion in the interpretation of ages quoted in the literature, for the exact method used to derive the calendar age from the original ^{14}C activity is not always described.

The general trend of the change in atmospheric ^{14}C level over the last 8000 years corresponds well with that of changes in the Earth's magnetic field (Bucha, 1970). The variation of the latter seems to have a periodicity of about 8900 years. Superimposed on this long-term trend is evidence of additional ^{14}C fluctuations on a time scale of a hundred years. The precision of the ^{14}C measurements, of the order of 0.5%, as well as possible small differences in calibration among the different laboratories, however, have virtually precluded a quantitative analysis of the small-amplitude, short-term fluctuations.

Over the past few years, high-precision ^{14}C counters have been constructed (Pearson, 1979; Schoch et al., 1980;Stuiver et al., 1979; Tans and Mook, 1979). As a consequence, tree-ring ^{14}C calibration measurements with a precision of 0.2% to 0.3% have now become available (Bruns et al., 1980; de Jong and Mook, 1980; Pearson, 1980; Stuiver, 1978a; Stuiver and Quay, 1980a, 1980b). At this level of precision, the 100-year fluctuations showing an amplitude peak to peak of about 2% to 2.5% are well defined. A comparison of the Quaternary Isotope Laboratory's results on Douglas-fir from the Pacific Northwest with those of the Belfast Laboratory for Irish and Scottish oaks shows excellent agreement within the 0.2% to 0.3% measuring precision (Pearson, 1980). This proves that measured ^{14}C levels in tree rings are independent of species or longitude and are unaffected by the measuring technique (carbon dioxide counters versus liquid-scintillation counters). In addition, no significant latitude or

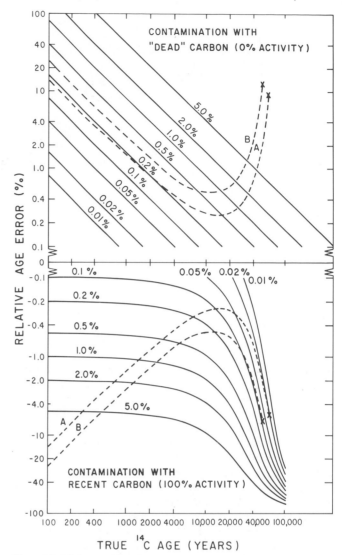

Figure 7-3. Relative age errors due to contamination as a function of the true ^{14}C age of a sample for different levels of contamination. "Dead" carbon gives positive age errors (samples measured are too old), whereas "recent" contaminants cause negative deviations. Included for comparison are the measuring precisions that are obtained in 4000 minutes (4 days) counting for two different counters. Counter A: recent count rate 85 cpm, background 1.5 cpm (Stuiver et al., 1979). Counter B: recent count rate 30 cpm, background 2.5 cpm (hypothetical), X-age limit of counter following the 2σ criterion. The relative age error caused by contamination should be less than the uncertainty caused by the activity measurement.

altitude effect has been identified, and the amplitude of real short-time, yearly fluctuations is expected to be small (Stuiver and Quay, 1981). Therefore, a calibration curve for converting radiocarbon to dendrochronologic (calendar) ages covering the period since the birth of Christ and based on tree-ring measurements in the Pacific Northwest (Stuiver, 1982) can be used worldwide (Figure 7-5). Although the uncertainty in the measured ^{14}C age is small (typically about 15 years), the uncertainty in the calendar-age obtained may be much larger for periods during which a rapid drop in atmospheric ^{14}C activity occurred.

A better record of the 100-year fluctuations has stimulated the

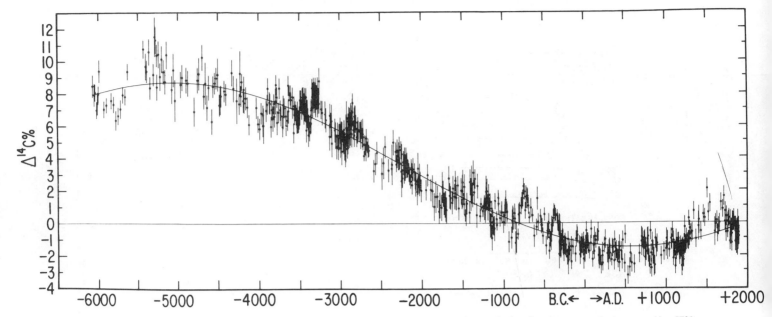

Figure 7·4. Difference Δ¹⁴C, in percentage of atmospheric content, between standard and measured atmospheric radiocarbon content in the past, taking 5730 years for the half-life of ¹⁴C and based on La Jolla bristlecone pine measurements in relation to a standard 19th century wood up to fall of 1979. The curve drawn through the points is a best-fit sine wave. (From Suess, 1981.)

search for their cause. Using a four-reservoir-box-diffusion model (Oeschger et al., 1975), Stuiver and Quay (1980a) have corrected their measured atmospheric ¹⁴C concentrations for the long-term geomagnetic field changes and calculated the variations in the ¹⁴C production rate required to produce the observed values. A good correlation between solar variability, as derived from sunspots and auroral observations, and ¹⁴C-production-rate variations has been obtained. It indicates that the modulation of the cosmic-ray intensity reaching the Earth by the solar wind may be the cause of the observed variations in ¹⁴C concentration. Stuiver and Quay (1980b) are able to explain two-thirds of the observed variability amplitude with solar modulation alone. The atmospheric ¹⁴C record derived from tree rings can, therefore, be used to generate a record of fluctuations of solar activity in the past; the record is important for solar studies (Stuiver and Grootes, 1980). So far, no significant correlation between ¹⁴C variations and climate has been found (Stuiver, 1980).

Varve Chronology

Like tree rings, varved sediments preserved in some lakes offer an opportunity to calibrate the ¹⁴C time scale. A problem in varve chronology is that in some cases the varves do not continue to the present, and so the connection has to be made by correlation. Alternatively, one can work with a floating chronology. A second difficulty is ensuring that a continuous record of annual layers is observed. Stuiver (1970) has measured the ¹⁴C activity in a varved core covering the last 10,000 years from Lake of the Clouds, a small lake in northeastern Minnesota (Figure 7-6). The oldest 2000 varves are of particular interest, for their ¹⁴C activity indicates that ¹⁴C activities beyond the range of the dendrochronological calibration do not return to the standard atmospheric ¹⁴C activity but rather continue to show an excess of about 10% for the next 2000 years. This means that the sinusoidal curve fitting into the dendrochronological ¹⁴C fluctuation curve may not be extrapolated back in time, either because the oscillation of the Earth's magnetic dipole moment, held responsible for the long-term change, is not symmetrical or because

other factors (for example, the changes associated with the end of the last glaciation) come into play. Using the varve calibration of Lake of the Clouds, Stuiver is able to show that apparent changes in sedimentation rate observed in a number of lakes (Rogers, Connecticut; Jacobson, Minnesota; Victoria, Africa; Yueh Tan and Jih Tan, Formosa) were partially caused by changes in atmospheric ¹⁴C activity.

A core from the Saanich Inlet, British Columbia, contains varved sediments covering the last 9000 years overlying a layer of unvarved material (Yang and Fairhall, 1973). Carbon-14 activity variations for terrestrial material embedded in the sediments, as well as for marine organic sediment, are in agreement with the general ¹⁴C trend as derived from the dating of tree rings. The oldest terrestrial sample, at about 8600 yr B.P., still shows a ¹⁴C excess of about 6%. Before 6000 yr B.P., the ¹⁴C activity of the sediment samples is much lower than that of terrestrial material; after that date, however, the differences are not statistically significant. Yang and Fairhall assume a time lag between terrestrial- and sediment-sample activity in order to explain the results, and they conclude that atmospheric ¹⁴C levels were near their standard value around 10,000 yr B.P. This contradicts the continued 10% ¹⁴C excess in the varves of Lake of the Clouds. An alternative explanation involving a change in the water circulation and upwelling regime causing the observed changes before 6000 yr B.P. cannot be excluded in interpretations of the Saanich Inlet sediment.

RESERVOIR EFFECTS

Another basic assumption in radiocarbon dating is that specific ¹⁴C activity is the same everywhere in the atmosphere, biosphere, and hydrosphere. Since the transfer of carbon from one reservoir to another is generally accompanied by isotopic fractionation, this assumption is unfounded. Moreover, processes like the addition of dead carbonate carbon dioxide to groundwater and the mixing of old oceanic deepwater with surface waters in upwelling zones can cause aquatic organisms living in this environment to have a reduced specific ¹⁴C activity and an older apparent age. A mean value of 750 ± 50 years has been reported for the top of the sediments near

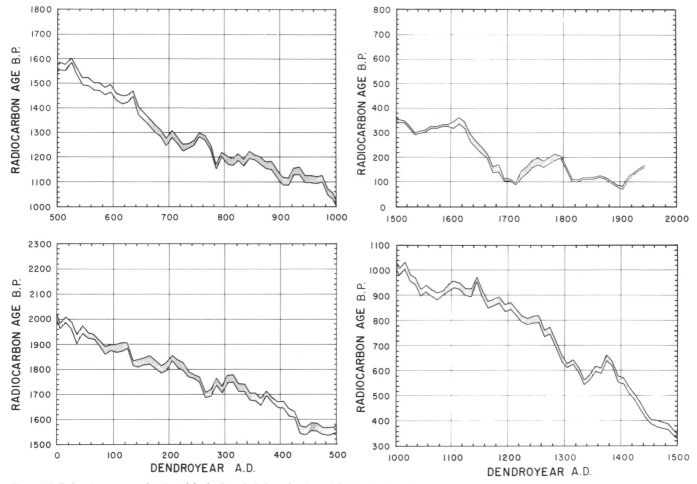

Figure 7-5. Radiocarbon age as a function of dendrochronological age for the period since the birth of Christ. (From Stuiver, 1982.)

Ellesmere Island (Canada) by Mangerud and Gulliksen (1975). Stuiver (1975) reports apparent ages for the top of the sediments in different lakes ranging from 200 years (Lake Jacobson, Minnesota) to 2500 years (Queechy Lake, New York).

The errors resulting from different initial ^{14}C concentrations in different reservoirs can be corrected, at least partially, when the abundance of the stable isotope ^{13}C is determined mass-spectrometrically. Fractionation occurs for molecules containing ^{13}C and ^{14}C roughly in the ratio one to two. Thus, ^{13}C is an indicator for what happened to the ^{14}C sample. Alternatively, it can be used to estimate the relative contributions of terrestrial organic material and carbonate material to the sediment carbon, because these have a distinctly different ^{13}C isotopic abundance ratio. For complicated systems such as a lake that receives an important contribution of bicarbonate-rich groundwater (hard-water effect) or a coastal area of strong and varying upwelling, no simple correction to ^{14}C ages can be made. This results in added uncertainty in dates for these systems of the order of 1000 years. For Holocene samples, this uncertainty is much larger than that of the ^{14}C date itself, which results mainly from counting statistics.

APPLICATIONS OF RADIOCARBON DATING

Such extensive use has been made of radiocarbon dating in many areas of Holocene studies—ranging from geology and hydrology to paleoecology and archaeology—that a review even of recent research

is inappropriate. In addition, such a review is unnecessary because applications of radiocarbon dating to various disciplines are illustrated in many of the other chapters in this volume.

Accelerator Mass Spectrometry

The detection of rare radioactive isotopes is generally based on the observation of their decay. The radiation emitted in the decay process is often highly characteristic of a specific isotope and makes it possible to identify the radioactive isotope in the presence of large numbers of other isotopes. Although this method is very sensitive, its efficiency decreases with the increasing half-life of the isotope because the fraction of the atoms decaying over a given time interval becomes small (Table 7-3). Furthermore, when those isotopes have a low natural abundance, the decomposition rate of a sample of practical size is very small, and extremely low background plus high stability are required for the counting setup. This has been a limiting factor for the use of ^{10}Be, ^{26}Al, and ^{36}Cl.

Directly detecting the isotopes of interest, instead of waiting for their (slow) decay, obviously has great advantages. The use of an accelerator and the particle-detection techniques of nuclear physics to measure directly the abundance of a natural radioactive isotope was first reported by Muller (1977), who used the 88-in. cyclotron at Berkeley to measure the T/H ratio in a water sample one mean-life

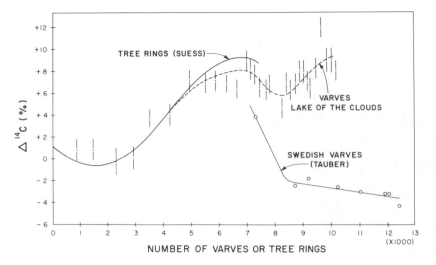

Figure 7-6. $\Delta^{14}C$, in percentage of atmospheric content, plotted against Lake of the Clouds varve years (dashed line and experimental points), varve years of the Swedish varve chronology (solid line at bottom), and number of tree rings (solid line at left). The lines are only approximate (From Stuiver, 1971; copyright 1971 Yale University Press; used with permission of author, editor, and publisher.)

old. In his report, he discusses the more general usefulness of the technique. The use of a Van de Graaff accelerator to detect directly ^{14}C in carbon derived from a piece of recent wood has been reported by Nelson and others (1977) and Bennett and others (1977). The latter group also reports a low background counting rate for a petroleum-based graphite sample. Since the late 1970s, the use of accelerators as ultrasensitive mass spectrometers has developed rapidly, both in the number of investigators working on it and in the number of isotopes and applications studied. In 1981, almost 20 accelerator groups reported their research at a symposium on accelerator mass spectrometry (AMS) at the Argonne Laboratory in Illinois (1981). Research reports include diverse topics such as the search for quarks and super-heavy elements, the measurement of cross sections for low yield reactions such as $^{26}Mg(p,n)^{26}Al$, the determination of half-lives of cosmogenic isotopes such as ^{32}Si, the application of cosmogenic isotopes to geologic and hydrologic studies and the trace-element analysis of stable isotopes.

A schematic diagram of a Van de Graaff accelerator mass spectrometer is shown in Figure 7-7. It contains the basic elements of a normal mass spectrometer, that is, an ion source CS, an accelerating stage with focusing elements for producing a well-defined beam of energetic ions (between CS and M_a), an analyzing magnet M_a for

selecting the desired ionic species, and a detector (either a Faraday cup DEC to measure the ion beam current or a solid-state detector ID to count and to identify individual ions). The advantages of the accelerator high-energy mass-spectrometry system over its low-energy counterpart originate from: (1) the initial use of negative ions that are later stripped of their electrons at the terminal and (2) the identification of individual particles that is made possible by high particle energy (typically more than 10 MeV). The use of negative ions, produced by bombarding a solid sample with a beam of positive cesium ions, reduces background. With ^{14}C, for instance, it virtually eliminates ^{14}N, which in low-energy mass spectrometry is the most serious and inseparable contaminant of the ^{14}C beam because $^{14}N^-$ is unstable. Further elimination of ions of different mass and charge takes place at the inflection magnet M_I, the terminal stripper S_1, and the analyzing magnet M_a. M_I selects ions of the desired mass and charge, injects them into the accelerator, and eliminates most other ions. At the positive terminal in the center, the now high-energy negative ions pass through a carbon foil that strips them of several electrons so that they emerge with a spectrum of positive charge states. After acceleration from the positive terminal to ground potential, the most abundant charge state of the isotope to be measured is selected by the analyzing magnet M_a. Molecular ions such as those for carbon $^{12}CH_2^-$ and $^{13}CH^-$, which masquerade as ^{14}C up to the stripper, are broken up by the stripper foil. The resulting positive ^{12}C and ^{13}C ions are generally eliminated by the analyzing magnet (reduction factor of the order of 10^5). After deflection by the switching magnet M_s and additional focusing, the ion beam passes through a velocity selector VS, where particles of the selected mass, charge state, and energy pass along a straight line; others are deflected and so eliminated from the beam (reduction factor about 100). The cumulative elimination of unwanted ions from the beam reduces the number of particles arriving at the detector system ID enough so that it can detect and identify the individual ions. The detector system consists of a thin, solid-state detector or a gas counter, in which particles lose part of their energy, followed by a thick, solid-state detector, in which they are stopped and, therefore, in which their residual energy is measured. The amounts of energy lost in the thin and thick detectors are used to identify particles and thus to provide final discrimination between the isotope to be measured and the unwanted background.

The beam intensity (in particles per minute) of the rare radioactive

Table 7-3. Efficiency of Detection
by Radioactive Decay (Fraction
Decaying in 24 Hours)

Isotope	Half-life ($T_{1/2}$ in years)	Fraction
T	12.26	1.55×10^{-4}
^{210}Pb	22.3	8.52×10^{-5}
^{137}Cs	30	6.33×10^{-5}
^{32}Si	300	6.33×10^{-6}
^{14}C	5730	3.31×10^{-7}
^{41}Ca	1.3×10^5	1.27×10^{-8}
^{36}Cl	3.08×10^5	6.17×10^{-9}
^{26}Al	7.3×10^5	2.60×10^{-9}
^{10}Be	1.5×10^6	1.27×10^{-9}
^{129}I	1.6×10^7	1.19×10^{-10}

isotope is compared with that of a stable isotope beam from the same sample (in microamperes) to obtain an isotopic abundance ratio. As in conventional isotope mass spectrometry, absolute measurements are difficult. The isotopic abundance ratio of the sample is, therefore, often compared with that obtained for a standard in order to determine the decay that has taken place and thus the age of the sample. In place of a Van de Graaff accelerator, a cyclotron may be used to produce the high-energy ion beam necessary for particle identification. In a cyclotron, charge changes cannot be used to reduce background, but the high mass resolution of the cyclotron eliminates all particles that have a mass noticeably different from the one selected.

Some important advantages of accelerator mass spectrometry are these. (1) A small amount of sample material (typically about 1 mg) is required for an accelerator measurement. (2) Relatively high sample-counting rates and low backgrounds can be obtained. As a result, counting statistics do not seriously limit the precision of the measurements, and short counting times (hours) are possible. (3) The high selectivity of the accelerator system combined with the particle identification eliminates the need for a sample preparation that yields a radiochemically pure sample. This greatly facilitates preparation of ^{10}Be and ^{26}Al samples, for example.

Some disadvantages are these. (1) The cost, size, and complexity of the accelerator system limit the use of accelerator mass spectrometry to a relatively small number of highly specialized facilities. (2) Fluctuations in the ion source output, along with magnetic and electrical field instability affecting the transmission efficiency of ions from the ion source to the detector, influence the measured isotopic abundance ratio. Therefore, the stability of the system may limit the measurement's precision and range. Improved stability may be obtained in specially designed systems dedicated exclusively to accelerator mass spectrometry.

APPLICATIONS

So far, most of the publications on accelerator mass spectrometry describe feasibility studies. Substantial results have been obtained only for ^{10}Be and ^{36}Cl.

Tritium (^{3}H or T)

The first direct measurement of a radioisotope with the use of an

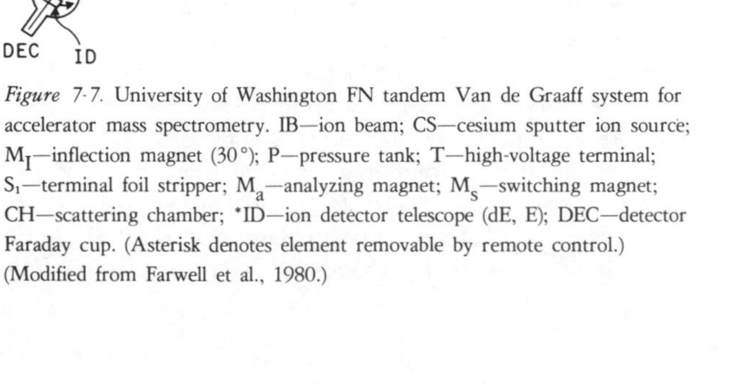

Figure 7-7. University of Washington FN tandem Van de Graaff system for accelerator mass spectrometry. IB—ion beam; CS—cesium sputter ion source; M_I—inflection magnet (30°); P—pressure tank; T—high-voltage terminal; S_1—terminal foil stripper; M_a—analyzing magnet; M_s—switching magnet; CH—scattering chamber; *ID—ion detector telescope (dE, E); DEC—detector Faraday cup. (Asterisk denotes element removable by remote control.) (Modified from Farwell et al., 1980.)

accelerator was the measurement of the T/H ratio in a water sample one mean-life old with the Berkeley 88-in. cyclotron (Muller, 1977). Because of its short half-life, conventional measuring techniques such as proportional counting and liquid-scintillation counting and especially the new technique of measuring T via its decay product ^{3}He are preferable.

Silicon-32 (^{32}Si)

Two groups recently measured the half-life of ^{32}Si with a Van de Graaff accelerator. The results, 108 ± 18 years (Elmore et al., 1980) and 101 ± 18 years (Kutschera, Henning, Paul, Smither, et al., 1980), agree well with each other but are considerably lower than the value of about 300 years (derived from sediment and ice studies) that is now commonly used. Using a different half-life would have important implications for the geochemical studies that have used ^{32}Si dating. A number of assumptions had to be made in order to arrive at the calculated half-life. Considering the discrepancy between the results from accelerator mass spectrometry and the half-lives obtained from ice and sediments by using beta decay counting of ^{32}P, a critical reevaluation of all half-life measurements appears to be needed. The prospect seems good that the new technique will lead to a more accurate half-life and that this combined with the small sample size required will lead to the increased use of ^{32}Si in geochemical studies.

Carbon-14 (^{14}C)

So far, radiocarbon has received most attention as an application of accelerator mass spectrometry. The first two papers describing the use of a Van de Graaff accelerator for high-sensitivity mass spectrometry report the detection of ^{14}C in sources prepared from recent wood (Bennett et al., 1977; Nelson et al., 1977) and a very low count rate of ^{14}C for a sample of petroleum-based graphite (Bennett et al., 1977). Since the late 1970s, several groups have developed the capacity to measure ^{14}C abundance ratios in natural samples. (See Argonne, 1981; Radiocarbon, 1980.) Radiocarbon dates have been reported by the Rochester-Toronto group (Bennett et al., 1978; Gove et al., 1980; Litherland, 1978) and by the Berkeley group (Muller, 1978; Muller et al., 1978). The results look promising. The reproducibility indicates a precision of 1% to 2% for the Rochester accelerator. A comparison of the accelerator-derived ages with those obtained by proportional counting shows in general a satisfactory agreement in relation to the quoted age uncertainty. This proves that accelerator mass spectrometry will be able to produce meaningful ages from samples containing 1 to 10 mg of carbon when all instrumental parameters have been carefully controlled and stabilized. For cases in which radiocarbon dating until now could not be used because not enough carbon was available or in which extraction of a sample for proportional counting would result in the destruction of a valuable specimen, it will be possible to obtain dates with a precision of a few percentage points from milligram quantities. It will probably be a few years, however, before the small accelerators dedicated to geochronology (Tucson, Arizona; Toronto, Canada; Oxford, England; Saclay, France; and Nagoya, Japan) will be able to accept samples on a routine basis or before other groups using existing accelerators will be able to provide dating services.

Chlorine-36 (^{36}Cl)

Chlorine-36 has been detected by three groups: Rochester/Toronto

(Elmore et al., 1979; Naylor et al., 1978; Nishiizumi et al., 1979), Berkeley (Mast, 1978), and Argonne (Kutschera, Henning, Paul, Stephenson, et al., 1980). Measurements of ^{36}Cl in water from different sources have yielded concentrations between 1950 and 120×10^{-15} $^{36}Cl/Cl$ (Elmore et al., 1979). The highest value (1950 ± 150) was found in groundwater from a well near the Rillito River in Arizona. The sample was expected to show the spike produced in the nuclear-weapons tests. The low value (120 ± 20) was from subsurface Lake Ontario water. Blank runs on zone-refined silver chloride established a lower detection limit of $(3 \pm 1) \times 10^{-15}$. A study of aquifer characteristics that uses ^{36}Cl has been reported by Bentley and Davis (1981).

Measurements of the ^{36}Cl content of a number of Antarctic meteorites yield activities from 4.6 ± 1.1 to 24.2 ± 2.5 dpm per kilogram (Fe + Co + Ni + 6 Ca) (Nishiizumi et al., 1979). The results indicate a two-stage irradiation history for two of the meteorites, a recent fall for one, and a long terrestrial age for the fourth. The ability of an accelerator mass spectrometer to obtain a ^{36}Cl activity from a sample of about 10 mg of separated magnetic phase from each meteorite made it possible to carry out the measurement. The amount of tens of grams needed for a decay measurement would have been prohibitively large. A sample of 5.8 kg of ice taken from a place close to the meteorite's recovery site in the Yamato Mountains gives $(58 \pm 6) \times 10^{-15}$ for the ratio $^{36}Cl/Cl$. In this measurement series, the silver chloride blank gives a ratio of $^{36}Cl/Cl$ less than 0.2×10^{-15}. These first results clearly demonstrate the possibilities of ^{36}Cl studies using the technique of accelerator mass spectrometry.

Calcium-41 (^{41}Ca)

Calcium-41's half-life (1.3×10^5 years) and the general abundance of calcium make it an isotope potentially useful in dating. Its small natural concentration (less than 10^{-14}) and its pure electron-capture decay preclude measurements by decay counting. Although no measurements have yet been attempted, Raisbeck and Yiou (1979) do discuss the feasibility of measuring it with accelerator technology, and they mention its potential use for dating carbonate sediments and bones.

Aluminum-26 (^{26}Al)

Exploratory measurements by three groups (Kilius, Beukens, Chang, Lee, Litherland, Elmore, Ferraro, and Gove, 1979; Kutschera Henning, Paul, Stephenson, et al., 1980; Raisbeck, Yiou, and Stephan, 1979) show that ^{26}Al can be counted directly. Acceleration to 200 MeV and complete stripping have been used to separate the ^{26}Al from its isobaric contaminant ^{26}Mg in the cyclotron (Raisbeck, Yiou, and Stephan, 1979). The use of negative ions for injection into the Van de Graaff accelerator eliminates most of the unwanted ^{26}Mg because the magnesium negative ion is highly unstable. The cyclotron measurements indicate the technique's capacity to measure 10^9 atoms of ^{26}Al, which correspond to about 40 g of fresh ocean sediment and $^{26}Al/Al$ ratios less than 10^{-12}. Kilius, Beukens, Chang, Lee, Litherland, Elmore, Ferraro, and Gove (1979) report a blank measurement indicating a ratio $^{26}Al/Al$ less than 10^{-14} and, on the basis of efficiency estimates, predict a considerably smaller minimum sample size. The measurement of ^{26}Al and ^{10}Be concentrations in sediments provides a possible dating tool independent of fluctuations in cosmic-ray intensity and sedimentation rate. Both isotopes have a similar production mechanism and chemistry and so probably will be affected in a similar way by fluctuations in the cosmic-ray intensity

and/or sedimentation rate. Therefore, their initial abundance ratio should be constant. Its change resulting from the different half-lives can then be used for dating.

Beryllium-10 (^{10}Be)

The first beryllium measurements were carried out with the Grenoble cyclotron by Raisbeck and his co-workers, whose results were published in 1978. Since then, the Grenoble group has made exploratory measurements of environmental ^{10}Be concentrations (Table 7-4). Measurements of monthly precipitation samples collected from four localities in France in 1966 and combined for ^{10}Be sampling show a distinct maximum of $4.5 \pm 0.5 \times 10^4$ atoms ^{10}Be per gram in precipitation for March and minimum values of $(1.6 \pm 0.3) \times 10^4$ atoms ^{10}Be per gram for late summer and fall (Raisbeck, Yiou, Fruneau, Loiseaux, Lieuvin, and Ravel, 1979). The average value for 6 months in 1966 is 2.33×10^4, which is in good agreement with a value of 2.49×10^4 atoms ^{10}Be per gram for a sample of the precipitation from April 1978 to April 1979 in Grenoble, France. These numbers lead to an estimated global deposition rate for ^{10}Be of 0.042 atoms per square centimeter per second, substantially larger than estimates of 0.015 to 0.018 based on earlier work.

Surface waters of the Pacific show concentrations of 530 ± 180 and 950 ± 390 atoms ^{10}Be per cubic centimeter (Raisbeck, Yiou, Fruneau, Loiseaux, and Lieuvin, 1979), and an Indian Ocean sample yields 1440 ± 240 (Raisbeck et al., 1980). At the Indian Ocean location, deep-ocean waters are higher in ^{10}Be by a factor of two (2500 ± 380 at 800 m and 2810 ± 420 at 3140 m depth), indicating a depletion of the ocean surface layer in ^{10}Be due to particle scavenging. The results indicate a mixed-layer residence time of about 10 years and a deep-ocean residence time of less than about 760 years. The residence times determine the time resolution that can be obtained for ^{10}Be concentration variations in the sediment.

In a North Pacific marine sediment core (V20-108), an average ^{10}Be concentration of $(3.68 \pm 0.35) \times 10^9$ atoms per gram of sediment has been found. The samples were spaced around a magnetic reversal at a depth of 1004 to 1006 cm in the core in order to determine whether a reversal of the Earth's magnetic field is accompanied by a decrease in the magnetic field's strength and, as a consequence, by an increase in the concentration of the cosmogenic isotopes. Although the concentration in the reversal sample is somewhat higher than in samples above and below it, the difference is less than two standard deviations and therefore is not statistically significant (Raisbeck, Yiou, Fruneau, Louiseaux, Lieuvin, Ravel, and Hays, 1979).

Beryllium-10 concentrations as a function of depth in a manganese nodule slab from the Rio Grande Ridge in the South Atlantic Ocean have been measured with a Van de Graaff accelerator (Lanford et al., 1980; Turekian et al., 1979) (Table 7-4). The measured concentrations imply a growth rate of about 4.5 mm per million years. Recent, more detailed studies using step sizes of 1 mm or less indicate variations in the growth rate of the nodules (Thomas et al., 1981). Similar results have been obtained by the Simon Fraser/McMaster group (E. Nelson, personal communication, 1981).

Samples from the 905-m-long Dome C ice core (latitude 74°39′S, 124°10′E, elevation 3240 m) show an increased ^{10}Be concentration in the core during the Maunder sunspot minimum (A.D. 1654 to 1714). Beryllium-10 concentrations change from about 10×10^4 atoms ^{10}Be per gram of ice below 500 m (estimated age about 15,000 years) to about 4×10^4 above this depth (Raisbeck et al., 1981). In this same depth interval, a marked change in the ^{18}O composition of the

Table 7-4. Environmental ^{10}Be Concentrations

Location	^{10}Be/^{9}Be[a]	Concentration ^{10}Be	Author(s)/Year
Precipitation, France, 1966		1.6 ± 0.3 to $4.50 \pm 0.55 \times 10^4$ at ^{10}Be/g water	Raisbeck, Yiou, Fruneau, Loiseaux, Lieuvin, and Ravel, 1979
Precipitation average, 1966		2.33×10^4 at ^{10}Be/g water	
Precipitation, Orsay, France, April 1978-April 1979		2.49×10^4 at ^{10}Be/g water	
Surface, Pacific Ocean	$7.0 \pm 2.1 \times 10^{-12}$	$5.3 \pm 1.8 \times 10^2$ at ^{10}Be/cm^3 water	Raisbeck, Yiou, Fruneau, Loiseaux, and Lieuvin, 1979
	$3.6 \pm 1.4 \times 10^{-12}$	$9.5 \pm 3.9 \times 10^2$ at ^{10}Be/cm^3 water	
Surface, Indian Ocean	$4.9 \pm 0.8 \times 10^{-13}$	$14.4 \pm 2.4 \times 10^2$ at ^{10}Be/g water	Raisbeck et al., 1980
Deep Indian Ocean 800 m	$8.5 \pm 1.3 \times 10^{-13}$	$25.0 \pm 3.8 \times 10^2$ at ^{10}Be/g water	
3140 m	$9.6 \pm 1.4 \times 10^{-13}$	$28.1 \pm 4.2 \times 10^2$ at ^{10}Be/g water	
Marine sediment, North Pacific V20-108			Raisbeck, Yiou, Fruneau, Loiseaux, Lieuvin, Ravel, and Hays, 1979
977-978 cm		$3.66 \pm 0.56 \times 10^9$ at ^{10}Be/g sediment	
990-991 cm		3.38 ± 0.51	
995-996 cm		3.66 ± 1.28	
1000-1001 cm		3.43 ± 0.60	
1004-1006 cm[b]		4.35 ± 0.88	
1029-1031 cm		3.62 ± 0.65	
average		3.68 ± 0.35	
Manganese nodule, South Atlantic			Turekian et al., 1979
0-5 mm		2.67×10^9 at ^{10}Be/g nodule	
5-11.6 mm		11.5×10^9 at ^{10}Be/g nodule	
19.4-29.4 mm		2.5×10^9 at ^{10}Be/g nodule	
Antarctic ice	$7.6 \pm 2.5 \times 10^{-13}$	$2.00 \pm 0.67 \times 10^4$ at ^{10}Be/g ice	Raisbeck, Yiou, Fruneau, Lieuvin, and Loiseaux, 1978b
	$3.6 \pm 1.4 \times 10^{-13}$	$9.5 \pm 3.9 \times 10^4$ at ^{10}Be/g ice	

[a]The ratio ^{10}Be/^{9}Be measured is determined by the amount of ^{10}Be in the sample and the amount of ^{9}Be carrier added.
[b]Position of geomagnetic reversal.

ice from δ^{18}O equals -50 ‰ above 400 m to about -55 ‰ below 500 m occurs (Lorius et al., 1979), indicating that this interval records the glacial/interglacial transition. Obviously, the ^{10}Be study of ice cores will reveal interesting data on the production rate of cosmogenic isotopes and the factors controlling their distribution and deposition.

Preliminary measurements of ^{10}Be/^{9}Be ratios in the 10^{-10} to 10^{-13} range with a Van de Graaff accelerator have been reported by Kilius and others (1980), Farwell and others (1980), Middleton and others (1981), and Kruse and others (1981). The advent of accelerator mass spectrometry has greatly increased the use of ^{10}Be in isotopic geochemical studies and will continue to do so because, in comparison to radioactive decay counting, its sample size is greatly reduced, the purification required is less stringent, and its measuring time is short.

Iodine-129 (^{129}I)

The Rochester, New York, Van de Graaff accelerator has been used to detect ^{129}I in two meteorite samples (Kilius, Beukens, Chang, Lee, Litherland, Elmore, Gove, and Finkel, 1979). The sensitivity for the ratio ^{129}I/^{127}I obtained is less than 10^{-12}. The limiting factor is resolution at the ion source insufficient to resolve ^{129}I and the stable ^{127}I. Development of a dating technique based on ^{129}I seems feasible.

TRACE-ELEMENT MASS SPECTROMETRY

The use of accelerator mass spectrometry, of course, is not limited to radioactive isotopes. It can also be used for high-sensitivity work at the parts-per-billion (ppb) level on stable isotopes. An example is the measurement of ^{195}Pt at the 59-ppb level in a platinum sample (the reference is in Litherland, 1980).

Conclusions

The use of natural and bomb-produced isotopes in environmental studies has increased over the past two decades and has now become well established. It has contributed significantly to a better understanding of natural processes, both active and past. Improved measurement techniques have made it feasible to measure accurately large numbers of ^{14}C samples, and, consequently, a detailed radiocarbon time control now exists for many regions and processes. Other cosmogenic and decay-produced isotopes are increasingly being used, and they provide complementary information. Bomb-produced isotopes such as T, ^{137}Cs, and ^{14}C have been and still are useful as natural tracers. The development of accelerator mass spectrometry promises to open up a whole range of radioactive and stable isotopes for use in environmental studies and to contribute to the growing field of isotopic geochemistry.

References

Andersen, L. J., and Sevel, T. (1974). Six years environmental tritium profiles in the unsaturated and saturated zones, Grønhøj, Denmark. In "Proceedings of the Symposium on Isotope Techniques in Groundwater Hydrology, Vienna, 1974," vol. 1, pp. 3-20, International Atomic Energy Agency, Vienna.

Appleby, P. G., and Oldfield, F. (1978). The calculation of lead-210 dates assuming a constant rate of supply of unsupported ^{210}Pb to the sediment. Catena 5, 1-8.

Appleby, P. G., Oldfield, F., Thompson, R., Huttunen, P., and Tolonen, K. (1979). ^{210}Pb dating of annually laminated lake sediments from Finland. Nature 280, 53-55.

Argonne (1981). "Proceedings of the Symposium on Accelerator Mass Spectrometry." Argonne National Laboratory, Argonne, Ill.

Armentano, T. V., and Woodwell, G. M. (1975). Sedimentation rates in a Long Island marsh determined by ^{210}Pb dating. *Limnology and Oceanography* 20, 452-56.

Atakan, Y., Roether, W., Münnich, K. -O., and Matthess, G. (1974). The Sandhausen shallow-groundwater tritium experiment. *In* "Proceedings of the Symposium on Isotope Techniques in Groundwater Hydrology, Vienna, 1974," vol. 1, pp. 21-43. International Atomic Energy Agency, Vienna.

Barnes, R. S., Birch, P. B., Spyridakis, D. E., and Schell, W. R. (1979). Changes in the sedimentation histories of lakes using lead-210 as a tracer of sinking particulate matter. *In* "Proceedings of the Symposium on Isotope Hydrology, Neuherberg, 1978," pp. 875-98. International Atomic Energy Agency, Vienna.

Begemann, F., and Libby, W. F. (1957). Continental water balances, groundwater inventory and storage times, surface ocean mixing rates and world wide circulation patterns from cosmic ray and bomb tritium. *Geochimica et Cosmochimica Acta* 12, 277-96.

Bennett, C. L.,Beukens, R. P., Clover, M. R., Elmore, D., Gove, H. E., Kilius, L., Litherland, A. E., and Purser, K. H. (1978). Radiocarbon dating with electrostatic accelerators: Dating of milligram samples. *Science* 201, 345-47.

Bennett, C. L., Beukens, R. P., Clover, M. R., Gove, H. E., Liebert, R. B., Litherland, A. E., Purser, K. H., and Sondheim, W. E. (1977). Radiocarbon dating using electrostatic accelerators: Negative ions provide the key. *Science* 198, 508-10.

Benninger, L. K. (1976). The uranium-series radionuclides as tracers of geochemical processes in Long Island Sound. Ph.D. dissertation, Yale University, New Haven, Conn.

Benninger, L. K. (1978). ^{210}Pb balance in Long Island Sound. *Geochimica et Cosmochimica Acta* 42, 1165-74.

Benninger, L. K., Aller, R. C., Cochran, J. K., and Turekian, K. K. (1979). Effects of biological sediment mixing on the ^{210}Pb chronology and trace metal distribution in a Long Island Sound sediment core. *Earth and Planetary Science Letters* 43, 241-59.

Bentley, H. W., and Davis, S. N. (1981). Application of AMS to hydrology. *In* "Proceedings of the Symposium on Accelerator Mass Spectrometry," pp. 193-227. Argonne National Laboratory, Argonne, Ill.

Broecker, W. S. (1965). Isotope geochemistry and the Pleistocene climatic record. *In* "The Quaternary of the United States" (H. E. Wright, Jr., and D. G. Frey, eds.), pp. 737-53. Princeton University Press, Princeton, N.J.

Brown, R. M. (1961). Hydrology of tritium in the Ottawa Valley. *Geochimica et Cosmochimica Acta* 21, 199-216.

Brown, R. M., and Grummitt, W. E. (1956). The determination of tritium in natural waters. *Canadian Journal of Chemistry* 34, 220-26.

Brugam, R. B. (1978). Pollen indicators of land-use change in southern Connecticut. *Quaternary Research* 9, 349-62.

Bruland, K. W., Bertine, K., Koide, M., and Goldberg, E. D. (1974). History of metal pollution in southern California coastal zone. *Environmental Science and Technology* 8, 425-32.

Bruland, K. W., Franks, R. P., and Landing, W. M. (1981). Southern California inner basin sediment trap calibration. *Earth and Planetary Science Letters* 53, 400-408.

Bruland, K. W., Koide, M., Bowser, C., Maher, L. J., and Goldberg, E. D. (1975). ^{210}Lead and pollen geochronologies on Lake Superior sediments. *Quaternary Research* 5, 89-98.

Bruns, M., Münnich, K. O., and Becker, B. (1980). Natural radiocarbon variations from AD 200 to 800. *In* "Proceedings of the Tenth International Radiocarbon Conference, Bern/Heidelberg, 1979 (M. Stuiver and R. Kra, eds.). *Radiocarbon* 22, 273-77.

Bucha, V. (1970). Influence of the Earth's magnetic field on radiocarbon dating. *In* "Proceedings of the Twelfth Nobel Symposium on Radiocarbon Variations and Absolute Chronology, Uppsala, 1969" (I. U. Olsson, ed.), pp. 501-11. Almqvist and Wiksell, Förlag AB, Stockholm.

Clarke, W. B., Jenkins, W. J., and Top, Z. (1976). Determination of tritium by mass spectrometric measurement of ^{3}He. *International Journal of Applied Radiation and Isotopes* 27, 515-22.

Clausen, H. B. (1973). Dating of polar ice by ^{32}Si. *Journal of Glaciology* 12, 411-16.

Crozaz, G., and Langway, C. C., Jr. (1966). Dating Greenland firn-ice cores with Pb-210. *Earth and Planetary Science Letters* 1, 194-96.

Crozaz, G., Picciotto, E., and Breuck, W. de (1964). Antarctic snow chronology with ^{210}Pb. *Journal of Geophysical Research* 69, 2597-2604.

Damon, P. E., Long, A., and Grey, D. C. (1970). Arizona radiocarbon dates for dendrochronologically dated samples. *In* "Proceedings of the Twelfth Nobel Symposium on Radiocarbon Variations and Absolute Chronology, Uppsala, 1969" (I. U. Olsson, ed.), pp. 615-18. Almqvist and Wiksell, Förlag AB, Stockholm.

Damon, P. E., Long, A., and Wallick, E. J. (1973). Dendrochronologic calibration of the carbon-14 time scale. *In* "Proceedings of the Eighth International Conference on Radiocarbon Dating, Lower Hutt, 1972," pp. A28-43. Royal Society of New Zealand, Wellington.

Dansgaard, W., Clausen, H. B., and Aarkrog, A. (1966). The ^{32}Si fallout in Scandinavia. *Tellus* 18, 187-91.

de Jong, A. F. M., and Mook, W. G. (1980). Medium-term atmospheric ^{14}C variations. *In* Proceedings of the Tenth International Radiocarbon Conference, Bern/Heidelberg, 1979 (M. Stuiver and R. Kra, eds.). *Radiocarbon* 22, 267-72.

DeMaster, D. J. (1980). The half-life of ^{32}Si determined from a varved Gulf of California sediment core. *Earth and Planetary Science Letters* 48, 209-17.

de Vries, Hl. (1958). Variations in concentration of radiocarbon with time and location on Earth. *Koninklijke Nederlandse Akademie van Wetenschappen Proceedings* B61, 94-102.

Durham, R. W., and Joshi, S. R. (1980). Recent sedimentation rates, ^{210}Pb fluxes and particle settling velocities in Lake Huron, Laurentian Great Lakes. *Chemical Geology* 31, 53-66.

Eichinger, L., Rauert, W., Salvamoser, J., and Wolf, M. (1980). Large-volume liquid scintillation counting of carbon-14. *In* Proceedings of the Tenth International Radiocarbon Conference, Bern/Heidelberg, 1979 (M. Stuiver and R. Kra, eds.) *Radiocarbon* 22, 417-27.

Elmore, D., Anantaraman, N., Fulbright, H. W., Gove, H. E., Hans, H. S., Nishiizumi, K., Murrell, M. T., and Honda, M. (1980). Half-life of ^{32}Si from tandem-accelerator mass spectrometry. *Physical Review Letters* 45, 589-92.

Elmore, D., Fulton, B. R., Clover, M. R., Marsden, J. R., Gove, H. E., Naylor, H., Purser, K. H., Kilius, L. R., Beukens, R. P., and Litherland, A. E. (1979). Analysis of ^{36}Cl in environmental water samples using an electrostatic accelerator. *Nature* 277, 22-25.

Faltings, V. Von., and Harteck, P. (1950). Der Tritiumgehalt der Atmosphäre. *Zeitschrift für Naturforschung* 5a, 438-39.

Farwell, G. W., Schaad, T. P., Schmidt, F. H., Tsang, M.-Y. B. Grootes, P. M., and Stuiver, M. (1980). Radiometric dating with the University of Washington tandem Van de Graaff accelerator. *In* Proceedings of the Tenth International Radiocarbon Conference, Bern/Heidelberg, 1979 (M. Stuiver and R. Kra, eds.). *Radiocarbon* 22, 838-49.

Farmer, J. G. (1978). The determination of sedimentation rates in Lake Ontario using the ^{210}Pb dating method. *Canadian Journal of Earth Sciences* 15, 431-37.

Ferguson, C. W. (1970). Dendrochronology of bristlecone pine, *Pinus Aristata*: Establishment of a 7484-year chronology in the White Mountains of eastern-central California, U.S.A. *In* "Proceedings of the Twelfth Nobel Symposium on Radiocarbon Variations and Absolute Chronology, Uppsala, 1969" (I. U. Olsson, ed.), pp. 237-59. Almqvist and Wiksell, Förlag AB, Stockholm.

Ferguson, C. W. (1973). Dendrochronology of bristlecone pine prior to 4000 B.C. *In* "Proceedings of the Eighth International Conference on Radiocarbon Dating, Lower Hutt, 1972," pp. A1-10. Royal Society of New Zealand, Wellington.

Friedman, I. (1983). Paleoclimatic evidence from stable isotopes. *In* "Late Quaternary Environments of the United States," vol. 1, "The Late Pleistocene" (S. C. Porter, ed.), pp. 385-89. University of Minnesota Press, Minneapolis.

Goldberg, E. D. (1963). Geochronology with ^{210}Pb. *In* "Proceedings of the Symposium on Radioactive Dating, Athens, 1962," pp. 121-31. International Atomic Energy Agency, Vienna.

Gove, H. E., Elmore, D., Ferraro, R. D., Beukens, R. P., Chang, K. H., Kilius, L. R., Lee, H. W., and Litherland, A. E. (1980). Radiocarbon dating with tandem electrostatic accelerators. *In* Proceedings of the Tenth International Radiocarbon Conference, Bern/Heidelberg, 1979 (M. Stuiver and R. Kra, eds.), *Radiocarbon* 22, 785-93.

Grootes, P. M. (1977). Thermal diffusion isotopic enrichment and radiocarbon dating beyond 50,000 years BP. Ph.D. dissertation, University of Groningen, Groningen, The Netherlands.

Grootes, P. M. (1978). Carbon-14 time scale extended: Comparison of chronologies. *Science* 200, 11-15.

Grootes, P. M., Mook, W. G., Vogel, J. C., de Vries, A. E., Haring, A., and Kistemaker, J. (1975). Enrichment of radiocarbon for dating samples up to 75,000 years. *Zeitschrift für Naturforschung* 30a, 1-14.

Grootes, P. M., and Stuiver, M. (1979). The Quaternary Isotope Laboratory thermal diffusion enrichment system: Description and performance. *Radiocarbon* 21, 139-64,

Grosse, A. V., Johnston, W. M., Wolfgang, R. L., and Libby, W. F. (1951). Tritium in nature. *Science* 113, 1-2.

Guizerix, J., Margrita, R., Launay, M., and Ruby, P. (1967). Tritium et hydrogeologie: Etudes et mesures effectuées au Centre d'études nucléaires de Grenobles. *In* "Proceedings of the Symposium on Isotopes in Hydrology, Vienna, 1966," pp. 433-50. International Atomic Energy Agency, Vienna.

Haring, A., de Vries, A. E., and de Vries, Hl. (1958). Radiocarbon dating up to 70,000 years by isotopic enrichment. *Science* 128, 472-73.

Imboden, D. M., Weiss, R. F., Craig, H., Michel, R. L., and Goldman, C. R. (1977). Lake Tahoe geochemical study: I. Lake chemistry and tritium mixing study. *Limnology and Oceanography* 22, 1039-51.

International Atomic Energy Agency (1967). "Proceedings of the Symposium on Isotopes in Hydrology, Vienna, 1966." International Atomic Energy Agency, Vienna.

International Atomic Energy Agency (1970). "Proceedings of the Symposium on Isotope Hydrology, Vienna, 1970." International Atomic Energy Agency, Vienna.

International Atomic Energy Agency (1974). "Proceedings of the Symposium on Isotope Techniques in Groundwater Hydrology, Vienna, 1974." Vols. 1 and 2. International Atomic Energy Agency, Vienna.

International Atomic Energy Agency (1979). "Proceedings of the Symposium on Isotope Hydrology, Neuherberg, 1978." International Atomic Energy Agency, Vienna.

International Atomic Energy Agency (1981). Low-level tritium measurement. *In* "Proceedings of I.A.E.A. Consultants Group Meeting, 1979." International Atomic Energy Agency Technical Document 246.

Jantsch, K. (1967). Kernreaktionen mit Tritonen beim ^{30}Si: Bestimmung der Halbwertzeit von ^{32}Si. *Kernenergie* 10, 89.

Kaufman, S., and Libby, W. F. (1954). The natural distribution of tritium. *Physical Review* 93, 1337-44.

Kharkar, D. P., Lal, D., and Somayajulu, B. L. K. (1963). Investigations in marine environments using radioisotopes produced by cosmic rays. *In* "Proceedings of the Symposium on Radioactive Dating, Athens, 1962," pp. 175-87. International Atomic Energy Agency, Vienna.

Kharkar, D. P., Turekian, K. K., and Scott, M. R. (1969). Comparison of sedimentation rates obtained by ^{32}Si and uranium decay series determinations in some siliceous Antarctic cores. *Earth and Planetary Science Letters* 6, 61-68.

Kilius, L. R., Beukens, R. P., Chang, K. H., Lee, H. W., Litherland, A. E., Elmore, D., Ferraro, R., and Gove, H. E. (1979). Separation of ^{26}Al and ^{26}Mg isobars by negative ion mass spectrometry. *Nature* 282, 488-89.

Kilius, L. R., Beukens, R. P., Chang, K. H., Lee, H. W., Litherland, A. E., Elmore, D., Ferraro, R., Gove, H. E., and Purser, K. H. (1980). Measurement of ^{10}Be/^{9}Be ratios using an electrostatic tamdem accelerator. *Nuclear Instruments and Methods* 171, 355-60.

Kilius, L. R., Beukens, R. P., Chang, K. H., Lee, H. W., Litherland, A. E., Elmore, D., Gove, H. E., and Finkel, R. C. (1979). The detection of ^{129}I by atom counting. *Bulletin of the American Physical Society* 24, 1186.

Kipphut, G. W. (1978). An investigation of sedimentary processes in lakes. Ph.D. dissertation, Columbia University, New York.

Kirkov, P., Kačurkov, D., Tolev, M., and Anovski, T. (1974). Determination of the origin of water in springs from the simultaneous application of natural and artificial isotopes: Some aspects of the origin of water in the Rashche Spring. *In* "Proceedings of the Symposium on Isotope Techniques in Groundwater Hydrology, Vienna, 1974," vol. 1, pp. 465-79. International Atomic Energy Agency, Vienna.

Koide, M., Bruland, K. W., and Goldberg, E. D. (1973). Th-228/Th-232 and lead-210 geochronologies in marine and lake sediments. *Geochimica et Cosmochimica Acta* 37, 1171-87.

Koide, M., Soutar, A., and Goldberg, E. D. (1972). Marine geochronology with lead-210. *Earth and Planetary Science Letters* 14, 442-46.

Krishnaswami, S., Benninger, L. K., Aller, R. C., and Von Damm, K. L. (1980). Atmospherically-derived radionuclides as tracers of sediment mixing and accumulation in near shore marine and lake sediments: Evidence from ^{7}Be, ^{210}Pb, and ^{239,240}Pu. *Earth and Planetary Science Letters* 47, 307-18.

Krishnaswami, S., Lal, D., Martin, J., and Meybeck, M. (1971). Geochronology of lake sediments. *Earth and Planetary Science Letters* 11, 407-14.

Krishnaswami, S., Lal, D., Somayajulu, B. L. K., Dixon, F. S., Stonecipher, S. A., and Craig, H. (1972). Silicon, radium, thorium, and lead in seawater: In-situ extraction by synthetic fibre. *Earth and Planetary Science Letters* 16, 84-90.

Kruse, T. H., Moniot, R., Herzog, G., Milazzo, T., Hall, G., and Savin, W. (1981). Tandem Van de Graaff measurement of ^{10}Be in meteorites. *In* "Proceedings of the Symposium on Accelerator Mass Spectrometry," pp. 277-84. Argonne National Laboratory, Argonne, Ill.

Kutschera, W., Henning, W., Paul, M., Smither, R. K., Stephenson, E. J., Yntema, J. L., Alburger, D. E., Cumming, J. B., and Harbottle, G. (1980). Measurement of the ^{32}Si half-life via accelerator mass spectrometry. *Physical Review Letters* 45, 592-96.

Kutschera, W., Henning, W., Paul, M., Stephenson, E. J., and Yntema, J. L. (1980). Radioisotope detection with the Argonne FN tandem accelerator. *In* Proceedings of the Tenth International Radiocarbon Conference, Bern/Heidelberg, 1979 (M. Stuiver and R. Kra, eds.). *Radiocarbon* 22, 807-15.

Lal, D., Arnold, J. R., and Somayajulu, B. L. K. (1964). A method for the extraction of trace elements from sea water. *Geochimica et Cosmochimica Acta* 28, 1111-17.

Lal, D., Goldberg, E. D., and Koide, M. (1960). Cosmic-ray-produced silicon-32 in nature. *Science* 131, 332-37.

Lal, D., Nijampurkar, V. N., and Rama, S. (1970). Silicon-32 hydrology. *In* "Proceedings of the Symposium on Isotope Hydrology, Vienna, 1970," pp. 847-68. International Atomic Energy Agency, Vienna.

Lal, D. Nijampurkar, V. N., Somayajulu, B. L. K., Koide, M., and Goldberg, E. D. (1976). Silicon-32 specific activities in coastal waters of the world oceans. *Limnology and Oceanography* 21, 285-93.

Lanford, W. A., Parker, P. D., Bauer, K., Turekian, K. K., Cochran, J. K., and Krishnaswami, S. (1980). Measurements of ^{10}Be distributions using a tandem Van de Graaff accelerator. *Nuclear Instruments and Methods* 168, 505-10.

Leventhal, J. S., and Libby, W. F. (1970). Tritium fallout in the Pacific United States. *Journal of Geophysical Research* 75, 7628-33.

Lindner, M. (1953). New nuclides produced in chlorine spallation reactions. *Physical Review* 91, 642-44.

Litherland, A. E. (1978). Radiocarbon dating with accelerators: Results from Rochester-Toronto-General Ionex Corporation. *In* "Proceedings of the First Conference on Radiocarbon Dating with Accelerators" (H. E. Gove, ed.), pp. 70-113. University of Rochester, Rochester, N.Y.

Litherland, A. E. (1980). Ultrasensitive mass spectrometry with accelerators. *Annual Review of Nuclear and Particle Science* 30, 437-73.

Livingston, H. D., and Bowen, V. T., (1979). Pu and ^{137}Cs in coastal sediments. *Earth and Planetary Science Letters* 43, 29-45.

Loosli, H. H., and Oeschger, H. (1979). Argon-39, carbon-14 and krypton-85 measurements in groundwater samples. *In* "Proceedings of the Symposium on Isotope Hydrology, Neuherberg, 1978," pp. 931-47. International Atomic Energy Agency, Vienna.

Loosli, H. H., and Oeschger, H. (1980). Use of ^{39}Ar and ^{14}C for groundwater dating. *In* Proceedings of the Tenth International Radiocarbon Conference, Bern/Heidelberg, 1979 (M. Stuiver and R. Kra, eds.). *Radiocarbon* 22, 863-70.

Lorius, C., Merlivat, L., Jouzel, J., and Pourchet, M. (1979). A 30,000 yr isotope climatic record from Antarctic ice. *Nature* 280, 644-48.

McCaffrey, R. J. (1977). A record of the accumulation of sediment and trace metals in a Connecticut, U.S.A., salt marsh. Ph.D. dissertation, Yale University, New Haven, Conn.

Mangerud, J., and Gulliksen, S. (1975). Apparent radiocarbon ages of recent marine shells from Norway, Spitsbergen, and Arctic Canada. *Quaternary Research* 5, 263-73.

Mast, T. S. (1978). Plans for radioisotope dating with the Lawrence Berkeley Laboratory 88-inch cyclotron. *In* "Proceedings of the First Conference on Radiocarbon Dating with Accelerators" (H. E. Gove, ed.), pp. 239-44. University of Rochester, Rochester, N.Y.

Michael, H. N., and Ralph, E. K. (1970). Correction factors applied to Egyptian radiocarbon dates from the era before Christ. *In* "Proceedings of the Twelfth Nobel Symposium on Radiocarbon Variations and Absolute Chronology, Uppsala, 1969" (I. U. Olsson, ed.), pp. 109-20. Almqvist and Wiksell, Förlag AB, Stockholm.

Michael, H. N., and Ralph, E. K. (1973). Discussion of radiocarbon dates obtained from precisely dated sequoia and bristlecone pine samples. *In* "Proceedings of the Eighth International Conference on Radiocarbon Dating, Lower Hutt, 1972," pp. A11-27. Royal Society of New Zealand, Wellington.

Michel, R. L., and Suess, H. E. (1975). Bomb tritium in the Pacific. *Journal of Geophysical Research* 80, 4139-52.

Middleton, R., Klein, J., and Tang, H. (1981). Instrumentation of an FN tandem for the detection of ¹⁰Be. *In* "Proceedings of the Symposium on Accelerator Mass Spectrometry," pp. 57-86. Argonne National Laboratory, Argonne, Ill.

Muller, R. A. (1977). Radioisotope dating with a cyclotron. *Science* 196, 489-94.

Muller, R. A. (1978). Radioisotope dating with the LBL 88″cyclotron. *In* "Proceedings of the First Conference on Radiocarbon Dating with Accelerators" (H. E. Gove, ed.), pp. 33-37. University of Rochester, Rochester, N.Y.

Muller, R. A., Stephenson, E. J., and Mast, T. S. (1978). Radioisotope dating with an accelerator: A blind measurement. *Science* 201, 347-48.

Münnich, K. O., Roether, W., and Thilo, L. (1967). Dating of groundwater with tritium and ¹⁴C. *In* "Proceedings of the Symposium on Isotopes in Hydrology, Vienna, 1966," pp. 305-20. International Atomic Energy Agency, Vienna.

Naylor, H., Elmore, D., Clover, M. R., Kilius, L. R., Beukens, R. P., Fulton, B. R., Gove, H. E., Litherland, A. E., and Purser, K. H. (1978). Determination of ³⁶Cl isotopic ratios. *In* "Proceedings of the First Conference on Radiocarbon Dating with Accelerators" (H. E. Gove, ed.), pp. 360-71. University of Rochester, Rochester, N.Y.

Nelson, D. E., Korteling, R. G., and Stott, W. R. (1977). Carbon-14: Direct detection at natural concentrations. *Science* 198, 507-8.

Nijampurkar, V. N., Amin, B. S., Kharkar, D. P., and Lal, D. (1966). "Dating" groundwaters of ages younger than 1,000-1,500 years using natural silicon-32. *Nature* 210, 478-80.

Nishiizumi, K., Arnold, J. R., Elmore, D., Ferraro, R. D., Gove, H. E., Finkel, R. C., Beukens, R. P., Chang, K. H., and Kilius, L. R. (1979). Measurements of ³⁶Cl in Antarctic meteorites and Antarctic ice using a Van de Graaff accelerator. *Earth and Planetary Science Letters* 45, 285-92.

Oeschger, H., Siegenthaler, U., Schotterer, U., and Gugelman, A. (1975). A box diffusion model to study the carbon dioxide exchange in nature. *Tellus* 27, 168-92.

Oldfield, F., Appleby, P. G., and Battarbee, R. W. (1978). Alternative ²¹⁰Pb dating: Results from the New Guinea Highlands and Lough Erne. *Nature* 271, 334-42.

Östlund, H. G., Dorsey, H. A., and Brescher, R. (1976). "Geosecs Atlantic Radiocarbon and Tritium Results (Miami)." Rosenstiel School of Marine Atmospheric Science, University of Miami, Tritium Laboratory Report 5.

Östlund, H. G., Dorsey, H. A., and Rooth, C. G. (1974). Geosecs North Atlantic radiocarbon and tritium results. *Earth and Planetary Science Letters* 23, 69-86.

Östlund, H. G., and Werner, E. (1962). The electrolytic enrichment of tritium and deuterium for natural tritium measurements. *In* "Proceedings of the Symposium on Tritium in the Physical and Biological Sciences, Vienna, 1961," vol. 1, pp. 95-104. International Atomic Energy Agency, Vienna.

Pearson, G. W. (1979). Precise ¹⁴C measurement by liquid scintillation counting. *Radiocarbon* 21, 1-21.

Pearson, G. W. (1980). High precision radiocarbon dating by liquid scintillation counting applied to radiocarbon time-scale calibration. *In* Proceedings of the Tenth International Radiocarbon Conference, Bern/Heidelberg, 1979 (M. Stuiver and R. Kra, eds.). *Radiocarbon* 22, 337-45.

Pearson, G. W., Pilcher, J. R., Baillie, M. G. L., and Hillam, J. (1977). Absolute radiocarbon dating using a low altitude European tree-ring calibration. *Nature* 270, 25-28.

Pennington, W., Cambray, R. S., and Fisher, E. M. (1973). Observations on lake sediments using fallout ¹³⁷Cs as a tracer. *Nature* 242, 324-26.

Picciotto, E., Cameron, R., Crozaz, G., Deutsch, S., and Wilgain, S. (1968). Determination of the rate of snow accumulation at the pole of relative inaccessibility, eastern Antarctica: A comparison of glaciological and isotopic methods. *Journal of Glaciology* 7, 273-87.

Picciotto, E., Crozaz, G., Ambach, W., and Eisner, H. (1967). ²¹⁰Pb and ⁹⁰Sr in an Alpine glacier. *Earth and Planetary Science Letters* 3, 237-42.

Radiocarbon (1980). Proceedings of the Tenth International Radiocarbon Conference,

Bern/Heidelberg, 1979 (M. Stuiver and R. Kra, eds.). *Radiocarbon* 22, 131-1016.

Raisbeck, G. M., and Yiou, F. (1979). Possible use of ⁴¹Ca for radioactive dating. *Nature* 277, 42-44.

Raisbeck, G. M., Yiou, F., Fruneau, M., Lieuvin, M., and Loiseaux, J. M. (1978a). ¹⁰Be detection using a cyclotron equipped with an external ion source. *In* "Proceedings of the First Conference on Radiocarbon Dating with Accelerators" (H. E. Gove, ed.), pp. 38-46. University of Rochester, Rochester, N.Y.

Raisbeck, G. M., Yiou, F., Fruneau, M., Lieuvin, M., and Loiseaux, J. M. (1978b). Measurement of ¹⁰Be in 1,000- and 5,000-year-old Antarctic ice. *Nature* 275, 731-33.

Raisbeck, G. M., Yiou, F., Fruneau, M., and Loiseaux, J. M. (1978). Beryllium-10 mass spectrometry with a cyclotron. *Science* 202, 215-17.

Raisbeck, G. M., Yiou, F., Fruneau, M., Loiseaux, J. M., and Lieuvin, M. (1979). ¹⁰Be concentration and residence time in the ocean surface layer. *Earth and Planetary Science Letters* 43, 237-40.

Raisbeck, G. M., Yiou, F., Fruneau, M., Loiseaux, J. M., Lieuvin, M., and Ravel, J. C. (1979). Deposition rate and seasonal variations in precipitation of cosmogenic ¹⁰Be. *Nature* 282, 279-80.

Raisbeck, G. M., Yiou, F., Fruneau, M., Loiseaux, J. M., Lieuvin, M., Ravel, J. C., and Hays, J. D. (1979). A search in a marine sediment core for ¹⁰Be concentration variations during a geomagnetic field reversal. *Geophysical Research Letters* 6, 717-19.

Raisbeck, G. M., Yiou, F., Fruneau, M., Loiseaux, J. M., Lieuvin, M., Ravel, J. C., Reyss, J. L., and Guichard, F. (1980). ¹⁰Be concentration and residence time in the deep ocean. *Laboratoire René Bernas* 80-02, 1-12.

Raisbeck, G. M., Yiou, F., Lieuvin, M., Ravel, J. C., Fruneau, M., and Loiseaux, J. M. (1981). ¹⁰Be in the environment: Some recent results and their applications. *In* "Proceedings of the Symposium on Accelerator Mass Spectrometry," pp. 228-43. Argonne National Laboratory, Argonne, Ill.

Raisbeck, G. M., Yiou, F., and Stephan, C. (1979). ²⁶Al measurement with a cyclotron. *Le Journal de Physique-Lettres* 40, 241-44.

Ralph, E. K., and Michael, H. N. (1970). Masca radiocarbon dates for sequoia and bristlecone-pine samples. *In* "Proceedings of the Twelfth Nobel Symposium on Radiocarbon Variations and Absolute Chronology, Uppsala, 1969" pp. 619-23. (I. U. Olsson, ed.), Almqvist and Wiksell, Förlag AB, Stockholm.

Ralph, E. K., Michael, H. N., and Han, M. C. (1973). Radiocarbon dates and reality. *Masca Newsletter* 9, 1-20.

Ritchie, J. C., McHenry, J. R., and Gill, A. C. (1974). Fallout cesium-137 in the soils and sediments of three watersheds. *Ecology* 55, 887-90.

Robbins, J. A. (1978). Geochemical and geophysical applications of radioactive lead. *In* "The Biogeochemistry of Lead in the Environment" (J. O. Nriagu, ed.), pp. 285-393. Elsevier, Amsterdam.

Robbins, J. A., and Edgington, D. N. (1975). Determination of recent sedimentation rates in Lake Michigan using ²¹⁰Pb and ¹³⁷Cs. *Geochimica et Cosmochimica Acta* 39, 285-304.

Robbins, J. A., Edgington, D. N., and Kemp, A. L. W. (1978). Comparative ²¹⁰Pb, ¹³⁷Cs, and pollen geochronologies of sediments from Lakes Ontario and Erie. *Quaternary Research* 10, 256-78.

Robbins, J. A., Krezoski, J. R., and Mosley, S. C. (1977). Radioactivity in sediments of the Great Lakes: Post-depositional redistribution by deposit-feeding organisms. *Earth and Planetary Science Letters* 36, 325-33.

Robbins, J. A., McCall, P. L., Fisher, J. B., and Krezoski, J. R. (1979). Effect of deposit feeders on migration of ¹³⁷Cs in lake sediments. *Earth and Planetary Science Letters* 42, 277-87.

Roether, W. (1974). The tritium and carbon-14 profiles of the Geosecs I (1969) and Gogo I (1971) North Pacific stations. *Earth and Planetary Science Letters* 23, 108-15.

Roether, W., Münnich, K. O., and Schoch, H. (1980). On the ¹⁴C to tritium relationship in the North Atlantic Ocean. *In* Proceedings of the Tenth International Radiocarbon Conference, Bern/Heidelberg, 1979 (M. Stuiver and R. Kra, eds.). *Radiocarbon* 22, 636-46.

Rooth, C. G., and Östlund, H. G. (1972). Penetration of tritium into the Atlantic thermocline. *Deep-sea Research* 19, 481-92.

Rózański, K., and Florkowski, T. (1979). Krypton-85 dating of groundwater. *In* "Proceedings of the Symposium on Isotope Hydrology, Neuherberg, 1978,'' pp. 949-61. International Atomic Energy Agency, Vienna.

Santschi, P. H., Li, Y-H., and Bell, J. (1979). Natural radionuclides in the water of Narragansett Bay. *Earth and Planetary Science Letters* 45, 201-13.

Schell, W. R. (1977). Concentrations, physico-chemical states and mean residence times of ^{210}Pb and ^{210}Po in marine and estuarine waters. *Geochimica et Cosmochimica Acta* 41, 1019-31.

Schell, W. R., Sauzay, G., and Payne, B. R. (1974). World distribution of environmental tritium. *In* ''Physical Behaviour of Radioactive Contaminants in the Atmosphere,'' pp. 375-400. International Atomic Energy Agency, Vienna.

Schink, D. R. (1962). The measurement of dissolved silica in sea water. Ph.D. dissertation, Scripps Institution of Oceanography, University of California, San Diego.

Schoch, H., Bruns, M., Münnich, K. O., and Münnich, M. (1980). A multi-counter system for high precision carbon-14 measurements. *In* Proceedings of the Tenth International Radiocarbon Conference, Bern/Heidelberg, 1979 (M. Stuiver and R. Kra, eds.). *Radiocarbon* 22, 442-47.

Schotterer, U., and Oeschger, H. (1980). Low-level liquid scintillation counting in an underground laboratory. *In* Proceedings of the Tenth International Radiocarbon Conference, Bern/Heidelberg, 1979 (M. Stuiver and R. Kra, eds.). *Radiocarbon* 22, 505-11.

Shokes, R. F. (1976). Rate-dependent distributions of lead-210 and interstitial sulfate in sediments of the Mississippi River delta. Ph.D. dissertation, Texas A & M University, College Station, Tex.

Simpson, H. J. (1970). Tritium in Crater Lake, Oregon. *Journal of Geophysical Research* 75, 5195-5207.

Smith, D. B., Wearn, P. L., Richards, H. J., and Rowe, P. C. (1970). Water movement in the unsaturated zone of high and low permeability strata by measuring natural tritium. *In* ''Proceedings of the Symposium on Isotope Hydrology, Vienna, 1970,'' pp. 73-87. International Atomic Energy Agency, Vienna.

Smith, J. N., and Walton, A. (1980). Sediment accumulation rates and geochronologies measured in the Saguenay Fjord using the Pb-210 dating method. *Geochimica et Cosmochimica Acta* 44, 225-40.

Somayajulu, B. L. K., Lal, D., and Craig, H. (1973). Silicon-32 profiles in the South Pacific. *Earth and Planetary Science Letters* 18, 181-88.

Stuiver, M. (1970). Long-term C-14 variations. *In* ''Proceedings of the Twelfth Nobel Symposium on Radiocarbon Variations and Absolute Chronology, Uppsala, 1969'' (I. U. Olsson, ed.), pp. 197-213. Almqvist and Wiksell, Förlag AB, Stockholm.

Stuiver, M. (1971). Evidence for the variation of atmospheric C^{14} content in the late Quaternary. *In* ''Late Cenozoic Glacial Ages'' (K. K. Turekian, ed.), pp. 57-70. Yale University Press, New Haven, Conn.

Stuiver, M. (1975). Climate versus changes in ^{13}C content of the organic component of lake sediments during the late Quaternary. *Quaternary Research* 5, 251-62.

Stuiver, M. (1978a). Radiocarbon timescale tested against magnetic and other dating methods. *Nature* 273, 271-74.

Stuiver, M. (1978b). The ultimate precision of ^{14}C dating is determined only by counting statistics. *In* ''Proceedings of the First Conference on Radiocarbon Dating with Accelerators'' (H. E. Gove, ed.), pp. 353-59. University of Rochester, Rochester, N.Y.

Stuiver, M. (1980). Solar variability and climatic change during the current millennium. *Nature* 286, 868-71.

Stuiver, M. (1982). A high-precision calibration of the AD radiocarbon time scale. *Radiocarbon* 24, 1-26.

Stuiver, M., and Grootes, P. M. (1980). Trees and the ancient record of heliomagnetic cosmic ray flux modulation. *In* Proceedings of the Conference on the Ancient Sun, Boulder, 1979 (R. O. Pepin, J. A. Eddy, and R. B. Merrill, eds.). *Geochimica et Cosmochimica Acta* (supplement) 13, 165-73.

Stuiver, M., Heusser, C. J., and Yang, I. C. (1978). North American glacial history extended to 75,000 years ago. *Science* 200, 16-21.

Stuiver, M., and Quay, P. D. (1980a). Changes in atmospheric ^{14}C attributed to a variable sun. *Science* 207, 11-19.

Stuiver, M., and Quay, P. D. (1980b). Patterns of atmospheric ^{14}C changes. *In* Proceedings of the Tenth International Radiocarbon Conference, Bern/Heidelberg, 1979 (M. Stuiver and R. Kra, eds.). *Radiocarbon* 22, 166-76.

Stuiver, M., and Quay, P. D. (1981). Atmospheric ^{14}C changes resulting from fossil fuel CO$_2$ release and cosmic ray flux variability. *Earth and Planetary Science Letters* 53, 349-62.

Stuiver, M., Robinson, S. W., and Yang, I. C. (1979). ^{14}C dating to 60,000 years BP with proportional counters. *In* ''Proceedings of the Ninth International Radiocarbon Dating Conference, Los Angeles/La Jolla, 1976'' (R. Berger and H. E. Suess, eds.), pp. 202-15. University of California Press, Los Angeles.

Suess, H. E. (1967). Bristlecone-pine calibration of the radiocarbon timescale from 4100 B.C. to 1500 B.C. *In* ''Proceedings of the Symposium on Radioactive Dating and Methods of Low-Level Counting, Monaco, 1967,'' pp. 143-51. International Atomic Energy Agency, Vienna.

Suess, H. E. (1970). Bristlecone-pine calibration of the radiocarbon timescale 5200 B.C. to the present. *In* ''Proceedings of the Twelfth Nobel Symposium on Radiocarbon Variations and Absolute Chronology, Uppsala, 1969'' (I. U. Olsson, ed.), pp. 303-11. Almqvist and Wiksell, Förlag AB, Stockholm.

Suess, H. E. (1978). La Jolla measurements of radiocarbon in tree-ring dated wood. *Radiocarbon* 20, 1-18.

Suess, H. E. (1981). Solar activity, cosmic ray-produced carbon-14, and the terrestrial climate. *In* ''Proceedings of the Conference on Sun and Climate, Toulouse, 1980,'' pp. 307-10. Centre Nationale d'Etude Spatiale, 31055 Toulouse, France.

Tans, P. P., and Mook, W. G. (1979). Design, construction and calibration of a high accuracy carbon-14 counting set-up. *Radiocarbon* 21, 22-40.

Taylor, C. B. (1981). Current status and trends in electrolytic enrichment of low-level tritium in water. *In* ''Proceedings of the Symposium on Methods of Low-Level Counting and Spectrometry, Berlin, 1981,'' pp. 303-23. International Atomic Energy Agency, Vienna.

Thomas, J., Parker, P., Mangini, A., Cochran, K., Turekian, K., Krishnaswami, S., and Sharma, P. (1981). ^{10}Be in manganese nodules. *In* ''Proceedings of the Symposium on Accelerator Mass Spectrometry,'' pp. 244-54. Argonne National Laboratory, Argonne, Ill.

Torgersen, T., Clarke, W. B., and Jenkins, W. J. (1979). The tritium/helium-3 method in hydrology. *In* ''Proceedings of the Symposium on Isotope Hydrology, Neuherberg, 1978,'' pp. 917-30. International Atomic Energy Agency, Vienna.

Torgersen, T., Top, Z., Clarke, W. B., Jenkins, W. J., and Broecker, W. S. (1977). A new method for physical limnology, tritium-helium-3 age-results for Lakes Erie, Huron and Ontario. *Limnology and Oceanography* 22, 181-93.

Turekian, K. K., Cochran, J. K., Krishnaswami, S., Lanford, W. A., Parker, P. D., and Bauer, K. A. (1979). The measurement of ^{10}Be in manganese nodules using a tandem Van de Graaff accelerator. *Geophysical Research Letters* 6, 417-20.

Turkevich, A., and Samuels, A. (1954). Evidence for ^{32}Si: A long-lived beta emitter. *Physical Review* series 2, 94, 364.

Unterweger, M. P., Coursey, B. M., Schima, F. J., and Mann, W. B. (1980). Preparation and calibration of the 1978 National Bureau of Standards tritiated-water standards. *International Journal of Applied Radiation and Isotopes* 31, 611-14.

Vogel, J. C., and Waterbolk, H. T. (1972). Groningen radiocarbon dates X. *Radiocarbon* 14, 6-110.

Von Buttlar, H. H., and Libby, W. F. (1955). Natural distribution of cosmic-ray produced tritium. *Journal of Inorganic Nuclear Chemistry* 1, 75-91.

Wahlgren, M. A., and Nelson, D. M. (1974). ''Residence Times for Plutonium-239 and Cesium-137 in Lake Michigan Water.'' Radiological and Environmental Research Division Annual Report, Ecology, January-December 1973, ANL-3060, Part III, pp. 85-89. Argonne National Laboratory, Argonne, Ill.

Yang, A. I. C., and Fairhall, A. W. (1973). Variations of natural radiocarbon during the last 11 millennia and geophysical mechanisms for producing them. *In* ''Proceedings of the Eighth International Conference on Radiocarbon Dating, Lower Hutt, 1972,'' pp. A44-57. Royal Society of New Zealand, Wellington.

Paleoecology

CHAPTER 8

Holocene Vegetational History of the Western United States

Richard G. Baker

Introduction

As valley glaciers and ice caps melted and pluvial lakes shrank at the end of the Pleistocene, substantial changes also occurred in the vegetation of the western United States. A new environmental setting prevailed. The macroclimate changed markedly during late-glacial and postglacial times. The microclimate also changed as large pluvial lakes disappeared. Fresh, unweathered soils were exposed as glaciers and pluvial lakes contracted. These new environmental conditions rendered old associations and distributions of plants out of equilibrium with their new environment. How was the vegetation affected by these changes? Did large latitudinal migrations like those in the eastern United States occur, or did species merely change their altitudinal distributions up and down mountainsides? The late-glacial-to-postglacial changes in vegetation were substantial. Are smaller Holocene changes detectable? Was there an ''Altithermal''? These questions form the basis of this chapter.

This review covers about the last 12,000 years in order to include the Pleistocene-Holocene transition. The area extends from the eastern edge of the Rocky Mountains to the Pacific Coast and from the Canadian border to the Mexican border. The diversity in climate, topography, and vegetation is enormous, ranging from maritime climates to deserts, from the highest peaks in the conterminous states to lowest points in North America, and from desert vegetation to forests to tundra. Although many sites in this area contain some data on Holocene vegetation, very few have continuous sections with adequate dating, and these few are not evenly distributed. The purpose of this chapter is to provide a brief summary of the vegetational changes for the western United States during the last 10,000 years; the summary is based on pollen and plant-macrofossil analyses.

I have divided the region into western Washington, eastern Washington, southwestern Montana-northern Wyoming, California, Utah-Colorado, and the desert Southwest. Pollen and/or plant-macrofossil sequences have been selected from each area. The general patterns of the modern vegetation are described for each area, and modern-pollen studies are summarized when they are available. The general pollen sequence from each area is summarized and the vegeta-tion from late-glacial time to the present is reconstructed. Late glacial is defined here as the time immediately preceding a major shift in pollen or plant macrofossils that signals the warming to Holocene conditions. The shift occurred between about 14,000 and 10,000 years ago.

The only previous attempts to summarize the vegetational history of this area were three papers prepared for the International Association for Quaternary Research Congress in Boulder, Colorado, in 1965. Martin and Mehringer (1965) summarized the pollen data for the Southwest deserts. Since 1965, few new pollen studies have been completed to alter Martin and Mehringer's picture, but a major new source of data has arisen in the form of plant macrofossils preserved in packrat (*Neotoma*) middens. Weber (1965) summarized the plant geography of the Rocky Mountains, but Maher's (1961) pollen work was his only source of Holocene vegetational data. New work in the Rocky Mountains has provided a basis for synthesis. Heusser (1965, 1966) summarized a considerable body of older work in the Pacific Northwest. Subsequent work in Washington provides new information about pollen influx and plant macrofossils and better-documented identifications. In all areas, the new sites have only begun to fill in a very sparse network of well-studied localities.

Western Washington

The western edge of Washington, a lush forested region, receives more precipitation than any other place in the conterminous United States. This area includes the Olympic Mountains, the Puget lowland, and the Cascade Range. The mountains once supported valley glaciers, and much of the lowland was invaded by piedmont glaciers as well as a lobe of the Cordilleran ice sheet during the last (Fraser) glaciation (Armstrong, 1975; Heusser, 1974; Thorson, 1980).

I want to thank C. S. Barnosky, H. E. Wright, Jr., W. G. Spaulding, and T. A. Ager for reviewing the manuscript for this chapter; Thompson Webb and Jonathan Overpeck for using their computer programs to draw the pollen diagrams; and Connie Reasoner and George Fisher for typing the manuscript.

Discrete altitudinal belts of vegetation are present today (Figure 8-1). Lowland forests are dominated by *Tsuga heterophylla* and *Thuja plicata*, accompanied by *Pseudotsuga menziesii*, *Abies* species, and the seral *Alnus rubra* and *Pinus contorta* (Table 8-1), up to 600 m along the coast and 900 m inland (Franklin and Dyrness, 1973). *Picea sitchensis* is codominant with *Tsuga heterophylla* and *Thuja* along the Pacific slope, and *Pinus contorta* is locally abundant along the coast. *P. ponderosa* is also locally present in dry prairie areas in the Puget lowland. The montane forests extend upward to 1000 m along the Pacific Ocean and 1500 m along the inland mountains. The forests are dominated by *Abies amabilis* and *Tsuga heterophylla*, along with *Abies procera*, *Thuja*, *Pseudotsuga*, and *Pinus monticola*. The subalpine forest extends upward to 1700 m in the Olympic Mountains and 2000 m in the Cascade Range. *Tsuga mertensiana* is prevalent in mature stands, along with *Abies amabilis*, *A. lasiocarpa*, *Pinus contorta*, and *Chamaecyparis nootkatensis*. The lower subalpine communities are usually closed forests, and upper communities are parklands and shrublands. Alpine vegetation occurs from the treeline to the snow line. Important taxa include Cyperaceae, Gramineae, Compositae, Ericaceae, Caryophyllaceae, and Rosaceae (Franklin and Dyrness, 1973; Heusser, 1974).

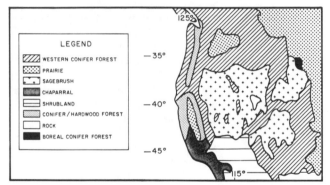

Figure 8-1. Generalized vegetational map of western United States. (From Wright, 1971; copyright 1971 Yale University Press; used with permission of publisher and author.)

Analyses of the modern pollen rain (Barnosky, 1981; C. J. Heusser, 1978a, 1978b, 1978c) show that *Alnus* is the most abundant pollen type in most areas. Coastal lowland forests contain high pollen percentages of *Picea* (representing *P. sitchensis*), Cupressaceae (representing *Thuja plicata*), and *Tsuga heterophylla*. Interior lowland forests show a dominance of *T. heterophylla* pollen, along with *Pseudotsuga* type and Cupressaceae (Hansen and Easterbrook, 1974; C. J. Heusser, 1978b). In the montane forest, pollen of *Abies* and *Tsuga heterophylla* occur together. In the upper montane and subalpine belts, percentages of *T. mertensiana* pollen are low but consistent, and *Abies* (probably *A. lasiocarpa*) and *Tsuga heterophylla* pollen percentages remain high. In the eastern Cascade Range, *Pinus* pollen prevails in the subalpine belt. Pollen of herbs and shrubs is subdominant in the upper subalpine belt and dominant in the alpine belt.

Three sites in western Washington (Figure 8-2 and Table 8-2) illustrate the Holocene vegetational sequence. The Hoh Valley peat bog on the western flank of the Olympic Mountains records the history of the coastal forest (Heusser, 1974). The site was ice covered until 17,000 yr B.P. and is now surrounded by *Tsuga heterophylla* forest with accessory *Picea sitchensis* and *Pseudotsuga*. Davis Lake

Table 8-1. Scientific Names, Families, and Common Names of Taxa Mentioned in the Text

Scientific Name	Family	Common Name
Abies amabilis (Doug.) Forbes	Pinaceae	Silver fir
A. concolor (Gord. & Glerd.) Lindl.	Pinaceae	White fir
A. grandis (Doug.) Forbes	Pinaceae	Grand fir
A. lasiocarpa (Hook.) Nutt.	Pinaceae	Subalpine fir
A. magnifica A. Murr.	Pinaceae	California red fir
A. procera Rehder	Pinaceae	Noble fir
Agave lechuguilla Torr.	Amaryllidaceae	Lechuguilla
Alnus rubra Bong.	Betulaceae	Red alder
Ambrosia dumosa Payne	Compositae	Bur-sage
Anacardiaceae	Anacardiaceae	Sumac family
Artemisia tridentata Nutt.	Compositae	Big sagebrush
Atriplex confertifolia Torr. & Frem.	Chenopodiaceae	Shadscale saltbush
Betula glandulosa Michx.	Betulaceae	Ground birch
B. occidentalis Hook.	Betulaceae	River birch
Caryophyllaceae	Caryophyllaceae	Pink family
Cercidium microphyllum (Torr.) Rose & Jtn.	Leguminosae	Paloverde
Chamaecyparis nootkatensis D. Don	Cupressaceae	Alaska cedar
Chenopodiineae		Goosefoot superfamily
Coleogyne	Rosaceae	Blackbush
Compositae	Compositae	Aster family
Cupressaceae	Cupressaceae	Cypress family
Cupressus	Cupressaceae	Cypress
Cyperaceae	Cyperaceae	Sedge family
Dasylirion	Liliaceae	Sotol
Encelia farinosa Gray	Compositae	Brittle-bush
Ericaceae	Ericaceae	Heath family
Filicales		Fern order
Flourensia cernua DC	Compositae	Tarbush
Fouquieria splendens Englem.	Fouqueriaceae	Ocotillo
Gramineae	Gramineae	Grass
Juniperus communis L.	Cupressaceae	Common juniper
J. deppeana Steud.	Cupressaceae	Alligator juniper
J. monosperma Engelm.	Cupressaceae	One-seeded juniper
J. occidentalis Hook.	Cupressaceae	Western juniper
J. osteosperma (Torr.) Little	Cupressaceae	Utah juniper
J. scopulorum Sarg.	Cupressaceae	Rocky Mountain juniper
Kochia	Chenopodiaceae	Red sage
Koenigia islandica L.	Polygonaceae	—
Larix	Pinaceae	Larch
Larrea divaricata Cav.	Zygophyllaceae	Creosote bush
Libocedrus	Cupressaceae	Incense cedar
Opuntia	Cactaceae	Cactus
Phlox	Polemoniaceae	Phlox
Picea engelmannii Parry	Pinaceae	Engelmann spruce
P. pungens Engelm.	Pinaceae	Blue spruce
P. sitchensis (Bong.) Carr.	Pinaceae	Sitka spruce
Pinus albicaulis Engelm.	Pinaceae	Whitebark pine
P. contorta Doug.	Pinaceae	Lodgepole pine
P. edulis Engelm.	Pinaceae	Pinyon pine
P. flexilis James	Pinaceae	Limber pine
P. longaeva Bailey	Pinaceae	Bristlecone pine
P. monophylla Torr. & Frem.	Pinaceae	Singleleaf pinyon
P. monticola Dougl.	Pinaceae	Western white pine
P. ponderosa Dougl.	Pinaceae	Ponderosa pine
P. sabiniana Dougl.	Pinaceae	Digger pine
P. strobiformis Engelm.	Pinaceae	Southwestern white pine
Polemonium	Polemoniaceae	Jacob's ladder, sky pilot
Polygonum bistortoides Pursh	Polygonaceae	American bistort
Polygonum viviparum L.	Polygonaceae	Viviparous bistort
Pseudotsuga menziesii (Mirb.) Franco	Pinaceae	Douglas-fir
Pteridium	Polypodiaceae	Bracken fern
Quercus douglasii H. & A.	Fagaceae	Blue oak
Q. garryana Doug.	Fagaceae	Oregon white oak
Q. kelloggii Newb.	Fagaceae	California black oak
Rhamnaceae	Rhamnaceae	Buck-thorn family
Rosaceae	Rosaceae	Rose family
Salicornia	Chenopodiaceae	Pickleweed
Salvia	Labiatae	Sage
Sarcobatus vermiculatus (Hook.) Torr.	Chenopodiaceae	Greasewood
Selaginella densa Rydb.	Selaginellaceae	Rock selaginella
Sequoia	Taxodiaceae	Redwood
Suaeda	Chenopodiaceae	Seablite
Taxaceae	Taxaceae	Yew family
Taxodiaceae	Taxodiaceae	Baldcypress family
Taxus	Taxaceae	Yew family
Thuja plicata Donn.	Cupressaceae	Western red cedar
Tsuga heterophylla (Raf.) Sarg.	Pinaceae	Western hemlock
T. mertensiana (Bong.) Carr.	Pinaceae	Mountain hemlock
Yucca	Liliaceae	Yucca

was beyond the limit of the Fraser Glaciation, and it chronicles the history of the interior lowland forests. The vegetation today is predominantly *Pseudotsuga, Tsuga heterophylla,* and *Thuja* (Barnosky, 1981). Pangborn Bog (Hansen and Easterbrook, 1974) is on an outwash plain that dates from about 11,500 yr B.P. (Armstrong, 1975), and it contains only a Holocene record. The site was heavily logged but is within the *Tsuga heterophylla* belt (Franklin and Dyrness, 1973).

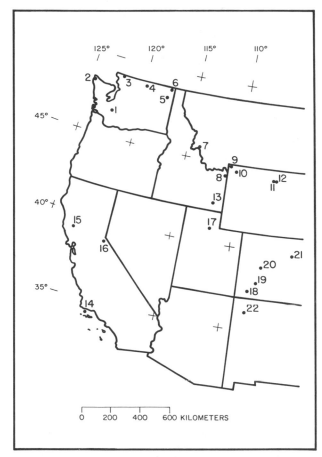

Figure 8-2. Map of pollen sites shown in figures and summarized in Table 8-2. (1) Davis Lake, Washington, (2) Hoh River valley, Washington, (3) Pangborn Bog, Washington, (4) Bonaparte Meadows, Washington, (5) Waits Lake, Washington, (6) Big Meadow, Washington, (7) Lost Trail Pass Bog, Montana, (8) Cub Lake, Idaho, (9) Gardiners Hole, Wyoming, (10) Cub Creek Pond, Wyoming, (11) "Beaver Lake," Wyoming, (12) Sherd Lake, Wyoming, (13) Swan Lake, Idaho, (14) Santa Barbara Basin, Pacific Ocean, (15) Clear Lake, California, (16) Osgood Swamp, California, (17) Snowbird Bog, Utah, (18) Twin Lakes, Colorado, (19) Hurricane Basin, Colorado, (20) Alkali Creek Basin, Colorado, (21) Redrock Lake, Colorado, (22) Chaco Canyon, New Mexico.

The basal portions of the Davis Lake and Hoh Valley diagrams are full glacial (Figure 8-3). They contain very high percentages of herb pollen and low pollen influx and are interpreted as representing tundra-parkland environments (Barnosky, 1981; Heusser, 1974). By 12,500 yr B.P., the pollen rain was dominated by *Pinus, Picea, Abies,* and *Tsuga mertensiana.* Needles of *Pinus contorta, Picea engelmannii,* and *Abies lasiocarpa* from Davis Lake indicate their local presence at this time (Barnosky, 1981). The late-glacial pollen

spectra resemble the modern pollen rain in the subalpine *A. lasiocarpa* belt east of the Cascade Range, except that *Picea engelmannii* and perhaps *Tsuga mertensiana* were more important in late-glacial forests than in modern ones. Mixed woodlands of *Pinus contorta, Picea engelmannii, Abies lasiocarpa,* and *Tsuga mertensiana* became established in western Washington during late-glacial time as the treeline migrated up the mountain slopes.

Pseudotsuga menziesii was probably the chief constituent of the early-Holocene communities in the Puget lowland from about 10,000 to 6000 yr B.P. (Figure 8-3). The occurrence of high pollen percentages of *Pseudotsuga* with low peaks of *Pteridium* and *Quercus* suggests that an open woodland was present (Barnosky, 1981; Franklin and Dyrness, 1973; Hansen, 1943; Hibbert, 1979). *Pseudotsuga's* dominance over *Tsuga heterophylla* suggests warmer, drier summers than at present (Heusser, 1977). Early-Holocene warmth and dryness are also indicated in a profile from Nisqually Lake, located in present-day prairies south of Puget Sound (Hibbert, 1979). High percentages of Gramineae, *Artemisia,* and Chenopodiineae in the early Holocene suggest that prairies expanded between about 10,000 and 7000 yr B.P. Prairie herbs declined as forests expanded again after 7000 yr B.P., signaling the end of the warm, dry interval.

Alnus and *Abies* were other components of the early-Holocene forest (Figure 8-3). *Alnus rubra* was apparently a persistent seral tree throughout the Holocene. *Abies* probably was represented by *A. lasiocarpa* during the earliest Holocene, and *A. grandis* seems to have appeared about 8000 yr B.P. (according to needle identifications in Barnosky, 1981). The replacement of open *Pseudotsuga* forests with closed *Pseudotsuga, Tsuga heterophylla,* and *Thuja plicata* forests about 6000 yr B.P. suggests a cooler, moister climate than that of the previous interval. The warmest part of the postglacial in this area seems to have been the early Holocene.

Coastal sites may have had a different vegetational history. For the Hoh Valley, Heusser (1974) reports that *Pinus contorta, Picea sitchensis,* and *Tsuga mertensiana* prevailed in early-Holocene forests that resembled modern subalpine forests. However, the dates for this period are out of sequence (Figure 8-3) (Heusser, 1974), and this pollen assemblage could be late glacial rather than early Holocene. The nearby section at Bogachiel contains no early-Holocene peaks of *Pinus* or *Tsuga mertensiana* (C. J. Heusser, 1978d). All sites in the Olympic Peninsula show that *Picea* (presumably *P. sitchensis*) and *Tsuga heterophylla* became established at the beginning of the Holocene. *Picea sitchensis* continues to the present as an important forest constitutent, but *Tsuga heterophylla* did not become dominant until about 8000 yr B.P. *T. heterophylla* requires moderately acid, well-drained, and highly organic soils in order to reproduce, and its slow rise to prominence in the Pacific Northwest may have been controlled by edaphic rather than climatic factors (Heusser, 1974).

The closed forests of the end of the early Holocene persisted throughout the late Holocene. *T. heterophylla* was apparently more abundant in the northern Puget lowland (Hansen and Easterbrook, 1974), whereas *Pseudotsuga* remained more important south of Puget Sound (Barnosky, 1981). *Thuja plicata* was apparently a dominant throughout western Washington, although its pollen (Cupressaceae) may not be equally recognized by all investigators. *Picea sitchensis* and *Abies* (probably *A. grandis*) were codominants with *Tsuga* and *Thuja* on the Olympic Peninsula (Heusser, 1974).

Pollen records for the last few hundred years show a decrease in conifer pollen percentages and an increase in *Alnus,* Gramineae, and *Pteridium* pollen (Figure 8-3) (Barnosky, 1981; Davis, 1973). This se-

Table 8-2. Data Available for Cores and Core Localities Mentioned in the Text.

Site (Reference[s])	Vegetational Map or Study	Surface-Pollen Study	Pollen Percentage	Pollen Concentration	Pollen Influx	Plant Macrofossils	Radiocarbon Dates (Number in Holocene)	Tephra	Elevation of Site (m)	Elevation of Treeline— Upper/Lower (m)
Davis Lake, Washington (Barnosky, 1981)	No	Yes	Yes	No	Yes	Selected	11	5-8	282	NA/NA
Hoh Valley, Washington (Heusser, 1974)	No	Yes	Yes	No	Yes	No	12	M	146	1500/NA
Pangborn Bog, Washington (Hansen and Easterbrook, 1974)	No	Yes	Yes	No	No	No	2	M	< 500	1700/NA
Bonaparte Meadows, Washington (Mack et al., 1979)	Yes	Yes	Yes	No	No	No	14	M, S	1021	NA/NA
Waits Lake, Washington (Mack et al., 1978e)	No	Yes	Yes	No	No	No	8	G, M	540	NA/NA
Big Meadow, Washington (Mack et al., 1978b)	No	Yes	Yes	No	Yes	No	8	M	1040	NA/NA
Lost Trail Pass Bog, Montana (Mehringer et al., 1977; Bright, unpublished data)	Yes	No	Yes	Yes	Zonal	Yes	16	G, M	2152	2800/NA
Cub Lake, Idaho (Baker, 1970)	No	Yes	Yes	No	No	None found	2	—	1840	3000/1750
Gardiner's Hole, Wyoming (Baker, 1970)	No	Yes	Yes	No	No	None found	1	M	2215	3000/1750
Cub Creek Pond, Wyoming (Waddington and Wright, 1974)	No	Yes	Yes	Yes	Zonal	No	2	G(?), M	2485	3000/1750
"Beaver Lake," Wyoming (Burkart, 1976)	Yes	Yes	Yes	Yes	No	No	1	—	2560	3000/2000
Sherd Lake, Wyoming (Burkart, 1976)	Yes	Yes	Yes	Yes	No	No	2	—	2660	3000/1500
Swan Lake, Idaho (Bright, 1966)	Yes	Yes	Yes	No	No	Yes	3	—	1450	NA/2000
Santa Barbara Basin, Pacific Ocean (L. E. Heusser, 1978)	NA	NA	Yes	Yes	Yes	No	0	—	− 550	NA/NA
Clear Lake, California (Adam, 1979b)	No	No(?)	Yes*	Yes*	Yes*	No	0	—	404	NA/NA
Osgood Swamp, California (Adam, 1967)	No	Yes	Yes	No	No	No	2	M	1980	> 3000
Snowbird Bog, Utah (Madsen and Currey, 1979)	No	One	Yes	No	No	Selected	5	—	2470	3200/1850
Twin Lakes, Colorado (Peterson and Mehringer, 1976)	No	Yes	Yes.	No	No	No	11	—	3290	3540/?
Hurricane Basin, Colorado (Andrews et al., 1975)	Yes	Yes	Yes	No	No	Selected	4	—	3650	3600/?
Alkali Creek Basin, Colorado (Markgraf and Scott, 1981)	No	One	Yes	Yes**	No	No	3	—	2800	3300/3000
Redrock Lake, Colorado (Maher, 1972)	No	Ratios	Yes	Yes**	Yes	No	7	—	3095	3500/1800
Chaco Canyon, New Mexico (Hall, 1977, 1981; Betancourt and Van Devender, 1980)	No	Yes	Yes	Yes**	No	No	18	—	1980	NA/2040

Abbreviations: NA, not applicable; G, Glacier Peak; M, Mazama ash; S, Mount St. Helens ash.
*Not fully published, but data are available.
**Available for total pollen only.

quence documents the destruction of the native forests by logging and the importance of *Alnus rubra* as a successional tree.

North-Central and Northeastern Washington

The pollen records at Waits Lake, Bonaparte Meadows, and Big Meadow are representative of the region (Figures 8-2 and 8-4 and Table 8-2). Fraser (Pinedale) glaciers covered these sites and withdrew about 12,500 yr B.P. The sites are in valleys cut in mountainous terrain. Waits Lake and Bonaparte Meadows are in a *Pseudo-tsuga menziesii* forest (Mack et al., 1978e, 1979), and Big Meadow is in the *Tsuga heterophylla* series (Daubenmire and Daubenmire, 1968; Mack et al., 1978b). Nearby highlands support forests of *Abies*

grandis, *Tsuga heterophylla*, and *Abies lasiocarpa* at successively higher elevations (Mack et al., 1979). South of the area, forests give way to grasslands and steppe (Küchler, 1964). Several studies of the modern pollen rain provide a basis for interpreting the diagrams (C. J. Heusser, 1978b; Mack and Bryant, 1974; Mack et al., 1978a).

The late-glacial sequence from 12,500 to 10,000 yr B.P. is dominated by pollen of *Artemisia*, Gramineae, other herbs, and haploxylon *Pinus* (Figure 8-4). No modern analogues exist for these pollen assemblages (Mack et al., 1978b, 1979), but they probably represent open tundralike environments, perhaps with scattered *P. albicaulis*.

During early-Holocene time, *Pinus* became more dominant in the pollen rain and changed to the diploxylon type. At Bonaparte

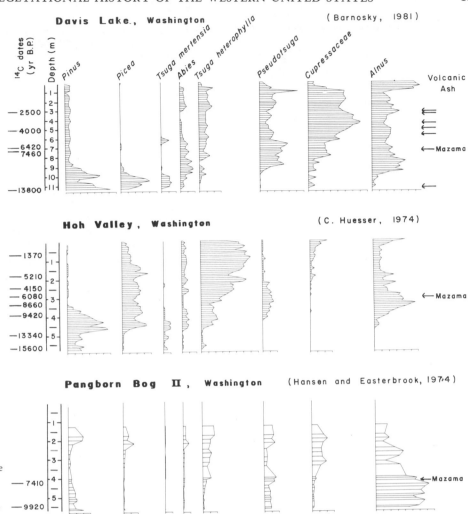

Figure 8-3. Pollen diagrams from Davis Lake, Hoh River valley, and Pangborn Bog, western Washington. (See Figure 8-2 for locations of the sites and Table 8-2 for data on the sites.) (Redrawn from Barnosky, 1981; C. Heusser, 1974; and Hansen and Easterbrook, 1974. [The latter two are from Geological Society of America publications].)

Meadows (Mack et al., 1979) and especially at sites near present grasslands (Nickmann, 1979), pollen percentages of Gramineae and *Artemisia* suddenly increase to a middle-Holocene peak. Even Big Meadow (Figures 8-2 and 8-4) and other sites within the forest (Mack et al., 1978b, 1978c) show a peak in Gramineae pollen between 9000 and 7000 yr B.P. More mesic sites reverted to forest diploxylon *Pinus* about 7000 yr B.P., but the more xeric sites maintained a predominantly herbaceous pollen rain until about 4000 yr B.P. (Mack et al., 1978b; Nickmann, 1979).

Apparently, *P. ponderosa* and/or *P. contorta* immigrated to northeastern Washington about 10,000 yr B.P. and remained at more mesic sites. However, grasses and *Artemisia* expanded their ranges shortly after 10,000 yr B.P., presumably moving northward from sources in central Washington. By middle-Holocene time the grassland/forest border was displaced northward some tens of kilometers. Even where *Pinus* remained, forests probably changed to open woodlands and savannas. Evidence seems strong in eastern Washington for an early-to-middle-Holocene interval warmer and drier than the present climate.

Diploxylon *Pinus* pollen has reached maximum values during the last 5000 years, but *Abies*, *Picea*, *Pseudotsuga-Larix*, and locally *Tsuga heterophylla* pollen replace Gramineae and other nonarboreal pollen (NAP) as important minor constituents of the pollen spectra (Figure 8-4) (Mack et al., 1978b, 1978c, 1978d, 1979). The replacement of steppes and grasslands by forests began in mesic habitats about 5000 yr B.P., but it did not occur at the present steppe/forest border in central Washington until about 4000 yr B.P. (Nickmann, 1979).

At Waits Lake and Bonaparte Meadows, the present *Pseudotsuga* forest became established about 5000 yr B.P. (Mack et al., 1978e, 1979). *Pseudotsuga* is notorious as an underproducer of pollen (Baker, 1976; Mack et al., 1978a), and its more-or-less continuous presence in the pollen diagram probably means that it was a dominant in the forest. *Pinus ponderosa* and *P. contorta* were probably present as well. No fluctuations in forest composition at these sites can be seen in the late Holocene.

In far-northeastern Washington and adjacent Idaho, the replacement of forest elements took place in two stages. About 4000 yr B.P., *Picea* and *Abies*, also pollen underproducers (Mack et al., 1978a), replaced the open pine forest of the middle Holocene. Although species identification is not possible, Mack and others (1978b) suggest on ecologic grounds that *Picea engelmannii* and *Abies lasiocarpa* shifted downslope; such a shift implies a decidedly cooler and moister climatic regime than is characteristic of the modern climate.

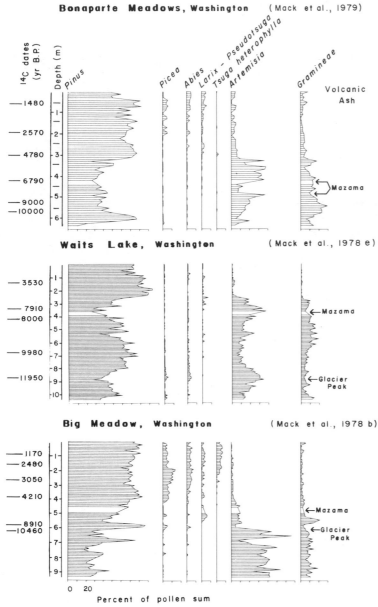

Figure 8-4. Pollen diagrams from Bonaparte Meadows, Waits Lake, and Big Meadow, eastern Washington. (See Figure 8-2 for locations of the sites and Table 8-2 for data on the sites.) (Redrawn from Mack et al., 1979; Mack et al., 1978e [copyright 1978 University of Chicago Press]; and Mack et al., 1978b [copyright 1978 Ecology Society of America].)

The modern *Tsuga heterophylla* forest of northeastern Washington and northern Idaho is often considered a Tertiary relict (Daubenmire, 1975). *T. heterophylla* pollen appeared in northeastern Washington about 2500 yr B.P. (Mack et al., 1978b) and in northern Idaho about 1500 yr B.P. (Mack et al., 1978c). Peaks in haploxylon *Pinus* and *Abies* pollen accompanying the *Tsuga* peak in northern Idaho suggest that *Pinus monticola* and *Abies grandis* were also late migrants to this community. Further documentation is needed, but this forest seems to have been a late-Holocene migrant, presumably from the northwest. The *Tsuga heterophylla* forest is more typically a coastal forest, and it requires a more equitable climate than the preceding *Picea-Abies* forest; warmer mean

temperatures and less extreme temperatures probably prevailed during the last 2500 years (Mack et al., 1978b).

Wyoming, Southwestern Montana, and Southeastern Idaho

Several sites from comparable elevations and vegetational belts are available from Wyoming and southwestern Montana (Figures 8-2 and 8-5 and Table 8-2). All sites were covered by Pinedale ice; the Bitterroot Range and the Bighorn Mountains supported valley glaciers (Mehringer et al., 1977; Nelson, 1970), and the Yellowstone Park area was covered by an ice cap (Pierce, 1979; Richmond, 1969; U.S. Geological Survey, 1972).

The sites in the Bighorn Mountains and Cub Lake and Gardiners Hole in Yellowstone Park are within the *Pinus contorta* belt, although Gardiners Hole is within a large, open sagebrush meadow (Figure 8-2). Cub Creek Pond and Lost Trail Pass Bog are in mixed coniferous forest of *Picea*, *Abies*, and *Pinus contorta*. From lowlands to uplands, the regional vegetational belts consist successively of chenopod steppe, *Artemisia* steppe, *Pinus flexilis* parkland, *Pseudotsuga* forest, *Pinus contorta* forest (the dominant cover in Yellowstone Park and the Bighorn Mountains), *Picea engelmannii-Abies lasiocarpa* forest, and alpine tundra (Arno, 1979; Despain, 1973). *Pinus albicaulis* is a locally important timberline tree in Yellowstone Park and the Bitterroot Range.

Modern pollen rain from the Bighorn Mountains and nearby basins (Burkart, 1976; McAndrews and Wright, 1969), the Beartooth Plateau (R. G. Baker, unpublished data), and Yellowstone Park (Baker, 1976) indicates that most vegetational belts have a distinctive pollen signature. The pollen rain in the basins is dominated by *Artemisia*, Gramineae, and Chenopodiineae pollen, with smaller maxima of Compositae, *Ambrosia*, and *Sarcobatus*. Rocky Mountain *Pseudotsuga* forests usually produce only low percentages of *Pseudotsuga* pollen. *Pinus contorta*-type pollen often comprises 80% to 90% of the pollen rain in the montane pine forests, and it may remain at levels of 50% to 80% in mixed *Picea-Abies-Pinus* forests. *Picea* and *Abies* pollen reach low peaks in the subalpine forests in which they dominate, and a variety of herb-pollen types, including Gramineae, Cyperaceae, and Compositae, reach peaks in the tundra.

The late-glacial sequence at all sites is marked by extremely low *Pinus* (10% to 20%) and high *Artemisia* pollen percentages (Figure 8-5). Gramineae and other herb pollen also register high values. Surface-pollen spectra from forest into tundra show trends toward low *Pinus* and high *Artemisia* percentages but do not match the late-glacial extremes. Pollen taxa characteristic of modern tundra are also found in late-glacial deposits. These include *Koenigia islandia*, *Polygonum bistortoides* or *P. viviparum*, *Polemonium*, *Caryophyllaceae*, *Phlox*-type, *Selaginella densa*, and others (Baker, 1976; Burkart, 1976). This pollen assemblage lasted from 14,000 to 11,500 yr B.P. in most areas, but it evidently lingered on until 10,000 yr B.P. in the western Bighorn Mountains (Figure 8-5) (Burkart, 1976). A widespread tundralike environment was apparently present at all sites across the entire Yellowstone Plateau and on surrounding mountain flanks. Macrofossil evidence suggests that scattered individuals of *Picea engelmannii*, *Pinus albicaulis*, and *Betula glandulosa* were present (Baker, 1976). If the treeline (presently about 3050 m) was below the lowest site at Cub Lake (elevation 1840 m),

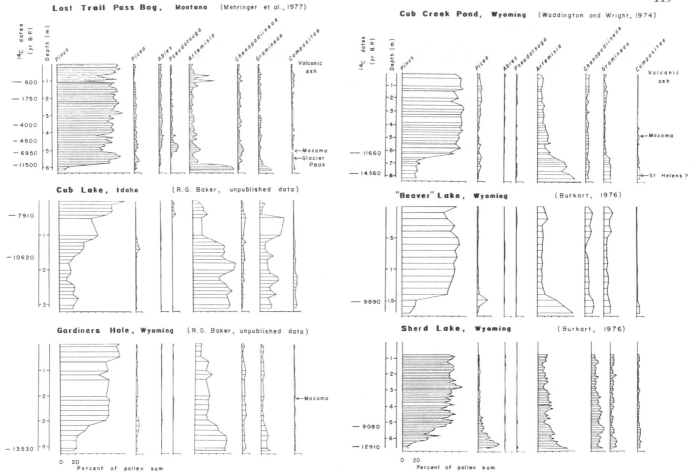

Figure 8-5. Pollen diagrams from Lost Trail Pass Bog, Montana; Cub Lake, Idaho; and Gardiners Hole, Cub Creek Pond, "Beaver" Lake, and Sherd Lake, Wyoming. (See Figure 8-2 for locations of the sites and Table 8-2 for data on the sites.) The pollen sum does not include Cyperaceae at Lost Trail Pass Bog, at Cub Lake, and from 0 to 50 cm at Gardiners Hole. (Lost Trail Pass Bog diagram redrawn from Mehringer et al., 1977; Cub Creek Pond, Waddington and Wright, 1974; Sherd Lake and "Beaver" Lake, Burkart, 1976)

a lowering of at least 1200 m is suggested. A drop in the treeline of 650 m is a minimum estimate for the Bighorn Mountains and the Bitterroot Range, because sites at low elevations are not known. If the upper treeline was 1200 m lower throughout the area, it would have been close to the base of most mountains. Yet, the nearby basins apparently were not covered by *Pinus* forests at that time, because *Pinus* pollen percentages were very low (Figure 8-5). Markgraf (1980) shows that the transport of pollen upslope is common in mountain regions. If *Pinus* were growing in the basins, its pollen percentages should be high in these mountain sites. Late-glacial *Pinus* forests were probably compressed into narrow belts along the base of the mountains. Ice- and sand-wedge casts from the Wyoming basins suggest that a cold, open environment was present in these lowlands (Mears, 1981), and perhaps the basins supported *Artemisia*, the dominant pollen type in the late glacial.

The early-Holocene record is characterized by a rise in *Picea* pollen, followed shortly upsection by a marked rise in *Pinus* pollen percentages. *Picea engelmannii* was evidently the first tree to expand its areal extent during the Holocene, followed by *Pinus albicaulis*. This mixed forest lasted until about 8800 yr B.P. at Cub Creek Pond and Lost Trail Pass Bog (Mehringer et al., 1977; Waddington and Wright, 1974). In the Bighorn Mountains, *P. albicaulis* is not present, and the montane *P. contorta* is not common at the present

treeline. For the early Holocene, *Picea* percentages are very high, and *Pinus* pollen rises to its modern values much later. Because *P. contorta* is not well adapted to near-timberline conditions, it was much slower to immigrate upslope and become prominent than was *P. albicaulis* farther west, leaving *Picea* to play a much more dominant role in the early-Holocene vegetation of the Bighorn Mountains. From about 9000 to 8000 yr B.P., *Picea* and *Pinus contorta* grew in mixed forests, and for the last 8000 years no observable change has occurred in the *P. contorta* forests that dominated the Bighorns.

In the Yellowstone Park area, the direction of middle- and late-Holocene changes depends on site location and elevation. At higher sites such as Cub Creek Pond, presently within subalpine forests, *Picea*, *Abies*, and *Pinus albicaulis* decline to minima in pollen percentages, and *P. contorta* reaches a maximum between the 8000 and 4000 yr B.P. These changes suggest that the former three trees retreated upslope during warmer, drier conditions (Baker, 1976; Waddington and Wright, 1974). During the last 4000 years, these species have returned to form mixed forests with *P. contorta*. Plant-macrofossil evidence supports this reconstruction (Baker, 1976).

Xeric sites such as Gardiners Hole and nearby Blacktail Pond (Gennett, 1977) are in large, open *Artemisia* steppe areas surrounded by forest or parkland. At these sites, *Pinus* (mainly *P. contorta*) percentages level off between 11,000 and 7000 yr B.P. but then fall to a Holocene minimum as *Artemisia* rises between 7000 and 1500 yr B.P. (Figure 8-5) (Gennett, 1977). Evidently, the forest boundary retreated and the forest thinned out during this interval. It seems likely that the warmest and driest part of the Holocene occurred later at low elevations than at high elevations. After 1500 yr B.P., percentages of *Artemisia* pollen decline and *Pinus* rises to its Holocene high; this implies denser forests and a closer forest border.

At Cub Lake (Figure 8-5), a mesic site at low elevation, *Pinus* pollen percentages atypically increase very gradually as *Artemisia* and Gramineae percentages decrease in the early Holocene. The implications of the substantial peaks in Gramineae pollen as late as 8000 yr B.P. are unknown. The site is only 10 km east of the extensive steppes of the Snake River Plain, and forest establishment may have been slower there than elsewhere. Thick loessial soils and dry winds off the Snake River Plain at Cub Lake may have inhibited rapid forest development. The early-Holocene vegetation is tentatively interpreted as an open savanna with *Artemisia*, *Gramineae*, *Pinus flexilis* or *P. albicaulis*. *Pseudotsuga* percentages range from 2% to 5% between 8000 and about 1500 yr B.P. (Baker, 1970). On the basis of surface-sample studies (Mack et al., 1978a), such percentages indicate a *Pseudotsuga* forest at the site. This type of forest is now found at lower elevations, indicating warmer conditions than at present. The present *Pinus contorta* forest became established about 1500 years ago; this suggests a downward migration of *P. contorta* and a cooler climate. However, fire can be important in the distribution of *P. contorta* and *Pseudotsuga* in Yellowstone Park (Houston, 1973; Taylor, 1973), and its role in the past is uncertain.

At Lost Trail Pass Bog (Figure 8-5), both needles and pollen indicate that *Pseudotsuga* formed dense forests with *Pinus contorta* from 7000 to 4000 yr B.P. (Bright, in preparation; Mehringer et al., 1977). *Pseudotsuga* is presently restricted to warm microhabitats and is at its upper limit at the pass. Its upward extension was probably caused by a warmer middle-Holocene climate. During the late Holo-

cene, *Pseudotsuga* became rare and the forests were dominated by *Pinus contorta*, *Picea engelmannii*, and *Abies lasiocarpa*. *Pinus albicualis* became more abundant in these forests about 1750 yr B.P. The elimination of *Pseudotsuga* and the increase in abundance of subalpine elements during the last 4000 years demonstrates that Neoglacial climates were cooler than those of the middle Holocene (Bright, in preparation).

Other pollen sequences from the Wind River Range, central Wyoming, show fluctuations similar to both low- and high-altitude sites in Yellowstone Park (Bright, unpublished data), and a recently completed pollen sequence from a site in the mountains of west-central Montana resembles the high-altitude sections (Brant, 1980).

Southeastern Idaho

Swan Lake, in the outlet of Lake Bonneville (Figures 8-2 and 8-6 and Table 8-2) is one of the few sites available below the lower treeline. It is surrounded by *Artemisia* steppe, with chenopod steppe nearby and shrublands on adjacent slopes (Bright, 1966). The nearest large stand of *Pinus contorta* is 40 km distant. *Pseudotsuga*, *Picea engelmannii*, *P. pungens*, *Abies concolor*, and rare *Abies lasiocarpa* occur in the mountains nearby. The area was unglaciated, but spill waters from Lake Bonneville scoured the outlet sometime before 12,000 yr B.P.

The late-glacial sequence is characterized by peaks in *Pinus* (including both *P. contorta* and *P. flexilis* types) and *Picea* pollen percentages (Figure 8-6). Nonarboreal pollen percentages are generally low (Bright, 1966). This assemblage represents a mixed forest of *Pinus contorta*, *Picea* (probably *P. engelmannii*), and *Pinus flexilis* or *P. albicaulis*. Vegetational zones must have been at least 700 m lower than they are at present in order to have brought coniferous forests to the site, and a more likely estimate is 1100 m.

The transition to early-Holocene vegetation began about 10,800 yr B.P. when Chenopodiineae, Compositae, *Ambrosia*, and Gramineae

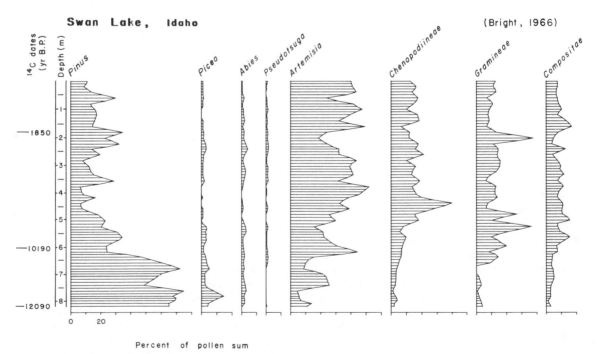

Figure 8-6. Pollen diagram from Swan Lake, southeastern Idaho. (See Figure 8-2 for location of the site and Table 8-2 for data on the site.) The pollen sum does not include Cyperaceae. (Redrawn from Bright, 1966.)

pollen became abundant and an *Artemisia*-Gramineae steppe replaced the forest. By middle-Holocene time, an *Artemisia*-chenopod steppe prevailed. The soil alkalinity may have increased since alkali-loving chenopods such as *Atriplex*, *Kochia*, *Salicornia*, and *Suaeda* became abundant (Bright, 1966). Peaks in the percentages of *Pinus* pollen in the late Holocene probably indicate that the lower treeline descended during the period from 3100 to 1700 yr B.P., but the vegetation at Swan Lake was probably still *Artemisia* steppe, as it is at present. After about 1700 yr B.P., the treeline ascended to its modern level, and the present vegetational pattern was established.

California

California's Holocene vegetational history is still relatively obscure. Not only is the diversity of trees greater than elsewhere in the West, but many important forest species have indistinctive pollen grains.

These include *Pinus*, *Quercus*, Cupressaceae, Taxodiaceae, and Taxaceae. There are also few carefully worked sites in the state. Three sites are reviewed here (Figure 8-7 and Table 8-2): Osgood Swamp on the eastern slope of the Sierra Nevada (Adam, 1967), Clear Lake in the Coast Range (Adam, 1979a, 1979b; Adam and Sims, 1977; Adam et al., 1981; Casteel et al., 1977), and the Santa Barbara Basin off the coast of California (L. E. Heusser, 1978).

The Santa Barbara Basin is a closed depression over 550 m deep off the coast of southern California (Table 8-2). Its varved sediments contain a 12,000-year pollen record of paleoecologic trends in the upland and coastal vegetation (L. E. Heusser, 1978). The validity of correlating the pollen in marine sediments with terrestrial vegetation history has been demonstrated in Washington State, where such pollen sequences correlate well with continental sequences (Heusser and Balsam, 1977; Heusser and Florer, 1973). The modern vegetation includes scrub vegetation on coastal terraces that is dominated by

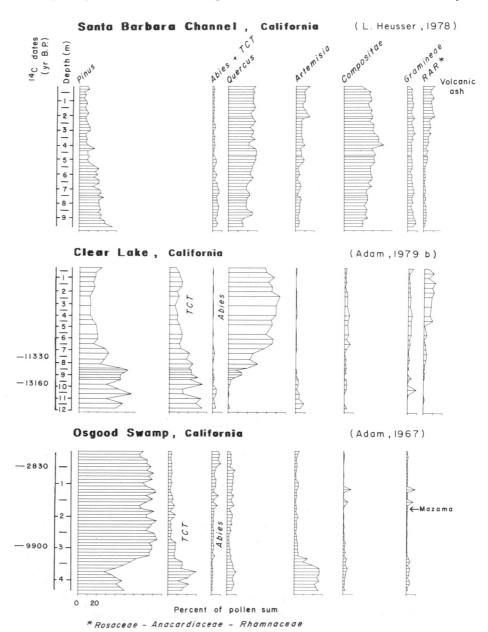

Figure 8-7. Pollen diagrams from Santa Barbara Channel, Clear Lake, and Osgood Swamp, California. (See Figure 8-2 for locations of the sites and Table 8-2 for data on the sites.) The pollen sum for *Pinus* at Osgood Swamp is computed in the usual way, but all other pollen percentages at that site are based on a pollen sum excluding *Pinus* pollen. (Redrawn from L. Heusser, 1978 [from a Geological Society of America publication]; Adam, 1979b; and Adam, 1967.)

Artemisia and *Salvia*. Chaparral dominated by Rhamnaceae, Anacardiaceae, and Rosaceae covers much of the foothills and Coast Range (L. E. Heusser, 1978). Higher mountains along the coast support a forest of mixed conifers and oak.

Despite problems in interpreting what pollen came from different vegetational belts, L. E. Heusser (1978) deduces a vegetational history consistent with subsequent work at Clear Lake. (See the discussion below.) *Pinus* and other conifer pollen percentages are predominant in the late-glacial and early-Holocene sediments from 12,000 to about 7800 yr B.P. (Figure 8-7). Filicales (fern) spores are also abundant, and nonarboreal pollen percentages are low. Mixed conifer-oak forests like those of the high parts of the Coast Range at present apparently occupied large areas at lower elevations during the early Holocene. About 7800 yr B.P., *Pinus* pollen percentages dropped and *Quercus*, Rosaceae-Anacardiaceae-Rhamnaceae, *Artemisia*, and other Compositae percentages rose. Middle- and late-Holocene vegetational changes are reflected by high pollen percentages of *Quercus* at about 5700 yr B.P., Compositae between 5700 and 2300 yr B.P., and Rosaceae-Anacardiaceae-Rhamnaceae and *Artemisia* from about 2300 yr B.P. to the present. The vegetation apparently was approaching its modern condition 7800 years ago. The progressive increase in chaparral pollen types and *Artemisia* pollen suggests increasing dominance by coastal sage scrub and chaparral communities, respectively, in the middle and late Holocene.

Clear Lake (Table 8-2), a tectonic basin on the northern Coast Range, was unglaciated. The lake has yielded long cores, including one that contains sediments of the last interglacial (Adam and Sims, 1977; Adam et al., 1981). Mixed forests of *Quercus douglasii* and *Pinus sabiniana* along with *Arbutus* and *Pseudotsuga* surround the lake. A chaparral community dominated by genera in the Rosaceae, Ericaceae, Anacardiaceae, and Rhamnaceae covers the lower mountains (L. E. Heusser, 1978); the higher mountains in the region support a mixed coniferous forest of *Pinus*, *Abies*, and *Pseudotsuga* (Küchler, 1964).

The late-glacial pollen spectra are typified by high percentages of *Pinus* and TCT (Taxodiaceae, Cupressaceae, Taxaceae), along with smaller maxima of *Abies* and *Artemisia* (Figure 8-7). *Quercus* pollen percentages gradually rise and *Artemisia* percentages decline beginning about 13,000 yr B.P. (Adam, 1979b). Closed coniferous forests with increasing proportions of *Quercus* must have dominated in the region. Small peaks in *Artemisia* suggest that some open areas may have been present at times.

About 10,000 yr B.P., percentages of *Pinus* and TCT decline and those of *Quercus*, Compositae, and Gramineae increase to their Holocene levels. Pollen of RAR (Rosaceae-Anacardiaceae-Rhamnaceae) become important between about 7000 and 6500 yr B.P. Mixed oak forests like those currently growing near the lake must have become established at the beginning of the Holocene. The appearance of RAR pollen indicates that the chaparral vegetation presently dominant in the Coast Range arrived about 7000 yr B.P. Differential plant migration rates may have affected the time these communities appeared. Barring such a nonclimatic effect, the climatic trend suggested by this sequence is increasing warmth and dryness throughout the Holocene.

The Clear Lake and Santa Barbara Basin pollen samples are the first obtained from lowland California. Although they provide a fairly consistent sequence of pollen assemblages, both sites occur in large catchment basins, and they integrate the pollen rain from a large region of high topographic relief. Such sites are often less sensitive to climatic shifts than sites from smaller basins. This sequence should, therefore, be considered tentative until small lake or bog sites are found.

Osgood Swamp (Table 8-2) was glaciated in Tioga (Pinedale?) time. The modern vegetation of the Sierra Nevada includes *Quercus* in the Central Valley west of the mountains; *Quercus* and *Pinus* species in the western foothills; *Pinus* species, *Abies concolor*, *Pseudotsuga*, *Libocedrus*, and *Quercus kelloggii* in the yellow pine belt; *Pinus* species, *Abies magnifica*, *Juniperus occidentalis*, and *Quercus* in the lodgepole pine/red fir belt; *Pinus* species and *Tsuga mertensiana* in the subalpine belt; and Gramineae, Cyperaceae, and other herbs in the alpine belt (Storer and Usinger, 1963). East of the mountains is the sagebrush belt, with *Artemisia tridentata*, *Pinus monophylla*, and *Juniperus osteosperma*, although *Artemisia tridentata* extends up to some mountaintop communities (Adam, 1967).

Late-glacial pollen spectra (Figure 8-7) are marked by low *Pinus* and high *Artemisia* and TCT pollen percentages between about 15,000 and 10,000 yr B.P. (Adam, 1967; Batchelder, 1980; Dorland et al., 1980). The vegetation is interpreted as tundra or tundralike, and the treeline was at least 650 m lower than it is today.

The early Holocene was distinguished by high percentages of *Pinus* and TCT pollen, by the steady presence of *Abies* pollen, and by low percentages of *Artemisia* pollen (Adam, 1967; Dorland et al., 1980). A mixed *Pinus-Abies*-Cupressaceae (probably *Juniperus*) forest apparently inhabited the area during early-Holocene time. *Pinus* pollen percentages remain high and TCT pollen percentages drop to low levels for the remainder of the Holocene. *Pinus-Abies* forests probably prevailed during the middle-Holocene time until about 2800 yr B.P., and the climate was somewhat warmer than it is at present. A rise in *Abies* pollen percentages about 2800 yr B.P. indicates that the late Holocene was somewhat cooler, and *Abies*, probably *A. magnifica* (Adam, 1967), was able to grow at slightly lower elevations than it did during middle-Holocene time.

On the basis of present forest distribution, Adam (1967) has interpreted an increase in Cupressaceae pollen in the Hodgdon Ranch section on the western side of the Sierra Nevada as representative of an upward movement of the *Libocedrus* forest in middle-Holocene time. Remnants of trees above the present treeline in the White Mountains of California have been dated as middle-Holocene by dendrochronology (LaMarche, 1973), indicating a higher treeline there as well. These data support the hypothesis of a warmer middle-Holocene climate in the mountains of California.

Colorado and Utah

Despite several recently published pollen records from Colorado and Utah, the Holocene vegetational history is relatively unsettled because some sites seem to be located far from ecotones or are otherwise ecologically insensitive and other records are difficult to interpret. Sites reviewed here include Redrock Lake (Maher, 1972) in the Colorado Front Range, Alkali Basin (Markgraf and Scott, 1981) in central Colorado, Hurricane Basin (Andrews et al., 1975) in the San Juan Mountains, Twin Lakes (Petersen and Mehringer, 1976) in the La Plata Mountains, and Snowbird Bog (Madsen and Currey, 1979) in the Wasatch Range of Utah (Figure 8-2 and 8-8 and Table 8-2).

Redrock Lake and Twin Lakes are within the *Picea-Abies* forest belt, and Hurricane Basin lies within tundra about 50 m above the treeline. Alkali Basin is 200 m below the treeline in an *Artemisia* steppe; forests above the site consist of *Picea engelmannii*, *Abies lasiocarpa*, and *Pseudotsuga menziesii*. Snowbird Bog lies in the bottom of a deep canyon amongst forests of *Picea engelmannii*, *Abies lasiocarpa*, and *Pseudotsuga*. *Pinus flexilis* occurs near the timberline in the Wasatch Range. Modern pollen-rain studies by Maher (1961, 1963, 1972) from central and southwestern Colorado are useful in interpreting these localities.

All sites but Alkali Basin were glaciated. All sequences but Snowbird Bog are from lake sediments; the peculiar deposits at Snowbird Bog consist of peat representing bog growth interbedded with forest-floor debris derived from snowslides coursing down the steep valley walls (Madsen and Currey, 1979).

Late-glacial pollen sequences are available from three sites (Figure 8-8)). Redrock Lake and Snowbird Bog show typical late-glacial profiles for Rocky Mountain sites, with high percentages of *Artemisia*, Gramineae, and Compositae pollen and very low values of *Pinus* prior to 10,000 yr B.P. (Legg and Baker, 1980; Madsen and Currey, 1979; Maher, 1972). Compared to the modern treeline, the treeline was a minimum of 500 m lower at Redrock Lake and 700 m lower at Snowbird Bog, and the sites were probably covered by alpine tundra. At Alkali Basin, the pollen spectra suggest a mixed *Pinus-Picea-Abies* forest during late-glacial time. Herb pollen percentages are low; this implies that a closed subalpine forest was present. At this site, the maximum lowering of the upper treeline must have been 500 m because any larger figure would place the site above the treeline, where herb pollen percentages are generally high.

During the Holocene, the sequence of changes at the sites varies because of the different topographic settings. At Redrock Lake, the early-Holocene sequence began about 7600 yr B.P., with peaks in *Picea* and *Abies* pollen (Figure 8-8). *Picea* percentages decline about 7600 yr B.P., and *Pinus* pollen gradually increases until about 5000 yr B.P. and remains at high levels until the present. *Betula* pollen percentages increase at about 2000 yr B.P. Maher (1972) interprets this sequence mainly on the basis of ratios of *Picea* to *Pinus*. In the modern pollen rain of the San Juan Mountains (Maher, 1961, 1963) and the Front Range (Maher, 1972), these ratios increase systematically with elevation up to the timberline and then decrease above the timberline. The ratios can, therefore, be used to predict the "apparent elevation" at any stratigraphic level on a pollen diagram (Maher, 1963). Using this method, Maher (1972) concludes that the timberline was lower and the climate was cooler and moister than at present from 10,000 to 7600 yr B.P. and from 6700 to 3000 yr B.P. and that conditions were much like modern ones from 7600 to 6700 yr B.P. and for the last 3000 years (Figure 8-9).

Phytogeographic considerations suggest a hypothesis different from Maher's. The high percentages and influx of *Picea* in the early Holocene indicate that this tree was more important locally then than it is now. *Picea* today is common above Redrock Lake. *Artemisia* percentages also are relatively high; this suggests that there were open areas nearby. The vegetation in the early Holocene was probably an open *Picea-Pinus* forest, a finding suggesting a climate cooler than the present's. The middle-Holocene dominance of *Pinus* pollen suggests a *Pinus* forest, which could represent a warm, dry middle Holocene. The rise of *Betula* pollen percentages and influx about 2400 yr B.P. probably marks the appearance of *B. glandulosa*, a shrub now present at the lake and most abundant near the treeline in the subalpine belt. (*B. occidentalis*, a riparian tree of mainly lower elevations, is a less likely species.) The *Betula* pollen peaks suggest a cooler Neoglacial climate at Redrock Lake. This hypothesis fits better

with other sites in Colorado (Figures 8-8 and 8-9) and elsewhere in the Rocky Mountains. Nevertheless, Maher's (1972) interpretation may be correct; the Holocene record of glacial fluctuations in the Front Range does not fit well with other areas (Benedict, 1981), and a late-Holocene sequence from the western side of the Front Range (Nelson et al., 1979) agrees with Maher's interpretation that the last 4000 years were similar to the present. More detailed work is needed to settle this problem.

At Hurricane Basin in the San Juan Mountains, low *Picea* and *Pinus* pollen percentages and high *Artemisia* occur for the period before about 8400 yr B.P. (Figure 8-8). *Picea* and *Pinus* dominated the pollen rain of the middle Holocene until about 3500 yr B.P., when *Pinus* and nonarboreal pollen became predominant. Andrews and others (1975) conclude that prior to 8400 yr B.P. the climate was somewhat cooler than the present (based on Maher's pollen ratios) but that the treeline was at the site (based on *Picea* macrofossils). From 8400 to 3000 yr B.P., a *Picea* forest was present at the site, *Pinus* was probably abundant at slightly lower elevations, and the climate was warmer and/or drier than it is now. During the last 3000 years, the site has been above the treeline (Figure 8-9). *Picea* macrofossils, present throughout the middle-Holocene deposits, are missing from the late-Holocene deposits, and so cooler conditions are implied.

At Twin Lakes in the La Plata Mountains, a *Picea-Pinus* pollen rain has been predominant throughout the last 10,000 years (Figure 8-8). *Picea-Pinus* pollen ratios indicate that the climate was cooler and the treeline lower from 10,000 to 8600 and from 8000 to 6900 yr B.P., for short periods at 5600, 4500, 3600, 2800, 1400 yr B.P., and from 750 to 150 yr B.P. (Petersen, 1981; Petersen and Mehringer, 1976). Warmer periods during which the treeline advanced upslope are suggested from 8600 to 8300, 6700 to 5900, and 4300 to 4000 yr B.P. Parts of the record correlate well with the nearby Hurricane Basin record (Figure 8-9), but it is unfortunate that other taxa are not more useful in interpreting the Twin Lakes sequence. Apparently, the lake is in an ecologically insensitive location.

At Snowbird Bog (Figure 8-8), late-glacial conditions evidently persisted until about 8000 yr B.P., when *Picea* and *Pinus* pollen percentages increase to maximum levels until about 5000 yr B.P. (Madsen and Currey, 1979). Little herb pollen is present for this middle-Holocene period of arboreal dominance. During the last 5000 years, however, percentages of *Picea* and *Pinus* have been lower, but they fluctuate substantially and nonarboreal pollen percentages increase. The unusual topographical conditions and depositional history of this site make the interpretation of this pollen sequence difficult. Madsen and Currey (1979) hypothesize that abundant conifer pollen reflects warm conditions, when the entire valley was covered with *Picea*, *Pinus*, and *Abies* forests. For periods when conifer pollen percentages are low, they reason that the treeline had fallen below the tops of the valley walls, limiting the conifers to the valley floors. Using a ratio of conifer pollen to total pollen, they derive a climatic curve for the site (Figure 8-9). From 8000 to 5000 yr B.P., the valley supported a coniferous forest from top to bottom, and the climate was warmer than present. During the last 5000 years, the treeline has been depressed, and conifers have had a more limited distribution than during the preceding warm phase. The general curve for this site correlates well with the trends from southwestern Colorado (Figure 8-9).

Alkali Basin, the only site from below the lower treeline in this

120

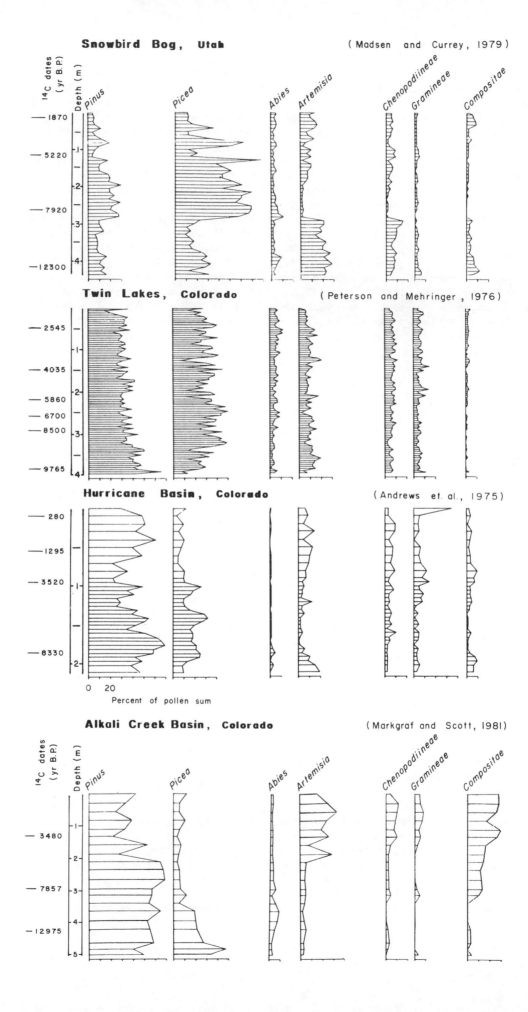

Snowbird Bog, Utah (Madsen and Currey, 1979)

Twin Lakes, Colorado (Peterson and Mehringer, 1976)

Hurricane Basin, Colorado (Andrews et. al., 1975)

Alkali Creek Basin, Colorado (Markgraf and Scott, 1981)

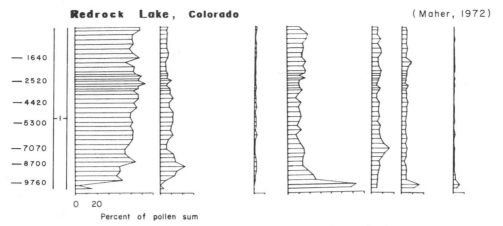

Percent of pollen sum

Figure 8-8. Pollen diagrams from Snowbird Bog, Utah; and Twin Lakes, Hurricane basin, Alkali Creek Basin, and Redrock Lake, Colorado. (See Figure 8-2 for locations of the sites and Table 8-2 for data on the sites.) The pollen sum at Snowbird Bog, Twin Lakes, and Hurricane Basin does not include Cyperaceae; the pollen sum at Redrock Lake does not include other Compositae. (Redrawn from Madsen and Currey, 1979; Petersen and Mehringer, 1976; Andrews et al., 1975; Markgraf and Scott, 1981 [from a Geological Society of America publication]; and Maher, 1972.)

region, presents an interesting contrast with the other sites. The early-Holocene spectra are dominated by *Pinus* pollen, with gradually rising percentages of *Artemisia* and Compositae pollen from 10,000 to 5000 yr B.P. The vegetation represented by this assemblage was an open montane *Pinus* forest growing about 200 m lower than it is now. The last 5000 years have been characterized by lower *Pinus* pollen percentages and higher *Artemisia* and Compositae percentages. Chenopodiineae and Gramineae also have reached maximum percentages during late-Holocene time. These sharp changes in pollen rain signal the establishment of the modern steppe vegetation about 5000 yr B.P. (Figures 8-8 and 8-9). This entire sequence from late glacial to present suggests a progressively warmer, drier climate. No hint is found of an ''Altithermal'' climate warmer than the present during the middle Holocene nor of a Neoglacial cooling in the late Holocene. Markgraf and Scott (1981) suggest that Neoglacial cooling could have occurred at high altitudes at the same time that warmer, drier conditions were forcing the lower treeline to rise if a stronger pattern of summer monsoonal rainfall existed. Apparently, their reasoning is that the warm, relatively moist summer air would have resulted in more rain at higher elevations, where more convective storms presently occur, whereas the main effect at lower elevations would have been warmer air.

Desert Southwest

The vegetational record of the Southwest differs in kind from records elsewhere in the West. Glaciated terrain with lakes and bogs is rare or absent. Many pluvial lakes contain pollen-bearing sediments of Wisconsin age but not of Holocene age. Alluvium and spring deposits have yielded several pollen sequences (Martin, 1963; Martin and Mehringer, 1965; Mehringer, 1965). Pollen has even been extracted from pack-rat middens (King and Van Devender, 1977), cave fills, animal dung (Martin et al., 1961), and human dung (Bryant and Williams-Dean, 1975). A few pollen sequences have been published since Martin and Mehringer's (1965) review; these are briefly summarized below.

Pollen records provide useful data, but the analysis of plant macrofossils preserved in packrat (*Neotoma*) middens has revolutionized southwestern paleoecology. The packrat collects a representative sample of the surrounding vegetation (Wells, 1976) and cements the various plant fragments together with urine. Each mid-

den must be radiocarbon dated because it represents only a short interval of time. Middens are common in certain areas, and they span much of late-Quaternary time. The preservation of plant macrofossils is excellent, and most plants can be identified to species. Midden analysis was pioneered by Wells (1966, 1976, 1979; Wells and Berger, 1967; Wells and Hunziker, 1977; Wells and Jorgensen, 1964) and further developed by Van Devender and his colleagues in Arizona (Spaulding et al., 1983; Van Devender 1977, 1980; Van Devender and Mead, 1976; Van Devender and Riskind, 1979; Van Devender and Spaulding, 1979). The rich plant assemblages contained in these middens are only briefly mentioned here because two recent review papers are available (Van Devender, 1977; Van Devender and Spaulding, 1979).

The southwestern United States is predominantly arid and semiarid; four separate deserts are recognized, each with a distinct vegetation. Mountain ranges provide areas of lower temperature and higher precipitation, and they support woodlands and forests separated by the broad desert basins. The Mohave Desert communities of

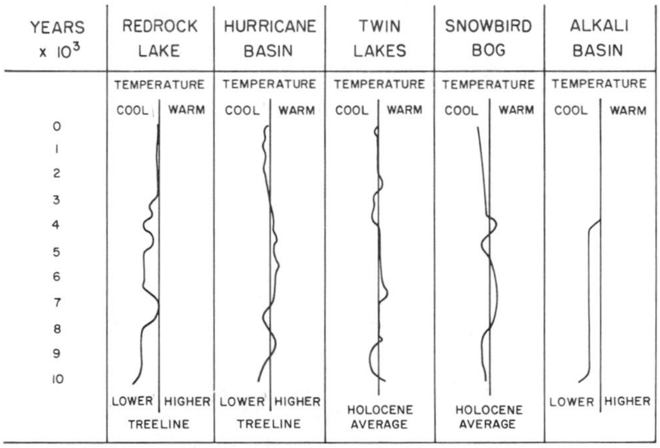

YEARS x 10³	REDROCK LAKE		HURRICANE BASIN		TWIN LAKES		SNOWBIRD BOG		ALKALI BASIN	
	TEMPERATURE		TEMPERATURE		TEMPERATURE		TEMPERATURE		TEMPERATURE	
	COOL	WARM	COOL	WARM	COOL	WARM	COOL	WARM	COOL	WARM
0 1 2 3 4 5 6 7 8 9 10										
	LOWER	HIGHER	LOWER	HIGHER	HOLOCENE AVERAGE		HOLOCENE AVERAGE		LOWER	HIGHER
	TREELINE		TREELINE							

Figure 8-9. Climatic curves for Redrock Lake, Hurricane Basin, Twin Lakes, Snowbird Bog, and Alkali Basin. The first four curves are the original authors' estimates of upper treeline based on pollen ratios; the curve for Alkali Basin is based on the original authors' verbal description of lower treeline fluctuations. The baselines for Redrock Lake, Hurricane Basin, and Alkali Basin are the modern treelines. The baselines for Twin Lakes and Snowbird Bog are the mean Holocene pollen ratios. (See Table 8-2 for other data on the sites.)

southern California and adjacent Nevada and northwestern Arizona are usually composed of *Larrea divaricata, Ambrosia dumosa*, and *Encelia farinosa*, along with several cacti and *Yucca* species (Küchler, 1964; Mehringer, 1965; Van Devender and Mead, 1976). The lower mountain slopes support *Juniperus osteosperma*, joined at slightly higher elevations by *Pinus monophylla. P. ponderosa, Abies concolor*, and *Pinus longaeva* occur at successively higher altitudes.

The cooler, moister Great Basin Desert of Nevada and Utah is dominated by *Atriplex confertifolia, Artemisia tridentata*, and *Sarcobatus vermiculatus* (Küchler, 1964; Van Devender and Mead, 1976), with *Juniperus osteosperma*, Pinyon pine (*Pinus monophylla* in the west and *P. edulis* in the east), and *Quercus* species (on the eastern and western margins) at higher elevations. Mountain slopes are covered by forests of *Pinus longaeva* and *P. flexilis*. In the eastern Great Basin ranges, these trees are joined by *P. ponderosa, Abies concolor, Pseudotsuga menziesii, Picea engelmannii*, and *Abies lasiocarpa* (Küchler, 1964).

The lowest, hottest desert in North America is the Sonoran Desert of southern Arizona. It is dominated by *Larrea divaricata, Ambrosia dumosa, Cercidium microphyllum*, and several taxa of cacti (Küchler, 1964; Van Devender, 1977). Pinyon-juniper woodlands include *Pinus monophylla, Juniperus osteosperma, J. deppeana*, and *Quercus* species. The Chihuahuan Desert of southern New Mexico and Texas is distinguished by *Larrea divaricata*, and *Flourensia cernua*, with *Agave lechugilla, Opuntia* species, *Dasylirion* species, *Fourquieria splendens*, and various grasses. Three Pinyon pines are present, though geographically separated from one another, in the next-higher vegetational zone, along with *Juniperus* species and *Quercus* species. Mixed coniferous forests of *Pinus ponderosa, P. strobiformis*, and *Pseudotsuga menziesii* occur atop many of the low mountains.

The late-glacial fossil record from the Southwest clearly indicates that a woodland prevailed in most areas that are now desert (Spaulding et al., 1983). The pollen record is strongly dominated by *Pinus* (50% to 85%), and the proportion of *Picea* pollen rises to over 8%, even at localities below 2000 m (Martin and Mehringer, 1965). These sites, which are now cool grasslands, were interpreted as supporting *Pinus* and *Picea* forests. Sites such as Searles Lake in the Mohave Desert contain a preponderance of *Pinus* and *Juniperus* pollen; this suggests that Pinyon-juniper woodlands prevailed (Leopold, 1967; Spaulding et al., 1983). Two mountain lakes above 2300 m in New Mexico and Arizona were surrounded by alpine tundra during late-glacial times (Whiteside, 1965; Wright et al., 1973). These lakes presently lie amidst *Pinus ponderosa* forests. Wright and others (1973) infer a depression of the upper treeline of about 900 m during late-glacial time in northwestern New Mexico. All vegetational belts apparently were not depressed equally, however; Wright and others (1973) give evidence that the adjacent basins supported Pinyon pine, and they conclude that the *P. ponderosa* belt was depressed only 500 m (*Pinus* species were identified by pollen-morphological studies). Wells and Berger (1967) indicate that Pinyon-juniper woodlands descended 600 m below their present limits in southern California. Montane and subalpine forests in the Southwest must have been compressed during late-glacial time. Van Devender and Spaulding (1979) suggest a similar compression of subalpine forests from packrat-midden evidence in southwestern Texas. Spaulding and others (1983) document several other examples of unequal movements of species from vegetational belts at different altitudes.

The record from packrat middens also indicates that Pinyon pines and juniper were dominant at low to middle elevations (550 to 1525 m) now occupied by desert-scrub communities (Van Devender and Spaulding, 1979). *Juniperus* woodlands without *Pinus* extended to as low as 425 m. Desert-scrub communities may have existed below about 400 m along the lower Colorado River, but no middens of Late Wisconsin age have been found with a desert flora (Van Devender and Spaulding, 1979). Many sites document woodland elements mixed with Mohave and Great Basin desert plants at lower altitudes and farther south than they occur today. Mixed forests of *Abies concolor, Pseudotsuga menziesii*, and *Picea* are known (but rare) from localities ranging from 1500 to about 1900 m (Van Devender and Spaulding, 1979). Above 1900 m, *Pinus longaeva* and *P. flexilis* grew together in southern Nevada (Van Devender and Spaulding, 1979) and in the central Great Basin (Thompson and Mead, 1982). Clearly, both latitudinal and altitudinal shifts in distributions occurred during late-glacial time (Spaulding et al., 1983). Like plants in the East and Midwest, species in the Southwest moved independently, not as groups in discrete communities.

The *Pinus*-dominated pollen profiles of the late glacial changed to NAP-dominated spectra with Chenopodiineae and Compositae about 12,000 yr B.P. (Clisby and Sears, 1956; Martin and Mehringer, 1965; Mehringer, 1965; Whiteside, 1965). Holocene pollen diagrams at low elevations show low *Pinus* percentages and irregular fluctuations in Chenopodiineae and Compositae, but there are few other changes (Haury et al., 1959; Martin, 1963; Mehringer, 1965). Martin and Mehringer (1965) suggest that these changes may reflect local edaphic conditions accompanying arroyo erosion and floodplain alluviation.

Hall (1977) presents a different interpretation for an alluvial sequence spanning the last 7000 years from Chaco Canyon, New Mexico (Figures 8-2 and 8-10 and Table 8-2). He suggests that *Pinus edulis* formed woodlands in that area from 7000 to about 5800 yr B.P. and that *P. ponderosa* forests were also more widespread. From 5800 to 600 yr B.P., *Pinus* woodlands were greatly diminished, and *Atriplex* and other Chenopodiineae became dominant. During the past 600 years, *Pinus edulis* forests expanded again to their present extent. Betancourt and Van Devender (1981) have found packrat middens dating from late-glacial and Holocene time in Chaco Canyon. They report that *Pseudotsuga menziesii, Juniperus scoplorum*, and *Pinus* cf. *flexilis* were dominant during Late Wisconsin time. These trees are all now many kilometers distant from the area. Their Holocene middens range in age from 5550 to 490 yr B.P. Both *P. edulis* and *Juniperus monosperma* were present in middens dating from 4480 to 1940 yr B.P. but absent by 490 yr B.P. Their data contradict the alluvial pollen record (Hall, 1977). Pollen analysis of the packrat middens suggests to Hall (1981) that scattered individuals of Pinyon pine and juniper were present, rather than a woodland. The pollen record generally reflects regional trends and macrofossils indicate local vegetation, so these records may be compatible. Further work in progress may clarify this discrepancy.

Changes in the record from packrat middens began about 11,000 yr B.P., when pinyon pines disappeared from middle-elevation woodlands, and xeric *Juniperus* or *Juniperus-Quercus* woodlands were widely prevalent between 11,000 and 8000 yr B.P. (Van Devender and Spaulding, 1979). The xeric *Ambrosia dumosa-Larrea divaricata* community was present below 300 m by about 10,000 yr B.P. in southeastern California and southwestern Arizona (Van Devender, 1977).

Forests in middle elevations changed to essentially modern communities during the early Holocene. In southern Nevada, mesic juniper communities between 1585 and 1860 m changed to xeric juniper woodlands before 9500 yr B.P. (Van Devender and Spaulding, 1979). These woodlands were, in turn, replaced by mixed desert scrub or *Coleogyne* scrub similar to the current vegetation by 7500 yr B.P. At altitudes between 1900 and 2100 m in the same area, Late Wisconsin forests were replaced before 9500 yr B.P. by the Pinyon pine-juniper woodland that still exists at these sites. Late-glacial forests at 2400 m contained *Pinus longaeva* and *P. flexilis* and *Juniperus communis*. By 10,000 yr B.P., forests of *Pinus longaeva* and *P. flexilis* (trees of high elevations) also contained some *P. ponderosa* and *Juniperus scopulorum* (trees of middle elevations). The modern forests contain only *Abies concolor*, *Pinus ponderosa*, and *Juniperus scopulorum* (Van Devender and Spaulding, 1979). Pollen and plant macrofossils from cave deposits in Utah show that montane conifers, including *Picea* cf. *engelmannii* and *Pseudotsuga*, were present during the late glacial (Spaulding and Petersen, 1980). By about 8500 yr B.P. the modern vegetation was established (Harper and Alder, 1970; Spaulding and Petersen, 1980).

Packrat middens younger than 8000 years that contain modern desert plants are common, but few have been studied. Those few suggest that, as the xeric woodlands had moved northward and upslope by 8000 yr B.P., desert-adapted species bacame dominant. *Larrea divaricata*-*Ambrosia dumosa* vegetation greatly expanded into the lowland areas of the Mohave and Sonoran Deserts. The Chihuahuan Desert was dominated by grasslands at elevations of 1200 and 2000 m until about 4000 yr B.P., when modern desert scrub became established (Van Devender and Spaulding, 1979). Martin (1963) and Van Devender and Wiseman (1977) argue convincingly that these grasslands indicate increased summer precipitation from a stronger monsoon during the middle-Holocene in the Chihuahuan Desert.

Summary

Late-glacial conditions differed substantially from the present environment in the western United States. In western Washington, the treeline was rising, and subalpine forests prevailed in the lowlands and along the coast. Tundra or tundralike environments were the rule at middle and upper altitudes from New Mexico to Montana. The floristic character of these environments has yet to be worked out, but some sites contain several taxa common in modern alpine tundra. One minimum estimate is that the late-glacial treeline may have been 1200 m lower in the Rocky Mountains than the present treeline, but many sites give a minimum estimate of only 500 to 900 m for the extent of treeline lowering. Pierce suggests, on the basis of equilibrium-line altitudes, the relationship of snowpack to altitude, and the presence of permafrost features in Wyoming basins, that Late Wisconsin temperatures in the Rocky Mountains may have been 12°C cooler than present temperatures (Porter et al., 1983). Such a large temperature difference would result in a 2000-m drop in the treeline. Subalpine and montane forests also grew at lower elevations, but few sites are available at lower elevations for determining the altitudes of these communities. In the Southwest, a lowering of 500 to 600 m has been suggested. Fossil packrat middens indicate that trees characteristic of these forests also grew south of their present distributional limits. Deserts of the Southwest were replaced during late-glacial time by woodlands of Pinyon pines and juniper. Spaulding and others (1983) suggest that Late Wisconsin climates in the deserts were characterized by cool, dry summers. California supported

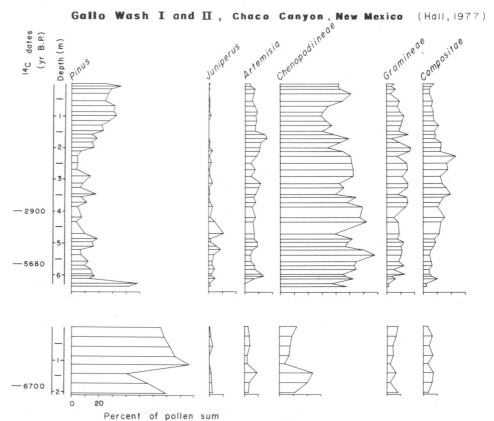

Figure 8-10. Pollen diagram from Gallo Wash, Chaco Canyon, New Mexico. (See Figure 8-2 for location of the site and Table 8-2 for data on the site.) (Redrawn from Hall, 1977 [from a Geological Society of America publication].)

woodlands of *Pinus* and one or more genera in the Taxodiaceae, Cupressaceae, or Taxaceae.

The early Holocene is sharply delimited from the late glacial. In western Washington, *Pseudotsuga* woodlands dominated. This fact suggests that the Holocene temperature maximum there occurred between 10,000 and 7000 yr B.P. Closed forests of *Tsuga heterophylla* in the northern part of the Puget lowland and *Pseudotsuga menziesii* in the southern part prevailed during the rest of the Holocene. *Picea sitchensis* and *Tsuga heterophylla* were important elements of the northwestern coastal forest throughout the Holocene. In eastern Washington, *Pinus ponderosa* and/or *P. contorta* became established in the early Holocene and remained on mesic sites. On xeric sites, grassland and steppe invaded from central Washington and prevailed from 10,000 to 5000 yr B.P., but forests returned in the late Holocene. Inliers of Pacific coastal forest reached northern Idaho in the last few thousand years. The montane and subalpine vegetational belts in Yellowstone Park shifted upward about 8000 yr B.P., as *P. contorta* forests expanded, but *Picea* and *Abies* returned to lower altitudes after 4000 yr B.P. At lower elevations in Yellowstone Park, an upward shift of steppe into forest occurred about 7000 yr B.P., and the appearance of modern forests occurred about 1500 yr B.P. Similar vegetational shifts occurred at Swan Lake in Idaho at 8500 and 3000 yr B.P. In the Bighorn Mountains of Wyoming, no shifts are detected on the pollen diagrams. Pollen ratios from Hurricane Basin, Twin Lakes, and Snowbird Bog, all subalpine sites in southwestern Colorado and Utah, show a warm interval from about 7000 to 4000 yr B.P. Alkali Basin, at a lower altitude in south-central Colorado, shows progressive Holocene warming up to the present. Sites in coastal California also show uninterrupted warming in the Holocene, as *Pinus* forests gave way to *Pinus-Quercus* forests by 7800 yr B.P. and chaparral and coastal scrub developed during the late Holocene. The mountain sites in the Sierra Nevada record a slight upward migration of *Juniperus* and *Libocedrus* during the middle Holocene, followed by a lowering of *Abies* during the late Holocene. The southwestern deserts were established about 8000 yr B.P. Pollen and packrat-midden sites in the Chihuahuan Desert indicate that grasslands prevailed during the middle Holocene, suggesting a stronger summer-monsoonal rainfall. The grasslands were replaced in the late Holocene by desert scrub. Throughout the Southwest, data from sites in the mountains seem to suggest a middle-Holocene warming, whereas the lowlands do not.

Conclusions

The geography of the western United States poses some interesting problems and possibilities for reconstructing Holocene vegetational history. Many areas have relatively undisturbed vegetation, and so modern pollen rain may be related to present vegetation. The high relief and large mass of many mountain ranges causes distinctive altitudinal vegetational belts. Vertical adjustments in these belts, especially in the upper and lower treeline, can be clearly demonstrated on many pollen diagrams. However, this altitudinal zonation can be blurred in the pollen record, because pollen easily blows from one zone to another. Furthermore, it may be difficult to distinguish vertical from latitudinal shifts in vegetation. The diversity of vegetation over the entire region is enormous. In a given area, the vegetational diversity may be high, as in coastal regions, but many species of important forest trees cannot be distinguished on the basis of their pollen morphology. Even so, several different arboreal pollen taxa are commonly important contributors to the pollen rain. In contrast, diversity is low in the Rocky Mountain region, where *Pinus* may contribute 70% to 95% of the total pollen during much of the Holocene. Despite these provisos, a pattern of vegetational changes has emerged.

The questions posed in the introduction to this chapter are still difficult to answer. The change from late-glacial to postglacial conditions caused marked shifts in plant populations. In the Rocky Mountains, areas of tundra or tundralike vegetation retreated to about their present position in the high mountains at this time, and modern forest belts at lower elevations probably became established. Only altitudinal shifts in plants can presently be demonstrated in most of the Rocky Mountains, although latitudinal shifts may well have occurred. In fact, the alpine-tundra flora in the Rocky Mountains consists of many circumboreal/arctic species (Weber, 1965) that must have immigrated during glacial times. However, no early-Quaternary pollen records exist to confirm this idea. In western Washington, both altitudinal and latitudinal shifts are suggested. *Pinus contorta*, *Tsuga mertensiana*, and *Picea engelmannii* all grew in lowland areas during the late glacial and migrated upslope during the Holocene. Current lowland species such as *Picea sitchensis*, *Tsuga mertensiana*, and *Thuja plicata* arrived during the Holocene and must have come from more southerly glacial refugia. Apparently, these trees did not reach as far south as central California, however, because their pollen is not seen in the lowland California profiles. The clearest case of both altitudinal and latitudinal plant migration can be made in the Southwest, for which Spaulding and others (1983) have given several convincing examples from the packrat-midden records.

Evidence for Holocene vegetational changes is more subtle, and only sites near ecotones show the changes clearly. Some of these changes are clearly latitudinal shifts, as in eastern Washington, where grasslands advanced into forests during Altithermal time (Nickmann, 1979). In many areas of high relief, only altitudinal changes are apparent. Individual species shifts in the Holocene confirm the loose character of community structure. For example, in southwestern Montana, *Pseudotsuga menziesii* and *Pinus albicaulis* seem to have moved independently from their respective communities during middle- and late-Holocene time (Bright, in preparation). Differential migration on this small scale was probably common during the Holocene, but large-scale migration lags like those that occurred in the eastern United States (Cushing, 1965; Davis, 1976; Delcourt, 1979) are not clear in the Holocene record of the West. One possible candidate for slow migration is *Pseudotsuga menziesii*. The species is known to have survived during Late Wisconsin time in uplands of the Southwest, and it may have had other refugia. It did not arrive in most sequences from the northern Rocky Mountain and Washington sites until about 9000 to 8000 yr B.P., although it probably could have survived climatically 2000 to 1000 years earlier. Migratory pathways for this and other taxa are still unknown.

A warmer Altithermal climate is strongly suggested in western and eastern Washington, southwestern Montana, Yellowstone Park, and southeastern Idaho. This climatic warming seems to have occurred later and lasted longer in sites at low versus high elevations. Sites in the Bighorn Mountains and in southwestern Colorado show little evidence for change in Holocene vegetation and climate, perhaps because they are not from ecologically sensitive areas. Progressive warming from the late glacial through the entire postglacial seems to characterize lowland sites in Colorado and California, and a wetter Altithermal is indicated for the Chihuahuan Desert of western Texas and New Mexico.

Quaternary paleoenvironmental reconstruction for the West lags behind that for the East by about 15 to 20 years. We need more detailed work on well-dated, carefully selected sites to fill in the large gaps in areal coverage. Especially large unstudied areas include much of Montana, Idaho, Oregon, Nevada, and California. Sites at low altitudes in the Rocky Mountains are needed to discover the nature of lowland vegetation and the extent of unequal compression of vegetational belts during the late glacial. We need to expand systematically our surface-sample coverage in the West. Pollen studies need to be combined with plant-macrofossil analyses, in small lakes and bogs as well as in packrat middens, so that we can learn more of floristics and regional and local vegetation. Studies of packrat middens need to be expanded to areas beyond the Southwest.

References

Adam, D. P. (1967). Late-Pleistocene and recent palynology in the central Sierra Nevada, California. In "Quaternary Paleoecology" (E. J. Cushing and H. E. Wright, Jr., eds.), pp. 275-301. Yale University Press, New Haven, Conn.

Adam, D. P. (1979a). "Raw Pollen Counts from Core 4, Clear Lake County, California." U.S. Geological Survey Open-File Report 79-663.

Adam, D. P. (1979b). "Raw Pollen Counts from Core 7, Clear Lake County, California." U.S. Geological Survey Open-File Report 79-1085.

Adam, D. P., and Sims, J. D. (1977). A continuous pollen record through the last glacial/interglacial cycle from Clear Lake, California. In "International Association for Quaternary Research Tenth Congress Abstracts," p. 3. Birmingham, England.

Adam, D. P., Sims, J. D., and Throckmorton, C. K. (1981). 130,000-yr continuous pollen record from Clear Lake, Lake County, California. Geology 9, 373-77.

Andrews, J. T., Carrara, P. E., King, F. B., and Stuckenrath, R. (1975). Holocene environmental changes in the alpine zone, northern San Juan Mountains, Colorado: Evidence from bog stratigraphy and palynology. Quaternary Research 5, 173-98.

Armstrong, J. E. (1975). "Quaternary Geology, Stratigraphic Studies and Reevaluation of Terrain Inventory Maps, Fraser Lowland, British Columbia." Canada Geological Survey Paper 75-1, pp. 377-80.

Arno, S. F. (1979). "Forest Regions of Montana." U.S. Department of Agriculture, Forest Service Research Paper INT-218.

Baker, R. G. (1970). Late Quaternary vegetation in Yellowstone Park. In "American Quaternary Association, First Biennial Meeting, Abstracts," pp. 4-5. Yellowstone Park and Montana State University, Bozeman.

Baker, R. G. (1976). "Late Quaternary Vegetation History of the Yellowstone Lake Basin, Wyoming." U.S. Geological Survey Professional Paper 729-E.

Barnosky, C. W. (1981). A record of late Quaternary vegetation from Davis Lake, southern Puget lowland, Washington. Quaternary Research 16, 221-39.

Batchelder, G. L. (1980). A Late-Wisconsinan and early Holocene lacustrine stratigraphy and pollen record from the west slope of the Sierra Nevada, California. In "American Quaternary Association, Sixth Biennial Meeting, Abstracts and Program, 18-20 August 1980," p. 13. Institute for Quaternary Studies, University of Maine, Orono.

Benedict, J. B. (1981). "The Fourth of July Valley: Glacial Geology and Archeology of the Timberline Ecotone." Center for Mountain Archeology Research Report 2.

Betancourt, J. L., and Van Devender, T. R. (1981). Holocene vegetation in Chaco Canyon, New Mexico. Science 214, 656-58.

Brant, L. A. (1980). A palynological investigation of postglacial sediments at two locations along the Continental Divide near Helena, Montana. Ph.D. dissertation, Pennsylvania State University, University Park.

Bright, R. C. (1966). Pollen and seed stratigraphy of Swan Lake, southeastern Idaho: Its relation to regional vegetational history and to Lake Bonneville history. Tebiwa 9, 1-47.

Bright, R. C. (in preparation). Postglacial history of Lost Trail Pass Bog, Montana: The "macrofossil" record.

Bryant, V. M., Jr., and Williams-Dean, G. (1975). The coprolites of man. Scientific American 63, 100-109.

Burkart, M. R. (1976). Biostratigraphy and late Quaternary vegetation history of the Bighorn Mountains, Wyoming. Ph.D. dissertation, University of Iowa, Iowa City.

Casteel, R. W., Adam, D. P., and Sims, J. D. (1977). Late-Pleistocene and Holocene remains of Hysterocarpus traski (tule perch) from Clear Lake, California, and inferred Holocene temperature fluctuations. Quaternary Research 7, 133-43.

Clisby, K. H., and Sears, P. B. (1956). San Augustin Plains: Pleistocene climatic changes. Science 124, 537-39.

Cushing, E. J. (1965). Problems in the Quaternary phytogeography of the Great Lakes region. In "The Quaternary of the United States" (H. E. Wright, Jr., and D. G. Frey, eds.), pp. 403-16. Princeton University Press, Princeton, N.J.

Daubenmire, R. (1975). Floristic plant geography of eastern Washington and northern Idaho. Journal of Biogeography 2, 1-18.

Daubenmire, R., and Daubenmire, J. B. (1968). "Forest Vegetation of Eastern Washington and Northern Idaho." Washington Agricultural Experiment Station Technical Bulletin 60.

Davis, M. B. (1973). Pollen evidence of changing land use around the shores of Lake Washington. Northwest Science 47, 133-48.

Davis, M. B. (1976). Pleistocene biography of temperate deciduous forests. Geoscience and Man 13, 13-26.

Delcourt, H. R. (1979). Late Quaternary vegetation history of the eastern Highland Rim and adjacent Cumberland Plateau of Tennessee. Ecological Monographs 49, 255-80.

Despain, D. G. (1973). Vegetation of the Bighorn Mountains, Wyoming, in relation to substrate and climate. Ecological Monographs 43, 329-55.

Dorland, D., Adam, D. P., and Batchelder, G. L. (1980). Two Holocene pollen records from Meyers Grade Marsh and Grass Lake, El Dorado County, California. In "American Quaternary Association, Sixth Biennial Meeting, Abstracts and Program, 18-20 August 1980," p. 64. Institute for Quaternary Studies, University of Maine, Orono.

Franklin, J. F., and Dyrness, C. T. (1973). "Natural Vegetation of Oregon and Washington." U.S. Department of Agriculture Forest Service General Technical Report PNW-8.

Gennett, J. A. (1977). Palynology and paleoecology of sediments from Blacktail Pond, northern Yellowstone Park, Wyoming. M.S. thesis, University of Iowa, Iowa City.

Hall, S. A. (1977). Late Quaternary sedimentation and paleoecologic history of Chaco Canyon, New Mexico. Geological Society of America Bulletin 88, 1593-1618.

Hall, S. A. (1981). Holocene vegetation at Chaco Canyon: Pollen evidence from alluvium and packrat middens. In "Program and Abstracts, Society for American Archeology, 46th Annual Meeting," p. 61. San Diego, Calif.

Hansen, B. S., and Easterbrook, D. J. (1974). Stratigraphy and palynology of late Quaternary sediments in the Puget lowland, Washington. Geological Society of America Bulletin 85, 587-602.

Hansen, H. P. (1943). A pollen study of two bogs on Orcas Island, of the San Juan Islands, Washington. Bulletin of the Torrey Botanical Club 70, 236-43.

Harper, K. T., and Alder, G. M. (1970). The macroscopic plant remains of the deposits of Hogup Cave, Utah, and their paleoclimatic implications. In "Hogup Cave" (C. M. Aikens, ed.), pp. 215-40. University of Utah Anthropological Papers 93.

Haury, E. W., Sayles, E. B., and Wasley, W. W. (1959). The Lehner mammoth site, southeastern Arizona. American Antiquity 25, 2-30.

Heusser, C. J. (1965). A Pleistocene phytogeographical sketch of the Pacific Northwest and Alaska. In "The Quaternary of the United States" (H. E. Wright, Jr., and D. G. Frey, eds.), pp. 469-83. Princeton University Press, Princeton, N.J.

Heusser, C. J. (1966). Pleistocene climatic variations in the western United States. In "Pleistocene and Post-Pleistocene Climatic Variations in the Pacific Area" (D. I. Blumenstock, ed.), pp. 9-36. Bernice P. Bishop Museum Press, Honolulu.

Heusser, C. J. (1974). Quaternary vegetation, climate, and glaciation of the Hoh River valley, Washington. Geological Society of America Bulletin 85, 1547-60.

Heusser, C. J. (1977). Quaternary palynology of the Pacific slope of Washington. Quaternary Research 8, 282-306.

Heusser, C. J. (1978a). Modern pollen rain in the Puget lowland of Washington. Bulletin of the Torrey Botanical Club 105, 296-305.

Heusser, C. J. (1978b). Modern pollen rain of Washington. Canadian Journal of Botany 56, 1510-17.

Heusser, C. J. (1978c). Modern pollen spectra from western Oregon. *Bulletin of the Torrey Botanical Club* 105, 14-17.

Heusser, C. J. (1978d). Palynology of Quaternary deposits of the lower Bogachiel River area, Olympic Peninsula, Washington. *Canadian Journal of Earth Sciences* 15, 1568-78.

Heusser, C. J., and Florer, L. E. (1973). Correlation of marine and continental Quaternary pollen records from the Northwest Pacific and western Washington. *Quaternary Research* 3, 661-70.

Heusser, L. E. (1978). Pollen in Santa Barbara Basin, California: A 12,000 year record. *Geological Society of America Bulletin* 89, 673-78.

Heusser, L. E. and Balsam, W. L. (1977). Pollen distribution in the Northeast Pacific Ocean. *Quaternary Research* 7, 45-62.

Hibbert, D. M. (1979). Pollen analysis of late-Quaternary sediments from two lakes in the southern Puget lowland, Washington. M.S. thesis, University of Washington, Seattle.

Houston, D. L. (1973). Wildfires in Yellowstone National Park. *Ecology* 54, 1111-17.

King, J. E., and Van Devender, T. R. (1977). Pollen analysis of fossil packrat middens from the Sonoran Desert. *Quaternary Research* 8, 191-204.

Küchler, A. W. (1964). "Potential Natural Vegetation of the Conterminous United States." American Geographical Society Special Publication 36.

LaMarche, V. C., Jr., (1973). Holocene climatic variations inferred from treeline fluctuations in the White Mountains, California. *Quaternary Research* 3, 632-60.

Legg, T. E., and Baker, R. G. (1980). Palynology of Pinedale sediments, Devlins Park, Boulder County, Colorado. *Arctic and Alpine Research* 12, 319-33.

Leopold, E. B. (1967). Summary of palynological data from Searles Lake. *In* "Pleistocene Geology and Palynology: Searles Valley, California." Guidebook for Friends of the Pleistocene, Pacific Coast Section (G. I. Smith, compiler), pp. 52-66.

McAndrews, J. H., and Wright, H. E., Jr. (1969). Modern pollen rain across the Wyoming basins and the northern Great Plains (U.S.A.). *Review of Palaeobotany and Palynology* 9, 17-43.

Mack, R. N., and Bryant, V. M., Jr. (1974). Modern pollen spectra from the Columbia Basin, Washington. *Northwest Science* 48, 183-94.

Mack, R. N., Bryant, V. M., Jr., and Pell, W. (1978a). Modern forest pollen spectra from eastern Washington and northern Idaho. *Botanical Gazette* 139, 249-55.

Mack, R. N., Rutter, N. W., Bryant, V. M., Jr., and Valastro, S. (1978b). Late Quaternary pollen record from Big Meadow, Pend Oreille County, Washington. *Ecology* 59, 956-66.

Mack, R. N., Rutter, N. W., Bryant, V. M., Jr., and Valastro, S. (1978c). Reexamination of postglacial vegetation history in northern Idaho: Hager Pond, Bonner County. *Quaternary Research* 10, 241-55.

Mack, R. N., Rutter, N. W., and Valastro, S. (1978d). Late Quaternary pollen record from the Sanpoil River valley, Washington. *Canadian Journal of Botany* 56, 1642-50.

Mack, R. N., Rutter, N. W., Valastro, S., and Bryant, V. M., Jr. (1978e). Late Quaternary vegetation history at Waits Lake, Colville River valley, Washington. *Botanical Gazette* 139, 499-506.

Mack, R. N., Rutter, N. W., and Valastro, S. (1979). Holocene vegetation history of the Okanogan Valley, Washington. *Quaternary Research* 12, 212-25.

Madsen, D. B., and Currey, D. R. (1979). Late Quaternary glacial and vegetation changes, Little Cottonwood Canyon area, Wasatch Mountains Utah. *Quaternary Research* 12, 254-70.

Maher, L. J., Jr. (1961). Pollen analysis and post-glacial vegetation history in the Animas Valley. Ph.D. dissertation, University of Minnesota, Minneapolis.

Maher, L. J., Jr. (1963). Pollen analysis of surface materials from the southern San Juan Mountains, Colorado. *Geological Society of America Bulletin* 74, 1484-1504.

Maher, L. J., Jr. (1972). Absolute pollen diagram of Redrock Lake, Boulder County, Colorado. *Quaternary Research* 2, 531-53.

Markgraf, V. (1980). Pollen dispersal in a mountain area. *Grana* 19, 127-46.

Markgraf, V., and Scott, L. (1981). Lower timberline in central Colorado during the past 15,000 years. *Geology* 9, 231-34.

Martin, P. S. (1963). "The Last 10,000 Years: A Fossil Pollen Record of the American Southwest." University of Arizona Press, Tucson.

Martin, P. S., and Mehringer, P. J., Jr. (1965). Pleistocene pollen analysis and biogeography of the Southwest. *In* "The Quaternary of the United States" (H. E.

Wright, Jr., and D. G. Frey, eds.), pp. 433-51. Princeton University Press, Princeton, N.J.

Martin, P. S., Sabels, B. E., and Shutler, D., Jr. (1961). Rampart Cave coprolite and ecology of the Shasta ground sloth. *American Journal of Science* 259, 102-27.

Mears, B., Jr. (1981). Periglacial wedges and the late Pleistocene environment of Wyoming's intermontane basins. *Quaternary Research* 15, 171-98.

Mehringer, P. J., Jr. (1965). Late Pleistocene vegetation in the Mohave Desert of southern Nevada. *Journal of Arizona Academy of Science* 3, 172-87.

Mehringer, P. J., Jr., Arno, S. F., and Petersen, K. L. (1977). Postglacial history of Lost Trail Pass Bog, Bitterroot Mountains, Montana. *Arctic and Alpine Research* 9, 345-68.

Nelson, A. R., Millington, A. C., Andrews, J. T., and Nichols, H. (1979). Radiocarbon-dated upper Pleistocene glacial sequence, Fraser Valley, Colorado Front Range. *Geology* 7, 410-14.

Nelson, R. S., Jr. (1970). The glaciation of the headwaters area of Clear Creek, Bighorn Mountains, Wyoming. Ph.D. dissertation, University of Iowa, Iowa City.

Nickmann, R. (1979). The palynology of Williams Lake Fen, Spokane County, Washington. M.S. thesis, Eastern Washington University, Cheney.

Petersen, K. L. (1981). 10,000 years of climatic change reconstructed from fossil pollen, La Plata Mountains, southwestern Colorado. Ph.D. dissertation, Washington State University, Pullman.

Petersen, K. L., and Mehringer, P. J., Jr. (1976). Postglacial timberline fluctuations, La Plata Mountains, southwestern Colorado. *Arctic and Alpine Research* 8, 275-88.

Pierce, K. L. (1979). "History and Dynamics of Glaciation in the Northern Yellowstone National Park Area." U.S. Geological Survey Professional Paper 729-F.

Porter, S. C., Pierce, K. L., and Hamilton, T. D. (1983). Late Wisconsin mountain glaciation in the western United States. *In* "Late Quaternary Environments of the United States," vol. 1, "The Late Pleistocene" (S. C. Porter, ed.), pp. 71-111. University of Minnesota Press, Minneapolis.

Richmond, G. M. (1969). Development and stagnation of the last Pleistocene icecap in the Yellowstone Lake Basin, Yellowstone National Park, U.S.A. *Eiszeitalter und Gegenwart* 20, 196-203.

Spaulding, W. G., Leopold, E. B., and Van Devender, T. R. (1983). Late Wisconsin paleoecology of the American Southwest. *In* "Late-Quaternary Environments of the United States," vol. 1, "The Late Pleistocene" (S. C. Porter, ed.), pp. 259-93. University of Minnesota Press, Minneapolis.

Spaulding, W. G., and Petersen, K. L. (1980). Late Pleistocene and early Holocene paleoecology of Cowboy Cave. *In* "Cowboy Cave" (J. D. Jennings, ed.), pp. 163-77. University of Utah Anthropological Papers 104.

Storer, T. I., and Usinger, R. L. (1963). "Sierra Nevada Natural History." University of California Press, Berkeley.

Taylor, D. L. (1973). Some ecological implications of forest fire control in Yellowstone National Park, Wyoming. *Ecology* 54, 1394-96.

Thompson, R. S., and Mead, J. I. (1982). Late Quaternary environments and biogeography in the Great Basin. *Quaternary Research* 17, 39-55.

Thorson, R. M. (1980). Ice-sheet glaciation of the Puget lowland, Washington, during the Vashon stade (late Pleistocene). *Quaternary Research* 13, 303-21.

United States Geological Survey. (1972). Surficial geologic map of Yellowstone National Park. U.S. Geological Survey Miscellaneous Geologic Investigations Map I-710.

Van Devender, T. R. (1977). Holocene woodlands in the southwestern deserts. *Science* 198, 189-92.

Van Devender, T. R. (1980). Holocene plant remains from Rocky Arroyo and Last Chance Canyon, Eddy County, New Mexico. *Southwest Naturalist* 35, 361-72.

Van Devender, T. R., and Mead, J. I. (1976). Late Pleistocene and modern plant communities of Shinumo Creek and Peack Springs Wash, lower Grand Canyon, Arizona. *Journal of the Arizona Academy of Science* 11, 16-22.

Van Devender, T. R., and Riskind, D. H. (1979). Late Pleistocene and early Holocene plant remains from Hueco Tanks State Historical Park: The development of a refugium. *Southwestern Naturalist* 24, 127-40.

Van Devender, T. R., and Spaulding, W. G. (1979). Development of vegetation and climate in the southwestern United States. *Science* 204, 701-10.

Van Devender, T. R., and Wiseman, F. M. (1977). A preliminary chronology of bioenvironmental changes during the paleoindian period in the monsoonal Southwest.

In "Paleoindian Lifeways" (E. Johnson, ed.), pp. 13-26. Museum Journal 17, West Texas Museum Association.

Waddington, J. C. B., and Wright, H. E., Jr. (1974). Late Quaternary vegetational changes on the east side of Yellowstone Park, Wyoming. *Quaternary Research* 4, 175-84.

Weber, W. A. (1965). Plant geography in the southern Rocky Mountains. *In* "The Quaternary of the United States" (H. E. Wright, Jr., and D. G. Frey, eds.), pp. 453-68. Princeton University Press, Princeton, N.J.

Wells, P. V. (1966). Late Pleistocene vegetation and degree of pluvial climatic change in the Chihuahuan Desert. *Science* 153, 970-75.

Wells, P. V. (1976). Macrofossil analysis of wood rat (*Neotoma*) middens as a key to the Quaternary vegetational history of arid America. *Quaternary Research* 6, 223-48.

Wells, P. V. (1979). An equable glaciopluvial in the West: Pleniglacial evidence of increased precipitation on a gradient from the Great Basin to the Sonoran and Chihuahuan Deserts. *Quaternary Research* 12, 311-25.

Wells, P. V., and Berger, R. (1967). Late Pleistocene history of coniferous woodlands in the Mohave Desert. *Science* 155, 1640-47.

Wells, P. V., and Hunziker, J. H. (1977). Origin of the creosote-bush (*Larrea*) deserts of southwestern North America. *Annals of the Missouri Botanical Garden* 63, 843-61.

Wells, P. V., and Jorgensen, C. D. (1964). Pleistocene wood rat middens and climatic change in Mohave Desert: A record of juniper woodlands. *Science* 143, 1171-74.

Whiteside, M. C. (1965). Paleoecological studies of Potato Lake and its environs. *Ecology* 46, 807-16.

Wright, H. E., Jr. (1971). Late Quaternary vegetational history of North America. *In* "The Late Cenozoic Glacial Ages" (K. K. Turekian, ed.), pp. 425-64. Yale University Press, New Haven, Conn.

Wright, H. E., Jr., Bent, A. M., Hansen, B. S., and Maher, L. J., Jr. (1973). Present and past vegetation of the Chuska Mountains, northwestern New Mexico. *Bulletin of the Geological Society of America* 84, 1155-80.

CHAPTER 9

Holocene Vegetational History of Alaska

Thomas A. Ager

Introduction

At the time of the Seventh Congress of the International Association for Quaternary Research in 1965 at Boulder, Colorado, few investigations of Quaternary vegetational history in Alaska had been undertaken. Reviews published at the time of the congress or soon thereafter show that the regions of the state having the best-known vegetational history were south-central and southeastern coastal Alaska (Heusser, 1965) and Arctic Alaska. Arctic history was based on a few widely scattered sites in the central Brooks Range, the Arctic Foothills and Coastal Plain, the Seward Peninsula, and some of the islands of the Bering Sea (Colinvaux, 1967b). The little that was known of the Holocene vegetational history of interior Alaska was based primarily upon undated shallow peat cores analyzed only for a few arboreal pollen types (Hansen, 1953; Heusser, 1965).

Since the mid-1960s, considerable progress has been made toward the goal of establishing a network of carefully investigated sites across Alaska, particularly in the many areas where no previous studies had been attempted. Much of this work was initiated during the past decade; a considerable part of it is currently in progress. Several large areas remain for which no Quaternary palynologic and plant-macrofossil studies have yet been initiated (e.g., the Kuskokwim River drainage basin, most of the Alaska Peninsula, and northeastern Alaska north of about latitude 65°N and east of longitude 148°W. Many opportunities for true pioneering research remain in Alaska.

Past and present Quaternary palynologic studies in Alaska have focused on several problems of paleoenvironmental reconstruction: (1) the postglacial history of the Pacific coastal forest (e.g., Heusser, 1960), (2) the history of interior Alaskan boreal forests (e.g., Hansen, 1953; Matthews, 1974b; Ager, 1975); and (3) the environment of the Bering Land Bridge and adjacent unglaciated regions (Beringia) (e.g., Colinvaux, 1964, 1967a; Matthews, 1974a, 1974b; Ager, 1975, 1982c). In addition, several researchers have conducted work related to archaeological sites, particularly early man sites in Alaska (e.g., Ager, 1975; Schweger, 1976).

For the purposes of this chapter, I have focused upon what I judge to be key sites from each of several geographic regions of Alaska from which radiocarbon-dated Holocene pollen records are available. This permits the direct comparison of vegetational histories from different geographic settings, climates, and vegetational types. The selected sites include some from regions of Alaska that were glaciated during the late Pleistocene as well as some from unglaciated regions. Wherever possible, I have selected sites that have been investigated since the time earlier reviews of Alaskan vegetational history (e.g., Heusser, 1965; Colinvaux, 1967b) were published.

In the interest of presenting readily comparable pollen diagrams, I have chosen records obtained primarily from nonthermokarst lakes. A number of important investigations of Alaskan vegetational history have been based on pollen and macrofossils preserved in peat deposits, alluvial sediments, and organic colluvial silt; many such deposits in northern, western, and interior Alaska are perennially frozen. Such sites have contributed a great deal to our current understanding of Alaska's vegetational history, but they are not emphasized here because these deposits are less likely than lacustrine deposits to have preserved continuous records. Poor preservation and reworked pollen from older deposits are more serious problems in studying alluvial and colluvial deposits and thermokarst lakes than for the sites considered here.

The time interval of interest in this volume is the Holocene. To fully understand the history of the development of vegetation in Alaska during the Holocene, however, some discussion of Late Wisconsin vegetational changes must be included. For the purposes of this chapter only the time interval between around 16,000 to 10,000 yr B.P. is discussed for the Wisconsin. A number of other papers discuss aspects of the Wisconsin vegetational history in Alaska (e.g., Colinvaux, 1964, 1967b; Matthews, 1974a, 1974b, 1979; Ager, 1975, 1982c; Schweger, 1976; Heusser, 1983).

Unpublished data included in this chapter were generously provided by C. E. Schweger, L. B. Brubaker, and P. M. Anderson. Permission to include the author's unpublished data from the Denali National Park/North Alaska Range region was granted by the supporting agencies, the U.S. National Park Service, and the National Geographic Society. The manuscript for this chapter was reviewed by L. B. Brubaker, N. L. Frederiksen, L. E. Edwards, D. M. Hopkins, and R. G. Baker.

Present Vegetation and Climate of Alaska

The topographical complexity of Alaska (Figure 9-1), the state's broad latitudinal range (51°16′N to 71°23.5′N), the history of multiple glaciation of Alaskan mountainous regions, and the repeated land-bridge connections with Siberia all contribute to the complexity of present-day floral and vegetational patterns (Hopkins, 1967, 1972; Hultén, 1937, 1968; Péwé, 1975). Three broadly defined vegetational types are recognized (Figure 9-2): (1) coastal spruce-hemlock forest, (2) interior spruce forest, and (3) tundra. Alaskan vegetational types have been discussed by Viereck and Little (1972), Aleksandrova (1980), Larsen (1980), Bailey (1980), and Heusser (1960).

Spruce-hemlock forests are restricted to the coastal areas of south-central and southeastern Alaska. Dominant trees are Sitka spruce (*Picea sitchensis*) and western hemlock (*Tsuga heterophylla*). Other important trees within the coastal forests include mountain hemlock (*Tsuga mertensiana*), black cottonwood (*Populus trichocarpa*), and Alaska-cedar (*Chamaecyparis nootkatensis*). Within southeastern Alaska, western red cedar (*Thuja plicata*) and lodgepole pine (*Pinus contorta*) also occur. The climate of this coastal region is maritime, and abundant annual precipitation ranges from about 600 mm in the southern Kenai Peninsula in south-central Alaska to 5200 mm in the southern end of the panhandle near Ketchikan. Mean annual temperatures over this region range from 2°C at Homer to 8°C at Ketchikan (U.S. National Oceanic and Atmospheric Administration, 1980).

Interior forests are composed of white spruce (*Picea glauca*), black spruce (*P. mariana*), paper birch (*Betula papyrifera*), quaking aspen

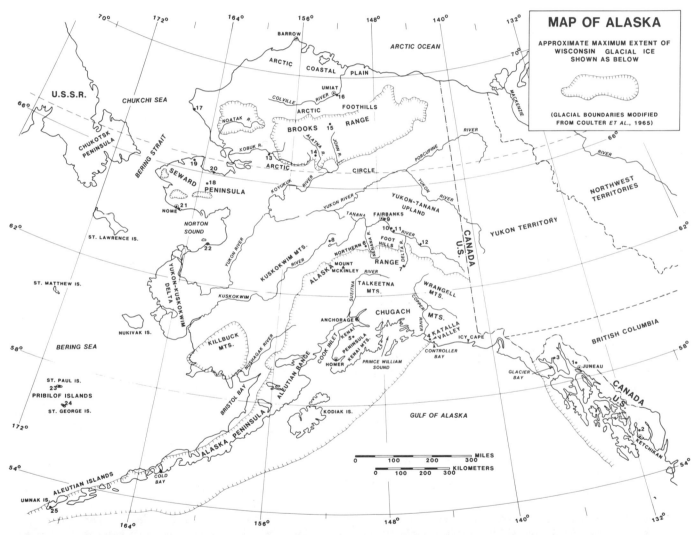

Figure 9-1. Map of Alaska showing the geographic features and localities mentioned in the text. Numbered sites are localities from which pollen data were collected. (1) Montana Creek (Heusser, 1960, 1965), (2) Ward Creek (Heusser, 1965), (3) Adams Inlet (McKenzie, 1970), (4) Controller Bay (Heusser, 1960; Sirkin and Tuthill, 1969), (5) Hidden Lake (Ager and Sims, 1981b, unpublished data), (6) Eightmile Lake (Ager, 1982b, unpublished data), (7) Tangle Lakes (Ager and Sims, 1981a; C. E. Schweger, unpublished data), (8) Lake Minchumina (Ager, 1976, unpublished data), (9) Isabella Basin (Matthews, 1974b), (10) Harding Lake (Ager, 1982a, unpublished data; Nakao et al., 1980), (11) Birch Lake (Ager, 1975), (12) Lake George (Ager, 1975), (13) Epiguruk (Schweger, 1976), (14) Ranger Lake (L. B. Brubaker, unpublished data), (15) Chandler Lake (Livingstone, 1955), (16) Umiat (Livingstone, 1957), (17) Ogotoruk Creek (Heusser, 1963), (18) Imuruk Lake (Colbaugh, 1968; Colinvaux, 1964; Shackleton, 1979), (19) Whitefish Lake (Shackleton, 1979), (20) Cape Deceit (Matthews, 1974a), (21) Nome (Hopkins et al., 1960), (22) St. Michael Island (Ager, 1980, 1982c, unpublished data), (23) St. Paul Island (Colinvaux, 1967a, 1967b), (24) St. George Island (Parrish, 1980), and (25) Umnak Island (Heusser, 1973).

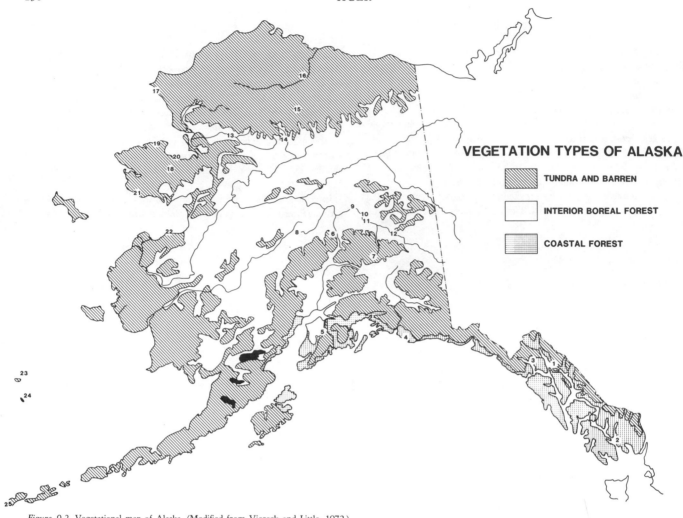

Figure 9-2. Vegetational map of Alaska. (Modified from Viereck and Little, 1972.)

(*Populus tremuloides*), balsam poplar (*P. balsamifera*), and, in some areas, larch (*Larix laricina*). Well-developed stands of *Picea glauca* and *Populus balsamifera* occur on fluvial terraces along the major rivers of interior Alaska. Extensive areas of interior lowlands and north-facing slopes of uplands have a vegetative cover of *Picea mariana* muskeg. South of the Arctic Circle, *Larix laricina* is a component of some of these muskegs, but the species also occurs on alluvial fans with gravel substrates. Upland vegetation consists of a mosaic of mixed and pure stands of *Picea*, *Betula papyrifera*, and *Populus tremuloides*. This mosaic reflects a history of frequent fires. The altitudinal treeline in the interior is variable but rarely exceeds 1000 m.

The climate within the interior is continental; summers are warm and winters are long and severely cold. Interior Alaska's mean annual temperatures are below freezing (e.g., – 4 °C at Fairbanks), and, consequently, much of the interior is underlain by permafrost (Hartman and Johnson, 1978; U.S. National Oceanic and Atmospheric Administration, 1980). Most of the annual precipitation falls as rain in August and September; annual precipitation is low (e.g., 292 mm at Fairbanks).

Tundra includes three types of treeless vegetation in Alaska: wet lowland tundra, moist tundra, and dry (alpine) tundra. Wet lowland tundra like that present over most of the Yukon-Kuskokwim deltaic lowland and the Arctic Coastal Plain includes a number of wetland Cyperaceae, particularly *Carex* species and non-tussock-forming *Eriophorum* species. Gramineae and a variety of forbs (e.g., *Petasites* spp., *Pedicularis* spp., *Polemonium acutiflorum*) are also common. Some woody shrubs also occur (e.g., *Salix fuscesens*, *S. pulchra*, *Empetrum nigrum*, *Betula nana*), but some of these are excluded from the Coastal Plain of northernmost Alaska by severe climate (e.g., *B. nana*, *Empetrum nigrum*).

Moist tundra includes the tussocky shrub tundra that covers much of the northern slope of the Brooks Range, the foothills of the Alaska Range, the central Seward peninsula, the Bristol Bay-Alaska Peninsula region, and the Aleutians. Tussock-forming *Eriophorum* (e.g., *E. vaginatum*), *Carex* species, and other Cyperaceae, along with Gramineae, Ericaceae, *Empetrum*, *Salix* species, *Betula nana*, and *B. glandulosa*, are often common elements. In some areas, *Alnus* and *Salix* shrubs form thickets in ravines and sheltered slopes and hollows. Most of the woody species are absent from the Aleutian Islands.

Dry tundra (alpine tundra) occurs in the higher elevations of the uplands and the mountainous regions of Alaska. Alpine tundra consists of discontinuous and commonly sparse vegetation composed primarily of hardy herbaceous plants (e.g., *Oxytropis*, Saxifragaceae, Caryophyllaceae), shrublets (e.g., *Dryas*), prostrate shrubs such as *Empetrum nigrum*, *Salix reticulata*, *S. arctica*, and ericaceous species (e.g., *Arctostaphylos* spp., *Vaccinium* spp.). Shrub *Betula* occurs in some alpine tundra communities.

The climate within the tundra regions of Alaska is quite variable, ranging from maritime and transitional maritime-continental in southwestern Alaska to arctic in northwestern and northernmost Alaska (Hartman and Johnson, 1978). Some areas of alpine tundra occur within the continental climatic regime. At Barrow in the arctic climate zone of northern Alaska, the mean annual temperature is about − 12 °C and the mean annual precipitation is about 127 mm. Permafrost is continuous. By contrast, Cold Bay in the maritime-transitional zone on the southern Alaska Peninsula has a mean annual temperature of 3 °C and a mean annual precipitation of 864 mm (U.S. National Oceanic and Atmospheric Administration, 1980). Permafrost is absent at Cold Bay. Most tundra areas are characterized by long, severe winters and short, cool summers.

Modern Pollen-Rain Studies in Alaska

Our knowledge of the present-day relationships between vegetation and modern pollen rain in Alaska remains at a rather primitive level. Although surface pollen spectra have been described from several sites in Alaska (e.g., Livingstone, 1955; Colinvaux, 1964; Matthews, 1970; Ager, 1975; Moriya, 1976; Nelson, 1979; Ager and Sims, 1981a), the total number of described samples is small. In addition, the samples represent a wide variety of sediment types, including moss polsters, shallow-water pond sediments, and deep-water core tops, and it is difficult to compare spectra from site to site. Only a general knowledge of the relationships between vegetation and the pollen spectra they produce is now available. No quantitative comparisons between the percentage of ground cover for individual plant taxa and the local pollen rain have yet been completed in Alaska.

Some of the qualitative relationships between modern vegetation and pollen rain have been discussed by Colinvaux (1967b), Matthews (1970), and Ager (1975). To some degree, the gaps in our knowledge of these vegetation/pollen-rain relationships are filled by more detailed pollen-rain studies in Canada (e.g., Ritchie and Lichti-Federovich, 1963, 1967; Rampton, 1971; Birks, 1973, 1977; Cwynar, 1980). Some of these studies are not entirely relevant to Alaska, however, because of differences in regional floras and topography. In the relatively flat regions of central Canada, a rather distinct latitudinal "zonation" of vegetational types has developed during Holocene time. In Alaska, however, mountain ranges and uplands cover much of the state, and lowland forests exist in close proximity to shrub tundra and alpine tundra in many mountainous regions. Therefore, the latitudinal "zonation" of vegetation is less distinct in Alaska, and trees and shrubs such as *Picea*, *Alnus*, and *Betula*, which produce abundant wind-borne pollen, contribute substantially to the pollen rain far beyond the areas where those plants now grow. This indistinct latitudinal zonation and the abundance of wind-borne pollen carried into tundra regions add to the difficulties of interpreting fossil-pollen spectra in Alaska.

During the past decade, virtually hundreds of additional surface pollen samples have been collected in Alaska, particularly in the northern, western, and interior regions of the state. The analysis of these samples should substantially improve our understanding of modern vegetation/pollen-rain relationships in Alaska.

Pollen Records from Southeastern Alaska

Our existing knowledge of the vegetational history of southeastern Alaska is based mostly on the work of Heusser (1957, 1960, 1965), who cored numerous peat bogs and muskegs in the region during the 1950s. Only a few palynologic samples have been described from this region since 1965 (McKenzie, 1970).

In brief summary, the pollen records of Heusser suggest that the earliest vegetation known to have developed after deglaciation consisted of *Pinus contorta* parkland with Cyperaceae and shrubs such as *Alnus* and *Salix*. The oldest dated record from Montana Creek near Juneau spans the last 10,300 years (Figures 9-1 and 9-3). As Wisconsin glaciers receded, the newly exposed terrain in southeastern Alaska was invaded by trees, shrubs, and herbaceous plants that had survived in unglaciated regions of British Columbia, Washington, and Oregon. Other refugia may have existed in southeastern Alaska as well, serving as centers from which some plants could reexpand onto newly deglaciated terrain. In addition, some elements of the coastal flora may have invaded from interior Alaskan refugia (Heusser, 1960, 1965). During early to middle Holocene time, *Picea sitchensis*, *Tsuga heterophylla*, and *T. mertensiana* invaded, forming forests that approach the composition of modern forests in the region (Table 9-1). Since middle Holocene time, *Tsuga* appears to have increased in importance in the coastal forests of southeastern Alaska.

Future palynologic research in southeastern Alaska may extend the record beyond the 10,300 years obtained near Juneau, particularly if lacustrine records are sought as coring sites. A significant length of time may have elapsed between deglaciation and the onset of peat deposition. This lag is suggested by pollen data from exposures of the Forest Creek Formation in Adams Inlet, Glacier Bay. The oldest pollen spectrum suggests that vegetation consisting of *Pinus contor-*

UPPER MONTANA CREEK

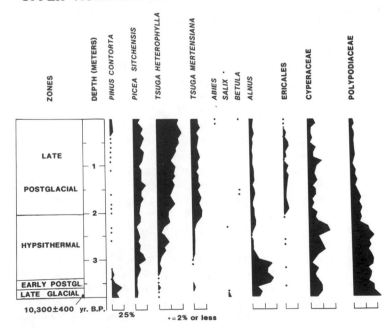

Figure 9-3. Pollen diagram (summary) from upper Montana Creek, near Juneau, southeastern Alaska. (Adapted from Calvin J. Heusser, *Late-Pleistocene Environments of North Pacific North America* [1960], with permission of the American Geographical Society.)

Table 9-1. Late-Quaternary Vegetational Histories from Selected Localities in Alaska

		SOUTHWESTERN		NORTHERN ALASKA		INTERIOR	ALASKA RANGE	SOUTH-CENTRAL		SOUTH-EASTERN
14C YEARS BEFORE PRESENT		ALEUTIAN ISLANDS —— UMNAK ISLAND (HEUSSER,1973)	YUKON DELTA-NORTON SOUND (AGER,1980, 1982a)	NORTH-CENTRAL BROOKS RANGE & ARCTIC FOOTHILLS (LIVINGSTONE, 1955,1957)	SOUTH-CENTRAL BROOKS RANGE (BRUBAKER, UNPUBLISHED)	TANANA VALLEY (AGER,1975,1982b; NAKAO et.al., 1980)	NORTHERN FOOTHILLS (AGER, 1982b, UNPUBLISHED)	CENTRAL KENAI PENINSULA (AGER & SIMS, 1981b)	CONTROLLER BAY-ICY CAPE (HEUSSER, 1960, 1965; SIRKIN & TUTHILL, 1969)	MONTANA-WARD CREEKS-ADAMS INLET (MODIFIED FROM HEUSSER, 1960, 1965; McKENZIE,1970)

HOLOCENE	LATE WISCONSIN									
— 1000		GRAMINEAE - CYPERACEAE HERBACEOUS TUNDRA	ALNUS – BETULA – ERICACEAE SHRUB TUNDRA with long-distance-transported PICEA pollen		ALNUS – BETULA SHRUB TUNDRA w/ PICEA GALLERY FOREST	PICEA – ALNUS – BETULA BOREAL FOREST	PICEA – ALNUS GALLERY FOREST and FOREST-TUNDRA w/ BETULA SHRUB TUNDRA	PICEA SITCHENSIS- TSUGA COASTAL FOREST (KENAI MTS.)	PICEA SITCHENSIS- TSUGA HETEROPHYLLA – T. MERTENSIANA – ALNUS COASTAL FOREST	TSUGA SPP. – PINUS CONTORTA- PICEA SITCHENSIS PICEA SITCHENSIS – TSUGA HETEROPHYLLA – T. MERTENSIANA COASTAL FOREST
— 2000										
— 3000			ALNUS–BETULA SHRUB TUNDRA							
— 4000		SALIX –								
— 5000		GRAMINEAE- CYPERACEAE HERBACEOUS TUNDRA	ALNUS – BETULA – ERICACEAE SHRUB TUNDRA	BETULA – ERICACEAE SHRUB TUNDRA	ALNUS– BETULA SHRUB TUNDRA	PICEA DECLINE		PICEA – ALNUS–BETULA BOREAL FOREST IN LOWLANDS	PICEA SITCHENSIS – ALNUS PARKLAND	PICEA SITCHENSIS – TSUGA HETEROPHYLLA – ALNUS COASTAL FOREST
— 6000										
— 7000										
— 8000				— ?–?–?–?–?		PICEA – ALNUS– BETULA FOREST				
— 9000		CYPERACEAE– GRAMINEAE – POLYPODIACEAE HERB TUNDRA	BETULA – ERICACEAE SHRUB TUNDRA	HERBACEOUS TUNDRA	BETULA SHRUB TUNDRA	PICEA – BETULA GALLERY FOREST and/or PARKLAND	BETULA – ERICACEAE SHRUB TUNDRA	ALNUS – BETULA SHRUB TUNDRA	ALNUS-CYPERACEAE- GRAMINEAE SHRUB TUNDRA w/ FERNS	PINUS CONTORTA – ALNUS PARKLAND
— 10,000						POPULUS-SALIX SCRUB FOREST and SHRUB TUNDRA	POPULUS-SALIX w/ SHRUB TUNDRA	POPULUS-SALIX w/ SHRUB TUNDRA	CYPERACEAE – ERICACEAE TUNDRA	PINUS CONTORTA PARKLAND
— 11,000			POPULUS-SALIX w/ SHRUB TUNDRA							
— 12,000			BETULA – ERICACEAE SHRUB TUNDRA			BETULA – ERICACEAE SHRUB TUNDRA	BETULA – ERICACEAE SHRUB TUNDRA	BETULA – ERICACEAE SHRUB TUNDRA		
— 13,000					HERBACEOUS TUNDRA					
— 14,000			HERBACEOUS TUNDRA			HERBACEOUS TUNDRA ("STEPPE-TUNDRA")	HERBACEOUS TUNDRA with abundant CYPERACEAE	HERBACEOUS TUNDRA		
— 15,000			TUNDRA		("STEPPE-TUNDRA")					

Note: Stippled sections show intervals for which no record is available. Horizontal lines within columns mark pollen-zone boundaries. Dashed lines indicated that the chronology is uncertain.

ta, Cyperaceae, Gramineae, Ericaceae, several forbs, and ferns was established as early as 10,940 ± 155 yr B.P. (I-2395). A cone identified as that of Picea was also recovered from these deposits in Adams Inlet. The radiocarbon date from the cone was 11,170 ± 225 yr B.P. (I-2396) (McKenzie, 1970: 40-41). If the identification of the cone was correct, its presence indicates a surprisingly early appearance of Picea as far north as Glacier Bay. The available pollen evidence from southeastern Alaska (Heusser, 1960) suggests that, if Picea was present at all in latest-Wisconsin and early-Holocene time, it was rare, perhaps colonizing the coast in scattered pockets as Picea sitchensis is now doing on the Alaska Peninsula and Kodiak Island.

Pollen Records from South-Central Alaska

Most of the palynologic research in south-central Alaska is that of Heusser (summarized in Heusser, 1957, 1960, 1965), who cored many peat deposits from Icy Cape to Kodiak Island (Figure 9-1) and published pollen diagrams from 28 of these localities (Heusser, 1960). One limitation of these data is the lack of radiocarbon control at many sites. Others have only basal dates. For example, Heusser produced pollen profiles from eight localities on the Kenai Peninsula, but none has radiocarbon control. Nevertheless, the diagrams, combined with Heusser's correlations with adjacent regions where some radiocarbon

dates are available, imply that forest vegetation was established in these coastal regions late in the Holocene. The oldest pollen spectra from Heusser's sites suggest that the earliest known postglacial vegetation was moist tundra with ferns, shrubs including Salix, Ericaceae, and dwarf birch (Betula nana), and herbs (Umbelliferae, Cyperaceae, Compositae, and others).

A recently obtained core from Hidden Lake in the central Kenai Peninsula (Figures 9-1 and 9-4) provides a pollen record spanning 14,000 years or more (Ager and Sims, 1981b; unpublished data). This moraine-dammed lake is near the western edge of the Kenai Mountains. The valleys and lower slopes of the Kenai Mountains support coastal forests, primarily of Sitka spruce, although some mountain hemlock also occurs. The western edge of this coastal forest extends to the eastern shore of Hidden Lake (Figure 9-2) (Viereck and Little, 1972). Most of the lowlands surrounding the lake are mantled with interior-type boreal forest and muskeg vegetation. Important local trees include Picea glauca, P. mariana, Betula papyrifera, Populus trichocarpa, P. balsamifera, and P. tremuloides. Alders (Alnus sinuata, A. tenuifolia) and a variety of willows (Salix spp.) are common local large shrubs.

The Hidden Lake pollen data and radiocarbon dates (Figure 9-4 and Table 9-1) suggest that vegetation first invaded the area during deglaciation, which began in some areas of Alaska more than 14,000 years ago (e.g., Denton, 1974). Pioneers were mostly herbs but included a few shrubs (*Salix* and Ericaceae). Herbs included Cyperaceae, Gramineae, *Artemisia*, *Ambrosia*, other Compositae (mostly Tubuliflorae), *Campanula*, and others. Ferns were also represented. This assemblage probably represents a discontinuous vegetative cover—a mosaic of tundra communities. By about 13,700 yr B.P., dwarf-*Betula* shrub tundra began to develop. The abrupt expansion of *Betula* probably coincided with or soon followed a climatic shift to warmer, moister summers about 14,000 yr B.P. (Ager, 1975, 1982c).

About 10,300 yr B.P., the vegetation changed from moist dwarf-*Betula* shrub tundra to a mixture of shrub tundra and deciduous scrub forest communities. The scrub forest included *Salix* and *Populus*, ericaceous shrubs, and abundant ferns. This shrub tundra-scrub forest phase of vegetational development has been recognized only recently in Alaska on the basis of pollen evidence (Nakao et al., 1980; Ager, 1982c) and plant-macrofossil evidence (Hopkins et al., 1981) from several widely scattered sites. By about 9500 yr B.P., *Alnus* shrubs began to invade the central Kenai Peninsula and probably many other areas around Cook Inlet. By about 9000 yr B.P., *Alnus* was a very important element of the regional pollen rain and, presumably, of the vegetation. The first conifer to reach the central Kenai Peninsula was *Picea*, which appeared abruptly about 8000 yr B.P. The invading spruce trees were probably *Picea glauca* and perhaps *P. mariana* from interior Alaska to the north (Figure 9-1). Traces of *Tsuga mertensiana* pollen appear in the Hidden Lake core, but apparently this tree was not present until about 5000 to 4000 yr B.P. *T. heterophylla* is present but uncommon in the Cook Inlet area today, and its pollen is rare or absent in most Kenai Peninsula pollen records. Pollen records from the east coast of the Kenai Peninsula and other sites in western Prince William Sound indicate that *Picea sitchensis* and *Tsuga mertensiana* may have arrived in the region at about the same time during the middle Holocene (Heusser, 1960). The record of *Tsuga* suggests that the coastal elements of the Kenai Peninsula's forests (e.g., *Picea sitchensis*, *Tsuga* spp.) may not have reached the western edge of the Kenai Mountains until the middle to late Holocene. This suggestion is consistent with interpretations of undated pollen records from coastal peat bogs and muskegs described by Heusser (1960) for the southern and northeastern Kenai Peninsula.

Other records from south-central Alaska deserve to be mentioned here. The 11 pollen records from Heusser's peat cores along the coast from Prince William Sound to Icy Cape (Figure 9-1) range in basal age from about 3800 to 10,800 years (Heusser, 1960: 149-52). The only published pollen work reported from that region since Heusser's (1960, 1965) is that of Sirkin and Tuthill (1969), who describe pollen profiles from exposures in the Controller Bay region east of the Copper River Delta (Figure 9-1). Fourteen radiocarbon dates associated with these profiles somewhat refine the chronology of vegetational changes in the region. One profile from the Katalla Valley has a radiocarbon date near the base of the section of 14,430 ± 890 yr B.P. (I-3082). This early date is associated with high percentages of *Alnus* pollen. No significant amounts of *Alnus* pollen have yet been reported from any other Wisconsin deposits in south-central or southeastern Alaska. The date is based on a sample of marine clay that could have incorporated older carbon, so the date may not be valid.

The pollen records from the Controller Bay area can be used to reconstruct the local vegetational history for the past 10,000 years or so (Table 9-1). The earliest known Holocene vegetation was *Alnus*-shrub tundra with Cyperaceae, Gramineae, *Salix*, herbs, and ferns. *Alnus* invaded during the early Holocene, perhaps as early as 10,000 yr B.P. Most profiles suggest that coastal *Picea sitchensis-Tsuga* forests developed during the middle to late Holocene, although one pollen profile (i.e., Ragged Mountain, *in* Sirkin and Tuthill, 1969: 202) shows *Picea* pollen in significant amounts as early as 8430 yr B.P. If the date is valid, that pollen profile may be a record of the invasion of *Picea glauca* from interior Alaska into the lower Copper River area during early-Holocene time rather than a record of invading coastal forest vegetation.

Heusser's pollen profile from Munday Creek near Icy Cape, east of Controller Bay (Figure 9-1), suggests that the earliest known vegetation along this section of the south-central Alaskan coast was tundra in which Cyperaceae and Ericaceae were important components. Tundra appeared there at least as early as 10,800 yr B.P. (Heusser, 1960).

Pollen Records from Interior Alaska

Within unglaciated interior Alaska, the most numerous and detailed records of late-Quaternary vegetational history are from the Tanana Valley and adjacent parts of the Yukon-Tanana Upland. Pollen analyses have been completed from sediment cores from 7 lakes (Ager, 1975; Anderson, 1975; Nakao et al., 1980; Ager, 1982a, unpublished data) and 15 peat bogs (Hansen, 1953; Ager, 1975; Anderson, 1975). Pollen studies, some of which have been supplemented by plant-macrofossil studies, have also been carried out on perennially frozen organic silts from exposures and a core from a borehole drilled through a thick, frozen valley fill (Matthews, 1970; 1974b; Péwé, 1975).

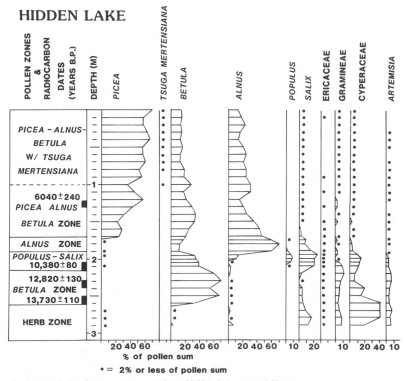

Figure 9-4. Pollen diagram (summary) from Hidden Lake, central Kenai Peninsula. (From Ager and Sims, 1981b, and unpublished data.)

Until the mid-1960s, Quaternary scientists assumed that unglaciated interior Alaska served as a refugium for *Picea* and other boreal forest taxa during past intervals of glacial climate (e.g., Heusser, 1965; Hopkins, 1967: 462). The pollen and macrofossil records now available, however, suggest that *Picea* and perhaps *Betula papyrifera*, *Larix*, and some species of *Alnus* may have been eliminated from Alaska during the last glacial interval (e.g., Matthews, 1974b; Ager, 1975; Hopkins et al., 1981). Probably, other species of *Alnus* as well as *Populus balsamifera* and perhaps *P. tremuloides* survived within unglaciated Alaska during the Wisconsin glacial interval (Hopkins et al., 1981; Ager, 1982c).

Pollen evidence from Birch Lake (Figures 9-1 and 9-5A), Lake George (Figure 9-1) (Ager, 1975), Harding Lake (Figures 9-1 and 9-5B) in the Tanana Valley (Nakao et al., 1980; Ager, 1982a, unpublished data), and Isabella Basin (Figure 9-1) in the Yukon-Tanana Upland near Fairbanks (Matthews, 1974b) indicate that the full-glacial vegetation of interior Alaska was treeless or nearly treeless. Shrubs such as Ericaceae, *Betula nana*, *B. glandulosa*, and *Alnus* were apparently uncommon. Only willow shrubs were well represented; herbaceous plants dominated.

Pollen types common in this herb zone include Gramineae, Cyperaceae, *Salix*, and *Artemisia*, along with Caryophyllaceae, *Plantago*, Cruciferae, Chenopodiaceae, Compositae (Tubuliflorae), *Thalictrum*, and other Ranunculaceae. Ferns, *Sphagnum*, and other mosses are poorly represented. This full-glacial pollen assemblage does not appear to have any totally convincing modern analogue in Alaska or elsewhere, although some alpine fell-field communities (Cwynar and Ritchie, 1980) and some dry-meadow communities with *Populus* on warm south-facing slopes in the interior (Young, 1976) have been suggested.

Some investigators interpret these full-glacial herbaceous pollen assemblages as representatives of a steppelike herb tundra characterized by discontinuous ground cover dominated by "pioneer" herbs tolerant of disturbances from frost heave, frost-boil formation, strong sand- and silt-bearing winds, and aridity. This herb tundra vegetation, sometimes referred to as steppe-tundra, tundra-steppe, or arctic steppe (e.g., Matthews, 1974a, 1976; Ager, 1975; Hopkins, 1979), appears to have been the predominant vegetation in Beringia during full-glacial times. Other coexisting vegetational types similar to their modern counterparts probably included alpine, wetland, and shrub tundra. The distribution of the latter two types was greatly restricted, however.

About 14,000 yr B.P., an abrupt climatic change to warmer, somewhat moister summers triggered the onset of a major vegetational change in many areas of Beringia. A dramatic change from herbaceous tundra or tundralike communities to shrub tundra characterized by dwarf *Betula*, Ericaceae, *Salix*, Cyperaceae, Gramineae, and other herbs spread throughout the Tanana Valley about 14,000 yr B.P. (Ager, 1975). In some areas of Alaska, this transition was apparently delayed for as much as several thousand years, perhaps as a result of local climatic or other environmental differences or possibly because of a lack of local seed sources of dwarf *Betula* and other shrub elements (Schweger, 1976; Ager, 1982c).

By about 11,000 yr B.P., the shrub-tundra vegetation of interior Alaska was invaded by deciduous scrub-forest vegetation, probably restricted to well-drained alluvial soils and south-facing slopes in uplands. This scrub forest included *Populus*, probably *P. balsamifera* and perhaps *P. tremuloides*, as well as various species of *Salix*. In most lakes cored thus far in interior Alaska, *Populus* pollen is rarely preserved in more than trace amounts. However, in one core from Harding Lake (Figure 9-5B) in the Tanana Valley, abundant, well-preserved *Populus* pollen occurs, bracketed by radiocarbon dates of 11,300 and 9860 yr B.P. (Nakao et al., 1980; Ager, 1982a, unpublished data). This and evidence from other sites indicates that *Populus* was the first tree to expand throughout much of Alaska about 11,000 yr B.P. (Hopkins et al, 1981; Ager, 1982a, 1982c). *Populus* was also a pioneer invading tree throughout much of Canada after the retreat of the last continental ice sheet (Mott, 1978).

Picea appears to have invaded the interior of Alaska from northwestern Canada via the Porcupine and Yukon Rivers (Hopkins et al., 1981). *Picea* began to invade the middle Tanana Valley about 9500 yr B.P. (Ager, 1975) and quickly spread from there eastward into the southwestern Yukon Territory by about 8700 yr B.P. (Rampton, 1971). It also spread southward from the Tanana Valley into some valleys of the Alaska Range as early as about 9100 yr B.P. (Ager and Sims, 1981a; Hopkins et al., 1981; C. E. Schweger, unpublished data). From the Alaska Range, *Picea* spread southward, reaching Hidden Lake on the Kenai Peninsula about 8000 yr B.P.

The next major vegetational change was an *Alnus* expansion about 8400 yr B.P. in the Tanana Valley. *Alnus* quickly became more abundant than at any time since perhaps as long ago as the previous interglacial; this finding is based on very limited evidence of pre-middle-Wisconsin vegetation in interior Alaska (e.g., Matthews, 1970; Péwé, 1975).

In some lacustrine pollen profiles from the Tanana Valley (e.g., Birch Lake, Figure 9-5A; Harding Lake, Figure 9-5B), there is evidence of a regional decline in *Picea* pollen percentages about 8000 to 6000 yr B.P. (Ager, 1975). This decline may be related to an episode of warm, dry climate in interior Alaska (Hypsithermal) that would perhaps have been unfavorable for *Picea*. Other factors could be responsible for the *Picea* decline, however (Ager, 1975; Hopkins et al., 1981).

By 6000 yr B.P., the boreal forest appears to have attained its present composition and perhaps its approximate present-day distribution at least in the Tanana Valley. The pollen records from lowland sites in interior Alaska show no clear evidence of significant vegetational change during the past 6000 years (Ager, 1975) (Table 9-1).

Pollen-Influx Studies

Only a few investigators in Alaska have attempted to calculate pollen influx (accumulation rate) in late-Quaternary deposits (e.g., Colinvaux, 1967a; Heusser, 1973; Ager, 1975). Several projects in progress should add considerably to our knowledge of present and past influxes in Alaska. At present, the evidence from interior Alaska suggests that pollen-accumulation rates were very low (about 100 to 400 grains per square centimeter per year) during the Late Wisconsin, when the vegetation of unglaciated regions of Alaska was sparse, herbaceous tundra. In latest-Wisconsin time (about 14,000 to 10,000 yr B.P.), when shrub tundra covered much of Alaska, pollen-accumulation rates increased substantially (to about 1000 to 1500 grains per square centimeter per year). During the early Holocene, when boreal forest began to develop in interior Alaskan lowlands, pollen accumulation rates again increased significantly, to about 5000 to 7000 grains per square centimeter per year. Later, during the middle to late Holocene, these accumulation rates dropped to about 3000 to 4000 grains per square centimeter per year (Ager, 1975).

Pollen Records from the Alaska Range

Since the mid-1970s, pollen profiles from five lakes have been analyzed from the Nenana Valley-Denali (Mount McKinley) National Park area (Ager, 1981b, unpublished data) and from a lacustrine core and an exposure of pond sediments from Tangle Lakes in the headwaters of the Delta River in the Central Alaska Range (Ager and Sims, 1981a; C. E. Schweger, unpublished data). The longest continuous section obtained thus far is a record of about 15,000 years from Eightmile Lake in the Northern Foothills of the Alaska Range (Figures 9-1 and 9-6). Eightmile Lake lies beyond the limit of late-Pleistocene glacial ice. Its pollen record is interesting because the lake is at a higher altitude (648 m) than the previously studied Tanana Valley lakes (200 to 400 m). The earliest recorded vegetation in the foothills was herbaceous tundra, rather similar to the herb-zone vegetation of the Tanana Valley. Cyperaceae seems to have been more abundant in the Alaska Range during full-glacial times than in the Tanana Valley, perhaps because of the effects of moisture from late-lasting snow, and *Artemisia* was less abundant.

By about 13,000 yr B.P., dwarf *Betula* began to invade at least the interfluves in the Northern Foothills. Herbaceous tundra communities may have persisted along glacial outwash deposits for several thousand years after shrub tundra began to develop on interfluves (T. A. Ager, unpublished data). By about 10,000 yr B.P., *Populus* had invaded the valleys of the Alaska Range, and *Salix* had increased in importance (Ager and Sims, 1981a; Ager, 1982b, unpublished data; C. E. Schweger, unpublished data).

Picea glauca began to invade the Alaska Range via the upper Delta River valley as early as 9100 yr B.P. (C. E. Schweger, unpublished data), but it did not reach the Northern Foothills in the Denali (Mount McKinley) National Park area until about 7500 yr B.P. or possibly later. Whereas in the Tanana Valley *Picea* preceded *Alnus* by about 1100 years, in the Nenana Valley-Northern Foothills area *Picea* and *Alnus* populations expanded at about the same time.

Lake Minchumina is located in the Tanana-Kuskokwim Lowland north of the Denali National Park segment of the central Alaska Range. A pollen record spanning about 8700 years was obtained from the lake (Ager, 1976, unpublished data). Early-Holocene vegetation at Lake Minchumina was shrub tundra. *Alnus* invaded the area about 7500 yr B.P. The low percentages of *Picea* pollen in the early to middle Holocene may be attributed to long-distance transport from the lowlands to the east, perhaps as far east as the Tanana Valley. *Picea* pollen percentages increased dramatically at the lake about 5700 yr B.P.; this suggests that spruce had finally invaded locally. *Picea* may, therefore, have been quite late (middle Holocene) in invading the adjacent upper Kuskokwim River drainage basin via this lowland migration route.

Pollen Records from Northern Alaska

When the last review of the Quaternary vegetational history of Arctic Alaska was published (Colinvaux, 1967b), few pollen sites were available from northern Alaska. Pioneer studies of pollen records at Chandler Lake (Table 9-1) in the north-central Brooks Range and near Umiat, in the Arctic Foothills north of the Brooks Range, by Livingstone (1955, 1957) provided the basis for a tentative pollen zonation and chronology across northern Alaska, and this chronology was applied to undated sites as far away as Ogotoruk Creek on the coast of northwestern Alaska (Heusser, 1963). Livingstone

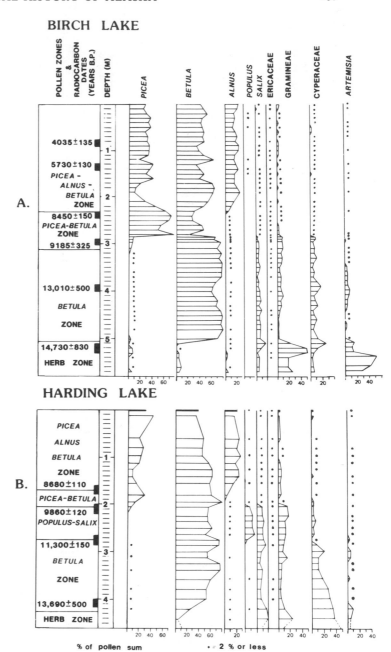

Figure 9-5. (A) Pollen diagram (summary) from Birch Lake (Core II), Tanana Valley. (From Ager, 1975.) (B) Pollen diagram (summary) from Harding Lake, Tanana Valley. (From Ager, 1982a, and unpublished data; and Nakao et al., 1980.)

recognized a three-part sequence of pollen zones, beginning with an early-Holocene (?) herb zone (Zone I), followed by a *Betula* zone (Zone II) when shrub tundra with dwarf *Betula* predominated, and an *Alnus* zone (Zone III) characterized by high percentages of *Alnus* pollen. The Chandler Lake pollen record displaying these zones lacked radiocarbon control, but a perennially frozen deposit of organic sediments near Umiat in the Arctic Foothills yielded a somewhat comparable pollen record supported by radiocarbon dates (Figure 9-1). Several problems exist in extending the Umiat chronology to Chandler Lake or throughout the rest of northern Alaska, however. First, the two sites are 125 km apart and their altitudes dif-

EIGHTMILE LAKE

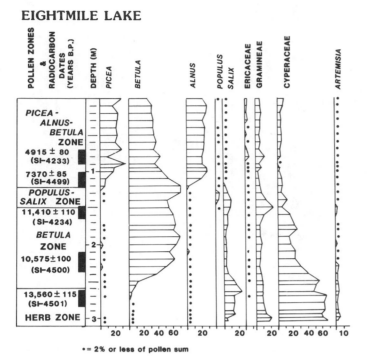

• = 2% or less of pollen sum

Figure 9-6. Pollen diagram (summary) from Eightmile Lake, Northern Foothills, Alaska Range. (From Ager, 1982b, and unpublished data.)

fer by 750 m. Second, the boundary between pollen Zones I and II at Umiat is less distinct than that at Chandler Lake. Although Zone I at Umiat does contain a significant herbaceous assemblage, it also contains a large amount of *Betula* pollen, more typical of Zone II. Third, the pollen record at Umiat may be interrupted by unconformities. Approximately the last 5000 years of the record is known to be missing from the Umiat section (Livingstone, 1957).

Later research in northern and western Alaska by Heusser (1963) and Colinvaux (1964, 1967b) showed that the three-zone sequence of Livingstone can be applied generally over extensive areas of Arctic Alaska, although chronologies and pollen assemblages vary somewhat. Research in progress in the north-central Brooks Range and areas to the north by P. A. Colinvaux and his colleagues at Ohio State University and on the Arctic Coastal Plain by R. Nelson from the University of Washington is likely to provide a much clearer record and a more firmly established chronology for Alaska north of the crest of the Brooks Range.

During the past decade, several projects have been undertaken on the southern side of the Brooks Range and adjacent lowlands by C. E. Schweger (1976, 1982), L. B. Brubaker of the University of Washington (unpublished data), and P. M. Anderson of Brown University (unpublished data). Schweger's pollen records are based on exposures of alluvial sediments along the Kobuk, John, and Koyukuk Rivers. Research by Brubaker and Anderson involves pollen analysis of lacustrine sediment from lakes in the south-central and southwestern Brooks Range and adjacent areas.

Ranger Lake, 9 km west of the Alatna River on the southern edge of the Brooks Range at lat. 67°08′ long. 153°38′ is one of the lakes cored by Brubaker (Figure 9-1). The lake is surrounded by shrub tundra, but a *Picea* gallery forest grows within a few kilometers. A summary pollen diagram (Figure 9-7) shows the major pollen types found in the upper 5 m of the 6-m core. No radiocarbon dates are yet avail-

able for this core, but dates from nearby lakes provide a basis for a preliminary regional chronology (Table 9-1). The pollen diagram shows a zonation quite similar to that from the Tanana Valley in interior Alaska (Figure 9-5A) (Ager, 1975), except that *Alnus* preceded the arrival of *Picea* at Ranger Lake. The Late Wisconsin vegetation was herbaceous tundra (steppe-tundra?) that was replaced by shrub tundra with dwarf *Betula* about 12,000 yr B.P. The *Alnus* rise began about 7800 yr B.P., and the *Picea* invasion appears to have begun roughly 5500 yr B.P. (L. B. Brubaker, personal communication, 1981).

The abrupt shift from herb tundra to dwarf-birch tundra seen in the Ranger Lake pollen profile (Figure 9-7) appears to have taken place about 2000 years earlier than a similar shift at Epiguruk (Figure 9-1) on the Kobuk River, 200 km to the west (Schweger, 1976). The two dates for this event may reflect real differences in the habitats and vegetational histories of these sites. However, the alluvial sections at Epiguruk may contain undetected hiatuses that could explain the differences in ages of the *Betula* rise at these two localities. Pollen studies under way involving analysis of cores from lakes west and northwest of Epiguruk also suggest the early expansion of *Betula* shrub tundra. P. M. Anderson has analyzed cores from Squirrel River Lake, about 100 km west of Epiguruk, and Kaiyak Lake, about 150 km northwest of Epiguruk. These cores also penetrate to deposits of Wisconsin age. Full-glacial herbaceous tundra was replaced by dwarf *Betula* tundra at these localities about 14,500 to 14,300 yr B.P. (P. M. Anderson, personal communication, 1980).

Pollen Records from the Seward Peninsula

Several major investigations of Quaternary vegetational history have been carried out on the Seward Peninsula (Hopkins et al., 1960; Colinvaux, 1964; Colbaugh, 1968; Matthews, 1974a; Shackleton, 1979). Some studies concentrated on reconstructing extraordinarily long-term histories of changing environments in Beringia, such as the records from Imuruk Lake (Figure 9-1) on the central Seward Peninsula (Colinvaux, 1964; Colbaugh, 1968; Shackleton, 1979) and the coastal exposures at Cape Deceit on the northeastern coast (Matthews, 1974a). In spite of the great importance of these investigations, they have not yielded complete, well-dated Holocene vegetational records because of the nature of the deposits. Therefore, at present, the late-Quaternary vegetational history of the Seward Peninsula must be pieced together from fragmentary records from several widely separated localities.

The full-glacial vegetation of the Seward Peninsula was herbaceous tundra or steppe-tundra (Colinvaux, 1964, 1967b; Matthews, 1974a). The full-glacial (Late Wisconsin) pollen assemblages are quite similar to those from the Yukon Delta-Norton Sound area to the south (Ager, 1980, 1982c).

Herbaceous tundra was replaced by dwarf *Betula* tundra late in the Wisconsin. Radiocarbon dates from Imuruk Lake suggest that the shift to shrub tundra took place about 13,000 to 12,000 yr B.P. (Colinvaux, 1964; Colbaugh, 1968). Other sites having radiocarbon dates (e.g., exposures near the southwestern coast of the Seward Peninsula at Nome) suggest that the transition may have taken place later in some areas but probably before 10,000 yr B.P. (Hopkins et al., 1960).

The pollen records from Nome on the southwestern coast and from Whitefish Lake near the northern coast of the Seward Peninsula (Shackleton, 1979) both suggest vegetational stability through what may be all or most of the Holocene. The Holocene profile from

RANGER LAKE

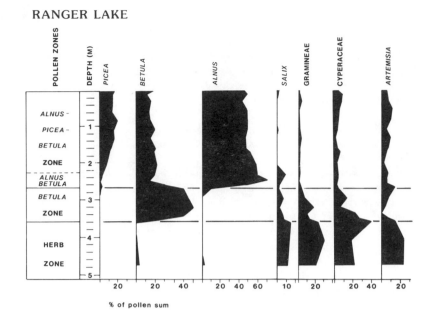

Figure 9-7. Pollen diagram (summary) from Ranger Lake, south-central Brooks Range, (Based on L. B. Brubaker, unpublished data.)

% of pollen sum

Nome records an undated history of shrub tundra, with a slight increase in *Alnus* pollen in the upper half of the pollen diagram (Hopkins et al., 1960). No dramatic *Alnus* rise is found like that in St. Michael Island sites on the southern side of the Norton Sound (Figures 9-1 and 9-8). The pollen diagram for Whitefish Lake, which is a maar lake, shows only minor changes. Two widely separated radiocarbon samples from the lower part of the core are dated at about 10,000 yr B.P. (Shackleton, 1979). Although the age of the base of the core is uncertain, the pollen record indicates that throughout most of the Holocene the vegetation of the northern coast of the Seward Peninsula probably has been shrub tundra with *Alnus*. The oldest radiocarbon date, if valid, indicates that *Alnus* appeared surprisingly early on the northern Seward Peninsula. *Alnus* apparently grows well on soils composed of porous volcanic ash, and, therefore, it may have been able to persist in that region when suitable habitats were rare elsewhere (Shackleton, 1979).

Pollen Records from Southwestern Alaska

For the purpose of this summary, southwestern Alaska includes the region from the southern side of Norton Sound to the Alaska Peninsula (Figure 9-1). Late-Quaternary pollen evidence from this region has been reviewed recently (Ager, 1982c) and will not be repeated here in detail. A previously unpublished pollen diagram from Zagoskin Lake on St. Michael Island, Norton Sound, is presented here (Figure 9-8) to serve as a key regional vegetational record. The chronology of regional vegetational changes is based on radiocarbon-dated segments of lacustrine sediment cores from several lakes in the Yukon Delta-Norton Sound (St. Michael Island) region (Ager, 1980, 1982c).

Pollen evidence from several sites indicates that the full-glacial vegetation was herbaceous tundra. *Artemisia* was present but less abundant than in full-glacial sites in the Tanana Valley (Ager, 1975, 1982c). Cyperaceae, Gramineae, *Salix*, and herbs such as *Thalictrum*, other Ranunculaceae, Caryophyllaceae, Cruciferae, *Plantago*, and Compositae are common elements of the Herb Zone pollen spectra.

The *Betula* rise marking the base of the *Betula* Zone occurs in

ZAGOSKIN LAKE

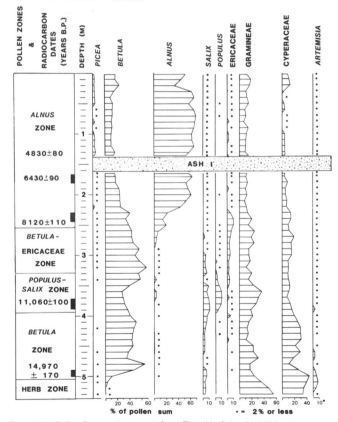

% of pollen sum   • = 2% or less

Figure 9-8. Pollen diagram (summary) from Zagoskin Lake, St. Michael Island, Norton Sound. (Based on T. A. Ager, unpublished data.)

several sites in this region at the level dated at about 14,000 yr B.P. (Ager, 1980, 1982c). This zone spans a long time interval (14,000 to about 7000 yr B.P.) during which *B. nana* shrub tundra with *Salix*, Gramineae, Cyperaceae, and a rich assemblage of herbs predominated over southwestern Alaska. At Puyuk Lake and Zagoskin Lake on St.

Michael Island, there are clear records of the brief development of *Populus-Salix* communities in well-drained sheltered sites in the region (Ager, 1982c) (Figure 9-8). These communities existed approximately at 11,000 yr B.P. in the Norton Sound area and may reflect an interval of warmer, drier climate. A slight increase in the proportion of Ericaceae pollen and *Sphagnum* spores during the early Holocene suggests an increase in moisture during the growing season.

Alnus crispa underwent a dramatic, abrupt regional expansion in southwestern Alaska about 7500 or 7000 yr B.P. Some pollen evidence indicates that *Alnus* may have been present in the region prior to that time, perhaps as early as about 11,000 yr B.P., but, if present, it was apparently rare and producing little pollen (Ager, 1982c). In south-central Alaska, *Alnus* invaded about 9500 yr B.P. or possibly earlier in some sites. In interior Alaska, it expanded rapidly after 8400 yr B.P. Thus, the rapid expansion of *Alnus* about 7500 yr B.P. in southwestern Alaska did not necessarily coincide with a climatic change.

Picea began to invade the eastern edge of southwestern Alaska about 5500 yr B.P. Today, the edge of the *Picea* forest reaches as far west as eastern Norton Sound, the east-central part of the Yukon-Kuskokwim Delta, and parts of the eastern edge of Bristol Bay (Figure 9-2) (Viereck and Little, 1972; Ager, 1982c). No palynologic investigations have yet been conducted in the region within or close to the modern *Picea* forest/tundra transition of southwestern Alaska, and, therefore, nothing is yet known about possible fluctuations in the position of that boundary during the past 5500 years.

Aside from the *Picea* invasion, the pollen records from southwestern Alaska do not suggest any dramatic vegetational changes since about 7500 yr B.P., when *Alnus* shrub tundra became established (Table 9-1). As elsewhere in Alaska, *Alnus* populations apparently increased abruptly once invasion began (or once conditions permitted small local populations, long suppressed, to expand). Just as abruptly, stabilization occurred.

Pollen Records from the Aleutian Islands

Pollen samples from late-Quaternary peats from a few scattered localities in the Aleutian chain have been described (Judson, 1946; Anderson and Bank, 1952; Everett, 1971; Heusser, 1973, 1978). Of those studies, only the peat profiles analyzed by Heusser (1973, 1978) are sufficiently continuous to permit detailed vegetational reconstruction.

Important components of the tundra vegetation of the Aleutians today include Cyperaceae, Gramineae, *Salix*, *Empetrum*, Umbelliferae, Ranunculaceae, Compositae and other herbs, and Polypodiaceae.

The Aleutian Islands (Figure 9-1) were extensively glaciated during the Wisconsin glacial interval (Coulter et al., 1965). The time of deglaciation is uncertain, but the oldest radiocarbon dates from the Aleutians suggest that peat first began to accumulate about 10,500 yr B.P. (Everett, 1971).

Peat profiles from Umnak Island (Figure 9-1) in the eastern Aleutians (Heusser, 1973) and Adak Island in the central Aleutians (Heusser, 1978) both record histories of tundra vegetation spanning about the past 10,000 years. Adak, located 560 km west of Umnak, yielded a pollen record that reflects its more depauperate flora; as the distance from the mainland seed sources increases, the island floras tend to become smaller (Hultén, 1968). In addition, islands close to the mainland receive larger amounts of long-distance pollen from trees and shrubs that do not grow in the Aleutians (e.g., *Picea*, *Betula*). *Alnus* grows only on the easternmost Aleutian Islands, according to Hultén (1968).

The pollen diagram from Umnak Island is reproduced in Figure 9-9. Heusser (1973) recognized three pollen zones representing (1) an early-Holocene Cyperaceae-Gramineae tundra, (2) a middle-Holocene *Salix*-Gramineae-Cyperaceae tundra, and (3) a late-Holocene Gramineae-Cyperaceae tundra (Table 9-1). Heusser inferred climatic changes from an early-Holocene interval of cool, moist climate (10,000 to 8500 yr B.P.), to a middle-Holocene interval of warmer, somewhat drier, and probably less windy climate (about 8500 to 3000 yr B.P.), to a late-Holocene interval of cooler, moist climate (3000 yr B.P. to the present).

The pollen record from Adak Island (Heusser, 1978) displays a basal Cyperaceae-*Salix*-*Empetrum*-*Lycopodium* pollen zone that represents the earliest recorded Quaternary vegetation on the island (about 10,250 to 10,000 yr B.P.) This is overlain by a sequence covering about 10,000 to 3000 yr B.P., during which Cyperaceae was predominant. No clear indication of a middle-Holocene vegetational change comparable to that seen on Umnak is in evidence. In the upper part of the section, however, a dramatic increase in *Empetrum* pollen percentages takes place. Heusser (1978) inferred from this increase a shift to cooler, stormier climatic conditions associated with Neoglaciation during the past 3000 years.

Pollen Records from the Pribilof Islands

Pollen profiles have been described from several sites in the Pribilof Islands. The record from St. Paul Island suggests that the present-day tundra vegetation, which is composed of a rich assemblage of herbs (Cyperaceae, Gramineae, Umbelliferae, *Artemisia*, Ranunculaceae, and others) and few low shrubs (*Empetrum nigrum*, *Salix*), has undergone very little change during the Holocene (Colinvaux, 1967a, 1967b). Some shifts in pollen percentages in profiles from St. George Island may, however, reflect slight vegetational changes during the middle to late Holocene. Percentages of Umbelliferae pollen increase, and *Artemisia* percentages decrease. These changes may record a shift from Hypsithermal to Neoglacial climatic regimes in the southern Bering Sea region (Parrish, 1980).

Summary and Conclusions

Palynologic investigations in Alaska during the past decade have greatly improved our knowledge of the vegetational history for the late Quaternary, although much remains to be done. Table 9-1 shows a comparison of some regional vegetational histories. It now appears that the vegetation of essentially all of unglaciated Alaska from which we have data for full-glacial time was herbaceous tundra, of which Cyperaceae, Gramineae, *Salix*, *Artemisia* and assorted forbs were important elements. Vegetative cover was probably patchy, and the mossy, tussocky, and shrubby tundra vegetation that characterizes the tundra in much of Alaska today was apparently rare. Some arctic and alpine tundra communities that exist today in Alaska and adjacent Canada probably serve as reasonable analogues for some full-glacial herb-tundra communities. However, other communities probably existed and can be described, at least tentatively, by terms such as ''steppe-tundra'' that differentiate them from modern tundra types. Regardless of how those herbaceous communities are characterized and interpreted, variations of herb tundra extended unbroken from the Bering Land Bridge (Hopkins, 1967, 1979) to unglaciated northwestern Canada during the Wisconsin glacial interval and from sites near present sea level to altitudes approaching glacial-age snowfields in the mountains. No dated tree-size wood has yet been found in full-glacial deposits (older than 12,000 and younger than 26,000 yr B.P.), and even evidence of woody shrubs, except *Salix*, is rare (Hopkins et al., 1981; Schweger, 1982). In spite of the sparse evidence, it still seems likely that eastern Beringia and interior Alaska served as a refugium for some species of trees (e.g., *Populus*) and shrubs (e.g., *Alnus*, but the long-held view that the interior was a major forest refugium for *Picea*, *Betula papyrifera*, and some other boreal plants is unsupported by the evidence presently available.

The abrupt replacement of herbaceous tundra by birch shrub tundra that began in several areas of Alaska as early as 14,500 to 13,700 yr B.P. appears to have been triggered by a climatic change to warmer, moister summers and perhaps deeper winter snows. This shrub tundra may have a modern analogue or near-analogue in tussocky dwarf-*Betula* shrub-tundra communities in the central Seward Peninsula and many other areas of Alaska, but most *Betula* Zone pollen spectra have higher percentages of *Betula* pollen than do surface pollen samples from modern shrub-tundra communities. It is not yet known whether significant floristic differences existed between the *Betula* Zone vegetational communities and their presumed modern counterparts.

Alaskan vegetation changed dramatically during the Holocene, particularly in interior, south-central, and southeastern Alaska, when successive invasions of trees transformed the treeless or nearly treeless landscape into a mosaic of forest, muskeg, tundra, and forest-tundra. An altitudinal "zonation" of vegetation developed that appears to have been minimal during full-glacial time. In interior Alaska before the population expansion of *Alnus*, *Picea* invaded the Tanana Valley from the northern Yukon Territory via the Porcupine and Yukon Rivers roughly 9500 yr B.P., and it spread rapidly into adjacent areas in the southwestern Yukon Territory.

In the Delta River valley, *Picea glauca* invaded the Alaska Range as early as 9100 yr B.P. and may have spread quickly southward to upper Cook Inlet and the central Kenai Peninsula by about 8000 yr B.P. In some valleys of the Alaska Range, *Alnus* and *Picea* seem to have spread simultaneously about 7500 yr B.P.

In southwestern and northwestern Alaska, the *Alnus* invasion or expansion preceded the *Picea* invasion. *Alnus* spread to those regions around 7800 to 7000 yr B.P. and was followed by *Picea* about 5500 yr B.P. In southeastern Alaska, where most of the terrain was covered by glacier ice during the Wisconsin glacial interval, coastal forests began to develop soon after deglaciation. Pioneer invaders such as *Alnus* and *Pinus contorta* apparently arrived first. Tree species that dominate the modern forests of the coastal regions (e.g., *Picea sitchensis*, *Tsuga* spp.) spread northward from British Columbia and Washington, eventually expanding northwestward to Cook Inlet during the middle Holocene.

The reconstruction of Alaskan climatic history for the past 16,000 years can be outlined only roughly on the basis of existing pollen data. A more detailed climatic record probably will eventually be reconstructed from less ambiguous indicators of climatic change (e.g., freshwater ostracodes, diatoms, insects, and tree rings). Some past vegetational changes do seem to coincide with climatic changes, how-

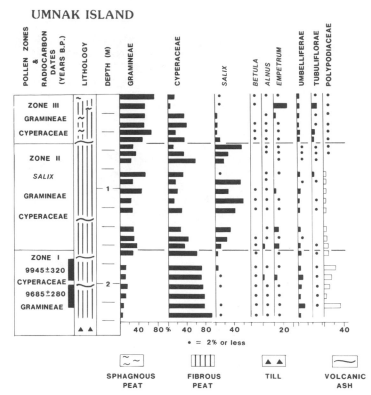

Figure 9-9. Pollen diagram (summary) from Umnak Island. (Adapted from Heusser, 1973, with the permission of Elsevier Scientific Publishing Company.)

ever. The *Betula* rise about 14,000 yr B.P. and the *Populus* expansion about 11,000 yr B.P. both were probably triggered by climatic changes. Shrub *Betula* and probably *Populus* were already present in scattered marginal habitats and then spread rapidly throughout most of the state as soon as local conditions became suitable and as rapidly as their migration rates permitted. The shrub *Betula* expansion probably was favored when summers became moister and warmer and winter snows perhaps deepened about 14,000 yr B.P. *Populus* expanded during an interval when climate appears to have been characterized by very warm, dry summers.

Climatic changes within the Holocene are less clearly defined, however. Some major vegetational changes such as the establishment of interior *Picea* forests are time transgressive, reflecting the progressive invasion of *Picea* westward and southward across Alaska during an interval of several thousand years. No one knows whether a significant interval elapsed between the time when the climate became suitable and the time when *Picea* invaded a given region. The gradual westward expansion of coastal forests in southeastern and south-central Alaska presents similar interpretational problems.

At present, an early-Holocene interval of warm, moist summer climate is recognized. Indications of a middle-Holocene interval of warmer, drier climate (Hypsithermal) is recognized at some but by no means all localities where paleoecologic studies have been done. Data from some sites suggest that a shift to cooler, moister neoglacial climate took place in late-Holocene time beginning about 3500 yr B.P. Most lowland sites in interior and western Alaska do not record significant changes in vegetation or climate during the past 6000 years or so, however.

References

Ager, T. A. (1975). "Late Quaternary Environmental History of the Tanana Valley, Alaska." Ohio State University, Institute of Polar Studies Report 54.

Ager, T. A. (1976). Holocene vegetational history of the Lake Minchumina area, Mt. McKinley quadrangle, Alaska. *In* "American Quaternary Association, Fourth Biennial Meeting, Abstracts," p. 96. Arizona State University, Tempe.

Ager, T. A. (1980). A 16,000 year pollen record from St. Michael Island, Norton Sound, western Alaska. *In* "American Quaternary Association, Sixth Biennial Meeting, Abstracts and Program, 18-20 August 1980," p. 3. Institute for Quaternary Studies, University of Maine, Orono.

Ager, T. A. (1982a). Pollen studies of Quaternary-age sediments in the Tanana Valley (Alaska). *In* "United States Geological Survey in Alaska: Accomplishments during 1980" (W. Coonrad, ed.), pp. 66-67. U.S. Geological Survey Circular 844.

Ager, T. A. (1982b). Quaternary history of vegetation in the North Alaska Range. *In* "United States Geological Survey in Alaska: Accomplishments during 1980" (W. Coonrad, ed.), pp. 109-11. U.S. Geological Survey Circular 844.

Ager, T. A. (1982c). Vegetational history of western Alaska during the Wisconsin glacial interval and Holocene. *In* "Paleoecology of Beringia" (D. M. Hopkins, J. V. Matthews, Jr., C. E. Schweger, and S. B. Young, eds.), pp. 75-93. Academic Press, New York.

Ager, T. A., and Sims, J. D. (1981a). Holocene pollen and sediment record from the Tangle Lakes area, central Alaska. *Palynology* 5, 85-98.

Ager, T. A., and Sims, J. D. (1981b). Late Quaternary pollen record from Hidden Lake, Kenai Peninsula, Alaska. *In* "American Association of Stratigraphic Palynologists, Program and Abstracts of the Fourteenth Annual Meeting," pp. 8-9.

Aleksandrova, V. D. (1980). "The Arctic and Antarctic: Their Division in Geobotanical Areas." Cambridge University Press, Cambridge.

Anderson, J. H. (1975). A palynological study of late Holocene vegetation and climate in the Healy Lake area, Alaska. *Arctic* 28, 29-62.

Anderson, S. T., and Bank, T. P., II (1952). Pollen and radiocarbon studies of Aleutian soil profiles. *Science* 116, 84-86.

Bailey, R. G. (1980). "Description of the Ecoregions of the United States." U.S. Department of Agriculture Miscellaneous Publication 1391.

Birks, H. J. B. (1973). Modern pollen rain studies in some arctic and alpine environments. *In* "Quaternary Plant Ecology" (H. J. B. Birks and R. G. West, eds.), pp. 143-68. John Wiley and Sons, New York.

Birks, H. J. B. (1977). Modern pollen rain and vegetation of the St. Elias Mountains, Yukon Territory. *Canadian Journal of Botany* 55, 2367-82.

Colbaugh, P. R. (1968). The environment of the Imuruk Lake area, Seward Peninsula, during Wisconsin time. M.S. thesis, Ohio State University, Columbus.

Colinvaux, P. A. (1964). The environment of the Bering Land Bridge. *Ecological Monographs* 34, 297-329.

Colinvaux, P. A. (1967a). Bering Land Bridge: Evidence of spruce in Late Wisconsin times. *Science* 156, 380-83.

Colinvaux, P. A. (1967b). Quaternary vegetational history of Arctic Alaska. *In* "The Bering Land Bridge" (D. M. Hopkins, ed.), pp. 207-31. Stanford University Press, Stanford, Calif.

Coulter, H. W., Hopkins, D. M., Karlstrom, T. N. V., Péwé, T. L., Wahrhaftig, C., and Williams, J. R. (1965). Map showing extent of glaciations in Alaska. U.S. Geological Survey Miscellaneous Geologic Investigations Map I-415.

Cwynar, L. C. (1980). A late-Quaternary vegetation history from Hanging Lake, northern Yukon. Ph.D. dissertation, University of Toronto, Toronto.

Cwynar, L. C., and Ritchie, J. C. (1980). Arctic steppe-tundra: A Yukon perspective. *Science* 208, 1375-77.

Denton, G. H. (1974). Quaternary glaciations of the White River valley, Alaska, with a regional synthesis for the north St. Elias Mountains, Alaska and Yukon Territory. *Geological Society of America Bulletin* 85, 871-92.

Everett, K. R. (1971). Composition and genesis of the organic soils of Amchitka Island, Aleutian Islands, Alaska. *Arctic and Alpine Research* 3, 1-16.

Hansen, H. P. (1953). Postglacial forests in the Yukon Territory and Alaska. *American Journal of Science* 251, 505-42.

Hartman, C. W., and Johnson, P. R. (1978). "Environmental Atlas of Alaska." 2nd ed. University of Alaska Institute of Water Resources, Fairbanks.

Heusser, C. J. (1957). Pleistocene and postglacial vegetation of Alaska and Yukon Territory. *In* "Arctic Biology" (H. P. Hansen, ed.), pp. 131-51. Oregon State College, Corvallis.

Heusser, C. J. (1960). "Late-Pleistocene Environments of North Pacific North America." American Geographical Society Special Publication 35.

Heusser, C. J. (1963). Pollen diagrams from Ogotoruk Creek, Cape Thompson, Alaska. *Grana Palynologica* 4, 149-59.

Heusser, C. J. (1965). A Pleistocene phytogeographical sketch of the Pacific Northwest and Alaska. *In* "The Quaternary of the United States" (H. E. Wright, Jr., and D. G. Frey, eds.), pp. 469-83. Princeton University Press, Princeton, N.J.

Heusser, C. J. (1973). Postglacial vegetation on Umnak Island, Aleutian Islands, Alaska. *Review of Palaeobotany and Palynology* 15, 277-85.

Heusser, C. J. (1978). Postglacial vegetation on Adak Island, Aleutian Islands, Alaska. *Bulletin of the Torrey Botanical Club* 105, 18-23.

Heusser, C. J. (1983). Vegetational History of the northwestern United States including Alaska. *In* "Late-Quaternary Environments of the United States," vol. 1, "The Late Pleistocene" (S. C. Porter, ed.), pp. 239-58. University of Minnesota Press, Minneapolis.

Hopkins, D. M. (ed.) (1967). "The Bering Land Bridge." Stanford University Press, Stanford, Calif.

Hopkins, D. M. (1972). The paleogeography and climatic history of Beringia during late Cenozoic time. *Internord* 12, 121-50.

Hopkins, D. M. (1979). Landscape and climate of Beringia during late Pleistocene and Holocene time. *In* "The First Americans: Affinities and Adaptations" (W. S. Laughlin and A. B. Harper, eds.), pp. 15-41. Gustav Fischer, New York.

Hopkins, D. M., MacNeil, F. S., and Leopold, E. B. (1960). The Coastal Plain at Nome, Alaska: A late Cenozoic type section for the Bering Strait region. *In* "Report of the 21st International Geological Congress (Copenhagen)," part 4, pp. 44-57.

Hopkins, D. M., Matthews, J. V., Jr., Schweger, C. E., and Young, S. B. (eds.) (1982). "Paleoecology of Beringia." Academic Press, New York.

Hopkins, D. M., Smith, P. A., and Matthews, J. V., Jr. (1981). Dated wood from Alaska: Implications for forest refugia in Beringia. *Quaternary Research* 15, 217-49.

Hultén, E. (1937). "Outline of the History of Arctic and Boreal Biota during the Quaternary Period." Bokforlags Aktiebolaget Thule, Stockholm.

Hultén, E. (1968). "Flora of Alaska and Neighboring Territories." Stanford University Press, Stanford, Calif.

Judson, S. (1946). Late glacial and postglacial chronology on Adak. *Journal of Geology* 54, 376-85.

Larsen, J. A. (1980). "The Boreal Ecosystem." Academic Press, New York.

Livingstone, D. A. (1955). Some pollen profiles from Arctic Alaska. *Ecology* 36, 587-600.

Livingstone, D. A. (1957). Pollen analysis of a valley fill near Umiat, Alaska. *American Journal of Science* 255, 254-60.

McKenzie, G. D. (1970). "Glacial Geology of Adams Inlet, Southeastern Alaska." Ohio State University, Institute of Polar Studies Report 25.

Matthews, J. V., Jr. (1970). Quaternary environmental history of interior Alaska: Pollen samples from organic colluvium and peats. *Arctic and Alpine Research* 2, 241-51.

Matthews, J. V., Jr. (1974a). Quaternary environments at Cape Deceit (Seward Peninsula, Alaska): Evolution of a tundra ecosystem. *Geological Society of America Bulletin* 85, 1353-84.

Matthews, J. V., Jr. (1974b). Wisconsin environment of interior Alaska: Pollen and macrofossil analysis of a 27-meter core from the Isabella Basin (Fairbanks, Alaska). *Canadian Journal of Earth Sciences* 11, 828-41.

Matthews, J. V., Jr. (1976). Arctic steppe: An extinct biome. *In* "American Quaternary Association, Fourth Biennial Meeting, Abstracts," pp. 73-77. Arizona State University, Tempe.

Matthews, J. V., Jr. (1979). "Beringia during the Late Pleistocene: Arctic-Steppe or Discontinuous Herb-Tundra? A Review of the Paleontological Evidence." Geological Survey of Canada Open-File Report 649.

Moriya, K. (1976). "Flora and Palynomorphs of Alaska." Kodansha Company, Tokyo. (In Japanese.)

Mott, R. J. (1978). *Populus* pollen in late Pleistocene pollen spectra. *Canadian Journal of Botany* 56, 1021-31.

Nakao, K., La Perriere, J., and Ager, T. A. (1980). Climatic changes in interior Alaska. *In* "Climatic Changes in Interior Alaska" (K. Nakao, ed.), pp. 16-23. Hokkaido University Department of Geophysics, Sapporo, Japan.

Nelson, R. E. (1979). Quaternary environments of the Arctic Slope of Alaska. M.S. thesis, University of Washington, Seattle.

Parrish, L. (1980). "A Record of Holocene Climatic Changes from St. George Island, Pribilofs, Alaska." Ohio State University, Institute of Polar Studies Report 75.

Péwé, T. L. (1975). "Quaternary Geology of Alaska." U.S. Geological Survey Professional Paper 835.

Rampton, V. (1971). Late Quaternary vegetational and climatic history of the Snag-Klutlan area, southwestern Yukon Territory, Canada. *Geological Society of America Bulletin* 82, 959-78.

Ritchie, J. C., and Lichti-Federovich, S. (1963). Contemporary pollen spectra in central Canada: I. Atmospheric samples at Winnipeg, Manitoba. *Pollen et Spores* 5, 95-114.

Ritchie, J. C., and Lichti-Federovich, S. (1967). Pollen dispersal phenomena in Arctic-Subarctic Canada. *Review of Palaeobotany and Palynology* 3, 255-66.

Schweger, C. E. (1976). Late Quaternary paleoecology of the Onion Portage region, northwestern Alaska. Ph.D. dissertation, University of Alberta, Edmonton.

Schweger, C. E. (1982). Late Pleistocene vegetation of eastern Beringia: Pollen analysis of dated alluvium. *In* "Paleoecology of Beringia" (D. M. Hopkins, J. V. Matthews, Jr., C. E. Schweger, and S. B. Young, eds.), pp. 95-112. Academic Press, New York.

Shackleton, J. (1979). Paleoenvironmental histories from Whitefish and Imuruk Lakes, Seward Peninsula, Alaska. M.S. thesis, Ohio State University, Columbus.

Sirkin, L. A., and Tuthill, S. (1969). Late Pleistocene palynology and stratigraphy of Controller Bay region, Gulf of Alaska. *In* "Etudes sur le Quaternaire dans le Monde," pp. 197-208. Eighth Congress of the International Association for Quaternary Research, Paris.

United States National Oceanic and Atmospheric Administration (1980). "Climates of the States." Vol. 1, 2nd ed. Gale Research Company, Detroit.

Viereck, L. A., and Little, E. L., Jr. (1972). "Alaska Trees and Shrubs." U.S. Department of Agriculture Handbook 410.

Young, S. B. (1976). Is steppe-tundra alive and well in Alaska? *In* "American Quaternary Association Fourth Biennial Meetings, Abstracts," pp. 84-88. Arizona State University, Tempe.

CHAPTER 10

Holocene Changes in the Vegetation of the Midwest

T. Webb, III, E. J. Cushing, and H. E. Wright, Jr.

Introduction

Describing the Holocene vegetational history of the midwestern United States has long challenged palynologists and ecologists. Early work from the 1930s to the 1950s revealed that the vegetation had changed, but only since 1960 has a data base accumulated that contains enough radiocarbon dates to permit study of the timing and spatial patterns of the changes. In the northern Midwest, the network of radiocarbon-dated pollen diagrams is now sufficiently dense that the pollen data can be mapped in a form that reveals the patterns of past vegetation.

Previous discussions of the Holocene vegetational history within the Midwest (Potzger, 1946; Curtis, 1959; Cushing, 1965; Wright, 1968a, 1968b, 1977; Bernabo and Webb, 1977; Amundson and Wright, 1979) have identified one or more of the following major events: (1) an initial eastward movement of the prairie from South Dakota into Wisconsin and a subsequent westward retreat after 7000 yr B.P.; (2) an early northward retreat of *Picea* and other boreal taxa and then, much later, a southward readvance; (3) a sequence of changes in the location and composition of the mixed conifer/hardwood forest beginning with increased abundances of *Betula*, *Alnus*, and *Pinus banksiana*, followed next by the migration and rise to dominance of *P. strobus* and finally by the spread of *Fagus*, *Tsuga*, and *Betula lutea*; and (4) the initial appearance and later changes in location and composition of the *Quercus*-dominated deciduous forests. By mapping the pollen data from 34 diagrams for the complete Holocene and from 15 other sites in the northern Midwest, we add spatial resolution to the previous descriptions of these events and try to demonstrate how they all are part of a combined response by the flora to changing Holocene environmental conditions. We also discuss the occurrence of certain assemblages of Late Wisconsin and early-Holocene pollen for which no modern analogues exist.

Modern Vegetation and Pollen

MAPPING THE PRESENT AND FORMER VEGETATION

Maps of the vegetational patterns that existed just before the arrival of Euro-Americans in the 19th century are fundamental to understanding the vegetational changes that occurred during the Holocene because these patterns provide a baseline against which the previous and subsequent vegetational patterns can be compared. Such maps (as they are normally drawn [Figure 10-1]) have two important limitations, however. First, the small scale of the map permits the display of only a fraction of the spatial complexity of the vegetation. When data are widespread and abundant, as they are for the period the map represents, the spatial generalization required for the map can be made with confidence. When data come only from scattered points, as they do for most of the Holocene, distinguishing between local and regional phenomena can be difficult at selected sites, and this difficulty can complicate the efforts to describe the composition of past vegetational types and to map their location.

We have, therefore, chosen to map directly the fossil data, in the form of pollen percentages, and then to reconstruct in narrative form the total vegetation on the basis of the patterns that emerge on the pollen maps. With this procedure, the small scale of the map can be turned to an advantage: it directs attention to those ecologic processes whose effects are more apparent at that small map scale, and a series of such maps at intervals through the Holocene should yield knowledge of the dynamics of those processes and should point the way to consideration of the regional climatic controls on those processes. The patterns on the map series should also help in assessing the regional climatic controls of the ecologic processes.

For the northern Midwest (Figure 10-1), we judge that the available data are sufficient to support several regional generalizations of data directly, and these we present and discuss. For this reason, we restrict Figure 10-1 to an area smaller than the usual definition of the Midwest. Data from farther south or from the Great Plains to the

NSF grants (ATM79-16234, ATM79-16580, ATM81-11870, and ATM81-12840) to COHMAP (Climates of the Holocene: Mapping Based on Pollen Data) supported this research. We thank M. Anderson, R. Arigo, D. C. Gaudreau, R. M. Mellor, S. Suter, and C. Thrall for technical assistance; R. E. Bailey, R. O. Kapp, and J. H. McAndrews for permission to use unpublished data; R. E. Bailey, K. M. Heide, R. Lawrenz, and L. C. K. Shane for permission to use data from unpublished theses; and M. B. Davis, H. J. B. Birks, and W. A. Watts for reviews of the manuscript.

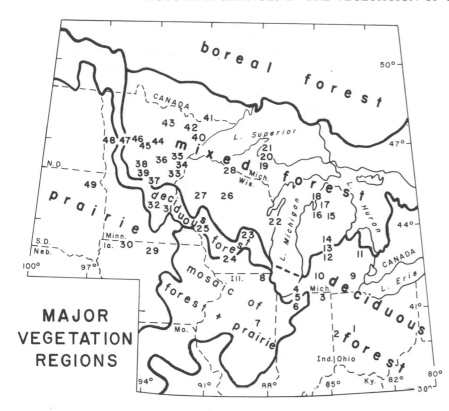

MAJOR
VEGETATION
REGIONS

Figure 10-1. Map showing the pollen sites and the major vegetational regions in the Midwest just before Euro-American settlement. (Table 10-3 provides a key to the site numbers.)

west are still few because of the scarcity of suitable Holocene deposits (E. Grüger, 1972; J. Grüger, 1973; King and Allen, 1979).

The second limitation of Figure 10-1 is that it can display only a single aspect of the vegetation, and that aspect may not be one readily recognized in fossil deposits or one highly responsive to Holocene changes in regional processes. We have, therefore, chosen traditional units for the map because their familiarity permits easy comparisons with other areas and because they are based largely on the characteristics of the dominant overstory species, which are also those most abundant and reliable in the fossil data.

Just to the north of our region is the boreal forest, dominated by conifers, especially *Picea glauca* and *P. mariana*; *Pinus banksiana*, *P. resinosa*, and *P. strobus*; *Abies balsamea*; and *Larix laricina* (plant names follow Gleason and Cronquist, 1963; English equivalents for trees are listed in Table 10-1). The important broad-leaved deciduous trees are *Betula papyrifera*, *Populus tremuloides*, and *P. balsamifera*. South of the boreal forest lies a mixed forest of coniferous and broad-leaved deciduous species. All of the species of the boreal forest occur here, joined by the conifer *Tsuga canadensis* and many angiosperms, including *Acer saccharum*, *A. rubrum*, *Betula lutea*, *Fagus grandifolia*, *Tilia americana*, *Quercus borealis*, *Q. macrocarpa*, *Ulmus americana*, *U. rubra*, *Fraxinus nigra*, and *F. pennsylvanica*. This group of angiosperms is often called the northern hardwoods. South of the mixed forest is deciduous forest in which conifers are rare, although a few species, notably *Pinus strobus*, occur in special habitats. The northern hardwoods are here joined by many additional deciduous trees, including species of *Quercus*, *Carya*, *Juglans*, *Prunus*, *Fraxinus*, and, to the southeast, *Platanus* and *Liriodendron*. South and west of the forest is the prairie. Its rich flora is dominated by the grass genera *Andropogon*, *Sorghastrum*, *Panicum*, *Spartina*, and *Stipa* and includes the forb genera *Ambrosia*, *Artemisia*, and

Chenopodium, which are wind pollinated and hence important components of fossil pollen assemblages.

These four vegetational regions are distinguished chiefly by their physiognomy, but the boundaries between some of the regions can also be defined by changes in flora. The location of the boundaries can be understood by reference not only to flora and structure but also to certain climatic variables. The boundary between the boreal forest and the mixed forest is drawn near the northern range limits of the northern hardwoods. These limits correspond well with the line marking an average annual minimum temperature of $-40\,°C$, and all the species (except *Tilia americana*) have been shown to survive subfreezing winter temperatures in their ray-parenchyma cells by the mechanism of deep supercooling of intracellular water, a mechanism that is ineffective below about $-41\,°C$ (Burke et al., 1976; George et al., 1974). The trees of the boreal forest, in contrast, tolerate these and lower temperatures by the mechanism of extracellular ice formation, and this mechanism has no effective lower limit. Although the details vary from species to species and are not clear, the correlation between species distribution and a plausible and sufficient physiologic mechanism is sufficiently strong to favor the hypothesis of a causal climatic link between the two. As pointed out by Bryson (1966), this vegetational boundary is correlated with the mean winter position of the arctic frontal zone.

The boundary between the mixed forest and the deciduous forest is also a floristic boundary, the ''tension zone'' described for Wisconsin by Curtis (1959) and for Michigan by Potzger (1948). A recent analysis of the boundary in Michigan by McCann (1979) verifies that a significant number of southern species of vascular plants have their northern range limits concentrated near the boundary. A physiologic mechanism common to these species is unknown, although McCann argues that the environmental variable correlating best with the

Table 10-1. English Equivalents for Latin Names
of Trees Mentioned in the Text

Latin Name	English Name
Abies balsamea	Balsam fir
Acer saccharum	Sugar maple
A. rubrum	Red maple
Alnus rugosa	Speckled alder
Betula lutea	Yellow birch
B. papyrifera	White birch (paper birch)
Carya	Hickory
Corylus	Hazel
Fagus grandifolia	Beech
Fraxinus nigra	Black ash
F. pennsylvanica	Green ash
Juglans	Walnut
Larix laricina	Tamarack
Liriodendron	Tulip-poplar
Ostrya/Carpinus	Ironwood/hornbeam
Picea glauca	White spruce
P. mariana	Black spruce
Pinus banksiana	Jack pine
P. resinosa	Red pine
P. strobus	White pine
Platanus	Sycamore
Populus balsamifera	Balsam poplar
P. tremuloides	Trembling aspen
Prunus	Cherry
Quercus borealis	Northern red oak
Q. macrocarpa	Bur oak
Tilia americana	Basswood
Tsuga canadensis	Eastern hemlock
Ulmus americana	American elm
U. rubra	Slippery elm

boundary and physiologically most plausible is a steepening in the gradient of growing-degree-days.

Other floristic gradients exist within the mixed and deciduous forests. Most notable is a decrease in the number of species from east to west. In particular, *Fagus grandifolia*, *Tsuga canadensis*, and *Betula lutea* are more abundant in the eastern part of the region and are rare or absent westward into Minnesota. The range boundary for *Fagus grandifolia* is in eastern Wisconsin and central upper Michigan, and that for *Tsuga canadensis* is in northeastern Minnesota. The western limit of these species may be controlled physiologically, perhaps by mechanisms related to the decreased soil moisture and precipitation to the west. On the other hand, a decrease in the relative abundance of the northern hardwoods as a group from Wisconsin into Minnesota and their replacement by *Betula papyrifera* and *Populus tremuloides* may be related to an increase in the frequency of fire in the forests of Minnesota (Heinselman, 1973). The increase in fire frequency may itself be related to the westward-increasing moisture stress.

The boundary between deciduous forest and prairie is more strongly physiognomic than floristic. Most of the species of the prairie occur in favorable habitats well within the forested region, and the reverse is true of the forest flora. Unlike the other boundaries, the change from forest to prairie is commonly quite abrupt. In detail, the position of the boundary is highly correlated with the local topography and the location of natural barriers to the spread of prairie fires, and little

doubt remains that fire was the dominant factor in determining the position and nature of the boundary on the scale of a county or a section of a state. In areas of rough morainic topography in Minnesota, the transition from prairie to forest was marked by a woodland of scrubby trees and shrubs, by thickets and groves of trees, and rarely by savanna (Grimm, 1981). In the driftless area of southwestern Wisconsin and adjacent parts of Minnesota and Iowa, the landscape is deeply dissected by stream valleys; these were occupied by deciduous forest, whereas the broad interfluves supported prairie. On Figure 10-1, this area is shown as a mosaic of the two vegetational units, and both the scarcity of fossil deposits in the area and the local heterogeneity of habitats make reconstruction of past vegetation there exceptionally difficult.

GEOGRAPHIC DISTRIBUTION OF CONTEMPORARY
POLLEN IN SEDIMENTS

Studies by McAndrews (1966), Davis and others (1971), Webb (1974a, 1974b), Webb and McAndrews (1976), and Peterson (1978) illustrate the patterns of modern pollen data within the Midwest. Davis and others (1973) and Webb and others (1978, 1981) investigated the relations between the pollen-accumulation rate and tree basal area and between pollen percentage and tree percentage. Within Michigan and Wisconsin, the pollen percentages (based on total tree pollen in lakes) for *Pinus*, *Betula*, and *Tsuga* are greater than the tree percentages (i.e., they overrepresent the tree percentages); *Quercus* is equitably represented; *Ulmus*, *Fagus*, and *Picea* are moderately underrepresented; and *Fraxinus* and *Acer* are strongly underrepresented. Of these pollen types, *Pinus* and *Quercus* are dispersed the farthest, but *Acer* is dispersed hardly at all (Webb et al., 1981).

When compared to the maps of forest regions, the distribution of contemporary pollen shows that the 20% contour for *Picea* pollen (based on total upland pollen) lies near the southern border of the boreal forest (Webb and McAndrews, 1976), that the 5% contour for *Picea* pollen matches the 5% contour for *Picea* trees (Bernabo and Webb, 1977), and that the 10% contour for *Betula* pollen and the 20% contours for *Pinus* and *Quercus* pollen lie near the mixed forest/deciduous forest border in central Minnesota, Wisconsin, and Michigan. The 1% contour for *Acer* pollen corresponds roughly to the distribution of extensive stands of northern hardwoods and other forests dominated by *Acer saccharum*. The 2% contour for *Fagus* pollen lies near the range limit for the trees, as does the 3% contour for *Tsuga* pollen.

The match between herb types and the prairie/forest ecotone is confused by the high abundances of *Ambrosia*, Gramineae, and other herb pollen types in areas of agricultural and logging disturbance. Bernabo and Webb (1977), however, show that a strong gradient in total herb pollen (excluding aquatics and Cyperaceae) marks this ecotone today and that the 30%-herb contour lay near the ecotone 500 years ago, prior to agricultural disturbance.

Webb and Bernabo (1977) averaged together the modern climatic data at all the sites with contemporary percentages of pollen above 5%, 10%, 20%, and 30% for each of the mapped pollen types. A summary of their data (Table 10-2) shows that the order of arboreal types from low to high temperatures is *Picea*, *Betula*, *Pinus*, *Tsuga*, *Acer*, *Fagus*, *Ulmus*, and *Quercus* and that among the major types in the mixed forest region the order from low to high annual precipitation is *Pinus*, *Betula*, *Ulmus*, *Tsuga*, *Fagus*, and *Acer*. The precipitation values associated with the prairie-herb types are much lower than the values associated with the arboreal types.

Table 10-2. Climatic Values Within Areas of Abundance for the Major Pollen Types

Pollen	Type	Values Greater Than (%)	Mean July Temperature (°C)	Mean Annual Temperature (°C)	Mean Annual Precipitation (mm)	Number of Sites
Picea		5	16.4	−0.7	622	321
		10	16.1	−1.3	605	266
		20	15.7	−2.0	597	178
		30	15.3	−2.3	638	101
Betula		5	17.9	2.1	693	476
		10	17.6	1.7	704	339
		20	17.3	1.8	752	172
		30	17.5	2.3	792	66
Pinus (North of 42°N)		5	18.3	2.3	630	606
		10	18.1	2.1	615	524
		20	17.9	1.8	602	363
		30	17.9	1.8	602	253
Tsuga		5	20.0	6.4	841	58
		10	20.5	6.8	833	12
Fagus		5	21.2	8.1	874	24
		10	21.9	8.6	861	3
Acer		5	20.6	8.1	963	33
		10	20.4	9.1	1125	5
Fagus, Tsuga, and *Acer*		5	20.5	7.2	874	143
		10	20.5	7.3	881	69
		20	20.8	7.9	909	9
Ulmus		5	21.6	7.6	803	74
		10	21.8	7.7	782	7
		20	22.2	8.8	803	2
Quercus		5	21.6	8.0	828	294
		10	21.9	8.4	831	214
		20	22.3	9.1	859	113
		30	22.3	8.8	836	50
Gramineae		5	20.9	6.4	650	292
		10	20.7	6.1	582	146
		20	21.1	6.6	551	45
		30	22.8	8.1	605	14
Ambrosia		5	21.2	7.2	792	316
		10	21.8	8.0	818	218
		20	22.4	8.5	808	119
		30	22.2	8.1	790	39
Artemisia		5	19.7	4.0	399	121
		10	19.7	4.2	373	72
		20	20.2	4.8	351	38
		30	20.8	5.8	315	21
Other Compositae		5	20.8	6.3	546	35
		10	21.9	8.1	638	14
Chenopodiineae		5	21.5	6.3	475	100
		10	21.6	6.3	445	76
		20	22.7	7.5	414	33
		30	23.7	8.7	429	21

Source: Webb and Bernabo, 1977: Table 3.

Data and Methods

POLLEN DATA

Extending from Ohio to South Dakota and from Illinois to upper Michigan, our study area includes pollen data from 47 radiocarbon-dated records and 2 undated records (Figure 10-1). Of the total, 34 sites cover the entire period from 11,000 (or at minimum 9000) to 500 yr B.P., 3 sites have records ending before 5000 yr B.P., 3 other sites have records ending before 1000 yr B.P., 7 sites contain records from 4500 to 500 yr B.P., and 2 sites have records for 2500 to 3000 years in the middle Holocene (Table 10-3).

DATING THE POLLEN SPECTRA

At 45 sites, dates were estimated for each sample depth by linear interpolation between radiocarbon, corrected radiocarbon, or stratigraphically assigned dates at various depths in the cores (Table 10-3). At the other 4 sites, quadratic regression analysis was used to estimate the dates. The dates and their depths are listed in computer files that are stored with the pollen data from each site. Table 10-3 lists the types of corrections and adjustments that were made to the dates at 25 sites. (At Weber Lake, Jacobson Lake, Pickerel Lake, and Vestaburg Bog, cores from two different locations at each site were used in obtaining the pollen counts and radiocarbon dates, and dates had to be assigned pollen stratigraphically within the core[s] with the pollen counts [Table 10-3]. Corrections for ancient carbonate contamination of 520 to 1200 years were required for all radiocarbon dates at Billys Lake, Ondris Lake, Rutz Lake, Jones Lake, Lake 27, and Marion Lake; for all but the bottommost date on wood at Pretty Lake; and for the bottom two dates at Lake West Okoboji, Green Lake, and Bog D. The date for the decline in the abundance of *Picea* pollen from a nearby site [Shay, 1971] was used to set the date for this event at Bog D. Dates at Lost Lake were assigned by pollen-stratigraphic correlation from Yellow Dog Pond and Camp 11 Lake [Brubaker, 1975], and dates from Hudson, Clear, and Round Lakes were adjusted to match those at Pretty Lake. Dates from Bog D were used at Terhell, Martin, and Thompson Ponds, and dates prior to 4000 yr B.P. at Crystal Lake were assigned from Vestaburg Bog and Demont Lake. One date from Demont Lake was added to the set of dates at Vestaburg Bog. Two pairs of dates were averaged at Silver Lake, and two dates were deleted at Wolf Creek in order to eliminate dating reversals. At Lake of the Clouds, the radiocarbon dates as corrected by Stuiver [1971] were used.) The presence of carbonate bedrock and calcareous till in parts of Minnesota, Wisconsin, Michigan, Indiana, and Ohio necessitates careful selection of the dates used at each site in those areas.

The oldest date for which data were mapped at each site was no more than 500 years older than the oldest (corrected) radiocarbon or pollen-stratigraphic date at the site. The youngest date mapped at each site depended on the date estimated for the topmost pollen sample. For instance, the data from Wolf Creek were truncated at 9000 yr B.P., those from Seidel Lake at 8500 yr B.P., and those from Stotzel-Leis at 6000 yr B.P. (Table 10-3).

The pollen percentages were calculated from a sum of all tree, shrub, and herb pollen. At two sites where Cyperaceae pollen percentages were consistently high because of local bog development (Blue Mounds Creek and Tamarack Creek Bog), Cyperaceae was also excluded from the pollen sum.

Tables of the pollen percentages interpolated for each 500-year in-terval from 11,000 yr B.P. to the present were then produced. The values in this table were linearly interpolated from the values at the two samples whose estimated ages most immediately bracket a given 500-year date (e.g., the percentages for 7000 yr B.P. at Bog D are interpolated from the samples estimated to be 6872 and 7114 yr B.P.) Maps were then produced from these interpolated values.

THE MAPS

Maps were prepared for 1000-year intervals in the pollen percentages of *Picea*, *Betula*, *Pinus*, *Quercus*, *Fraxinus*, *Ulmus*, *Acer*, *Fagus*, and *Tsuga* and of the sum of *Ambrosia*, *Artemisia*, other Compositae, and Chenopodiaceae/Amaranthaceae (here called prairie-forb pollen). The percentage maps were contoured (with isopolls) and used to produce both difference maps and isochrone maps for selected time intervals (Bernabo and Webb, 1977). The difference maps illustrate the pattern of change in pollen percentages between two dates, and the isochrone maps display the changing position of selected isopolls during the past 11,000 years.

We have chosen a limited number of maps for discussion in order to depict the major changes in the vegetation during the past 11,000 years. The main set of maps (Figures 10-3 through 10-10) illustrate the patterns of change for selected pollen types and emphasize the individualistic behavior of each type. A final set of maps shows the joint distribution of several types of single maps for selected dates (Figures 10-11 and 10-12). These maps provide "snapshots" of the vegetational patterns at 11,000, 10,000, 9000, 6000, 3000, and 500 yr B.P. The map for 500 yr B.P. shows the patterns just prior to major disturbance by Euro-Americans and illustrates how well the patterns in the pollen data match the patterns on Figure 10-1.

We illustrate the history of the prairie/forest border by presenting a series of isopoll and difference maps for the sum of the percentages of prairie-forb pollen. The isochrone map for the 20% contour of prairie-forb pollen summarizes the main trends in this series of maps (Figure 10-5). For mapping prairie vegetation, we exclude Gramineae as well as Cyperaceae pollen because increases in these two types can reflect local changes at a site as it shallows and as peat begins to form, as at Kirchner Marsh (Wright et al., 1963), Disterhaft Farm Bog (Baker, 1970; Webb and Bryson, 1972; West, 1961) and Chatsworth Bog (Figure 10-2).

To illustrate the changing position and composition of the forests, we chose isochrone maps for *Picea* (20% and 5% contours), *Betula* (10%), *Pinus* (20%), *Quercus* (20%), *Acer* (1% and 2%), *Fagus* (2%), *Tsuga* (3%), *Fraxinus* (5%), and *Ulmus* (10%). A pair of difference maps for *Tsuga*, *Acer*, and *Fagus* combined illustrate the late-Holocene vegetational changes of these pollen types. The mapping intervals on the isochrone maps are 11,000, 10,000, 9000, 8000, 7000, 6000, 3000, and 500 yr B.P.

Mapped Summary of Vegetational Change

PRAIRIE/FOREST BORDER

At 10,000 yr B.P., the percentages of prairie-forb pollen were low, and a weak west-to-east gradient of 10 percentage units existed across Minnesota (Figure 10-3a). Woodland vegetation of varying composition grew in this area. By 9000 yr B.P., prairie-forb values had increased dramatically throughout all but northeastern Minnesota, and the 10% contour had moved from western Minnesota and Iowa into southwestern Wisconsin and northern Illinois (Figure 10-3b). Prairie vegetation dominated in western Minnesota. This trend of major increases in herb pollen continued until 8000 yr B.P. and resulted in a steep west-to-east gradient of prairie-forb pollen from South Dakota to central Wisconsin. The 20% contour at 8000 yr B.P. extended from northwestern Minnesota to northeastern Illinois (Figure 10-3c), and prairie vegetation grew in all of Iowa, most of Minnesota, and central Illinois. Between 8000 and 7000 yr B.P., the eastward increase in prairie-forb pollen continued in north-central Minnesota and southwestern Wisconsin (Figure 10-3d), but data at Chatsworth Bog (Figure 10-2) (King, 1981) show that this group of pollen types decreased and that the 20% contour moved westward in Illinois. The isopolls suggest that the gradient was steeper at 7000 yr B.P. than earlier. By 6000 yr B.P., this contour moved to eastern Iowa, and in northern Minnesota it also moved westward about 50 km (Figure 10-4a). It remained at its extreme eastward extent in southeastern Minnesota and southwestern Wisconsin, however. The difference map between 10,000 and 6000 yr B.P. show that the major increase in prairie-forb pollen over this period was west of a line from northcentral Minnesota to southwestern Wisconsin and central Illinois (Figure 10-4b).

From 6000 to 3000 yr B.P. (Figure 10-4c), prairie-forb pollen decreased westward across northwestern Minnesota but advanced in central Illinois to its former values at 8000 yr B.P. The difference map for this period (Figure 10-4d) shows a broad decrease in the values of prairie-forb pollen north and west from central Minnesota but an increase in Iowa, Illinois, and northwestern Indiana. The wedge of the Prairie Peninsula thus extended eastward in the central Midwest, while the prairie retreated westward in northern Minnesota.

By 500 yr B.P., just prior to Euro-American settlement, the 20% contour for prairie-forb pollen had extended slightly eastward in northwestern Minnesota but had retreated westward into eastern Iowa (Figure 10-4e). The main decrease in prairie-forb pollen occurred in central Minnesota (Figure 10-4f); a gradient of 35 percentage units developed from southwest to northeast from western Iowa to east-central Minnesota and Wisconsin (Figure 10-4e).

The isochrones for the 20% contour of prairie-forb pollen neatly summarize the above trends (Figure 10-5). In broad time intervals, the prairie moved eastward from 9000 to 7000 yr B.P. and then retreated westward to its position at 500 yr B.P. Intermediate times show the movement to have been more complex. Before 9000 yr B.P., no broad area of prairie vegetation existed in the study area. From 9000 to 8000 yr B.P., the prairie moved eastward along its entire margin and extended farthest eastward in central Illinois. The eastward movement continued in Minnesota and Wisconsin from 8000 to 7000 yr B.P., but in Illinois the prairie moved westward. By 7000 yr B.P., the prairie had reached its most eastward position in north-central Minnesota. From 7000 to 6000 yr B.P., the prairie border was stable except for a westward retreat in north-central Minnesota. By 3000 yr B.P., the prairie was far west in northern Minnesota but had again extended eastward in central Illinois. After 3000 yr B.P., the prairie border moved westward in the south but was stable in the north.

This isochrone map adds detail to the isochrone map of herb pollen in Bernabo and Webb (1977). Their map shows only that the prairie/forest border moved eastward until 7000 yr B.P. and then retreated westward. The addition of data for Lake West Okoboji, Chatsworth Bog, Volo Bog, and Stotzel-Leis (Table 10-3) reveals, however, differences in the motion north and south along the border.

BOREAL VEGETATION

The movement of the 20% contour for *Picea* pollen shows that boreal spruce-dominated vegetation had retreated northward to a position outside the Midwest by 8000 yr B.P. (Figure 10-6a) and that by 500 yr B.P. it had returned to only one site. The most rapid movement northward, from 11,000 to 10,000 yr B.P., may indicate the rapid warming in the Midwest at that time. The 5% contour for *Picea* pollen shows the same trend and suggests that few *Picea* trees grew in northern Minnesota from 8000 to 3000 yr B.P. (Figure 10-6b). The rise in abundance of *Picea* trees is a relatively recent event in the extreme northern Midwest.

MIXED FOREST/DECIDUOUS FOREST ECOTONE

As the main populations of *Picea* trees moved northward after 11,000 yr B.P., *Pinus* trees extended their range into Minnesota and increased in abundance in Michigan and Wisconsin (Figure 10-6c). After 10,000 yr B.P. in Michigan, the abundance of *Pinus* pollen decreased below 20% first in the south and later in the north. The trend ended at 6000 yr B.P., and the relative abundance of *Pinus* pollen increased southward in central Michigan from 6000 to 500 yr B.P. (Figure 10-6d). In Minnesota the trend in decreasing *Pinus* pollen after 10,000 yr B.P. was northeastward, and the trend reversed after 7000 yr B.P., when the prairie was at its eastern extreme there. *Pinus* values then increased westward and southwestward in Minnesota as the prairie/forest border retreated westward (Jacobson, 1979).

The pattern for *Quercus* pollen is similar to that for *Pinus* pollen, but *Quercus* percentages increased northward in Ohio and Michigan from 11,000 to 6000 yr B.P. (Figure 10-7a). After that time, its values decreased in the north (Figure 10-7b). This pattern suggests that the conifer-hardwood/deciduous forest ecotone moved northward in Michigan until 6000 yr B.P. and then retreated southward. (Figure 10-7b).

In Minnesota, the pattern for *Quercus* is more complex because *Quercus* trees competed both with prairie to the south and west and with *Pinus* and *Betula* trees to the north. The pattern of its isochrones in Minnesota shows the development of this dual competition between 10,000 and 8000 yr B.P.; the extension to the northwest of a narrow buffer zone of *Quercus*, probably *Q. macrocarpa*, at the prairie/forest border from 7000 to 3000 yr B.P.; and the reduction of *Quercus* in the northwest after 6000 yr B.P.

Betula pollen in Minnesota shows similar trends; it decreased in abundance northeastward from 10,000 to 7000 yr B.P. and then increased in abundance to the south and west after 7000 yr B.P. (Figure 10-7c). *Betula* trees in Minnesota before 6000 yr B.P. were probably mostly *B. papyrifera*. Since the mixed-forest/deciduous-forest ecotone moved southward after 6000 yr B.P., the isochrones for *Betula* pollen open up to the east, perhaps because *B. lutea* became an important component in the conifer-hardwood forests (Figure 10-7d). This compositional change in the mixed forest was accentuated by the simultaneous rises in abundance of *Acer*, *Fagus*, and *Tsuga* pollen after 6000 yr B.P.

Acer trees had moved into the Midwest by 9000 yr B.P., and they then moved north in Wisconsin and Michigan while being pushed out of Iowa by the advancing prairie (Figure 10-8a and b). Up to 6000

Table 10-3. Location, Dating Information, and References for Sites with Pollen Data (Key for Figure 10-1)

Site	Latitude (N)	Longitude (W)	Elevation (m)	¹⁴C Dates (N)	Adjustments to Dates	Youngest Date (yr B.P.)	Oldest Date (yr B.P.)	Youngest Date Mapped (yr B.P.)	Oldest Date Mapped (yr B.P.)	Reference(s)
1. Silver Lake, Ohio	40°21'	83°48'	361	10	980-year carbonate correction	170 C	9800 C	0	10,000	Ogden, 1966
2. Stotzel-Leis site, Ohio	40°13'	84°41'	320	11	—	5830	24,110	6000	11,000	Shane, 1976
3. Pretty Lake, Ind.	41°35'	85°15'	293	15	920-year carbonate correction	0 C	13,295	0	11,000	Williams, 1974
4. Hudson Lake, Ind.	41°40'	86°32'	239	6	Match Pretty Lake	2060	12,100 A	0	11,000	Bailey, 1972
5. Clear Lake, Ind.	41°39'	86°32'	244	5	Match Pretty Lake	2000 A	12,100 A	0	11,000	Bailey, 1972
6. Round Lake, Ind.	41°14'	86°38'	216	5	Added two bottom dates from Pretty Lake	0	12,100 A	0	11,000	R. Bailey, unpublished data
7. Chatsworth Bog, Ill.	40°40'	88°20'	219	8	—	3370	14,380	3000	11,000	King, 1981
8. Volo Bog, Ill.	42°21'	88°11'	229	6	—	1050	11,070	0	11,000	King, 1981
9. Frains Lake, Mich.	42°20'	83°38'	271	7	—	1240	12,570	0	11,000	Kerfoot, 1974
10. Wintergreen Lake, Mich.	42°24'	85°23'	283	8	—	2450	13,195	0	11,000	Manny et al., 1978
11. Chippewa Lake, Mich.	42°07'	83°15'		4	—	1200	9540	100	9500	Bailey and Ahearn, 1981
12. Crystal Lake, Mich.	43°15'	84°55'	260	2	Three dates added from Demont	1455	10,800 A	0	10,500	R. O. Kapp, unpublished data
13. Vestaburg Bog, Mich.	43°25'	84°53'	255	3	One date added from Demont	3150	10,300	0	11,000	Gilliam et al., 1967
14. Demont Lake, Mich.	43°29'	85°00'	248	5	—	640	11,140 A	0	10,500	R. O. Kapp, unpublished data
15. Jones Lake, Mich.	44°46'	84°36'	365	3	1040-year carbonate correction	90 C	1115 C	500	1000	Bernabo, 1981
16. Green Lake, Mich.	44°53'	85°07'	350	3	1200-year carbonate correction for two based dates only	4215	14,015 C	500	11,000	Lawrenz, 1975
17. Lake 27, Mich.	45°04'	84°47'	378	3	790-year carbonate correction	115 C	3350 C	500	3000	Bernabo, 1981
18. Marion Lake, Mich.	45°14'	85°15'	223	3	520-year carbonate correction	126 C	2775 C	500	2500	Bernabo, 1981
19. Camp 11 Lake, Mich.	46°40'	88°01'	549	11	—	1055	10,200	0	10,000	Brubaker, 1975
20. Lost Lake, Mich.	46°43'	87°58'	500	8	Match Camp 11	3000 A	9000 A	0	9500	Brubaker, 1975
21. Yellow Dog Pond, Mich.	46°45'	87°57'	445	9	—	2170	9100 A	0	9000	Brubaker, 1975
22. Seidel Lake, Wis.	44°27'	87°31'	219	3	—	8680	10,440	8500	11,000	J. C. B. Waddington, unpublished data
23. Disterhaft Farm Bog, Wis.	43°55'	89°10'	329	6	—	2850	15,560	500	11,000	Baker, 1970; Webb and Bryson, 1972
24. Blue Mounds Creek, Wis.	43°05'	89°52'	335	4	—	4235	10,485	0	10,000	Davis, 1977
25. Tamarack Creek Bog, Wis.	44°09'	91°27'	244	4	—	1190	4410	0	4500	Davis, 1979

Table 10-3. (Continued)

Site	Latitude (N)	Longitude (W)	Elevation (m)	14C Dates (N)	Adjustments to Dates	Youngest Date (yr B.P.)	Oldest Date (yr B.P.)	Youngest Date Mapped (yr B.P.)	Oldest Date Mapped (yr B.P.)	Reference(s)
26. Wood Lake, Wis.	45°20'	90°05'	350	6	—	1040	13,000	0	11,000	Heide, 1981
27. Stewart's Dark Lake, Wis.	45°18'	91°27'	335	7	Pollen and dates from two cores	1370	10,570	0	10,500	Peters and Webb, 1979
28. Lake Mary, Wis.	46°15'	89°54'	488	3	—	3650	9460	500	9500	Webb, 1974b
29. Woden Bog, Iowa	43°14'	93°55'	381	6	—	2830	11,570	3000	11,000	Durkee, 1971
30. Lake West Okoboji, Iowa	43°22'	95°11'	415	10	1000-year carbonate correction for two basal dates only	390	12,990 C	0	11,000	Van Zant, 1979
31. Kirchner Marsh, Minn.	44°50'	93°07'	254	6	—	1660	13,270	500	11,000	Wright et al., 1963
32. Rutz Lake, Minn.	44°52'	93°52'	314	8	1000-year carbonate correction	100 C	11,000 C	0	11,000	Waddington, 1969
33. Willow River Pond, Minn.	46°18'	92°47'	314	2	—	5160	7890	5000	8000	Jacobson, 1979
34. Nelson Pond, Minn.	46°24'	92°41'	335	2	—	5540	7245	5500	7500	Jacobson, 1979
35. Jacobson Lake, Minn.	46°25'	92°43'	324	4	Pollen and dates from two cores	3920	10,800	0	10,500	Wright and Watts, 1969
36. Rossburg Bog, Minn.	46°35'	93°36'	372	4	—		10,100	2000	10,000	Wright and Watts, 1969
37. Wolf Creek, Minn.	46°07'	94°07'	375	15	Two dates deleted	9150	20,500	9000	11,000	Birks, 1976
38. Ondris Lake, Minn.	46°21'	94°25'	363	2	560-year carbonate correction	100 C	975 C	0	1500	Jacobson, 1979
39. Billys Lake, Minn.	40°16'	94°33'	383	2	890-year carbonate correction	100 C	1110 C	0	1500	Jacobson, 1979
40. Weber Lake, Minn.	47°28'	91°40'	567	5	Pollen and dates from two cores	7300	14,690	0	11,000	Fries, 1962
41. Lake of the Clouds, Minn.	48°09'	91°07'	453	30	See Stuiver (1971)	780	9490	500	10,000	Craig, 1972
42. Shagawa Lake, Minn.	47°55'	91°52'	492	1	—		1935	0	2000	Bradbury and Waddington, 1973
43. Myrtle Lake, Minn.	47°58'	93°23'	393	5	—	2680	11,120	0	11,000	Janssen, 1968
44. Portage Lake, Minn.	47°12'	94°09'	396	3	—	3685	9780	0	10,000	J. H. McAndrews, unpublished data
45. Martin Pond, Minn.	47°11'	94°56'	429	0	—	2700 A	10,000 A	0	10,500	McAndrews, 1966
46. Bog D, Minn.	47°11'	95°10'	457	4	1000-year carbonate correction for two basal dates only	2730	10,000 C	0	10,500	McAndrews, 1966
47. Terhell Pond, Minn.	47°12'	95°47'	442	1	Bottom date from Bog D	4270	10,000 A	0	10,000	McAndrews, 1966
48. Thompson Pond, Minn.	47°12'	96°05'	370	0	Four dates from Bog D	2500 A	10,000 A	0	10,000	McAndrews, 1966
49. Pickerel Lake, S.D.	45°30'	97°20'	395	4	Dates from separate cores	2700	10,670	0	11,000	Watts and Bright, 1968

Abbreviations: A = adjusted or added date; C = corrected value of the date.

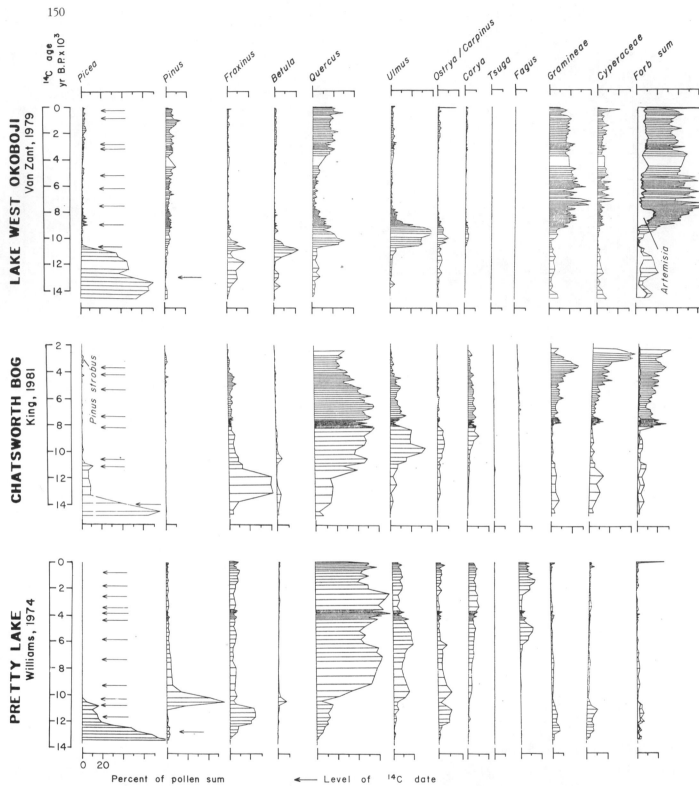

Figure 10-2. Pollen diagrams for seven sites in the Midwest. The pollen percentages are plotted against a chronology estimated from the available radiocarbon dates. (See Table 10-3 for corrections to dates Lake West Okoboji, Chatsworth Bog, and Green Lake.) The diagrams are ordered from west to east and then from south to north.

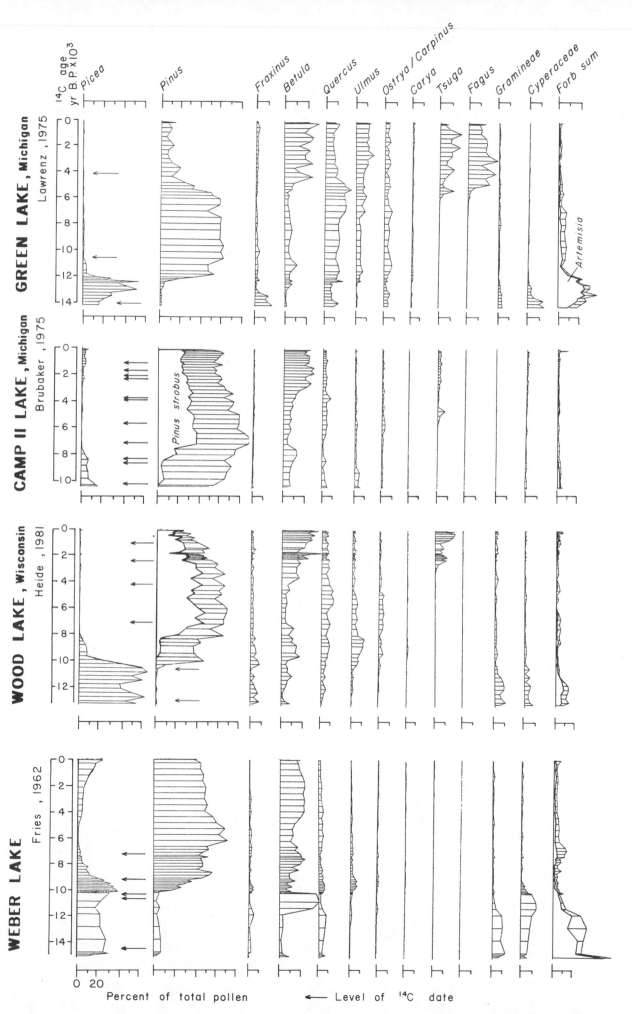

^{14}C age
yr B.P x10^3

Picea   *Pinus*   *Fraxinus*   *Betula*   *Quercus*   *Ulmus*   *Ostrya/Carpinus*   *Carya*   *Tsuga*   *Fagus*   *Gramineae*   *Cyperaceae*   *Forb sum*

GREEN LAKE, Michigan
Lawrenz, 1975

Artemisia

CAMP II LAKE, Michigan
Brubaker, 1975

Pinus strobus

WOOD LAKE, Wisconsin
Heide, 1981

WEBER LAKE
Fries, 1962

0 20

Percent of total pollen ⟵ Level of ^{14}C date

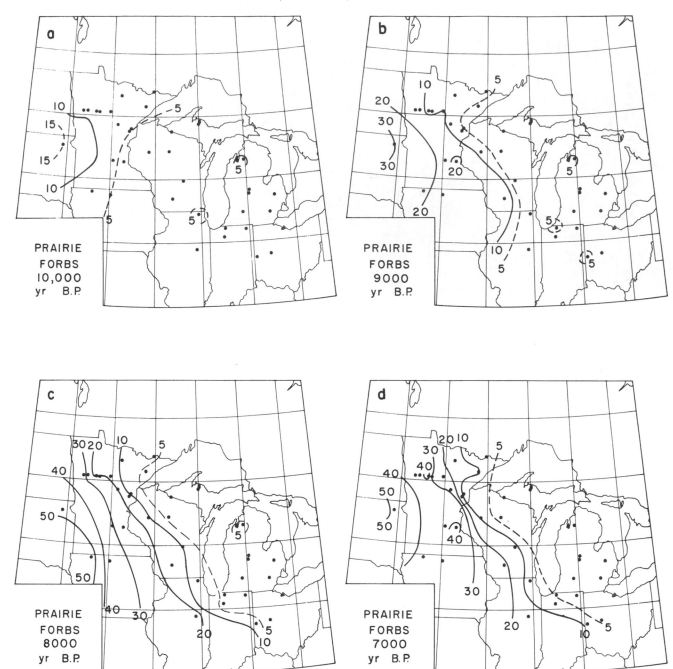

Figure 10-3. Maps with isopolls (isofrequency contours) for prairie-forb pollen (*Artemisia*, *Ambrosia*, other Compositae, and Chenopodiaceae/Amaranthaceae) for (a) 10,000 yr B.P., (b) 9000 yr B.P., (c) 8000 yr B.P., and (d) 7000 yr B.P.

yr B.P., *Acer* trees grew throughout both Michigan and Wisconsin but had high abundances mainly in southern Michigan, Indiana, and Ohio, where *Fagus* pollen and trees were also present and abundant (Figure 10-8c and d). *Fagus* trees first appeared in eastern Michigan about 8500 yr B.P. (Figure 10-9a).

After 6000 yr B.P., with the prairie moving westward in northern and central Minnesota, *Acer* pollen increased in abundance in northwestern lower Michigan, in upper Michigan, and in northern Wisconsin (Figure 10-8d). *Tsuga* appeared for the first time in northwestern lower Michigan, and its range boundary moved westward into northern Wisconsin (Figure 10-9), where the abundance of *Acer*

and *Betula* trees had increased within the forests. The range of *Fagus* also extended into northwestern lower Michigan and eastern upper Michigan (Figure 10-9).

The difference map from 6000 to 3000 year B.P. for *Tsuga*, *Fagus*, and *Acer* pollen (Figure 10-9c) shows a major rise in abundance of these types in northern lower Michigan but a decrease to the south in Ohio, Indiana, and southwestern Michigan. This decrease parallels the increase in herbs in central Illinois and Indiana, indicating the eastward extension of the prairie in Illinois during this time interval (Figure 10-5). After 3000 year B.P., *Tsuga* and *Acer* pollen continued to rise in abundance in northern Wisconsin, and *Fagus* and

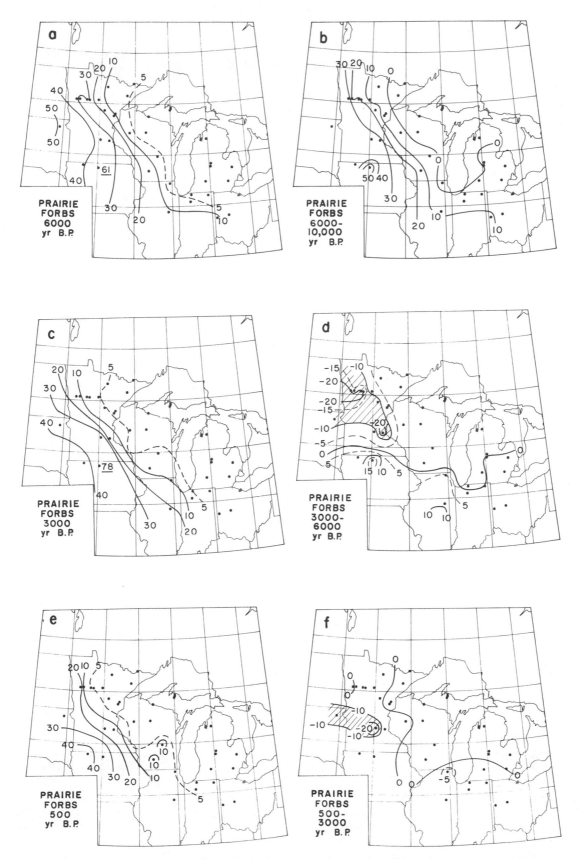

Figure 10-4. Isopoll maps showing the regional patterns for prairie-forb pollen for (a) 6000 yr B.P., (c) 3000 yr B.P., and (e) 500 yr B.P. and difference maps showing the patterns of change (b) from 10,000 to 6000 yr B.P., (d) from 6000 to 3000 yr B.P., and (f) from 3000 to 500 yr B.P. Positive values on the difference maps indicate a net increase in the percentages of prairie-forb pollen during the time interval.

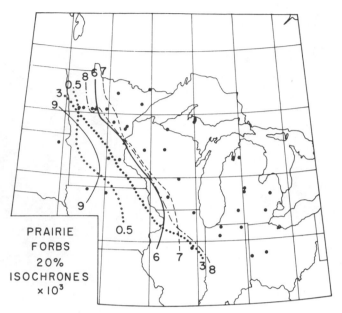

PRAIRIE
FORBS
20%
ISOCHRONES
×10³

Figure 10-5. Isochrone map showing the position of the 20% isopoll for prairie-forb pollen from 9000 to 500 yr B.P., with higher percentages to the west. We interpret the changes in position as indications of the changes in the prairie/forest border.

Acer pollen became more prominent in northeastern Indiana (Figure 10-9d). These abundance changes fit into the regional pattern of eastward-moving prairie in the southwestern Midwest and the southward increase in the abundances of *Picea* pollen and trees in northern Minnesota and upper Michigan (Figures 10-5 and 10-6b).

FRAXINUS AND *ULMUS*

In contrast to the other types that are mapped, high percentages of *Fraxinus* and *Ulmus* pollen were most widespread at 10,000 yr B.P. and occupied their smallest area at 500 yr B.P. and today (Figure 10-10). The highest values occurred in the Late Wisconsin for *Fraxinus* and in the early Holocene for *Ulmus* (Figure 10-2), and no samples of contemporary pollen contain assemblages analogous to those fossil assemblages with peak values of *Fraxinus* and *Ulmus* pollen. This situation creates problems in interpreting the Late Wisconsin and early-Holocene vegetation from central Minnesota to western Ohio, but the broad-scale patterns on the isochrone maps indicate that these tree genera were responding in some degree to environmental changes on a comparable scale.

By 11,000 yr B.P., the peak phase for *Fraxinus* pollen was already past in southern Michigan and central Illinois, but high values still existed from Illinois to western Ohio (Figure 10-10a). *Ulmus* values had just risen above 10% at Chatsworth Bog (Figures 10-2 and 10-10c). By 10,000 yr B.P., the values for both types had increased northward, and they covered their largest extent from southern Michigan into Minnesota. *Ulmus* values were above 10% even in eastern South Dakota and northwestern Minnesota. After this time, their regions of moderate values contract southeastward, with *Fraxinus* values above 5% becoming immediately confined to Indiana and east-central Michigan. The changes for *Ulmus* pollen were gradual, but at 7000 yr B.P. the mapped isopolls of both types have a similar distribution and match the patterns of *Fagus* (above 2%) (Figure 10-9a) and *Acer* (above 2%) (Figure 10-8c). After 6000 yr B.P., the

contraction continued and at 500 yr B.P. *Ulmus* was above 10% only in northwestern lower Michigan and *Fraxinus* was above 5% in central Michigan and north Indiana (Figure 10-10b and d). Since their distributions after 10,000 yr B.P. lay mainly in the area dominated by *Quercus* pollen, the changes in these types resulted from compositional changes within the deciduous forest during the Holocene. *Carya, Ostrya/Carpinus, Acer,* and *Fagus* are other types that changed in abundance within this forest region (Figures 10-2, 10-8, 10-9, 10-11, and 10-12).

VEGETATIONAL PATTERNS AT SELECTED DATES; 11,000 TO 500 YR B.P.

A picture of the changing composition and distribution of the vegetation can be gained by plotting the joint distribution of several pollen types for 11,000, 10,000, 9000, 6000, 3000, and 500 yr B.P. (Figures 10-11 and 10-12). These maps show the distribution of the pollen types and the position of ecotones within the Midwest at each of the time periods. At 11,000 yr B.P., *Picea*-dominated forest and woodland grew within almost all of the Midwest except central Illinois and western Ohio, where an ecotone existed between boreal forest and deciduous forest (Figure 10-11a). South of this ecotone *Quercus* and *Ulmus* trees were prominent in the forests, together with *Fraxinus, Ostrya/Carpinus,* and some *Picea* trees (Figure 10-2).

In eastern Michigan, *Pinus* trees (mostly *P. banksiana*) had entered the *Picea* forest, and in northeastern Minnesota high percentages of forb pollen (mostly *Artemisia*) and Cyperaceae (Figure 10-2) suggest an open *Picea* woodland with *Betula* shrubs and trees. *Fraxinus* trees grew within the southern part of the *Picea* forest from northern Indiana into central Wisconsin and Minnesota and across to northeastern Iowa.

By 10,000 yr B.P., forest covered all of the Midwest except for west-central Minnesota and eastern South Dakota, where woodlands grew (Figure 10-11b). The boreal-forest/deciduous-forest ecotone disappeared and was replaced by two new ecotones as a result of the immigration of *Pinus*. The first lay across central Minnesota, northern Wisconsin, and upper Michigan; to its north were *Picea*-rich forests with *Betula* (probably *B. papyrifera*) and *Pinus* (mostly *P. banksiana*) trees. The second ecotone was located in extreme northern Iowa, Illinois, Indiana, and Ohio; to its south were deciduous forests composed of *Quercus, Acer, Ulmus, Fraxinus,* and *Ostrya/Carpinus.* In the extreme southeast, *Fraxinus* values decreased as *Carya* values increased. The second ecotone was sharply defined from southern Michigan into Illinois, where the 20% isopolls for *Quercus* and *Pinus* lie close together, but it was more diffuse to the west. There appreciable numbers of *Ulmus* and *Fraxinus* trees grew in central Minnesota and Wisconsin and separated the region between *Quercus* forests and conifer forests. In Michigan and into Wisconsin, *Pinus*-dominated forests grew between the two ecotones. The composition of these forests became more boreal northwestward into Minnesota. *Acer* trees were almost entirely confined to the region of the deciduous forest.

At 9000 yr B.P., the *Picea* and *Quercus* forests had both moved farther north, and prairie had begun to develop in the southwest (Figure 10-12a). A boreal forest of *Picea, Pinus* (mostly *P. banksiana*), and *Betula* grew in northeastern Minnesota, and elements of this forest grew in northwestern Wisconsin and western upper Michigan. The deciduous forest extended northward almost into central Wisconsin and lower Michigan. *Acer* trees were still confined

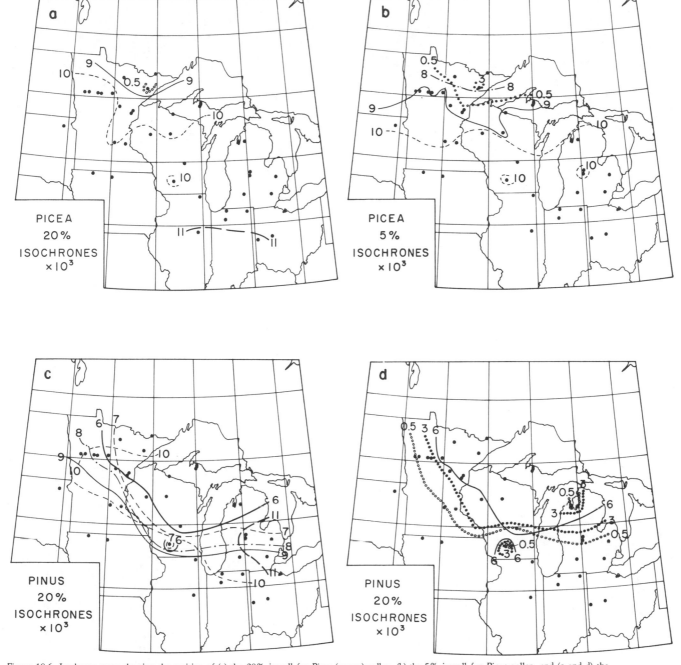

Figure 10-6. Isochrone maps showing the position of (a) the 20% isopoll for *Picea* (spruce) pollen, (b) the 5% isopoll for *Picea* pollen, and (c and d) the 20% isopoll for *Pinus* (pine) pollen from 11,000 to 500 yr B.P. In a, c, and d, the values above 20% lie to the north of the isochrones except at 11,000 and 10,000 yr B.P. for *Pinus*, when the values above 20% lie between the isochrones. In b the values above 5% are north of the isochrones.

mainly to the deciduous forest in the east and south, and the mapped populations of *Fraxinus* trees grew only in northern Indiana and southeastern lower Michigan. In northern lower Michigan and Wisconsin, a *Pinus*-dominated forest grew between the deciduous and boreal forests. The prairie/forest ecotone had developed but was diffuse in comparison to its later development. In Iowa and central Minnesota, the open woodland included trees of *Acer* and *Ulmus* as well as *Quercus*.

The deciduous forest had moved to its farthest northern extent in lower Michigan and Wisconsin by 6000 yr B.P., and few if any *Picea*

trees grew in northern Minnesota at that time (Figure 10-12b). The close proximity of the two isopolls for prairie-forb pollen indicates that the prairie/forest ecotone was then sharply defined and prairie grew from northwestern Minnesota to Illinois. This vegetational pattern contrasts markedly with the almost uniform boreal vegetation at 11,000 yr B.P. and the almost entirely forested landscape at 10,000 yr B.P. In lower Michigan, deciduous forests included *Quercus* trees with moderate numbers of *Acer*, *Fagus*, and *Ulmus* trees. *Fraxinus* and then *Carya* trees became more abundant from southern Michigan into western Ohio. An east-west compositional gradient ex-

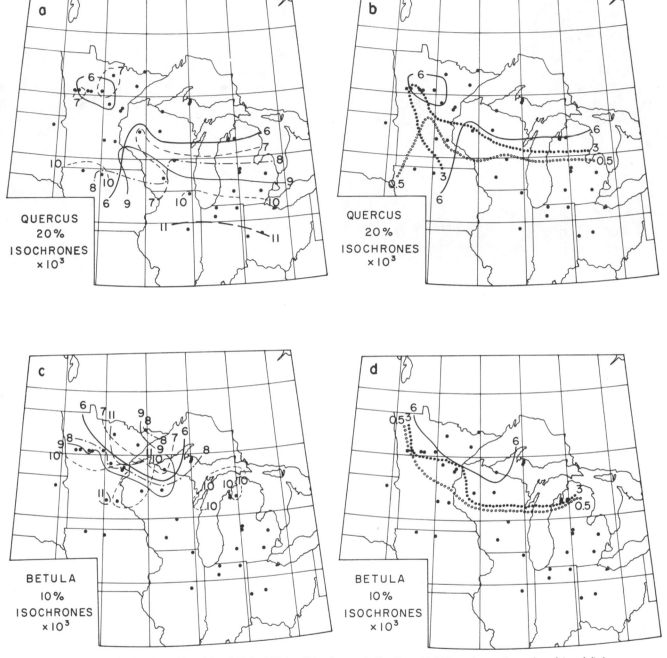

Figure 10-7. Isochrone map showing the position of (a and b) the 20% isopoll for *Quercus* (oak) pollen, with higher values to the south, and (c and d) the 10% isopoll of *Betula* (birch) pollen with higher values to the north from 11,000 to 500 yr B.P.

isted in the deciduous forests between lower Michigan and southeastern Wisconsin. The map shows no *Fagus*, and fewer *Carya* and *Fraxinus* trees growing in Wisconsin than to the east. The east-west gradient in the north is from a mixed forest of *Pinus* with some *Acer* trees in the east to forests of *Pinus* with increased numbers of *Betula* trees in the west (northeastern Minnesota). This is the first time that the pollen data show evidence of *Acer* trees extending beyond the region of the deciduous forest. *Quercus*-shrub vegetation grew along the prairie/forest border and extended northward to northwestern Minnesota.

A reversal in many of the early-Holocene trends is evident by 3000

yr B.P., when the ecotone between deciduous forest and mixed forest had moved southward into central lower Michigan and Wisconsin, the prairie/forest border had retreated westward in central and northern Minnesota, and a few *Picea* trees were again growing in northern Minnesota (Figure 10-12c). The gradient across the prairie/forest border continued to be well defined, and this ecotone extended eastward across Illinois almost into northern Indiana, where *Fagus* trees became less abundant in the deciduous forests. Within the deciduous forests, *Ulmus* trees had also decreased in abundance, and the number of *Carya* trees had increased. A major new development at this time was the clear pattern of an east-west compositional gra-

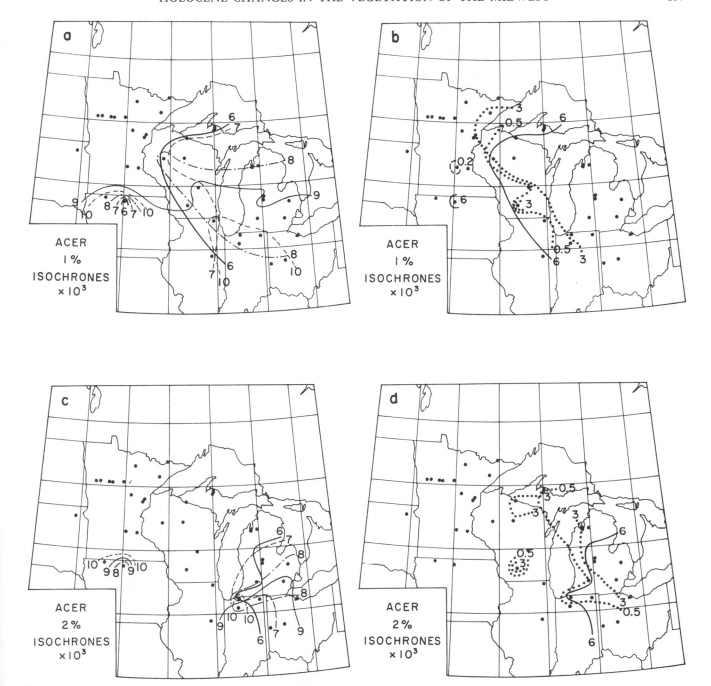

Figure 10-8. Isochrone maps showing *Acer* (maple) pollen for (a and b) the position of the 1% isopoll and (c and d) the 2% isopoll. Higher percentages are to the east or to the south.

dient among northern hardwood trees across northern Michigan and Wisconsin into Minnesota. *Tsuga* populations were still rebuilding from the *Tsuga* decline at 4800 yr B.P. (Davis, 1981), but a gradient in the pollen percentages existed along which first *Fagus grandifolia* became scarce, then *Tsuga canadensis*, then *Acer*, then *Pinus*, and finally *Betula* before *Quercus* scrub and prairie were reached in northwestern Minnesota.

By 500 yr B.P., some of the patterns at 3000 yr B.P. had become more pronounced (Figure 10-12d). Detectable numbers of *Picea* trees grew over a wider area of northern Minnesota. The deciduous forest of southern Michigan had more *Carya* and fewer *Fraxinus* trees, and

the east-west compositional gradient was well developed in both the deciduous forest of southern lower Michigan and Wisconsin and the mixed forests of northern Michigan, Wisconsin, and Minnesota. The prairie, however, was less extensive in the southwest, and the prairie/forest ecotone across Iowa and southern Minnesota was more diffuse than at 3000 yr B.P. These patterns in the pollen fit well with the mapped presettlement vegetation (Figure 10-1).

This series of maps from 11,000 to 500 yr B.P. illustrates the sequence of vegetational changes leading to the development of the major patterns in the vegetation just prior to Euro-American settlement. Several of these patterns developed early in the Holocene, while

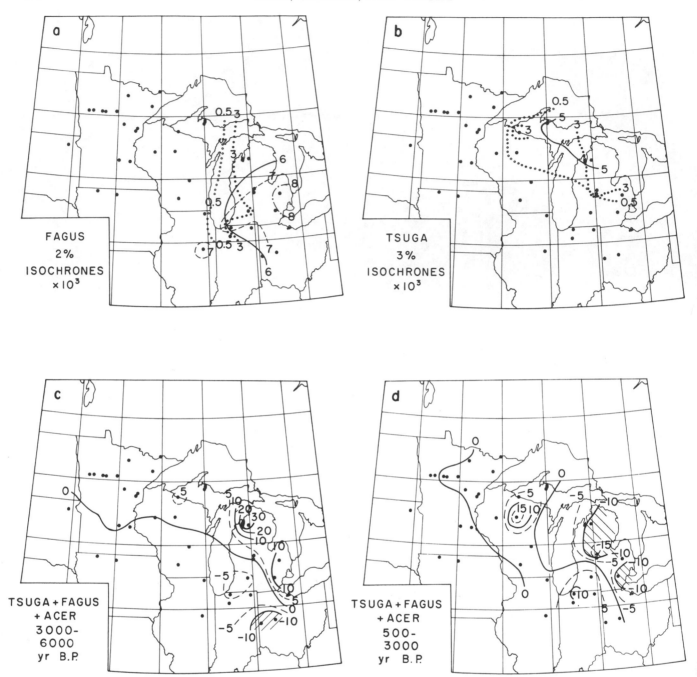

Figure 10-9. Isochrone maps showing the position of (a) the 2% isopoll for *Fagus* (beech) pollen and (b) the 3% isopoll for *Tsuga* (hemlock) pollen and difference maps showing the sum of *Tsuga*, *Fagus*, and *Acer* pollen and the patterns of change (c and d) from 6000 to 500 yr B.P. Positive values in c and d indicate areas in which the pollen percentages were higher at 3000 than at 6000 yr B.P. In a and b, percentages above 2% or 3% are to the east.

others only appeared relatively recently within the Midwest. The *Pinus-Quercus* ecotone had developed by 10,000 yr B.P., and the north-south gradient from *Quercus* to *Pinus* to *Picea* also dates from this time. Prairie first appeared in the study area at 9000 yr B.P., but the steepness of the prairie/forest ecotone varied from diffuse at 9000 and 500 yr B.P. to abrupt at 6000 and 3000 yr B.P. The composition within the deciduous forest has been diverse since at least 10,000 yr B.P., but the current compositional diversity within the mixed forest, and perhaps the floristic tension zone at its southern edge, dates from after 6000 yr B.P. The current east-west compositional gradients in

the Midwest are, therefore, younger than the north-south gradients.

Discussion

To the extent that causal links can be established between regional climate and regional vegetation and between regional vegetation and pollen assemblages, certain climatic patterns of the past can be inferred. Plausible reconstructions are further constrained by a requirement for spatial and temporal consistency. That is, the patterns inferred over space and time must be explicable by the dynamic pro-

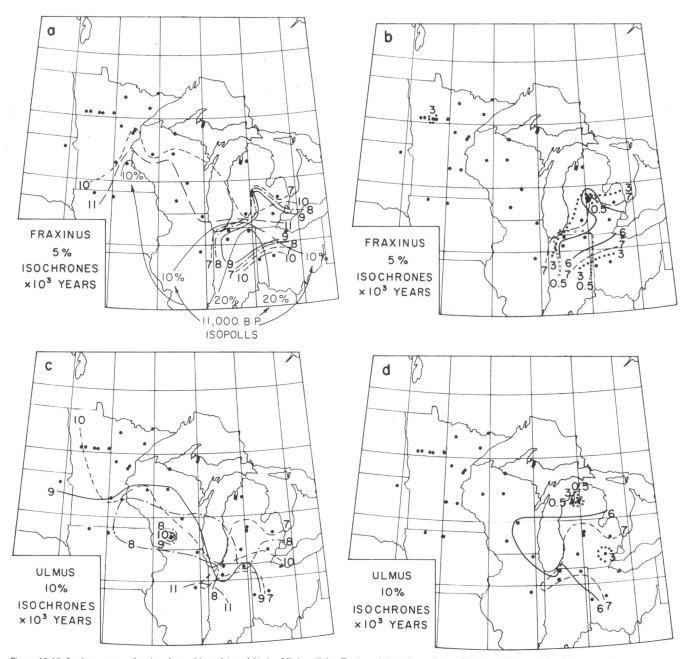

Figure 10-10. Isochrone maps showing the position of (a and b) the 5% isopoll for *Fraxinus* (ash) pollen and (c and d) the 10% isopoll for *Ulmus* (elm)
pollen. Values above 5% or 10% are to the south or between the isochrones. The 10% and 20% isopolls for 11,000 yr B.P. also appear on a.

cesses of climate and vegetation as they are currently understood.
Because the broad-scale and long-term dynamics of climate are more
closely linked to physical processes and hence better understood at
present than those of vegetation, our reconstructions of vegetation
have been aided by considerations of climatic implications, and these
we present below for several critical periods.

PICEA DECLINE AND ASSOCIATED MIXED FOREST; 14,000 TO 10,000 YR B.P.

About 14,000 yr B.P., after a relatively stable *Picea* forest had existed for several thousand years during the Late Wisconsin glacial
maximum, the vegetation began to change. The ice sheet was melting

rapidly. At Chatsworth Bog in north-central Illinois (Figure 10-2),
the *Picea* forest disappeared about this time. The *Picea* forest then
shifted northward with the margin of the ice sheet until it reached
its present distribution in northern Minnesota about 10,000 years
ago. *Pinus* trees did not reach Wisconsin and Minnesota from the
Appalachians until about 10,000 yr B.P., and a mixed forest of *Fraxinus*, *Betula*, *Ulmus*, *Picea*, and other taxa developed to the south
of the *Picea* forest as the latter shifted to the north.

The forest composition at certain sites during this time interval
seems anomalous in comparison with present-day forests, and the
succession after the *Picea* forest varied from place to place. At
Chatsworth Bog (Figure 10-2), the pollen assemblage following the
Picea zone is dominated by *Fraxinus* (exceeding 40%), along with

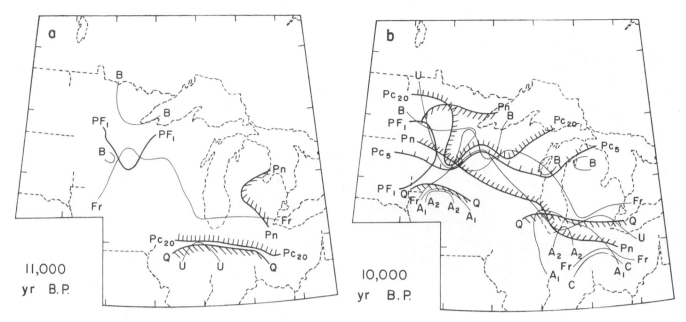

Figure 10-11. Maps of selected isopolls for (a) 11,000 and (b) 10,000; the isopolls are 20% (Pc$_{20}$) and 5% (Pc$_{5}$) *Picea*, 20% *Pinus* (P$_{n}$), 20% *Quercus* (Q), 20% (PF$_{2}$) and 10% (PF$_{1}$) prairie forbs, 10% *Betula* (B), 5% *Fraxinus* (Fr), 10% *Ulmus* (U), 3% *Tsuga* (T), 2% *Fagus* (Fg), 2% (A$_{2}$) and 1% (A$_{1}$) *Acer*, and 5% *Carya* (C).

minor amounts of *Quercus*, *Picea*, and graminoids. This is indeed a strange assemblage. Because of the low pollen influx, King (1981) interprets it as derived from *Fraxinus nigra* in the lowlands and tundra on the uplands, with *Quercus* pollen blown in from the south. This assemblage might also represent *Fraxinus nigra* over most of the landscape, for this tree can readily colonize uplands as well as lowlands when competition from other taxa is weak. No modern analogue exists for the pollen assemblage involved in either of these hypothetical vegetational types. A somewhat similar pollen assemblage accumulated at Pretty Lake between 12,000 and 10,800 yr B.P. (Figures 10-2 and 10-10a).

About 11,600 yr B.P., the *Fraxinus* pollen zone at Chatsworth was followed by a second although minor maximum of *Picea* pollen, along with an increase in *Alnus*, *Quercus*, and then *Ulmus* pollen (Figure 10-10c). Such a second maximum of *Picea* pollen is much more strongly expressed farther east in eastern Indiana and western Ohio, for which Shane (1980) interprets it as the product of a climatic oscillation that once more brought cold to the area—an oscillation roughly contemporaneous with the post-Two Creeks advance of the Lake Michigan ice lobe. Although a modern analogue for this assemblage is not available, *Picea* is almost uniquely a boreal genus and so provides convincing evidence in Ohio and Illinois for a climate colder than today's.

Farther northwest, the decline in the percentages of *Picea* pollen was also followed by a brief interval of rapid succession of forest types before the early-Holocene deciduous forest became established (in Iowa [Figure 10-2], eastern South Dakota, and southern Minnesota) or before the immigration of pine (in central Minnesota). *Alnus*, *Betula*, *Fraxinus nigra*, and *Abies balsamea* were the principal components of these successional forests. Peak values of these types are also associated with the *Picea* decline in New England pollen diagrams (Davis, 1981). In the Midwest, *Betula* and *Fraxinus nigra* were already present in the *Picea* forest that preceded, and *Alnus* and *Abies balsamea* readily colonized. Again, no satisfactory modern

analogues for communities of these types have been identified, but the very low values for other northern conifers (*Picea*, *Larix*) imply distinctly temperate conditions.

At Chatsworth Bog (Figure 10-2), the mixed forest lasted from 13,800 to 11,000 yr B.P. (if the second *Picea* pollen peak is included), terminating with the increase in percentages of *Ulmus* and *Quercus* pollen to a maximum (Figure 10-11a). Farther north and west, the *Picea* decline that initiated the vegetational changes occurred later (from about 11,300 yr B.P. at Lake West Okoboji in northwestern Iowa to about 10,500 yr B.P. in central Minnesota [Figure 10-6a]). The major increase in *Ulmus* and *Quercus* trees terminating this interval at about 10,000 yr B.P. also came somewhat later here than farther south (Figures 10-7a and 10-10c), and in central Minnesota and Wisconsin *Pinus banksiana* invaded from the east at about this time (Figures 10-2, 10-6c, and 10-11b). The *Betula-Fraxinus-Abies* forest in central Minnesota is thus barely identifiable, lasting only a few hundred years during the decline of *Picea* but before the *Pinus* invasion (Amundson and Wright, 1979). In Wisconsin and Michigan, *Pinus* was present when *Picea* declined, but farther south a mixed forest without *Pinus* lasted 1000 to 3000 years (Figure 10-2).

The lack of modern analogues makes it difficult to make a direct quantitative paleoclimatic interpretation of this mixed assemblage in which *Pinus* pollen is absent. We suppose that the vegetation changed rapidly as the climate became more and more unfavorable for the regeneration of *Picea* trees. The trees that replaced the *Picea* forest were those that were already at hand or were good colonizers on space vacated by *Picea*. These included *Betula papyrifera*, *Fraxinus nigra*, *Populus tremuloides*, *Abies balsamea*, and *Alnus rugosa*. All of these trees are common today in at least the southern part of the boreal forest. Because *Picea* trees were decreasing in abundance, however, the climate may well have been temperate like that of the present-day Great Lakes forest, which is dominated by three species of *Pinus* as well as by *Tsuga* and various northern hardwoods. The mixed forest probably experienced a different combination of

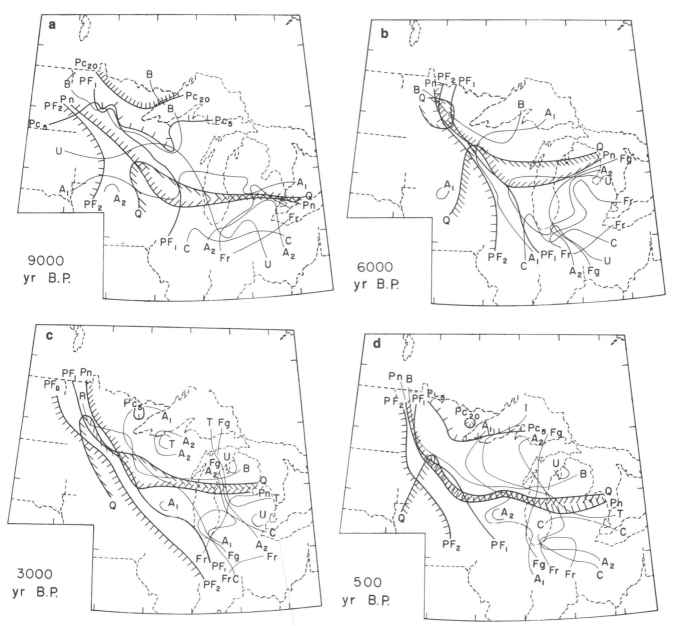

Figure 10-12. Maps of selected isopolls for (a) 9000, (b) 6000, (c) 3000, and (d) 500 yr B.P. (See legend for Figure 10-11 for an explanation of the abbreviations.)

temperature and moisture conditions than those in the region of the Great Lakes today, however.

The climate in the Midwest during this time at the end of the Late Wisconsin (14,000 to 10,000 yr B.P.) can be reconstructed in broad pattern by focusing on the more familiar pollen assemblages to the north and south of this anomalous mixed assemblage. At 11,000 yr B.P., *Picea* dominated pollen assemblages existed in northern Minnesota, Wisconsin, and Michigan, and *Quercus*-dominated pollen assemblages existed in central Illinois (Figure 10-11a). By interpolating the estimates to the north and south, one may be able to gain climatic estimates for the mixed assemblages between.

EARLY HOLOCENE; 10,000 TO 8000 YR B.P.

The increase in *Ulmus* and *Quercus* pollen (in the south) and *Pinus* (in the north) marked the termination of the mixed assemblage by about 10,000 yr B.P. (Figures 10-6c, 10-7a, 10-10c). The early-Holocene forest in Iowa and southern Minnesota apparently contained much more *Ulmus* and commonly more *Ostrya/Carpinus*, *Fraxinus pennsylvanica*, *Carya*, *Acer saccharum*, *Corylus*, and other mesic deciduous trees than occur in deciduous forests today (Figures 10-8a, 10-10c, 10-11b). The contrast is clearly seen at Kirchner Marsh, which stands now in *Quercus* forest in southern Minnesota, and also at Lake West Okoboji (Figure 10-2) and Pickerel Lake (in northeastern South Dakota), which are now in the prairie but are locally fringed with deciduous woodland. The presence of *Acer saccharum* during the early Holocene at the latter two sites is especially striking. Apparently, the *Ulmus*-dominated deciduous forest spread rapidly west and north to replace the *Picea* forest in the area where prairie was later to develop. This relation implies that the early Holocene was slightly cooler and more moist than the late Holocene in Minnesota, Iowa, and South Dakota. This early-Holocene climate was certainly influenced by the wasting ice sheet, which still existed in the Hudson Bay region until 7500 years ago. The influence may have accounted for the cooler summers and more mesic forest in the western Midwest.

HOLOCENE HISTORY OF *PINUS* SPECIES

One significant detail not represented on the pollen maps (Figures 10-6, 10-11, and 10-12) concerns the role of the species of *Pinus* in the Holocene history. Although only a few pollen diagrams show the separation of *P. strobus* from *P. banksiana/P. resinosa* and only one (Birks, 1981a) so far distinguishes the latter two species, an interesting pattern is emerging. The first species of *Pinus* to immigrate from the east was probably *P. banksiana.* It may have been slowed in its westward migration from the Appalachian Highlands by late-lasting ice lobes and associated proglacial lakes in the Great Lakes basins (Wright, 1968a). *P. banksiana/P. resinosa* did reach western Ohio, however, at the time of the second *Picea* pollen maximum described above (Shane, 1980). *Pinus* entered eastern Minnesota about 10,000 years ago and rapidly spread to Manitoba, where it encountered populations of *P. banksiana* thought to have immigrated from the west (Ritchie, 1976) (Figure 10-6c). In northeastern Minnesota, *P. banksiana* arrived at about the same time (10,700 yr B.P.) as shrub tundra gave way to *Picea* forest (Birks, 1981a). (Today in the boreal forest, *Pinus banksiana* nowhere reaches the tundra border.) About 1000 years later, *P. resinosa* arrived in northeastern Minnesota, likewise from the Appalachians, causing a steep increase in the *Pinus* pollen percentage at Kylen Lake in northeastern Minnesota (Birks, 1981a).

P. strobus migrated rapidly from the Appalachian Highlands to western Indiana, where it joined *P. banksiana/P. resinosa* and other conifers in the second *Picea* maximum (Shane, 1980). It moved more slowly northwestward, however, reaching Wood Lake in north-central Wisconsin (Figure 10-2) at 8000 yr B.P. and Jacobson Lake in northeastern Minnesota by 7200 yr B.P. For the next 3000 years it expanded little more while the prairie was farthest east. *P. strobus* then resumed its westward migration and was still expanding at the time of logging a hundred years ago (Jacobson, 1979). It thus caught up with the other *Pinus* species and in fact overtook them in northwestern Minnesota in its expansion after 4000 yr B.P. During the last 1000 years, however, *P. banksiana* or *P. resinosa* has invaded the area of *P. strobus,* which is presumably continuing to spread west into the narrow band of deciduous forest bordering the prairie in northwestern Minnesota (McAndrews, 1966).

In northern Michigan and Wisconsin, *P. strobus* was the dominant pollen type in the middle Holocene, and then it decreased in abundance during the late Holocene as *Tsuga*-northern hardwood forests developed (Figures 10-2 and 10-9). This change is discussed below.

MIDDLE AND LATE HOLOCENE; 8000 YR B.P. TO THE PRESENT

The maps (Figures 10-5 through 10-7 and 10-11 and 10-12) show systematic shifts in the major midwestern ecotones during the last 10,000 years, in particular the changing alignment of the prairie/forest border, which today extends from northwestern Minnesota to Illinois as the northeastern flank of the Prairie Peninsula (Figures 10-1 and 10-5). During the time of the Late Wisconsin glaciation, when boreal *Picea* forest was widespread throughout the

Midwest, treeless openings were apparently common, as indicated by the relatively high pollen percentages of *Artemisia,* *Ambrosia,* and other herbs (Figures 10-2 and 10-11a) (Bernabo and Webb, 1977). But there is no evidence for extensive prairie anywhere on the Great Plains at this time, although the scarcity of pollen sites in this region provides little confidence in any reconstruction.

Regardless of the Pleistocene history of the prairie, in the heart of the Great Plains (e.g., northern Nebraska at Rosebud [Watts and Wright, 1966]), the Holocene opened with an abrupt transformation from spruce forest to the prairie, with no substantial intermediate phase of deciduous forest or woodland such as occurred farther east (e.g., in northwestern Iowa) (Figure 10-2). This transformation is also recorded in the northern Great Plains in Manitoba and Saskatchewan (Ritchie, 1976), where today the prairie meets the spruce forest with only the aspen parkland between. In these western areas, the prairie was thus fully established in its modern expanse by 10,000 years ago, even though the Laurentide ice sheet had scarcely withdrawn across the Canadian border (Denton and Hughes, 1981; Prest, 1969). By this time, the Great Plains were probably dominated in summer by the dry, warm Pacific air that crossed the western cordillera, and the cool summers previously engendered by the nearby ice sheet at its maximum no longer prevailed.

The maps (Figures 10-3a and 10-5) show that the Prairie Peninsula, which was nonexistent 10,000 and 9000 years ago, was fully developed by 8000 years ago. Its shape subsequently changed as its northeastern flank advanced far into eastern Minnesota and even into western Wisconsin, while its point retreated from eastern to western Illinois. (The lack of pollen sites south of the Prairie Peninsula prevents delineation of its southern flank.) At this time (7000 to 6000 years ago), the greater strength of the dry westerlies apparently affected a broader latitudinal belt in the Midwest, but the moist air mass from the Gulf of Mexico dominated the East, thus blunting the point of the Prairie Peninsula. Still later (3000 years ago), the shape reversed as the northeastern flank retreated and the point was reestablished.

The late-Holocene climatic trends, as manifested by the migration of the prairie/forest border and by the history of *Pinus* species, can also be detected in the development of the extensive patterned peatlands of northern Minnesota. These date largely after 4000 yr B.P., with the inception of a cooler, wetter climate that inhibited the decomposition of plant material on the poorly drained bed of glacial Lake Agassiz and other lake plains (Griffin, 1975). *Picea* populations also increased in abundance in northern Minnesota, Wisconsin, and Michigan (Figure 10-6b), and developing peatlands were one habitat where the *Picea* trees grew.

FIRE AND THE PRAIRIE/FOREST BORDER IN MINNESOTA

The prairie/forest border today is clearly related to climate and must have been so in the past. This generalization, however, is valid only at a scale appropriate to major vegetational formations and to macro-climatic variables, such as air-mass frequency. The scale of our maps is appropriate to demonstrate this relationship.

The problem of scale (Webb, 1981) arises from the fact that a pollen assemblage is related differently to local and regional vegetation, depending on the size of the site, the pollen production and dispersal of the local and regional vegetation, the complexity of the vegetation, the steepness of the vegetational gradients, and other factors (Jacobson and Bradshaw, 1981; Janssen, 1973). Recognizing regional trends in pollen diagrams over broad areas was one of the major contributions of Von Post (1946), who introduced the method of stratigraphic pollen analysis for paleoclimatic reconstructions. Mapping the pollen data over a broad area with wide contour intervals is one method of employing a low-pass spatial filter in order to emphasize the broad regional patterns.

In the case of the prairie/forest border, a larger map scale (showing counties rather than states) might reveal so many local complexities that it could fail to depict those broad shifts in vegetational formations and their ecotones that result from changes in climate. On the other hand, knowledge of the nonclimatic factors controlling the location of an ecotone or of plant distributions is needed to guide paleoclimatic reconstructions (Birks, 1981b).

The shift in controls with scale is especially well illustrated by the prairie/forest border in south-central Minnesota, where the detailed study of the Big Woods by Grimm (1981) has elucidated the factors controlling the transition westward from mesic deciduous forest (*Ulmus, Acer, Tilia, Quercus*) through *Quercus macrocarpa/Populus tremuloides* forest and brushland to prairie. The dominating local controls on the distribution of these vegetational types are soils, topography, and especially fire, which itself is affected by such topographic elements as lakes, streams, areas of high relief, and other features that provide fire breaks. On a county scale, the mosaic of vegetational types is depicted on maps compiled from the records of presettlement land surveys of the 19th century (Grimm, 1981; Marschner, 1974). Such maps show the effectiveness of fire and other ecologic factors in building the mosaic: the almost annual late-summer fires (many apparently set by Indians) were blown by dry southwesterly winds. Prairie, therefore, commonly terminated against streams and lakes or against rough morainic topography, and the forest border was dominated by *Quercus macrocarpa*, which has thick, fire-resistant bark, or by *Populus tremuloides*, which sprouts readily after fire. The forest fringe provided by these two tree types was commonly burned by ground fires, but the sweep of the wind was greatly reduced within the woodland fringe, and the fires did not carry easily into the mesic forest beyond, which was composed of species not adapted to repeated fires. This situation was especially prevalent where rough topography reduced the wind fetch and thus the spread of fire.

The topography and soils in the glaciated terrain of the Big Woods area were such, however, that broad areas of low relief with sandy soils and prairie openings existed east of forest-covered morainic ridges. After the middle-Holocene prairie expansion in this area, the forest advance in the late Holocene presumably involved the filling in of such prairie openings by *Quercus-Populus* brush prairie or woodland, and the *Quercus* woodland on the morainic ridges to the west became sufficiently dense and extensive to protect the area in its lee from fire and to permit the development there of mesic forest. The forest seems to have moved westward in jumps and stages, and the elements of the prairie/forest mosaic formed a continually changing pattern under the influence of broad-scale climatic changes. This detailed study of the prairie/forest ecotone shows that climatic change may be the ultimate cause for regional vegetational change but that topography, fire, and soil are proximal factors controlling the exact timing and local expression of the vegetational change.

COMPOSITIONAL CHANGES IN THE MIXED FOREST OF THE MIDDLE AND LATE HOLOCENE

After its arrival, *Pinus strobus* became the dominant pollen type at sites from northern Michigan into northeastern Minnesota during the middle Holocene. The record at Wood Lake in northern Wisconsin (Figure 10-2) is a good example of this pattern, and the pollen evidence from a small hollow, Kelly's Hollow, within 10 km of Wood Lake verifies that *P. strobus* trees grew on morainic soils (Heide, 1981).

After 6000 yr B.P., as *P. strobus* was extending its range in northern Minnesota (Figure 10-6d), its populations decreased in northern Michigan and Wisconsin and were replaced by *Acer, Betula*, and *Tsuga* trees. In Michigan, *Fagus* trees also increased in abundance. The development of cooler and especially moister conditions in the northern Great Lakes region seems to have favored this major change in the forest composition. Both the southward movement of *Picea* populations (Figure 10-6b) and the westward movement of the prairie/forest border (Figure 10-5) support this climatic interpretation.

An alternative hypothesis for the late development of the northern hardwoods is that biologic factors delayed the migration of *Tsuga* and *Fagus* populations into the Midwest and that such factors are more important for these taxa than are climatic factors. Our maps certainly show that the migration of *Tsuga* and *Fagus* lagged behind that of *Pinus, Acer, Quercus*, and other types, but that fact does not necessarily imply that biologic factors such as seed dispersal or competition limited the migration of *Tsuga* and *Fagus*. Had climatic conditions suitable for *Tsuga* and *Fagus* existed in the northern Great Lakes region prior to 3000 yr B.P., then populations of *Acer* should have increased in the region. *Acer* trees were present in the Midwest before either *Fagus* or *Tsuga* trees and were confined to the deciduous forest in the early Holocene. *Acer* populations then expanded within the northern Great Lakes area only after 6000 yr B.P. (Figure 10-8d), and the distribution of *Acer* pollen (above 2%) at 7000 and 6000 yr B.P. is similar to that of *Fagus* pollen (Figures 10-9a and 10-12b), which only occurred above 2% at sites from central Michigan southward.

The behavior of *Fagus* also creates problems for the hypothesis that biologic factors limited its migration. *Fagus* arrived in the Midwest much earlier than *Tsuga* but remained in southern Michigan until after 6000 yr B.P. (Figures 10-9a and b). *Fagus* then moved into eastern upper Michigan after *Tsuga*. If the rate of seed dispersal controlled the timing of migration for *Fagus* relative to *Tsuga*, why did *Fagus* arrive in southern Michigan first but then lag behind *Tsuga* in entering upper Michigan? Some element of climatic influence seems a likely answer to this question, and ample evidence of climatic change exists in the Midwest for this time period. The late-Holocene range extensions of *Tsuga, Fagus*, and *Pinus strobus* and the population expansions of *Tsuga, Fagus, Acer*, and *Betula* are all consistent with the climatic changes associated with the westward movement of the prairie-forest border and the southward movement of the populations of *Picea* trees.

Conclusions

The Holocene vegetational history of the Midwest reveals a changing pattern of north-south and east-west compositional gradients. These in part result from the changing duration and characteristics of the air masses within the study areas.

In the Late Wisconsin and early Holocene, several pollen assemblages existed for which no analogues are found today among the samples of contemporary pollen. As the climate warmed, populations of some taxa expanded, other taxa entered the Midwest for the first time in many thousands of years, and still others decreased in abundance or extent of range. The isochrone maps show that north-south trends dominate for certain taxa (e.g., *Picea* and *Quercus*), whereas east-west trends dominate for other taxa (e.g., *Pinus*, *Tsuga*, *Fagus*, and prairie forbs). A deciduous forest of diverse composition existed in the southeast by 10,000 yr B.P., and extensive prairie first grew in the study area by 9000 yr B.P. *Fagus* appeared in southern Michigan by 8000 yr B.P., but *Tsuga*-northern hardwood forests began growing in the mixed forests of the Midwest only after 6000 yr B.P.

These broad-scale changes indicate the long-term response of the vegetation to variations in temperature, moisture, and air-mass patterns. The maps are an effective means of emphasizing this macroscale response, which is often obscured on individual pollen diagrams by the mesoscale influence of topography and fire breaks and by the microscale influence of the development of local wetland communities as sediment accumulates at a site (Delcourt et al., 1983).

References

Amundson, D. C., and Wright, H. E., Jr. (1979). Forest changes in Minnesota at the end of the Pleistocene. *Ecological Monographs* 49, 1-16.

Bailey, R. E. (1972). Late and postglacial environmental changes in northeastern Indiana. Ph.D. dissertation, Indiana University, Bloomington.

Bailey, R. E., and Ahearn, P. J. (1981). A late and postglacial pollen record from Chippewa Bog, Lapeer Co., MI: Further examination of white pine and beech immigration into the central Great Lakes region. *In* "Geobotany II" (R. C. Romans, ed.), pp. 53-74. Plenum Press, New York.

Baker, R. G. (1970). A radiocarbon-dated pollen chronology for Wisconsin: Disterhaft Farm Bog revisited. *Geological Society of America Abstracts* 2, 488.

Bernabo, J. C. (1981). Quantitative estimates of temperature changes over the last 2700 years in Michigan based on pollen data. *Quaternary Research* 15, 143-59.

Bernabo, J. C., and Webb, T., III (1977). Changing patterns in the Holocene pollen record of northeastern North America: A mapped summary. *Quaternary Research* 8, 64-96.

Birks, H. J. B. (1976). Late-Wisconsinan vegetational history at Wolf Creek, central Minnesota. *Ecological Monographs* 46, 395-429.

Birks, H. J. B. (1981a). Late Wisconsin vegetational and climatic history at Kylen Lake, northeastern Minnesota. *Quaternary Research* 16, 322-55.

Birks, H. J. B. (1981b). The use of pollen analysis in the reconstruction of past climates: A review. *In* "Climate and History" (T. M. L. Wigley, M. J. Ingram, and G. Farmer, eds.), pp. 111-38. Cambridge University Press, Cambridge.

Bradbury, J. P., and Waddington, J. C. B. (1973). The impact of European settlement on Lake Shagawa, northeastern Minnesota, U.S.A. *In* "Quaternary Plant Ecology" (H. J. B. Birks and R. B. West, eds.), pp. 289-307. Blackwell Scientific Publishing, Oxford.

Brubaker, L. B. (1975). Postglacial forest patterns associated with till and outwash in northcentral upper Michigan. *Quaternary Research* 5, 499-527.

Bryson, R. A. (1966). Air masses, streamlines, and the boreal forest. *Geographical Bulletin* 8, 228-69.

Burke, M. J., Gusta, L. V., Quamme, H. A., Weiser, C. J., and Li, P. H. (1976). Freezing injury in plants. *Annual Review of Plant Physiology* 27, 507-28.

Craig, A. J. (1972). Pollen influx to laminated sediments: A pollen diagram from northeastern Minnesota. *Ecology* 53, 46-57.

Curtis, J. T. (1959). "The Vegetation of Wisconsin." University of Wisconsin Press, Madison.

Cushing, E. J. (1965). Problems in the Quaternary phytogeography of the Great Lakes region. *In* "The Quaternary of the United States" (H. E. Wright, Jr., and D. G. Frey, eds.), pp. 403-16. Princeton University Press, Princeton, N.J.

Davis, A. M. (1977). The prairie-deciduous forest ecotone in the upper Middle West. *Annals of the Association of American Geographers* 67, 204-13.

Davis, A. M. (1979). Wetland succession, fire, and the pollen record: A Midwestern example. *American Midland Naturalist* 102, 86-94.

Davis, M. B. (1981). Outbreaks of forest pathogens in Quaternary History. *In* "Proceedings IV International Palynological Conference, Lucknow (1976-77)," 3, pp. 216-27.

Davis, M. B., Brubaker, L. B., and Beiswenger, J. (1971). Pollen grains in lake sediments: Pollen percentages in surface sediments from southern Michigan. *Quaternary Research* 1, 450-67.

Davis, M. B., Brubaker, L. B., and Webb, T., III (1973). Calibration of absolute pollen influx. *In* "Quaternary Plant Ecology" (H. J. B. Birks and R. G. West, eds.), pp. 9-25. Blackwell Scientific Publishing, Oxford.

Delcourt, H. R., Delcourt, P., and Webb, T., III (1983). Dynamic plant ecology: The spectrum of vegetational change in space and time. *Quaternary Science Reviews*, 1, 153-75.

Denton, G. H., and Hughes, T. J. (eds.) (1981). "The Last Great Ice Sheets." Wiley, New York.

Durkee, L. H. (1971). A pollen profile from Woden Bog, Hancock County, Iowa. *Ecology* 52, 835-44.

Fries, M. (1962). Pollen profiles of late Pleistocene and recent sediments at Weber Lake, northeastern Minnesota. *Ecology* 43, 295-308.

George, M. F., Burke, M. J., Pellett, H. M., and Johnson, A. G. (1974). Low temperature exotherms and woody plant distribution. *Horticultural Science* 9, 519-22.

Gilliam, J. A., Kapp, R. O., and Bogue, R. D. (1967). A post-Wisconsin pollen sequence from Vestaburg Bog. Montcairn County, Michigan. *Michigan Academy of Science, Arts and Letters* 52, 3-17.

Gleason, H. A., and Cronquist, A. (1963). "Manual of Vascular Plants of Northeastern United States and Adjacent Canada." Van Nostrand, New York.

Griffin, K. O. (1975). Vegetation studies and modern pollen spectra from Red Lake peatland, northern Minnesota. *Ecology* 56, 531-46.

Grimm, E. C. (1981). An ecological and paleoecological study of the vegetation in the Big Woods region of Minnesota. Ph.D. dissertation, University of Minnesota, Minneapolis.

Grüger, E. (1972). Pollen and seed studies of Wisconsinan vegetation in Illinois, U.S.A. *Geological Society of America Bulletin* 83, 2715-34.

Grüger, J. (1973). Studies on the late Quaternary vegetation history of northeastern Kansas. *Geological Society of America Bulletin* 84, 239-50.

Haworth, E. Y. (1972). Diatom succession in a core from Pickerel Lake, northeastern South Dakota. *Geological Society of America Bulletin* 83, 157-72.

Heide, K. M. (1981). Late Quaternary vegetational history of northcentral Wisconsin, U.S.A.: Estimating forest composition from pollen data. Ph.D. dissertation, Brown University, Providence, R.I.

Heinselman, M. L. (1973). Fire in the virgin forests of the Boundary Waters Canoe Area, Minnesota. *Quaternary Research* 3, 329-82.

Jacobson, G. L. (1979). The paleoecology of white pine in Minnesota. *Journal of Ecology* 67, 699-728.

Jacobson, G. L., and Bradshaw, R. H. W. (1981). The selection of sites for paleoenvironmental studies. *Quaternary Research* 16, 80-96.

Janssen, C. R. (1968). Myrtle Lake: A late and postglacial pollen diagram from northern Minnesota. *Canadian Journal of Botany* 46, 1397-1410.

Janssen, C. R. (1973). Local and regional pollen deposition. *In* "Quaternary Plant Ecology" (H. J. B. Birks and R. G. West, eds.), pp. 31-42. Blackwell Scientific Publishing, Oxford.

Kerfoot, W. C. (1974). Net accumulation rates and the history of cladoceran communities. *Ecology* 55, 51-61.

King, J. E. (1981). Late Quaternary vegetational history of Illinois. *Ecological Monographs* 51, 43-62.

King, J. E., and Allen, W. H., Jr. (1979). A Holocene vegetation record from the Mississippi River valley, southeastern Missouri. *Quaternary Research* 8, 307-23.

Lawrenz, R. (1975). Biostratigraphic study of Green Lake Michigan. M.S. thesis, Central Michigan University, Mount Pleasant.

McAndrews, J. H. (1966). "Postglacial History of Prairie, Savanna and Forest in Northwestern Minnesota." Torrey Botanical Club Memoir 22 (2).

McCann, M. T. (1979). The plant tension zone in Michigan. M.S. thesis, Western Michigan University, Kalamazoo.

Manny, B. A., Wetzel, R. G., and Bailey, R. E. (1978). Paleolimnological sedimentation of organic carbon, nitrogen, phosphorus, fossil pigments, pollen, and diatoms in a hypereutrophic, hardwater lake: A case history of eutrophication. *Polskie Archiwum Hydrobiologii* 25, 243-67.

Marschner, F. G. (1974). Original vegetation of Minnesota (map). U.S. Forest Service, North Central Forest Experiment Station, St. Paul.

Ogden, J. G., III (1966). Forest history of Ohio: Radiocarbon dates and pollen stratigraphy of Silver Lake, Ohio. *Ohio Journal of Science* 66, 387-400.

Peters, A., and Webb, T., III (1979). A radiocarbon-dated pollen diagram from west-central Wisconsin. *Bulletin of the Ecological Society of America* 60, 102.

Peterson, G. M. (1978). Pollen spectra from surface sediments of lakes and ponds in Kentucky, Illinois, and Missouri. *American Midland Naturalist* 100, 333-40.

Potzger, J. E. (1946). Phytosociology of the primeval forests in central-northern Wisconsin and upper Michigan, and a brief postglacial history of the Lake Forest formation. *Ecological Monographs* 16, 211-50.

Potzger, J. E. (1948). A pollen study in the tension zone of lower Michigan. *Butler University Botanical Studies* 8, 161-77.

Prest, V. K. (1969). Retreat of Wisconsin and recent ice in North America. Geological Society of Canada Map 1257A.

Ritchie, J. C. (1976). The late-Quaternary vegetational history of the western interior of Canada. *Canadian Journal of Botany* 54, 1793-1818.

Shane, L. C. K. (1976). Late-glacial and postglacial palynology and chronology of Drake County, west-central Ohio. Ph.D. dissertation, Kent State University, Kent, Ohio.

Shane, L. C. K. (1980). Detection of a late-glacial climatic shift in central mid-western pollen diagrams. *In* "American Quaternary Association, Sixth Biennial Meeting, Abstracts and Program, 18-20 August 1980," p. 171. Institute for Quaternary Studies, University of Maine, Orono.

Shay, C. T. (1971). "The Itasca Bison Kill Site: An Ecological Analysis." Minnesota Historical Society, St. Paul.

Stuiver, M. (1971). Evidence for variation of atmospheric C-14 content in the late Quaternary. *In* "The Late Cenozoic Glacial Ages" (K. K. Turekian, ed.), pp. 57-70. Yale University Press, New Haven, Conn.

VanZant, K. (1979). Late-glacial and postglacial pollen and plant macrofossils from Lake West Okoboji, northwestern Iowa. *Quaternary Research* 12, 358-80.

Von Post, L. (1946). The prospect for pollen analysis in the study of the Earth's climatic history. *New Phytologist* 45, 193-217.

Waddington, J. C. B. (1969). "A Stratigraphic Record of Pollen Influx to a Lake in the Big Woods of Minnesota." Geological Society of America Special Paper 123, pp. 263-82.

Watts, W. A., and Bright, R. C. (1968). Pollen, seed, and mollusk analysis of a sediment core from Pickerel Lake, northeastern South Dakota. *Geological Society of America Bulletin* 79, 855-76.

Watts, W. A., and Wright, H. E., Jr. (1966). Late Wisconsin pollen and seed analysis from the Nebraska sandhills. *Ecology* 47, 202-10.

Webb, T., III (1974a). Corresponding distributions of modern pollen and vegetation in lower Michigan. *Ecology* 55, 17-28.

Webb, T., III (1974b). A vegetational history from northern Wisconsin: Evidence from modern and fossil pollen. *American Midland Naturalist* 92, 12-34.

Webb, T., III (1981). The past 11,000 years of vegetational change in eastern North America. *Bioscience* 31, 501-6.

Webb, T., III, and Bernabo, J. C. (1977). The contemporary distribution and Holocene stratigraphy of pollen types in eastern North America. *In* "Contributions of Stratigraphic Palynology," vol. 1, "Cenozoic Palynology" (W. C. Elsik, ed.), pp. 130-46. American Association of Stratigraphic Palynologists, Contribution Series 5A.

Webb, T., III, and Bryson, R. A. (1972). Late- and postglacial climatic change in the northern Midwest, U.S.A.: Quantitative estimates derived from fossil pollen spectra by multivariate statistical analysis. *Quaternary Research* 2, 70-115.

Webb, T., III, and McAndrews, J. H. (1976). Corresponding patterns of contemporary pollen and vegetation in central North America. *Geological Society of America Memoir* 145, 267-99.

Webb, T., III, Howe, S. E., Bradshaw, R. H. W., and Heide, K. M. (1981). Estimating plant abundances from pollen percentages: The use of regression analysis. *Review of Paleobotany and Palynology* 34, 269-300.

Webb, T., III, Laseski, R. A. and Bernabo, J. C. (1978). Sensing vegetation patterns with pollen data: choosing the data. *Ecology* 59, 1151-63.

West, R. G. (1961). Late and postglacial vegetational history in Wisconsin, particularly changes associated with the Valders readvance. *American Journal of Science* 259, 766-83.

Williams, A. S. (1974). "Late-Glacial-Postglacial Vegetational History of the Pretty Lake Region, Northeastern Indiana." U.S. Geological Survey Professional Paper 686, pp. D1-D23.

Wright, H. E., Jr. (1968a). History of the Prairie Peninsula. *In* "The Quaternary of Illinois" (R. E. Bergstrom, ed.), pp. 78-88. Special Report 14, College of Agriculture, University of Illinois, Urbana.

Wright, H. E., Jr. (1968b). The roles of pine and spruce in the forest history of Minnesota and adjacent areas. *Ecology* 49, 937-55.

Wright, H. E., Jr. (1977). Quaternary vegetation history: Some comparisons between Europe and America. *Annual Review of Earth and Planetary Sciences* 5, 123-58.

Wright, H. E., Jr., and Watts, W. A. (1969). "Glacial and Vegetational History of Northeastern Minnesota." Minnesota Geological Survey Special Paper SP-11, pp. 1-59.

Wright, H. E., Jr., Winter, T. C., and Patten, H. L. (1963). Two pollen diagrams from southeastern Minnesota: Problems in the late- and postglacial vegetational history. *Geological Society of America Bulletin* 74, 1371-96.

CHAPTER 11

Holocene Vegetational History of the Eastern United States

Margaret Bryan Davis

Introduction

The eastern United States (Figure 11-1), now farmland and secondary forest, was largely a region of deciduous forest at the time of the European settlement. Today, spruce and fir (Table 11-1) occur in the boreal forest (region 1, Figure 11-2) and also at high elevations throughout the Appalachian mountain chain, and tundra is found on the highest peaks in the North. Throughout the northernmost United States (region 2, Figure 11-2), the deciduous forests are dominated by sugar maple, beech, and yellow birch. Hemlock and white pine are also common except at high elevations, and oaks occur in warmer areas. The deciduous forests are more complex and diverse in the central Appalachians (region 5, mesophytic forest). Several investigators distinguish "cove" forests in steep first-order valleys in the mountains. Cove forests contain tulip poplar, hemlock, and many hardwood species. Steep side slopes are dominated by oak, and dry ridges support pine or chestnut-oak (*Q. prinus*) (Hack and Goodlett, 1960). Chestnut (*Castanea*) was once abundant in such sites, which occur abundantly in region 3, oak-chestnut forest. Oak, red maple, and hickory are now the dominant species in region 3. Oak and pine form forests on the Coastal Plain (region 8). Farther south, pine often grows among open, fire-prone grassy vegetation, and swamps are common (southern mixed forest). Palmettos are often found in association with pine in open grassy vegetation in northern Florida. Cypress swamps cover large areas, and broad-leaved evergreens such as live oak grow along the coast. Tropical mangrove swamps can be found at the southern end of the peninsula (subtropical vegetation). The deciduous forests become less diverse west of the Appalachians (regions 4 and 5). They are dominated by oak and hickory and finally by oak, in a mosaic of patches of forest and scrub interspersed with areas of prairie (region 6).

The vegetation of the eastern United States has been mapped and described in detail by Braun (1950) and Küchler (1964). Figure 11-2 is a modification and simplification of their maps. In fact, however, few quantitative data on forest composition are available, and regional maps of "natural" or "potential" vegetation are generalizations expanded from detailed studies of tiny areas of old-growth forest thought to be representative of primeval vegetation. Relatively little

is known about the way in which the modern, second- or third-growth forests differ from the original forests. From central Pennsylvania westward, surveyor's records from the early 19th century provide a census of presettlement forest; these records have not yet been compiled and analyzed for the entire region, however. More detailed descriptions of modern forests are provided in regional treatments by Bormann and others (1970), Hack and Goodlett (1960), Goodlett (1954), Siccama (1974), Brush and others (1980), Whittaker (1956), and others. The problems in reconstructing the forest history of a region whose modern vegetation is so poorly understood are discussed at length in another review (Davis, 1965).

Wisconsin-Holocene Transition (Late-Glacial) North of the Glacial Boundary

As the ice sheets began retreating from their terminal positions about 15,000 years ago, tundra plants colonized the landscape. Glacial retreat occurred rapidly. In New England, the ice retreated from Long Island to Quebec in less than 3000 years, a retreat that averaged 150 km per 1000 years. In the high mountains of New Hampshire, where relief is almost 2000 m, radiocarbon dates from basal sediments in lakes at high elevations, now above the treeline, are similar to those from sites at low elevations, in the valleys; this suggests the downwasting of a 2000-m thickness of ice within a few hundred years (Davis et al., 1980).

The late-glacial tundra included a number of species that grow in the Arctic. At Cambridge, Massachusetts, a late-glacial site includes macrofossils from *Salix herbacea*, *Dryas integrifolia*, and *Vaccinium uliginosum* ssp. *alpinum* (Argus and Davis, 1962). A similar fossil flora has been found in northeastern New York (D. R. Whitehead, personal communication, 1982). *Dryas* has also been identified from leaves in the sediments of Mirror Lake, 250 km farther north. *Dryas* pollen occurs at Moulton Pond in Maine (Figure 11-3A). *Selaginella selaginoides* spores have been found in 12,600-year-old sediment at

This work was supported by the National Science Foundation. I am grateful to Donald R. Whitehead and Stephen Jackson for their critical comments.

Figure 11-1. Outline map of the eastern United States showing the localities referred to in the text.

several sites in western New York State (Miller, 1973; Spear and Miller, 1976). A rich macrofossil deposit at Columbia Bridge in northernmost Vermont contains many plant species characteristic of tundra (Miller and Thompson, 1979). Miller and Thompson demonstrate conclusively that several plants that combine broad distribution in western North America with disjunct stations in eastern North America were members of a late-glacial flora that moved northward following the ice's retreat, colonizing the eastern Arctic from the south. Fernald (1925) noted these disjunctions many years ago, postulating that the eastern populations were relicts that survived glaciation in situ on nunataks protruding above the ice. The failure of these species to spread out during the Holocene from their putative

refuges was attributed to their loss of genetic variability, or their "senescence." Their fossil history (Anderson, 1954; Miller and Thompson, 1979), however, suggests that they are now restricted to isolated islands of favorable habitat.

Tundra pollen assemblages from late-glacial deposits in the eastern United States contain high (10% to 40%) percentages of sedge pollen. Grass is less frequent, its pollen usually comprising less than 10% of the assemblage. *Artemisia* contributes 1% to 6%, and willow 2% to 8%. In many cases, the percentages of willow pollen are lowest where *Artemisia* percentages are high (Maine, New Hampshire) and highest where *Artemisia* percentages are low (southern Connecticut). This relationship may indicate heavy

Table 11-1. Colloquial Names and Latin Equivalents (after Fernald, 1950) of Plant Species and Genera

Colloquial Name	Latin Name
Alder	*Alnus* spp.
Ash	*Fraxinus* spp.
Beech	*Fagus grandifolia*
Birch	*Betula* spp.
dwarf	*B. glandulosa*
yellow	*B. lutea*
Chestnut	*Castanea dentata*
Cypress	*Taxodium*
Elm	*Ulmus* spp.
Fir, balsam	*Abies balsamea*
Fir, fraser	*Abies fraseri*
Grass	Gramineae
Gum, sweet	*Liquidambar styraciflua*
Hemlock	*Tsuga canadensis*
Hickory	*Carya* spp.
(shagbark, pignut, mockernut)	
Ironwood (hornbeam)	*Ostrya virginiana* and *Carpinus caroliniana*
Larch (tamarack)	*Larix laricina*
Mangrove	*Rhizophora* spp.
Maple	*Acer* spp.
red	*A. rubrum*
sugar	*A. saccharum*
Oak	*Quercus* spp.
chestnut	*Q. prinus*
Palmetto	Palmae
Pine	*Pinus* spp.
jack	*P. banksiana*
white	*P. strobus*
Poplar	
balsam	*Populus balsamifera*
tulip	*Liriodendron tulipifera*
Sedge	Cyperaceae
Legumes	Leguminosae
Spruce	
black	*Picea mariana*
red	*P. rubens*
white	*P. glauca*
Willow	*Salix* spp.

Figure 11-2. Vegetation map of the eastern United States. (Modified from Braun [1950] and Küchler [1964].)

VEGETATION REGIONS
1. Boreal Forest
2. Hemlock–White Pine– Northern Hardwood Forest
3. Oak–Chestnut Forest
4. Beech–Maple Forest
5. Mesophytic Forest
6. Deciduous Forest–Prairie Mosaic
7. Prairie
8. Oak–Pine Forest
9. Southern Mixed Forest
10. Subtropical Vegetation

snowfall in regions where willow was abundant and drier, more continental climate where *Artemisia* was more abundant. Soils, however, were also important. Watts (1979) shows that the tundra assemblage at Longswamp, Pennsylvania, outside the glacial boundary, contained a much higher ratio of grass pollen to sedge pollen than pioneer tundra a few kilometers away on deglaciated landscape. Both *Artemisia* and willow were uncommon at Longswamp.

Although dwarf birch is a common constituent of present-day Arctic tundra, it was rare during the late glacial. Similarly, alder was absent, arriving at most sites after spruce had become established (Figures 11-3A, 11-3B, and 11-3C). The macrofossil flora at Columbia Bridge, Vermont, is important in showing that spruce, balsam poplar, and dwarf birch were present although rare 11,500 years ago. *Juniperus communis* was also present, but alder was not (at least not

as a macrofossil). At most pollen sites in northern New Hampshire, the oldest tundra samples contain pollen of herbs and what is presumably far-traveled pollen of spruce and pine. After about 1000 years, willow, Ericaceae, and juniper pollen became more abundant. This assemblage is similar to the Columbia Bridge flora. Macrofossils of spruce occurred at 11,300 yr B.P., about 2000 years after the glacial retreat (Davis et al., 1980). But 250 km farther south in Massachusetts, a fossil beetle was found in a late-glacial tundra assemblage. This species, *Deronectes griseostriatus* (Argus and Davis, 1962), occurs today in low-arctic tundra or within the boreal forest. Its occurrence suggests that climate in Massachusetts was not limiting to the growth of trees but rather that factors such as availability of seed source were (Davis, 1961). The presence of a few fossil spruce and balsam poplar remains at Columbia Bridge in northern Vermont, however, suggests that trees were present in low abundance there, coexisting with tundra plants, perhaps on different microsites. Their failure to expand must have reflected an unfavorable environment. Columbia Bridge and Moulton Pond (Davis et al., 1975) may be similar in this regard (Figure 11-3A). The more-southern sites may represent treeless tundra from which trees were absent because of migration lag (Figure 11-3C).

Whether or not trees were limited by temperature, the presence of many species of herbs that now characterize alpine, arctic, and subarctic regions suggests that temperatures at that time were lower than present temperatures. Frost action was probably important, producing an unstable substrate for plant growth. Massive frost features, now inactive, found above the treeline in the high mountains of New Hampshire (Goldthwait, 1976) probably formed just after the glacier melted from the summits 13,000 years ago. The pollen record of high-elevation sites indicates that between 12,000 and 11,750 years ago the high peaks were a barren, periglacial desert (Spear, 1981). Such extensive fossil frost features do not occur at lower elevations; this fact suggests that frost action there was less intense and that perennially frozen ground was not extensive, if it existed at all. However, solifluction and frost heaving may have accelerated the rate of landscape denudation. The rate of the loss of inorganic particulates from the catchment of Mirror Lake has been estimated from the sediment mass that accumulated in the lake during late-glacial time (Davis and Ford, 1981). At that time, the loss rate was 650 kg per hectare per year, more than 30 times the modern rate of particulate export from nearby forested watersheds (Likens et al., 1977). Exports of particulate inorganics from the lake catchment declined dramatically at the time spruce appeared in the local flora between 12,000 and 11,000 years ago (Davis and Ford, 1981).

Soils were poorly developed during late-glacial time as rapid erosion and solifluction removed incipient soil profiles. Miller and Thompson (1979) note that the flora at Columbia Bridge is dominated by calcicolous species, although a few species characteristic of acid habitats are also present. Several of the species present are capable of fixing nitrogen and, therefore, appear to be well adapted to the nitrogen-poor, sterile soils left by the melting ice sheet.

Spruce was the first tree to colonize the late-glacial landscape. Its northward movement is time transgressive. A sharp rise in spruce-pollen influx occurs at 12,300 yr B.P. (uncorrected date) in Connecticut (Figure 11-3C), at 12,500 yr B.P. in western New York State, at 12,500 to 12,000 yr B.P. in western Massachusetts (Whitehead, 1979), and at 11,000 yr B.P. in New Hampshire (Figure 11-3B) (Bernabo and Webb, 1977). The spruce pollen influx never rises sharply in Maine, which leads R. B. Davis and others (1975) to conclude that a spruce forest never developed there (Figure 11-3A), although spruce was present as scattered trees on a tundra landscape. Even at sites where the pollen influx indicates the presence of spruce trees, the continuing presence of herb pollen in high percentages suggests a partially open vegetation, not a closed forest like the modern boreal forest in Canada (Davis, 1967; Whitehead, 1979). In Connecticut, southern New Hampshire, and western and northeastern New York State, fir and jack pine arrived soon after spruce and coexisted with it (Figure 11-3). Farther north in New Hampshire and Maine, fir and larch lagged behind spruce, arriving and reaching population maxima after spruce had begun to decline. The modern boreal forest of Canada, a species-poor formation, is made up of species that reacted in an individualistic manner to the rapidly changing climate and the sudden appearance of new habitat at the beginning of the Holocene. Maps of migration patterns (Figure 11-4) (Davis, 1981b) show that jack pine moved northward and westward, whereas larch, which was already present on the Great Plains, moved northward and eastward. The map for larch migration follows Watts's conclusions (1979) in showing larch absent east of the Appalachians; more recent data indicate that it was present there (Figure 11-5) (Whitehead, 1981). The presence of larch on the Coastal Plain still leaves intact a difference in the distributions of jack pine and larch during the full glacial, as well as differences in their rates of migration; these factors affected their times of arrival at sites north of the glacial boundary and thus the species composition of late-glacial woodlands.

Three species of spruce and two species of fir now grow in the eastern United States. *Abies fraseri* is endemic in the southern Appalachians, while *A. balsamea* occurs throughout the northern United States and Canada. I presume that the fir that moved northward to colonize deglaciated territory (Figure 11-4) was *A. balsamea*, but the distributions of the two species during the Wisconsin remain problematical. Spruce is more complicated. *Picea glauca* and *P. mariana* occur throughout the northern United States and Canada. Frequently, *P. glauca* occurs on drier sites, and *P. mariana* grows in muskegs and bogs. However, *P. mariana* dominates both kinds of habitat in Labrador and Quebec, and it is the only species in the

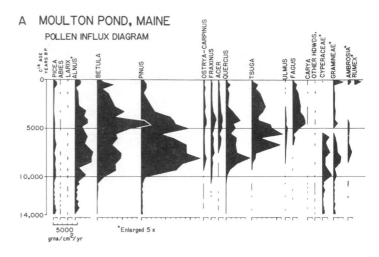

A MOULTON POND, MAINE
POLLEN INFLUX DIAGRAM

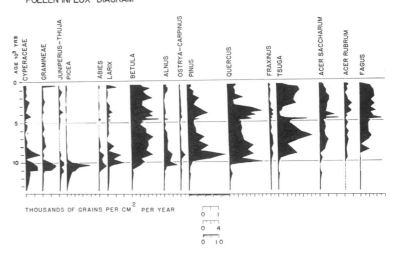

B SANDOGARDY POND, N.H.
POLLEN INFLUX DIAGRAM

THOUSANDS OF GRAINS PER CM^2 PER YEAR

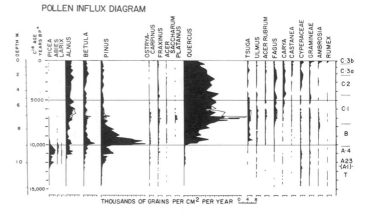

C ROGERS LAKE, CT
POLLEN INFLUX DIAGRAM

THOUSANDS OF GRAINS PER CM^2 PER YEAR

*Not corrected. Note change from previously published dates.

Figure 11-3. Pollen influx diagrams from three sites in the northern part of the eastern United States. (A) Moulton Pond, Maine. (Redrawn from Davis et al., 1975.) (B) Sandogardy Pond, New Hampshire. (C) Roger Lake, Connecticut.

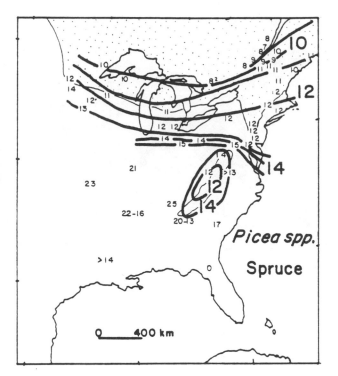

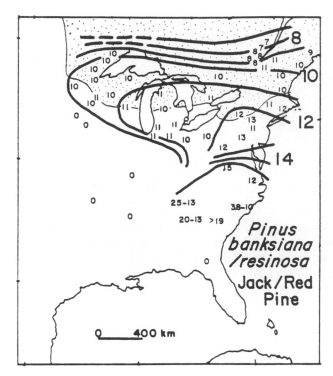

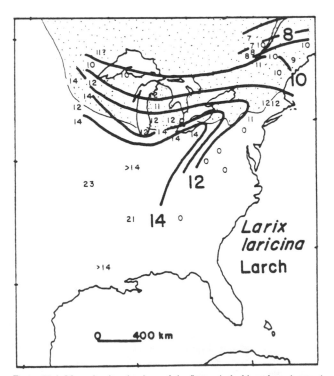

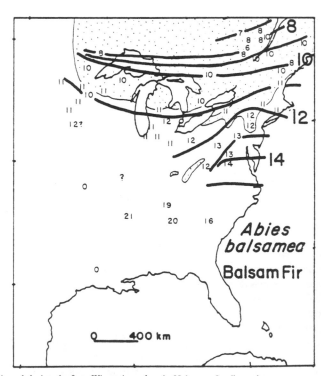

Figure 11·4. Maps showing the times of the first arrival of boreal species moving northward during the Late Wisconsin and early Holocene. Small numbers indicate arrival times in thousands of years at individual sites; isopleths connect points of similar age and represent the frontier for the species at 1000-year intervals. Stippled areas are modern ranges for the species. (Redrawn from Davis, 1981b.)

ROCKYHOCK BAY, N.C.

POLLEN PERCENTAGE DIAGRAM (ARBOREAL POLLEN)

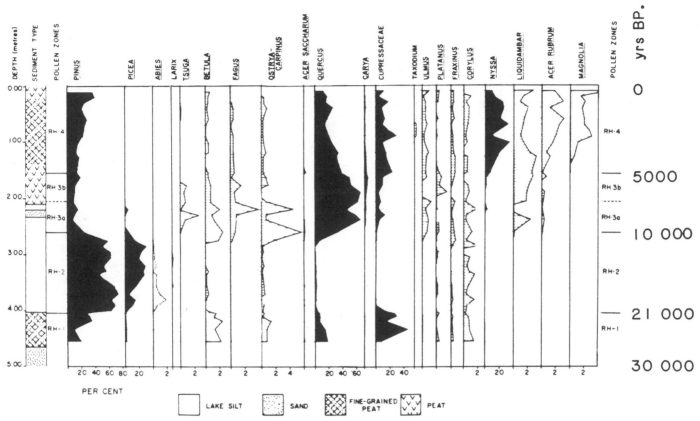

Figure 11 5. Pollen diagram from Rockyhock Bay, North Carolina. Only tree pollen is shown. (Redrawn from Whitehead, 1981.)

subarctic woodland near the arctic treeline. It grows as krummholz at the alpine treeline in New Hampshire and New York. At the southern limit of its range (e.g., northern New Hampshire and Vermont), *P. glauca* grows only in regions with nutrient-rich soil. *P. rubens*, a third species, occurs with *P. glauca* on rich soils and alone on upland sites in regions of acid bedrock. It grows both in subalpine forests with *Abies balsamea* and in hardwood forests at lower elevations. *Picea rubens* does not grow north of latitude 68° or west of longitude 80°, and its range extends southward into the Appalachians beyond the other two species.

P. rubens is similar morphologically to *P. mariana*; many of the morphological characteristics that separate the two species are described in relative terms (Fernald, 1950). In certain kinds of habitats, *P. rubens* and *P. glauca* resemble one another. All three species have the same number of chromosomes. Under the circumstances, it seems remarkable that the pollen of all three species can be separated, according to several investigators (Birks and Peglar, 1980; Richard, 1970; Watts, 1979). These workers separate *P. rubens* subjectively on the basis of morphological traits, primarily shape and wall sculpturing. *P. mariana* and *P. glauca* are distinguished from one another by size. Several of the preparations of *P. rubens* that I have seen meet the morphological criteria established by the investigators. Other preparations, however, seem less easily

separated from *P. glauca* and *P. mariana*. Birks and Peglar (1980) cite only four collections of *P. rubens* pollen. Therefore, I am slightly skeptical about identifications made from fossil material. Most of the preparations of *P. mariana* and *P. glauca* that Birks and Peglar studied were from Minnesota. I do not question their identifications for this region, although I feel that additional material should be studied before the identification methods they have developed are extended to materials from other regions.

Early measurements of the gross size of spruce pollen (Davis, 1958) suggest that a large-grained species (*P. glauca* or *P. rubens*) was the first to appear in Massachusetts, followed by a species with smaller-sized pollen (*P. mariana*). Watts (1979) suggests from pollen morphology that outside the glacial limit in Pennsylvania *P. glauca* dominated from 15,000 to 11,000 yr B.P., when *P. rubens* invaded and quickly replaced it. However, at high elevations close to the modern treeline in the mountains of New Hampshire, *P. mariana* seems to have been the only spruce species to have grown as krummholz (Spear, 1981). The present distribution of spruce in New England suggests that the neutral or calcareous soils of the late-glacial period (suggested by the presence of herbaceous plants such as *Dryas*) could have been conducive to the growth of *P. glauca* in regions like New Hampshire, where it does not grow today. The definitive identification of the spruce species that grew as woodland during late-glacial time is, therefore, of primary importance in distinguishing the effects of soil and climate on the growth of trees during the late-glacial interval.

Jack pine (*P. banksiana*) moved northward quickly and cooccurred with spruce in Connecticut and New York (Figure 11-3C). Its pollen never occurred commonly in sediments at Mirror Lake in northern New Hampshire, surprising because the species grows today a few kilometers away as an isolated stand on the granite ledges of a nearby mountain. Presumably, this disjunct stand is a relict of a putative late-glacial population, rather than a recent immigrant. Jack pine is not common today in eastern Canada. There are small stands in Maine and New Hampshire and along the St. Lawrence River in Canada, but they do not approach the extent of the forests of jack pine that grow on sand and gravel soils in the Great Lakes region (Brubaker, 1975).

Development of Forest

The opening of the Holocene 10,000 years ago marked the development of forest throughout the deglaciated region and at high elevations south of the glacial boundary where there had previously been tundra. At low elevations in the Southeast, significant changes took place in the relative abundances of species within the existing forests.

In the North, the composition of Holocene forest communities varied from region to region, depending largely on which species were available to form forests. Spruce, however, was not an important component of forests anywhere in the eastern United States. After 10,000 years ago, conditions apparently became less favorable for spruce than for other species (Solomon et al., 1981), and it declined in abundance everywhere, even in regions such as Maine, where it had been growing as scattered trees in a tundra environment (Bernabo and Webb, 1977; Davis et al., 1975). The percentages of spruce pollen dropped steeply, but the pollen influx shows that the decline of spruce abundance occurred slowly, beginning about 10,500 years ago and continuing over the next millennium. At some sites, significant populations of spruce persisted until about 9000 yr B.P.

Alder began to increase while spruce was still abundant, but it peaked after spruce began to decline. Poplar pollen percentages sometimes appear to reach a maximum before spruce pollen percentages do (Davis, 1958; Mott, 1978), but influx diagrams show that poplar increased in the spruce zone but actually reached its maximum abundance after spruce began to decline. Poplar became abundant in the early-Holocene forests of northern New England. A plot of poplar pollen influx is shown in Davis (1981b). Poplar is not shown in Figure 11-3.

In Connecticut, a number of temperate forest species were present 10,000 years ago, and these trees all expanded at the opening of the Holocene. The earliest Holocene forests included white pine, oak, elm, ash, birch, ironwood, and sugar maple (Figure 11-3C). In several important respects, however, these early forests differed from late-Holocene vegetation. Chestnut, hickory, and red maple, which later became forest dominants, were all absent, as were hemlock and beech. In most cases, maps showing the progressive northward movement of these species from various directions indicate that their absence from Connecticut at this time was due to migration lag, not unfavorable climate. Each arrived in Connecticut at a different time, and much of the Holocene pollen stratigraphy reflects the successive arrivals and expansions of hemlock, beech, hickory, and chestnut. Poplar, birch, alder, balsam fir, and spruce were all more abundant in

early-Holocene forests than in more modern forests of that region. The pollen assemblage resembled modern pollen assemblages from the northern Great Lakes region (M. B. Davis, 1967). Ironwood pollen was more abundant than at any time later during the Holocene. Ironwood pollen is very common in surface assemblages from the prairie-forest margin in Minnesota (Grimm, 1983); the tree grows as an understory tree in open oak forests on bluffs along river valleys. The abundance of this species in the early Holocene suggests a forest with diffuse canopy and well-lighted forest floor.

Farther north, in New Hampshire, the transition to forest 10,000 years ago involved an expansion of balsam fir, poplar, and birch. White pine, oak, and sugar maple did not arrive in the Far North at Mirror Lake until about 9000 years ago, at which time they also expanded. In central Maine, forest developed 9700 years ago with an expansion of pine, birch, and alder. Oak expanded a little later, reaching a maximum 8000 years ago (Figure 11-3).

In western New York State, the early-Holocene forest was more diverse and more similar to the forest in Connecticut. Fir, white pine, oak, hemlock, alder, and birch all expanded 10,000 years ago (Spear and Miller, 1976). In northeastern New York State, alder expansion preceded white pine and hemlock (Whitehead et al., 1981).

South of the glacial boundary, forest had been established much earlier, and in most regions it was already dominated by deciduous species. Late Wisconsin forests in Tennessee contained ironwood, which contributed 20% of the arboreal pollen 12,500 to 9000 years ago. The pollen diagram from Delcourt (1979) is shown in Watts (1983). Fifty percent of the tree pollen assemblage was oak. White ash, maple, and elm are represented by small amounts of pollen. Spruce was still present, but pine and fir were absent. Farther north, at Cranberry Glades in the mountains of West Virginia, a major expansion of deciduous forest, including oak, ironwood, and hemlock, occurred some time after 12,000 years ago (Watts, 1979). Unfortunately, either a sedimentary hiatus or a period of extremely slow sedimentation occurred at the site at that time, and the exact date of the expansion is unclear. By analogy with Buckle's Bog in western Maryland (Maxwell and Davis, 1972), the expansion of white pine and birch can be dated at 10,000 yr B.P., with hemlock increasing a few hundred years later. At this time, oak, hemlock, and hickory replaced pine and spruce in the valleys of Virginia (Craig, 1969). Chestnut was certainly well established in the Appalachians by at least 8000 yr B.P. (Craig, 1969; Delcourt, 1979; Watts, 1979), although it was slow to move northward and did not reach Connecticut until 2000 years ago (M. B. Davis, 1976, 1981b)(Figure 11-7).

Along the Coastal Plain, deciduous forest replaced full-glacial conifer woodland during the Late Wisconsin. On the southern Coastal Plain and in northern Florida, a mesophytic forest including beech grew throughout the late-glacial interval, when tundra and spruce woodland were developing on the deglaciated landscape farther north. These southern forests experienced changes in composition around 10,000 years ago. Farther north on the Coastal Plain of North Carolina, pine and spruce forests were replaced at this time by deciduous forest with white pine, hemlock, and beech (Whitehead, 1981) (Figure 11-5). Since migration lags were not involved or were not extensive in these latitudes, the changes in forest composition suggest that the climate was altered in important ways over a wide area at the opening of the Holocene 10,000 years ago. In the northern part of the United States, spruce declined and closed-canopy forests (of variable composition) developed; on the central Coastal Plain, pine and spruce were replaced by deciduous forest; in the southern part of the United States mesic forests were replaced by a more xeric woodland dominated by oak and pine (Table 11-2).

Migrations of Forest Species

The arrival of deciduous forest species throughout the eastern United States can be mapped by noting the radiocarbon age of the first sharp increase in pollen influx. First occurrences of macrofossils have been used at some sites, and increases in pollen percentages have been used where influx data are not available. The absolute age of the arrival (to the nearest thousand years) has been plotted on maps, and isoclines have been drawn to indicate the postulated position of the species frontier at various times in the past (Figures 11-4, 11-6, and 11-7) (M. B. Davis, 1976, 1981b). The maps show migration routes and migration rates for various species. The differences among the maps show that the various species of trees moved northward individualistically (Gleason, 1939). Forest communities have represented fortuitous combinations of species during the Holocene. Many modern communities are very young; they include dominant species that have grown locally for only a few thousand years. For example, communities that are mapped as a single unit (e.g., the oak-chestnut communities in Connecticut and Pennsylvania) have contained chestnut for different lengths of time: 5000 years in Pennsylvania and 2000 years in Connecticut. Forests in Ohio acquired hickory 10,000 years ago and beech 6000 years ago, but these genera entered forest communities in Connecticut in reverse order, beech 8000 years ago and hickory 5000 years ago. Evolutionary processes that may have adapted cooccurring species to one another have had very little time in which to take effect (Davis, 1981b).

Both white pine and hemlock appear to have spread out from Late Wisconsin refuges in the Appalachian foothills, the Coastal Plain, or the continental shelf (Figure 11-6). Beech grew in northern Florida (Watts, 1980) and spread northward east of the Appalachian mountain chain, moving westward through New York State and southern Canada and then southward into Ohio and Indiana and northward into Michigan (M. B. Davis, 1976, 1981b; Kapp, 1977) (Figure 11-7). Its movement into Wisconsin and upper Michigan is presently under study by M. B. Davis. A full-glacial occurrence of beech is reported from a site in the Mississippi River valley near Memphis (Delcourt et al., 1980), but beech failed to expand northward up the Mississippi River valley into Ohio during the early Holocene. Many other deciduous species did move northward very early through the central Plains. Sugar maple, elm, oak, and hickory all show this pattern (Figures 11-6 and 11-7). Chestnut seems to have moved northward along the Appalachian mountain chain. The slowness of its expansion is of interest because it contrasts with most other species. Chestnut is self-sterile, so its chance of successfully establishing itself from seeds may be limited. Other heavy-fruited species, such as oak, however, spread rapidly; this suggests that seed weight alone is not a predictor of the speed of range extension.

Speculation about the biogeographic effect of Pleistocene climatic changes often assumes that present-day communities were displaced southward during each glaciation. Biogeographers have mapped putative Pleistocene distributions of modern forest types and assumed that they provided habitat for animal species presently associated with boreal forest, deciduous forest, et cetera (Martin, 1958; Mengel, 1964). New data suggest, however, that the long glacial intervals caused extensive range displacement for individual species. The communities we recognize today in the northern United States did not exist during the Wisconsin, although subsets of species that presently characterize deciduous forest grew in pockets of favorable habitat in various parts of the southern United States. As warming occurred at the termination of the Wisconsin glaciation, plants spread northward rapidly, each species being limited only by its intrinsic capacity to expand its population and disperse northward. Because of the enormous distances involved, many plant species have been out of equilibrium with the climate over a part of their range during much of the Holocene, reaching climatic limits only after the warmest part of the Holocene had ended. Hickory is an example. Although it probably was limited by climate in the Great Lakes region during the early Holocene, having arrived there 10,000 years ago, it continued to expand slowly eastward, reaching Connecticut only 5000 years ago, after the warmest part of the Holocene had ended.

These extensive contractions and expansions of species ranges occurred during each glacial-interglacial cycle, that is, 16 to 18 times during the Quaternary. Chance differences in climate and geography from one interglacial to another, and loss or gain of biotypes through time, have resulted in different geographic distributions and thus dif-

Table 11-2. Summary of the Holocene Pollen Stratigraphy
of the Eastern United States

Age (years)	Northern New England	Southern New England	Central Appalachians	Coastal Plain	Central Florida
2000	Expansion of boreal elements		Oak, hickory, sweetgum, pine	Pine, oak, hickory	Sand-pine scrub
	Deciduous forest				
5000	Hemlock decline		Hemlock decline		
	Expansion of temperate elements		Oak, hickory, sweetgum, chestnut	Oak, sweetgum, hickory, pine	
	Mixed forest				
10,000		Mixed forest			Oak scrub
	Tundra	Open spruce Woodland	Oak, hornbeam, spruce	Mesic deciduous forest	
12,500					
15,000	Ice	Tundra	Spruce forest with pine and deciduous trees	Pine and spruce forest	
		Ice	Jack pine, spruce		?

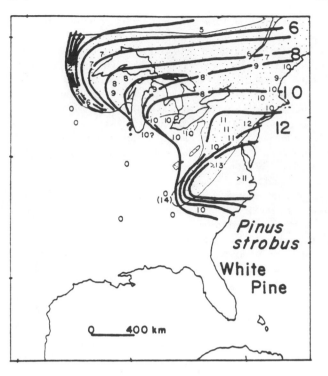

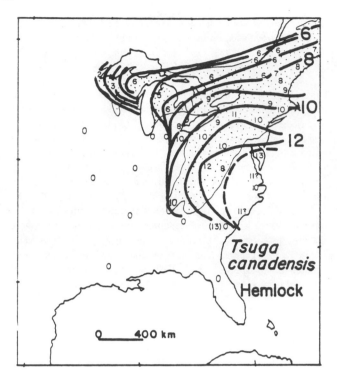

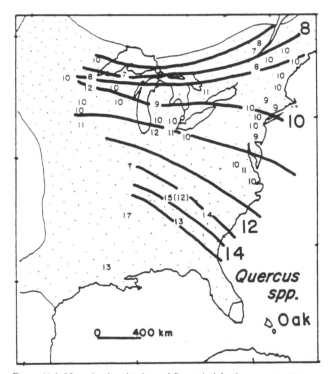

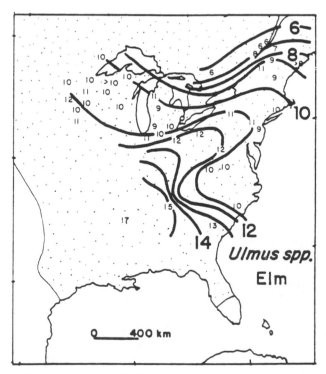

Figure 11-6. Maps showing the times of first arrival for four tree species/genera. (For an explanation, see the caption for Figure 11-4.) (Redrawn from Davis, 1981b.)

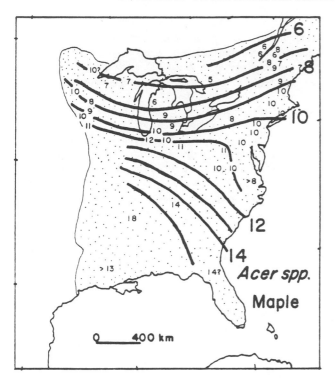

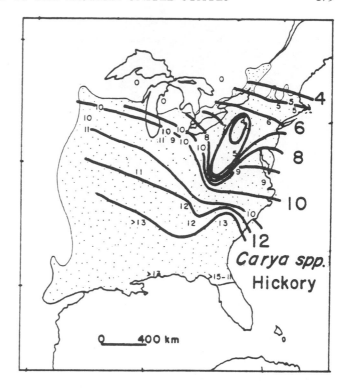

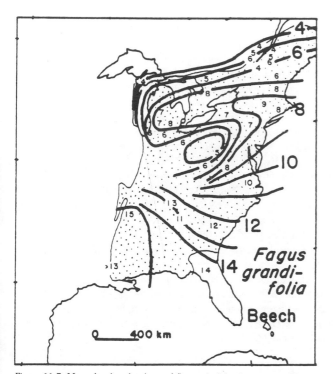

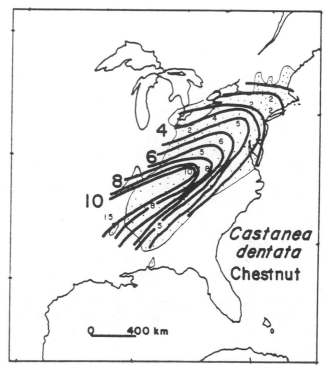

Figure 11-7. Maps showing the times of first arrival for four tree species/genera. (For an explanation, see the caption for Figure 11-4.) (Redrawn from Davis, 1981b.)

ferent communities during each interglacial. This effect can be seen very clearly in Europe, where interglacial floras have been studied intensively (West, 1970).

Climatic Changes

Evidence for climatic change during the Holocene in the northeastern United States is difficult to deduce from changes in pollen abundances (Davis, 1965). Many of the conspicuous stratigraphic changes in pollen percentages, such as the sharp rise of hickory pollen 5000 years ago and the increase in chestnut pollen 2000 years ago (zones C2 and C3 in Connecticut [Deevey, 1939]), reflect the delayed arrival of forest species (M. B. Davis, 1976). These events are time transgressive from south to north, east to west. Another major pollen event is the middle-Holocene decline and subsequent recovery of hemlock, which can be identified throughout the northern half of the region. This appears to be the result of a biotic event rather than a climatic change (Davis, 1965, 1981a). Accordingly, the discussion that follows focuses on changes in distribution as evidence for climatic change, and it places very little emphasis on changes in abundance.

TEMPERATURE CHANGES

The Appalachian Mountains in New York, Vermont, New Hampshire, and Maine have well-developed alpine zones extending from about 1500 m elevation to the summits (maximum elevation 1962 m in New Hampshire). The upper limit of trees is diffuse, with an extensive zone of krummholz composed of balsam fir and black spruce. Winds are exceptionally strong on the mountains of New Hampshire, averaging 60 km per hour at the highest summit. The windiness is largely an orographic effect that should not have changed very much through time, even if the prevailing climate changed. In a study of the history of alpine and subalpine vegetation in New Hampshire, Spear (1981) has found that the effects of wind exposure, snow depth, and atmospheric moisture appear to be more important than the effects of temperature on vegetation at the treeline. His data from pollen and macrofossils show that 10,300 years ago krummholz replaced tundra at elevations where krummholz occurs today. This suggests a climatic regime similar to today's at the beginning of the Holocene. The krummholz was more extensive than now during the first half of the Holocene; the treeline may have risen slightly at that time, but the changes were not extensive.

In contrast, two conifer species now confined to low elevations in the mountains of New Hampshire and New York expanded their ranges greatly during the early Holocene (Davis et al., 1980; Jackson, 1981; Whitehead et al., 1981). White pine is now abundant in New Hampshire only below 450 m, and hemlock is now abundant only below 650 m (Bormann et al., 1970). After white pine invaded the region 9000 years ago, pine needles occurred not only at low elevations, where the tree grows today, but at sites up to 800 m elevation, 350 m above the present-day limit for the species. About 9000 years ago, pine pollen percentages were also much higher than they were later. Because pine is now confined to low elevations, its expansion during the early Holocene suggests higher temperatures and lower precipitation, similar to the conditions prevailing in the mountain valleys today. Oak, which has a distribution similar to that of white pine on the modern landscape of New Hampshire, was also well represented by pollen during this period. Unfortunately, no macro-

fossils of oak have been found. Needles of white pine occurred at high-elevation sites throughout most of the Holocene until the last few hundred years before settlement. Apparently, the climatic changes that ushered in the Little Ice Age (A.D. 1450-1850) were sufficient to cause species ranges to contract downward, limiting them to elevations where we find them today.

Fossil hemlock needles also occur at sites in both New Hampshire and New York well above the present limits for the species. Again, the highest elevations at which they are found in New Hampshire are 350 m above the present limit. At sites up to 1000 m elevation, hemlock needles occur from the time hemlock arrived in the region 7000 years ago until 5000 years ago. The population shrank in size about 5000 years ago, apparently as a result of biotic events. High-elevation populations were never reestablished, although hemlock needles persisted at sites a few meters above their present limit until the time of the Little Ice Age a few hundred years ago.

At the present time, the mean annual temperature at the elevations where the fossil needles occur in New Hampshire is about 2 °C colder than the temperature at the modern limits for white pine and hemlock, and the precipitation is about 125 mm higher (as estimated from lapse rates). The expansion in range suggests that the temperature during the early Holocene was about 2 °C warmer and the precipitation was about 125 mm lower than now. The temperature estimate is similar to estimates for the Holocene of northwestern Europe (Andersson, 1902; Iversen, 1944).

Changes in forest composition throughout the East at around 10,000 yr B.P., including the development of krummholz at the treeline (Spear, 1981), suggest that the opening of the Holocene was marked by a change of climate to essentially modern conditions. Soon afterward, at least by 9000 yr B.P., the climate became warmer than today. Temperatures warmer than present appear to have persisted until the time of the Little Ice Age (A.D. 1450-1850). The occurrence of widespread hemlock, the greater development of krummholz above the limit of forest, and the increased pollen influx for pine and oak at northern sites (Figure 11-3A), however, suggests that the most extreme warmth occurred before 5000 years ago. It is not easy to pinpoint the time this warm phase began. White pine serves as a temperature indicator at montane sites, but prior to 9000 yr B.P. it was absent from the flora. Hemlock also appears to have been absent before 7000 yr B.P., when fossil needles appear at many sites at a wide range of elevations. However, one hemlock needle found in older sediment (Davis et al., 1980) might mean (if contamination can be ruled out) that hemlock was present earlier but was rare because of unfavorable climate. The possible occurrence of several macrofossils from hemlock prior to 7000 yr B.P. in the mountains of New York State may support this view (S. T. Jackson, personal communication, 1982). This interpretation would suggest that temperatures rose gradually during the early Holocene and permitted the expansion of white pine 9000 years ago and the expansion of hemlock 7000 years ago, with the maximum warmth occurring between 7000 and 5000 years ago. The treeline history, however, suggests maximum warmth 9000 years ago.

PRECIPITATION CHANGES

Many forest trees are sensitive to both temperature and moisture, and the effects of the two parameters are difficult to separate. Lapse rates, for example, suggest that the changes in elevation distribution of hemlock and white pine resulted from a drop in precipitation as well as a rise in temperature during the early Holocene.

Charcoal counts provide a means for measuring the frequency of fires. Fires are more frequent in regions of dry, continental climate (such as Minnesota) than in regions of moist, relatively maritime climate (such as New Hampshire). Charcoal was counted at several sites in Minnesota, where it was related to known fire histories. An example is Lake of the Clouds in the Boundary Waters Canoe Area (Heinselman, 1973; Swain, 1973). Charcoal occurs at much lower concentrations in the sediment of Mirror Lake, New Hampshire, indicating a much lower regional fire frequency than northern Minnesota throughout the Late Wisconsin (Davis, 1983) and Holocene. In a core from the center of the lake, charcoal is almost entirely absent, but it is consistently present in a nearshore core of coarser-grained sediment. Even there, where the record is biased toward charcoal, the influx to sediment is only one-tenth the rate measured in a core from the center of the Minnesota lake. Charcoal influx to Mirror Lake changed through time, however. The overall influx was higher between 11,000 and 7000 yr B.P. than it was later in the Holocene, with a peak abundance at about 7500 yr B.P. These data suggest higher fire frequencies when coniferous trees were abundant and when drier, more continental climatic conditions prevailed during the Late Wisconsin and early Holocene (Davis, 1983).

The abundance of ironwood pollen in the early Holocene of New England is compatible with drier climate and higher fire frequency. Ironwood today is characteristic of woodlands in Minnesota along the prairie margin, where frequent fires are an important force controlling the distribution of vegetation (Grimm, 1983). However, Delcourt (1979) interprets high percentages for ironwood pollen in Tennessee sediments 12,000 to 9000 years old as evidence for mesic conditions, at least relative to the present climate of Tennessee, which is much drier and warmer than New Hampshire and much warmer than Minnesota. She points out that ironwood pollen reaches 8% in surface samples from cove hardwood forests in the Appalachian Mountains. The assemblage 12,000 to 9000 years old contains spruce, oak, hickory, sugar maple, and many mesic taxa. It was replaced 9000 years ago by a more xeric assemblage dominated by oak and sweet gum. Delcourt feels that the Late Wisconsin/Holocene transition assemblage represents a diverse deciduous-coniferous forest that was characteristic of the mid-South 12,000 to 9000 years ago (an interval termed Holocene in her paper), whereas contemporaneous forest of the southern Atlantic Coastal Plain and Gulf Coastal Plain were more xeric and were dominated by oak and hickory (Delcourt, 1979). By 5000 yr B.P., forests throughout the South were dominated by oak, hickory, and ash, and mesic genera such as beech and sugar maple were restricted to moist microhabitats such as coves in the Applachian Mountains. Pollen percentages at Rockyhock Bay in North Carolina (Figure 11-5) indicate that these mesic genera were rare on the Coastal Plain (relative to oak). The restriction of mesic genera appears to have occurred 9000 years ago in Tennessee (Delcourt, 1979), whereas it was delayed until 7200 years ago on the Coastal Plain of North Carolina (Whitehead, 1981).

Biotic Events

Biotic events have had a major impact on the forests of the eastern United States. At the present time, Dutch elm disease, caused by an insect-vectored fungal pathogen, is taking a heavy toll on American elm trees. The disease was introduced to Pennsylvania during the 1930s, apparently from Europe. Since then, it has spread throughout the United States. Almost all of the elms planted as ornamentals in American cities have been affected. Losses have also been heavy in natural forest stands.

Even more dramatic was the loss of chestnut trees during the early decades of the 20th century. Chestnut was a far more valuable tree than elm since it was used for timber and nuts, which were harvested commercially. Chestnut was the dominant tree over a large part of the Appalachian Mountains, especially on dry ridge tops, where it grew with oak, commonly making up 50% or more of the stand. *Endothia parasitica*, another insect-vectored fungal parasite, was first observed in New York City in 1904. It spread rapidly throughout the range of chestnut, decimating all the commercially valuable stands by the 1930s. The root stock still persists in many locations, but its sprouts contract the disease by the time they reach 10 to 20 cm in diameter, usually before they are large enough to bear fruit. Except for these sprouts, the species is now virtually absent from the forest communities in which it was once so important. The fall in chestnut pollen percentages is recorded in subsurface sediments at many sites (Anderson, 1974). It is particularly clearly shown at a depth of 25 cm in the pollen diagram from Linsley Pond, Connecticut (Brugam, 1978) (Figure 11-8). Also visible in the diagram is the secondary succession set in motion by the demise of chestnut. At a depth of 20 cm, there is a rise in the pollen percentage of birch, which was increased by seeding into the large openings left by the death of groups of chestnut trees. This is followed (at 15 cm) by an increase in the percentage of oak pollen; oak succeeded birch and eventually occupied most of the space given up by chestnut in this region.

Pollen diagrams from the eastern United States record an earlier abrupt decrease in hemlock pollen closely resembling the chestnut decline (Davis, 1981a). Sediments deposited 4800 radiocarbon years ago all show a rapid decline in both the percentage and the concentration of hemlock pollen (Figure 11-3). The pollen influx decreased by 80% within a 30-year interval at Mirror Lake in northern New Hampshire (M. B. Davis, unpublished data). A similar decline is seen everywhere within what was then the range of hemlock, extending westward from New Brunswick (Mott, 1975) and Quebec (Richard, 1970, 1971, 1973) to central northern Michigan (Brubaker, 1975) and southward to the central Appalachians (Watts, 1979). Macrofossils from hemlock also declined in abundance (Davis et al., 1980). Hemlock continued to be represented in low abundance for about 1000 years, when it began to increase, returning to its former abundance 2000 years after the initial decline (Davis, 1981a).

All sites show a characteristic successional sequence following the hemlock decline. Birch pollen usually shows an immediate sharp increase in both percentage and influx. A short-lived birch maximum is followed by increases in the dominants that grow together with hemlock in the modern forest: beech and maple in western New York and northern New Hampshire; pine and oak in southern New Hampshire (Figure 11-3); pine, oak, and red maple in Massachusetts (Whitehead, 1979); birch, oak, and hickory in Virginia (Watts, 1979); oak in Tennessee (Delcourt, 1979); and white pine in northern Michigan (Davis, 1981a). The regional differences in the secondary succession that follows the hemlock decline demonstrate that hemlock competes with a different spectrum of species in different parts of its range. As hemlock declined, these competitors increased and replaced it. As it expanded again 2000 years later, they were gradually overcome and replaced by hemlock.

It seems unlikely that a climatic change could explain the unilateral decrease in abundance for one species over such a wide geographic

LINSLEY POND, CT

POLLEN PERCENTAGE DIAGRAM

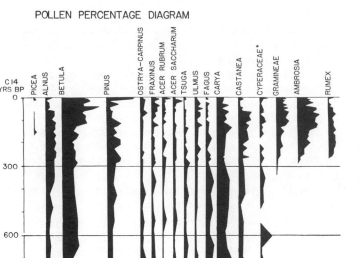

Figure 11-8. Pollen diagram from near-surface sediments from Linsley Pond, Connecticut, showing changes associated with human disturbance in the vicinity of the site. (From Brugam, 1978.)

and climatic range. In view of the similarity of the pollen stratigraphy to the changes that record the chestnut blight, it seems possible that the hemlock decline also represents the outbreak of a fungal pathogen. An insect pest also may have been responsible, but that seems less likely because the decline is a single event rather than a series of periodic outbreaks and recoveries, as is often the case with losses caused by forest insects. There seems to have been a single sharp decline, nearly synchronous throughout the range of hemlock, followed by a long period when the hemlock population was held low. There was then an expansion, presumably after some kind of resistance to the pathogen (or insect pest) had developed through evolutionary processes. Still unexplained is the origin of the putative pathogen. Both western hemlock (*Tsuga heterophylla*) and eastern hemlock are attacked by the hemlock looper (*Lamdina fiscellaria*), but neither species is affected seriously by fungi at the present time. It is conceivable that the forest trees that survived as small populations in refuges during the long Wisconsin glacial period lost some of their insect and fungal predators and that the rapid Holocene expansion of populations involved some that were free of endemic pathogens and lacked defenses. The later expansion of the pathogen, which might have survived in other refuges, could have resulted in the massive die-off of the host. A simpler hypothesis involves the origin 5000 years ago of a lethal strain of pathogen through evolutionary processes. Neither hypothesis is easily testable.

Late-Holocene Events

Spruce pollen shows an increase in abundance at most northeastern and Appalachian sites during the late Holocene. After declining 10,000 years ago at low and middle elevations throughout the East and after declining 9000 years ago at the treeline in the high mountains of the

Northeast, spruce seems to have remained rare at all elevations for about 8000 years (Davis et al., 1980; Spear, 1981). About 2000 years ago, spruce began to increase again in the coniferous forests at 750 to 1000 m elevation in the mountains at the expense of balsam fir and in competition with hardwood species below 750 m elevation. Farther south, in Massachusetts and northern Connecticut, where the resurgence of spruce pollen is of lesser magnitude (Deevey, 1943), spruce may have increased locally in bogs. At Rogers Lake in southern Connecticut, the reappearance is hardly noticable; it cannot be detected farther south along the Coastal Plain or in Tennessee (Delcourt, 1979). In the central Appalachians, the resurgence of spruce is more conspicuous. At Cranberry Glades in West Virginia (elevation 1029 m) and Tannersville Bog in Pennsylvania (277 m), the return of spruce shows clearly, with pollen percentages reaching from 2% to 5% of the total (Watts, 1979). The rise begins about 5000 years ago and presumably represents the growth of *Picea rubens* in bogs (Watts, 1979). (See the pollen diagrams in Watts, 1983). Maps of Holocene pollen percentages show increasing spruce percentages over a wide region in the northern United States, starting in the North at least 4000 years ago and moving southward until the present (Bernabo and Webb, 1977). More detailed maps including data from eastern Canada are now being processed (Webb, 1981; Webb et al., 1983).

The meaning of the late-Holocene increase in abundance of spruce is, of course, very different from the meaning of the late-glacial spruce maximum. During the late Holocene, red spruce is the principal species involved. It has increased on well-developed soils in competition with established forest trees. Deevey and Flint (1957) regard the return of spruce as a stratigraphic marker for the end of the Holocene hypsithermal interval. I agree that it has been caused by a decrease in temperature or an increase in moisture or by both.

R. B. Davis (1976) has called attention to a series of late-Holocene changes in pollen abundances. In some regions of northern New England, the resurgence of spruce pollen within the last two millennia has been accompanied by an increase in fir and alder. At the same time, pollen of beech and hemlock decline in abundance. R. B. Davis (1976) identifies these changes as a "boreal trend" that has continued into the period of European settlement. Swain (1980) notes similar changes in Maine lakes. His charcoal counts indicate that the incidence of fires has increased during the last 600 years, and he believes the decline in shade-tolerant tree species (hemlock, beech) and the increase in charcoal denotes cooler and drier conditions. Other sites across the northern United States show evidence for increased moisture (Swain, 1980). Futyma (1980) has found evidence for a regional rise in the water table beginning as early as 6000 to 5000 years ago in northern Michigan.

SETTLEMENT HORIZON

Changes in pollen percentages caused by European settlement during the 17th and 18th centuries have been studied in detail at several sites in the East (among others Brugam, 1978; R. B. Davis, 1967; R. E. Moeller, unpublished data). At Linsley Pond, Connecticut (Figure 11-8), the pollen stratigraphy can be correlated with changes in the mineral content of the sediment caused by local human disturbance. As a result, the pollen changes can be dated from historical records. The sediments have also been dated radiometrically (Brugam, 1978). The establishment of a farm on the shore of the pond in A.D. 1700 caused changes in the pollen percentages, most notably a decrease in hemlock and an increase in herbs such as grass, *Rumex*, and *Am-*

brosia. Plantago lanceolata pollen appears in more-recent sediment, but in general this species is poorly represented in American material. *Zea* is also poorly represented; Brugam (1978) reports finding only two grains in the entire profile. The landscape near Linsley Pond was almost completely deforested during the peak of agricultural activity in the early 19th century. At that time, herb pollen percentages were 35% of the total tree pollen. Successional forests that have developed in the last 100 years as farms have been abandoned are reflected by increased percentages of birch and red maple pollen. The very recent peak in pine pollen is from recent plantings. The decline in chestnut pollen at a depth of 25 cm reflects the chestnut blight, which reached this region in A.D. 1913. The rapid sedimentation in Linsley Pond and the paucity of bottom-dwelling fauna have left an exceptionally clear record of recent changes in pollen input.

Culture horizons have been described from other parts of the eastern United States (Davis et al., 1983; Delcourt, 1979; Maxwell and Davis, 1972; R. E. Moeller, unpublished data; and many others). The increased frequency of ragweed pollen is the universal indicator of agriculture. In regions where agriculture was less intensive, the rise in herbs from local open land may be diluted by pollen blown into the region from agricultural land elsewhere. Even lakes from relatively undisturbed environments above the treeline in the mountains of New Hampshire, for example, show a clear rise in herb pollen, especially ragweed, near the top of the profile, indicating that pollen can be carried from the valleys up to the mountain summits. Ragweed pollen might have been blown into the eastern United States from agricultural areas in the Midwest (R. E. Moeller, personal communication, 1981). The presence of significant numbers of ragweed pollen grains in subsurface sediments in Greenland indicates that very long distance transport is possible (Fredskild, 1980). In southern Ontario, however, ragweed pollen appears for the first time at the European settlement horizon; this fact suggests that the plant was introduced by early farmers (McAndrews, 1976). Ragweed seeds have been found in 3000-year-old Connecticut River floodplain sediments (B. Casper, unpublished data), however, indicating that ragweed is native to Connecticut.

Many native Indians occupied the eastern United States at the time of European colonization. Indians cultivated *Zea mays*, legumes, and squash and maintained fairly extensive fields in some areas. Although they did not keep cattle and so may not have had an impact on the landscape comparable to that of prehistoric peoples in Europe, they did use fire very extensively to modify the forests in order to clear underbrush and to improve the fodder for deer (Day, 1953). The impact of prehistoric burning has not yet been detected in the pollen record. Changes in the frequency of charcoal in sediment have been attributed to climatic changes rather than to human influences (Swain, 1980). McAndrew's (1976) remarkable pollen diagram from Crawford Lake, Ontario, remains a unique example of pollen evidence for prehistoric settlement, or "landnam" (Iverson, 1941), which has not been replicated in the United States. However, ecologists have speculated that forest communities at the time of European settlement had already been modified by prehistoric burning over the course of many centuries (Niering and Goodwin, 1962). This hypothesis can be tested by pollen analysis, charcoal counts, and archaeological investigations.

Conclusions

The region discussed here includes latitudes 25° to 47° N and comprises the Coastal Plain, the Piedmont, and the Appalachian Mountains. It includes regions of sedimentary, metamorphic, and igneous bedrock. The relief ranges from near sea level in parts of Florida to 2000 m in the northern mountains. The climate ranges from tropical to cool temperate, with alpine conditions on mountain summits. The climate of different parts of the region is controlled by different weather systems, and it may have changed differentially in the past (Wahl, 1968). The South is affected predominantly by tropical air masses moving north from the Gulf of Mexico, and the North is affected both by air masses traveling up the Atlantic coast and by colder air moving in from the Northwest.

Nevertheless, certain general features of the pollen stratigraphy stand out as common features for the entire region (Table 11-2). (1) At the time the ice began to retreat 16,000 to 15,000 years ago, forests of spruce and pine in the South were replaced by deciduous trees. (2) Around 12,000 years ago, spruce woodland replaced tundra in western Maryland, western New York, and southern New England. The appearance of spruce is time transgressive from south to north in New England, reflecting the migratory speed of spruce, at least within small areas. However, the nearly coincidental timing of the development of spruce woodland over a large area suggests that a climatic amelioration 12,000 years ago allowed spruce to grow in regions where it was previously limited by climate. Changes 12,000 years ago also occurred much farther south outside the glaciated region. Oak scrub developed in south-central Florida, where the full-glacial vegetation is still unknown (Watts, 1982). Mesic deciduous trees replaced jack pine along the southern Coastal Plain; this suggests greater moisture and perhaps elevated temperature. In Tennessee, a mixed forest of spruce, oak, ironwood, and other deciduous elements replaced a forest of spruce, pine, fir, and oak; this suggests a rise in temperature. (3) About 10,000 years ago, forests of variable composition developed in the North, and the forests underwent a series of changes as new species migrated northward. A series of changes in species composition also occurred in forests in many parts of the South. Boreal species were replaced by deciduous forest on the central Coastal Plain, while farther south mesic trees were replaced by species that now occur in greater abundance in warmer regions or in drier habitats. (4) Five thousand years ago, vegetation throughout the South changed from oak forest to forest dominated by pine. Even in Tennessee, where pine was never abundant, pine pollen percentages began to increase at that time. In the North, a number of indicators of warmth and dryness declined in abundance or contracted in range 5000 years ago; this suggests that the warmest and driest part of the Holocene had ended, although temperatures 5000 years ago were still higher than they are today. About 2000 years ago, although the time varied (from 5000 to 1000 years ago) from region to region, boreal elements began to increase in abundance; this suggests cooling. The trend has continued to the present.

A useful stratigraphic marker is provided by the sharp drop in hemlock pollen, which is synchronous within the error of radiocarbon dating at 4800 yr B.P.

Much of the vegetational history of the eastern United States involves the expansion northward of boreal and temperate trees. The individualistic manner in which trees expanded into new territory and the different directions and rates of movement affected the flora available for colonization or for expansion when climatic changes occurred. As a result, a variety of forest communities developed, many of them ephemeral (Whitehead, 1981). Many changes in abundance are time transgressive from one region to another. Presumably, herbs and shrubs now recognized as "characteristic" of particular deciduous forest communities were also individualistic in their history. Certain herbaceous species (e.g., xeric herbs) have probably existed in the southeastern United States through many glacial and interglacial cycles; while other species, especially mesic herbs that require shady, dark forest environments with humic soils, have been displaced many times. Some species may have persisted in very small populations in particular habitats, from which they have expanded slightly during the interglacial or glacial periods. Others may have undergone the extraordinary geographic displacements documented for most of the boreal or temperate trees. The flora of the eastern United States is a fortuitous grab bag of species, the survivors of 16 to 18 glacial-interglacial cycles. Each interglacial probably has been as complex in its development as the Holocene. Although biogeographers have long noted the richness of the flora of the eastern United States by comparison to Europe (Watts, 1982), it can be supposed that many genetic lineages have been lost through extinction at times of low population size or through failure to expand quickly enough to exploit favorable habitats as they became available.

The southern Appalachian Mountains contain the most diverse tree flora of the eastern region. So many species and genera occur there that the region was long considered a center from which the flora had spread out to colonize other areas, a refugium that had maintained species richness through a long history of stable climate during which evolutionary processes had caused diversification (Watts, 1982). It now appears that the southern Appalachian region has received almost all of its arboreal flora within the last 15,000 years. However, the southern Appalachians obtained their complement of deciduous trees thousands of years earlier then the northern Appalachians because of the southern mountains' proximity to refuge areas. In this sense, a time hypothesis is relevant to diversity after all. Perhaps the fossil record, when it is explored in greater detail, may indicate how many plant species are still moving northward to occupy their potential habitats at this time, near the end of the Holocene interglacial.

References

Andersen, S. T. (1954). A late-glacial pollen diagram from southern Michigan, U.S.A. *Danmarks Geologiske Undersögelse* 2 (80), 140-55.

Anderson, T. W. (1974). The chestnut pollen decline as a time horizon in lake sediments in eastern North America. *Canadian Journal of Earth Sciences* 11, 678-85.

Andersson, G. (1902). Hasseln i Sverige dordom och nil: 1902. *Sveriges Geologiska Undersokning* Series Ca, 3, 1-168.

Argus, G. W., and Davis, M. B. (1962). Macrofossils from a late-glacial deposit at Cambridge, Massachusetts. *American Midland Naturalist* 67, 106-17.

Bernabo, J. C., and Webb, T., III (1977). Changing patterns in the Holocene pollen record from northeastern North America: A mapped summary. *Quaternary Research* 8, 64-96.

Birks, H. J. B., and Peglar, S. M. (1980). Identification of *Picea* pollen of late Quaternary age in eastern North America: A numerical approach. *Canadian Journal of Botany* 58, 2043-58.

Bormann, F. H., Siccama, T. G., Likens, G. E., and Whittaker, R. H. (1970). The Hubbard Brook ecosystem study: Composition and dynamics of the tree stratum. *Ecological Monographs* 40, 373-88.

Braun, E. L. (1950). "Deciduous Forests of Eastern North America." Blakeston Company, Philadelphia.

Brubaker, L. B. (1975). Postglacial forest patterns associated with till and outwash in northcentral upper Michigan. *Quaternary Research* 5, 499-527.

Brugam, R. B. (1978). Pollen indicators of land-use change in southern Connecticut. *Quaternary Research* 9, 349-62.

Brush, G. S., Lenk, C., and Smith, J. (1980). The natural forests of Maryland: An explanation of the vegetation map of Maryland. *Ecological Monographs* 50, 77-92.

Craig, A. J. (1969). Vegetational History of the Shenandoah Valley, Virginia. Geological Society of America Special Paper 123, pp. 283-96.

Davis, M. B. (1958). Three pollen diagrams from central Massachusetts. *American Journal of Science* 256, 540-70.

Davis, M. B. (1961). Pollen diagrams as evidence of late-glacial climatic change in southern New England. *New York Academy of Science Annals* 95, 623-31.

Davis, M. B. (1965). Phytogeography and palynology of northeastern United States. *In* "The Quaternary of the United States" (H. E. Wright, Jr., and D. G. Frey, eds.), pp. 377-401. Princeton University Press, Princeton, N.J.

Davis, M. B. (1967). Late-glacial climate in northern United States: A comparison of New England and the Great Lakes region. *In* "Quaternary Paleoecology" (E. J. Cushing and H. E. Wright, Jr., eds.), pp. 11-44. Yale University Press, New Haven, Conn.

Davis, M. B. (1976). Pleistocene biogeography of temperate deciduous forests. *Geoscience and Man* 13, 13-26.

Davis, M. B. (1981a). Outbreaks of forest pathogens in Quaternary history. *In* "Proceedings of the Fourth International Palynological Conference," vol. 3, pp. 216-27. Lucknow, India, 1976-1977.

Davis, M. B. (1981b). Quaternary history and the stability of deciduous forests. *In* "Forest Succession" (D. C. West, H. H. Shugart and D. B. Botkin, eds.), pp. 132-77. Springer-Verlag, New York.

Davis, M. B. (1983). History of the vegetation on the watershed. *In* "An Ecosystem Approach to Aquatic Ecology: Mirror Lake and its Watershed" (G. E. Likens, ed.). Springer-Verlag, New York, in press.

Davis, M. B., and Ford, M. S. (1981). Sediment focusing in Mirror Lake, New Hampshire. *Limnology and Oceanography* 27, 137-50.

Davis, M. B., Heathcote, I. W., and Ogg, J. A. (1983). Surface and presettlement pollen assemblages from New England. *Canadian Journal of Botany* (in press).

Davis, M. B., Spear, R. W., and Shane, L. C. K. (1980). Holocene climate of New England. *Quaternary Research* 14, 240-50.

Davis, R. B. (1967). Pollen studies of near-surface sediments in Maine Lakes. *In* "Quaternary Paleoecology" (E. J. Cushing and H. E. Wright, Jr., eds.), pp. 144-73. Yale University Press, New Haven, Conn.

Davis, R. B. (1976). Late and postglacial vegetational history of northern New England. *American Quaternary Association Abstracts*, p. 136.

Davis, R. B., Bradstreet, T. E., Stuckenrath, R., Jr., and Borns, H. W., Jr. (1975). Vegetation and associated environments during the past 14,000 years near Moulton Pond, Maine. *Quaternary Research* 5, 435-65.

Day, G. M. (1953). The Indian as an ecological factor in the northeastern forest. *Ecology* 34, 329-46.

Deevey, E. S., Jr. (1939). Studies on Connecticut lake sediments: I. A post-glacial climatic chronology for southern New England. *American Journal of Science* 237, 691-724.

Deevey, E. S., Jr. (1943). Additional pollen analyses from southern New England. *American Journal of Science* 241, 717-52.

Deevey, E. S., Jr., and Flint, R. F. (1957). Postglacial hypsithermal interval. *Science* 125, 182-84.

Delcourt, H. R. (1979). Late Quaternary vegetation history of the eastern highland rim and adjacent Cumberland Plateau of Tennessee. *Ecological Monographs* 49, 255-80.

Delcourt, P. A., Delcourt, H. R., Brister, R. C., and Lackey, L. E. (1980). Quaternary vegetation history of the Mississippi Embayment. *Quaternary Research* 13, 111-32.

Fernald, M. L. (1925). Persistence of plants in unglaciated areas of boreal America. *American Academy of Arts and Sciences Memoir* 15, 238-42.

Fernald, M. L. (1950). "Gray's Manual of Botany." 8th ed. American Book Company, New York.

Fredskild, B. (1980). The Holocene vegetational history of Greenland. *In* "Fifth International Palynological Conference Abstracts," p. 139. Cambridge University, Cambridge.

Futyma, R. P. (1980). Postglacial vegetation history of eastern upper Michigan, USA.

In "Fifth International Conference Abstracts," p. 144. Cambridge University, Cambridge.

Gleason, H. A. (1939). The individualistic concept of the plant association. *American Midland Naturalist* 21, 92-110.

Goldthwait, R. P. (1976). Past climates on "The Hill": Part 1. When glaciers were here. Part 2. Permafrost fluctuations. *Mt. Washington Observatory Bulletin* March 12-16; June, 38-41.

Goodlett, J. C. (1954). "Vegetation Adjacent to the Border of the Wisconsin Drift in Potter County, Pennsylvania." Harvard Forest Bulletin 25.

Grimm, E. C. (1983). Chronology and dynamics of vegetation change in the prairie-woodland region of southern Minnesota, U.S.A. *New Phytologist* 93, 311-50.

Hack, J. T., and Goodlett, J. C. (1960). "Geomorphology and Forest Ecology of a Mountain Region in the Central Appalachians." U.S. Geological Survey Professional Paper 347, pp. 1-66.

Heinselman, M. L. (1973). Fire in the virgin forests of the Boundary Waters Canoe Area, Minnesota. *Quaternary Research* 3, 329-82.

Iversen, J. (1941). Landnam i Danmarks stenalder (Land occupation in Denmark's Stone Age). *Danmarks Geologiske Undersogelse* 2 (66), 1-68.

Iversen, J. (1944). *Viscum, Hedera,* and *Ilex* as climatic indicators. *Geologiska Forenigens i Stockholm* Förhandlingar 66, 403-83.

Jackson, S. T. (1981). Late-glacial and postglacial local vegetational changes in the Adirondack Mountains, New York: The macrofossil record. *Ecological Society of America Bulletin* 62, 126.

Kapp, R. O. (1977). Late Pleistocene and postglacial plant communities of the Great Lakes region. *In* "Geobotany" (R. C. Romans, ed.), pp. 1-27. Plenum Publishing, New York.

Küchler, A. W. (1964). "Potential Natural Vegetation of the Conterminous United States (Map and Manual)." American Geographical Society Special Publication 36.

Likens, G. E., Bormann, F. H., Pierce, R. S., Eaton, J. S., and Johnson, N. M. (1977). "Biogeochemistry of a Forested Ecosystem" Springer-Verlag, New York, N.Y.

McAndrews, J. H. (1976). Fossil history of man's impact on the Canadian flora: An example from southern Ontario. *Canadian Botanical Association Bulletin* 9, 1-6.

Martin, P. S. (1958). Pleistocene ecology and biogeography of North America. *In* "Zoogeography" (C. Hubbs, ed.), pp. 375-420. American Association for the Advancement of Science Publication 15.

Maxwell, J. A., and Davis, M. B. (1972). Pollen evidence of Pleistocene and Holocene vegetation on the Allegheny Plateau, Maryland. *Quaternary Research* 2, 506-30.

Mengel, R. M. (1964). The probable history of species formation in some northern wood warblers (Parulidae). *Living Bird* 3, 9-43.

Miller, N. G. (1973). "Late-Glacial and Postglacial Vegetation Change in Southwestern New York State." New York State Museum of Science Service Bulletin.

Miller, N. G., and Thompson, G. G. (1979). Boreal and western North American plants in the late Pleistocene of Vermont. *Journal of the Arnold Arboretum* 60, 167-218.

Mott, R. J. (1975). Palynological studies of the lake sediment profiles from southwestern New Brunswick. *Canadian Journal of Earth Sciences* 12, 273-88.

Mott, R. J. (1978). *Populus* in late-Pleistocene pollen spectra. *Canadian Journal of Botany* 56, 1021-31.

Niering, W. A., and Goodwin, R. H. (1962). Ecological studies in the Connecticut Arboretum Natural Area: Introduction and a survey of vegetation types. *Ecology* 43, 41-54.

Richard, P. (1970). Atlas pollinique des arbres et de quelque arbustes indigénes du Québec: I. Introduction generale. II. Gymnospermes. *Naturaliste Canadien* 97, 1-34.

Richard, P. (1971). Two pollen diagrams from the Quebec City area, Canada. *Pollen et Spores* 13, 523-59.

Richard, P. (1973). Histoire postglaciare de la végétation dans la région du Saint-Raymond de Portneuf, telle que révélée par l'analyse pollinique d'une tourbiere. *Naturaliste Canadien* 100, 561-75.

Richard, P. (1975). Contribution à l'histoire postglaciare de la végétation dans les Catons-de-l'Est: Etude des sites de Weedon et Albion. *Cahiers de Geographie de Quebec* 19, 267-84.

Siccama, T. G. (1974). Vegetation, soil and climate on the Green Mountains of Vermont. *Ecological Monographs* 44, 325-49.

Solomon, A. M., West, D. C., and Solomon, J. A. (1981). The role of climate change and species immigration in forest succession. *In* "Forest Succession" (D. C. West, H. H. Shugart and D. B. Botkin, eds.), pp. 154-77. Springer-Verlag, New York.

Spear, R. W. (1981). History of alpine and subalpine vegetation in the White Mountains, New Hampshire. Ph.D. dissertation, University of Minnesota, Minneapolis.

Spear, R. W., and Miller, N. G. (1976). A radiocarbon pollen diagram from the Allegheny Plateau of New York State. *Journal of the Arnold Arboretum* 57, 369-403.

Swain, A. M. (1973). A history of fire and vegetation in northeastern Minnesota as recorded in lake sediments. *Quaternary Research* 3, 383-96.

Swain, A. M. (1980). Environmental changes between 600-100 B.P. in mid-latitudes of North America: Pollen and charcoal evidence from annually laminated lake sediments. *In* "Fifth International Palynological Conference Abstracts, p. 381. Cambridge University, Cambridge.

Wahl, E. W. (1968). A comparison of the climate of eastern United States during the 1830's with the current normals. *Monthly Weather Review* 96, 73-82.

Watts, W. A. (1979). Late Quaternary vegetation of central Appalachia and the New Jersey Coastal Plain. *Ecological Monographs* 49, 427-69.

Watts, W. A. (1980). Late Quaternary vegetation history at White Pond on the inner Coastal Plain of South Carolina. *Quaternary Research* 13, 187-99.

Watts, W. A. (1983). Vegetational history of the eastern United States 25,000 to 10,000 years ago. *In* "Late-Quaternary Environments of the United States," vol. 1. "The Late Pleistocene" (S. C. Porter, ed.), pp. 294-310. University of Minnesota Press, Minneapolis.

Webb, T., III (1981). The past 11,000 years of vegetational change in eastern North America. *Bioscience* 31, 501-6.

Webb, T., III, Richard, P., and Mott, R. J. (1983). Holocene pollen maps from southern Quebec. *In* "Climatic Change in Canada" (C. R. Harrington, ed.), vol. 3, *in press*. Syllogeus, National Museums of Canada, Ottawa.

West, R. G. (1970). Pleistocene history of the British flora. *In* "Studies in the Vegetational History of the British Isles" (D. Walker and R. G. West, eds.), pp. 1-11. Cambridge University Press, New York.

Whitehead, D. R. (1981). Late-Pleistocene vegetational changes in northeastern North Carolina. *Ecology* 51, 451-71.

Whitehead, D. R., Sheehan, M. C., and Jackson, S. T. (1981). Late-glacial and postglacial regional vegetational changes in the Adirondack Mountains, New York: The pollen record. *Ecological Society of America Bulletin* 62, 126.

Whittaker, R. H. (1956). Vegetation of the Great Smokey Mountains. *Ecological Monographs* 26, 1-80.

CHAPTER 12

Holocene Mammalian Biogeography and Climatic Change in the Eastern and Central United States

Holmes A. Semken, Jr.

Introduction

Paleontologic, as opposed to archeological, localities representing the last 10,000 years seem relatively scarce when the abundance of literature is used as a measure. Kurtén and Anderson (1980: 39) record 15 early-Holocene paleontologic local faunas "that have yielded information on the appearance or extinction of species," compared to 149 significant Wisconsin local faunas. Although several Holocene sites have been reported since that book was prepared, the ratio of Holocene paleontologic sites to Wisconsin paleontologic sites has not been altered.

The paucity of paleontologic sites assigned to the Holocene is not a result of their scarcity; it reflects a lack of interest on the part of vertebrate paleontologists. Their attitude can be traced at least to Osborn (1910), who divided the Quaternary into four faunal zones. The latest zone, or *Cervus* Zone, which Osborn (1910: 440) related to the Holocene, is defined as the prehistoric (archaeological) fauna of the forests, prairies, great plains, and arid regions of the United States and includes all of the mammals reported by early settlers. Because vertebrate paleontologists are interested primarily in extinct fauna, evolution, taxonomic relationships, or biostratigraphy, those fossils representative of extant species and communities are rarely studied. Also, Holocene sites often contain archaeological material, and paleontologists are reluctant to disturb potential cultural associations; archaeologists, on the other hand, do not excavate a site unless artifacts are present. Thus, Holocene vertebrate research has been confined primarily to fossils recovered from archaeological sites. The major publications on Holocene bison, deer, and wapiti over the last three decades have been prepared by archaeologists (e.g., Davis and Wilson, 1978). A third factor relates to the problem of identifying Holocene sites: a local fauna with an extinct species is automatically assigned to the Pleistocene, where as one with modern taxa only is assigned to the Holocene, a radiocarbon date being unnecessary by biostratigraphic definition except for studies dealing with the chronology of extinction.

Archaeological sites are a potential reservoir of information on Holocene vertebrates, but archaeologists paid little attention to associated fauna until Theodore White (1952a, 1952b, 1953a, 1953b,

1954, 1955, 1956) discussed problems of species abundance, dietary composition, and butchering techniques as inferred from zoo-archaeological collections. After White's contributions, "Faunal reports began to appear with increasing regularity, though the organization and quality of data presented were highly variable, usually consisting of a brief narrative summary or a species list" (Falk, 1977: 153). A survey of published faunal lists associated with archaeological sites in North Dakota, South Dakota, and Iowa (Semken and Falk, 1980) reveals 59 localities with 11 or more associated mammals. Of these lists, only 22 relatively recent ones include identifications of the more ecologically significant rodent and insectivore remains. In contrast, site reports with counts of major subsistence-base animals (e.g., bison, deer) number in the hundreds. The number of sites with extensive faunal lists may be more abundant in the northern Plains than in other areas because of White's pioneering efforts in this region. However, Cleland (1965) reviews 13 sites from northwestern Arkansas alone, and Purdue and Styles (in preparation) catalogue 45 significant sites from Illinois and Missouri. Although these numbers may not be representative of those for all regions of the United States, a review of two zooarchaeological bibliographies (Bogan and Robison, 1978; Lyman, 1979) demonstrates that the potential for obtaining faunal data from archaeological sites is immense. Lyman (1979) cites 2318 titles for the United States, and Bogan and Robison (1978) note approximately 1400 for eastern North America. Both papers reference applicable neontologic literature (skull keys, etc.) as well as

Excellent critical reviews of the manuscript for this chapter were provided by John E. Guilday, Carnegie Museum of Natural History; Herbert E. Wright, Jr., University of Minnesota; Russell W. Graham, Illinois State Museum; and Ernest L. Lundelius, Jr., University of Texas. In addition, Guilday, Graham, Carl R. Falk (University of North Dakota), and John Ludwickson (University of Nebraska) supplied unpublished material for inclusion in the manuscript. Marsha Satorius-Fox deserves special recognition for the uncounted hours she spent searching library stacks for faunal lists and for reviewing this manuscript during its preparation. The figures were prepared by Joyce E. Chrisinger, and the manuscript was typed by George Fisher, Debra Curnan, Ellen Haman, and Connie Reasoner. Partial funding for the preparation of the illustrations was provided both by the University of Iowa Graduate College and the Mobile Foundation Fund to the Department of Geology, University of Iowa.

zooarchaeological papers, but it is clear that a definitive review of Holocene mammalian biogeography is beyond the limit of this chapter. This review, therefore, is organized to illustrate selected aspects and sites (Figure 12-1) of the Holocene mammalian record that exemplify interpretations of Holocene stratigraphy, paleoecology, and biogeography. Examples are biased toward the north-central Plains region of the United States because of the author's experience.

ZOOARCHAEOLOGICAL SAMPLES

Both archaeologists and paleontologists view faunal remains from archaeological sites with varying degrees of skepticism. Many regard zooarchaeological material as so strongly biased by the "cultural filter" that the zoologic and paleoecologic significance of the remains is minimal. Others attribute the small-mammal component (rodents and insectivores) of an archaeological site to burrowing activity and assume that it is an admixture accumulating from the time of occupation until the present. There is no question that some sites have been bioturbated by rodents and that the period of faunal accumulation is questionable. The distribution of small mammals within other sites, however, suggests that postoccupational burrowing is not a major factor, that the association is contemporaneous with occupation, and that the remains represent a relatively precise interval of time (Semken and Foley, 1979).

The effect of the "cultural filter" on small mammals associated with archaeological sites is difficult to evaluate. A zooarchaeological sample, just as a paleontologic sample, is biased by a variety of factors. The type of site is a major factor. A temporary camp would have little effect on the local biota, whereas a village with permanent structures and agriculture would have a major impact on the surroundings (Guilday and Parmalee, 1965). A permanent village creates an artificial environment and generates a micromammalian community with proportions different from those seen in natural situations.

Another bias is introduced by the subsistence value of an animal. Bison, deer, and wapiti were selected for food, and large numbers of

Figure 12-1. Localities of Holocene mammal remains in the eastern and central United States discussed in the text. *Northeastern Forest:* (1) Lamoka Lake (Guilday, 1965); (2) Kipp Island (Guilday and Tanner, 1965); (3) New Paris Sinkholes No. 2 and 4, Unit A (Guilday, et al., 1964); (4) Bootlegger Sink (Guilday et al., 1966); (5) Sheep Rock Shelter (Guilday and Parmalee, 1965); (6) Hosterman's Pit (Guilday, 1967a); (7) Meadowcroft Rockshelter (Guilday et al., in press); (8) Chesser Cave (Cleland and Kearney, 1967); (9) Blain Village (Parmalee and Shane, 1970); (10) Eau Claire (Wilson, 1967); (11) Lenawee (Wilson, 1967); (12) Sleeping Bear (Pruitt, 1954); (13) Milwaukee (West and Dallman, 1980); (14) Raddatz Rockshelter (Cleland, 1966); (15) Lost River Sink (Theiling, 1973); (16) Meyer Cave (Parmalee, 1967); (17) Koster (Butzer, 1977); and (18) Modoc Rockshelter (Fowler, 1959). *Northern Plains:* (19) Callahan-Thompson (Lewis, 1974); (20) Zebree (Guilday and Parmalee, 1971); (21) Peccary Cave (Semken, 1969); (22) Rodgers Shelter (Parmalee et al., 1976); (23) Brynjulfson Cave No. 2 (Parmalee and Oesch, 1972); (24) Tick Creek Cave (Parmalee, 1965); (25) Arnold Research Cave (Falk, 1970); (26) Graham Cave (Klippel, 1971); (27) Mud Creek (Kramer, 1972); (28) Willard Cave (Eshelman, 1971); (29) Hadfield's Cave (Benn, 1980); (30) Garrett Farm (Fay, 1980); (31) Pleasant Ridge (Fay, 1980); (32) Cherokee Sewer Site (Semken, 1980); (33) Itasca (Shay, 1971); (34) Helb (Falk and Calabrese, 1973); (35) Lower Grand (Semken, 1976); (36) Walth Bay (Ahler et al., 1974); (37) White Buffalo Robe (Semken and Satorius, 1980); (38) Hudson-Meng (Agenbroad, 1977); (39) Agate Basin (Walker, 1982); (40) Casper (Frison, 1974); and (41) Medicine Lodge Creek (Walker, 1975). *Southeastern Forest:* (42) Nichol's Hammock (Hirschfield, 1968); (43) Little Salt Spring (Clausen et al., 1979); (44) Vero (Weigel, 1962); (45) Melbourne (Sellards, 1952); (46) Devil's Den (Martin and Webb, 1974); (47) Russell Cave (Weigel et al., 1974); (48) Stanfield-Worley Bluff Shelters (Parmalee, 1963a); (49) Baker Bluff Cave (Guilday et al., 1978); (50) Salts Cave (Watson, 1969); and (51) Avery Island (Gagliano, 1967). *Southern Plains:* (52) Ben Franklin (Slaughter and Hoover, 1963); (53) Friesenhahn Cave (Graham, 1976b); (54) Longhorn Cavern (Semken, 1961); (55) Wunderlich (Lundelius, 1967); (56) Kline Cave (Roth, 1972); (57) Rattlesnake Cave (Semken, 1967); (58) Miller's Cave (Patton, 1963); (59) Schultze Cave (Dalquest et al., 1969); (60) Centipede (Lundelius, 1963); (61) Damp Cave (Lundelius, 1963); (62) Bonfire Shelter (Frank, 1968); (63) Dye Creek (Dalquest, 1965); (64) Lubbock Lake (Johnson and Holliday, 1979); (65) Rex Rodgers site (Schultz, 1978); (66) Deadman's Shelter (Schultz and Rawn, 1978); (67) Kyle (Lundelius, 1962); and (68) Birch Creek (Henry et al., 1979). *Palynologic Sites:* (a) West Lake Okoboji (Baker and Van Zant, 1980), (b) Old Field (King and Allen, 1977), and (c) Anderson Pond (Delcourt, 1980).

these animals in a site could be disproportionate to their relative biomass in the immediate area of the site if they were secured during extended foraging activities. Small animals also were consumed. Watson (1969: 55) records both cranial and postcranial elements of *Peromyscus* and *Microtus* in human feces from Salts Cave, Kentucky. Butcher marks on moles (Parmalee, 1975b) and gophers (Dallman, in press) confirm selection of small mammals as food sources. Moreover, Walker (1975) summarizes ethnographic accounts on the hunting, processing, and consumption of rodents by native North Americans. But small mammals, whether commensal to the village or harvested, reflect to some degree the local biota at the time of occupation. Zooarchaeological remains are influenced by a variety of selective factors but perhaps no more than any other mode of fossil accumulation. Except for specimens that may have been involved in trade, the presence of a species in an archaeological site is as valid as that procured by any other carnivore over large geographic areas (e.g. bear claws with a pelt).

PALEONTOLOGIC SAMPLES

Guilday (1962) notes the difference in faunal composition between the late-Pleistocene Natural Chimneys and New Paris No. 4 local faunas in Pennsylvania with respect to their mode of accumulation. The Natural Chimneys local fauna, an owl-pellet accumulation, contained rabbit-size or smaller mammals, a high percentage of nocturnal flying squirrels, and low numbers of diurnal chipmunks. New Paris No. 4, a natural trap, produced some larger mammals and a high percentage of chipmunks in proportion to flying squirrels.

Mellett (1974) describes the appearance of recent specimens passed through the digestive tracts of carnivores and relates their altered nature to certain Tertiary local faunas. Mayhew (1977) contrasts the characteristics of bone consumed and regurgitated by owls as opposed to hawks and notes differences related to diet, digestive processes, and manner of feeding. Moreover, Dodson and Wexlar (1979) describe species-specific patterns of bone destruction among three species of owl. The combination of food preference and differential digestion among predators biases paleontologic samples in the same manner that food selection and processing influence zooarchaelogical counts. Reworking and fluvial transport, which frequently bias paleontologic samples, usually do not affect archaeological samples.

The relative abundance of taxa in a site is influenced by both the kind of site, its availability, and preferences of any predator involved. Thus, comparisons of relative abundance should be restricted to those between similar types of sites, for example, owl roost to owl roost (Guilday et al., 1978), processing site to processing site (Semken, 1980), earth-lodge village to earth-lodge village (Sammis, 1978), or sequence to sequence within a depositional continuum (Parmalee, et al., 1976). The method of faunal accumulation definitely affects the quantitative measures derived from a site. The counting methodology used, for example, the number of individuals versus the number of specimens (Grayson, 1978), also strongly influence data summaries.

CHRONOLOGIC FRAMEWORK FOR HOLOCENE FAUNAS

Isotope dates provide the basis for Holocene mammalian chronology, but cultural units established by archaeologists also constitute useful divisions. Much of the available faunal data has an archaeological perspective, and temporal control is enhanced by dates associated with similar cultural material at other sites. Cultural units usually are geographically limited, but they can be time transgressive between regions (See Aikens, 1983, and Stoltman and Baerreis, 1983). For this reason, an age range assigned to a given culture may differ from site to site, and different ages are occasionally assigned to a given culture in this chapter. Though not exact, the archaeological chronology is more precise than most biostratigraphic schemes and is an asset that is not available in most North American pre-Holocene sites. Utilizing both isotopic and archaeological control, Wendland (1978) updates a system of climatic episodes originally defined in North America by Bryson and others (1970). This terminology also is included here because it provides a suitable framework for collating the stratigraphically and geographically disjunct nature of Holocene local faunas for climatic interpretation. Sites discussed in this chapter (Figure 12-1) generally are limited to those 10,000 years old or less; older faunas are reviewed by Lundelius and others (1983).

Holocene Mammalian Distributions

SURVIVAL OF EXTINCT "PLEISTOCENE" TAXA

The possible survival of some now extinct glacial taxa past the 10,000 yr B.P. date is relevant to this chapter since Holocene faunas are often identified by the absence of Wisconsin taxa. Conversely, when a local fauna contains one or more species normally associated with the Wisconsin megafauna, it is assigned to the Pleistocene. This biostratigraphic philosophy, combined with a paucity of local faunas dated between 10,000 and 7000 yr B.P., has resulted in an open question on the issue of megafaunal survival into the Holocene.

Two major theoretical models have been proposed for the extinction problem, (1) overkill and (2) environmental change. Those favoring the overkill hypothesis (Martin, 1967, 1973) reject all post-Pleistocene isotope dates on extinct fauna and point to the nearly synchronous appearance of Clovis hunters and the loss of the North American megafauna. The environmentalists, championed by Guilday (1967b), recognize the sudden change in megafaunal composition but relate the event to an abrupt or square-wave (Wendland, 1978) change in the environment about 10,000 yr B.P. Environmentalists generally permit survival of relict populations past the 10,000 yr B.P. date in selected areas (Martin and Webb, 1974). Details regarding the late-Pleistocene extinction are reviewed by Lundelius and others (1983).

Hester (1960, 1967) provided the first comprehensive summary of radiocarbon dates on extinct North American megafauna and graphed the records for 32 taxa; 12 phena are associated with dates younger then 9000 yr B.P. Subsequent lists were collated by Martin and Webb (1974: 136) and Kurtén and Anderson (1980: Table 19.6). Holocene survivors recorded in the eastern United States include *Mylohyus*, Russell Cave, Alabama, 8500 to 7565 yr B.P. (Weigel et al., 1974); *Dasypus bellus*, Miller's Cave, 7200 yr B.P. (Patton, 1963); *Mammuthus*, Eau Clair, Michigan, 8200 yr B.P. (Wilson, 1967); *Mammut*, Lapeer, Michigan, 5950 yr B.P. (Wilson, 1967); and *Castoroides*, *Mylohyus*, and *Equus*, Ben Franklin, Texas, 9550 yr B.P. (Slaughter and Hoover, 1963). Martin and Webb (1974) listed *Platygonus*, *Smilodon*, *Megalonyx*, *Tremarctos*, *Canis dirus*, *Equus*, and *Mammut* from the postglacial Devil's Den local fauna, which subsequently has been radiocarbon dated between 8000 and 7000 yr B.P. Kurtén and Anderson (1980) noted 17 taxa, some of which are based on the same determination, associated with dates younger than 10,000 yr B.P.; but they suggested, as did Martin (1967), that the uncritical acceptance of radiocarbon dates or provenience is not wise. Land and others (1980) demonstrated that radiocarbon dates derived from either bone apatite or collagen can be controlled by environmental factors. Long and Martin (1974) published over 20 radiocarbon dates associated with the Shasta ground sloth. Five post-10,000 yr B.P. dates on this animal were rejected for cause by Long and Martin (1974). Since 1974, Thompson and others (1980) have reported 15 newer dates on sloth dung, and none is younger than 10,000 yr B.P. The bulk of the isotope dates available on extinct megavertebrates are pre-Holocene, and their rare association with Archaic or younger archaeological sites (e.g., Adams, 1941) probably reflects the curiosity of the sites' inhabitants.

The range of radiocarbon dates published on extinct Pleistocene megafauna generally follows that shown in Figure 12-2. This chart is a compilation of all dates reported on North American proboscidians in *Radiocarbon* prior to 1974. None has been deleted for cause. The distribution reflects three factors attributed to the extinction of the Pleistocene megafauna: (1) most of the dates are pre-10,000 yr B.P., and 81% of them predate 9000 yr B.P.; (2) the slope of the curve is the greatest during the period assigned to the extinction event, 11,000 to 10,000 yr B.P.; and (3) bone dates, which frequently are rejected because they ordinarily are younger than any associated botanical material, generally are in the latest portion of the curve. When these dates are removed from the scatter, only dates based on carbon from gyttja remain, the termination of the curve is more abrupt, and the overkill hypothesis is supported. The curve also supports the argument of those who believe that some elements of the Pleistocene megafauna survived in substantially reduced and progressively decreasing numbers after the Wisconsin because (1) its shape is that expected for a diminishing population after a rapid

decline in numbers and (2) the curve is not random but skewed toward early-Holocene time. If contamination were the controlling factor, more than one date younger than 5500 yr B.P. would be expected. The graph can be interpreted to support either model, depending on which, if any, dates are rejected.

POST-ALTITHERMAL RANGE FLUCTUATIONS OF SELECTED SMALL MAMMALS

The Holocene is generally regarded as a period of time characterized by modern community structure. Even though there is abundant evidence for change in climatic regimes over the last 10,000 years (Wendland, 1978), these changes are measured by readjustment of existing ecotones or by comparison with modern climatic conditions in the area (Baerreis, 1980). This is in direct contrast to the late Pleistocene, in which the biotic provinces are totally unlike those of today (Graham, 1979; Lundelius et al., 1983; Martin and Neuner, 1978; Semken, 1974) and the climates inferred have no modern analogues. However, there is evidence for major changes in the range of some mammals during the Holocene. The 11 examples given here were selected because they postdated the Altithermal and occurred too late in the Holocene to have been an immediate result of deglaciation. These changes in distribution must reflect other causes.

The dramatic range reduction of the Holocene megafauna over the past three centuries is common knowledge. The near "Extermination of American Bison" (Hornaday, 1887) and the sharp reduction in the range of the wapiti, bear, bighorn sheep, wolf, and pronghorn, as well as the extinction of the passenger pigeon, are documented events that can be directly attributed to the expansion of European

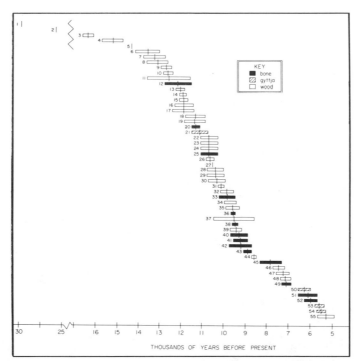

Figure 12-2. Distribution of radiocarbon dates on proboscidian remains in the United States. (1) L-440, (2) I-1559, (3) I-3922, (4) OWU-140, (5) W-1571, (6) OWU-260C, (7) M-1971, (8) OWU-220, (9) OWU-260A, (10) OWU-260B, (11) M-139, (12) M-507, (13) I-3929, (14) I-3930, (15) S-172, (16) I-586, (17) S-30, (18) Y-460, (19) S-29, (20) GSC-611, (21) M-1743, (22) M-1781, (23) M-1780, (24) M-1746, (25) M-1254, (26) OWU-126, (27) W-1358, (28) W-1038, (29) M-1745, (30) M-1782, (31) OWU-141, (32) M-1739, (33) M-1778, (34) W-325, (35) M-1744, (36) WIS-267, (37) M-282, (38) WIS-265, (39) OWU-194, (40) M-694, (41) M-1783, (42) M-490, (43) GSC-614, (44) UCLA-1325, (45) M-281, (46) M-1742, (47) M-1741, (48) M-1790, (49) M-280, (50) S-16, (51) M-67, (52) M-347, (53) OWU-224B, (54) OWU-224A, (55) M-138, and (56) GX-107.

settlement across North America. The most distinct and best-documented reduction is that of the American bison. Hornaday's (1887) map is redrawn here (Figure 12-3) in order to show not only the rapidity of population decline but also the extensive pre-1800 range of these animals, which were not confined to the Plains. Changes in the distribution of smaller mammals over the past 5000 years are equally dramatic but are familiar only to those directly concerned with Holocene biogeography.

Pre-Columbian Distribution of Northern Species

Both the fisher (*Martes pennanti*) and the porcupine (*Erethizon dorsatum*) presently are confined to either the Canadian or boreal biomes in eastern North America or the mountainous regions (Burt and Grossenheider, 1976). Both animals are known from late-Holocene localities up to 1200 km south of their present limits (Figures 12-4 and 12-5). Because fishers are regarded as the primary predators of porcupines (Banfield, 1974), the similarity of their past distributions could be a function of this relationship. However, the two are not common to any site south of latitude 40°N, and relatively few sites (e.g., Meadowcroft Rockshelter) have produced the two animals in the same horizon (Guilday et al., in press). Moreover, in the southern United States, the fisher is present primarily in sites younger than 1000 years old, whereas the porcupine appears to be most common in sites older than 1000 years. Both animals are associated with sites dated in the 1600s, and better chronologic control will be necessary to affirm or to reject this generalization.

Another animal of boreal affinities, the red-backed vole (*Clethrionomys gapperi*), also is generally associated with the Canadian and boreal forests of North America (Figure 12-6). Specimens of this taxon have been found in an earth lodge of the Central Plains Tradition (Fulmer, 1974) dated at about A.D. 1215 ± 95 (Hotopp, 1978) in southwestern Iowa. A lower right first molar of this vole, reported

here for the first time, was found in association with a Woodland occupation (A.D. 1040) at Rock Run Shelter in Cedar County, southeastern Iowa. The southwestern Iowa locality is approximately 300 km southwest of the southernmost occurrence of this taxon in the Plains region. The Cedar County locality is almost 190 km southeast of the modern southern record in Iowa. The red-backed vole recently has been recovered from an Early or Middle Archaic horizon at the Barnhart site (R. W. Graham, personal communication, 1981), which is equivalent to the post-Clovis horizon D at Kimmswick (Graham, 1981). It also is known throughout the Peccary Cave sequence, which ranges in age from 16,700 to 2230 yr B.P. (Quinn, 1972). The red squirrel (*Tamiasciurus hudsonicus*) has been recovered from the Archaic (2000-800 yr B.P.) Westmoreland-Barber site in Marion County, Tennessee, as well as from other southeastern archaeological sites (Parmalee and Guilday, 1966). It also is identified in Cultural Horizon II (7300 yr B.P.) at Cherokee, Iowa (Semken, 1980). Both establish a more widespread distribution of the species during the Holocene. Similarly, a major post-3000 yr B.P. reduction in the range of the prairie vole (*Microtus ochrogaster*) from the southern Plains is recorded by Schultz and Rawn (1978).

Pre-Columbian Distribution of Eastern Species

The woodchuck, or groundhog (*Marmota monax*), is a common inhabitant of the eastern deciduous forest, and its present range extends west to the prairie/forest ecotone. Specimens have been found west of its present range (Figure 12-7) in three archaeological sites along the Missouri River (Semken and Falk, in preparation). Two of these sites, Tony Glas and Crow Creek, fall within or close to the Neo-Atlantic climatic episode (A.D. 900-1150) and may support the relatively moist condition predicted for that interval (Wendland, 1978). Gallery forests, developed to varying degrees and expected near each of these sites, may have provided the necessary habitat for

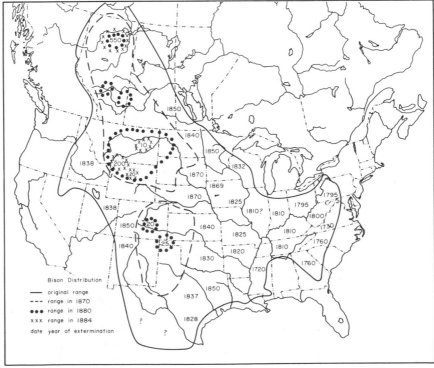

Figure 12-3. Isochron map depicting the extermination of the American bison. (Modified from Hornaday, 1887.)

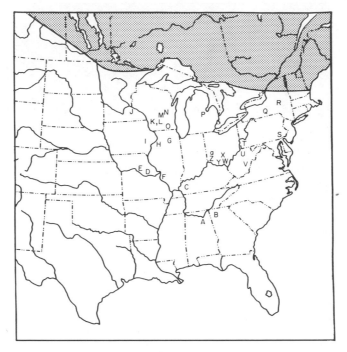

Figure 12-4. Representative Holocene localities and modern distribution (stippled) of the fisher (*Martes pennanti*). (A) Laws, A.D. 1700 (Barkalow, 1961); (B) Etowah, A.D. 1100-1500 (Parmalee, 1959c); (C) Indian Knoll, 5950-2950 yr B.P. (Cleland, 1966); (D) Graham Cave, after 7600 yr B.P. (Parmalee, 1971); (E) Arnold Research Cave, A.D. 500 (Parmalee, 1971); (F) Cahokia, A.D. 1200-1550 (Parmalee, 1958); (G) Zimmerman, A.D. 900-1790 (Rogers, 1975); (H) Heins Creek, A.D. 732 (Cleland, 1966); (I) Rock Run, 2560 yr B.P. (Semken, this report); (J) Bryan, A.D. 1200-1600 (Lunkens, 1963); (K) Raddatz Rockshelter, 9000-3492 yr. B.P. (Parmalee, 1959b); (L) Durst Rockshelter, A.D. 1200 (Parmalee, 1960b); (M) Bornick, A.D. 1290 (Gibbon, 1971); (N) Bell, A.D. 1680-1730 (Parmalee, 1963b); (O) Atztalan, mean date A.D. 1200 (Parmalee, 1960a); (P) Schuitz, 50 B.C. to A.D. 500 (Cleland, 1966); (Q) Kipp Island, A.D. 310-630 (Cleland, 1966); (R) Wilson Sand Hill, A.D. 1400+ (Cleland, 1966); (S) Eschelman, A.D. 1400+ (Cleland, 1966); (T) Meadowcroft, 11,300+-0 yr B.P. (Guilday et al., in press); (U) Globe Hill, 2950-3950 yr B.P. (Guilday, 1956); (V) Johnston, A.D. 1400+ (Cleland, 1966); (W) Feurt, A.D. 1400+ (Cleland, 1966); (X) Cramer Village, A.D. 1000-1400 (Cleland, 1966); (Y) Baum, A.D. 1000-1400 (Cleland, 1966); (Z) Madisonville, A.D. 1400+ (Cleland, 1966); and (a) Anderson, A.D. 1000-1400 (Cleland, 1966).

the woodchuck to survive into the drier post-A.D. 1150 Pacific episode at the Black Partizan site.

A more dramatic reduction in western range is recorded by *Blarina brevicauda kirtlandi* (a medium-sized phenon of short-tailed shrew), a taxon that is regarded as specifically distinct by Graham and Semken (1976). This phenon has been recorded west of its present range (Figure 12-8) in both the Mud Creek local fauna of eastern Iowa at about 6200 yr B.P. (Kramer, 1972) and the 3400 yr B.P. Garrett Farm local fauna of western Iowa (Fay, 1980). It also is known from the uppermost level (about 2230 yr B.P.) in Peccary Cave, northwestern Arkansas (Graham and Semken, 1976; Quinn, 1972; Semken, 1982). An unusual living population of *Blarina* with a larger body size than expected for its geographic location is recorded by Jones and Glass (1960) from Douglas County, Kansas. It compares in size with the more eastern *B. b. kirtlandi* and may represent a relict of its more widespread former range.

Pre-Columbian Distribution of Western Species

Plains species, including both the thirteen-lined ground squirrel (*Spermophilus tridecemlineatus*) and the prairie chicken (*Tympanuchus cupido*), have been recorded east of their present ranges during both Pleistocene and Holocene time. Figure 12-9 illustrates the present and Wisconsin range of the thirteen-lined ground squirrel and some Holocene sites from which it has been recovered. The ground squirrel is associated with post-9000 yr B.P. Archaic artifacts

in Baker Bluff Cave in eastern Tennessee (Guilday et al., 1978), the A.D. 1380 to 1470 Callahan-Thompson site in southeastern Missouri (Lewis, 1974), and the circa A.D. 900 to 1300 Zebree site in northeastern Arkansas (Guilday and Parmalee, 1971). It is difficult to establish whether these records reflect a continuing range reduction from late-Wisconsin time or whether this ground squirrel's range expanded periodically from more western limits. Thirteen-lined ground squirrels, as well as prairie chickens, are open-country forms, and so an eastern expansion during the theoretically moister Neo-Atlantic episode at the Zebree site is difficult to reconcile. The ground squirrel's presence in the Callahan site of the more xeric Pacific episode is more compatible with the proposed climatic regimen. Alternatively, clearing for agriculture may be responsible for the distribution of the thirteen-lined ground squirrel in these sites. Its widespread range during the Pleistocene, through most of the eastern United States, must have been more ecologically controlled. The plains pocket gopher (*Geomys* cf. *bursarius*) also lived east and south of its present range during both the Pleistocene and the Holocene (Parmalee and Klippel, 1981).

Pre-Columbian Distribution of Southern Species

The significance of the rice rat (*Oryzomys palustris*) in archaeological sites (Guilday and Mayer-Oakes, 1952) north and west of its present range (Figure 12-10) was first discussed in 1961, when Guilday suggested that *Oryzomys* was a commensal pest living in the

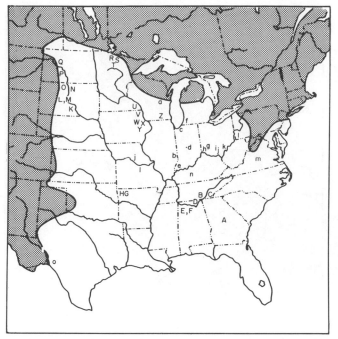

Figure 12-5. Representative Holocene localities and recent distribution of the porcupine (*Erethizon dorsatum*). (A) unidentified site, no age given (Ray and Lipps, 1970); (B) Westmoreland-Barber, Late Archaic/Early Woodland (Parmalee and Guilday, 1966); (C) Bible, Late Archaic/Early Woodland (Parmalee and Guilday, 1966); (D) Russell Cave, 1500-1165 yr B.P. (Weigel et al., 1974); (E) Little Bear Creek, 3950-2950 yr B.P. (Barkalow, 1961); Stanfield-Worley, 9256 yr B.P. (Parmalee, 1963a); (G) Peccary Cave, Holocene (Semken, 1969); (H) Ten Mile Rockshelter, Holocene (Sealander, 1979); (I) Tick Creek Cave, Woodland (Parmalee, 1971); (J) Brynjulfson Cave, 2460-1400 yr B.P. (Parmalee and Oesch, 1972); (K) Black Partizan, A.D. 1450-1685 (Semken and Falk, in preparation); (L) Buffalo Pasture, A.D. 1675-1845 (Semken and Falk, in preparation); (M) Gillette, multicomponent (Semken and Falk, in preparation); (N) Walth Bay, A.D. 1515 (Semken and Falk, in preparation); (O) Lower Grand, A.D. 1354 (Semken and Falk, in preparation); (P) Upper Sanger, A.D. 1675-1780 (Semken and Falk, in preparation); (Q) Kipps Post, A.D. 1830 (Semken and Falk, in preparation); (R) Smith, A.D. 600-1200 (Lunkens, 1963); (S) McKinstry, A.D. 1000-1400 (Lunkens, 1963); (T) White Oak, A.D. 1000-1650 (Lunkens, 1963); (U) La Moille, 3600-3000 yr B.P. (Lunkens, 1963); (V) Lane Enclosure, A.D. 1460-1740 (Jenkins and Semken, 1972); (W) Willard Cave, 3500 yr B.P. (Eshelman, 1971); (X) Schmidt Shelter, A.D. 100-600 (Eshelman, 1972); (Y) Hadfields Cave, A.D. 300-800 (Benn, 1980); (Z) Cooper's Shore, Middle Woodland (Lippold, 1973); (a) Bell, A.D. 1680-1730 (Parmalee, 1963b); (b) Riverton, Swan Island, and Robeson, A.D. 3490-3110 yr B.P. (Parmalee, 1969); (c) Fifield, A.D. 1000-1400 (Parmalee, 1972); (d) Bowen, A.D. 800-1300 (Dorwin, 1971); (e) Angel, A.D. 1000-1400 (Black, 1967); (f) Moccasin Bluff, A.D. 1000-1400 (Cleland, 1966); (g) Anderson, A.D. 1000-1400 (Cleland, 1966); (h) Madisonville, A.D. 1400+ (Cleland, 1966); (i) Baum, A.D. 1000-1400 (Cleland, 1966); (j) Feurt, A.D. 1400+ (Cleland, 1966); (k) Canter Caves, no date (Cleland, 1966); (l) Meadowcroft, 11,300⁺-0 yr B.P. (Guilday et al., in press); (m) three unidentified localities, late Quaternary (Ray and Lipps, 1970); (n) Mammoth Cave, early Woodland (Watson, 1969); and (o) Eagle Cave and Coontail Spin (Raun and Eck, 1967).

artificial environment of an Indian settlement. In the early 1950s, all records of *Oryzomys* north of its present range were associated with Indian villages of late-prehistoric age. Johnson (1972) extended this extralimital village/rice rat association to the southwestern Iowa Glenwood Village at about A.D. 1100, which was recently reviewed by Bardwell (1981). Moreover, Pyle (1981) has recorded *Oryzomys* in the Late Woodland MAD site (A.D. 250) in western Iowa, and Satorius-Fox (1981) has identified rice rats in the Schmidt and surrounding central Nebraska earth-lodge villages dating from the 1100s. The latter is the westernmost known Holocene population, but specimens are known from southwestern Kansas during Late-Illinoian and Wisconsin time (Hibbard, 1970). Rice rats were replaced by black rats (*Rattus* sp.) in eastern Indian villages after European contact (Guilday, 1971a). Guilday regards this as an example of competitive exclusion.

In 1972, rice-rat remains were reported northwest of their present range from two cave deposits in Missouri, one of which (Brynjulfson No. 2) was, at most, infrequently occupied (Parmalee and Oesch, 1972). Parmalee and Oesch (1972: 35) add that, "although there may be some relationship, direct or indirect, between the prehistoric distribution of *Oryzomys* and man in certain localities, the Boone (Brynjulfson) and Montgomery (Graham Cave) records lend little credence to this theory." They regard climatic change as the most likely explanation for the restriction of the rice rat's distribution. Three noncultural sites with the rice rat recently have been reported from southern Indiana (Richards, 1980). These localities occur north of the modern range of the rice rat, when the distribution of Hall (1981) is used as a basis, but they are within the rice rat's range as mapped by Burt and Grossenheider (1976). These localities are too close to the modern range or *Oryzomys* to offer strong support for the climatic model when the latter distribution is used, but they may support climatic change if Hall's (1981) distribution is more accurate. The differences between the two maps are interpretive, since the same occurrences were used for both. Purdue and Styles (in preparation) note that there are now five premaize sites with *Oryzomys* in Illinois, and Styles (1981: 139) regards the circa A.D. 400 to 750 Newbridge specimens as noncultural. Guilday and others (in press) counter this argument, noting that the presence of only two *Oryzomys* bones in the historic to 11,300 yr B.P. Meadowcroft Cave cultural sequence is in strong contrast with the high numbers of these

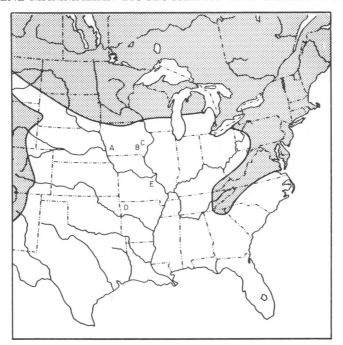

Figure 12-6. Representative Holocene localities and modern distribution (stippled) of the red-backed vole (*Clethrionomys gapperi*). (A) 13 ML 124, A.D. 1215 (Fulmer, 1974); (B) Rock Run Shelter, level 3, A.D. 1040, and level 12, 4180 yr. B.P. (Semken, this report); (C) Mud Creek, 6265 yr B.P. (Kramer, 1972); (D) Peccary Cave 16,700-2230 yr B.P. (Semken, this report); and (E) Barnhart, Early to Middle Archaic (R. W. Graham, personal communication, 1981).

animals he reports from 16 late-prehistoric villages in the upper Ohio Valley. Lodges, rather than stored corn or people per se, may be the limiting factor. The paucity of noncultural Holocene sites precludes speculation on nonarchaeological associations of these animals, but both culture (particularly villages [Guilday, 1971a]) and environment (Richards, 1980) may be involved.

Although there has been no attempt to catalogue all mammals or sites having specimens that are extralimital to 20th-century distributions, a review of the above demonstrates substantial changes in the distribution of small mammals over the last 5000 years. Like the bison, each of the species mentioned apparently had more extensive ranges until a few hundred years ago. The latest dates associated with extralimital records are about A.D. 1700 for the fisher in Alabama, about A.D. 1100 to 1500 for the porcupine in Georgia, about A.D. 1450-1685 for the woodchuck in the Dakotas, about A.D. 1380 to 1470 for the thirteen-lined ground squirrel in Missouri, about A.D. 1600 for the rice rat in Pennsylvania and A.D. 1155 in Nebraska, about A.D. 1275 for the red-backed vole in Iowa, about A.D. 1143 for Kirtland's short-tailed shrew in southwestern Iowa, and about 3000 yr B.P. for the prairie vole in central Texas.

Post-Columbian Micromammal Distributional Changes

Distributional changes for other small mammals also appear to be directly correlated with settlement by Europeans. Guilday (1972) notes the presence of the eastern mole (*Scalopus aquaticus*) in five archaeological sites in the upper Ohio Valley. The species does not inhabit the area today, but some of the records are from sites dating in the 1600s. Guilday attributes the loss of the eastern mole in this region to "significant ecological changes brought about by industry and engineering in the valley" (Guilday, 1972: 905). Their presence in the region originally may have been a result of the prairie's expan-

sion into Pennsylvania before the development of the late-Holocene closed-canopy forest (Guilday, 1961).

Postcontact distributional changes are better substantiated in central Texas. Semken (1961, 1967) relates the black fill of both Longhorn Cavern and Rattlesnake Cave to the period of European settlement on the basis of the occurrence of the house mouse (*Mus musculus*), a Eurasian native, in both deposits. The plains pocket gopher (*Geomys bursarius*), now no closer than 65 to 120 km to either locality, also was present in both black fills. Since these two Edwards Plateau caves are over 255 km apart, the pocket gopher must have been widespread in the region before the 1800s. Its regional demise apparently correlates with the loss of habitat (soil) resulting from the overgrazing and resultant erosion of the black soil, now found only as residuum in pockets on the surface and in the internal drainage system of the denuded limestone plateau.

The occurrence of northern taxa, particularly the fisher and the porcupine, south of their present range in some archaeological sites "may have reflected cultural exchange and transport rather than environmental change" (Purdue and Styles, in preparation). The presence of a fisher skull in an early-historic Indian grave (Barkalow, 1961), the dominance of skull elements of both species in the sites, the evidence of depelting, the known utilization of porcupine quills in garments, and the general absence of other boreal species in each site support this hypothesis.

The fisher and the porcupine may represent trade items, but trade of red-backed voles, woodchucks, rice rats, prairie voles, thirteen-lined ground squirrels, gophers, short-tailed shrews, and eastern moles appears less likely. Gifford (1967) records subsistence utilization of ground squirrels, field mice, gophers, red squirrels, moles, gray squirrels, chipmunks, and woodrats among the Pomo Indians of California. Such observations demonstrate that mammals from archaeological sites represent a harvesting bias. However, harvested

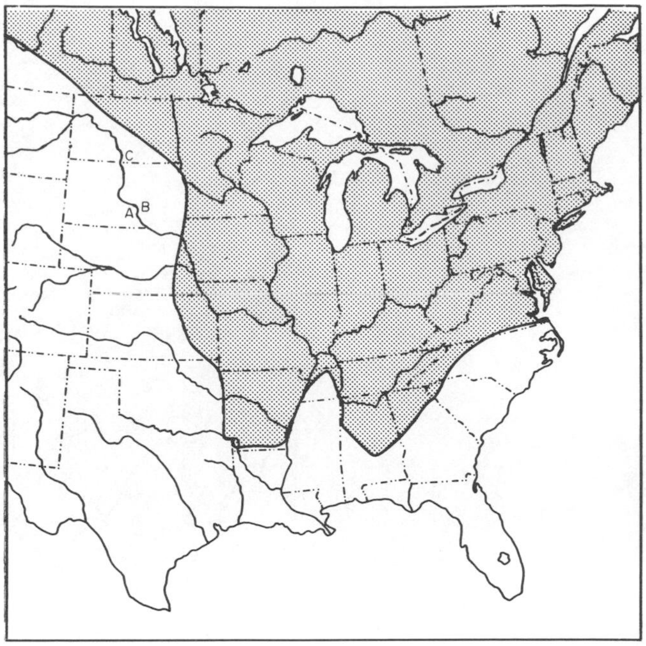

Figure 12-7. Representative Holocene localities and modern distribution (stippled) of the groundhog (*Marmota monax*). (A) Black Partizan, A.D. 1450-1685; (B) Crow Creek, A.D. 1140; and (C) Tony Glas, A.D. 1275. (all dates from Falk and Semken, in preparation).

species are available locally and therefore reflect nearby environments.

The ranges of many species are expanding and contracting at the present time irrespective of their primary community associations. Undoubtedly, this process could typify changing Holocene faunal structure and so could explain the presence of remnant species from the boreal component of the late-Pleistocene fauna south of their present limits. Examples include the expansion of the opossum (*Didelphis marsupialis*) to the northeast (Guilday, 1958) and to the northwest (Hoffman and Jones, 1970) and the northern expansion of the range of the armadillo (*Dasypus novemcinctus*) (Buchanan and Talmage, 1954). Numerous other examples of the changing nature in the ranges of modern species are given by Udvardy (1969). Dynamic faunal processes were in effect during the Holocene, and modern distributions cannot be wholly attributed to climatic change associated with deglaciation. In summary, the modern distributions of many small mammals, like that of the bison and the pronghorn, have been strongly affected by (1) European settlement, (2) pre-Columbian cultural patterns, (3) climatic change during the Holocene, and (4) climatic alteration associated with the glacial retreat.

SIZE CLINES IN HOLOCENE SPECIES

Clines, that is, progressive changes in one or more physical characteristics through a series of contiguous populations (Mayr, 1970: 215), are useful but underutilized factors for interpreting Holocene environments. Character gradients can be progressive through time (chronoclines), they can reflect gradual variation in response to changing environments across a geographic area (topoclines), or they can be a combination of both. Chronoclines probably will play an increasing role in dating Holocene sites, and the application of modal parameters of a Holocene sample to an established topocline may provide additional paleoecologic evidence for environmental reconstruction. Since nearly every well-studied species exhibits clinal variation (Mayr, 1970), a systematic search for those trends in the osteology of any taxon common in Holocene sites would be fruitful.

The *Bison* chronocline of Wilson (1978: Figure 12), reproduced here as Figure 12-11, documents a gradual size diminution of the distance between horn-cone tips in northern Plains *Bison* over the past 10,000 years and illustrates the transition from *Bison bison occidentalis* and *B. b. antiquus* to modern *B. b. bison*. This reduction in horn-cone size is commensurate with previous indications via metapodial parameters that the bulk and stature of the American bison decreased over this period (Bedord, 1974). Examination of Figure 12-11 also reveals a broad size range in specimens representing any period of time along the chronocline. All unusually large individuals reported from isolated Holocene specimens can be accounted for in this distribution "simply as population extremes in a single evolving biospecies" (Wilson, 1978: 13). It is also clear that the rate of decrease was greater for the larger animals representative of each sample (42 mm per 1000 years) than for those representative of the mode (32 mm per 1000 years). Dwarfing took place continually throughout the Holocene, although there may have been a "slight" increase in the rate between 6500 and 4500 yr B.P. (Wilson, 1978: 15). The replacement of a large subspecies by a smaller one during the Holocene also has been documented by Wright and Lundelius (1963), who note the presence of a robust raccoon in cave deposits and archaeological sites in central Texas. This Holocene raccoon population, which is distinctly larger than the recent central Texas population, was characterized by more massive skulls and mandibles, larger canines, more laterally directed third upper incisors, and increased sexual dimorphism. Originally described as *Procyon simus*, this animal resembles living populations of raccoons currently confined to Idaho, eastern Oregon, and the Pacific Northwest. Wright and Lundelius (1963) note the subspecific relationship of the northwestern population to other North American raccoons and reduce the fossil taxon to subspecific status. They apply *P. lotor simus* to the northwestern raccoons and suggest that this massive morphotype "had a much wider distribution during the Pleistocene, probably throughout the western half of North America" (Wright and Lundelius, 1963: 20-21). The large taxon ranged widely in Texas

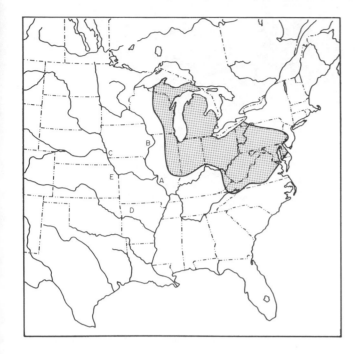

Figure 12-8. Representative Holocene localities and modern distribution (stippled) of Kirtland's short-tailed shrew (*Blarina b. kirtlandi*). (A) Meyer Cave, Holocene, (Parmalee, 1967); (B) Mud Creek, 6100 yr B.P. (Kramer, 1972); (C) Garett Farm, 3400 yr B.P. (Fay, 1980); (D) Peccary Cave, Holocene (Semken, this report); and (E) Douglas County, recent (Graham and Semken, 1976).

about 9300 yr B.P. but persisted along the edge of the Edwards Plateau until about A.D. 1300.

East-west clinal variation in the size of the recent eastern cottontail (*Sylvilagus floridanus*), with smaller populations toward the west, was identified by Purdue (1980), who extracted a single factor from nine osteological variables to account for 90.7% of the variation. When the factor is applied to the Holocene sequence at Rodgers Shelter (Wood and McMillan, 1976), a record (Figure 12-12) of decreasing size from 8600 to 5200 yr B.P. followed by an increase to modern proportions at the top of the section becomes evident. Since smaller forms are characteristic of the more arid western portion of the modern transect, a relatively dry interval culminating between 6700 and 5200 yr B.P. (the Altithermal) is indicated for the Rodgers Shelter sequence.

Purdue (1980) using revised dates for Rodgers Shelter, also identifies an inverse clinal relationship between the fox squirrel (*Sciurus niger*) and the gray squirrel (*S. carolinensis*) over their essentially sympatric range. The gray squirrel increases in size toward the west, whereas the fox squirrel decreases in size in that direction. The two are most similar in size at their common western extreme (Purdue, 1980: Figure 4). When the clinal factors are applied to the gray squirrel samples from Rodgers Shelter and Graham Cave, an increase in size is apparent between 9500 and 7000 yr B.P. in Graham Cave and between 8100 and 5200 yr B.P. in Rodgers Shelter. Since the modern sample from the Rodgers vicinity represents a smaller population than the sample reflecting the period between 6700 and 5200 yr B.P., a decrease in effective moisture is inferred for middle-Holocene time at these sites.

Nelson and Semken (1970: Figure 4) record a north-south topocline for the length/width ratio of the first lower molar for Plains populations of the muskrat, *Ondatra zibethicus*; specimens from more-southern latitudes have a lower ratio than those to the north. This ratio is confirmed in part for the eastern United States by Martin and Tedesco (1977) with a modern sample from New Jersey. Muskrat remains are common in Plains archaeological sites, but they have

not been used in conjunction with this cline for paleoecologic interpretations.

Clinal and specific/subspecific relationships among three phena of the short-tailed shrew, *Blarina*, (Graham and Semken, 1976) have been used to interpret Pleistocene environments (Lundelius et al., 1983). These distinctions also have value for the Holocene. Kramer (1972) records the taxon *B. b. kirtlandi* 210 km west of its present range in the Mud Creek local fauna of eastern Iowa at about 6200 yr B.P. Fay (1980) extends this range 560 km west to the southwestern corner of the state with its identification in the circa 3400 yr B.P. Garrett Farm local fauna. It may exist as a relict as far west as Douglas County, Kansas (Jones and Glass, 1960), and its association with post-middle-Holocene faunas suggests a late-Holocene reduction in range (Figure 12-8). Moreover, the coexistence of *B. b. kirtlandi* and *B. b. carolinensis* in the Garrett Farm local fauna (Fay, 1980) and of both *B. b. brevicauda* and *B. b. carolinensis* in the A.D. 980 Thurman local fauna (Jenkins, 1972) supports the contention of Graham and Semken (1976) that the above phena are of specific rather than subspecific rank. Major range fluctuations of these phena took place during the Holocene, and the cooccurrence of two taxa of *Blarina* in a single local fauna may illustrate an "edge effect" along the prairie/forest border.

Clines also have been established within the microtine rodents. Kurtén and Anderson (1980) reiterate the reverse Bergman's response noted by Hibbard (1963) in modern populations of *Synaptomys cooperi*, the southern bog lemming. Guilday and others (1964) use the bog lemming to interpret climatic change in the late Pleistocene of Pennsylvania (New Paris No. 4) and postglacial climatic stability (Guilday et al., in press) in the Holocene Meadowcroft sequence. Semken (1966) has observed an increase from east (Massachusetts) to west (Dakotas) in the number of closed triangles in the first lower molar of the meadow vole (*Microtus pennsylvanicus*). Davis (1975) has contoured this effect (Figure 12-13) as well as the progressive size variation of the first lower molar within the taxon. The tight size gradient, with smaller sizes in forested

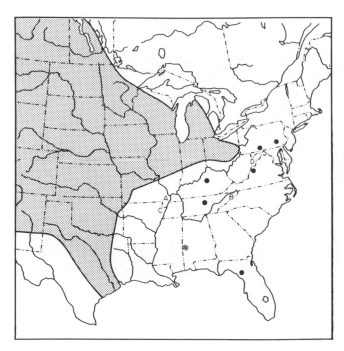

Figure 12-9. Representative Holocene localities and modern distribution (stippled) of the thirteen-lined ground squirrel (*Spermophilus tridecemlineatus*). (A) Callahan-Thompson, A.D. 1380-1470 (Lewis, 1974); (B) Baker Bluff, 9000-8000 yr B.P. (Guilday et al., 1978); (C) Zebree, A.D. 900-1300 (Guilday and Parmalee, 1971); and (D) Eagle Cave, early Holocene (Kurtén and Anderson, 1980). Solid circles denote representative late Wisconsinan localities: Bootlegger Sink and New Paris No. 4, Pennsylvania; Clark's Cave, Virginia; Welsh Cave, Kentucky; Robinson Cave, Tennessee; and Haile XIVA, Florida (all dates from Guilday et al., 1978).

regions, should be a useful tool in determining the extent of environmental change. Satorius-Fox (1981) identifies the large morph of *M. pennsylvanicus* (*M. xanthognathus* size) now characteristic of south-central Nebraska (Davis, 1975) in the circa A.D. 1150 Schmidt site. This discovery indicates that this large form was capable of withstanding Pacific climates in that region.

An example of a chronocline also is provided by the geologic record of the meadow vole. There was a tendency for an increase in the number of closed triangles in the first lower molar through time (Semken, 1966) as well as between forest and plains environments. The chronoclinal nature within *M. pennsylvanicus* has been used by McMullen (1974) to order Illinoian local faunas by increasing triangle counts. Thus, it appears that specimens of *M. pennsylvanicus* with five closed triangles can represent populations from either earlier Pleistocene time or a forested situation. Those specimens with six, seven, or eight closed triangles should be both late in time and in association with either the prairie or the prairie/forest-edge environment. Care must be used in utilizing triangle counts unless the age or the paleoenvironmental association of the deposits is known.

Holocene Mammalian Faunas

COMMUNITY STRUCTURE

The concept of a biologic community as it is applicable to the vertebrate paleontologic/archaeological record has been recently reviewed by King and Graham (1981). They point out that the numerous working definitions of community can be divided into two basic categories: organismic or individualistic. The organismic approach contends that communities have discrete boundaries and that the sum of the species in an area behaves as an organism with both structure and function. The individualistic concept regards communities as collections of species requiring similar environmental conditions. "Because the ecological requirements of the individual species differ, sharp boundaries between 'communities' do not exist"

(King and Graham, 1981: 129). Investigators who prefer the organismic approach expect an entire community or biome to respond as a unit and to relocate as climatic conditions change. Pleistocene biome migration in response to multiple glaciation, the accordian effect, is a classic example of this model (Blair, 1958). The individualist expects each species experiencing similar climatic changes to respond independently and, thus, the community composition of an area to change via both immigration and emigration of some individual taxa while others remain in the area. Communities are not stable under this model but reorganize in response to changing local conditions.

The Pleistocene fauna of the United States was characterized by a combination of (1) extinct megavertebrates and (2) extant temperate megavertebrates and microvertebrates in association with (3) now disjunct large and small northern species. The Holocene fauna of temperate regions generally is composed of the second category only. This reduction in the number of species has led Martin (1967: 110) to regard the Holocene fauna as "depauperate" and Semken (1974) and Martin and Webb (1974) to consider it "impoverished" with respect to the high species densities characteristic of the late Pleistocene (Graham, 1976a, 1979). This faunal change is used to define the Pleistocene/Holocene vertebrate biostratigraphic boundary, which has been placed by various investigators at 18,000, 14,500, 14,000, 11,800, 10,500, 10,350, 10,030, 8500, 5000, and 4000 yr B.P. (Mercer, 1972; Reider, 1980; Wendland and Bryson, 1974); but some (Morrison, 1965) intended theirs to be strictly regional in application. Breaks in the vertebrate record appear at about 12,500, 10,500, 8500, and 5000 yr B.P.

The response time of various components of the biota to deglaciation trails climatic change to varying degrees (Bryson et al., 1970). In the pollen record for many areas, there is an abrupt shift in dominance from coniferous to deciduous pollen that, when present, may be used to define the Pleistocene/Holocene boundary. In many cores, this transition, which varies geographically between 12,000 and 10,500 yr B.P., took place within a few hundred years (Baker, 1983). The same concept appears to be valid for the vertebrate transition.

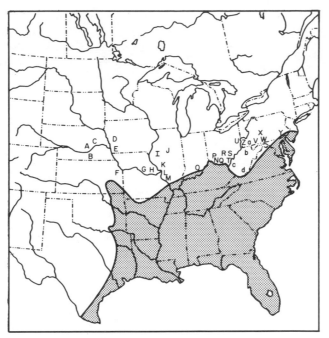

Figure 12-10. Representative Holocene localities and modern distribution (stippled) of the rice rat (*Oryzomys palustris*). (Some letters refer to more than one locality.) (A) Schmidt, A.D. 1150 (Satorius-Fox, 1982); (B) Shipman, A.D. 1250-1450 (Ludwickson, 1978; personal communication 1981); (C) Fullerton, A.D. 1250-1450 (Ludwickson, personal communication, 1981; Satorius-Fox, 1982); (D) MAD, A.D. 250 (Pyle, 1980); (E) Glenwood, A.D. 1100 (Johnson, 1972; Bardwell, 1981); (F) Trowbridge, A.D. 130-400 (Reid, 1976); (G) Brynjulfson Cave, 2460-1400 yr B.P. (Parmalee and Oesch, 1972); (H) Graham Cave, after 7600 yr B.P. (Parmalee, 1971); (I) Scovill, Middle Woodland (Munson et al., 1971); (J) Kingston, A.D. 1000-1400 (Cleland, 1966); (K) Schild, A.D. 1000-1150 (Perino, 1971); Newbridge and Carlin, both A.D. 400-750 (Styles, 1981); (L) Cahokia, A.D. 1200-1550, and Kane, A.D. 1000 (Parmalee, 1973); (M) Meyer Cave, Holocene (Parmalee, 1967); (N) Madisonville, A.D. 1400+ (Cleland, 1966); (O) Anderson Pit, early Holocene (Richards, 1980); (P) Anderson, A.D. 1000-1400 (Cleland, 1966); (Q) Feurt, A.D. 1400+ (Cleland, 1966); (R) Baum and Cramer, both A.D. 1000-1400 (Cleland, 1966); (S) Blaine, A.D. 970-1221 (Parmalee and Shane, 1970); (T) McCune, A.D. 1320, and Gabriel, A.D. 1450 (Murphy, 1975); (U) Speidel, A.D. 1200-1600 (Guilday and Mayer-Oakes, 1952); (V) Martin, A.D. 1200-1600 (Guilday and Mayer-Oakes, 1952); (W) Fort Hill, A.D. 1200-1600 (Guilday and Mayer-Oakes, 1952); (X) Johnson, A.D. 1400+ (Cleland, 1966); (Y) Eschelman, A.D. 1625 (Guilday, 1971a); (Z) Meadowcroft, 11,300 yr B.P. to present (Guilday et al., in press); (a) Varner, A.D. 1000-1400 (Cleland, 1966); (b) Fairchance, A.D. 1530 (Guilday, 1971a); (c) Buffalo, A.D. 1650 (Guilday, 1971a); and (d) Mount Carbon, A.D. 1500-1600 (Guilday, 1971a).

NORTHEASTERN UNITED STATES

The mammalian fauna from Hosterman's Pit, Pennsylvania, dated at 9290 yr B.P., is modern in every aspect (Guilday, 1967a). This is in direct contrast to the 11,300 yr B.P. faunule from Unit B of New Paris Sinkhole No. 4, approximately 80 km to the north, which contains a strong boreal component. The turnover to modern community structure, therefore, must have occurred within this 2000-year interval in central Pennsylvania (Guilday, 1971b). The Unit A faunule at New Paris, which overlies that of Unit B, is not dated but contains a greater proportion of temperate elements than Unit B and suggests that the change to Holocene associations, contoured by Guilday and others (1964: Maps A and B), was transitional but clear with respect to variation in relative species abundance during glacial time.

The stratum-IIa faunule from Meadowcroft Rockshelter (Adovasio et al., 1978; Guilday et al., in press), also dated at 11,300 yr B.P., is totally modern in aspect. The error limits of 700 years on the Meadowcroft date and 1000 years on the New Paris date restrict this transition to a probable 1700-year interval. However, if the 11,300 yr B.P. dates on both Meadowcroft and New Paris No. 4 approach their central tendency, the Pleistocene/Holocene faunal transition in the Northeast must be compared with the approximate 500-year interval characteristic of the pollen record.

Guilday and others (in press) find little evidence for climatic change

between 11,300 yr B.P. and A.D. 1265 in the 11 Meadowcroft strata. Guilday (1967a: 232) previously noted that "*all* Indian archaeofaunas from the East within the time span of at least the last 6000 years contain faunas which are essentially modern." This comment is based on evidence from a number of sites, and the longevity of the Holocene woodlands is confirmed by both the 11 superimposed units (A.D. 1265 to 11,300 yr B.P.) at Meadowcroft (Guilday et al., in press), and the Archaic (circa 8920 yr B.P.) to modern (circa 490 yr B.P.) faunal succession in Sheep Rock Shelter, Pennsylvania, in which all species recovered from all levels are typical of the area at present (Guilday and Parmalee, 1965). A faunal shift of any magnitude should have been detected from either stratigraphically controlled sample. The influence of the "climatic optimum" apparently was negligible in the well-forested East (Guilday and Parmalee, 1965: 48). The only change in the Sheep Rock record probably is of nonclimatic origin and can be related to the activity of humans. Forest-dwelling flying squirrels (*Glaucomys volans*) predominate in the lower and upper parts of the section but are rare in the middle levels; open-ground rabbits (*Sylvilagus* sp.) show a reciprocal pattern (Figure 12-14). The obvious interpretation suggests that the site was initially forested, that open country then predominated, and that this was followed by reforestation. Since the dominance of rabbits is contemporary with the period of heaviest occupation in the shelter, Guilday and Parmalee (1965) relate the change to the clearing of the original forest by

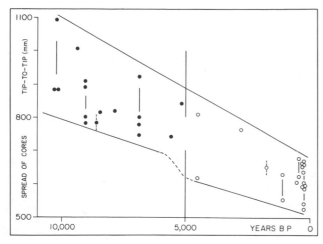

Figure 12-11. Holocene chronocline for bison horn cores. (From Wilson, 1978.)

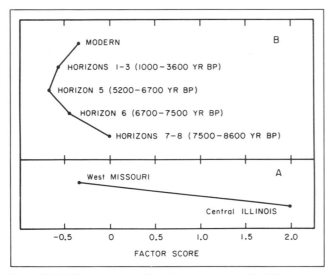

Figure 12-12. Clinal variation in (A) modern eastern cottontail rabbits (*Sylvilagus floridanus*) and (B) prehistoric and modern specimens from Rodgers Shelter. (From Purdue, 1980.)

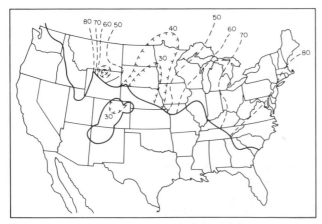

Figure 12-13. Clinal variation for the triangle counts on first lower molars of the meadow vole (*Microtus pennsylvanicus*). (From Davis, 1975).

Indians for agriculture. The reversion to forests reflects the area's abandonment by the Indians before the period of European colonization, which is indicated by the presence of domestic sheep, chickens, and European rats in the uppermost deposits. The only evidence of the climatic optimum between 5500 and 3500 yr B.P. in the eastern forest is noted by Guilday (1965) at the Lamoka Lake site, New York, on the basis of the fox squirrel (*Sciurus niger*) and the box-turtle (*Terrapene carolina*), each of which reflects a warming trend and perhaps a reduction of the closed-canopy deciduous forest at that time.

Evidence of Holocene climatic change is apparent but subtle in the Great Lakes region. Cleland (1966: Appendix I) located over 160 eastern zooarchaeological lists and organized the faunal data from the Great Lakes region chronologically with respect to both the climatic episodes of Baerreis and Bryson (1965) and the paleoclimatologic periods of Griffin (1961). The most nearly complete faunal sequence was obtained from the Raddatz Rock shelter, Wisconsin (Cleland, 1966; Parmalee, 1959). Cleland (1966: Figure 2) arranged the Raddatz species by habitat preference and plotted the relative abundance of each group (Figure 12-15). Deciduous-forest forms are dominant throughout the Holocene, as noted by Guilday in more-eastern sites. However, coniferous and aquatic forms are strongly represented (but not dominant) in the lowest levels, and they suggest that the earliest deposits probably accumulated during a cool, wet pre-Altithermal climate. Both the decrease of coniferous and aquatic species between levels 12 and 8 and the sudden appearance of grassland forms in level 11 may represent the onset of the Altithermal, an event that also is reflected by an increase in prairie forms through level 8. A radiocarbon date of 5241 yr B.P. on level 10 is consistent with other dates on Altithermal deposits. A circa 7750 ± 150 yr B.P. fish fauna, which also represents a prairie association, was collected from an excavation in Milwaukee, Wisconsin, and further documents the presence of prairie in eastern Wisconsin during the Altithermal (West and Dallman, 1980). Above level 8 in Raddatz, an increase in coniferous and aquatic forms probably reflects a return to cooler and moister post-Altithermal climatic conditions. Short-term changes in relative abundance within the above pattern are attributed to smaller-scale climatic stages and episodes postulated by both Griffin (1961) and Baerreis and Bryson (1965).

At Lost River Sink, changes in the relative abundance of the 20 species present imply a slight reduction in forest cover to a grassy parkland and a subsequent period of reforestation between 3970 and 2720 yr B.P. (within the Sub-Boreal episode) in southwestern Wisconsin (Theiling, 1973). This change is subtle and equal in intensity to post-Altithermal fluctuations at Raddatz. Climatic fluctuations may be better represented at Raddatz and Lost River than elsewhere in the eastern United States because the sites are strategically located near the southern margin of the coniferous forest, the northern border of the deciduous forest, and the southwestern Wisconsin prairie (Cleland, 1966). Also, the effect of increased aridity on the Plains would have become progressively muted toward the east, being noticeable in the Great Lakes region but having little impact on the more-maritime areas (King and Graham, 1981).

Small-mammal remains from archaeological sites along the Illinois and Mississippi Valleys also provide evidence of environmental change near the Prairie Peninsula (Parmalee, 1968). Parmalee suggests that a cool-moist boreal environment prevailed in the northern Mississippi Valley following glacial retreat. Northern indicator species retreated from the area during a subsequent warm-dry period,

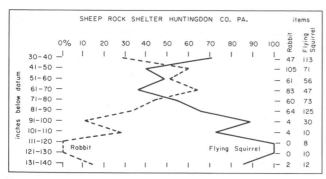

Figure 12-14. Fluctuation in the relative abundance of rabbits (*Sylvilagus* sp.) and southern flying squirrels (*Glaucomys volans*) recovered from Sheep Rock Shelter (Guilday and Parmalee, 1965)

and dry upland ecotypes entered the region at that time. Cooler climates with more moisture followed and resulted in both the extirpation of the xeric species from the region and the establishment of the modern fauna. Purdue and Styles (in preparation) and McMillan and Klippel (1981) both confirm Parmalee's (1968) interpretations from an expanded data base. They have collated data from more than 25 fauna-bearing sites and, like Cleland (1966) in Michigan, note that the upper Mississippi Valley Holocene was mainly affiliated with the eastern deciduous forest but that the mammalian communities were modified during the middle-Holocene time toward warmer and drier associations. The Prairie Peninsula extended eastward, but the effects were not as dramatic as they were farther west. The Koster site, which has one of the most complete fauna-bearing Holocene stratigraphic sequences in North America (Butzer, 1977; Hajic, 1981), undoubtedly will provide a superb climatic history for southern Illinois. The climatic cycles and gradients of the upper Mississippi Valley region are discussed in the section that follows.

NORTHERN PLAINS AND PRAIRIE-FOREST ECOTONE

The fauna of both the Casper site (Frison, 1974; Wilson, 1974) and the Medicine Lodge Creek site (Walker, 1975) indicate that the western Great Plains (Wyoming) achieved a faunal assemblage similar to that of today by 10,000 yr B.P. The transition from typical Pleistocene climate apparently took place in a series of sharp changes beginning sometime before 11,000 yr B.P. (Walker, 1982). First, tundra members of the Pleistocene fauna were lost. It is apparent that coniferous forests of a post-tundra mesic-montane association still predominated in eastern Wyoming at 10,700 yr B.P. because the area of sympatry constructed for the Folsom faunule of the Agate Basin site (Walker, 1982) lies in northwestern Wyoming. The area of sympatry representing the overlying 10,300 yr B.P. Hell Gap faunule at Agate Basin reflects a shift to conditions similar to those of modern south-central Wyoming—primarily a sagebrush/tall-grass/short-grass prairie with remnants of the coniferous forest. Short-grass prairie was common by 10,000 yr B.P. in central Wyoming. These changes in community structure followed an individualistic pattern, with the addition and subtraction of two or three species between each of the approximately 400-year-long time planes. Snails and pollen associated with the Hudson-Meng site (Agenbroad, 1977) imply that conditions were still relatively mesic in western Nebraska at about 9000 yr B.P. and that cool-moist grassland existed there at that time. Conditions of increasing aridity then developed in the Hudson-Meng area through 4000 yr B.P. and possibly until 2000 yr B.P.; faunal and floral re-

mains indicate a semiarid climate for the vicinity. Modern relationships, with increased numbers of mesic species, were established after 2000 yr B.P. Data from the Hawken site (Frison et al., 1976) suggest that the Black Hills may have been an oasis-like feature during the Altithermal.

Excavations at the Cherokee Sewer site in northwestern Iowa (Anderson and Semken, 1980) have revealed three cultural horizons dated at 8400, 7300, and 6350 yr B.P. within a stratified alluvial fan. Each horizon has produced a micromammalian fauna sufficient to construct an area of sympatry (Figure 12-16). For Cultural Horizon III (8400 yr B.P.), the faunule most resembles the community presently living in eastern Iowa. More moderate temperatures, increased moisture, and a longer growing season compared to that at present are inferred for western Iowa for that time. For Cultural Horizon II (7300 yr B.P.), the faunule is more difficult to interpret because it is not harmonious; the modern range of the hispid pocket mouse (*Perognathus hispidus*) is disjunct from the common range of the other species. This disharmonious arrangement suggests a community structure somewhat different from any modern one. Otherwise, the sympatry for Horizon II reflects a drier environment with cooler summers than in the area at about 8400 yr B.P. Prairie species were more common, and greater temperature extremes may have prevailed. Climatic conditions then changed markedly. For Cultural Horizon I (6350 yr B.P.), the southern boundary of the sympatry is north of Cherokee and its eastern boundary is 440 km west of its Cultural Horizon II position. The mean annual temperature appears to have decreased sharply, and the climate was substantially more arid than during either previous interval or at present. This evidence for increasing aridity at Cherokee between 8400 and 6350 yr B.P. is in agreement with the concept of an ''Altithermal'' on the eastern Plains margin. The sharply increased deposition of clastic sediment at West Lake Okoboji (Baker and Van Zant, 1980) between 7300 and 6300 yr B.P. implies increased erosion and supports an Altithermal in northwestern Iowa. However, for eastern Iowa at 6200 yr B.P., the Mud Creek local fauna sympatry (Kramer, 1972) occurs in eastern Wisconsin (Figure 12-17) and indicates that effective moisture was greater in eastern Iowa at that time than at present. Therefore, the climatic gradient between western and eastern Iowa may have been more pronounced at about 6200 yr B.P. than it is today. The Prairie

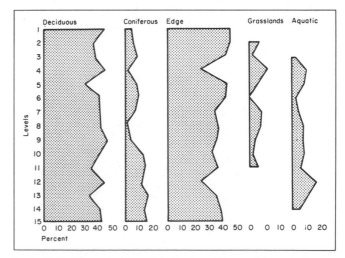

Figure 12-15. Relative abundance of deciduous, coniferous, edge, grassland, and aquatic mammals in Raddatz Rockshelter. (From Cleland, 1966.)

Peninsula expanded into Illinois sometime around Mud Creek time (King and Allen, 1977), but it is not as evident in Illinois as in western Iowa. Mud Creek reinforces a reduction in the effect of Altithermal climates from west to east and a cessation of the Altithermal in eastern Iowa approximately 200 years after the deposition of Cultural Horizon I at Cherokee.

A muted Altithermal also is apparent to the southeast between 8000 and 6300 yr B.P. in the faunal sequence at Rodgers Shelter in central Missouri (Wood and McMillan, 1976). The earliest deposits indicate that from about 10,000 yr B.P. until approximately 8400 yr B.P. a forest-edge community typical of the area today was present around the site. Prairie species then appear (Figure 12-18) and become most numerous, with a concomitant decline in forest species, between 8500 and 6500 yr B.P. Mammalian remains from 12 arbitrary levels excavated in nearby Graham Cave (Klippel, 1971) also indicate that central Missouri was cool, moist, and largely forested between 10,800 and 8000 yr B.P. At about 8000 yr B.P., warmer and drier conditions resulted in the forest community being replaced by a forest-edge or parkland environment. Faunal associations from the upper levels of Graham Cave indicate increasingly moist climatic conditions after 5000 yr B.P. Size gradients in both tree squirrels and rabbits from Rodgers Shelter and Graham Cave also point to a middle-Holocene prairie expansion (Purdue, 1980). Unfortunately, deposits dating between, 5200 and 3600 yr B.P. (revised from 6300 and 3000 by Kay [in Purdue, 1980]) at Rodgers are sterile, but prairie species were substantially reduced in diversity and abundance by 3600 yr B.P. and disappeared by 2000 yr B.P. That "the landscape again supported forest after about 3000 B.P. is evidenced by the in-

creased frequencies of deer, turkey, and raccoon, and the accompanying disappearance of all grassland vertebrate species'' (McMillan, 1976: 228).

Areas of sympatry for Atlantic- (Altithermal-) age faunas in Iowa and Missouri also illustrate the geographic difference in severity of climate during middle-Holocene time. Sympatries for about 8400 yr B.P. at Cherokee and Rodgers (Figure 12-17A) show that woodland or forest-edge communities were characteristic of both sites. By 6300 yr B.P., the Rodgers sympatry (Figure 12-17B) lay on the mixed forest/prairie ecotone of central Iowa; the Cherokee sympatry reflected basic prairie typical of western Minnesota and the Dakotas today. The climatic gradient was clearly greater than that at present between the two sites during the Altithermal, but it was not as great as the gradient between Cherokee Cultural Horizon I and Mud Creek.

Modoc Rockshelter (Fowler, 1959), Randolph County, Illinois, approximately 320 km due east of Rodgers Shelter, shows a climatic relationship to Rodgers similar to that of the Mud Creek local fauna to Cherokee Cultural Horizon I. The Modoc deposits span the period between 10,000 and 4000 yr B.P. and contain vertebrate fossils throughout. Parmalee (1959a: 63) infers uniform ecologic conditions from 10,000 to 4000 yr B.P. except for an initial slightly cooler and damper period. Fowler (1959: 41), however, states that after 8000 yr B.P. fish rather than deer became highly important in the diet of the Modoc inhabitants until 5500 yr B.P., when deer and waterfowl again became important. Perhaps the shift in subsistence patterns resulted from forest reduction because of drier middle-Holocene climates, as interpreted by McMillan (1976) at Rodgers Shelter. King

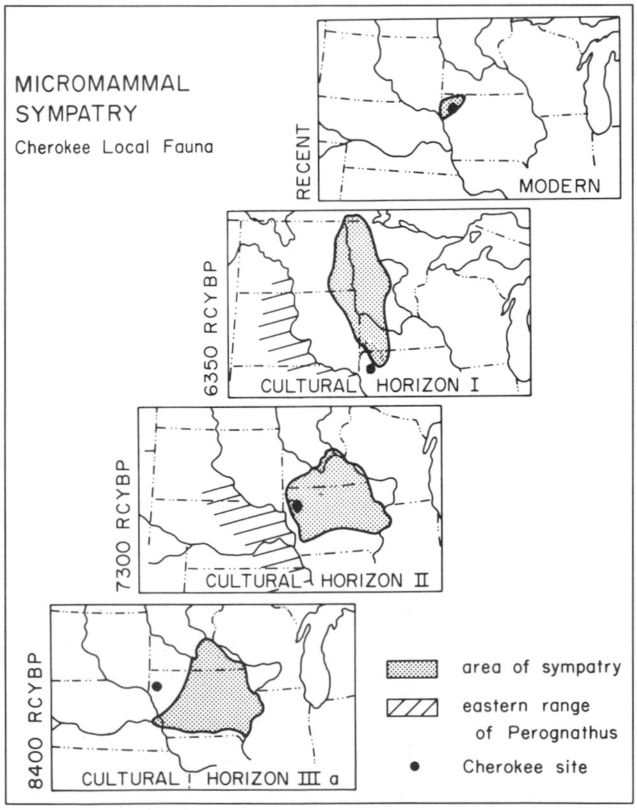

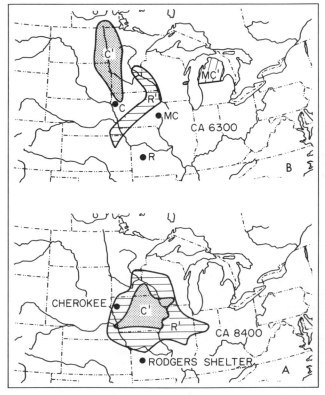

Figure 12-16. Changing areas of sympatry at the Cherokee Sewer site between 8400 and 6350 yr B.P. The modern mammal community of Cherokee County, Iowa, is included for comparison. (From Semken, 1980.)

Figure 12-17. Area of micromammal sympatry for faunules of the Cherokee (C), Mud Creek (MC), and Rodgers Shelter (R) local faunas. (A) about 8400 yr B.P.; (B) About 6300 yr B.P. The woodrat (*Neotoma floridana*) is not included within the 8400 yr B.P. Rodgers sympatry.

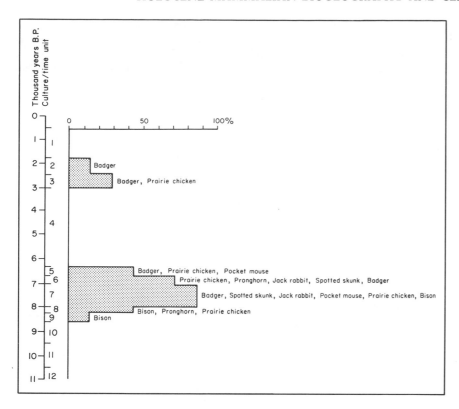

Figure 12-18. Relative abundance of prairie mammals at Rodgers Shelter. (From McMillan, 1976.)

and Allen (1977) record a replacement of arboreal pollen with grasses and other nonarboreal pollen between 8700 and 5000 yr B.P. in the Old Field site, a presently dry Mississippi Valley swamp near Advance, in eastern Missouri. Apparently, this also reflects maximum expansion of the Prairie Peninsula during this time. Thus, the Cherokee, Modoc, and Rodgers records suggest that the Altithermal had an effect on micromammalian populations to the southeast of Cherokee but that directly to the east at Mud Creek, Raddatz, Meadowcroft, and Sheep Rock this effect was reduced.

Butzer (1977), in his analysis of geomorphic features associated with the Koster site and the Illinois Valley, also depicts a series of Holocene climatic cycles for the upper Mississippi Valley, with three relatively arid periods occurring between A.D. 705 and 1000, 2100 and 1900 yr B.P., and 9700 and 5000 yr B.P., with the most vigorous erosion taking place between 8500 and 7700 yr B.P. within the earliest arid period. Cyclic deposition also is evident in the Koster alluvial fan where Hajic (1981) recorded depositional breaks at 8700, 8450, 7000, 5825, 4880, and 3950 yr B.P., with the strongest corresponding with the Altithermal (8400 to 4880 yr B.P.) boundaries. The Koster cycles also correlate remarkably with those proposed by Knox (1972) for depositional cycles elsewhere in the upper Mississippi Valley and the boundaries of the climatic episodes proposed by Bryson and others (Wendland, 1978). They also conform well with the initiation of fan deposition at Cherokee at about 8400 yr B.P. (Hoyer, 1980) and with evidence of increasing aridity over this period in northwestern Iowa, as suggested by pollen (Baker and Van Zant, 1980), mollusks (Baerreis, 1980), and microvertebrates (Semken, 1980). Butzer's (1977) record of a relatively more mesic interval sometime after 7700 yr B.P. within the arid period between 9700 and 5000 yr B.P. is substantiated by the mesic 6200 yr B.P. Mud Creek local fauna (Kramer, 1972). Thus, Butzer's data support the concept of an increased middle-Holocene climatic gradient between the

Mississippi and Missouri Valleys as predicted from the Mud Creek and Cherokee local faunas.

Two local faunas in southwestern Iowa, Garrett Farm (circa 3400 yr B.P.) and Pleasant Ridge (circa 1540 yr B.P.), indicate that western Iowa was progressing toward modern climatic conditions over that period. A post-Altithermal decline of the prairie is demonstrated by steppe species that progressively decline from 32% at the Garrett Farm to 27% at Pleasant Ridge and finally to 18% at present (Fay, 1980). Both local faunas are disharmonious and add to the growing body of evidence that Holocene climates were more variable than previously thought.

The Brynjulfson Caves (Parmalee and Oesch, 1972) of Boone County, Missouri, reveal two periods of deposition. Cave No. 1, dated between 9440 ± 760 and more than 27,000 yr B.P. (Coleman and Liu, 1975), contains nine extinct large mammals and an associated fauna that best reflects the boreal aspect typical of Pleistocene faunas. Cave No. 2, with Late Woodland artifacts in the upper levels and with randomly collected radiocarbon dates of 1400 ± 200 and 2460 yr B.P., contains both the pronghorn (*Antilocapra americana*) and the plains pocket gopher, (*Geomys bursarius*) which attest to a remnant of "prevailing arid conditions at or near the cave locale prior to reestablishment of the deciduous forests" (Parmalee and Oesch, 1972: 49). Cave No. 2 also contains three extinct taxa, *Platygonus compressus* (flat-headed peccary), *Dasypus bellus* (giant armadillo), and *Canis dirus* (dire wolf). These specimens may support the opinion of J. H. Quinn (personal communication, 1972) that the Ozark Highlands were a Holocene refugium for the extinct Pleistocene megafauna, but Parmalee and Oesch (1972) also note that some of the matrix in Brynjulfson Cave No. 2 may have accumulated before the Altithermal. Unfortunately, no stratigraphic information has been preserved with the specimens (R. W. Graham, personal communication, 1981). The Tick Creek Cave (Parmalee, 1965) and Arnold

Research Cave sequences (Falk, 1970) of Missouri have extensive faunas that only have been analyzed from the standpoint of subsistence and should be examined for paleoecologic significance.

Prairies invaded Minnesota at the same time they invaded western Iowa. Shay (1971: 73) notes from the Itasca site that by 8000 yr B.P. "plant and animal associations in the Middle West approximated those of recent times except that prairie and oak savanna were more widespread." By this time, two major cultural/ecologic adaptations had developed, based on bison in the North and West and deer in the South and East. This division of bison or deer subsistence patterns was later graphically displayed by Shay (1978), who plotted percentages of bison and deer bone recovered from post-3500 yr B.P. archaeological sites along an east-west transect across the Prairie Peninsula (Figure 12-19). Fluctuations in this sharp interface probably could be used to plot the expansion and contraction of prime bison range and to test the climatic-episode model (Wendland, 1978) when a sufficient number of sites representative of each episode become known. Shay's (1978) initial attempts at this with the Neo-Atlantic, Pacific, and Neo-Boreal episodes are encouraging.

Extensive excavations of earth-lodge villages along the Missouri River in South and North Dakota have yielded an array of faunal remains dated from A.D. 900 until the present. Although the collecting strategies of archaeologists have varied widely (Falk, 1977), several sites have been intensively waterscreened for total recovery of biologic and lithologic debris. Sympatries (Figure 12-20) constructed from the microfauna of three juxtaposed sites in South Dakota (the Helb, Lower Grand, and Walth Bay villages), all of which have similar topographic positions on high terraces, illustrate differences in the biogeographic associations of each village.

The circa A.D. 1050 Helb site (Falk and Calabrese, 1973) was occupied during the Neo-Atlantic episode, a relatively mesic period on the northern Plains (Wendland, 1978). The Helb sympatry (Figure 12-20A) reflects a gallery-forest environment typical of the valley floors in the region today. It is probable that this forest extended up to the high-terrace position when the site was occupied. The Lower Grand village, occupied at about A.D. 1350, can be related to the Pacific episode, a dry phase on the northern Plains that began at about A.D. 1100 and continued into the 1600s. A sympatry (Figure 12-20B) for Lower Grand indicates that an upland prairie typical of the region today surrounded the site and suggests that the gallery forest was limited to lower elevations along the Missouri Valley than

during Neo-Atlantic time. This interpretation is compatible with a "dry" Pacific episode. Judging from the sympatry for Walth Bay (Figure 12-20 C) (Ahler et al., 1974), the drought was more severe around A.D. 1550, for the prime area of sympatry is associated with the short-grass prairie of central Colorado. The latter sympatry may be an overstatement; the Walth Bay taxa are almost sympatric everywhere within the dashed lines on Figure 12-20 C, and the differences may result from our incomplete knowledge (within 50 km) of the range extremes of a few species. Nonetheless, the sympatry reflects drier conditions in the area than prevailed earlier in the Pacific episode. Cooler temperatures are suggested, since the configuration of the Walth Bay sympatry is either to the north of, or at higher elevations than, the Walth Bay village. This interpretation supports the concept that the Pacific episode can be subdivided into an early warm Pacific I interval and a later cool Pacific II interval. A similar sequence of increasing severity in climatic conditions is demonstrated by Semken and Satorius (1980) from the multicomponent White Buffalo Robe site. There, the Nailati phase (between about A.D. 1200 and 1400) was associated with a typical well-balanced prairie fauna, and the middle component faunule (Heart River phase, A.D. 1420 to 1660) reflected a decrease in grassy ground cover, as shown by an increase in browsing rodents. The post-1780 Knife River phase faunule, with a single browsing species comprising half of a faunule with low species diversity, indicates that the area may have been denuded. The mammal association of this faunule is typical of a pioneer plant community (Voight and Glenn-Lewin, 1979).

SOUTHEASTERN FOREST

Except for the Florida peninsula, evidence for faunal change in the southeastern deciduous forest region over the Holocene is sparse. The absence of faunal change is supported in part by the pollen spectrum from Anderson Pond, White County, Tennessee, where Delcourt (1980: 270) notes that, "from 9500 yr B.P. to the present, as far as is possible to detect from pollen analysis, the arboreal flora of the eastern Highland River and adjacent Cumberland Plateau of Tennessee has changed little." Delcourt (1980: 271) adds that "between 8000 yr B.P. and 5000 yr B.P., oak, ash, hickory, birch, alder, button bush, and Virginia willow became more important" and that "a mid-Holocene warming and drying trend is inferred for Middle Tennessee

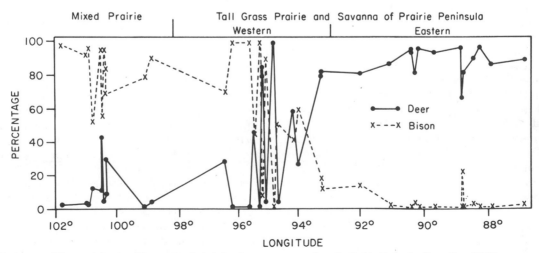

Figure 12-19. Percentages of bison and deer bone from archaeological sites from west to east along the Prairie Peninsula. (From Shay, 1978.)

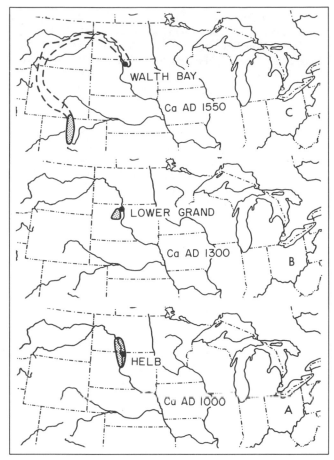

Figure 12-20. Area of sympatry for three earth-lodge village sites on the
northern plains. (A) Helb, about A.D. 1000, Neo-Atlantic site; (B)
Lower Grand, about A.D. 1300, Pacific I site; and (C) Walth Bay, about
A.D. 1550, Pacific II site (Semken and Falk, in preparation). Dashed line
represents possible extension of the sympatry if some allopatric species
actually came into contact.

from 8000 yr B.P. to 5000 yr B.P.'' After 5000 yr B.P., mesic
deciduous forest taxa are represented at Anderson Pond.

Baker Bluff Cave (Guilday et al., 1978) in eastern Tennessee contains a stratified late-Pleistocene faunal sequence, but deposits
representing the Holocene/Pleistocene transition are absent. Above
this hiatus, 11 artifacts date the overlying meter of section as Archaic
to Late Woodland in age. Disturbance by successive occupations pro-
hibited analysis by level, but 19 species of mammals are recorded
from the Holocene unit, one of which (*Castoroides ohioensis*, the
giant beaver) is extinct. All small mammals except the thirteen-lined
ground squirrel (*Spermophilus tridecemlineatus*) presently reside in
the area, and this fact suggests that there were no major changes
throughout the Holocene. The ground squirrel remains could be
either residual from its widespread eastern distribution during the
late Pleistocene (Figure 12-8) or reworked, perhaps along with
Castorides, from underlying deposits. Guilday and others (1978:
61-62) contrast the Pleistocene faunal gradient with that of the present and note the following.

A comparison of the late Wisconsinan faunal gradient of the
Ridge and Valley province with that of today shows significant differences. Based upon Recent small mammal ecological require-

ments, the late Pleistocene gradient, Pennsylvania (north) to Tennessee (south), was in the direction ''boreal'' to ''temperate''.
Today the same gradient runs ''temperate'' to ''austral''. This
reflects the regional postglacial rise in mean temperature. . . .The
faunal gradient is more extreme today than it was under late glacial
conditions although different combinations of species are involved.

The similarity of the mammalian fauna throughout the Holocene
of the southeastern United States is supported by deposits from both
Russell Cave and the Stanfield-Worley Bluff Shelters in northern
Alabama. Excavation of seven superimposed levels in Russell Cave,
which range in age from 8500 to 370 yr B.P. (A.D. 1580), produced
22 species of mammals. Faunal remains are not extensive in every
level, but ''all vertebrates (except the passenger pigeon, long-nosed
peccary, and porcupine) occur as residents or migrants in the area.
. . . There is nothing to indicate drastic ecological or climatic changes
during the period of cave occupancy.'' (Weigel et al., 1974: 85). Remains of mammals from the Stanfield-Worley Bluff Shelter are
sparse, but through the section similar frequencies of both the white-
tailed deer and the gray squirrel, which live in the area today, support
continuity of climate throughout Holocene time (Parmalee, 1962).
Judged from the two faunal lists, neither of which contains species
smaller than a chipmunk, it appears that excavation strategies were
not designed to recover smaller mammals. Thus, any evidence for
subtle changes within the deciduous-forest sequence detected by
Delcourt (1980) from pollen may have been lost.

This is confirmed by detailed excavations in Salts Cave, Kentucky
(Duffield, 1974). The Horizon III faunule, from the base of the
stratigraphically controlled sample, contains a more diverse microfauna than present in the two overlying horizons. Also, grassland
species (e.g., *Sigmodon*) only appear in Horizon III (circa 2940 to
2560 yr B.P.), indicating a period of climatic fluctuation during
which grasslands were introduced into the forested region. Climate
stabilized with the initiation of the Sub-Atlantic (circa 2500 yr B.P.),
and forest situations persisted after that date.

The Devil's Den local fauna of Florida has been dated by radiocarbon analysis of bone as representing a period between 8000 and 7000
yr B.P. (Martin and Webb, 1974). This fauna, which contains eight
extinct taxa, is regarded as either very latest Pleistocene or early
Holocene by Martin and Webb. Holocene age is suspected because of
the date and because most of the extant mammals reflect a xeric
parkland-savanna similar to that present in the region today. Martin
and Webb note that the forest in the vicinity was more mesic then
than now because woodrats (*Neotoma floridana*), flying squirrels
(*Glaucomys volans*), and gray squirrels (*Sciurus carolinensis*),
among other mammals, were in the fauna. Three other Devil's Den
taxa, the muskrat (*Ondatra zibethicus*), the meadow vole (*Microtus
pennsylvanicus*), and the gray myotis (*Myotis grisescens*), are now
found only north of the locality; this suggests cooler summers during
the period of deposition. The total Devil's Den association suggests
that early-Holocene conditions in Florida, as in the northeastern
United States, were similar to those of today but somewhat cooler and
moister.

A study of Florida archaeofaunas by Bullen (1965) reveals that all
mammals except the muskrat that are not present in Florida today
were gone by about 5000 yr B.P. The muskrat apparently survived in
the state until about 3000 yr B.P. ''Retraction to their present limits
of the ranges of such species as *Synaptomys cooperi*, *Microtus pennsylvanicus*, and *Ondatra zibethicus* apparently occurred after the

major times of megafaunal extinction'' (Martin and Webb, 1974: 144).

Little Salt Spring, in southwestern Florida, a freshwater cenote during dry intervals of the Holocene, has provided a record for climatic change since 12,030 yr B.P. via botanical, archaeological, eustatic, and vertebrate data. Clausen and others (1979) report a paleo-Indian occupation on a now submerged ledge 26 m below the lip of the sink. Remains of an extinct ground sloth (*Megalonyx*) and a slaughtered giant land tortoise (*Geochelone crassiscutata*) were found at this level along with a variety of other vertebrates. By 10,000 yr B.P., the water table in the sink had risen to 11 m below the surface and white-tailed deer (*Odocoileus virginianus*) remains became common in the deposits.

The rise in the groundwater table crested about 8500 yr B.P. 1 m below the rim. Paleobotanical evidence suggests a moist interval until that date. Between 8500 and 5500 yr B.P., the period recorded by Delcourt (1980) as being slightly warmer and drier in central Tennessee, the water table dropped 8 m. After 5500 yr B.P., more mesic conditions returned, the water table rose to its present position above the lip of the sink, and aquatic plants such as water lilies again became abundant. A relatively mesic climate can be presumed for the last 5000 years. Little Salt Spring demonstrates that in marginally arid regions of the Southeast, as on the High Plains of the United States, climatic change had greater apparent impact on the flora and fauna than it did in the eastern deciduous forest. The combination of mesic, and semiaquatic and xeric species that characterized the Holocene local faunas of northern and central Florida also was represented in the late-Holocene deposits of Nichol's Hammock natural trap in southern Florida (Hirschfield, 1968), as well as in the modern fauna of that region.

A sharply pointed wooden stake, the source of the 12,030 yr B.P. date on the Paleo-Indian horizon at Little Salt Spring, was taken from the body cavity of a giant tortoise. It apparently had been driven in between the carapace and the plastron. Moreover, the tortoise was overturned, resting on fire-hardened clay, and the internal skeletal elements appeared carbonized. From these observations, Clausen and others (1979: 613) conclude that Little Salt Spring contains the earliest evidence of human activity in Florida and that the site represents the first evidence in the Southeast for the association of humans and extinct vertebrates. The latter portion of the statement is not correct, however. Sellards (1952) reviewed the circumstances surrounding the human remains, known since 1917, associated with Pleistocene fauna at Vero (now Vero Beach), Florida, and Loomis (1924) recorded the contemporaneity of humans and extinct animals at Melbourne, Florida. Field reconnaissance, apparently without excavation, led Rouse (1950) to question the association at these sites, but chemical analyses of both human and extinct vertebrate remains (Heizer and Cook, 1952; Sellards, 1952) have supported the association. Weigel (1962) redescribed the Vero exposures in detail and reported a confusing series of radiocarbon dates from 30,000 to 1625 yr B.P., but he confirmed the association of artifacts with Pleistocene faunal remains in a test excavation 3 m deep. Although these sites should be regarded with suspicion because the stratigraphic relationships are questionable, they cannot be ignored as possible Holocene faunas. Human remains also were recorded with the Devil's Den local fauna (8000 to 7000 yr B.P.) discussed above. However, extinct mammals have not been recovered in any clearly defined Archaic site in Florida and only rarely from other prehistoric sites in the United States.

The complex stratigraphic deposits from Avery Island, Louisiana, which are similar to those at Vero-Melbourne, have yielded a variety of recent and extinct mammals over the past century. Gagliano (1967) reviews these deposits, the fossils, and a bewildering array of radiocarbon dates with determinations as young as 8390 yr B.P. on extinct taxa. The Quaternary sequence (Gagliano, 1967) records a change from an early-through-middle-Holocene prairie grassland with parklike stands of hardwoods around water holes to a seashore environment beginning at about 5000 yr B.P. By 3650 yr B.P., sea level reached its present level, and the landscape was transformed into an area of coastal swamps. These dates are similar to those marking ecologic changes elsewhere in North America. At present, Avery Island, a salt dome with topographic expression, serves as a refuge for coastal animals during storms.

SOUTHERN PLAINS

As in other regions, length of time involved in the transition from Pleistocene to Holocene conditions on the southern Plains is questionable. Although it is clear that megamammalian extinction and dissolution of disharmonious faunas began earlier, about 12,000 to 10,000 yr B.P. (Graham, 1976b), aspects of typical Pleistocene conditions appear to have persisted some 2000 years later on the southern Plains than the 10,000 yr B.P. date frequently assigned to the boundary. Slaughter (1967: 155) notes ''that extant animals whose geologic ranges extend back into the Pleistocene have modern ranges established about the same time as the last major period of extinction.'' He further maintains that the modern climate on the southern Plains developed during the extinction period, which he places between 9000 and 8000 yr B.P.

The Domebo local fauna of Oklahoma (circa 10,100 to 11,045 yr B.P.). collected on the southeastern side of the Plains (as defined by the western boundary of pedalfers), represents a climatic situation approaching the modern configuration but with milder winters and cooler summers (Slaughter, 1966). This interpretation, typical for many late-Pleistocene local faunas, was based on the joint occurrence of the southern bog lemming (*Synaptomys cooperi*), now found only north of the site, and the giant tortoise (*Geochelone*), a southern-temperate or subtropical taxon. The Rex Rodgers local fauna of northwestern Texas suggests that this ''equable'' climate extended into earliest Holocene time. The cooccurrence of the water snake (*Natrix sipedon*), the short-tailed shrew (*Blarina brevicauda*), and the meadow vole (*Microtus pennsylvanicus*), all of which now range at least 320 km to the north or east of the Texas panhandle, also indicate that summers were cooler and moister than at present between 10,500 and 9000 yr B.P. in northwestern Texas (Schultz, 1978).

Johnson (1976) suggests that this pluvial condition continued well into the Holocene and that ''essentially modern conditions did not prevail until Late Paleo-Indian times (7067 ± 80 yr B.P.) in the immediate vicinity of the Lubbock Lake site on the southwestern plains.'' Johnson and Holliday (1981: 185) describe an almost modern faunule associated with late Paleo-Indian Feature FA6-3 (8655 to 8095 yr B.P.), which ''probably reflects the onset of modern climatic conditions.'' Between 9000 and 8000 yr B.P., a widespread marsh existed at the now relatively arid locale (Johnson and Holliday, 1979). The continuation of Pleistocene-like ''pluvial'' climates into the early Holocene of central Texas, if the radiocarbon dates are correct (Land et al., 1980), is supported by mammal remains recovered from the circa 9680 to 9310 yr B.P. Layer C of Schultze Cave (Dalquest et al., 1969), the circa 9000 to 7000 yr B.P. Units 3C, 2C, and

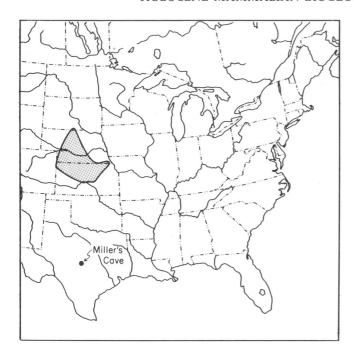

Figure 12-21. Area of sympatry for the travertine faunule, about 7200 yr B.P., Miller's Cave, Texas.

2D of Friesenhahn (Graham, 1976b), and the circa 8000 yr B.P. Kline Cave local fauna (Roth, 1972). Each of these units shares faunal combinations of the eastern chipmunk (*Tamias striatus*), the southern bog lemming (*Synaptomys cooperi*), the ermine (*Mustela erminea*), the short-tailed shrew (*Blarina brevicauda*), and either the prairie vole (*Microtus ochrogaster*) or the pine vole (*M. pinetorum*). All these species presently reside either north on the Plains or east in the deciduous forest, and their presence indicates a cooler and moister climate than that in the area today. Van Devender and Wiseman (1977: 22) reinforce the findings for the mesic nature of early-Holocene conditions in the Southwest with botanical evidence and report that the "last major charge in the vegetation of the southwest that could be interpreted as a shift from pluvial to a nonpluvial climate was 8000 years ago." The shift to a nonpluvial climate after about 8000 yr B.P. also is indicated by the circa 7200 yr B.P. Travertine faunule from Miller's Cave (Patton, 1963). A sympatry (Figure 12-21) based on rodent and insectivore remains recovered from this unit lies in the central Plains and indicates the breakdown of the "pluvial" aspect of the early Holocene. However, temperatures were still similar to those north of the site at the present time.

Lundelius (1967) divides the late-Quaternary fauna of Texas into three assemblages: (1) a late-Pleistocene unit consisting of large extinct forms (e.g., mammoths, sloths, and camels), (2) extant species that have retreated from the state (e.g., the ermine, short-tailed shrew, and southern bog lemming), and (3) species still present in the region (e.g., the coyote, white-tailed deer, and least shrew). Members of the above assemblages coexisted in central Texas until approximately 8000 years ago. By that time, most if not all members of group 1 disappeared, leaving species associated with groups 2 and 3 residing in the state. Elements of group 2 gradually withdrew, the last disappearing approximately 1000 years ago. This left a residuum of group 3 species, which, augmented by two late-Holocene immigrants (the collared peccary [*Tayassu tajacu*] and the armadillo [*Dasypus novemcinctus*]), make up the modern fauna of the state. Lundelius (1967) notes that *Sorex cinereus* (the masked shrew), *Microtus pennsylvanicus* (the meadow vole), and *Mustela erminea* (the ermine)

seem to have disappeared from central Texas about 8000 yr B.P., the time of final megafaunal extinction. *Synaptomys cooperi* (the southern bog lemming) and *Peromyscus nasutus*, now *P. difficilis* (the rock mouse), were still present about 7000 years ago. *Microtus ochrogaster* (the prairie vole) persisted in central Texas until 3000 years ago and did not disappear from northwestern Texas (Deadman's Shelter) until sometime after A.D. 120 to 710 (Schultz and Rawn, 1978). *Blarina brevicauda* (the short-tailed shrew), *Scalopus aquaticus* (the eastern mole), and *Procyon lotor simus* (a large species of raccoon) remained until at least 1000 years ago. *Geomys bursarius* (the plains pocket gopher), *Cynomys ludovicianus* (the prairie dog), and perhaps *Scalopus aquaticus* were decimated on the Edwards Plateau of central Texas during the past 200 years, as noted earlier in this review.

"There is no indication in any of these (central Texas) faunas that the climate during the interval 4,000-6,000 years ago was any drier or warmer than at present. In fact the occurrences of the otter (*Lutra canadensis*) in the Wunderlich site in Comal County 5,000 years ago indicates more humid conditions" at that time (Lundelius, 1967: 312). Both Lundelius (1967) and Frank (1965), the latter working with sediments, believe that Holocene warming and drying gradually progressed from west to east across the Edwards Plateau. Middens in Trans-Pecos Texas, faunal data from Howell's Ridge Cave, New Mexico, and pollen data from southwestern Arizona suggest that the summer monsoon was well-developed from 4,000 to 8,000 years ago and that present desert communities formed at some later date" (Van Devender and Wiseman, 1977: 23). Van Devender and Wiseman add that the limited data available do not support an Altithermal in the Southwest. Deposits of "Altithermal" age at the Lubbock Lake site (Johnson and Holliday, 1979) contain lacustrine marl and indicate that the cienega (marsh) associated with this western Texas site did not dry up until about 5000 yr B.P.

In their analysis of the circa 3800 yr B.P. Layer B of Schultze Cave in central Texas, Dalquest and others (1969) conclude that there is no evidence of a post-Altithermal mesic situation around the cavern at that time. Patton (1963: 37), in summarizing the circa 3000 yr B.P.

brown clay faunule from Miller's Cave, also concludes that "there appears to be an uninterrupted sequence of events leading gradually from the more moist, perhaps cooler, conditions during travertine formation [noted above] to the hotter, semiarid conditions of the present. There is no evidence for a return to any pre-existing conditions." However, Dillehay (1974) notes the absence of bison in southern High Plains sites dated between about 7500 and 4500 yr B.P. and again over the interval between about A.D. 500 and A.D. 1250. He attributes their absence to unfavorable climatic conditions during those periods.

Increased post-Altithermal desiccation is suggested by the 1350 yr B.P. Dye Creek local fauna (Dalquest, 1965). All 11 mammals in this northern Texas local fauna live in Montague County today, except *Perognathus merriami* (Merriam's pocket mouse) and *Onychomys leucogaster* (northern grasshopper mouse). Both of these species presently exist to the west of the site; this fact suggests that the prairie expanded farther into eastern Texas 1350 years ago than it had either before or after that date. Hall (1980), working in Oklahoma, substantiates this interpretation by recording a shift in pollen and land-snail frequencies, which indicate a change to drier conditions about 1200 yr B.P.

A slight revision to conditions cooler and/or more humid than today between A.D. 500 and 1291 is suggested by the central Texas Kyle site local fauna (Lundelius, 1962). Cooler environs for this interval also are supported by the presence of the prairie vole (*Microtus ochrogaster*) in the circa A.D. 120 to 710 Deadman's Shelter local fauna in northwestern Texas (Schultz and Rawn, 1978), for this plains animal is now found only to the north and east.

Arid conditions characteristic of western Texas today clearly have prevailed in southwestern Texas over the past 2000 years, at least until very recently. There is little or no evidence of faunal change in the stratified sequences of Bonfire Shelter (Frank, 1968), Centipede Cave (Lundelius, 1963), and Damp Cave (Lundelius, 1963), and any climatic fluctuations over this interval probably were minor. Since all species found in these sites presently inhabit the area, modern climates are projected into the late Holocene. Both climatic stability and similarity to modern conditions also are postulated for northern Oklahoma between A.D. 800 and 1500 (Henry et al., 1979). Remains of vertebrates, pollen, and mollusks recovered from four sites spanning the above interval along Birch Creek are essentially equivalent to the modern biota.

There is evidence of a marked change in climate at about A.D. 1200 in western Oklahoma and northwestern Texas. Bryson and others (1970) report a sudden influx of Panhandle Aspect villages in this region at that date and attribute this immigration of sedentary people to a marked increase in mesic conditions (sufficient for crop production) at that time. This sharp change in climate corresponds in time to increased aridity to both the north and the west. An increase in moisture for the late prehistoric of the region is confirmed by the Lubbock Lake sequence. Johnson and Holliday (1979) show that evidence for flowing streams is almost absent in the late Holocene of western Texas until the late prehistoric, when depositional systems at Lubbock Lake changed toward those present during the late Paleo-Indian period. Faunal changes associated with the historic period, a relatively mesic interval, on the southern Plains were discussed earlier in this chapter.

Conclusions

(1) The Holocene mammalian fauna of North America can be regard-

ed as an impoverished residuum of the late-Pleistocene fauna. However, analysis of local faunas younger than 10,000 years, primarily zooarchaeological associations, demonstrates that modern distributions cannot be wholly attributed to climatic change associated with deglaciation. The present distribution of many small mammals was strongly affected by (a) climatic change during the Holocene, (b) pre-Columbian cultural patterns, and (c) European settlement, as were such large mammals as bison and pronghorn.

(2) Holocene distributional changes involved all cardinal directions and included boreal, prairie, marsh, fossorial, and deciduous-forest ecotypes. These changes were individualistic rather than collective (organismic) in nature, except for those related to the introduction of European agriculture.

(3) A chronocline based on *Bison* horn core and metapodial dimensions is the best Holocene evidence for time-transgressive morphological variation. Other taxa should be examined for other useful indexes. Application of established modern topoclines to the fossil record has been successful in predicting both climatic fluctuation and stability during the Holocene.

(4) The Pleistocene/Holocene faunal transition in the northeastern United States apparently was complete within a few hundred years. After that event, modern deciduous and mixed forests dominated the region. These forests evidently were persistent in more maritime areas, since all Holocene local faunas were essentially modern in aspect. Evidence for a middle-Holocene xeric period, the Altithermal, between 5000 and 8000 years ago, as well as comparatively minor climatic cycles, is detectable to the west (Mississippi Valley), where the fluctuating prairie ecotone left its imprint on the fossil and depositional records. The evidence for these events becomes more subtle toward the east.

(5) The post-Pleistocene fauna of the western Great Plains achieved modern xeric configurations about 10,000 years ago, but this transition apparently occurred in a steplike fashion beginning 11,000 years ago. Desiccation progressed from west to east, with the first evidence of aridity affecting the eastern prairie/forest ecotone after 9000 years ago. Altithermal climates between 8000 and 6000 years ago were strongly diminished toward the east, but prairie species invaded both the Ozark Mountains of Missouri and the Illinois plains during the period. The climatic gradient across the prairie/forest border was stronger during the Altithermal than at present. Post-Altithermal environments closely resembled the relatively mesic pre-Altithermal climate, but post-Altithermal climatic fluctuations were sufficient to affect the life-styles of the prehistoric people of the Plains.

(6) In the Southeastern deciduous forest, mammalian evidence of Holocene climatic alteration other than range changes of individual species is sparse, and available data do not indicate any drastic ecologic or climatic changes. Similar frequencies of both deer and gray squirrel in the two major stratified sequences reported to date also support continuity of forest composition throughout the Holocene. The appearance of grassland species in a 2940 to 2560 yr B.P. horizon in Kentucky, palynologic data reported elsewhere in this volume, and marked changes in the distribution of gophers, porcupines, and fishers suggest that refined collecting techniques will produce vertebrate information supporting climatic fluctuation in the southeastern forest. Faunal modifications resulting from climatic change undoubtedly will be most apparent on the xeric Florida peninsula, where the water table dropped 8 m between 8500 and 5500 yr B.P., because of the greater number of ecotonal situations in that region.

(7) On the southern Plains, as in other regions, there is a question

regarding the length of time involved in the transition from Pleistocene to Holocene conditions. Although it is clear that megamammalian extinction and dissolution of disharmonious faunas began about 12,000 years ago, pluvial aspects of typical Pleistocene conditions appear to have persisted until about 8000 years ago, when modern climatic conditions first appeared. However, there is no evidence in any central Texas local fauna that the climate between 7000 and 4000 years ago (Altithermal time) was either drier or warmer than at present. In fact, the Holocene pattern, although erratic, appears to represent a period of increased aridity through time, with no overall post-Altithermal return to more mesic conditions prior to 1000 years ago. The absence of bison remains in southern High Plains sites from 7500 to 4500 years ago and again from about A.D. 500 to 1250 demonstrates that this pattern was not a continuum. Substantial changes in the distribution of small mammals on the southern Plains can be directly related to postcontact overgrazing and agriculture.

References

Adams, R. M. (1941). Archaeological investigations in Jefferson County, Missouri. *Transactions of the Academy of Science of Saint Louis* 5, 183-84.

Adovasio, J. M., Gunn, J. D., Donahue, J., Stuckenrath, R., Guilday, J., and Lord, K. (1978). Meadowcroft Rockshelter. *In* "Early Man in America" (A. L. Bryan, ed.), pp. 140-80. Department of Anthropology, University of Alberta Occasional Paper 1.

Agenbroad, L. D. (1977). Climatic change and early man in northwest Nebraska. *In* Paleoindian lifeways (E. Johnson, ed.). *The Museum Journal* 17, 117-25.

Ahler, S. A., Davies, D. K., Falk, C. R., and Madsen, D. B. (1974). Holocene stratigraphy and archaeology in the Middle Missouri River Trench, South Dakota. *Science* 184, 905-8.

Aikens, C. M. (1983). Environmental archaeology of the western United States. *In* "Late-Quaternary Environments of the United States," volume 2, "The Holocene" (H. E. Wright, Jr., ed.), pp. 239-51. University of Minnesota Press, Minneapolis.

Anderson, D. C., and Semken, H. A., Jr. (eds.) (1980). "The Cherokee Excavations: Holocene Ecology and Human Adaptations in Northwestern Iowa." Academic Press, New York.

Baerreis, D. A. (1980). Habitat and climatic interpretation from terrestrial gastropods at the Cherokee site. *In* "The Cherokee Excavations: Holocene Ecology and Human Adaptations in Northwestern Iowa" (D. C. Anderson and H. A. Semken, Jr., eds.), pp. 101-22. Academic Press, New York.

Baerreis, D. A., and Bryson, R. (1965). Climatic episodes and dating of Mississippian cultures. *Wisconsin Archaeologist* 46, 203-20.

Baker, R. G. (1983). Holocene vegetational history of the western United States. *In* "Late-Quaternary Environments of the United States," vol. 2, "The Holocene" (H. E. Wright, Jr.), pp. 109-27. University of Minnesota Press, Minneapolis.

Baker, R. G., and Van Zant, K. L. (1980). Holocene vegetational reconstruction in northwestern Iowa. *In* "The Cherokee Excavations: Holocene Ecology and Human Adaptations in Northwestern Iowa" (D. C. Anderson and H. A. Semken, Jr., eds.), pp. 123-38. Academic Press, New York.

Banfield, A. W. F. (1974). "The Mammals of Canada." University of Toronto Press, Toronto.

Bardwell, J. (1981). The paleoecological and social significance of the zooarchaeological remains from Central Plains Tradition earthlodges of the Glenwood locality, Mills County, Iowa. M.S. Thesis, University of Iowa, Iowa City.

Barkalow, F. S. (1961). The porcupine and fisher in Alabama archaeological sites. *Journal of Mammalogy* 42, 544-45.

Bedord, J. N. (1974). Morphological variation in bison metacarpals and metatarsals. *In* "The Casper Site: A Hell Gap Bison Kill on the High Plains" (G. C. Frison, ed.), pp. 199-240. Academic Press, New York.

Benn, D. W. (1980). "Hadfields Cave: A Perspective on Late Woodland Culture in Northeastern Iowa." University of Iowa, Office of the State Archaeologist Report 13.

Black, G. A. (1967). "Angel Site: An Archaeological, Historical, and Ethnological Study." Indiana Historical Society, Indianapolis.

Blair, W. F. (1958). Distributional patterns of vertebrates in the southern United States in relation to past and present environments. *In* "Zoogeography" (C. L. Hubbs, ed.), pp. 433-68. American Association for the Advancement of Science Publication 51.

Bogan, A. E., and Robison, N. D. (1978). "A History and Selected Bibliography of Zooarchaeology in Eastern North America." Tennessee Anthropological Association Miscellaneous Paper 2.

Bryson, R. A., Baerreis, D. A., and Wendland, W. M. (1970). The character of late-glacial and post-glacial climatic changes. *In* "Pleistocene and Recent Environments of the Central Great Plains" (W. Dort, Jr., and J. K. Jones, Jr., eds)., pp. 53-74. University Press of Kansas, Lawrence.

Buchanan, G. D., and Talmage, R. V. (1954). The geographic distribution of the armadillo in the United States. *Texas Journal of Science* 6, 142-50.

Bullen, R. P. (1965). Florida's prehistory. *In* "Florida from Indian Trail to Space Age" (C. W. Tabeau and R. L. Carson, eds.), pp. 305-16. Southern Publishing Company, Delray Beach, Fla.

Burt, W. H., and Grossenheider, R. P. (1976). "A Field Guide to the Mammals." 3rd ed. Houghton Mifflin, Boston.

Butzer, K. W. (1977). "Geomorphology of the Lower Illinois Valley as a Spatial-Temporal Context for the Koster Archaic Site." Illinois State Museum Reports of Investigations 34.

Clausen, C. J., Cohen, A. D., Emiliani, C., Holman, J. A., and Stipp, J. J. (1979). Little Salt Spring, Florida: A unique underwater site. *Science* 203, 609-14.

Cleland, C. E. (1965). Faunal remains from bluff shelters in northwestern Arkansas. *The Arkansas Archeologist* 6, 46.

Cleland, C. E. (1966). "The Prehistoric Animal Ecology and Ethnozoology of the Upper Great Lakes Region." University of Michigan Anthropological Paper 29.

Cleland, C. E., and Kearney, J. (1967). The vertebrate fauna of the Chesser Cave site, Athens County, Ohio. *In* "Studies in Ohio Archaeology" (O. H. Prufer and D. H. McKenzie, eds.), pp. 43-48. Western Reserve University Press, Cleveland.

Coleman, D. D., and Liu, C. L. (1975). Illinois State Geological Survey radiocarbon dates VI. *Radiocarbon* 17, 160-73.

Dallman, J. E. (in press). "Choice of Diet: Response to Climatic Change". Office of the State Archaeologist, University of Iowa, Report 16.

Dalquest, W. W. (1965). 1,350-year-old vertebrate remains from Montague County, Texas. *The Southwestern Naturalist* 10, 315-16.

Dalquest, W. W., Roth, E., and Judd, F. (1969). The mammal fauna of Schulze Cave, Edwards County, Texas. *Bulletin Florida State Museum* 13, 205-76.

Davis, L. B., and Wilson, M. (1978). "Bison Procurement and Utilization: A Symposium." Plains Anthropologist Memoir 14.

Davis, L. C. (1975). Late Pleistocene geology and paleontology of the Spring Valley Basin, Meade County, Kansas. Ph.D. dissertation, University of Iowa, Iowa City.

Delcourt, H. R. (1980). Late Quaternary vegetation history of the eastern highland rim and adjacent Cumberland Plateau of Tennessee. *Ecological Monographs* 49, 255-80.

Dillehay, T. D. (1974). Late Quaternary bison population changes on the southern plains. *Plains Anthropologist* 19, 180-96.

Dodson, P., and Wexlar, D. (1979). Taphonomic investigations of owl pellets. *Paleobiology* 5, 275-84.

Dorwin, J. T. (1971). "The Bowen Site: An Archaeological Study of Culture Process in the Late Prehistory of Central Indiana." Indiana Historical Society, Prehistory Research Series 4.

Duffield, L. F. (1974). Nonhuman vertebrate remains from Salts Cave Vestibule. *In* "Archaeology of the Mammoth Cave Area" (P. J. Watson, ed.), pp. 123-33. Academic Press, New York.

Dulian, J. J. (1975). Paleoecology of the Brayton local biota, Late Wisconsinan of southwestern Iowa. M.S. thesis, University of Iowa, Iowa City.

Eshelman, R. E. (1971). The paleoecology of Willard Cave, Delaware County, Iowa. M.S. thesis, University of Iowa, Iowa City.

Eshelman, R. E. (1972). Faunal analysis of the Schmitt site. *Proceedings of the Iowa Academy of Science* 79, 56-68.

Falk, C. R. (1970). The application of factor analysis in the interpretation of unmodified vertebrate remains from an archaeological cave deposit in central Missouri. M.S. thesis, University of Missouri, Columbia.

Falk, C. R. (1977). Analyses of unmodified vertebrate fauna from sites in the middle

Missouri subarea: A review. *In* "Trends in Middle Missouri Prehistory" (W. R. Wood, ed.), pp. 151-61. Plains Anthropologist Memoir 13.

Falk, C. R., and Calbrese, F. A. (1973). Helb: A preliminary statement. *Plains Anthropologist* 18, 336-43.

Fay, L. P. (1980). Mammals of the Garrett Farm and Pleasant Ridge local biotas (Holocene), Mills County, Iowa. *In* "American Quaternary Association, Sixth Biennial Meeting, Abstracts and Program, 18-20 August 1980," pp. 74-75. Institute for Quaternary Studies, University of Maine, Orono.

Fowler, M. L. (1959). "Summary Report of Modoc Rock Shelter." Illinois State Museum Report of Investigations 8.

Frank, R. M. (1965). Petrographic study of sediments from selected Texas caves. M.S. thesis, University of Texas, Austin.

Frank, R. M. (1968). Identification of miscellaneous faunal remains from Bonfire Shelter. *In* "Bonfire Shelter: A Stratified Bison Kill Site, Val Verde County, Texas" (D. S. Dibble and D. Lorrain, eds.), pp. 133-34. Texas Memorial Museum Miscellaneous Paper 1.

Frison, G. C. (1974). "The Casper Site: A Hell Gap Bison Kill on the High Plains." Academic Press, New York.

Frison, G. C., Wilson, M. C., and Wilson, D. J. (1976). Fossil bison and artifacts from an early Altithermal period arroyo trap in Wyoming. *American Antiquity* 41, 28-57.

Fulmer, D. W. (1974). A central Plains earthlodge: 13 ML 124. M.S. thesis, University of Iowa, Iowa City.

Gagliano, S. M. (1967). "Occupation Sequence at Avery Island." Louisiana State University Press, Baton Rouge.

Gibbon, G. (1971). The Bornick site: A Grand River phase Oneota site in Marquette County. *Wisconsin Archaeologist* 52, 85-137.

Gifford, E. W. (1967). Ethnographic notes on the southwestern Pomo. *University of California Anthropological Records* 25, 17.

Graham, R. W. (1976a). Late Wisconsin mammalian faunas and environmental gradients of the eastern United States. *Paleobiology* 2, 343-50.

Graham, R. W. (1976b). Pleistocene and Holocene mammals, taphonomy, and paleoecology of the Friesenhahn local fauna, Bexar County, Texas. Ph.D. dissertation, University of Texas, Austin.

Graham, R. W. (1979). Paleoclimates and late Pleistocene faunal provinces in North America. *In* "Pre-Llano Cultures of the Americas: Paradoxes and Possibilities" (R. L. Humphreys and D. Stanford, eds.), pp. 49-69. Anthropological Society of Washington, D.C., Washington, D.C.

Graham, R. W. (1981). Kimmswick: A Clovis-Mastodon association in eastern Missouri. *Science* 213, 1115-17.

Graham, R. W., and Semken, H. A., Jr. (1976). Paleoecological significance of the short-tailed shrew (*Blarina*), with a systematic discussion of *Blarina ozarkensis*. *Journal of Mammalogy* 57, 433-49.

Grayson, D. K. (1978). Minimum numbers and sample size in vertebrate faunal analysis. *American Antiquity* 43, 53-65.

Griffin, J. B. (1961). Post glacial ecology and culture changes in the Great Lakes area of North America. *University of Michigan Great Lakes Research Publication* 7, 147-55.

Guilday, J. E. (1956). Archaeological evidence of the fisher in West Virginia. *Journal of Mammalogy* 37, 287.

Guilday, J. E. (1958). The prehistoric distribution of the oppossum. *Journal of Mammalogy* 39, 39-43.

Guilday, J. E. (1961). Prehistoric record of *Scalopus* from western Pennsylvania. *Journal of Mammalogy* 42, 117-18.

Guilday, J. E. (1962). The Pleistocene local fauna of the Natural Chimneys, Augustana County, Virginia. *Annals of the Carnegie Museum* 36, 87-122.

Guilday, J. E. (1965). Bone refuse from the Lamoka Lake site. *In* "The Archaeology of New York State" (by W. A. Ritchie), pp. 56-58. Natural History Press, Garden City, N.Y.

Guilday, J. E. (1967a). The climatic significance of the Hosterman's Pit local fauna, Centre County, Pennsylvania. *American Antiquity* 32, 231-32.

Guilday, J. E. (1967b). Differential extinction during late Pleistocene and recent times. *In* "Pleistocene Extinctions: The Search for a Cause" (P. S. Martin and H. E. Wright, Jr., eds.), pp. 121-54. Yale University Press, New Haven, Conn.

Guilday, J. E. (1971a). "Biological and Archaeological Analysis of Bones from a 17th Century Indian Village (46 PU 31), Putnam County, West Virginia." West Virginia Geological and Economic Survey Report of Archaeological Investigations 4.

Guilday, J. E. (1971b). The Pleistocene history of the Appalachian mammal fauna. *In* "The Distributional History of the Biota of the Southern Appalachians: Part III. Vertebrates" (P. C. Holt, ed.). pp. 233-62. Research Monograph 4, Virginia Polytechnic Institute.

Guilday, J. E. (1972). Archaeological evidence of *Scalopus aquaticus* in the upper Ohio Valley. *Journal of Mammalogy* 53, 905-7.

Guilday, J. E., and Bender, M. S. (1957). A recent fissure deposit in Bedford County, Pennsylvania. *Annals of the Carnegie Museum* 35, 127-38.

Guilday, J. E., Hamilton, H. W., Anderson, E., and Parmalee, P. W. (1978). "The Baker Bluff Cave Deposit, Tennessee, and the Late Pleistocene Faunal Gradient." Carnegie Museum of Natural History Bulletin 11.

Guilday, J. E., Hamilton, H. W., and McCrady, A. D. (1966). The bone breccia of Bootlegger Sink, York County, Pa. *Annals of the Carnegie Museum* 8, 145-63.

Guilday, J. E., Martin, P. S., and McCrady, A. D. (1964). New Paris No. 4: A late Pleistocene cave deposit in Bedford County, Pennsylvania. *Bulletin of the National Speleological Society* 26, 121-94.

Guilday, J. E., and Mayer-Oakes, W. J. (1952). Occurrence of the rice rat (*Oryzomys*) in West Virginia. *Journal of Mammalogy* 33, 253-55.

Guilday, J. E., and Parmalee, P. W. (1965). Animal remains from the Sheep Rock Shelter (36 HU 1), Huntingdon County, Pennsylvania. *Pennsylvania Archaeologist* 35, 34-49.

Guilday, J. E., and Parmalee, P. W. (1971). Thirteen-lined ground squirrel, prairie chicken and other vertebrates from an archaeological site in northeastern Arkansas. *American Midland Naturalist* 86, 227-29.

Guilday, J. E. Parmalee, P. W., and Wilson, R. (in press). Vertebrate faunal remains from Meadowcroft Rockshelter (36 WH 297). Washington County, Pennsylvania. *In* "Meadowcroft Rockshelter" (by J. M. Adovasio et al., eds.). University of Pittsburgh Press, Pittsburgh.

Guilday, J. E., and Tanner, D. P. (1965). Vertebrate remains from Kipp Island site. *In* "The Archaeology of New York State" (by W. A. Ritchie), pp. 241-42. Natural History Press, Garden City, N.Y.

Hajic, E. R. (1981). Paleopedology ad microstratigraphy of the Koster archaeological site, Illinois. M.S. thesis, University of Iowa, Iowa City.

Hall, E. R. (1981). "The Mammals of North America." John Wiley and Sons, New York.

Hall, S. A. (1980). Corresponding geomorphic, archaeologic and climatic change in the southern plains: New evidence from Oklahoma. *In* "American Quaternary Association, Sixth Biennial Meeting, Abstracts and Program, 18-20 August 1980," p. 89. Institute for Quaternary Studies, University of Maine, Orono.

Harper, K. T., and Alder, G. M. (1970). The macroscopic plant remains of the deposits of Hogup Cave, Utah, and their paleoclimatic implications. *In* "Hogup Cave" (by C. M. Aikens), pp. 215-40. University of Utah Anthropological Paper 93.

Heizer, R. F., and Cook, S. F. (1952). Fluorine and other cheimical tests of some North American human and animal bones. *American Journal of Physical Anthropology* 10, 289.

Henry, D. O., Butler, B. H., and Hall, S. A. (1979). The late prehistoric human ecology of Birch Creek Valley, northeastern Oklahoma. *Plains Anthropologist* 24, 207-38.

Hester, J. J. (1960). Late Pleistocene extinction and radiocarbon dating. *American Antiquity* 26, 58-77.

Hester, J. J. (1967). The agency of man in animal extinctions. *In* "Pleistocene Extinctions: The Search for a Cause" (P. S. Martin and H. E. Wright, Jr., eds.), pp. 169-92. Yale University Press, New Haven, Conn.

Hibbard, C. W. (1963). A late Illinoian fauna from Kansas and its climatic significance. *Papers of the Michigan Academy of Science, Arts, and Letters* 48, 187-221.

Hibbard, C. W. (1970). Pleistocene mammalian local faunas from the Great Plains and central lowland provinces of the United States. *In* "Pleistocene and Recent Environments of the Central Great Plains" (W. Dort, Jr., and J. K. Jones, Jr., eds.), pp. 395-433. University Press of Kansas, Lawrence.

Hirschfield, S. E. (1968). Vertebrate fauna of Nichol's Hammock, a natural trap. *Quarterly Journal of the Florida Academy of Science* 31, 177-89.

Hoffman, R. S., and Jones, J. K., Jr. (1970). Influence of late-glacial and post-glacial events on the distribution of recent mammals on the northern Great Plains. *In* "Pleistocene and Recent Environments of the Central Great Plains" (W. Dort, Jr., and J. K. Jones, Jr., eds.), pp. 355-94. University Press of Kansas, Lawrence.

Hornaday, W. T. (1887). The extermination of the American bison with a sketch of its discovery and life history. *In* "Report of the U.S. National Museum for 1887," pp. 367-548. Washington, D.C.

Hotopp, J. (1978). Glenwood: A contemporary view. *In* "The Central Plains Tradition" (D. J. Blakeslee, ed.), pp. 109-33. Office of the State Archaeologist, University of Iowa, Report 11.

Hoyer, B. E. (1980). The geology of the Cherokee Sewer site. *In* "The Cherokee Excavations: Holocene Ecology and Human Adaptations in Northwestern Iowa" (D. C. Anderson and H. A. Semken, Jr., eds.), pp. 21-66. Academic Press, New York.

Jenkins, J. T., Jr. (1972). The Pleistocene geology and paleontology of the Fremont County quarry, Fremont County, Iowa. M.S. thesis, University of Iowa, Iowa City.

Jenkins, J. T., Jr., and Semken, H. A., Jr. (1972). Faunal analysis of the Lane Enclosure, Allamakee County, Iowa. *Proceedings of the Iowa Academy Science* 78, 76-78.

Johnson, E. (1976). Investigations into the zooarchaeology of the Lubbock Late site. Ph.D. dissertation, Texas Tech University, Lubbock.

Johnson, E., and Holliday, V. T. (1979). Prehistoric life on the southern High Plains of Texas. *Archaeology* 32, 60-61.

Johnson, E., and Holliday, V. T. (1981). Late paleo-Indian activity at the Lubbock Lake site. *Plains Anthropologist* 26, 173-93.

Johnson, P. C. (1972). Mammal remains associated with Nebraska phase earth lodges in Mills County, Iowa. M.S. thesis, University of Iowa, Iowa City.

Jones, J. K., and Glass, B. G. (1960). The short-tailed shrew, *Blarina brevicauda*, in Oklahoma. *Southwestern Naturalist* 5, 136-42.

King, F. B., and Graham, R. W. (1981). Effects of ecological and paleoecological patterns on subsistence and paleoenvironmental reconstructions. *American Antiquity* 46, 128-42.

King, J. E., and Allen, W. H. (1977). A Holocene vegetation record from the Mississippi River valley, southeastern Missouri. *Quaternary Research* 8, 307-23.

Klippel, W. E. (1971). Prehistory and environmental change along the southern border of the Prairie Peninsula during the Archaic period. Ph.D. dissertation, University of Missouri, Columbia.

Knox, J. C. (1972). Valley alluviation in southwestern Wisconsin. *Annals of the Association of American Geographers* 62, 401-10.

Kramer, T. L. (1972). The paleoecology of the post glacial Mud Creek biota, Cedar and Scott Counties, Iowa. M.S. thesis, University of Iowa, Iowa City.

Kurtén, B., and Anderson, E. (1980). "Pleistocene Mammals of North America." Columbia University Press, New York.

Land, L. S., Lundelius, E. L., Jr., and Valastro, S. (1980). Isotopic ecology of deer bones. *Palaeogeography, Palaeoclimatology, Palaeoecology* 32, 143-51.

Lewis, R. B. (1974). Ecological data from archaeological sites. *In* Mississippi exploitative strategies: A southeast Missouri example. *Missouri Archaeologist* 11, Appendix I.

Lippold, L. K. (1973). Animal resource utilization at the Coopers Shore site (47 RO 2), Rock County, Wisconsin. *Wisconsin Archaeologist* 54, 36-62.

Long, A., and Martin, P. S. (1974). Death of American ground sloths. *Science* 186, 638-40.

Loomis, F. B. (1924). Artifacts associated with the remains of a Columbian elephant at Melbourne, Florida. *American Journal of Science* 8, 503-8.

Ludwickson, J. (1978). Central Plains tradition settlements in the Loup River basin: The Loup River phase. *In* "The Central Plains Tradition" (D. J. Blakeslee, ed.), pp. 94-108. Office of the State Archaeologist, University of Iowa, Report 11.

Lundelius, E. L., Jr. (1962). Non-human skeletal material from the Kyle site. *In* "A Stratified Central Texas Aspect Site in Hill County, Texas" (by Edward Jelks), Appendix 2. University of Texas, Department of Anthropology, Archaeology Series 5.

Lundelius, E. L., Jr. (1963). Non-human skeletal material in Centipede and Damp Caves. *Bulletin of the Texas Archaeological Society* 33, 127-29.

Lundelius, E. L., Jr. (1967). Late-Pleistocene and Holocene faunal history of central Texas. *In* "Pleistocene Extinctions: The Search for a Cause" (P. S. Martin and H. E. Wright, Jr., eds.), pp. 287-336. Yale University Press, New Haven, Conn.

Lundelius, E. L., Jr., Graham, R. W., Anderson, E., Guilday, J., Holman, J. A. Steadman, D., and Webb, S. D. (1983). Terrestrial vertebrate faunas. *In* "Late-Quaternary Environments of the United States," volume 1, "The Late Pleistocene" (S. C. Porter, Jr., ed.) pp. 311-53. University of Minnesota Press, Minneapolis.

Lunkens, P. W. (1963). Some ethnozoological implications of mammalian faunas from Minnesota Archaeological sites. Ph.D. dissertation, University of Minnesota, Minneapolis.

Lyman, R. L. (1979). "Archaeological Faunal Analysis: A Bibliography." Occasional Paper of the Idaho Museum of Natural History 31.

McMillan, R. B. (1976). The dynamics of cultural and environmental change at Rodgers Shelter, Missouri. *In* "Prehistoric Man and His Environments" (W. R. Wood and R. B. McMillan, eds.), pp. 211-32. Academic Press, New York.

McMillan, R. B., and Klippel, W. E. (1981). Environmental changes and hunter-gatherer adaptations in the southern Prairie Peninsula. *Journal of Archaeological Science* 8, 215-45.

McMullen, T. L. (1974). The mammals of the Duck Creek local fauna, late Pleistocene of Kansas. M.S. thesis, Fort Hays Kansas State University, Fort Hays.

Martin, L. D., and Neuner, A. M. (1978). The end of the Pleistocene in North America. *Transactions of the Nebraska Academy of Science* 6, 117-26.

Martin, P. S. (1967). Prehistoric overkill. *In* "Pleistocene Extinctions: The Search for a Cause" (P. S. Martin and H. E. Wright, Jr., eds.), pp. 75-120. Yale University Press, New Haven, Conn.

Martin, P. S. (1973). The discovery of America. *Science* 179, 969-74.

Martin, R. A., and Tedesco, R. (1977). *Ondatra annectens* (Mammalia: Rodentia) from the Pleistocene Java local fauna of South Dakota. *Journal of Paleontology* 50, 846-50.

Martin, R. A., and Webb, S. D. (1974). Late Pleistocene mammals from the Devil's Den fauna, Levy County. *In* "Pleistocene Mammals of Florida" (S. D. Webb, ed.), pp. 114-45. University Presses of Florida, Gainesville.

Mayhew, D. F. (1977). Avian predators as accumulators of fossil mammal material. *Boreas* 6, 25-31.

Mayr, E. (1970). "Populations, Species, and Evolution." Harvard University Press, Cambridge, Mass.

Mellett, J. S. (1974). Scatological origin of microvertebrate fossil accumulations. *Science* 185, 349-50.

Mercer, J. H. (1972). The lower boundary of the Holocene. *Quaternary Research* 2, 15-24.

Morrison, R. B. (1965). Quaternary geology of the Great Basin. *In* "The Quaternary of the United States" (H. E. Wright, Jr., and D. G. Frey, eds.), pp. 265-85. Princeton University Press, Princeton, N.J.

Munson, P. J., Parmalee, P. W., and Yarnell, R. A. (1971). Subsistence ecology of Scoville, a terminal Middle Woodland village. *American Antiquity* 30, 410-31.

Murphy, J. L. (1975). "An Archaeological History of the Hocking Valley." Ohio University Press, Athens.

Nelson, R. S., and Semken, H. A., Jr. (1970). Paleoecological and stratigraphic significance of the muskrat in Pleistocene deposits. *Geological Society of America Bulletin* 81, 3733-38.

Osborn, H. F. (1910). "The Age of Mammals in Europe, Asia, and North America." Macmillan, New York.

Parmalee, P. W. (1958). Evidence for the fisher in central Illinois. *Journal of Mammalogy* 39, 153.

Parmalee, P. W. (1959a). Animal remains from the Modoc Rock Shelter site, Randolph County, Illinois. *In* "Summary Report of Modoc Rock Shelter" (by M. L. Fowler), pp. 61-65. Illinois State Museum, Reports of Investigations 8.

Parmalee, P. W. (1959b). Animal remains from Raddatz Rockshelter, SK 5, Wisconsin. *Wisconsin Archaeologist* 40, 83-90.

Parmalee, P. W. (1959c). A prehistoric record of the fisher in Georgia. *Journal of Mammalogy* 41, 409-10.

Parmalee, P. W. (1960a). Animal remains from the Aztalan site, Jefferson County, Wisconsin. *Wisconsin Archaeologist* 41, 1-10.

Parmalee, P. W. (1960b). Animal remains from the Durst Rock Shelter, Sauk County, Wisconsin. *Wisconsin Archaeologist* 41, 11-17.

Parmalee, P. W. (1962). Faunal remains from the Stanfield-Worley Bluff Shelter, Colbert Co., Alabama. *Journal of Alabama Archaeology* 8, 112-14.

Parmalee, P. W. (1963a). A prehistoric occurrence of porcupine in Alabama. *Journal of Mammalogy* 44, 267-68.

Parmalee, P. W. (1963b). Vertebrate remains from the Bell site, Winnebago County, Wisconsin. *Wisconsin Archaeologist* 44, 58-69.

Parmalee, P. W. (1965). The food economy of Archaic and Woodland peoples at the Tick Creek Cave site, Missouri. *Missouri Archaeologist* 27, 1-34.

Parmalee, P. W. (1967). A recent bone cave deposit in southwestern Illinois. *National Speleological Society Bulletin* 29, 119-85.

Parmalee, P. W. (1968). Cave and archaeological faunal deposits as indicators of post-Pleistocene animal populations and distribution in Illinois. *In* ''The Quaternary of Illinois'' (R. E. Bergstrom, ed.), pp. 104-13. University of Illinois College of Agriculture Special Publication 14.

Parmalee, P. W. (1969). Animal remains from the Archaic Riverton, Swan Island, and Robeson Hill sites, Illinois. *In* ''The Riverton Culture: A Second Millennium Occupation in the Central Wabash Valley'' (by H. D. Winters), Appendix I, Illinois State Museum Report of Investigations 13.

Parmalee, P. W. (1971). Fisher and porcupine remains from cave deposits in Missouri. *Transactions of the Illinois Academy of Science* 64, 225-29.

Parmalee, P. W. (1972). Vertebrate remains from the Fifield site, Porter Co., Indiana. *In* ''Late Prehistoric Occupation of Northwest Indiana: A Study of the Upper Mississippi Culture of the Kankakee Valley'' (by C. H. Faulkner), Appendix 4, Indiana Historical Society Prehistory Research Series 5.

Parmalee, P. W. (1973). Faunal remains from the Kane Village site (MS 194), Madison County, Illinois. *In* ''Late Woodland Site Archaeology in Illinois,'' Appendix A, Investigations in South Central Illinois, Illinois Archaeological Survey Bulletin 9.

Parmalee, P. W. (1975a). A general summary of the vertebrate fauna from Cahokia. *In* ''Perspectives of Cahokia Archaeology.'' Illinois Archaeological Survey Bulletin 10.

Parmalee, P. W. (1975b). Mole food? *Tennessee Archaeologist* 31, 37-40.

Parmalee, P. W., and Guilday, J. E. (1966). A recent record of porcupine from Tennessee. *Journal of the Tennessee Academy of Science* 41, 81-82.

Parmalee, P. W., and Klippel, W. E. (1981). A late Pleistocene population of the pocket gopher, *Geomys* cf. *bursarius*, in the Nashville Basin, Tennessee. *Journal of Mammalogy* 64, 831-35.

Parmalee, P. W., McMillan, R. B., and King, F. B. (1976). Changing subsistence patterns at Rodgers Shelter. *In* ''Prehistoric Man and His Environment: A Case Study in the Ozark Highland'' (W. R. Wood and R. B. McMillan, eds.), pp. 141-61. Academic Press, New York.

Parmalee, P. W., and Oesch, R. D. (1972). ''Pleistocene and Recent Faunas from Brynjulfson Caves, Missouri.'' Illinois State Museum Reports of Investigations 25.

Parmalee, P. W., and Shane, O. C., III (1970). The Blain site vertebrate fauna. *In* ''Blain Village and the Fort Ancient Tradition in Ohio'' (O. H. Prufer and O. C. Shane, eds.), pp. 185-206. Kent State University Press, Kent, Ohio.

Patton, T. H. (1963). ''Fossil Vertebrates from Miller's Cave, Llano County, Texas.'' Bulletin of the Texas Memorial Museum 7.

Perino, G. H. (1971). ''The Mississippian Component at the Schild Site (No. 4), Greene County, Illinois.'' Illinois Archaeological Survey Bulletin 8.

Pruitt, W. O., Jr. (1954). Additional animal remains from under Sleeping Bear Dune, Leelanau County, Michigan. *Papers of the Michigan Academy of Science, Arts, and Letters* 39, 253-56.

Purdue, J. R. (1980). Clinal variation of some mammals during the Holocene in Missouri. *Quaternary Research* 13, 242-58.

Purdue, J. R., and Styles, B. W. (in preparation). Changes in the mammalian fauna of Illinois and Missouri during the late Pleistocene and Holocene. *In* ''Late Pleistocene/Holocene Environmental Changes in the High Plains'' (R. W. Graham and H. A. Semken, Jr., eds.). Illinois State Museum Scientific Papers.

Pyle, K. B. (1981). ''Faunal Remains from the MAD Site, Crawford County, Iowa.'' Report to the State Historic Preservation Program, Iowa City.

Quinn, J. H. (1972). Extinct mammals in Arkansas and related C-14 dates circa 3000 years ago. *In* ''Quaternary Time Scale, Stratigraphy and Climate,'' section 12, pp. 89-96, International Geological Congress Papers 24, Montreal.

Raun, G. G., and Eck, L. J. (1967). Vertebrate remains from four archaeological sites in the Amistad Reservoir area, Val Verde County, Texas. *Texas Journal of Science* 19, 138-50.

Ray, C. E., and Lipps, L. (1970). Southerly distribution of porcupine in eastern United States during late Quaternary time (abstract). *Bulletin of the Georgia Academy of Science* 28, 24.

Reid, K. C. (1976). Prehistoric trade in the lower Missouri River valley: An analysis of Middle Woodland bladelets. *In* ''Hopewellian Archaeology in the Lower Missouri River Valley'' (A. E. Johnson, ed.), pp. 63-99, University of Kansas Publications in Anthropology 8.

Reider, R. G. (1980). Interpretation of climatic change at the Pleistocene-Holocene boundary using soils associated with paleoindian cultural levels at the Sheaman and Agate Basin (Brewster) sites in eastern Wyoming. *Plains Conference Abstracts* 38, 36.

Richards, R. L. (1980). Rice rat (*Oryzomys* cf. *palustris*) remains from south-Indiana caves. *Proceedings of the Indiana Academy of Science* 89, 425-31.

Rogers, K. (1975). Faunal remains from the Zimmerman site. *In* ''The Zimmerman Site: Further Excavations at the Grand Village of Kaskaskia'' (by M. Brown), Tables 35 and 36, Appendix I. Illinois State Museum Reports of Investigations 32.

Roth, E. L. (1972). Late Pleistocene mammals from Klein Cave, Kerr County, Texas. *Texas Journal of Science* 24, 75-84.

Rouse, I. (1950). Vero and Melbourne man: A cultural and chronological interpretation. *New York Academy of Science Transactions* 12, 220-24.

Sammis, C. G. (1978). Mammalian osteo-archaeology of the Mitchell site, (39 DV 2), Mitchell, South Dakota. M.S. thesis, University of Iowa, Iowa City.

Satorius-Fox, M. R. (1982). Paleoecological analysis of micromammals from the Schmidt site, a central Plains tradition village in Howard County, Nebraska. M.S. thesis, University of Iowa, Iowa City.

Schultz, G. E. (1978). Supplementary data on the Rex Rodgers site: Micromammals. *In* ''Archaeology at MacKenzie Reservoir'' (P. W. Willey, B. R. Harrison, and J. T. Hughes, eds.), pp. 113-14. Texas Historical Commission Archaeological Survey Report 24.

Schultz, G. E., and Rawn, V. M. (1978). Faunal remains from the Deadman's Shelter site. *In* ''Archaeology at MacKenzie Reservoir'' (P. S. Willey, B. P. Harrison, and J. T. Hughes, eds.), pp. 191-97. Texas Historical Commission Archaeological Survey Report 24.

Sealander, J. A. (1979). ''A Guide to Arkansas Mammals.'' River Road Press, Conway, Ark.

Sellards, E. H. (1952). ''Early Man in America.'' University of Texas Press, Austin.

Semken, H. A., Jr. (1961). Fossil vertebrates from Longhorn Cavern, Burnet County, Texas. *Texas Journal of Science* 13, 290-310.

Semken, H. A., Jr. (1966). ''Stratigraphy and Paleontology of the McPherson *Equus* Beds (Sandahl Local Fauna), McPherson County, Kansas.'' Contributions to the Museum Paleontology, University of Michigan 20, pp. 121-78.

Semken, H. A., Jr. (1967). ''Mammalian Remains from Rattlesnake Cave, Kinney County, Texas.'' Texas Memorial Museum Pearce-Sellards Series 7.

Semken, H. A., Jr. (1969). Paleoecological implications of micromammals from Peccary Cave, Newton County, Arkansas. *Geological Society America, South-Central Section, Abstracts for 1969*, 27.

Semken, H. A., Jr. (1974). Micromammal distribution and migration during the Holocene. *In* ''American Quaternary Association, Third Biennial Meeting, Abstracts,'' p. 25. University of Wisconsin, Madison.

Semken, H. A., Jr. (1976). ''Descriptive Analysis of the Rodent and Insectivore Cranial Elements Collected from the Helb, Lower Grand, and Walth Bay Sites, Oahe Reservoir, South Dakota.'' Report to the National Park Service, Midwest Region.

Semken, H. A., Jr. (1980). Holocene climatic reconstructions derived from the three micromammal bearing cultural horizons of the Cherokee Sewer site, northwestern Iowa. *In* ''The Cherokee Excavations: Holocene Ecology and Human Adaptations in Northwestern Iowa'' (D. C. Anderson and H. A. Semken, Jr., eds.), pp. 67-99. Academic Press, New York.

Semken, H. A., Jr. (1982). The Late Wisconsinan-Holocene micromammal sequence in Peccary Cave, northwestern Arkansas. Unpublished manuscript.

Semken, H. A., Jr., and Falk, C. R. (1980). Holocene climatic changes in the northern Plains of the United States: the Mammalian record. *In* ''Abstracts of the 38th

Plains Conference,'' p. 37. University of Iowa, Iowa City.

Semken, H. A., Jr., and Falk, C. R. (in preparation). Holocene climatic changes in the northern Plains of the United States. *In* ''Late Pleistocene/Holocene Environmental Changes in the High Plains'' (R. W. Graham and H. A. Semken, Jr., eds.). Illinois State Museum Scientific Papers.

Semken, H. A., Jr., and Foley, R. L. (1979). Small mammal distributions in four Plains village sites. *In* ''Abstracts of the 37th Plains Conference,'' p. 29. Kansas City.

Semken, H. A., Jr., and Satorius, M. R. (1980). Distribution and climatic significance of small mammals from the White Buffalo Robe site. *In* ''The Archaeology of the White Buffalo Robe Site'' (C. H. Lee, ed.), pp. 618-37. Department of Anthropology, University of North Dakota Report.

Shay, C. T. (1971). ''The Itasca Bison Kill Site: An Ecological Analysis.'' Minnesota Historical Society, St. Paul.

Shay, C. T. (1978). Late prehistoric bison and deer use in the eastern prairie-forest border. *In* ''Bison Procurement Utilization: A Symposium'' (L. B. Davis and M. Wilson, eds.), pp. 194-212. Plains Anthropologist Memoir 14.

Slaughter, B. H. (1966). The vertebrata of the Domebo local fauna, Pleistocene of Oklahoma. *In* ''Domebo: A Paleo-Indian Mammoth Kill in the Prairie-Plains'' (F. C. Leonhardy, ed.). pp. 31-35. Museum of the Great Plains Contribution 1.

Slaughter, B. H. (1967). Animal ranges as a clue to late-Pleistocene extinction. *In* ''Pleistocene Extinctions: The Search for a Cause'' (P. S. Martin and H. E. Wright, Jr., eds.), pp. 155-67. Yale University Press, New Haven, Conn.

Slaughter, B. H., and Hoover, B. R. (1963). Sulphur River Formation and the Pleistocene mammals of the Ben Franklin local fauna. *Journal of the Graduate Research Center* (SMU) 31, 132-48.

Smith, P. W. (1957). An analysis of post-Wisconsin biogeography of the Prairie Peninsula region based on distributional phenomena among terrestrial vertebrate populations. *Ecology* 38, 205-18.

Stoltman, J. E., and Baerreis, D. A. (1983). The evolution of human ecosystems in the eastern United States. *In* ''Late-Quaternary Environments of the United States,'' vol. 2, ''The Holocene'' (H. E. Wright, Jr., ed.), pp. 252-68. University of Minnesota Press, Minneapolis.

Styles, B. H. (1981). ''Faunal Exploitation and Resource Selection: Early Late Woodland Subsistence in the Lower Illinois Valley.'' Northwestern University Archaeological Program Scientific Papers 3.

Theiling, S. C. (1973). The Pleistocene fauna of Lost River Sink, Iowa County, Wisconsin. M.S. thesis, University of Iowa, Iowa City.

Thompson, R. S., Van Devender, T. R., Martin, P. S., Foppe, T., and Long, A. (1980). Shasta ground sloth (*Nothrotheriops shastense* Hoffstetter) at Shelter Cave, New Mexico: Environment, diet, and extinction. *Quaternary Research* 14, 360-76.

Udvardy, M. D. F. (1969). ''Dynamic Zoogeography.'' Van Nostrand Reinhold, New York.

Van Devender, T. R., and Wiseman, F. M. (1977). A preliminary chronology of bioenvironmental changes during the paleoindian period in the monsoonal southwest. *In* Paleoindian lifeways (E. Johnson, ed.), *The Museum Journal* 17, 13-27.

Voight, J. R., and Glenn-Lewin, D. C. (1979). Strip mining, *Peromyscus* and other small mammals in southern Iowa. *Proceedings of the Iowa Academy of Science* 86, 133-36.

Walker, D. N. (1975). A cultural and ecological analysis of the vertebrate fauna from the Medicine Lodge Creek site (48 BH 499). M.S. thesis, University of Wyoming, Laramie.

Walker, D. N. (1982). Early Holocene vertebrate fauna. *In* ''The Agate Basin Site: A Record of the Paleo-Indian Occupation of the Northwestern High Plains'' (G. C. Frison and D. J. Stanford, eds.), pp. 274-308. Academic Press, New York.

Watson, P. J. (1969). ''The Prehistory of Salts Cave, Kentucky.'' Illinois State Museum Reports of Investigations 16.

Webb, T., and Bryson, R. A. (1972). Late and postglacial climatic change in the northern Midwest, USA: Quantitative estimates derived from fossil pollen spectra by multivariate statistical analysis. *Quaternary Research* 2, 70-115.

Weigel, R. D. (1962). ''Fossil Vertebrates of Vero, Florida.'' Special Publication of the Florida Geological Survey 10.

Weigel, R. D., Holman, J. A., and Paloumpis, A. A. (1974). Vertebrates from Russell Cave. *In* ''Investigations in Russell Cave'' (by J. W. Griffen), pp. 81-85. U.S. National Park Service Publications in Archaeology 13.

Wendland, W. M. (1978). Holocene man in North America: The ecological setting and climatic background. *Plains Anthropologist* 23, 273-87.

Wendland, W. M., and Bryson, R. A. (1974). Dating climatic episodes of the Holocene. *Quaternary Research* 4, 9-24.

West, R. M., and Dallman, J. E. (1980). Late Pleistocene and Holocene vertebrate fossil record of Wisconsin. *Geoscience Wisconsin* 4, 25-45.

White, T. E. (1952a). Observations on the butchering techniques of some aboriginal peoples: I. *American Antiquity* 17, 337-38.

White, T. E. (1952b). Suggestions for facilitating the identification of animal bone from archaeological sites. *Plains Archeological Conference Newsletter* 5, 3-5.

White, T. E. (1953a). A method for calculating the dietary percentage of various food animals utilized by aboriginal peoples. *American Antiquity* 18, 396-98.

White, T. E. (1953b). Observations on the butchering techniques of some aboriginal peoples: No. 2. *American Antiquity* 19, 160-64.

White, T. E. (1954). Observations on the butchering techniques of some aboriginal peoples: Nos. 3, 4, 5, and 6. *American Antiquity* 19, 254-64.

White, T. E. (1955). Observations of the butchering techniques of some aboriginal peoples: Nos. 7, 8, and 9. *American Antiquity* 21, 170-78.

White, T. E. (1956). The study of osteological materials in the plains. *American Antiquity* 21, 401-4.

Wilson, M. (1974). The Casper local fauna and its fossil bison. *In* ''The Casper Site: A Hell Gap Bison Kill on the Great Plains'' (G. C. Frison, ed.), pp. 125-71. Academic Press, New York.

Wilson, M. (1978). Archaeological kill site populations and the Holocene evolution of the genus *Bison*. *In* ''Bison Procurement and Utilization: A Symposium'' (L. B. Davis and M. Wilson, eds.), pp. 9-22. Plains Anthropologist Memoir 14.

Wilson, R. L. (1967). The Pleistocene vertebrates of Michigan. *Papers of the Michigan Academy of Science, Arts, and Letters* 52, 197-234.

Wood, W. R., and McMillan, R. B. (eds.) (1976). ''Prehistoric Man and His Environments: A Case Study in the Ozark Highland.'' Academic Press, New York.

Wright, T., and Lundelius, E. L., Jr. (1963). ''Post-Pleistocene Raccoons from Central Texas and Their Zoogeographic Significance.'' Texas Memorial Museum, Pearce-Sellards Series, 2.

CHAPTER 13

Holocene Paleolimnology

Richard B. Brugam

Introduction

Although the pollen content of lake sediments mainly reflects climatic influences on upland vegetation, remains of lake organisms in sediments show changes in lake environments that may or may not be directly related to climatic change. Climatic influences on lake organisms generally occur by means of indirect mechanisms such as changes in evaporation and runoff or watershed vegetational type. The usefulness of the paleolimnological record lies not in its being a directly readable record of climatic change but in its being a record of long-term changes in populations of lake organisms and in lake environments. The record includes, along with the morphologically identifiable remains of organisms, evidence of the physical structure and composition of the sediment itself.

In this chapter, special emphasis is placed on the North American work of the 1970s and early 1980s, but consideration is also given to other works that help to illuminate problems encountered in North American paleolimnology. The chapter also emphasizes the rationale behind paleolimnological investigations and the validity of the field's techniques rather than a recounting of numerous fine stratigraphic investigations pursued over the last decade. This chapter is arranged in a manner similar to the form of an aquatic food chain, with chemistry considered first, then plant fossils and pigments, and finally animal fossils. Other reviews of paleolimnology treat portions of the science not emphasized here. The volume edited by Frey (1969) and volume 25 of the *Polish Archives of Hydrobiology* published in 1978 are examples. Crisman (1978b) and Frey (1976) have reviewed the importance of animal remains in paleolimnology. Birks and Birks (1980) provide another excellent overview of paleolimnology.

Sediment Chemistry

The source of lake sediment is a major problem in the interpretation of chemical profiles from cores. A particular sediment constituent may be transported without change from the watershed soils and thus reflect watershed conditions, or it may be precipitated directly from the lake water and reflect conditions in the water column.

Deevey (1942) argues that most organic carbon in a lake sediment

is autochthonous (internal) in origin and thus records past primary productivity. In contrast, Mackereth (1966) argues that most constituents of lake sediment have their origin in watershed soils, although biophil elements such as phosphorus and sulfur can originate from biologic uptake and precipitation. Likens and Davis (1975) suggest that the sediment chemistry more clearly represents the chemistry of both autochthonous and allochthonous particulate material carried in water, rather than dissolved substances. Such an argument implies that the reconstruction of past lake-water chemistry from chemical analyses of sediment will be difficult and that fossil remains of organisms may be better chemical indicators.

A primary goal in the study of lake-sediment chemistry is to develop a suitable chemical index of lake trophic status. Deevey (1942) argues that the carbon concentration of the sediment indicates trophic status because it represents the remains of autochthonous primary production. His suggestion is based on studies of Linsley Pond, Connecticut, in which he has found that late-glacial sediments have low carbon concentrations associated with midge remains characteristic of oligotrophic lakes. As carbon concentrations rose in postglacial times, midge remains characteristic of eutrophic lakes appeared. Livingstone (1957) suggests, however, that mineral matter washed in from sparsely vegetated late-glacial landscape diluted the organic material. Davis (1976) and Brugam (1978) indicate that much of the organic material preserved in lake sediment may be derived from watershed soils and, therefore, does not reflect trophic status.

Mackereth (1966) suggests that iron and manganese are reasonable indicators of trophic status because their ratio should vary with the redox potential of the lake hypolimnion. Manganese should be depleted relative to iron in sediments where redox potentials are only moderately low, whereas in highly anaerobic sediments both iron and manganese would be lost. Presumably, eutrophic lakes have highly anaerobic hypolimnia and sediments. Complications arise because variations in the redox potential of watershed soils influence rates of iron and manganese input into a lake. In several Wisconsin lakes, Bortleson (1971) has found that iron-manganese ratios changed with

H. E. Wright, Jr., H. H. Birks, and D. G. Frey reviewed the manuscript for this chapter, and their suggestions resulted in substantial improvements.

agricultural settlement (and eutrophication) in some but not all. In many cases, concentrations of both iron and manganese increased; this implies increased outputs of these elements resulting from soil erosion and not from changes in the redox potential of watershed soils or lake hypolimnia. At Linsley Pond, Brugam (1978) has found that, when the lake watershed was disturbed by the construction of homes, the iron-manganese ratios approached those of watershed soils and the inputs of both elements increased. These results suggest that iron and manganese are not only affected by oxygen conditions in watershed soils and the lake but also by erosion of watershed soils and that iron-manganese ratios should be used cautiously in reconstructing past lake trophic conditions.

Mackereth (1966) argues that precipitation of phosphorus occurs mainly through uptake and sedimentation by organisms and thus indicates levels of past biologic activity. Bortleson and Lee (1974), however, show a close statistical relationship among iron, manganese, and phosphorus concentrations in Wisconsin lake sediments and suggest that precipitation of phosphorous with ferric and manganese hydroxides is an important mechanism of phosphorus input to the sediment in contrast to biotic uptake and precipitation. An abiotic mechanism of phosphorus precipitation may not preclude a relationship between sediment phosphorus levels and lake trophic status when the amount of phosphorus removed from the lake water is always a constant fraction of the amount of phosphorus present.

A number of investigators have examined sediment cores from lakes of known eutrophication history to see how faithfully the sediment records lake trophic status. Shapiro and others (1971) have found that phosphorus, nitrogen, and organic matter increased near the surface in three cores from Lake Washington in Washington State taken in 1958 and 1959. In two cores taken in one of the same locations in 1968, however, the high surface concentrations of organic matter and nitrogen found earlier had declined after burial. Phosphorus concentration, on the other hand, remained high and may reflect past lake trophic status.

Bradbury and Megard (1972) have found in Shagawa Lake, Minnesota, that phosphorus levels from a short core accurately reflect changes in phosphorus inputs from the sewage of the growing town of Ely. Bortleson (1971) also has found increases in sediment phosphorus levels after European settlement in a group of Wisconsin Lakes, but he is reluctant to attribute all of the increase to eutrophication because increases in iron concentration caused the phosphorus-retentive capacity of the sediment to increase. Brugam (1978) has examined the influx (milligrams per square centimeter per year) of phosphorus to the sediment of Linsley Pond, Connecticut. Influx calculation corrects for any differences in the accumulation rate of the sediment matrix and gives the rate at which the phosphorus enters the sediment at the core site. Phosphorus influx only increased at Linsley Pond after human disturbance in the lake watershed drastically increased sedimentation rates and inputs of mineral matter. Similarly, in a core from Lake Harriet, Minnesota, Brugam and Speziale (1983) have found data on phosphorus and diatom content to be in disagreement as to the timing and extent of eutrophication. In contrast, Birch and others (1980) have examined phosphorus influxes to the sediment of four Washington State lakes and found them to correlate well with algal primary productivity.

Although phosphorus appears to be a good indicator of past trophic status in some lakes, it is not in others. The anomalous results at Linsley Pond and Lake Harriet are probably the result of significant amounts of phosphorus reaching the lake sorbed on mineral particles.

Although leaching within the water column might not be strong enough to dissolve such phosphorus, the sorbed phosphorus is measured when the sediment is analyzed. It should be possible to fractionate the sediment phosphorus to exclude the portion that is strongly bound to mineral particles.

Additional problems with measuring phosphorus levels are caused by sedimentary processes within lakes. Tessenow (1975) points out the possibility of significant postdepositional movements of phosphorus, iron, and manganese in the deep waters of lakes with anaerobic hypolimnia. Such movements could complicate interpretations based on phosphorus influx profiles. Lehman (1975) has argued that "sediment focusing" may alter the apparent influx of a sediment constituent without any change in the overall input of the constituent to the lake as a whole. Phosphorus profiles from only one core would then be suspect.

The extent of postdepositional movement and sediment focusing can be resolved by dating many sediment cores from a lake basin and calculating influx over the whole lake bottom, as Evans and Rigler (1980) have done for Bob Lake in Ontario. They have found that the deposition rate of phosphorus agreed very well with a phosphorus loading rate previously calculated from long-term data on stream inputs.

In contrast to phosphorus, which may undergo considerable post depositional movement within the sediments, concentrations of heavy metals seem to show a closer relation to the input history, probably because of their strong affinity for particulate matter in the sediment. Radionuclide dating techniques based on ^{210}Pb and ^{137}Cs depend on the strong affinity of these heavy-metal isotopes for organic materials (in the case of ^{210}Pb) and clays (^{137}Cs). If there were significant postdepositional movements of thes nuclides, dating techniques based on their abundance would be impossible. It is, therefore, not surprising that a number of investigators have been able to document the recent input of heavy metals to lakes using the paleolimnological record.

Norton and others (1978) have examined profiles of zinc from 12 northern New England lakes dated by pollen analysis. They have found a two-phase increase in zinc in the sediment correlated with the colonization of the area by European settlers and the introduction of galvanized iron to the region.

Allott (1978) has found a strong increase in copper influx to the sediment of Twin Lake, Minnesota, as a result of the addition of copper sulfate to kill algae. Galloway and Likens (1979) have examined the influx of a large number of elements to the sediment of Woodhull Lake, a wilderness lake in Adirondack State Park, New York, and have found large increases in silver, gold, cadmium, chromium, copper, lead, antimony, vanadium, and zinc influxes to the sediment over the last 30 years. They argue that increases in cadmium, copper, lead, and zinc probably resulted from increased atmospheric influxes of these elements.

It appears, then, that concentrations of heavy metals in lake sediment can be used as indicators of environmental pollution in recently deposited lake sediments. The data suggest that these elements are transported as aerosols even to remote areas.

Although recent investigations of lake-sediment chemistry emphasize the latest portion of the Holocene record, spanning the period of European colonization in North America, the results of these studies are applicable to the entire Holocene period. The changes wrought by human technology can be viewed as a great experiment that allows us to test our understanding of sediment chemistry as a paleoenvironmental indicator. Problems associated with the interpretation of the impact of human disturbance on sediment chemistry, however, suggest the need for caution in the interpretation of the longer postglacial record. For example, sediment chemistry may, as Likens and Davis (1975) suggest, represent the record of particulate matter carried to lakes rather than a record of past lake-water chemistry. In seeking a more definite record of water chemistry, we must, therefore, turn to microfossils of aquatic organisms. Presumably, these are more directly affected by dissolved chemicals.

Diatoms

Fossils of diatoms (class Bacillariophyceae) provide a means of reconstructing past lake history. These algae are especially well suited for paleolimnological studies because their siliceous cell walls are more or less well preserved in lake sediment. The diatom cells, or frustules, can generally be identified by species. Diatoms are useful paleoenvironmental indicators because they inhabit lakes of all types and because particular species apparently prefer particular environmental conditions.

Diatom fossils in lake sediments have played an important part in the study of Quaternary paleoecology since they have yielded information on the impacts of changing climate on lake chemistry and fertility. In an early study at Linsley Pond, Connecticut, Patrick (1943) found that the clear-water diatoms dominating the lake during late-glacial times were replaced by a planktonic flora at the same time the climate improved enough to permit the growth of oak-dominated forests around the lake.

Bradbury (1971) has tried to correlate the diatom stratigraphy of Lake Texcoco, Mexico, which probably covers 100,000 years, with the Northern Hemisphere's glacial/interglacial cycle, but he has found the correlations poor. The diatom stratigraphy indicates that early in the record the lake contained fresh water, but the freshwater species (such as *Stephanodiscus niagarae*) were quickly replaced by species suited to brackish water. Throughout Wisconsin time, conditions at the core site fluctuated from those of a brackish-water marsh to a brackish-water lake.

Florin and Wright (1969) have used diatom fossils to demonstrate that the "trash" layer found at the bottom of the sediment column in many Minnesota lakes represents the remains of a terrestrial community that grew on stagnant ice blocks immediately after glacial retreat. At Kirchner Marsh, Minnesota, they found diatoms from terrestrial environments in the trash layer.

For Pickerel Lake, South Dakota, Haworth (1972) relates changes in diatom assemblages to the effects of climatic changes on the lake and infers changes in water chemistry as the lake developed. She demonstrates that in late-glacial times the lake was relatively infertile, supporting many of the diatom species that now live in the modern lakes of northeastern Minnesota. At the same core levels in which the pollen data show that the late-glacial spruce forest was replaced by prairie, the diatom assemblages become dominated by alkaline-indicating species that are presently characteristic of modern prairie lakes. Both the pollen and the diatom changes resulted from the changing climate at the close of the Wisconsin glacial period.

Stoermer (1977) has examined changes in fossil diatom assemblages from Douglas Lake, Michigan, and has found a late-glacial flora similar to late-glacial floras from the northern European lakes around the Baltic. At the time this assemblage was deposited, Douglas Lake was apparently part of the complex of proglacial lakes that preceded the modern Laurentian Great Lakes. This assemblage was replaced by one dominated by benthic species, and this in turn by a group of species that have remained in the lake until modern times.

For Kirchner Marsh, Minnesota, Brugam (1980a) demonstrates the close correspondence between limnological conditions as indicated by diatom assemblages and climatic conditions as indicated by pollen and macrofossils of vascular plants. At this site, the diatoms confirm that lake levels were drastically reduced during the middle-Holocene warm period, which also permitted the advance of prairie vegetation to the lake watershed.

Figure 13-1 shows the striking changes that occurred in the diatom assemblages during the middle-Holocene warm period at Kirchner Marsh. The K pollen zone represents late-glacial tundra; the A zones, closed spruce forest; the B zone, pine forest; the C-a, oak-forest; the C-b, prairie; and the C-c, modern oak woodland. The diagram shows that in the C-a zone the clear-water benthic diatom assemblages dominated by *Fragilaria construens* and *F. construens venter*, which characterized the lake during late-glacial and early postglacial times, gave way to a brief pulse of the eutrophic plankter *Stephanodiscus hantzschii*. This pulse coincided with the establishment of a series of assemblages dominated by shallow-water epiphytic general such as *Gomphonema* and *Cocconeis*, assemblages that persisted through zones C-b and C-a. These changes in diatom assemblages expand and confirm Watts and Winter's (1966) conclusions based on macrofossils, which also document the lowering of water levels during the prairie period.

Interpretations of fossil diatom assemblages are usually based on the environmental preferences of species on the various diatom "spectra" (summarized by Foged, 1954). An alternative method of interpretation is to compare down-core assemblages with those from surface sediments of modern lakes whose physical and chemical conditions are well known. The assumption is made that the lakes with assemblages that are similar to ones from deep in a core represent similar environmental conditions. The application of this method can lead to precise estimates of past limnological conditions. Using recently deposited assemblages from the tops of cores, Imbrie and Kipp (1971) have employed similar methods to interpret fossil foraminiferal assemblages from ocean cores.

Florin (*in* Bright, 1968) has examined surface-sediment diatom assemblages from 18 Minnesota lakes for which chemical data are available. She concludes that the diatom assemblages vary with differences in lake-water chemistry across the state. Brugam (1983) has applied cluster analysis to fossil diatom samples from the surface sediment of an additional 105 Minnesota lakes whose environmental conditions are well known. Total alkalinity and water transparency show the greatest differences between clusters, indicating that these are the most important controlling factors in the species composition of the fossil assemblages.

In order to make comparisons between surface and down-core fossil assemblages, it is necessary to employ a multivariate statistical technique (especially when sample numbers are high). Brugam (1980) has found cluster analysis to be an appropriate technique for this purpose. He has applied cluster analysis to a data set that includes the down-core assemblages from Kirchner Marsh and surface samples from 100 lakes in Minnesota, the Dakotas, and Labrador. Figure 13-2 shows the resulting clusters, and Figure 13-3 shows the species composition of the surface samples that are most similar to the down-core samples and the water depths of the modern lakes from which the surface samples were taken. Brugam concludes that samples from the

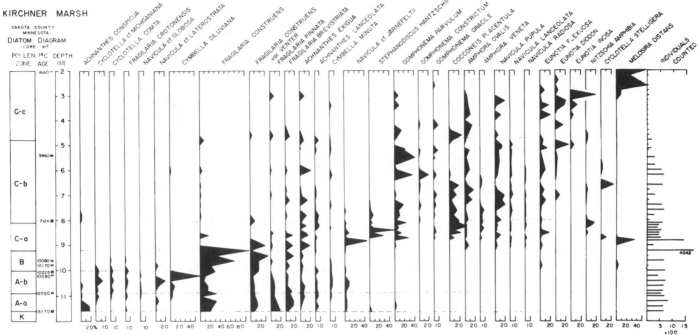

Figure 13-1. Diatom diagram from Kirchner Marsh, Dakota County, Minnesota. (From Brugam, 1980.)

middle-postglacial warm period at Kirchner Marsh (pollen zone C-a and diatom cluster V in Figure 13-3) are most similar to modern surface samples from shallow (less than 2.5 m) prairie lakes. Late-glacial samples (pollen zones A-a and A-b, diatom clusters I and II) cluster with modern samples from Labrador and Minnesota. The surface sample analogues of the down-core assemblages suggest that the lake at Kirchner Marsh was deep but became shallow as a result of the warmer and drier climate that prevailed during the middle-postglacial warm period.

Diatoms respond very strongly to human disturbance in lakes. In North America, in contrast to Europe, the period of human disturbance is relatively recent (less than 200 years in most locations) and is clearly marked by increases in weed pollen (mainly *Ambrosia*) in the sediment. Furthermore, much of the postdisturbance record can be dated with ^{210}Pb and ^{137}Ce.

Stockner and Benson (1967) and Stockner (1972) identify a repeatable sequence of diatom changes in the lakes they examined (Lake Washington, the English Lake District, and the Experimental Lakes Area, Ontario) in the form of a consistent shift from centric diatoms to araphidinate species as a result of disturbance. Stockner has formulated an index of eutrophication based on the ratio of these two diatom families (Stockner, 1972).

Although it has been widely applied, the Araphidineae/Centrales (A/C) ratio does not appear to be universally valid. In many lakes, the abundance of small *Stephanodiscus* species, rather than the Araphidineae, increase with disturbance. Bradbury (1975) has found a postcolonization increase in *Stephanodiscus* in 4 out of 8 lakes examined. Brugam (1979) has compared pre- and post-European-colonization diatom assemblages from 27 Minnesota lakes, finding no increase in the A/C index but large increases in small *Stephanodiscus* species instead. In surface samples from 100 Minnesota lakes, *S. hantzschii* is abundant at total phosphorus concentrations greater than 15 $\mu g/\ell$ (Figure 13-4), but the A/C index showed no good relationship to phosphorus levels (Figure 13-5). The primary difference between most Minnesota lakes and the lakes Stockner examined is in

their higher alkalinities (generally greater than 0.75 meq/ℓ HCO$_3^-$). Brugam (1979) speculates that the failure of the A/C index in Minnesota is a result of the presence there of a centric diatom that prefers alkaline eutrophic waters (i.e., *Stephanodiscus*). Other work on low-alkalinity lakes has suggested that this is true. Bradbury (1978) shows that the eutrophication of Burntside Lake (a low-alkalinity lake in Minnesota) causes an increase in *Asterionella formosa*, an araphidinate species. Similarly, Lily Lake, which is located on a small area of noncalcareous drift in eastern Minnesota, shows an increase in araphidinate species as cultural disturbance occurs (R. B. Brugam, unpublished data). It seems, then, that the Araphidineae may not be the only indicators of eutrophication among the diatoms.

Another possible index of trophic status is the influx (frustules per square centimeter per year) of diatoms to the sediment (Battarbee, 1974, 1978). Higher levels of nutrient inputs should result in a higher production of phytoplankton; so, if diatoms are an important component of the lake's phytoplankton, the influx of diatom remains to the sediment should increase with increasing trophic status. It is expected that, where production is particularly high, blue-green algae will become more abundant than diatoms, and the diatom influx may be reduced because of competition (Brugam, 1978). In moderately eutrophic lakes, however, diatom influx, in theory, reflect trophic status.

Recent geochemical evidence suggests, however, that the relationship between diatom production and accumulation in the sediment is not as simple as the process outlined above. Parker and Edgington (1976) have found in Lake Michigan that recognizable diatom fossils are concentrated in only the shallowest sediment levels. They argue that diatoms dissolve and disappear from older sediment layers. Parker and others (1977) confirm that considerable silica recycling occurs in the sediment of Lake Michigan. Battarbee (1978) calculates that the amount of silica fixed by diatoms during any year in Lough Neagh, Northern Ireland, is more than the silica inputs from the surrounding landscape could sustain; this implies that there is considerable recycling of silica within the lake. These results suggest that

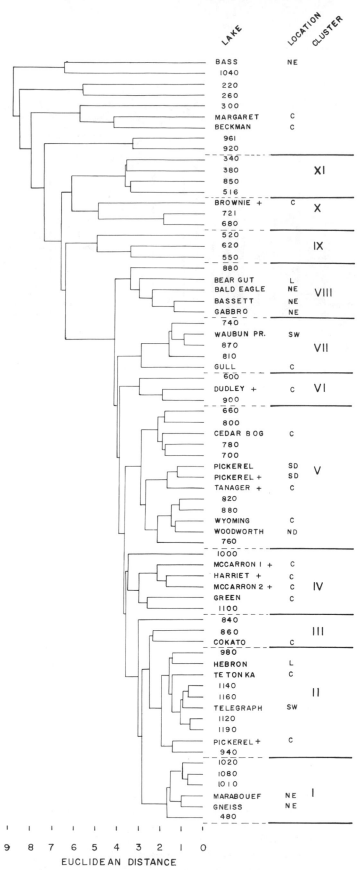

Figure 13-2. Cluster analysis of down-core diatom samples from Kirchner Marsh and surface samples from lakes in Minnesota, North and South Dakota, and Labrador. Plus signs designate presettlement levels in short (one m) cores. Numbers indicate depth in Kirchner Marsh core. Location codes are: northeastern (NE), central (C), and southwestern (SW) Minnesota, South Dakota (SD), North Dakota (ND), and Labrador (L). (From Brugam, 1980.)

the diatoms produced in a year's bloom are built from silica redissolved from the preceding year's bloom.

The geochemist's argument is difficult to reconcile with the paleoecologist's argument about the significance of diatom influxes as indicators of a lake's trophic status. It seems that the diatoms preserved in the sediment (and counted by the diatomist) are only a residue of the diatoms produced in a given time period. Therefore, diatom influx represents the difference between diatom production and dissolution and may not be a close reflection of a lake's trophic status.

The dissolution of diatoms may cause other difficulties; because it is undoubtedly species specific, more delicate species such as *Asterionella formosa* are lost first. Haworth (1976) and Parker and Edgington (1976) have commented on the preferential loss of delicate species. Comparisons between the species compositions of down-core and surficial samples may reduce the problems of interpretation caused by differential dissolution because both surface samples and down-core samples can be expected to have undergone similar amounts of dissolution. Surface-sample analogues may, therefore, be more valuable than influx in the reconstruction of past lake conditions.

The problem of imperfect preservation is not unique to diatoms; nearly every constituent of the sediment that can be examined by the paleolimnologist is susceptible to diagenetic change. In some constituents, such as plant pigments, the possibility of diagenetic loss is widely accepted (see discussion in Daley, 1973), whereas for others, such as cladoceran remains, this possibility is not acknowledged. In order to make diatoms a totally reliable tool for paleolimnological reconstructions, it will be necessary to consider the effect of dissolution on the fossil assemblages.

Other Algae

Remains of algae other than diatoms can be recovered from lake sediments. They include *Pediastrum*, *Botryococcus*, *Gloeotrichia*, and many species of desmids. These remains are best identified in samples that are deflocculated in potassium hydroxide, but some desmids such as *Pediastrum*, and dinoflagellates even survive acetolysis and can be counted in pollen samples. Another group of algal remains is siliceous and dissolves when treated with potassium hydroxide or hydrofluoric acid. These remains are best seen in diatom preparations, where organic material is oxidized away and siliceous remains are retained. The siliceous remains include synuracean scales and chrysophycean cysts.

Two major problems occur with using the remains of algae other than diatoms as paleoecologic indicators. The first problem is that many of the remains are resting cells, which are difficult to relate to the species that produced them. Leventhal (*in* Hutchinson, 1970) has followed Nygaard's (1959) practice by assigning temporary names (*nomina tempora*) to her chrysophycean material to be identified later. Similarly, Norris and McAndrews (1970) call the dinoflagellate cysts they find types B, C, and D, although they do tentatively relate their types to known dinoflagellate species.

The second problem with algae other than diatoms is that their environmental preferences are largely unknown. Crisman (1978a) has examined the distribution of *Gloeotrichia*, *Botryococcus*, *Stauraustrum*, *Peridinium*, and seven species of *Pediastrum* from the surficial sediment of 96 lakes in Minnesota whose water chemistry is well known. He notes that some taxa show strong

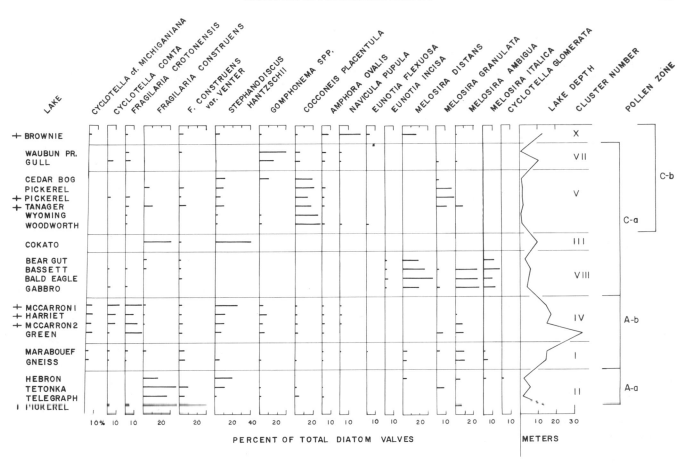

Figure 13-3. Species composition of surface samples that cluster most closely with the down-core Kirchner Marsh samples. Right panel shows the depths of the lakes from which the surface sample analogues were taken. Plus signs indicate a presettlement sample from a short core. (From Brugam, 1980).

preferences for particular chemical conditions. He also reports that, in the pollen profiles from Minnesota in which *Pediastrum* has been counted, the genus generally peaks during the late-glacial spruce zone. Furthermore, at Wolf Creek (Birks, 1976), where individual species of *Pediastrum* have been identified, most of the increase can be attributed to *Pediastrum boryanum*, which shows a decided preference for eutrophic lakes in the surface samples. Crisman's results suggest that during the late-glacial spruce maximum in Minnesota lakes may have been rather eutrophic.

Other investigators have attempted to use algae other than diatoms as stratigraphic indicators, with results that suggest interesting relationships. Problems of identifying and understanding environmental preferences, however, make firm conclusions difficult. Leventhal (*in* Hutchinson, 1970) has found that chrysophyte cysts declined in number in the core from the Lago di Monterosi when significant human disturbance (and eutrophication) began. Smol (1980) identifies major changes in the species composition of synuracean scale assemblages in Jake Lake and Found Lake, Ontario, that resulted from road building in the watershed, but he reports finding no change in nearly undisturbed Delano Lake assemblages. Munch (1980) also has found that Synuraceae responded strongly to the human disturbance of the watershed of Hall Lake, Washington. Brugam (1978) reports that *Gloeotrichia* remains declined in Linsley Pond with each increase in trophic status caused by human disturbance. In contrast, *Staurastrum gracile* remains only appeared in abundance during the most culturally disturbed and eutrophic phase of the lake's history.

In summary, the remains of algae other than diatoms in lake sediments show potential for expanding our knowledge of past lake environments. Unfortunately, however, this potential remains largely unrealized because of problems of taxonomy and of understanding environmental preferences. More work in these areas is clearly needed.

Pigments

The remains of plant pigments can be used to reconstruct past limnological conditions. Chlorophyll *a* is seldom preserved in lake sediments, but its degradation products—pheophytin *a*, pheophorbide *a*, and chlorophylide *a*—and the corresponding degradation products of chlorphylls *b* and *c* are found in lake sediments (Vallentyne, 1960) and are referred to as sediment chlorophyll degradation products (SCDPs). Carotenoids are also found in sediments and can be linked to particular organisms living in the lake. For example, Brown (1968) relates bacterial carotenoids in sediment from Little Round Lake, Ontario, to populations of *Rhodopseudomonas spheroides* in the lake. Griffiths and others (1969) use the fossilized carotenoid oscillaxanthin to reconstruct the history of *Oscillatoria* blooms in Lake Washington, Washington State.

The interpretation of pigment profiles from sediment cores is difficult and complex because of the pigments' extreme chemical lability. Pigments are subject to photooxidation in the lighted surface waters of the lake after the death of the plant cell in which they were

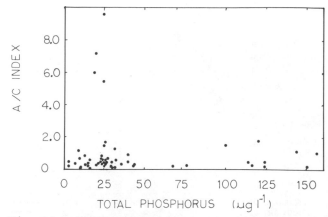

Figure 13·4. Percentage of *Stephanodiscus hantzschii* from surface-sediment samples versus water-column total phosphorus concentration in 48 Minnesota lakes. (From Brugam, 1979.)

Figure 13·5. Araphidineae/Centrales index from surface sediment samples versus water-column total phosphorus concentration in 48 Minnesota lakes. (From Brugam, 1979.)

produced (Daley and Brown, 1973). Furthermore, consumption by herbivores destroys some of the pigments more readily than others (Daley, 1973). In spite of these difficulties, pigment abundances seem to be at least generally correlated with a lake's trophic status. The reason for this correlation may be the greater algal production in eutrophic lakes, or it may be the more favorable conditions for preservation in the dark, cool, anaerobic hypolimnion of eutrophic lakes in contrast to the aerobic conditions prevailing in the deep waters of oligotrophic lakes.

Not only are pigment abundances useful in reconstructing past lake environments, but ratios and diversities of pigments may have paleoecologic significance. Ratios of SCDPs to carotenoids seem to indicate the relative importance of allochthonous sources in the origin of lake sediment. In decaying leaves on the terrestrial landscape, chlorophyll derivatives are preserved better than carotenoids, but the decay of algal cells in water favors the preservation of carotenoids. Thus, high SCDP/carotenoid ratios imply that the source of the sedimentary organic matter was mainly terrestrial, but lower ratios imply a predominately lacustrine source.

Sanger and Gorham (1973) suggest the use of pigment diversity (or chromatogram spot number) as an aid to paleoecologic reconstruction. Pigment diversity is the total number of pigment spots seen on a two-dimensional thin-layer chromatography plate and represents the number of colored chlorophyll derivatives and carotenoids present in the sediment. Pigment diversities from terrestrial detritus are low, presumably because of oxidative destruction. Lake-sediment pigment diversities may be high because algal decay produces a wide diversity of degradation products. It is more likely, however, that conditions for preservation are more favorable in an aquatic environment. Pigment diversities in oligotrophic lakes are generally lower (30 pigment spots on a chromatogram) than in eutrophic lakes (38 spots) (Sanger and Gorham, 1970), and diversities are lower still in dry terrestrial soils.

Plant pigments preserved in lake sediments have been useful in understanding lake development during the Holocene. Our advancing knowledge about the role of preservation in controlling pigment concentrations found in lake sediment promises to increase the value of pigment-based paleolimnological reconstructions.

Wetzel (1970) has found in Pretty Lake, a marl lake in Indiana, that chlorophyll-degradation products and carotenoids were low in the late-glacial sediments, that in the immediate postglacial sediments

pigments reached a peak, and that another minor peak occurred between 6500 and 5500 yr B.P. He postulates a maximum productivity between 10,000 and 9000 yr B.P., followed by lower productivity with a series of oscillations in trophic status. During the last 150 years, productivity has increased in the lake because of human disturbance. In general, pigment concentrations were inversely proportional to calcium carbonate deposition in the lake sediment, suggesting that in the past, as today, carbonate precipitation limited productivity by removing limiting nutrients from the water column.

Sanger and Gorham (1972) identify major changes in pigments at Kirchner Marsh, Minnesota, that may be correlated with changes in pollen and macrofossil assemblages (Figure 13-6). As in Pretty Lake (Wetzel, 1970), they report an early postglacial peak of pigment concentrations, which they interpret as a result of an increase in primary productivity. They suggest the possibility that the lake was meromictic at this time. From the time of lowered water levels in the middle Holocene, the pigment data indicate higher inputs of allochthonous organic matter, probably from the surrounding reed swamps, and perhaps also resulting from shoreline disturbance due to fluctuating water levels. Allochthonous organic inputs increased still more as the reed swamp grew toward the core site.

Whitehead and others (1973) and Whitehead and Crisman (1978) have examined pigments and a number of other sediment constituents at Berry Pond and North Pond in western Massachusetts and have found that all the sediment constituents showed a number of major ''trophic oscillations'' in the ponds' developmental histories. In both ponds, trophic status was low during late-glacial times and increased to a maximum during early-postglacial times, when first spruce forest and then pine/birch forest dominated the landscape. Productivity declined but rose again when hemlock pollen (and presumably hemlock trees) declined catastrophically. Berry Pond had a third trophic oscillation associated with the gradual filling of the lake basin and the oblinteration of the hypolimnion.

Whitehead and Crisman (1978) have found that the trophic status of their lakes was very sensitive to change in watershed vegetation. They believe, on the basis of models of chemical input and output (Likens and Bormann, 1977), that change in the surrounding watershed vegetation would be reflected in ecological change in nearby lakes. Because climatic change influences vegetational type and because vegetation controls the chemical quality of runoff waters, climatic change should be reflected in changing lake conditions.

Sanger and Crowl (1979) have examined pigment concentrations from the sediment of Browns Lake, Ohio, which lies on the edge of a large peatland and is heavily influenced by it. As with the lakes already mentioned, pigment levels are low in the late-glacial clays of Browns Lake. Maximum pigment concentrations occur immediately above the late-glacial clay in a zone of laminated sediment. Sanger and Crowl presume that the high pigment concentrations and laminated sediments indicate a period of meromixis and good pigment preservation rather than one of high lake productivity. At this core level, carotenoid concentrations are also high, supporting the argument that pigment preservation was especially good. After this period of presumed meromixis and high pigment concentration, levels decline to concentrations characteristic of oligotrophic lakes. The most recent strata of the core contain increasing amounts of reed-swamp peat, and chlorophyll derivative/carotenoid ratios indicate increasing allochthonous inputs as the lake became increasingly dominated by the surrounding peatlands.

In general, most pigment analyses have been carried out on sediments of eutrophic lakes or boggy ponds. Likens and Davis (1975) have examined an oligotrophic lake (Mirror Lake, New Hampshire) in a study comparing pigments, sediment chemistry, and pollen assemblages from a postglacial core. They report that pigment levels were low in late-glacial times, rose at the establishment of forest around the lake, and remained at the same levels to the present time. SCDP/carotenoid ratios were high from 9000 to 7000 yr B.P. and later declined but increased again after the colonization of the lake watershed by Europeans. Likens and Davis argue that the interpretation of fossil pigments at Mirror Lake is more complex than a simple allochthonous/autochthonous interpretation indicates; they point out that, in an oligotrophic lake, pigment production rates are low and conditions of preservation may be poor. Likens and Davis emphasize the necessity of examining many different indicators of past lake environments before making conclusions.

Pigment analyses have also been used to document the consequences in lakes of recent human disturbance. Gorham and Sanger (1976) have examined a short (1.5 m) core from Shagawa Lake, an extremely eutrophic lake in northeastern Minnesota, and have found large increases in pigment concentrations in the sediment levels corresponding to the period following the establishment of the town of Ely on the lakeshore.

Pigments can be useful indicators of past lake conditions, but, as Likens and Davis (1975) have indicated, they must be used cautiously. Problems of preservation are manifest enough (Daley, 1973; Daley and Brown, 1973) that it is now impossible to interpret fossil pigment concentrations simply as direct records of primary production, as some investigators have in the past. The realization that problems of preservation must be considered before interpretations can be made does not apply only to pigment analysis but extends to all fossils that are important in paleolimnological reconstructions. Investigators dealing with fossil pigments have been forced by the nature of their material to deal with the problem of diagenesis sooner than other paleolimnologists. The preservation problem is no less important to the interpretation of the remains of diatoms and crustaceans.

Chydrorid Cladocera

Chydorids are Crustacea adapted for life on submerged aquatic plants, rocks, and sediments (Fryer, 1968). They are valuable to the paleolimnologist because their remains (carapaces and head shields) are well preserved in lake sediment and are identifiable to species (Frey, 1959). Furthermore, like diatoms, the species composition of their fossil assemblages seems to reflect prevailing lake conditions.

D. G. Frey and his students have developed and used chydorid analysis to interpret past lake conditions in many parts of the world (Frey, 1960, 1964). Goulden (1964) relates changes in fossil chydorid assemblages to disturbances by prehistoric peoples at Esthwaite Water in England. He reports that littoral chydorids were less important than the planktonic nonchydorid Cladocera throughout the core. At nearby Blelham Tarn, Harmsworth (1968) has found that littoral species dominated before Sub-Atlantic time but after that period the assemblages became more like those at Esthwaite. He argues that this change reflected increasing lake productivity as a consequence of climatic amelioration. Throughout postglacial time, the succession of planktonic Cladocera indicated that the lake was becoming progressively more productive as it filled in.

The interpretation of chydorid remains from lake sediment suffered from lack of knowledge about the ecology and environmental preferences of the living organisms until Whiteside (1970) examined chydorid assemblages from surface muds of 80 Danish lakes whose water chemistry was well known. By applying discriminant analysis to the chemical and physical data for the lakes, he found that he was able to classify them into three types—clear-water lakes, pond and bog lakes, and culturally disturbed clear-water lakes. When the chydorid data were compared with this grouping, Whiteside found that there were certain species characteristic of certain lake types. He compared down-core chydorid samples from two Danish lakes, Grane Langso and Esromso, with his lake types to determine past lake conditions. Throughout most of its history, Esromso was classified by discriminant analysis as a clear-water lake. Analysis of the down-core samples from Grane Langso was not as successful because the dominant species in the down-core samples was relatively unimportant in the surface samples; that is, the surface samples were poor analogues of the Grane Langso core. Whiteside argues that Grane Langso was more sensitive than Esromso to the direct effects of climate because of its smaller size.

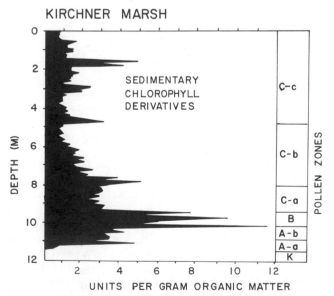

KIRCHNER MARSH

Figure 13-6. Sedimentary chlorophyll derivative profile from Kirchner Marsh, Minnesota. (After Sanger and Gorham, 1972.)

Synerholm (1979) has recently completed work similar to Whiteside's; Synerholm uses fossil assemblages from 32 Minnesota and North Dakota lakes analyzed by cluster analysis and principal-components analysis. She reports that the chydorid assemblages differ according to vegetational region in response to lake-water chemistry in ways similar to aquatic macrophytes (Birks, 1973). Moyle (1956) and Bright (1968) had earlier found that lakes in the prairie sections of Minnesota were alkaline and concentrated in ions, that those of the deciduous forest region were intermediate in pH and ionic concentration, and that those of the northeastern coniferous forest region were acidic and dilute. Synerholm reports that each lake type had its own characteristic recent assemblage of chydorids.

As with diatom influx, chydorid concentrations in lake sediment could be used as a direct indicator of a lake's trophic status. Such a relationship would offer a valuable alternative to the statistical techniques of Whiteside and Synerholm in estimating a lake's past trophic status. Unfortunately, Harmsworth and Whiteside (1968) have found no good relationship between lake productivity and chydorid concentrations in their examinations of data from 19 Danish and 14 Indiana lakes. They acknowledge that by considering only concentrations of fossil remains they have ignored differences in sedimentation rates among lakes. Certainly, influxes of fossil chydorids to the sediment would be more appropriately compared with lake productivity, but even influxes might not correlate very well with productivity if there were a significant loss of remains as a result of decay.

Still another approach to the problem of interpreting chydorid data applies modern theories of population ecology to the species composition of samples. Goulden (1969) applies MacArthur's (1957, 1960) broken-stick distribution to fossil chydorid samples from lakes. When the species randomly divide the available resources among themselves and are ranked from most abundant to least, a plot of abundance versus rank order should be a gently curving line. Goulden reports that this relationship holds in samples from undisturbed core levels; but for the Aguada de Santa Ana Vieja, a small pond in Guatemala, in periods of pre-Columbian human disturbance, the broken-stick distribution does not hold. The assemblage characterizing disturbed conditions is lower in species diversity than that from undisturbed conditions. Goulden has found similar responses to disturbance in the Lago di Monterosi, Italy, in response to Roman colonization and in Lake Lacawac, Pennsylvania, following glacial retreat. In each of these cases, the disturbance of the watershed was associated with lowered chydorid species diversity. Goulden argues that watershed disturbance increases silt and nutrient runoff and encourages the proliferation of one or a few eurytopic species that dominate the aquatic community. As watershed vegetation undergoes succession to a stable climax community, the chydorid communities respond by becoming more diverse.

Tsukada's (1972) work on Lake Nojiri in Japan supports Goulden's arguments. Lake Nojiri has been subject to volcanic ash falls throughout its history. Tsukada reports that each ash fall is associated with a reduction in chydorid species diversity. After each disturbance however, there is a slow return to the MacArthur broken-stick distribution. The recent eruptions of Mount St. Helens in Washington State and subsequent ash falls in the western United States offer a good opportunity for examining the mechanisms of these changes in modern lakes and for testing Goulden's argument relating watershed disturbance and chydorid species diversity.

Crisman and Whitehead (1978) compare the results of pigment studies at North Pond and Berry Pond, Massachusetts, with chydorid species diversities and report that at Berry Pond each increase in productivity (as indicated by pigments) coincides with a decrease in chydorid species diversity and equitability. In contrast, at North Pond diversity does not change much over the whole post glacial record.

It is evident from the research reviewed here that the recent emphasis in North American chydorid research has been on increasing our ability to interpret changes in fossil assemblages in cores. The main problem has been in understanding the environmental preferences of these organisms. Although knowledge on this score has advanced, more work needs to be done.

Planktonic Crustacea

A number of planktonic Crustacea leave recognizable remains in lake sediment similar to the littoral chydorids. Of the planktonic community, the Cladocera are the best preserved in the sediment. Copepods, though abundant in some lake-water columns, are preserved hardly at all in the sediment. Of the noncrustacean zooplankton, rotifers leave resting eggs in the sediment that are sometimes identifiable to species (Mueller, 1970).

Remains of planktonic animals are particularly interesting because of the large amount of recent research on their population ecology (summarized in O'Brien, 1980), which suggests that their numbers are primarily regulated by predation from other invertebrates and from fishes. In general, fishes select large, visible zooplankton and invertebrates select small, easily subdued individuals as prey. Such selectivity in predators leads to the evolution of morphological adaptions in prey that thwart predation. The paleolimnological record can be used to trace the historical development of such interactions over long periods of time.

A major unsolved problem in the paleolimnology of planktonic Cladocera is the shift from a variety of *Bosmina coregoni* to *B. longirostris* during postglacial time in most of the sediment cores that have been looked at (Crisman and Whitehead, 1978; Deevey, 1942, 1969; Goulden, 1964; Harmsworth, 1968). Figure 13-7 shows the dramatic shift from *B. coregoni* to *B. longirostris* that Crisman and Whitehead (1979) have found in North Pond, Massachusetts. Numerous arguments have been offered as explanations for this replacement. Goulden (1964) suggests that the shift reflects an increase in productivity due to increases in eutrophic midges at the same time. Deevey (1969) notes that the replacement occurred in Rogers Lake, Connecticut, at the time *Chaoborus* remains became abundant in the sediment. Brooks (1969) presents evidence that the introduction of the planktivorous fish *Alosa aestivalis* to Crystal Lake, Connecticut, caused the replacement of *Bosmina coregoni longispina* by *B. longirostris*. He suggests that the large *B. c. longispina* was more susceptible to fish predation than *B. longirostris* and that the shift Deevey notes in Rogers Lake resulted from the invasion of the lake by the fish *Alosa pseudoharengus* during early-postglacial times. Crisman and Whitehead's (1978) data from North Pond support the interpretation (Figure 13-7) because they show a sharp drop in average size of *Bosmina* carapaces as *B. longirostris* replaced *B. coregoni* (Figure 13-8). Kerfoot's (1975, 1978) recent results suggest that, in addition to fish predation, copepod predation is important in the life of bosminids. He shows that predatory copepods are better able to capture and to subdue *Bosmina* of particular morphologies.

It may be that changes in copepod predation lead to changes in the bosminid species and varieties present in a lake. The mechanisms suggested in the papers reviewed may represent parts of a causal chain, each link of which is important in determining the particular *Bosmina* species and morph that will be successful in a lake. A hypothetical causal chain might be as follows. A lake may become more productive as the climate improves or as the surrounding vegetation changes. In Linsley Pond and Rogers Lake (Deevey, 1942, 1969), for example, the replacement of *B. coregoni* by *B. longirostris* occurred as the late-glacial coniferous forest was replaced by deciduous forest. Increased productivity allowed the colonization of the lake by zooplanktivorous fishes. These fishes exterminated the large copepods that would usually have preyed on *B. longirostris*, allowing it to replace *B. coregoni*. The final step of the postulated causal chain is supported by Deevey's (1969) results that indicate that the replacement resulted in a decrease in bosminid size and by Kerfoot's demonstration that predatory copepods preferentially consume small bosminids. Proof of this mechanism can only come from the intensive study of bosminid population ecology in modern lakes.

In addition to *Bosmina*, *Daphnia* remains are also preserved in lake sediments. *Daphnia* is especially important for the role it plays in Brooks and Dodson's (1965) size-selective predation hypothesis, which suggests that zooplanktivorous fishes prefer larger and more visible individuals as food. Because *Daphnia* species range in size from the large *Daphnia pulex* and *D. magna* to the tiny *D. ambigua*, their remains can be important indicators of changing predation intensity with time. Hrbacek (1969) first suggested using measurements of *Daphnia* postabdominal claws as indicators of past individual size. Kerfoot (1974), however, has found that differential preservation makes postabdominal-claw size less useful than mandible size for this purpose. For Frains Lake, Michigan, he reports that *Bosmina* was abundant in late-glacial times (11,000 to 9000 yr B.P.). Between 9000 and 150 yr B.P., large-sized *Daphnia* declined drastically in numbers and size, to be replaced by *Bosmina* (Figure 13-9). Kerfoot links this change to eutrophication caused by agricultural runoff and disturbance of the watershed.

More recently, Kitchell and Kitchell (1980) have demonstrated that changes in zooplankton remains need not be associated with eutrophication. They studied two halves of a dystrophic lake that had been divided by a dam during the 1950s. Both halves contained the

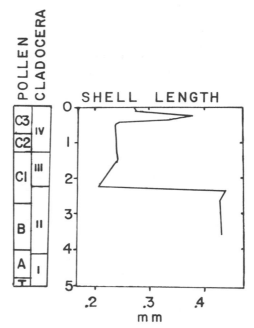

Figure 13-8. Changes of *Bosmina* shell (carapace) length with time in the core from North Pond, Massachusetts. (After Crisman and Whitehead, 1978.)

same fish stock (rainbow trout, *Salmo gardineri*). In 1951, liming was begun on half of the lake to precipitate humic colloids, and the treatment continues to the present day. Kitchell and Kitchell examined the change in zooplankton remains in the sediment caused by the experiment and found that the originally dominant *Daphnia pulex* was replaced first by *D. rosea* and then by *Bosmina* in the experimental half of the lake but that no changes occurred in the control half. They argue that these changes resulted from increases in the trout's efficiency of predation in the clearer water.

Kitchell and Kitchell's work suggests that zooplankton remains are subtle indicators of lake conditions. It may not be possible to use changes in zooplankton microfossils to reconstruct climatic change as such, but their potential usefulness may lie in reconstructions of the impact of climatic change on community interactions within a lake. In order to do this effectively, however, it is necessary to know more about mechanisms of change in modern zooplankton communities. As with the other sedimentary constituents discussed in this chapter, it is also necessary to know more about the preservation of zooplankton remains, particularly about whether preservation is size specific.

Insect Remains

The remains of insects, though not so abundant in lake sediments as those of diatoms and crustaceans, are abundant enough to be used as paleoenvironmental indicators. The most common insect remains are from midges—the head capsules of larval Chironomidae and the mandibles of larvae of the genus *Chaoborus*. Both groups dwell as larvae in lake sediments. *Chaoborus* also becomes planktonic during the evening and preys on zooplankton.

Midge remains can be important paleolimnological indicators because different species of the Chironomidae prefer different ambient oxygen concentrations. For this reason, the abundance of par-

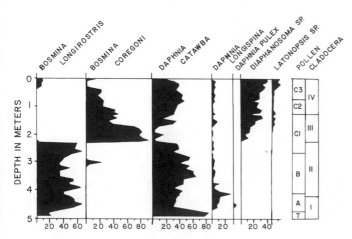

Figure 13-7. Abundance of planktonic cladocera from a core from North Pond, Massachusetts. Note the rapid replacement of *Bosmina coregoni* with *Bosmina longirostris*. (After Crisman and Whitehead, 1978.)

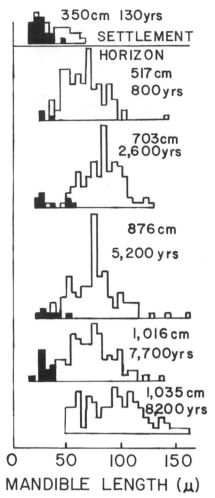

Figure 13-9. Changes in *Bosmina* and *Daphnia* mandible sizes in a core from Frains Lake, Michigan. Dark histograms represent *Bosmina* mandible sizes; open histograms represent *Daphnia* mandible sizes. (After Kerfoot, 1974; copyright 1974 Ecological Society of America.)

compound microscope. The time-consuming nature of the analysis results from the low concentrations of midge remains in a usual lake sediment. In order to obtain a statistically significant number of remains, one must examine a large amount of sediment.

Identification of chironomid remains also poses problems. In general, keys to the living organisms use soft parts to differentiate taxa. Hofmann's (1971) paper is a valuable taxonomic work on European species that takes into account the paleoecologist's problems of identification. Mason's (1973) work, though primarily oriented toward those working with living material, provides superb photomicrographs of major American genera.

Because of these difficulties, there has been little recent application of chronomid analysis to Holocene paleolimnological problems in North America. Stahl (1969) has reviewed the earlier work in the field.

Lawrenz's (1975) work on Green Lake, Antrum County, Michigan, is one notable exception, however. Like Douglas Lake (Stoermer, 1977), mentioned earlier, Green Lake underwent water-level fluctuations in response to changes in the levels of the nearby Laurentian Great Lakes. Lawrenz identifies five phases of chironomid development: (I) an initial late-glacial oligotrophic phase, (II) a moderately oligotrophic phase associated with the Valders ice re-advance, (III) an oligotrophic littoral phase, (IV), a moderately oligotrophic profundal phase, and (V) an oligotrophic profundal phase. The whole profile is dominated by oligotrophic-indicating Tanytarsini. Changes in chironomids coincided with changes in pollen zones and also with sediment type. Lawrenz suggests that the midge fauna from Green Lake may have been responding not to changes in the lake's trophic status but more likely to changes in water level as deglaciation occurred and levels of the surrounding Great Lakes changed.

More recently, Warwick (1980) has examined changes in the midge fauna of the Bay of Quinte, Ontario, over the past 2800 years. The Bay of Quinte is a narrow extension of Lake Ontario, which before European settlement was a center of American Indian activities because it offered a sheltered waterway leading to interior Ontario, the Georgian Bay, and the upper Great Lakes. Warwick dated a core from the area of the bay near Glenora, Ontario, using ^{210}Pb, ^{14}C, and pollen analysis. He documents changes in the bay's midge fauna caused by the slash-and-burn agriculture of the Indians and by the more intensive agriculture of the Europeans that followed. The impact of the native peoples was small compared to that of the European colonists. At the beginning of European settlement, the midge fauna became progressively more eutrophic. When large-scale deforestation began, however, oligotrophic-indicating species appeared. Warwick attributes this change to the dilution of organic nutrients in the sediment by silt rather than to declines in aquatic productivity. When sedimentation declined, the more eutrophic fauna returned. Warwick's study is particularly important because it is one of the few on temperate North America that seeks to relate changes in a lake to pre-Columbian agriculture. The Yucutan Peninsula of Central America is the only other place in the Western Hemisphere where extensive evidence of the impact of pre-Columbian agriculture on lakes has been found (Deevey et al., 1979).

In summary, analysis of midge remains shows promise, but the application of the technique is restricted by the necessarily laborious preparation techniques and by difficulties in identifying fossil remains. As the study of midge remains is developed, it will surely contribute to the accuracy of our paleoecologic reconstructions.

ticular fossils can be used to estimate past hypolimnetic oxygen levels and, therefore, to some degree a lake's past trophic status, although Hutchinson (1957) points out that hypolimnetic oxygen levels are a function of lake morphometry as well of trophic status. In general, the genus *Tanytarsus* has been taken to indicate oxygen-rich conditions and the genus *Chironomus* to indicate oxygen-poor ones. Furthermore, Hofmann (1979) in a recent review points out that all members of the genus *Tanytarsus* are not confined to the hypolimnion but also appear in the shallow water of eutrophic lakes and may be washed into deeper water. He reiterates Brundin's (1949) warning that not only *Tanytarsus* but all members of the European *Tanytarsus lugens*-dominated community, which is characteristic of oligotrophic lakes, should be used for paleoecologic reconstructions. The presence of some *Tanytarsus* remains does not necessarily imply oligotrophic conditions.

Preparation of samples for chironomid analysis is a laborious process (see Hofmann, 1979) because the sediment must be deflocculated, sieved, and scanned with a binocular microscope for head capsules. Each head capsule must be removed from the processed sample and separately mounted. Identifications are made under a

Other Remains

A number of other fossil remains found in sediment have contributed to our knowledge of Holocene environments at particular lakes.

Plant macrofossils found in lake sediments are generally derived from locally occurring marsh and aquatic taxa and may, therefore, be useful in paleolimnological reconstructions. Birks (1973) demonstrates that plant macrofossils are sensitive indicators of lake environments. Watts and Winter (1966) use plant macrofossils to reconstruct the history of water-level fluctuations at Kirchner Marsh, Minnesota. They have found seeds of plants that colonize the seasonally flooded margins of prairie lakes at core levels corresponding to the middle-postglacial warm period.

Ostracod, mollusk, and fish remains have also been used for paleolimnological reconstructions. Stark (1976) examines ostracod, mollusk, and chironomid remains from a series of cores from Elk Lake, Minnesota. She identifies cold-hardy ostracod species in the late-glacial and early-postglacial *Picea* pollen zone. Fossils from shallower sediment levels indicate declines in hypolimnetic oxygen levels and increases in lake productivity.

Casteel and others (1977) have found fish scales to be useful in reconstructing climatic changes in Clear Lake, California. At this site, they have been able to reconstruct the past growth rates of the Tule perch (*Hysterocarpus traski*). They relate growth rates of the fish to changes in temperature. From 10,000 to 9000 yr B.P., slow growth rates indicated lowered temperatures. Between 10,000 and 4000 yr B.P., temperatures increased to a peak between 4000 and 2800 yr B.P. Since that time, the declining growth rates of the fish imply declining temperatures.

Conclusions

The preceding discussion demonstrates the potential and the problems of paleolimnology in the Holocene. It is apparent that paleolimnological data cannot be used directly in paleoclimatic reconstructions. Lake organisms are more sensitive to lake environments than to the aspects of climate that humans usually consider important. Whiteside and others (1980) underscore the problem of using lake organisms as paleoclimatic indicators with their admission that most of the species they found in the lakes on the Klutlan glacier moraines in the Yukon Territory are also found in temperate-zone lakes.

What, then, is the value of paleolimnology to Quaternary studies? Insofar as climatic change is reflected in changes in lacustrine chemical and physical variables, one would expect changes in fossil assemblages. For example, a drier climate might result in declines in water levels, and the attendant evaporation might increase solute concentrations in the lake water. Such changes are profound ones for lake organisms and should result in major changes in fossil assemblages.

Other, more subtle effects may be important in determining the ecologic conditions of a particular lake. Recent geochemical investigations have emphasized the control that watershed vegetation exerts on the chemistry of watershed runoff (Likens and Bormann, 1977). One might expect that changes in watershed vegetation in response to climatic changes might cause changes in limnological conditions in a lake downstream. Whitehead and Crisman (1978) point out that controls of runoff chemistry by vegetation provide an important link between lake environments and climate (insofar as vegetation itself is controlled by climate). In no case will it be possible to reconstruct climatic change solely from paleolimnological data. Paleolimnological information does provide valuable additional information on the effects of climatic change on lakes, however. Paleolimnological techniques allow one to ask and to answer a wide variety of questions concerning lake development that may or may not be pertinent to the question of climatic change but that do enrich our understanding of the broader aspects of Holocene environmental change.

A more realistic goal of paleolimnological investigation is to determine the effects of climatic change on aquatic ecosystems. Examinations of the paleolimnological record represent one of the few methods of investigating long-term changes in populations of organisms. Given recent advances in our understanding of the population ecology of aquatic organisms, the sedimentary record can be used to test theories of population ecology against a long-term history of population change.

If paleolimnology is to reach the goal of interpreting the significance of this long-term record of population change, three areas for future research should be investigated: (1) extensive utilization of surface-sample analogues for down-core fossil assemblages, (2) increased awareness of the role diagenesis and dissolution play in shaping fossil assemblages, and (3) use of new groups of organisms as indicators of past environments.

A more extensive utilization of surface samples will help the paleolimnologist make more precise reconstructions of past environments. Multivariate statistical analysis can be employed with the surface-sample analogues and to reduce the effort involved in making comparisons between the surface and down-core samples. Ultimately, it may be possible to make fairly precise statements about past lake environments on the basis of fossil assemblages. Such statements would include information on chemical and physical properties of the lake in the past in addition to biologic factors such as intensity and type of predation in the planktonic environment.

An increased understanding of the role of diagenesis and dissolution in determining what species will appear in a given fossil assemblage may allow the use of fossil material in making precise estimates of past population numbers. Undoubtedly, there is a species-specific loss of microfossils from cores over time. Before precise paleolimnological reconstructions can be made, it will be necessary to quantify such losses. For some sediment constituents, such as pigments and diatoms, some progress has already been made on understanding the role of diagenesis in shaping what we see as fossils. For other constituents, such as sediment phosphorus and crustacean microfossils, very little is known about the effects and conditions of diagenesis.

It will always be necessary for the paleolimnologist to search for and to use new fossil remains and sediment characteristics for paleoecologic reconstructions. In recent years, it has become evident that to have a good understanding of environmental change in lakes it is necessary to use as many sediment constituents as possible. The development of new fossil indicators of past lake conditions can only increase the precision of paleolimnological reconstructions.

In summary, the paleolimnological study of North America is vigorous and developing. It is expected that paleolimnology will continue to contribute to our knowledge of the North American environment and its development over time.

References

Allott, N. (1978). Recent paleolimnology of Twin Lake near St. Paul, Minnesota, based on a transect of core. M.S. thesis, University of Minnesota, Minneapolis.

Battarbee, R. W. (1974). A new method for the estimation of absolute microfossil numbers with reference especially to diatoms. *Limnology and Oceanography* 18, 647-53.

Battarbee, R. W. (1978). Observations on the recent history of Lough Neagh and its drainage basin. *Philosophical Transactions of the Royal Society of London* 281, 303-45.

Birch, P. B., Barnes, R. S., and Spyridakis, D. E. (1980). Recent sedimentation and its relationship with primary productivity in four western Washington lakes. *Limnology and Oceanography* 25, 240-47.

Birks, H. H. (1973). Modern macrofossil assemblages from lake sediments in Minnesota. *In* "Quaternary Plant Ecology" (H. J. B. Birks and R. G. West, eds.), pp. 173-89. Blackwell, Oxford.

Birks, H. J. B. (1976). Late-Wisconsin vegetational history at Wolf Creek, central Minnesota. *Ecological Monographs* 46, 395-429.

Birks, H. J. B., and Birks, H. H. (1980). "Quaternary Palaeoecology." Edward Arnold, London.

Bortleson, G. C. (1971). "Chemical Investigation of lake sediments from Wisconsin Lakes and Their Interpretation." U.S. Environmental Protection Agency, Washington, D.C.

Bortleson, G. C., and Lee, G. F. (1974). Phosphorus, iron, and manganese distribution in sediment cores of six Wisconsin lakes. *Limnology and Oceanography* 19, 794-801.

Bradbury, J. P. (1971). Paleolimnology of Lake Texcoco, Mexico: Evidence from diatoms. *Limnology and Oceanography* 16, 180-200.

Bradbury, J. P. (1975). "Diatom Stratigraphy and Human Settlement in Minnesota." Geological Society of America Special Paper 171, pp. 1-74.

Bradbury, J. P. (1978). "A Paleolimnological Comparison of Burntside and Shagawa Lakes, Northeastern Minnesota." U.S. Environmental Protection Agency Ecological Research Series, EPA-600/3-78-004.

Bradbury, S. P., and Megard, R. O. (1972). Stratigraphic record of pollution in Shagawa Lake, northeastern Minnesota. *Bulletin of the Geological Society of America* 85, 2639-48.

Bright, R. C. (1968). "Surface Water Chemistry of Some Minnesota Lakes with Preliminary Notes on Diatoms." University of Minnesota Limnological Research Center Interim Report 3, pp. 1-58.

Brooks, J. L. (1969). Eutrophication and changes in the composition of the zooplankton. *In* "Eutrophication: Causes, Consequences, Correctives" (G. Rohlich, ed.), pp. 236-55. National Academy of Sciences, Washington, D.C.

Brooks, J. L., and Dodson, S. I. (1965). Predation, body size, and composition of the plankton. *Science* 150, 28-35.

Brown, S. R. (1968). Bacterial carotenoids from freshwater sediments. *Limnology and Oceanography* 8, 352-53.

Brugam, R. B. (1978). Human disturbance and the historical development of Linsley Pond. *Ecology* 59, 19-36.

Brugam, R. B. (1979). A re-evaluation of the Araphidineae/Centrales index as an indicator of lake trophic status. *Freshwater Biology* 9, 451-60.

Brugam, R. B. (1980). Post-glacial diatom stratigraphy of Kirchner Marsh, Minnesota. *Quaternary Research* 13, 133-46.

Brugam, R. B. (1983). The relationship between fossil diatom assemblages and limnological conditions. *Hydrobiologia* 98: 223-35.

Brugam, R. B., and Speziale, B. J. (1983). The eutrophication history of Lake Harriet, Minnesota. *Ecology* 64: 578-91.

Brundin, L. (1949). "Chironomiden und andere Bodentiere der Suedschwedischen Urgebirgseen." Reports Institute Freshwater Research Drottningholm 30.

Casteel, R. W., Adam, D. P., and Sims, J. D. (1977). Late-Pleistocene and Holocene remains of *Hystercarpus traski* (tule perch) from Clear Lake, California, and inferred Holocene temperature fluctuations. *Quaternary Research* 7, 133-43.

Crisman, T. L. (1978a). Algal remains in Minnesota lake types: A comparison of modern and late-glacial distributions. *Verhandlungen der Internationalen Vereinigung für Theoretische und Angewandte Limnologie* 20, 445-51.

Crisman, T. L. (1978b). Reconstruction of past lacustrine environments based on the remains of aquatic invertebrates. *In* "Biology and the Quaternary Environment" (D. Walker and J. C. Guppy, eds.), pp. 69-102. Australian Academy of Sciences, Canberra.

Crisman, T. L., and Whitehead, D. R. (1978). Paleolimnological studies of small New England (U.S.A) ponds. II: Cladoceran community responses to trophic oscillation. *Polish Archives of Hydrobiology* 25, 482-92.

Daley, R. J. (1973). Experimental characterization of lacustrine cholorphyll diagenesis. II: Bacterial, viral and herbivore grazing effects. *Archiv für Hydrobiologie* 72, 409-39.

Daley, R. J., and Brown, S. (1973). Experimental characterization of lacustrine chlorophyll diagenesis: I. Physiological and environmental effects. *Archiv für Hydrobiologie* 72, 277-304.

Davis, M. B. (1976). Erosion rates and land use-history in southern Michigan. *Environmental Conservation* 3, 139-48.

Deevey, E. S. (1942). Studies on Connecticut lake sediments: III: The biostratonomy of Linsley Pond. *American Journal of Science* 240, 233-64 and 313-24.

Deevey, E. S. (1969). Cladoceran populations of Rogers Lake, Connecticut, during late- and post-glacial time. *Mitteilungen der Internationalen Vereinigung für Theoretische und Angewandte Limnologie* 17, 56-63.

Deevey, E. S., Rice, D. S., Rice, P. M., Vaughn, H. H., Brenner, M., and Flannery, M. S. (1979). Mayan urbanism: Impact on a tropical karst environment. *Science* 206, 298-306.

Evans, R. D., and Rigler, F. H. (1980). Measurement of whole lake sediment accumulation and phosphorus retention using lead-210 dating. *Canadian Journal of Fisheries and Aquatic Science* 37, 817-22.

Florin, M. B., and Wright, H. E., Jr. (1969). Diatom evidence for the persistence of stagnant ice in Minnesota. *Geological Society of America Bulletin* 80, 695-704.

Foged, N. (1954). On the diatom flora of some Funen lakes. *Folia Limnologica Scandinavica* 6, 7-81.

Frey, D. G. (1959). The taxonomic and phylogenetic significance of the head pores of the Chydoridae (Cladocera). *International Revue der gesamten Hydrobiologie* 44, 27-50.

Frey, D. G. (1960). The ecological significance of cladoceran remains in lake sediments. *Ecology* 41, 684-99.

Frey, D. G. (1964). Remains of animals in Quaternary lake and bog sediments and their interpretation. *Archiv für Hydrobiologie, Beihefte, Ergebnisse der Limnologie* 2, 1-114.

Frey, D. G. (ed.) (1969). Symposium on paleolimnology. *Mitteilungen der Internationalen Vereinigung für Theoretische und Angewandte Limnologie* 17, 1-448.

Frey, D. G. (1976). Interpretation of Quaternary paleoecology from cladocera and midges. *Canadian Journal of Zoology* 54, 2208-26.

Fryer, G. (1968). Evolution and adaptive radiation in the Chydoridae (Crustacea: Cladocera): A study in comparative functional morphology and ecology. *Philosophical Transactions of the Royal Society* series B, 254, 221-385.

Galloway, J. N., and Likens, G. R. (1979). Atmospheric enhancement of metal deposition in Adirondack lake sediments. *Limnology and Oceanography* 24, 427-33.

Gorham, E., and Sanger, J. (1976). Fossil pigments as stratigraphic indicators of cultural eutrophication in Shagawa Lake, northeastern Minnesota. *Geological Society of America Bulletin* 87, 1638-42.

Goulden, C. E. (1964). The history of the cladoceran fauna of Esthwaite Water (England) and its limnological significance. *Archiv für Hydrobiologie* 60, 1-52.

Goulden, C. (1969). Temporal changes in diversity. *In* "Diversity and Stability in Ecological Systems" (G. M. Woodwell and H. H. Smith, eds.), pp. 96-102. Brookhaven Symposium in Biology 22.

Griffiths, M., Perrott, P. S., and Edmondson, W. T. (1969). Oscillaxanthin in the sediment of Lake Washington. *Limnology and Oceanography* 14, 317-26.

Harmsworth, R. V. (1968). The developmental history of Blelham Tarn (England) as shown by animal microfossils with special reference to the cladocera. *Ecological Monographs* 38, 223-41.

Harmsworth, R. V., and Whiteside, M. C. (1968). Relation of cladoceran remains in lake sediments to primary productivity of lakes. *Ecology* 49, 998-1000.

Haworth, E. Y. (1972). Diatom succession in a core from Pickerel Lake, northeastern South Dakota. *Geological Society of America Bulletin* 83, 157-72.

Haworth, E. Y. (1976). The changes in the composition of the diatom assemblages found in the surface sediments of Blelham Tarn in the English Lake District during 1973. *Annals of Botany* 40, 1195-1205.

Hofmann, W. (1971). Zur Taxonomie und Palökologie Sub-fossiler Chironomiden (Dipt.) in See Sedimenten. *Archiv für Hydrobiologie, Beihefte, Ergebnisse der Limnologie* 6, 1-50.

Hofmann, W. (1979). Chronomid Analysis. In "Paleohydrological Changes in the Temperate Zone in the Last 15,000 Years," Subproject B Lake and Mire Environments (B. B. Berglund, ed.), vol. 2, B.13, pp. 259-77. International Geological Correlation Program, Lund.

Hrbacek, J. (1969). On the possibility of estimating predation pressure and nutrition levels of populations of Daphnia from their remains in sediments. Mitteilungen der Internationalen Vereinigung für Theoretische und Angewandte Limnologie 17, 269-76.

Hutchinson, G. E. (1957). "A Treatise on Limnology." Vol. 1. John Wiley and Sons, New York.

Hutchinson, G. E. (1970). Ianula: An account of the history and development of the Lago di Monterosi, Latium, Italy. Transactions of the American Philosophical Society 60 (Part 4), 1-178.

Imbrie, J., and Kipp, N. (1971). A new micropaleontological method for quantitative paleoclimatology: Application to a late Pleistocene Caribbean core. In "The Late Cenozoic Glacial Ages" (K. K. Turekian, ed.), pp. 71-181. Yale University Press, New Haven, Conn.

Kerfoot, W. C. (1974). Net accumulation rates and the history of cladoceran communities. Ecology 55, 51-61.

Kerfoot, W. C. (1975). The divergence of adjacent populations. Ecology 56, 1298-1313.

Kerfoot, W. C. (1978). Combat between copepods and their prey: Cyclops, Epischura and Bosmina. Limnology and Oceanography 23, 1098-1103.

Kitchell, J. A., and Kitchell, J. F. (1980). Size-selective predation, light transmission and oxygen stratification: Evidence from the recent sediments of manipulated lakes. Limnology and Oceanography 25, 389-403.

Lawrenz, R. W. (1975). The developmental paleoecology of Green Lake, Antrim County, Michigan. M.S. thesis, Central Michigan University, Mount Pleasant.

Lehman, J. T. (1975). Reconstructing the rate of accumulation of lake sediment: The effect of sediment focusing. Quaternary Research 5, 541-50.

Likens, G. E., and Bormann, F. H. (1977). "Biogeochemistry of a Forested Ecosystem." Springer Verlag, Berlin.

Likens, G. E., and Davis, M. B. (1975). Post-glacial history of Mirror Lake and its watershed in New Hampshire, U.S.A.: An initial report. Verhandlungen der International Vereinigung für Theoretische und Angewandte Limnologie 19, 982-93.

Livingstone, D. A. (1957). On the sigmoid growth phase in the history of Linsley Pond. American Journal of Science 255, 304-73.

MacArthur, R. H. (1957). On the relative abundance of bird species. Proceedings of the National Academy of Science U.S.A. 43, 293-95.

MacArthur, R. H. (1960). On the relative abundance of species. American Naturalist 94, 25-36.

Mackereth, F. J. H. (1966). Some chemical observations on post-glacial lake sediments. Philosophical Transactions of the Royal Society series B, 250, 165-213.

Mason, W. T., Jr. (1973). "An Introduction to the Identification of Chironomid Larvae." U.S. Environmental Protection Agency, Cincinnati, Ohio.

Moyle, J. B. (1956). Relationships between chemistry of Minnesota surface waters and wildlife management. Journal of Wildlife Management 20, 303-20.

Mueller, H. (1970). Oekologische Veraenderungen im Otterstedter See im Laufe der Nacheiszeit. Berichte der Natur hisstorischen Gessell schaft zu Hannover 114, 33-47.

Munch, C. S. (1980). Fossil diatoms and scales of chysophyceae in the recent history of Hall Lake. Freshwater Biology 10, 61-66.

Norris, G., and McAndrews, J. H. (1970). Dinoflagellate cysts from post-glacial lake muds, Minnesota (U.S.A.) Review of Paleobotany and Palynology 10, 131-56.

Norton, S. A., Dubiel, R. F., Sasseville, D. R., and Davis, R. B. (1978). Paleolimnologic evidence for increased zinc loading in lakes of New England, U.S.A. Verhandlungen der Internationalen Vereinigung für Theoretische und Angewandte Limnologie 20, 538-45.

Nygaard, G. (1956). The ancient and recent flora of diatoms and chrysophyceae in Lake Gribso. Folia Limnologica Scandinavica 8, 32-93.

O'Brien, J. (1980). The predator-prey interaction of planktivorous fish and zooplankton. American Scientist 5, 572-81.

Parker, J. I., Conway, H. L., and Yaguchi, E. M. (1977). Dissolution of diatom frustules and recycling of amorphous silica in Lake Michigan. Journal of the Fisheries Research Board of Canada 34, 545-51.

Parker, J., and Edgington, D. (1976). Concentrations of diatom frustules in Lake Michigan cores. Limnology and Oceanography 21, 887-93.

Patrick, R. (1943). The diatoms of Linsley Pond, Connecticut. Proceedings of the Academy of Natural Sciences, Philadelphia 95, 53-110.

Sanger, J., and Crowl, G. H. (1979). Fossil pigments as a guide to the paleolimnology of Browns Lake, Ohio. Quaternary Research 11, 342-55.

Sanger, J., and Gorham, E. (1970). The diversity of pigments in lake sediments and its ecological significance. Limnology and Oceanography 15, 59-69.

Sanger, J., and Gorham, E. (1972). Stratigraphy of fossil pigments as a guide to the post-glacial history of Kirchner Marsh, Minnesota. Limnology and Oceanography 17, 840-54.

Sanger, J., and Gorham, E. (1973). A comparison of the abundance and diversity of fossil pigments in wetland peats and woodland humus layers. Ecology 54, 605-11.

Shapiro, J., Edmondson, W. T., and Allison, D. E. (1970). Changes in the chemical composition of sediments of Lake Washington (1958-1970). Limnology and Oceanography 6, 437-52.

Smol, J. (1980). Fossil synuracean (Chrysophyceae) scales in lake sediments: A new group of paleoindicators. Canadian Journal of Botany 58, 458-65.

Stahl, J. (1969). The uses of chironomids and other midges in interpreting lake histories. Mitteilungen der Internationalen Vereinigung für Theoretische und Angewandte Limnologie 17, 111-25.

Stark, D. M. (1976). Paleolimnology of Elk Lake, Itasca State Park, northwestern Minnesota. Archiv für Hydrobiologie Supplement Band 50, 208-74.

Stockner, J. G. (1972). Paleolimnology as a means of assessing eutrophication. Verhandlungen der Internationalen Vereinigung für Theoretische und Angewandte Limnologie 18, 1018-30.

Stockner, J. G., and Benson, W. W. (1967). The succession of diatom assemblages in the recent sediments of Lake Washington. Limnology and Oceanography 12, 513-32.

Stoermer, E. F. (1977). Post-Pleistocene diatoms succession in Douglas Lake, Michigan. Journal of Phycology 13, 73-80.

Synerholm, C. C. (1979). The chydorid cladocera from surface lake sediments in Minnesota and North Dakota. Archiv für Hydrobiologie 86, 137-51.

Tessenow, U. (1975). Accumulation processes in the maximum depths of lakes through post-sedimentary concentration migration. Verhandlungen der Internationalen Vereinigung für Theoretische und Angewandte Limnologie 19, 1251 62.

Tsukada, M. (1972). The history of Lake Nojiri, Japan. Transactions of the Connecticut Academy of Arts and Sciences 44, 337-65.

Vallentyne, J. R. (1960). Fossil pigments. In "Comparative Biochemistry of Photoreactive Systems" (M. B. Allen, ed.), pp. 83-105. Academic Press, New York.

Warwick, W. F. (1980). Paleolimnology of the Bay of Quinte, Lake Ontario: 2800 years of cultural influence. Canadian Bulletin of Fisheries and Aquatic Science 206, 1-117.

Watts, W. A., and Winter, T. (1966). Plant macrofossils from Kirchner Marsh, Minnesota: A paleoecological study. Geological Society of America Bulletin 77, 1339-60.

Wetzel, R. G. (1970). Recent and post-glacial production rates of a marl lake. Limnology and Oceanography 15, 49-503.

Whitehead, D. R., and Crisman, T. L. (1978). Paleolimnological studies of small New England (U.S.A.) ponds: I. Late-glacial and post-glacial trophic oscillations. Polish Archives of Hydrobiology 25, 471-81.

Whitehead, D. R., Rochester, H. R., Jr., Rissing, S. W., Douglass, C. B., and Sheehan, M. C. (1973). Late glacial and post-glacial productivity changes in a New England pond. Science 181, 744-46.

Whiteside, M. (1970). Danish chydorid cladocera, modern ecology and core studies. Ecological Monographs 40, 79-115.

Whiteside, M. C., Bradbury, J. P., and Tarapchak, S. J. (1980). Limnology of the Klutlan Moraines, Yukon Territory, Canada. Quaternary Research 14, 130-49.

CHAPTER 14

Tree-Ring Studies of Holocene Environments

Linda B. Brubaker and Edward R. Cook

Introduction

Concern over food, water, and energy resources has prompted a growing interest in recent climatic variations. Information on conditions during preceding centuries is essential for anticipating future conditions because the climate of the 20th century is thought to be somewhat anomalous (Bryson and Hare, 1974; Lamb, 1972, 1974). In North America, variations in the annual growth rings of trees constitute the most important source of information about climatic change during recent centuries (Fritts, 1976). Tree rings provide continuous and precisely dated records of annual and even seasonal climatic conditions. Because trees grow in a variety of climatic environments across North America, spatial arrays of growth records reveal large-scale patterns of diverse climatic variables. The temporal and spatial resolution of the climatic data derived from tree rings is, therefore, suitable for comparison with synoptic meteorologic data collected with modern instruments.

Emphasizing the analysis of environmental changes during the past three to four centuries, this chapter describes the techniques and results of tree-ring research in the United States. Conditions during this period can be studied in detail because old trees are relatively common in North America. Long tree-ring records can be obtained from forests throughout the western United States, since large portions of this region have never been cleared. Although land clearance has been extensive in the eastern United States, trees 300 to 400 years old still remain in widely scattered stands and provide a basis for reconstructing past conditions in this region (Cook, 1982).

History

Dendrochronology became an established scientific field through the work of Andrew Ellicot Douglass (1867-1959), an astronomer who was interested in the influence of sunspot activity on the Earth's climate (Robinson, 1976). His career started in 1904 at a time when climatic records were too short to be compared with records of sunspot activity. He noted that ring-width patterns varied synchronously among trees located tens of kilometers apart in Arizona and reasoned that only climate could influence tree growth over such long distances. In 1914, he was able to demonstrate a correlation between winter precipitation and the ring widths of trees in Arizona (Douglass, 1914).

During his long and active career, Douglass developed many important research techniques that are still used today. Perhaps his most important contribution was the dating of wood specimens from prehistoric Indian dwellings in the southwestern United States (Douglass, 1921). By overlapping the ring-width patterns of archaeological specimens with those of living trees, he established a regional dated sequence of ring widths extending back to A.D. 700. This chronology has since been used to date ruins throughout the Southwest.

The work of researchers following Douglass has demonstrated that tree rings can be used to study a wide range of late-Holocene events. For example, glacial advances and retreats, landslides, snow avalanches, fires, floods, earthquakes, and lake-level fluctuations have been investigated through the study of tree-rings. Since 1960, emphasis has been placed on developing objective numerical techniques for reconstructing past environmental variables from ring-width data. Very recently, investigators have examined the isotopic ratios of carbon ($^{13}C/^{12}C$), oxygen ($^{18}O/^{16}O$), and hydrogen ($^{2}H/^{1}H$) and the densitometric properties of annual rings. In some cases, these techniques may be capable of extracting environmental information that is not revealed by traditional ring-width data.

Methods

GENERAL PROCEDURE FOR ENVIRONMENTAL RECONSTRUCTIONS

The reconstruction of environmental records from tree-ring widths is carried out in several stages (Fritts, 1976). First is the careful selection of sites where ring widths are strongly correlated with measured environmental variables. Many trees must be sampled in order to identify ring-width patterns that are common to the site and free from variations caused by the unique growth patterns of individual trees. The rings of all samples must be accurately dated, because interpretations of ring sequences rely on year-by-year comparisons between ring widths and environmental data. The ring-width measurements are standardized in order to remove the effects of tree size, age, and competitive interactions on ring-width patterns. The standardized values, called ring-width indices, are compared with environmental data by regression techniques in order to obtain equations that express environmental variables in terms of tree-ring variables (Figure 14-1). Reconstructions of past conditions are derived by entering tree-ring data from early time periods into these equations, and the validity of these reconstructions is evaluated by comparing the estimated values with environmental data withheld from the original calibration.

Although these steps are a part of every quantitative reconstruction of past environments, the exact research procedures depend on the tree species, geographic region, environmental variables, and experimental design of the study. The following sections describe in more detail the techniques and objectives of each research step, with an emphasis on the study of climatic variables.

SITE SELECTION

The criteria for site selection depend on the particular problem being investigated (Fritts, 1976; LaMarche, 1982). For example, when a reconstruction of precipitation is sought, sites where soil moisture is limited should be selected. Sites in semiarid regions (Fritts et al., 1965) or well-drained sites in nonarid regions (Cook and Jacoby, 1977) are frequently sampled in such studies. In a similar fashion, when a reconstruction of summer temperature is made, sites are selected in areas where growing-season temperatures are frequently limiting. Altitudinal treeline sites (LaMarche, 1974) and high-latitudinal treeline sites (Cropper, 1981; Garfinkel and Brubaker, 1980; Jacoby, 1982; Jacoby and Cook, 1982) are appropriate for this purpose, but arid sites where high temperatures lead to low growth can also be used (Fritts et al., 1979). For the reconstruction of stream flow, tree-ring samples should come from dry sites within important runoff-producing regions of the watershed (Stockton, 1971, 1975; Stockton and Boggess, 1979). However, even when strict site-selection criteria are followed, tree-ring chronologies rarely contain information for only one factor and one season. Growth-ring sequences generally form complex records of different climatic factors for different seasons.

Early workers collected samples primarily from climatically stressed stands where tree densities were low (Stokes and Smiley, 1968). However, recent studies show that the ring-width sequences of trees growing on more favorable sites also contain significant

climatic information. In particular, closed forests of the northwestern (Brubaker, 1980) and eastern (Cleaveland, 1975; Conkey, 1979; Cook and Jacoby, 1977, 1979; Duvick and Blasing, 1981) United States have proven suitable for dendroclimatic research.

CROSS-DATING

Although trees in temperate and subarctic latitudes typically produce one growth layer per year, simple ring counts frequently lead to the incorrect dating of rings (Fritts, 1976; Stokes and Smiley, 1968). Dating errors can result from locally absent rings (growth rings that are discontinuous or patchy around the stem of a tree) and from intraannual latewood bands (layers of thick-walled cells witin earlywood portions of rings) (Stokes and Smiley, 1968). Locally absent and intraannual rings can usually be identified by cross-dating, the process of matching ring-width patterns of wood specimens (Fritts, 1976; Robinson, 1976; Stokes and Smiley, 1968). This technique is based on the principle that climatic variations produce synchronous (i.e., matching) ring-width patterns in all trees on a site. If the ring patterns of two specimens are offset by one year, it can be assumed that one of them contains a locally absent ring or an intraannual latewood band. It is almost always possible to discriminate between these sources of error by examining the cellular characteristics of rings at the point of disagreement (Stokes and Smiley, 1968). Only accurately dated specimens are included in the final tree-ring chronology.

STANDARDIZATION

Trees growing in upper and lower treeline environments are generally unaffected by competition because of the wide spacing between neighboring trees. The ring widths of trees in such areas generally decline exponentially with increasing age until a relatively constant average width is reached. This growth trend must be removed from ring-width measurements before growth data from different-aged trees can be combined and compared to records of climate (Fritts, 1976; Graybill, 1982). Simple linear regression and negative exponential curves (Fritts et al., 1969; Graybill, 1982) are commonly used to eliminate the growth trends of such open-grown trees.

In the closed-canopy forests of the eastern and northwestern United States, competition among trees has an increasingly important influence on ring-width variations (Cook, 1982; Fritts, 1976). Changes in the competitive status of a tree cause variable growth trends that must be removed by flexible curves such as cubic

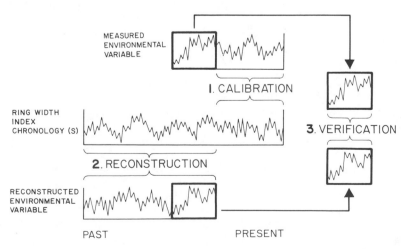

MEASURED
ENVIRONMENTAL
VARIABLE

I. CALIBRATION

RING WIDTH
INDEX
CHRONOLOGY (S)

3. VERIFICATION

2. RECONSTRUCTION

RECONSTRUCTED
ENVIRONMENTAL
VARIABLE

PAST   PRESENT

Figure 14-1. Steps in estimating past environmental variables from tree-ring measurements. The instrumental observations of modern environmental variables are used in three steps: (1) calibration, to establish the equations for estimating past variables; (2) reconstruction to estimate past variables from ring measurements for early periods; and (3) verification, to check the accuracy of those estimates.

smoothing splines (Cook and Peters, 1981) and orthogonal polynomials (Fritts, 1976). Unfortunately, these curves eliminate growth variations caused by long-term climatic variations as well as by competition among trees.

The values of the curve fitted to the ring-width measurements are considered to be the expected annual growth in a constant climatic environment. Ring-width indices are computed as ratios of actual growth to expected growth. The resulting index series has a mean of 1.0 and a variance more or less independent of the mean. A site chronology suitable for climatic analysis is produced by averaging the indices for all samples according to year (Fritts, 1976).

CALIBRATION

Dendroclimatic analyses of the site chronologies typically include three steps: 1) calibration, the identification of the climatic signal in site chronologies; 2) reconstruction, the estimation of past climatic variables from ring-width index series; and 3) verification, the comparison of the reconstructions with independent climatic data (Figure 14-1). The statistical techniques used in these steps are described by Fritts (1976).

As a first step in the calibration process, the climatic signal in tree-ring chronologies is identified by response-function analysis (Fritts, 1976; Fritts et al., 1971). In this procedure, ring-width indices are regressed on temperature and precipitation data from a nearby climatic station. The climatic data consist of monthly total precipitation and average temperature for a period starting at the beginning of the previous growth year and terminating at the end of the current growth season. Conditions before the growth season are included because they can influence the plant processes that affect the amount of wood formed when the cambium is active. Principal components of the climatic data are used as the predictor variables because they represent a concentration of the temperature and precipitation information into a smaller set of variables and satisfy the assumption of linear independence for the regression analysis. The regression coefficients of the principal components are transformed back to numerical weights corresponding to the monthly temperature and precipitation variables (Fritts, 1976). These weights (the response function) reflect the contribution of variations in each month's temperature and precipitation to variations in ring-width indices. Although response functions cannot be used directly to reconstruct climate, they can serve as a basis for making simple deductions about past climatic conditions from tree-ring chronologies.

A response function for a Douglas-fir (*Pseudotsuga menziesii*) chronology in the Olympic Mountains of Washington State is illustrated in Figure 14-2 (Brubaker, 1980). A positive value on such plots indicates a direct relationship (positive correlation) between growth and a given climatic variable, and a negative value indicates an inverse relationship (negative correlation). Response functions provide insight into the biologic processes limited by climate, and as such they are useful tools for studying the ecologic tolerances of tree species. By describing the nature of growth responses to climate, they also suggest which climatic variables can be quantitatively reconstructed from site chronologies.

A second calibration procedure, also based on regression techniques, derives equations for reconstructing climatic variables (Fritts, 1976; Lofgren and Hunt, 1982). These regression equations, called transfer functions, provide a set of weights that can be applied to ring-width data to estimate past values of climatic variables. Transfer functions thus "transfer" tree-ring data into estimates of climatic varia-

tions for time period in which climatic data are not directly available. The numerous transfer-function methods used to estimate paleoenvironmental data are all variants of a basic multiple linear regression model (Sachs et al., 1977). The simplest transfer functions are equations in which one environmental variable is estimated by one tree-ring variable, scaled by an appropriate constant. Duvick and Blasing (1981) use this approach to estimate annual precipitation in Iowa from an average of three oak chronologies.

Most investigators, however, have applied multiple regression equations (Cleaveland, 1975; Cook and Jacoby, 1977, 1979; Fritts, 1976; Garfinkel and Brubaker, 1980; Meko et al., 1980; Stockton, 1975), which predict one environmental variable from two or more tree-ring variables, usually growth records from two or more sites. These equations often include lagged tree-ring indices. Lagged growth data improve the ability of transfer functions to predict climate because trees act as integrators of environmental information. The widths of a ring can reflect the conditions of several seasons prior to the period of ring formation, since climatic variations are linked to growth during subsequent periods by many processes (e.g., food storage and bud formation) (Fritts, 1976, 1982). The most successful calibration equations, therefore, usually include growth-ring variables for a number of years surrounding the period of estimated climate. These studies typically use principal components of the tree-ring data as the predictor variables in the transfer function.

Transfer functions based on canonical regression techniques predict several climatic variables from several tree-ring variables. Blasing (1978) describes the statistical procedures used in this technique, and Glahn (1968) explains its mathematical basis. These techniques have been successfully used for reconstructing spatial anomalies of atmospheric pressure, precipitation, and temperature (Blasing and Fritts, 1975; Conkey, 1979; Fritts et al., 1979); hydrologic records (Stockton and Fritts, 1973); and droughts (Meko

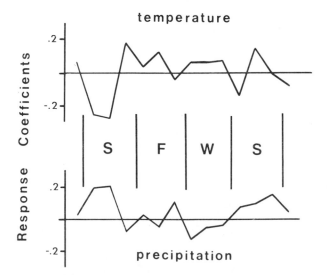

Figure 14-2. Response function for Douglas-fir from the Olympic Mountains, Washington. This response function includes 14 coefficients for average monthly temperature and 14 coefficients for total monthly precipitation from June of the year prior to the season of growth through the July concurrent with the growth season. A positive value indicates a positive correlation between growth and the climatic variable; a negative value indicates a negative correlation. S, F, W, S = summer, fall, winter, spring. (Redrawn from Brubaker, 1980.)

et al., 1980; Mitchell et al., 1979; Stockton and Meko, 1975) in several parts of North America.

RECONSTRUCTION AND VERIFICATION

Past conditions are estimated (reconstructed) by entering the ring-width data from early time periods into the transfer functions (Fritts, 1976; Lofgren and Hunt, 1982). Successful transfer functions typically reconstruct seasonal rather than annual climatic variables; this reflects the tendency of trees to respond differently to the same variable in different seasons (Fritts, 1982a). For example, in western Washington, above-average summer precipitation is favorable to growth, but above-average winter precipitation is unfavorable (Figure 14-2) (Brubaker, 1980). Because of such differences in seasonal responses, ring widths are often poorly correlated with climatic conditions averaged across several seasons. Seasonal variables are generally reconstructed for individual climatic stations. These values are averaged in different ways to display annual conditions, long-term trends, or conditions over broad areas (Fritts, 1982b; Fritts et al., 1979).

Finally, climatic reconstructions must be carefully verified by comparing reconstructed values with independent information about past conditions, such as written descriptions of weather and reports of crop failures or other events linked to climate. The most rigorous verification approach involves statistical comparisons between reconstructed values and meteorologic data withheld from the calibration procedure. Five statistical tests are commonly used to evaluate the similarity of the reconstructed and actual values: (1) correlation, (2) sign test, (3) reduction of error, (4) cross-product means, and (5) chi-square (Fritts, 1976; Gordon, 1982). When these tests indicate that the reconstructed and actual values are statistically the same, it can be concluded that the reconstructions provide accurate information about past conditions.

Results

DENDROCLIMATOLOGY

Since 1960, tree-ring studies in the United States have concentrated primarily on the interpretation of past climatic variability (Fritts, 1976). This section discusses the results of this research according to geographic regions—the western United States, the eastern United States, and Alaska—as defined in Figure 14-3. Dendroclimatic research has been most extensive in the western United States, where old trees and climatically stressed environments are relatively common (Brubaker, 1982). Open-grown trees from semiarid portions of this region have proven to be especially useful for climatic reconstructions (Fritts, 1976). Recent studies have shown, however, that trees in the closed forests of the eastern United States (Cook, 1982) and the boreal forests of Alaska (Cropper, 1981; Garfinkel and Brubaker, 1980; Jacoby, 1982) are also suitable for describing past climates. The climatic control over growth appears to be more subtle and complex in these mesic forests than in the forests of semiarid regions. Dendroclimatic research in mesic forest zones was not possible before the development of multivariate techniques for describing complex climate-growth relationships (Cook, 1982; Fritts, 1982a). In the past, research in the eastern United States has also been limited by the belief that few old-growth stands remain to provide adequate data for climatic reconstructions. Recent field surveys, however, have found a surprising number of old stands that have escaped land clearance (Cook, 1982). Although old trees are very common in Alaska, research in this region has been constrained primarily by the logistical difficulty of conducting fieldwork and by the lack of adequate climatic data for calibrating tree-ring sequences (Jacoby, 1982).

Western United States

Tree-ring chronologies from the western United States contain information about a wide range of climatic variables. This diversity of climatic information reflects the diversity of growth-limiting environments within the region, because trees experiencing different climatic limitations provide information about different aspects of past climate (Fritts, 1976, 1982b; LaMarche, 1982).

The most important accomplishments of dendroclimatic research in this region are the widespread investigation of climate-growth relationships by response-function techniques and the interpretation of past climatic variations at different spatial and temporal scales. Very long, qualitative climatic records have been inferred for southwestern parts of the region from the bristlecone pine chronologies at high-elevation sites (LaMarche, 1974; LaMarche and Stockton, 1974). These chronologies are a unique source of climatic information, for no other paleoenvironmental data, with the exception of annually laminated lake sediments, exhibit annual dating over such long periods. Quantitative climatic reconstructions that cover large portions of North America, the North Pacific, and even eastern Asia have been derived for shorter time periods from a large network of tree-ring sites in the western United States (Blasing, 1975; Blasing and Fritts, 1976; Fritts et. al., 1979). These reconstructions are especially valuable for evaluating models of climatic variation and for anticipating future conditions (LaMarche, 1978).

Tree-Ring Collections

Dendrochronologists have extensively sampled forests in the western United States since the mid-20th century (Drew, 1974, 1975, 1976; Stokes et al., 1973). Despite this effort, most of the existing samples come from low-elevation species in semiarid regions (Brubaker, 1982). This sampling bias reflects the initial emphasis of dendroclimatic research on drought-stressed trees. Recent collections (Brubaker, 1980) have extended the existing tree-ring data base into

Figure 14-3. Map showing the major geographic regions of the United States.

the mesic forest zones of the Pacific Northwest (Washington, Oregon, and Idaho).

Response Functions

Two major studies describe response functions from sites in the western United States. Fritts (1974) summarizes the response functions for 10 coniferous species at 127 semiarid sites in the region. Although most of these response functions are from low-elevation sites, several come from upper treeline areas. Overall, this study indicates that tree growth in semiarid regions is more strongly correlated with precipitation than with temperature variations.

At low-elevation and southern sites, growth tends to be positively correlated with precipitation and negatively correlated with temperature during all seasons. These results are consistent with the findings of studies on the physiology of trees from similar sites. Net photosynthesis at semiarid sites in northern Arizona (Fritts, 1976) and California (Rutter, 1977), for example, is severely reduced whenever soil moisture is low. Both experimental and response-function studies from such sites suggest, therefore, that droughts during any season can reduce radial growth by decreasing net photosynthesis.

Response functions from high elevations, high latitudes, and north-facing slopes indicate that high winter precipitation and low summer temperatures reduce radial growth. These results agree with the traditional view (Daubenmire, 1954; Hare, 1950; Hopkins, 1959; Koppen, 1936; Pearson, 1931) that the temperature and length of the growing season limit tree growth at altitudinal and latitudinal treeline sites. Low summer temperatures can reduce growth, as in the case of bristlecone pine (Fritts, 1969; LaMarche and Stockton, 1974), by directly reducing photosynthetic rates. High winter precipitation can reduce growth when snow cover lingers into summer and lowers the soil temperatures and decreases the ability of roots to take up water and nutrients (Kaufmann, 1977; Running and Reid, 1980).

Brubaker (1980) analyzes 38 response functions from mesic and semiarid sites in Washington, Oregon, and Idaho. This study suggests that late-spring/summer droughts constitute the most important factor limiting tree growth in the region. Spring/summer droughts apparently cause positive growth correlations among sites throughout the area. Winter precipitation has the second-greatest influence on radial growth, but its effect differs on opposite sides of the Cascade Mountain crest. West of the crest, winter precipitation restricts growth, probably through the direct effect of heavy cloud cover on winter photosynthesis and through the delayed effect of heavy snowpack on root growth and absorption during the summer. East of the Cascades, where the climate is much drier, high winter precipitation favors tree growth. Because the eastern areas are generally too cold for photosynthesis during the winter, winter precipitation may favor growth by recharging soil water levels, thereby allowing higher rates of photosynthesis during the spring and summer.

Although some species, for example, bristlecone pine (Fritts, 1974) and ponderosa pine (*Pinus ponderosa*) (Brubaker; 1980), show unique growth responses to climate, most growth responses do not appear to be species specific. Site factors such as slope, latitude, and elevation seem to be more important in determining the response of a given species to macroclimatic conditions.

Long Tree-Ring Records

Chronologies spanning several thousand years have been developed for bristlecone pine in the southwestern United States (Ferguson, 1968, 1970; LaMarche, 1974; LaMarche and Harlan, 1973; LaMarche and Stockton, 1974). Typically, bristlecone pine occupies dry, high-elevation sites, where climate severely limits growth (LaMarche, 1969). Since the oldest trees are generally found in the harshest environmental conditions, living trees can give very long and sensitive records of past climate. Such records can be extended even farther back in time because the wood of dead trees decays slowly and can be cross-dated with that of living trees. Ferguson has developed a ring-width chronology spanning more than 8200 years in this way (Ferguson, 1980). Wood pieces dated by this chronology have been analyzed in order to describe systematic errors in radiocarbon age determinations due to variations in atmospheric concentrations of ^{14}C (Grootes, 1983).

LaMarche and others (LaMarche and Mooney, 1967, 1972; LaMarche and Stockton, 1974) show that the ring patterns of these trees can be cross-dated among stands separated by more than 1000 km (LaMarche and Stockton, 1974). This suggests that the growth variations of bristlecone pine contain information about regional climatic variations. The climatic interpretation of these records is based on response-function analyses and physiologic studies. Response functions from several different upper treeline sites indicate that radial growth is favored by above-average summer temperatures and precipitation. Field experiments with bristlecone pine in the White Mountains of California (Fritts, 1969) indicate that warm growing season temperatures increase rates of photosynthesis and leaf development at the upper treeline. Precipitation, though less critical than temperature, favors growth by improving the availability of water at the upper treeline.

The paleoclimatic interpretation of bristlecone-pine ring patterns has centered on sites in the White Mountains. LaMarche (1974) establishes a 5405-year-long chronology from the upper treeline and concludes from physiologic and response-function studies that this chronology is a record of warm-season temperature. Interpreted in terms of temperature, this chronology indicates that relatively warm summers prevailed in the southwestern United States during the periods 5500 to 3300 yr B.P., 2000 to 1950 yr B.P., 800 to 600 yr B.P., and 100 yr B.P. to the present. Cool conditions were common during the periods 3300 to 2000 yr B.P., 1700 to 800 yr B.P., and 600 to 100 yr B.P.

On the basis of positions and ages of living and dead trees, LaMarche (1973) also describes past changes in the elevation of the upper treeline in the same area. The treeline was 150 km higher than it is at present 5500 to 3500 yr B.P., but it decreased in elevation 3500 to 2500 yr B.P. and again 900 to 500 yr B.P. Since warm-season temperatures presumably limit the position of the upper treeline in this area, these elevation changes corroborate the temperature interpretation of ring-width variations.

In the same study, LaMarche also examines the growth records of bristlecone pine at lower forest-border sites. Climatic limitations on the growth of these trees differ somewhat from those at high elevations. Drought stress is more severe at the lower treeline because annual precipitation and soil moisture levels are very low. Summer precipitation, as a result, favors growth more strongly at the lower treeline than at the upper treeline, and cool summers enhance rather than decrease growth rates. These differences in growth responses to climate make it possible to use the ring-width records from both environments to distinguish past temperature variations from past precipitation variations. For example, if growth at both treelines was above average, temperature and precipitation must have been above average. But, if growth was above average at the upper treeline and below average at the lower treeline, then temperatures must have been warm and precipitation low. Figure 14-4 shows the temperature/precipitation interpretation of growth records from bristlecone pine at upper and lower treeline sites for the period A.D. 800 to 1959.

Even though these paleoclimatic records are based on trees from a small part of the southwestern United States, they provide information for inferring large-scale circulation changes. LaMarche (1974) has been able to identify modern regional circulation features that correspond to inferred past temperature and precipitation conditions. Because these regional circulation features are a part of larger, global circulation patterns, it is possible to identify changes in large-scale atmospheric circulation that might have caused the inferred climatic variations in the White Mountains.

Synoptic Climatic Reconstruction

An extensive tree-ring data set from the western United States has been used to reconstruct spatial patterns of precipitation, temperature, and pressure anomalies for the past few centuries across much of North America (Blasing, 1975; Blasing and Fritts, 1975; Fritts et al., 1979). When plotted on maps, these climatic reconstructions resemble synoptic-scale weather maps. As such, they can be used to compare past circulation features (e.g., the location and strength of the Aleutian Low) with those occurring today.

The synoptic reconstructions produced so far (Blasing, 1975; Blasing and Fritts, 1975; Fritts et al., 1979) have been based on 65 high-quality chronologies from semiarid sites. These chronologies were originally selected by Fritts and Shatz (1975) as being well suited for reconstructing past climatic variables. The techniques of calibration, reconstruction, and verification were developed as part of this research, and many analyses are still in progress. In the most recent study, Fritts and others (1979) calibrate and verify the reconstructions for these chronologies against (1) seasonal temperature records from 77 weather stations in the western and eastern United States and southwestern Canada, (2) seasonal precipitation records over the same area, and (3) seasonal sea-level pressure at 96 points in the north Pacific sector from longitude 100°E to latitude 80°N and longitude

20°E to latitude 70°N. The seasons are defined as winter (December-February), spring (March-June), summer (July-August), and autumn (September-November). The chronologies are calibrated by canonical techniques against meteorologic data for 1901 to 1963. The reconstructed variables are analyzed directly or averaged in various ways (e.g., decade means, regional means) in order to simplify the results for climatic interpretation.

This study and others (Cropper, 1981; Duvick and Blasing, 1981; Garfinkel and Brubaker, 1980; Kutzbach and Guetter, 1980) show that accurate climatic reconstructions can be derived for an area that is much larger than that of the tree-ring network. Large-scale reconstructions are possible because the local conditions to which a tree responds are part of general circulation patterns that influence conditions over broad regions. This predictability of climate permits the reconstruction of climatic variables beyond the immediate range of the tree-ring sites. The accuracy of climatic reconstructions generally decreases, however, with increasing distance from the tree-ring sites (Kutzbach and Guetter, 1980).

The conditions reconstructed by Fritts and others (1979) demonstrate that climatic variations generally do not extend uniformly in sign or magnitude across the United States. Variation in precipitation and temperature are usually correlated. Low precipitation typically accompanies warm temperatures and high precipitation accompanies cool temperatures, but warm-wet and cool-dry anomalies have occurred on numerous occasions. The following paragraphs summarize the major patterns of reconstructed precipitation and temperature during the period A.D. 1600 to 1960 (H. C. Fritts, personal communication, 1982).

Contrasting temperature anomalies existed in the eastern and western United States during most of the 17th century. The West was warmer than the eastern areas except during the period 1610 to 1630. The first half of the 17th century was wetter than the 20th century in the eastern United States and drier than the 20th century only in the extreme Southwest. The second half of the 17th century was dry in much larger portions of the United States.

Temperatures during the 18th century were generally above the 20th-century mean in the western United States but below this mean in the eastern United States. Precipitation was often below 20th-

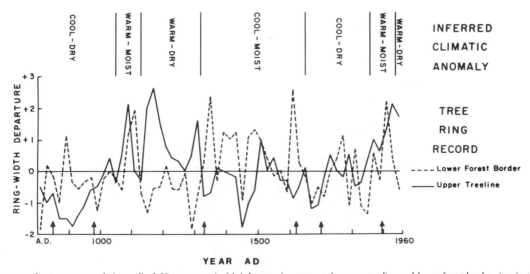

Figure 14-4. Departures from mean growth (normalized 20-year means) of bristlecone pine trees at the upper treeline and lower forest-border sites in the White Mountains, California, and inferred climatic anomalies. Arrows show dates of glacial moraines in nearby mountains. (Redrawn from LaMarche, 1974.)

century values for large areas, but the variations did not show well-defined temporal or spatial patterns.

Temperatures in most regions of the United States were above the 20th-century average during the periods 1800 to 1820 and 1860 to 1870 and below this average during the period 1840 to 1850. During the rest of the 19th century, colder-than-average temperatures prevailed in the eastern and central states and warmer-than-average temperatures prevailed in the western and southwestern states. With the exception of the period between 1830 and 1840, precipitation tended to be below the 20th-century average in the southwestern and south-central states. Between 1850 and 1870, the entire United States was drier than the 20th-century average.

During the 20th century, temperatures throughout the United States were warm during the periods 1920 to 1940 and 1950 to 1960. Temperature variations for other periods were not consistent across the country, however. Precipitation, mirroring temperature variations, was below average in most states between 1920 and 1940 and between 1950 and 1960. The country generally experienced above-average precipitation during the remaining periods, especially in the Southwest.

Meteorologic data indicate that average annual temperatures in many parts of the Northern Hemisphere increased between the 19th and 20th centuries (Brinkman, 1976; Mitchell, 1961). Bryson and Hare (1974) suggest that the 20th century may be the warmest century in the past 1000 years. Fritts and others (1979; H. C. Fritts, personal communication, 1982) compare 20th-century conditions with those of earlier centuries by subtracting 1901-to-1960 mean annual values from the 1602-to-1900 values and plotting the differences (Figure 14-5). These maps indicate that, if a hemisphere-wide warming did occur in the 20th century, there were large-scale variations from this pattern over North America. Average 1601-to-1900 temperatures were reconstructed as cooler than 20th-century temperatures everywhere except over the Rocky Mountains to the Pacific Coast. This area, particularly the Great Basin, was markedly warmer during past centuries. The reconstructed precipitation values indicate that the southwestern and south-central states were drier but that the Pacific Northwest and the eastern United States were wetter before 1900. These dendroclimatic reconstructions, therefore, suggest that hemisphere-wide average climatic shifts are not manifest everywhere. The western United States, for example, appears to have been out of phase with the average temperature patterns for the Northern Hemisphere during the last several centuries.

Eastern United States

Several severe droughts (1964-1966, 1980) and cold winters (1974-1976) during the past two decades have increased the demand for information about natural climatic variations in the eastern United States. Recent dendroclimatic studies from mesic sites in this region show that reconstructions based on eastern chronologies should increase the accuracy and spatial resolution of the broad-scale reconstructions derived from the western tree-ring network (Cook, 1982).

Tree-Ring Collections

Numerous tree-ring collections have been made in the eastern United States, but the spatial distribution of sites is extremely patchy and there are large unsampled areas in the Great Lakes region and in states bordering the Gulf of Mexico (Cook, 1982). Furthermore, many of the existing chronologies are based on the beams of historical structures or the collections of very early workers. These must be updated with new samples from living trees before they can be used in climatic reconstructions. Because of such problems, current research in the eastern United States emphasizes the collection of new sites to establish a suitable network for synopotic reconstructions.

Dendroclimatic Research

Several pioneering studies (Diller, 1935; Hawley, 1941; Lutz, 1944; Lyon, 1936; Schulman, 1942) firmly showed that the ring patterns of eastern tree species could be cross-dated. Lyon (1936) also demonstrated the feasibility of inferring past climatic events from tree-ring sequences of eastern hemlock (*Tsuga canadensis*). However, except for work by Fritts (1958, 1959, 1960, 1962), Phipps (1961, 1967), and Estes (1969), dendroclimatic research experienced little progress in the 1950s and 1960s.

The first quantitative reconstruction of climate from eastern ring sequences has been accomplished by Cleaveland (1975). Using tree-ring data for shortleaf pine (*Pinus echinata*) at a single site on the

RECONSTRUCTIONS
1602-1900
COMPARED TO
INSTRUMENTED RECORD
1901-1970

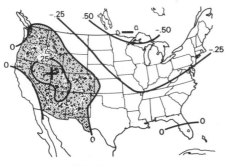

MEAN TEMPERATURE

TOTAL PRECIPITATION

Figure 14-5. Mean annual temperature and precipitation for 1602 to 1900 plotted as departures from a percentage of the 1901 to 1970 average values. Warm, dry anomalies are shaded. (Based on H. C. Fritts, personal communication, 1982.)

South Carolina piedmont, he estimates records of total June-to-August precipitation. These records verify well against independent data. His best reconstruction model, in terms of verification statistics, includes lagged tree-ring variables. Summer precipitation is estimated back to 1683. However, because of its limited sample size for early periods, this reconstruction is most reliable for times since 1831. Overall, the results of this study indicate that summer droughts have been relatively rare in South Carolina; only the summers of 1869, 1974, and 1925 stand out as unusually dry. The period between 1831 and 1868 is the longest interval without significantly dry years.

Duvick and Blasing (1981) have derived the first climatic reconstruction from ring patterns of a deciduous tree species (Figure 14-6). They estimate a 300-year record of annual precipitation in Iowa from a white oak (*Quercus alba*) chronology that is formed by averaging three separate site chronologies. The transfer function for this reconstruction is unusually simple in that it estimates annual, as opposed to seasonal, conditions from a single, unlagged tree-ring variable. The simplicity and annual resolution of the transfer function is consistent with the form of the response functions from these sites, which indicate that radial growth is positively correlated with precipitation during all months. Uniform growth responses to a single climatic variable should result in strong correlations of growth with annual values of that variable. The reconstructed precipitation record in this study indicates that numerous droughts have occurred in past centuries. Of the five severe droughts identified by Duvick and Blasing, three were even more severe than the one that caused widespread crop failures in the central United States between 1931 and 1940. This finding indicates that long-term agricultural planning in this region should include the possibility of serious droughts in the future.

Cook and Jacoby (1977, 1979) reconstructed the July Palmer Drought Severity Index (PDSI) (Palmer, 1965) in the Hudson Valley (New York State) from the chronologies of four coniferous and two oak species in that region (Figure 14-7). The transfer function for this reconstruction is based on a regression analysis using the principal components of lagged tree-ring variables. The PDSI reconstruction reveals that severe droughts have been rare but that moderate droughts have been relatively common in this area since 1694. This finding has implications for water-resource planning in this area because even moderate droughts can adversely affect the major metropolitan areas supplied by the Hudson River. Cook and Jacoby (1979) also examine oscillatory tendencies within the reconstructed PDSI values by using spectral analysis. They identify two statistically significantly quasi-periodic components with periods of 11.4 and 26 years. Although these periodicities are quite close to those associated

with the sunspot cycle and the moon's effect on zonal westerly air flow, no causal link between them is known.

Conkey (1979) examines the suitability of canonical regression techniques in reconstructions of spatial patterns of climate from seven chronologies in New York and New England. On the basis of the form of the response functions for these chronologies, she selected winter (November-March) temperature and spring (April-June) precipitation as climatic variables for reconstruction. The reconstructed winter temperatures for the period 1710 to 1750 and the middle 1800s were generally cooler than the 20th-century mean, whereas the temperatures for 1750 to 1790 were warmer. The 20th-century warming reconstructed by H. C. Fritts (personal communication, 1982) for the eastern United States is not strongly evident in Conkey's reconstruction, perhaps because the tree-ring widths in her study are standardized by polynomial functions, which can eliminate some of the long-term climatic signal. The spring precipitation reconstruction indicates dry intervals for 1715 to 1725, 1736 to 1745, and 1810 to 1822 and wet periods during the decades of 1730, 1750, 1780, and 1880. These wet and dry intervals agree in most cases with the reconstructed July PDSI series of Cook and Jacoby (1979). The reconstructions of these two studies are not entirely independent, however, for four of the chronologies in the tree-ring data sets are the same. Nevertheless, considering the differences in the transfer functions and tree-ring series in these studies, the agreement of their reconstructions encourages future research in this region.

Alaska

Since climatic fluctuations at high latitudes tend to be more pronounced than those at temperate latitudes (Kellogg, 1975), dendroclimatic studies from arctic and subarctic latitudes are especially important for understanding mechanisms of global climatic change (Cropper, 1981). Despite this potential, researchers have only recently begun to examine the climatic information recorded by Alaskan trees. One of the most serious problems in dendroclimatic research in Alaska is the scarcity of meteorologic stations with records long enough for carrying out the calibration and verification steps (Jacoby, 1982).

Tree-Ring Collections

Early tree-ring collections in Alaska were made by Giddings (1941, 1943, 1947) and Oswalt (1952, 1958). Although new sites have been sampled in recent years, the collections of these early workers still make up the majority of existing Alaskan chronologies. Cropper and Fritts (1981) show that the statistical properties of Alaskan

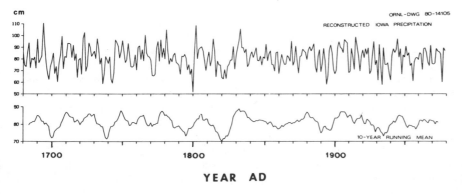

Figure 14-6. Reconstructed values of annual (August-July) precipitation for Iowa from 1680 through 1980 (top) and 10-year running means, plotted for the 5th year of each 10-year sequence (bottom). (Redrawn from Duvick and Blasing, 1981.)

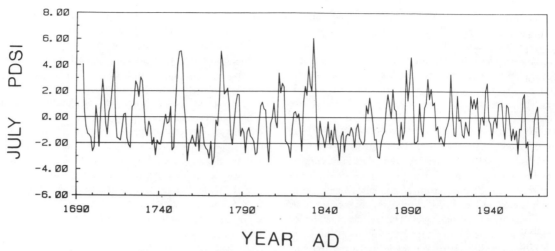

Figure 14-7. July PDSI reconstruction from the Hudson Valley, New York. Positive values indicate wetter-than-average conditions, and negative values indicate drier conditions. A PDSI of about −2.00 represents drought of at least moderate intensity. (Redrawn from Cook and Jacoby, 1979.)

chronologies contrast sharply with those of semiarid regions of the southwestern United States. Alaskan chronologies show many similarities to chronologies from the Pacific Northwest and the eastern United States, however. Growth records from each of these regions generally exhibit greater variability among individual trees and somewhat weaker climate-growth correlations than those from semiarid areas. In order to enhance the identification of common growth patterns among trees, investigators in mesic regions generally collect large samples at each site.

Growth Responses to Climate

Even though they did not have access to modern statistical techniques, Giddings (1941) and Oswalt (1958) were able to identify major climate-growth relationships for subarctic trees. They concluded that the mean growing season temperature was the overriding limiting factor at treeline sites. Recent statistical (Garfinkel and Brubaker, 1980; Haugen, 1967; Jacoby, 1982) and physiologic studies (Black and Bliss, 1980; Goldstein, 1981) have corroborated and amplified this understanding of the growth-limiting factors at the treeline. Although virtually all recent studies emphasize the importance of summer temperature, several (Black and Bliss, 1980; Garfinkel and Brubaker, 1980; Goldstein, 1981; Jacoby, 1982) also describe the importance of other variables, such as precipitation.

Garfinkel and Brubaker (1980) analyze the response functions of 14 white spruce (*Picea glauca*) chronologies from continuous forest and treeline sites in the central Brooks Range of Alaska. In addition to a positive correlation with current summer temperature, growth at these sites exhibits a strong positive correlation with previous autumn temperature and a negative correlation with spring temperature. Growth correlations with precipitation are generally positive, especially for summer months.

Certain aspects of these climate-growth relationships have been investigated in more detail through field and laboratory experiments with white spruce trees from the same field area (Goldstein, 1981). These studies indicate that midday temperatures during the summer usually exceed optimum values for photosynthesis. Stomata generally remain closed for extended periods during summer days due to moderate and high vapor-pressure deficits. This finding suggests that the positive correlations between summer temperature and growth of

subarctic trees do not occur because warm temperatures increase photosynthetic rates, as has been proposed by most workers. Goldstein (1981) suggests that the effect of warm temperatures may, instead, result from an increase in soil temperatures, which facilitates water uptake by roots. Soil temperature and transpiration measurements in his study indicate that resistance to water uptake and internal water stress decrease markedly with increased soil temperature. The effect of soil temperature is probably more important at the arctic treeline than at the alpine treeline because of the presence of permafrost at high latitudes.

Climatic Reconstructions

Two dendroclimatic studies of Alaskan chronologies have yielded reconstructions of summer climatic variables (Cropper, 1981; Garfinkel and Brubaker, 1980). (See Figure 14-8.) Garfinkel and Brubaker (1980) estimate average May-to-July temperatures for Fairbanks, 1828 to 1979, from the eigenvectors of 14 white spruce chronologies from the central Brooks Range. This record has been extended back to 1800 by Cropper (1981) by means of a correlated reconstruction of summer pressure anomalies in the North Pacific. Cropper's overall study reconstructs seasonal pressure-anomaly patterns from 56 Alaskan chronologies, the majority of which were assembled by Giddings and Oswalt (Cropper and Fritts, 1981). The temperature reconstructions of the two studies indicate that average May-to-July temperatures were approximately 3 °C colder between 1820 and 1860 than in the 20th century. Temperatures during the period 1800 to 1820 were nearly as warm as average temperatures during the 20th century. The cold period between 1820 and 1860 was characterized in Cropper's study by anomalously low pressures over Alaska. This pressure pattern would have resulted in cold conditions in the Fairbanks area. High pressures were reconstructed from 1870 to the end of Cropper's data period in 1938. Such conditions would have brought clear skies and warm temperatures to Fairbanks. Cropper suggests that these fluctuations in atmospheric pressure patterns are controlled by large-scale, hemispheric circulation systems. In support of this idea, he identifies similarities between the reconstructed Alaskan pressure records and inferred shifts in atmospheric pressure on Baffin Island and Greenland.

The reconstructed warming since 1860 in Fairbanks is consistent with information on age structures of treeline white spruce populations in the Brooks Range. Today, young trees established since the late 1800s are abundant above the treeline in these mountains. However, there is little evidence for tree establishment during the reconstructed cold period between 1820 and 1860 (Goldstein, 1981).

HYDROLOGIC STUDIES

Instrumental hydrologic records (e.g., stream flow, lake levels) are typically shorter than meteorologic records, and augmentations of the data currently available for describing hydrologic parameters (e.g., mean, variance) are greatly needed. Although several workers during the first half of this century (Hardman and Reil, 1936; Hawley, 1937; Kapteyn, 1914; Keen, 1937; Schulman, 1945a, 1945b, 1947, 1951) have described correlations between ring-width data and records of stream flow or surface runoff, the most extensive research in this field has been carried out since 1970 by Stockton and co-workers (Meko et al., 1980; Mitchell et al., 1979; Stockton, 1971, 1975; Stockton and Boggess, 1979; Stockton and Fritts, 1973; Stockton and Jacoby, 1976; Stockton and Meko, 1975). In general, the same techniques and principles used to reconstruct climatic histories are applied in studies of past hydrologic conditions.

Stockton's first major attempt to augment hydrologic data from tree-ring records was a reconstruction of runoff from three watersheds in the Colorado River drainage (Stockton, 1971, 1975). The upper Colorado River Basin, the largest watershed, was investigated in the most detail. Seventeen chronologies from major runoff-producing areas in this watershed were used to reconstruct annual river flow records for the period 1564 to 1961 at Lee Ferry, a gauging station at the mouth of the basin. The principal components of these chronologies were used in a multiple-regression calibration against a 66-year record of stream flow at Lee Ferry. Figure 14-9 shows the reconstructed runoff values, smoothed to emphasize long-term trends. The average for the reconstructed runoff over the entire 398-year period is approximately 16.0×10^9 metric tons as compared to 18.5×10^9 metric tons during the calibration period. There were two rather long periods of low flow, 1564 to 1600 and 1870 to 1890, and two extended period of high flow, 1601 to 1621 and 1905 to 1930.

The results of this study have important implications for water resource management in the Colorado River Basin. Because instrumental records of river flow do not span major drought periods, they overestimate the long-term availability of water in this system. The error is especially great because the instrumental record includes the longest period of high water flow during the entire 398 years. The reconstructed runoff records, therefore, suggest that management plans based on instrumental records can overestimate the water available from the Colorado River by as much as 2.5×10^9 metric tons per year.

In a similar study, Stockton and Fritts (1973) used tree rings of white spruce to lengthen the record of water levels of Lake Athabasca in northern British Columbia. The trees in this study were growing on floodplains and levees of rivers draining into the lake. The tree-ring data were calibrated by canonical regression techniques against 33-year records of water levels for 10-day intervals in May, July, and September. The resulting transfer function was used to reconstruct lake levels for the same intervals from 1810 through 1934. Although the means for the reconstruction and calibration periods were quite similar, the variance for the reconstructed period was greater than that for the calibration period, especially for May lake levels. These results are valuable for making ecologic decisions about the regulation of dams in the headwaters of rivers leading into Lake Athabasca.

Most recent water-resource studies have concentrated on assessing the frequency and magnitude of droughts in the agricultural regions of the United States. Stockton and Meko (1975) calibrate chronologies from sites in the western United States against July PDSI values for climatological divisions west of the Mississippi River. The calibration period in this study is 1930 to 1970 and the reconstruction period is 1700 to 1930. No verification, however, is made on independent data. Several droughts in the past were longer than those in the calibration period. The longest drought occurred during the early part of the 20th century, and the most extreme drought year in the entire record was 1936.

In a comparison of this reconstructed drought record with records of sunspots, Mitchell and others (1979) demonstrate that the extent of past droughts varied on the same 22-year periodicity as the sunspot cycle. Although the mechanism possibly linking sunspots and drought has not been identified, these investigations suggest that the risk of a widespread drought west of the Mississippi River is greater during the first few years following a sunspot minimum than at other times in the cycle.

STABLE-ISOTOPIC STUDIES

Several investigators have recently examined the possibility that stable-isotope ratios of carbon, oxygen, and hydrogen in tree rings contain information about past climate. Much of the work so far has centered on developing laboratory methods for quantifying isotopic ratios in rings, but many studies have also investigated relationships between isotopic ratios and climatic parameters. Jacoby (1980) and Long (1982) present an overview of this field of study.

A major objective of current research is to describe the fractiona-

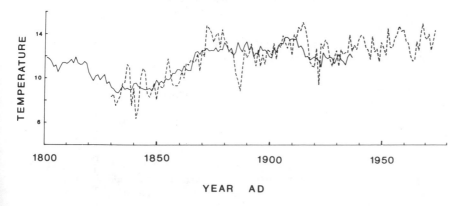

Figure 14-8. Summer temperature reconstruction for Fairbanks, Alaska (Garfinkel and Brubaker, 1980), indicated by solid line; and type 3 summer pressure reconstruction for Alaska (Cropper, 1981), indicated by dotted line.

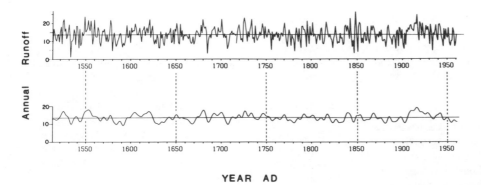

Figure 14-9. Reconstructed annual virgin runoff of the Colorado River at Lee Ferry (unfiltered) and 10-year running means (filtered), plotted for the 5th year of each 10-year sequence (bottom). (Based on C. W. Stockman, personal communication, 1982.)

tion and equilibrium processes important in determining isotopic ratios in tree rings. This is a difficult task because many factors can influence isotopic concentrations, including the rates of cellular reactions such as photosynthesis, the seasonal variations in such reactions, the effect of site conditions (e.g., latitude, exposure to wind) on concentrations of isotopes in the local environment of the tree, and the global concentrations of isotopes in the atmosphere.

Theoretical, expermental, and empirical approaches are being applied in tree-ring isotopic studies (Jacoby, 1980; Long, 1982). Models of global carbon reservoirs have been used to evaluate $^{13}C/^{12}C$ records in tree rings (Broecker, 1980; Freyer, 1980; Stuiver, 1978; Tans, 1980). These analyses indicate that carbon isotopes primarily provide information about changes in global carbon dioxide production. Global atmospheric concentrations of carbon isotopes have varied recently in response to the increased combustion of fossil fuel low in ^{13}C and the widespread removal of the forests that respire carbon dioxide low in ^{13}C. The effect of local carbon dioxide production within stands has also been examined through models and empirical studies (Lerman and Long, 1980). Oxygen-isotopic studies have modeled variations in isotopic concentrations in the atmosphere and isotopic exchange and equilibrium processes in the leaves and roots of individual plants (Burk and Stuiver, 1980; Ferhi et al., 1980). Models dealing with plant processes typically are developed from the results of experimental studies that carefully control environmental conditions and isotopic concentrations of each element available to the study plants (Long, 1982).

Empirical studies provide essential evidence for showing whether or not the processes controlling stable isotopes ultimately result in the observed statistical correlations between isotopic ratios and environmental variables. Burk and Stuiver (1980) demonstrate that ^{18}O in cellulose from trees in Alaska, Washington, and California shows latitudinal and elevational correlations with environmental water levels and mean annual temperature. Gray and Thompson (1980) also report a correlation between the ^{18}O content of cellulose and the mean annual temperature along a latitudinal transect in North America. Combined empirical and theoretical studies by Long (1982) suggest the ^{18}O and ^{2}H levels in cellulose may facilitate the reconstruction of temperature and humidity sequences. Although studies such as these suggest that stable isotopes in tree rings contain useful information about past conditions, scientists in this field caution that many studies are still needed for the full evaluation of the potential of isotopic ratios.

X-RAY DENSITY STUDIES

Several recent studies of conifer species have shown that the density of wood in annual rings can vary greatly from one year to the next even when the widths of rings remain relatively constant (Conkey, 1979; Parker and Henoch 1971; Schweingruber, 1982; Schweingruber et al., 1978). In the cases examined so far, latewood-density patterns are remarkably consistent among trees within a stand and among stands separated by long distances. These results suggest that there may be more climatic information in latewood-density data than in ring-width data in some trees.

Samples for densitometric studies are collected with standard-increment borers, and sections of uniform thickness are prepared for exposure to X rays. The intensity of the X rays passing through the sample is converted into estimates of wood density, and these values are standardized to remove age-related trends. The resulting series are averaged to form site chronologies. These wood-density chronologies are compared to climatic data with the same techniques as used for ring-width chronologies.

Conkey (1979) has measured wood densities for red spruce (*Picea rubens*) at three sites in Maine. The latewood-density patterns are strikingly similar among sites separated by more than 200 km (Figure 14-10). She also has compared response functions for latewood density and ring-width index data from one of the sites. Sixty-seven percent of the variation in density at that site could be attributed to climatic variations, but only 34% of the ring-width variation could be related to climatic variables. The form of these response functions suggests that latewood-density variations are more strongly correlated to summer conditions than are ring-width variations. The highest latewood-density values correspond to years having late-summer droughts. This study and others suggest that the climatic information in maximum wood density has finer seasonal resolution than that in ring widths. The ring widths appear to integrate the effects of conditions over several seasons (e.g., a year), while densities respond primarily to conditions during restricted periods (e.g., late summer). However, the higher resolution of latewood-density data apparently comes at the expense of low-frequency climatic information. Thus, density data alone may not be adequate to reconstruct a climatic series. Combined densitometric and ring-width studies should add significant precision to dendroclimatic research in the future (Schweingruber et al., 1978).

Conclusion

During the first half of the 20th century, Douglass and his co-workers demonstrated that ring-width sequences are a source of paleoecologic information. Subsequent research developed statistical techniques for describing the quality and type of environmental information in ring-width patterns and for estimating past conditions from ring sequences. A wide variety of variables has been reconstructed, and ex-

Yearly   Maximum   Density

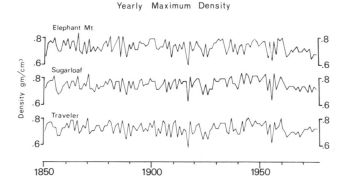

Figure 14-10. Yearly maximum density of samples of *Picea rubens* Sarg. at three sites in Maine. Each series is the average maximum density. (Redrawn from Conkey, 1979.)

isting techniques can be successfully applied to trees growing in mesic climates outside of the southwestern United States, where the research originated.

Future tree-ring research will probably emphasize the widespread reconstruction of specific environmental variables. Major tree-ring collections are being made in unsampled microsites and geographic regions of the United States. These new collections should increase the quality and diversity of existing reconstructions. The long-term environmental data base provided by such research will be essential for identifying mechanisms of natural environmental change, for evaluating the impact of humans on such variations, and for anticipating future conditions.

References

Black, R. A., and Bliss, L. C. (1980). Reproductive ecology of *Picea mariana* at treeline near Inuvick, Northwest Territories, Canada. *Ecological Monographs* 50, 331-54.

Blasing, T. J. (1975). Methods for analyzing climatic variations in the North Pacific sector and western North America for the last few centuries. Ph.D. dissertation, University of Wisconsin, Madison.

Blasing, T. J. (1978). Time series and multivariate analysis in paleoclimatology. *In* "Time Series and Ecological Processes" (H. H. Shugart, Jr., ed.), pp. 211-25. SIAM-SIMS Conference Series 5.

Blasing, T. J., and Fritts, H. C. (1975). Past climate of Alaska and northwestern Canada as reconstructed from tree-rings. *In* "Climate of the Arctic" (G. Weller and S. A. Bowling, eds.), pp. 48-58. Proceedings of the 24th Alaska Scientific Conference, August 15-17, 1973, Fairbanks, Alaska.

Blasing, T. J., and Fritts, H. C. (1976). Reconstruction of past climatic anomalies in the North Pacific and western North America from tree-ring data. *Quaternary Research* 6, 563-79.

Brinkman, W. A. R. (1976). Soil temperature influences on root resistance of *Pinus contorta* seedlings. *Plant Physiology* 65, 635-40.

Broecker, W. S. (1980). Status of C¹³/C¹² research on tree rings. *In* "Proceedings of the International Meeting on Stable Isotopes in Tree-Ring Research" (G. C. Jacoby, ed.), pp. 6-9. U.S. Department of Energy CONF-7905180, New Platz, N.Y.

Brubaker, L. B. (1980). Spatial patterns of tree growth anomalies in the Pacific Northwest. *Ecology* 61, 798-807.

Brubaker, L. B. (1982). Dendroclimatic research in western North America. *In* "Climate from Tree Rings" (M. K. Hughes, P. M. Kelly, J. Pilcher, and V. C. LaMarche, Jr., eds.), pp. 118-26. Cambridge University Press, Cambridge.

Bryson, R. A., and Hare, F. K. (1974). "Climates of North America." Elsevier, Amsterdam.

Burk, R. L., and Stuiver, M. (1980). Factors affecting ¹⁸O/¹⁶O ratios in cellulose. *In*

"Proceedings of the International Meeting on Stable Isotopes in Tree-Ring Research" (G. C. Jacoby, éd.), pp. 68-70. U.S. Department of Energy CONF-7907180, New Platz, N.Y.

Cleaveland, M. K. (1975). Dendroclimatic relationships of shortleaf pine (*Pinus schinata* Mill.) in the South Carolina Piedmont. M.S. thesis, Clemson University, Clemson, S.C.

Conkey, L. (1979). Dendroclimatology in the northwestern United States. M.S. thesis, University of Arizona, Tucson.

Cook, E. R. (1982). A prospectus on the development of a tree-ring network in eastern North America. *In* "Climate from Tree Rings" (M. K. Hughes, P. M. Kelly, J. Pilcher, and V. C. LaMarche, Jr., eds.), pp. 126-34. Cambridge University Press, Cambridge.

Cook, E. R., and Jacoby, G. C. (1977). Tree-ring-drought relationships in the Hudson Valley, New York. *Science* 198, 399-401.

Cook, E. R., and Jacoby, G. C. (1979). Evidence for quasi-periodic July drought in the Hudson Valley, New York. *Nature* 282, 390-92.

Cook, E. R., and Peters, K. (1981). The smoothing spline: A new approach to standardizing forest interior tree-ring width series for dendroclimatic studies. *Tree-Ring Bulletin* 41, 45-53.

Cropper, J. P. (1981). Reconstruction of North Pacific surface pressure anomaly types from Alaskan and western Canadian tree-ring data. M.S. thesis, University of Arizona, Tucson.

Cropper, J. P., and Fritts, H. C. (1981). Ring-width chronologies from the North American Arctic. *Arctic and Alpine Research* 13, 245-60.

Daubenmire, R. (1954). Alpine timberlines in the Americas and their interpretation. *Butler University Botanical Studies* 11, 119-36.

Dewitt, E., and Ames, M. (1978). "Tree-Ring Chronologies of Eastern North America," Vol. 1. University of Arizona Laboratory of Tree-Ring Research Chronology Series IV.

Diller, O. D. (1935). The relation of temperature and precipitation to the growth of beech in northern Indiana. *Ecology* 16, 72-81.

Douglass, A. E. (1914). A method of estimating rainfall by the growth of trees. *In* "The Climatic Factor" (E. Huntington, ed.), pp. 101-22. Carnegie Institute of Washington Publication 192.

Douglass, A. E. (1921). Dating our prehistoric ruins. *Natural History* 21, 27-30.

Drew, L. G. (ed.) (1974). "Tree-Ring Chronologies of Western America: Colorado, Utah, Nebraska and South Dakota." Vol. 4. University of Arizona Laboratory of Tree-Ring Research Chronology Series I.

Drew, L. G. (ed.) (1975). "Tree-Ring Chronologies of Western America: Washington, Oregon, Idaho, Montana and Wyoming." Vol. 5. University of Arizona Laboratory of Tree-Ring Research Chronology Series I.

Drew, L. G. (ed.) (1976). "Tree-Ring Chronologies for Dendroclimatic Analysis: An Expanded Western North American Grid." University of Arizona Laboratory of Tree-Ring Research Chronology Series II.

Duvick, D. N., and Blasing, T. J. (1981). A dendroclimatic reconstruction of annual precipitation amounts in Iowa since 1680. *Water Resources Research* 17, 1183-89.

Estes, E. T. (1969). The dendrochronology of three tree species in the central Mississippi Valley. Ph.D. dissertation, Southern Illinois University, Carbondale.

Ferguson, C. W. (1968). Bristlecone pine: Science and esthetics. *Science* 159, 839-46.

Ferguson, C. W. (1970). Dendrochronology of bristlecone pine, *Pinus aristata*: Establishment of a 7484-year chronology in the White Mountains of eastern-central California, USA. *In* "Radiocarbon Variations and Absolute Chronology" (I. U. Olsson, ed.), pp. 237-59. Nobel Symposium 12. Almqvist and Wiksell, Stockholm, and John Wiley and Sons, New York.

Ferguson, C. W. (1980). Four hundred years, and four thousand to go: Appreciation and study of bristlecone pine. *International Bonsai* 2 (4), 16-17.

Ferhi, A., Létolle, R., Long, A., and Lerman, J. C. (1980). Factors controlling the variations of oxygen-18 in plant cellulose. *In* "Proceedings of the International Meeting on Stable Isotopes in Tree-Ring Research" (G. C. Jacoby, ed.), pp. 82-84. U.S. Department of Energy CONF-7907180, New Platz, N.Y.

Freyer, H. D. (1980). Record of environmental variables by ¹³C measurements in tree rings. *In* "Proceedings of the International Meeting on Stable Isotopes in Tree-Ring Research" (G. C. Jacoby, ed.), pp. 13-21. U.S. Department of Energy CONF-7907180, New Platz, N.Y.

Fritts, H. C. (1958). An analysis of radial growth of beech in a central Ohio forest during 1954-1955. *Ecology* 39, 705-20.

Fritts, H. C. (1959). The relation of radial growth to maximum and minimum temperatures in three tree species. *Ecology* 40, 261-65.

Fritts, H. C. (1960). Multiple regression analysis of radial growth in individual trees. *Forest Science* 6, 334-49.

Fritts, H. C. (1962). The relation of growth ring widths in American beech and white oak to variations in climate. *Tree-Ring Bulletin* 25, 2-10.

Fritts, H. C. (1974). Relationships of ring widths in arid-site conifers to variations in monthly temperature and precipitation. *Ecological Monographs* 44, 411-40.

Fritts, H. C. (1976). "Tree Rings and Climate." Academic Press, New York.

Fritts, H. C. (1981). Statistical climatic reconstructions from tree-ring widths. *In* "Climate Variations and Variability: Facts and Theories" (A. Berger, ed.), pp. 135-53. Reidel, Dordrecht.

Fritts, H. C. (1982a). The climate-growth response in tree-ring chronologies. *In* "Climate from Tree Rings" (M. K. Hughes, P. M. Kelly, J. Pilcher, and V. C. LaMarche, Jr., eds.), pp. 33-37. Cambridge University Press, Cambridge.

Fritts, H. C. (1982b). An overview of dendroclimatic techniques, procedures, and prospects. *In* "Climate from Tree Rings" (M. K. Hughes, P. M. Kelly, J. Pilcher, and V. C. LaMarche, Jr., eds.), pp. 191-98. Cambridge University Press, Cambridge.

Fritts, H. C., Blasin, T. J., Hayden, B. P, and Kutzbach, J. E. (1971). Multivariate techniques for specifying tree-growth and climate relationships and for reconstructing anomalies in paleoclimate. *Journal of Applied Meteorology* 10, 845-64.

Fritts, H. C., and Gordon, C. A. (1982). Annual precipitation for California, U.S.A., since 1600 reconstructed from western North American tree rings. *In* "Climate from Tree Rings" (M. K. Hughes, P. M. Kelly, J. Pilcher, and V. C. LaMarche, Jr., eds.), pp. 185-90. Cambridge University Press, Cambridge.

Fritts, H. C., Lofgren, G. R., and Gordon, G. A. (1979). Variations in climate since 1602 as reconstructed from tree rings. *Quaternary Research* 12, 18-46.

Fritts, H. C., Mosimann, J. E., and Bottorff, C. P. (1969). A revised computer program for standardizing tree-ring series. *Tree-Ring Bulletin* 29, 15-20.

Fritts, H. C., and Shatz, D. J. (1975). Selecting and characterizing tree-ring chronologies for dendroclimatic analysis. *Tree-Ring Bulletin* 35, 31-40.

Fritts, H. C., Smith, D. G., Cardis, J. W., and Budelsky, C. A. (1965). Tree-ring characteristics along a vegetation gradient in northern Arizona. *Ecology* 46, 393-401.

Garfinkel, H. L., and Brubaker, L. B. (1980). Modern climate-tree-growth relationships and climatic reconstruction in sub-Arctic Alaska. *Nature* 286, 874.

Giddings, J. L. (1941). "Dendrochronology in Northern Alaska." University of Arizona Laboratory of Tree-Ring Research Bulletin 1.

Giddings, J. L. (1943). Some climatic aspects of tree growth in Alaska. *Tree-Ring Bulletin* 9, 4.

Giddings, J. L. (1947). Mackenzie River Delta chronology. *Tree-Ring Bulletin* 13, 4.

Glahn, H. R. (1968). Canonical correlation and its relationship to discriminant analysis and multiple regression. *Journal of Atmospheric Science* 25, 23-31.

Goldstein, G. H. (1981). Ecophysiological and demographic studies of white spruce (*Picea glauca* [Moench] Voss) at treeline in the central Brooks Range of Alaska. Ph.D. dissertation, University of Washington, Seattle.

Gordon, G. A. (1980). "Verification Tests for Dendroclimatic Reconstructions." Northern Hemisphere Climatic Reconstruction Group, University of Arizona Laboratory of Tree-Ring Research, Technical Note 19.

Gordon, G. A. (1982). Verification of dendroclimatic reconstructions. *In* "Climate from Tree Rings" (M. K. Hughes, P. M. Kelly, J. Pilcher, and V. C. LaMarche, Jr., eds.), pp. 58-62. Cambridge University Press, Cambridge.

Gray, J., and Thompson, P. (1980). Natural variations in the ^{18}O content of cellulose. *In* "Proceedings of the International Meeting on Stable Isotopes in Tree-Ring Research" (G. C. Jacoby, ed.), pp. 84-92. U.S. Department of Energy CONF-7907180, New Platz, N.Y.

Graybill, D. A. (1982). Chronology development and analysis. *In* "Climate from Tree Rings" (M. K. Hughes, P. M. Kelly, J. Pilcher, and V. C. LaMarche, Jr., eds.), pp. 21-28. Cambridge University Press, Cambridge.

Grootes, P. M. (1983). Radioactive isotopes in the Holocene. *In* "Late-Quaternary Environments of the United States." vol. 2, "The Holocene" (H. E. Wright, Jr., ed.), pp. 86-105. University of Minnesota Press, Minneapolis.

Hardman, G., and Reil, O. E. (1936). "The Relationship between Tree-Growth and Stream Runoff in the Truckee River Basin, California-Nevada." University of Nevada Agricultural Experimental Station Bulletin 141.

Hare, F. K. (1950). Climate and zonal divisions of the boreal formation in eastern Canada. *Geographical Review* 40, 615-35.

Haugen, R. K. (1967). Tree-ring indices: A circumpolar comparison. *Science* 158, 773-75.

Hawley, F. M. (1937). Relationship of southern cedar growth to precipitation and runoff. *Ecology* 18, 398-405.

Hawley, F. M. (1941). "Tree-Ring Analysis and Dating in the Mississippi Drainage." University of Chicago Press, Chicago.

Hopkins, D. M. (1959). Some characteristics of the climate in forest tundra regions in Alaska. *Arctic* 12, 215-20.

Jacoby, G. C. (ed.) (1980). "Proceedings of the International Meeting on Stable Isotopes in Tree-Ring Research" U.S. Department of Energy CONF-7905180, New Platz, N.Y.

Jacoby, G. C. (1982). Arctic dendroclimatology. *In* "Climate from Tree Rings" (M. K. Hughes, P. M. Kelly, J. Pilcher, and V. C. LaMarche, Jr., eds.), pp. 107-16. Cambridge University Press, Cambridge.

Jacoby, G. C., and Cook, E. R. (1981). Past temperature variations inferred from a 400-year tree-ring chronology from Yukon Territory, Canada. *Arctic and Alpine Research* 13, 409-18.

Kapteyn, J. C. (1914). Tree-growth and meteorological factors. *Recueil des Travaux Botaniques Neerlandais* 11, 71-93. (Cited in Schulman, 1945b.)

Kaufmann, M. R. (1977). Soil temperature and drying cycle effects on water relations of *Pinus radiata*. *Canadian Journal of Botany* 55, 2413-18.

Keen, F. P. (1937). Climatic cycle in eastern Oregon as indicated by tree rings. *Monthly Weather Review* 65, 175-88.

Kellogg, W. W. (1975). The poles: A key to climatic change. *In* "Climate of the Arctic" (G. Weller and S. A. Bowling, eds.), pp. v-vii. Proceedings of the 24th Alaska Scientific Conference, Auguest 15-17, 1973, Fairbanks, Alaska.

Koppen, W. (1936). Das geographische System der Klimate. *In* "Handbuchder Klimatologie" (W. Koppen and R. Geiger, eds.), vol. 1. Gebruder Borntrager, Berlin.

Kutzbach, J. E., and Guetter, P. J. (1980). On the design of paleoclimatic data networks for estimating large-scale circulation patterns. *Quaternary Research* 14, 169-87.

LaMarche, V. C., Jr. (1969). Environmental relation of bristlecone pines. *Ecology* 50, 53-59.

LaMarche, V. C., Jr. (1973). Holocene climatic variations inferred from treeline fluctuations in the White Mountains, California. *Quaternary Research* 3, 632-60.

LaMarche, V. C., Jr. (1974). Paleoclimatic inferences from long tree-ring records. *Science* 183, 1043-48.

LaMarche, V. C., Jr. (1978). Tree-ring evidence of past climatic variability. *Nature* 276, 334-38.

LaMarche, V. C., Jr. (1982). Sampling strategies in dendrochronology. *In* "Climate from Tree Rings" (M. K. Hughes, P. M. Kelly, J. Pilcher, and V. C. LaMarche, Jr., eds.), pp. 2-6. Cambridge University Press, Cambridge.

LaMarche, V. C., Jr., and Harlan, T. P. (1973). Accuracy of tree-ring dating of bristlecone pine for calibration of the radiocarbon time scale. *Journal of Geophysical Research* 78, 8849-58.

LaMarche, V. C., Jr., and Mooney, H. A., (1967). Altithermal timberline advance in western United States. *Nature* 218, 980-82.

LaMarche, V. C., Jr., and Mooney, H. A. (1972). Recent climatic change and development of the bristlecone pine (*P. longaeva* Bailey) Krummholz zone, Mt. Washington, Nevada. *Arctic and Alpine Research* 4, 61-72.

LaMarche, V. C., Jr., and Stockton, C. W. (1974). Chronologies from temperature-sensitive bristlecone pines at upper treeline in western United States. *Tree-Ring Bulletin* 34, 21-45.

Lamb, H. H. (1972). "Climate: Present, Past and Future: Fundamentals and Climate Now." Vol. 1. Methuen, London.

Lamb, H. H. (1974). "Climate: Present, Past and Future: Climatic History and the Future." Vol. 2. Methuen, London.

Lerman, J. C., and Long, A. (1980). Carbon-13 in tree rings: Local or canopy effects?

In "Proceedings of the International Meeting on Stable Isotopes in Tree-Ring Research" (G. C. Jacoby, ed.), pp. 22-34. U.S. Department of Energy CONF-7905180, New Platz, N.Y.

Lofgren, G. R., and Hunt, J. H. (1982). The transfer function. *In* "Climate from Tree Rings" (M. K. Hughes, P. M. Kelly, J. Pilcher, and V. C. LaMarche, Jr., eds.), pp. 50-56. Cambridge University Press, Cambridge.

Long, A. (1982). The Study of isotopic parameters. *In* "Climate from Tree Rings" (M. K. Hughes, P. M. Kelly, J. Pilcher, and V. C. LaMarche, Jr., eds.), pp. 12-17. Cambridge University Press, Cambridge.

Lutz, H. L. (1944). Swamp-grown eastern white pine and hemlock in Connecticut as dendrochronological material. *Tree-Ring Bulletin* 10, 26-28.

Lyon, C. J. (1936). Tree ring width as an index of physiological dryness in New England. *Ecology* 17, 457-78.

Meko, D. M., Stockton, C. W., and Boggess, W. R. (1980). A tree-ring reconstruction of drought in southern California. *Water Resources Bulletin* 16, 594-600.

Mitchell, J. M. (1961). Recent secular changes of global temperature. *Annals of the New York Academy of Science* 95, 235-50.

Mitchell, J. M., Jr., Stockton, C. W., and Meko, D. M. (1979). Evidence of a 22-year rhythm of drought in the western United States related to the Hale Solar Cycle since the 17th century. *In* "Proceedings of the Solar-Climate Conference, 24-28 August 1979," pp. 125-43. Ohio State University, Columbus.

Oswalt, W. H. (1952). Spruce samples from Copper River, Alaska. *Tree-Ring Bulletin* 19, 5-10.

Oswalt, W. H. (1958). Tree-ring chronologies in south central Alaska. *Tree-Ring Bulletin* 23, 16-22.

Palmer, W. C. (1965). "Meteorological Drought." U.S. Weather Bureau Research Paper 45.

Parker, M. L., and Henoch, W. E. S. (1971). The use of Engelmann spruce latewood density for dendrochronological purposes. *Canadian Journal of Forestry Research* 1, 90-98.

Pearson, G. A. (1931). "Forest Types of the Southwest as Determined by Climate and Soil." U.S. Department of Agriculture Technical Bulletin 247.

Phipps, R. L. (1961). "Analysis of Five Year Dendrometer Data Obtained within Three Deciduous Forest Communities of Neotoma." Ohio Agricultural Experimental Station Research Circular 105.

Phipps, R. L. (1967). "Annual Growth of Suppressed Chestnut Oak and Red Maple: A basis for Hydrologic Inference." U.S. Geological Survey Professional Paper 485-C.

Robinson, W. J. (1976). Tree-ring dating and archaeology in the American Southwest. *Tree-Ring Bulletin* 36, 9-20.

Running, S. W. (1976). Soil temperature influences on root resistance of *Pinus contorta* seedlings. *Plant Physiology* 65, 635-40.

Running, S. W., and Reid, C. P. (1980). Soil temperature influences on root resistance of *Pinus contorta* seedlings. *Plant Physiology* 65, 635-40.

Rutter, M. R. (1977). An ecophysical field study of three Sierra conifers. Ph.D. dissertation. University of California, Berkeley.

Sachs, H. M., Webb, T., III, and Clark, D. R. (1977). Paleoecological transfer functions. *Annual Review of Earth Planet Science* 5, 159-78.

Schulman, E. (1942). Dendrochronology in pines of Arkansas. *Ecology* 23, 309-18.

Schulman, E. (1945a). Runoff histories in tree-rings of the Pacific slope. *Geographical Review* 35, 59-73.

Schulman, E. (1945b). "Tree-Ring Hydrology of the Colorado River Basin." University of Arizona Laboratory of Tree-Ring Research Bulletin 2.

Schulman, E. (1947). "Tree-Ring Hydrology in Southern California." University of Arizona Laboratory of Tree-Ring Research Bulletin 4.

Schulman, E. (1951). Tree-ring indices of rainfall, temperature and river flow. *In* "Compendium of Meteorology" (American Meteorological Society editors, eds.), pp. 1024-29, American Meteorological Society, Boston.

Schweingruber, F. J. (1982). Measurement of densitometric properties of wood. *In* "Climate from Tree Rings" (M. K. Hughes, P. M. Kelly, J. Pilcher, and V. C. LaMarche, Jr., eds.), pp. 8-11. Cambridge University Press, Cambridge.

Schweingruber, F. J., Fritts, H. C., Braker, O. U., Drew, L. G., and Schä, E. (1978). The x-ray technique as applied to dendroclimatology. *Tree-Ring Bulletin* 38, 61-91.

Stockton, C. W. (1971). The feasibility of augmenting hydrologic records using tree-ring data. Ph.D. dissertation, University of Arizona, Tucson.

Stockton, C. W. (1975). "Long-term Streamflow Records Reconstructed from Tree Rings." Papers of the University of Arizona Laboratory of Tree-Ring Research 5.

Stockton, C. W., and Boggess, W. R. (1979). Augmentation of hydrologic records using tree rings. *In* "Improved Hydrologic Forecasting: How and Why," pp. 239-65. Proceedings of Engineering Foundation Conference, American Society of Civil Engineers, March 25-30, 1979, Pacific Grove, Calif.

Stockton, C. W., and Fritts, H. C., (1973). Long-term reconstruction of water level changes for Lake Athabasca by analysis of tree-rings. *Water Resources Bulletin* 9, 1006-27.

Stockton, C. W., and Jacoby, G. C., Jr. (1976). "Long-term Surface-Water Supply and Streamflow Trends in the Upper Colorado River Basin Based on Tree-Ring Analysis." Lake Powell Research Project Bulletin 18.

Stockton, C. W., and Meko, D. M. (1975). A long-term history of drought occurrences in western United States as inferred from tree rings. *Weatherwise* 28, 245-49.

Stokes, M. A., Drew, L. G., and Stockton, C. W. (eds.) (1973). "Tree-Ring chronologies of Western America: Selected Tree-Ring Stations." Vol. 1. University of Arizona Laboratory of Tree-Ring Research Chronology Series I.

Stokes, M. A., and Smiley, T. L. (1968). "An Introduction to Tree-Ring Dating." University of Chicago Press, Chicago.

Stuiver, M. (1978). Atmospheric carbon dioxide and carbon reservoir changes. *Science* 199, 253-58.

Tans, P. P. (1980). Some requirements for a ^{13}C tree-ring record to study the global carbon reserves. *In* "Proceedings of the International Meeting on Stable Isotopes in Tree-Ring Research" (G. C. Jacoby, ed.), pp. 35-39. U.S. Department of Energy CONF-7907180, New Platz, N.Y.

Environmental Archaeology

CHAPTER 15

Environmental Archaeology in the Western United States

C. Melvin Aikens

Introduction

The concept of the natural environment as a variable shaping human society has been a primary focus of archaeological study in the western United States for a long time, and significant progress has been made during the past two decades in delineating the culture-environment relationship. The work of Baumhoff and Heizer (1965) constitutes a baseline from which this chapter can start. The framework of their paper follows the Neothermal concept of Antevs (1948, 1955), which describes post-Pleistocene time in terms of the Anathermal period, about 9000 to 7000 yr B.P., with climatic conditions initially cool and moist but growing warmer, an Altithermal period, about 7000 to 4500 yr B.P., drier and warmer than at present, during which many Great Basin lakes disappeared; and the Medithermal period, 4500 yr B.P. to the present, which saw the rebirth of Great Basin lakes under a temperature-moisture regime similar to that of modern times.

At the time Baumhoff and Heizer were making their assessment, there was significant disagreement among archaeologists working in the region as to the character and dating of the temperature-moisture phases outlined by Antevs and particularly as to the relevance of such changes to human occupation in the desert West. Jennings (1957, 1964), for example, stressed the long-term stability of the Great Basin Desert Culture pattern and minimized the importance of environmental change as a variable significantly affecting that pattern. Baumhoff and Heizer (1965), on the other hand, argued that Altithermal aridity forced the abandonment of much of the interior desert West between about 7000 and 4500 yr B.P.

Accumulated paleoenvironmental data now make the general sequence of climatic shifts outlined by Antevs uncontestable. At the same time, discussions such as those of Aschmann (1958) and Bryan and Gruhn (1964), as well as many new, detailed paleoenvironmental sequences from a variety of settings (Grayson, 1982; Mehringer, 1977), make it clear that the severity of the Altithermal interval has been overstated and that the relationship between general climatic change and such on-the-ground change in local environments as might actually have affected specific human populations is more complex than previously realized. As a result, archaeologists today are

seldom tempted to indulge in the kinds of broad-scale interpretations that once imputed to an arid Altithermal climate the power to drive human occupants away from hundreds of thousands of square kilometers of latitudinally and altitudinally varied territory in the intermontane West and in the Great Plains as well (Mulloy, 1958; Wallace, 1962; Wedel, 1961).

Current approaches stress the development of detailed local and regional paleoenvironmental sequences as frameworks for examining cultural sequences of similarly local or regional scope. A number of recent studies are reviewed in this chapter in order to illustrate some of the most important current problems of environmental archaeology in western North America.

Climatic and Environmental Change

Lines of evidence bearing on past environment (Mehringer, 1977) are derived from studies of fossil pollen, plant macrofossils from archaeological and peat deposits as well as from fossil woodrat (*Neotoma*) middens, fluctuations in the elevation of upper and lower treelines, patterns of varying annual growth rings in several arboreal species, vertebrate remains (especially the bones of small mammals), modern biogeographic distributions, glacial moraines and pluvial-lake beaches, fluctuating Holocene lake strandlines, sand-dune activity, erosion, sedimentation, soil formation, and other biologic and geologic phenomena.

There seem to have been two chief junctures in late-Quaternary time at which significant climatic changes affected far-western environments. The first and more marked of these took place between about 12,000 and 10,000 years ago, at the close of the Pleistocene, when glaciers, woodlands, and basin lakes diminished and the Pleistocene megafauna disappeared.

The drying trend continued for several thousand years; the second climatic change, minor compared to the first, was accomplished around 7500 years ago, by which time desertic conditions much like those of the present prevailed. Climatic fluctuations of lesser magnitude characterized subsequent times. The details are not yet fully understood, but several lines of evidence suggest that effective precipitation reached its lowest level between about 7500 and 4500 yr

B.P. Even during this interval, however, there was apparently at least one brief period of increased effective moisture. By 4000 yr B.P. or so, there had developed an upward curve of increasing effective moisture that dipped once more toward aridity by 2000 yr B.P. Between about 1500 and 600 yr B.P. in the northeastern Great Basin and between about 1000 and 800 yr B.P. in the Southwest, a final prehistoric interval of increased effective moisture was recorded (Euler et al., 1979; Mehringer, 1977: 148-49; Van Devender and Spaulding, 1979).

In addition to these relatively major phenomena, local records give evidence of shorter-term variations less easily correlated with regional patterns. Throughout the western United States, postglacial climatic change mediated by local and regional factors of latitude, altitude, and topography repeatedly modified environmental mosaics in ways that probably had important effects on local food resources and, hence, on human societies. Paleoenvironmental records suggest even decennial or annual variability in local effective moisture and biotic productivity like that which characterizes the West today. Human societies that supported themselves by the systematic exploitation of environmental diversity were, no doubt, obliged to adjust their harvesting patterns continually to variations in the local and regional productivity of one or another resource.

Aschmann (1958) points out that the desert peoples' highly mobile and broadly based economic adaptation, necessary for occupying the region even under the best of conditions, effectively preadapted them to surviving the vicissitudes of the climate. The relative size and number of food-resource patches could grow or diminish with fluctuations in effective moisture, but the basic climatic and physiographic structure of the region guaranteed that the same general pattern of diverse and broadly distributed resource patches would persist there even if the overall rate of annual precipitation were doubled or halved by climatic fluctuations.

The studies summarized in this chapter discuss the possible and probable responses of the prehistoric societies of the western United States to climatically induced environmental changes, to biogeographic and physiographic circumstances, and to volcanism.

Environmental and Demographic Change

A series of correlations between environmental and demographic change in the southwestern United States have been developed by Irwin-Williams and Haynes (1970). Archaeological data from both surface surveys and excavations provide information on the duration and distribution of different cultural patterns. The basic paleoenvironmental framework is based on fossil-pollen records of fluctuations in the amount of effective moisture and on geologic records of erosional and depositional episodes and periods of soil formation. Botanical analysis of several hundred radiocarbon-dated woodrat (Neotoma) middens from throughout the Southwest has also contributed greatly to a clarification of the character and chronology of late-Pleistocene and Holocene environmental trends in the region (Van Devender and Spaulding, 1979).

The earliest cultural pattern well documented for the Southwest is that of the Clovis paleo-Indians, who were hunters of large Pleistocene animals, including mammoths and giant bison and possibly horses, sloths, and camels. Several radiocarbon-dated sites place the age of this occupation between about 11,500 and 11,000 yr B.P. The large, lanceolate Clovis fluted points diagnostic of this complex have been found throughout New Mexico, Arizona, and southern California, and more recent surface finds have extended this distribution northward throughout the Great Basin (Aikens, 1978: Figure 4.2). Pollen evidence indicates that there was more effective moisture in this time range than exists at present in the same region.

Later paleo-Indian cultures of the Southwest, represented by the Folsom, Agate Basin, Hell Gap, Alberta, Cody, and Frederick complexes, became progressively more restricted in geographic distribution throughout the period 10,500 to 7500 yr B.P. These complexes all represent essentially a continuation of the broad-ranging, big-game-hunting way of life of the Clovis period, but there were some important changes. The pedologic context of such finds suggests that this was a period of gradually decreasing soil moisture; pollen and other biotic evidence indicates a series of climatic fluctuations culminating in a long-term trend toward decreasing effective moisture. The mammoths had disappeared by Folsom times, and these folk preyed upon the early-postglacial bison. Folsom artifacts are rarely found west of central Arizona. Artifacts of the Agate Basin and Cody Complexes that followed are even more restricted in distribution, rarely occurring west of New Mexico. From the time of the terminal paleo-Indian Frederick complex, southwestrn remains assignable to the big-game-hunting tradition are found only in the plains grassland terrain of eastern New Mexico.

Thus, it appears from environmental and archaeological evidence that the early-postglacial decrease in the amounts of effective moisture available probably brought about the continuing shrinkage of the grasslands that harbored the bison on which later paleo-Indian hunters depended. As the Southwest more and more took on the desertic character it exhibits today, the paleo-Indian hunting pattern became restricted to the Great Plains, where optimum conditions for large herbivorous animals still existed.

In parts of western New Mexico and Arizona and probably in California and the Great Basin region, the Clovis paleo-Indian culture was succeeded locally by the Jay, Sulfur Springs Cochise, San Dieguito, and Lake Mohave complexes. Obsidian-hydration measurements show that Lake Mohave/San Dieguito-like specimens postdate Clovis fluted points in northern California (Meighan and Haynes, 1970), and typological similarities suggest that the western complexes were derived from Clovis just as were the Agate Basin and other late-paleo-Indian complexes of the Great Plains (Aikens, 1978: 148). These early Archaic complexes of the western deserts, essentially contemporary with the late-paleo-Indian Folsom-Frederick continuum farther east, may represent a transition between the Clovis way of life and a later hunting-foraging pattern, which in turn gave way after about 7500 yr B.P. to a broad-spectrum foraging economy of fully Archaic type.

The period between about 7500 and 5000 yr B.P. in the Southwest was characterized by a further reduction in effective moisture, although evidence from pollen and other sources also suggests some short-term reversals of this trend. The Jay and Bajada phases of northwestern New Mexico (Irwin-Williams, 1973) and the probably contemporaneous Silver Lake and early Pinto Basin and Little Lake complexes of southern California, which span this period, exemplify a foraging and hunting way of life under full desertic conditions.

Pollen and pedologic evidence suggest a marked increase in effective moisture in the Greater Southwest after about 5000 yr B.P. Dating to a period roughly 5000 to 2500 yr B.P. are the later Pinto Basin complex of southern California-Nevada and western Arizona, the San José complex of the northern Southwest, and the Chiricahua

Cochise complex of the southern Southwest. Plant gathering and foraging are believed to have been the dominant economic pursuits, with incipient maize horticulture in the Cochise and San José areas occurring perhaps as early as 4000 yr B.P. The relatively abundant remains of these highly similar and intergrading complexes, which have been grouped as the Picosa, or Elementary Southwestern, culture by Irwin-Williams (1967), suggest that populations expanded as effective moisture increased.

Archaeological evidence for the period since 2500 yr B.P. is increasingly abundant, suggesting continued population growth. The Pinto sector to the west gave rise to the nonhorticultural Rose Spring phase of western Nevada and eastern California, while the San José and Chiricahua Cochise phases of Arizona-New Mexico gave rise to full-blown horticultural developments of the classical Southwestern Basketmaster/Mogollon type during the succeeding En Medio and San Pedro Cochise phases.

The paleoenvironmental record of the last 2000 or so years in the Colorado Plateaus of the northern Southwest is more richly detailed than that for earlier periods. Because of the precision of dendrochronology and dendroclimatology, environmental fluctuations can be fixed in time and local sequences of climatically controlled events can be cross-correlated with one another within about 25 years. Archaeological data are also much more abundant for the last 2000 years than for earlier times, making possible a reliable assessment of settlement patterns, population distribution, and changes in these variables over relatively short periods. Euler and others (1979) show that among the village-farming populations of the region demographic changes again and again coincided closely with environmental fluctuations in a way that strongly implicates environmental change as a motive force.

Pollen and hydrologic evidence record a peak in effective moisture betweeen 1350 and 1250 years ago (A.D. 600 to 700). This coincides with a major expansion of the range of Basketmaker III horticulturists in adjacent parts of Arizona, New Mexico, Utah, and Colorado. A second moisture peak occurred between about 1000 and 800 years ago (A.D. 950-1150), when there was a further notable increase in the density and areal extent of human settlements. Several upland plateau areas in northeastern Arizona and Utah, before this interval only sparsely inhabited, experienced major population growth.

A downswing in the effective moisture curve began about 800 years ago (A.D. 1150), reaching its lowest point by about 500 years ago (A.D. 1450). Between these dates, most of the northern part of the Southwest was gradually abandoned by horticultural peoples, and populations became concentrated in large towns near the Hopi mesas in Arizona and the upper Rio Grande in New Mexico, where dependable sources of water remained. During this difficult interval, water-control devices such as check dams and irrigation works became increasingly common throughout the area. The evidence shows conclusively that moister climatic intervals fostered horticultural expansion into drier and/or warmer areas and that major droughts triggered movements toward cooler and wetter localities (Euler et al., 1979).

From the cultural and environmental sequence of more than 11,000 years so briefly sketched here for the southwestern United States, a simple and powerful correlation emerges between changes in the spatial/temporal abundance of effective moisture and the density and distribution of human populations. This human-environment correlation is clearly evident, whether the human populations involved

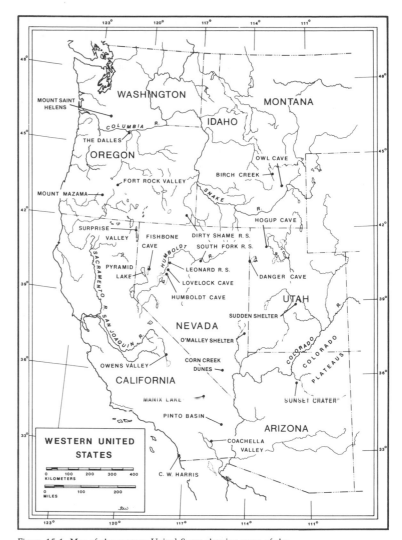

Figure 15-1. Map of the western United States showing some of the locations referred to in the text.

were big-game hunters, foragers, or horticulturists. The archaeological record of the Great Basin region to the north and west offers additional perspectives on climatically induced environmental change and human occupation.

Antevs (1948) speculated that much of the Great Basin may have been abandoned during the Altithermal phase of middle-postglacial time, when geologic and hydrologic indicators suggested that climatic warming and drying reduced effective moisture to a level significantly below that of modern times. This seemed a reasonable hypothesis, especially in view of Steward's (1938) ethnologic observations stressing the precariousness of aboriginal life in the Great Basin deserts, and it attracted a large following (Baumhoff and Heizer, 1965, and the references cited therein; Malde, 1964b).

By the mid-1960s, when the problem was reviewed by Baumhoff and Heizer (1965), the concept of an Altithermal abandonment had been invoked in both the southern California deserts and the western Great Basin. Although no long stratigraphic sequences exhibiting hiatuses or unconformities during the Altithermal interval were known in either area and although the chronology of surface finds and open sites was very poorly established, it was assumed nevertheless that none of the cultural materials then known could be dated to the Altithermal period, roughly 7000 to 5000 yr B.P. Cultural remains radiocarbon-dated after about 4000 yrs B.P. at Lovelock Cave in central Nevada and about 4500 yr B.P. at South Fork Shelter in east-central Nevada were interpreted as evidence of reoccupation after an Altithermal interval of depopulation (Baumhoff and Heizer, 1965: 704). (Dates of 7000 and 5700 yr B.P. from Leonard Rockshelter, a short distance from Lovelock Cave, might have been used in arguments for continuity of occupation throughout Altithermal times, but, instead, they were discounted.)

At Danger Cave, in the eastern Great Basin near the northern Utah-Nevada boundary, a deep cultural deposit radiocarbon-dated as older than 11,000 yr B.P. at its base and about 2000 yr B.P. near its top, seemed to span the Altithermal interval without a break (Jennings, 1957), but Baumhoff and Heizer considered this finding to be an anomaly. They pointed out that level DIII, which was bracketed by dates of about 9000 and about 5000 yr B.P., was not itself radiocarbon-dated. Although it was a very thick unit, level DIII might not represent the whole of the 4000-year interval implied by the bracketing dates.

This problem was resolved by subsequent research. At Danger Cave, radiocarbon dates of 8100, 6800, and 6600 yr B.P. were obtained from level DIII, showing that this rich cultural stratum had indeed accumulated during Altithermal time (Fry, 1976). Investigations at nearby Hogup Cave (Aikens, 1970) produced a record from about 8400 to 500 yr B.P. that showed clear evidence of a flourishing occupation throughout the Altithermal interval. In fact, it seems that the low middle-postglacial level of the Great Salt Lake, which apparently fostered a mosaic of shallow-water marshland vegetation on the lowlands near the site, contributed significantly to the richness of the lifeway attested to by the cave record for that period. Sudden Shelter (Jennings et al., 1980), in central Utah, also produced a series of middle-postglacial radiocarbon dates showing occupation throughout the interval 8000 to 3300 yr B.P.

In southwestern Nevada, radiocarbon dates from O'Malley Shelter (Fowler et al., 1973) indicate middle-postglacial occupation between 7000 and 3000 yr B.P., though with a possible hiatus between about 6500 and 4500 yr B.P. At Corn Creek Dunes, in the same region, radiocarbon-dated hearths document occupation between about 5200 and 4000 yr B.P. (Williams and Orlins, 1963).

Excavations at the Connley Caves in the Fort Rock Valley of south-central Oregon produced contrasting evidence. A series of 21 radiocarbon determinations showed human occupation from about 11,000 to after 3000 yr B.P., but with little or no occupation between 7000 and 5000 yr B.P. during the Altithermal interval (Bedwell, 1973; Bedwell and Cressman, 1971). Charcoal of pine (*Pinus* sp.) from campfires was not found after about 7000 yr B.P., whereas juniper (*Juniperus* sp.), which tolerates drier conditions than pine, appeared in the cave deposits at about this time and became increasingly abundant in later levels. Today, the Connley Caves are fronted by a healthy juniper forest.

A detailed analysis of the caves' fauna convincingly documents a significant drying of the local habitat after about 7000 yr B.P. (Grayson, 1979). The pika (*Ochotona princeps*), a small mammal of the cooler uplands today, was present at the Connley Caves between 11,000 and 7000 yr B.P. but absent thereafter. Decreased effective precipitation in the area after 7000 yr B.P. is also indicated by a progressive decline in the abundance of the white-tailed jackrabbit (*Lepus townsendii*), a creature that today occupies higher, cooler elevations and grassier habitats. Conversely, the black-tailed jackrabbit (*Lepus*

californicus), which currently dominates the now arid, shrubby Fort Rock Valley, is more common in the faunal assemblage dated after 7000 yr B.P. A marked decrease in the abundance of both marsh-dwelling and upland birds after 7000 yr B.P. is further evidence of increased aridity. Both types of birds would have been strongly affected by a reduction in the availability of free water in the area, and it is likely that this occurred with the shrinkage of Paulina Marsh, 1.5 km south of the Connley Caves. The largest body of water in the Fort Rock Valley today, it formerly was of still greater extent.

Dirty Shame Rockshelter, in extreme southeastern Oregon, has produced a comparable record (Aikens et al., 1977; Grayson, 1977). A series of 22 radiocarbon determinations span a period from 9500 to 500 yr B.P. and show, in conjunction with stratigraphic evidence, that there was a long period of nonoccupancy at the site between about 6000 and 3000 yr B.P. Shifts from level to level in the relative abundance of rodents of mesic and xeric habitat preference suggest greater effective moisture in the site's vicinity between 9500 and 6800 B.P. and decreased effective moisture since that time.

Archaeological evidence from Surprise Valley in northeastern California adds a further dimension to the picture for the northern Great Basin (O'Connell, 1975). Three sites show that human groups occupied large, substantial semisubterranean dwellings there between 6500 and 4500 yr B.P., after which these structures were abandoned for smaller, more lightly built brush shelters like those still in use by native people of the region in historic times. O'Connell suggests that an environmental change after 4500 yr B.P. may have reduced the abundance of local food resources enough to trigger this cultural change.

In summary, in the southern and western Great Basin, dated material assignable to the Altithermal interval is scarce. In the eastern Great Basin, Danger Cave, Hogup Cave, and Sudden Shelter were occupied throughout the time range of concern. In the northern Great Basin, the Connley Caves were abandoned between roughly 7000 and 5000 yr B.P., and Dirty Shame Rockshelter from roughly 6000 to 3000 yr B.P., and an adaptation to a sparser biota brought about by environmental change is postulated for Surprise Valley after 4500 yr B.P. Thus, the apparent effects of environmental change on human populations were varied according to both place and time. Obviously, local factors such as rain-shadow effects, drainage-basin size and pattern, and latitude and altitude all intervened to modify the effect of large-scale climatic change on specific local habitats. The arid climate of middle-postglacial times properly takes its place as simply one of many environmental factors conditioning the lives of Great Basin aboriginal people, rather than as the overriding, dominant, and catastrophic factor it was once pictured to be.

Lakeshores and Archaeology

Another topic of continuing interest is the relationship of human populations to the great lakes that exist or existed in the past throughout the intermontane desert region. Pleistocene Lake Bonneville in the eastern Great Basin (Gilbert, 1890) at its maximum reached a depth of over 300 m and covered an area of some 50,000 km². Modern Great Salt Lake, Utah Lake, and Sevier Lake in western Utah are its major remnants. Lake Lahontan was considerably shallower and smaller but still of vast extent (Russell, 1885). Pyramid Lake, Walker Lake, and Mono Lake in western Nevada/eastern California are its modern remnants. Other pluvial lake basins that have figured significantly in archaeological investigations will be identified in the following discussion.

Archaeologists have long recognized that the occurrence of artifactual remains on the ancient shorelines of now-vanished lakes offers possibilities for dating and cultural interpretation. Pioneering reconnaissance surveys in the southern California deserts by Rogers (1929, 1939) and the Campbells (Campbell and Campbell, 1935; Campbell et al., 1937) explored the possibilities of dating the open-surface sites that abounded there through the technique of horizontal stratigraphy. This approach proceeded from the knowledge that during the last glacial period many desert valleys were filled by great lakes that progressively dried after the end of glacial times. It could reasonably be expected that artifacts of the earliest people to live around such lakes would be found along the higher shorelines and that artifacts of the latest people would be found along the lower ones.

On this basis, Rogers sketched out an early Malpais and later Playa I and Playa II complexes, which he subsequently called San Dieguito I, II, and III (Rogers, 1939, 1958). The Campbells (Campbell et al., 1937) defined a Lake Mohave culture, which subsumed the three complexes named by Rogers, from the ancient shoreline of Pleistocene Lake Mohave. Lake Mohave was a freshwater lake; at its maximum, it covered an area of some 200 km² to a depth of up to 12 m. The lake maintained a high level during early postglacial times, sustained by runoff from the nearby mountains, but it subsequently dried completely (Ore and Warren, 1971). In a continuation of the earlier studies, Warren (1967) redefined the previously named assemblages as parts of a single San Dieguito complex, radiocarbon dated at the C. W. Harris site in San Diego at between 8500 and 9000 yr B.P. This complex, characterized by a variety of scraper types, leaf-shaped bifacial knives, large leaf-shaped and shouldered projectile points, crescent-shaped bifaces, and choppers, represented, Warren believed, an early hunting tradition that had existed in the Far West during the early postglacial waning of the pluvial lakes there. Not recognized as part of the complex are the manos and milling stones so characteristic of later cultures in the region.

The Pinto Basin remains, which follow the San Dieguito in time, are dated to a period of increased effective moisture after the drying of the pluvial lakes because Pinto Basin artifacts were found in abundance along the banks of a desert watercourse, now dry, which presumably held dependable water during the time of the human occupation (Campbell and Campbell, 1935). Wallace's (1962) estimates place the age of this complex between about 5000 and 2000 yr B.P. Lyneis (1982), however, believes that it probably emerged as early as 7000 yr B.P. and was succeeded by Gypsum and Elko periods after about 4000 yr B.P. Along with new types of flaked stone projectile points and other artifacts, manos and milling stones for the grinding of seeds appear in the Pinto Basin assemblage. This suggests that people were by then fully adapted to desertic conditions in which it was necessary to exploit food resources previously neglected. Seed-processing technology, involving both the mano and metate and the mortar and pestle, became increasingly abundant in remains from Gypsum and Elko times, and upland occupation is much more widely documented (Lyneis, 1982).

Sites of the succeeding Amargosa complex, which contain both large and small projectile points and occasional trade sherds of southwestern Anasazi pottery, as well as milling stones, are found on sandy terrain near water holes and playas where water can be obtained by digging. Most of these sites are said to have been small, apparently temporary camps, and they are loosely dated to the period 2000 to 1000 yr B.P. During the last half of this period, also, southwestern Anasazi horticulturists penetrated the lowland desert as far west as the Mojave Sink region, where they established small, scattered villages and farmed while engaging in turquoise mining and trade in marine shell beads from southern California.

The archaeological evidence of the past 1000 or so years in the southern California deserts seems quite clearly to represent the Yuman and Shoshonean peoples, who emerged into history there in the middle 1500s and who remain today. Archaeological sites are abundant throughout the area; pottery, small triangular arrowpoints, milling stones and manos, bedrock mortars, pestles, and a variety of flaked stone knives and scrapers attest to the peoples' use of a well-developed tool kit for the exploitation of a broad spectrum of desert plants and animals (Wallace, 1962). This period will be discussed further in the recounting of a fascinating recent study of lakeshore cultural adaptation in extreme southern California.

Besides the studies already mentioned, many others attempt to relate evidence of human occupation to the past stands of now-dry lakes. Pioneering work in the northern Great Basin sought to date early human occupation there by seeking correlations with the lake histories (Cressman, et al., 1940, 1942). Correlations were similarly sought for western Nevada by Carter (1958), for the Death Valley region by Clements and Clements (1953) and Hunt (1960), for Pleistocene Lake Manix in southern California by Simpson (1958, 1960), and for Pleistocene Lake Chapala in Baja California by Arnold (1957). (The much-publicized investigations at Calico Hills [Leakey et al., 1972], though an outgrowth of Simpson's concern with lake-level dating of specimens found along the strandline of Pleistocene Lake Manix, are not discussed here inasmuch as Haynes [1973] and Davall and Venner [1979] have raised convincing doubts about whether the Calico specimens are of human manufacture.)

There is no need to recount every attempt that has been made or to discuss all the frustrations that have vexed investigators carrying out the work. The sequence set out earlier for the southern California desert is illustrative. It is the product of many investigators' labor, and it enjoys a good measure of support. The residual ambiguities derive from the fact that, although surficial remains on a given beach terrace most probably date to the time when the lake washed that beach, they also could conceivably date to *any time after* the lake dropped below that terrace. There is the further complication that even relatively minor, short-term climatic shifts can cause lakes in closed basins to fluctuate in level, sometimes dramatically, so that the same beach terraces may have been occupied by the lake and by humans more than once in the past. These problems will not be easily resolved by further study of surface assemblages. Future excavation and radiometric dating of buried archaeological sites in strandline contexts will be necessary to refine the chronology of human occupation of lakeshore situations in this area.

It remains to be seen to what extent the excellent radiocarbon-dated sequences that have been developed for Pleistocene Lake Bonneville in the eastern Great Basin and for Pleistocene Lake Lahontan in the western Great Basin (Broecker and Kaufman, 1965; Gwynn, 1980; Morrison, 1965) may prove useful in understanding early cultural chronology. To date, archaeological study around the shores of these lakes has been confined almost wholly to their lower, more recent levels.

LAKESIDE CULTURAL ADAPTATIONS

Both archaeological and ethnographic evidence show that people of the desert West have long exploited waterside habitats. The evidence already reviewed bears equally on the question of lakeshore cultural adaptation. The San Dieguito complex, which is at least 9000 years old in southern California, is well represented along the shoreline of Pleistocene Lake Mohave. Points of the Clovis fluted type, dated from sites in Arizona and New Mexico at between 11,500 and 11,000 yr B.P. (Haynes, 1969), have been found on the same land surfaces (Campbell et al., 1937). A northern Great Basin analogue of the San Dieguito, called the Western Pluvial Lakes Tradition by Bedwell (1973), is dated to at least 11,000 yr B.P. and may be as old as 13,000 yr B.P. Clovis fluted points have also been found along low standlines of Pleistocene lakes in this region (Aikens, 1978). These are the earliest well-established human occupations known in the desert West. At the near end of the time scale, lacustrine adaptations flourished among the historic Klamath and Pyramid Lake Paiute of the northern Great Basin (Barrett, 1910; Loud and Harrington, 1929: 156-57; Spier, 1930), the Timpanogos or Utah Lake Utes of the eastern Great Basin (Warner, 1976: 60-61), and the Yokuts of the Tulare Lake region in southern California (Gayton, 1948).

Archaeological sites in open locations have left only a very thin record of what the earliest lakeshore occupations were like. Large stone points found on sites of the early San Dieguito complex or the Western Pluvial Lakes Tradition suggest the hunting of relatively large game animals more than the taking of fishes, small mammals, waterfowl, and many kinds of vegetable foods that are characteristic of marshlands. These earliest people may have been primarily hunters, as suggested by Warren (1967). Or it may be that definitive gathering implements such as bone or shell fishhooks, wooden fish spears, basketry traps, and spun-fiber fish nets and other indicators such as fish and bird bones and edible plant remains have long since decomposed in exposed open sites where they were once used and discarded.

Dry cave sites where organic materials are preserved offer a larger data base from which to approach these questions. Danger Cave and Hogup Cave in the Great Salt Lake region of the eastern Great Basin occur at the western and eastern edges, respectively, of a vast playa once flooded by the waters of a late, low stage of Pleistocene Lake Bonneville. Danger Cave was first occupied about 11,000 yr B.P. by people who built their campfires on the clean sand left in the cave after lake waters dropped below the level of its portal (Jennings, 1957). Hogup Cave, at a considerably higher elevation, had long been free of lake waters when people first occupied its rocky floor about 8400 yr B.P. (Aikens, 1970). Both caves yielded bones of water birds, ungulates, small rodents, rabbits, and hares in some abundance. Specimens of bulrush (*Scirpus*) and other water-loving plants were common, and the seeds and chaff of pickleweed (*Allenrolfea*), a low-growing halophyte characteristic of saline playa edges, were superabundant in the earlier levels. These remains show clearly that the lakeshore biota was being exploited, along with that of the nearby uplands, although no fishbones or fishing gear were recovered from either site.

It has been suggested that the early Archaic occupation in the Great Salt Lake region between 10,000 and 5500 yr B.P. was primarily adapted to the exploitation of lake-periphery ecosystems (Madsen and Berry, 1975). There is evidence of substantial upland settlement in the region only after about 5500 yr B.P., by which time there may have been an adaptive shift from a relatively specialized and semisedentary lakeshore-adaptive system to one that involved a broader range of exploitation in both lake-periphery and upland settings. At Hogup Cave, bones of water birds and chaff of pickleweed ceased to be deposited in significant quantity after about 3000 yr B.P.; a sediment core from a nearby playa-edge spring indicates that rising lake waters drowned the marshland resource base at about the same period. Available survey and excavation data from a variety of sources show that sites in the uplands became increasingly common in later times. The later strata of Hogup Cave in particular contained evidence of plants and animals native to distant upland locations. The postulated change to a more extensive and less sedentary pattern of settlement may represent the inception in this area of the cyclical, broad-ranging annual round that was characteristic of many Great Basin societies in historical times. (As part of this same scenario, it has been suggested that the rising waters of the neoglacial climatic interval in the Great Salt Lake Basin drowned many of the rich and productive lakeshore marshes after about 3500 yr B.P., leading to a total abandonment of the region by Archaic peoples by 2500 yr B.P., with the area then remaining unoccupied until Fremont culture farmers entered it about 1500 years ago [Madsen and Berry, 1975]. Additional evidence is needed, but the interpretation offered for the period 10,000 to 5500 yr B.P. is plausible and is reasonably congruent with such data as are currently available. [see also Green, 1972, and Sims, 1977.] The hypothesis of a total abandonment after 2500 yr B.P., however, is neither well supported nor plausible, and it is unlikely to become an accepted part of our understanding of the prehistory of the eastern Great Basin [Aikens, 1976].)

The richest picture of lakeside cultural adaptation in the Great Basin is afforded by Lovelock Cave, which overlooks the Humboldt-Carson Sink at the terminus of the Humboldt River in western Nevada (Loud and Harrington, 1929). The Lovelock culture manifested at this site has not been closely dated, but radiocarbon dates obtained long after the primary excavations indicate human activity there from as early as 4500 to as late as 150 yr B.P., with the most intensive occupation between about 3500 and 1500 yr B.P. (Heizer and Napton, 1970: 37-44).

Fishbones were found in all parts of the cave. Bone fishhooks, a fishlike effigy carved of horn, a possible model of a fish trap, and nets perhaps used in fishing show the importance of this food resource at the site. Remarkably lifelike duck decoys made of tule, some of them painted, some covered with fully feathered skins removed whole from real animals, were part of an assemblage that also included the heads and necks of water birds stuffed with tule; a possible wooden effigy of a water bird; loose feathers or bundles of feathers from pelicans (*Pelicanus*), herons (*Ardea*), ducks (*Anas*, *Aythya*), and geese (*Anser*); and the remains of nets that might have been used for taking low-flying water birds. These varied finds clearly indicate the importance of avian marshland species to the occupants of Lovelock Cave. Basketry, matting, and heavy cord made of marsh plants such as bulrush (*Scirpus*) and cattail (*Typha*) were relatively abundant. Very common were chewed masses of plant fiber, the inedible remnants of the shoots and rhizomes of bulrushes that had been masticated and then discarded (Loud and Harrington, 1929).

Analysis of some 300 human coprolites from Lovelock Cave indicated that over 90% of the foods represented were obtained from the lacustrine biota of the adjacent Humboldt-Carson Sink. Seeds of the aquatic bulrush and cattail and of the wetland grasses *Elymus* and *Panicum* made up the bulk of the food remains found in the coprolites. Fish remains, especially bones of the Tui chub (*Gila bicolor*), were common, and bones of ducks and mudhens (*Fulica americana*) were also identified (Heizer and Napton, 1970: 107-9).

An essentially sedentary occupation of the locality for most of the annual round is suggested; many of the foods identified from the coprolites, harvestable in the autumn months, were suitable for storage as winter staples. Wild fowl could have been obtained during the winter to supplement stored foods, as could the rhizomes of water plants such as *Scirpus* and *Typha* (Heizer and Napton, 1970).

Other archaeological sites in western Nevada have produced related evidence, most notably Humboldt Cave (Heizer and Krieger, 1956) and a series of sites near Pyramid Lake not yet reported in detail (D. R. Tuohy, personal communication, 1980).

The date of the inception of this lacustrine tradition in the western Great Basin is not clear. It is believed that the Lovelock Cave complex reflects a reestablishment in the Humboldt-Carson Sink area of a productive marshland environment after middle-postglacial times but that the cultural tradition of lacustrine adaptation may well be much older. This possibility is hinted at in Fishbone Cave (Orr, 1956) and Leonard Rockshelter (Heizer, 1951), both sites that overlook lakebeds and contain evidence of occupation perhaps as old as 11,000 years. A comparable recent tradition of lacustrine adaptation was maintained in historical times in the same region by the Northern Paiute, who exploited a major fishery at Pyramid Lake, about 90 km west of the Lovelock Cave area. The art of fashioning duck decoys of tule and covering them with feathered skins is still known to the modern native inhabitants (Wheat, 1967).

A recent study of late prehistoric human ecology at Lake Cahuilla in the Colorado Desert of extreme southern California offers important new insights into the adaptive and demographic effects of the sudden collapse of a flourishing lacustrine cultural specialization (Wilke, 1978). The Lake Cahuilla Basin, which encompasses the Coachella Valley, the Salton Sea, and the Imperial Valley, has many times in the geologic past been filled by a great freshwater lake about 185 km long and up to 95 m deep. Archaeological data and radiocarbon determinations indicate one major lacustral interval, which may have included one or more lake stands, between about 1100 and 800 years ago and a second interval between about 700 and 500 years ago.

Remarkably, the historical Cahuilla who occupy the region today have preserved an oral tradition detailing clearly the favorable conditions of existence around the ancient lake, said to have been extant "as far back as the lives of four or five very old men." The tales relate the abundance of fish and waterfowl then available. Archaeological middens rich in fishbones and other remains, scattered mile after mile along the old lakeshore, as well as hundreds of stone fishing weirs, attest to the accuracy of the traditional accounts. Remains of seeds and bones in human coprolites from such sites also demonstrate the consumption of aquatic plants and animals (Aschmann, 1959; Wilke, 1978; Wilke and Lawton, 1976). The numerous sites dated within the last few hundred years either by radiocarbon determinations or by their temporally diagnostic pottery, shell beads, and projectile points, suggest the existence of a very considerable population along the now-barren lakeshore during the last time that the waters stood at that level. This prosperity, however, was at the mercy of the natural mechanisms that controlled the existence of the lake itself, and environmental and archaeological evidence documents a major change just prior to the dawn of historical times in the region.

The primary mechanism controlling the rise and fall of Lake Cahuilla was not climatic but geologic and hydrographic. The Salton Trough in which the lake lay is a northward continuation of the great tectonic rift that forms the Gulf of California. The Colorado River, which before recent damming carried an enormous load of silt, has over geologic time built up a vast delta across the seaward end of the trough at the point where the river enters the Gulf of California. The river constantly changes its channel in this delta area as its natural levees are flooded by the huge spring freshet carried down from the river's Rocky Mountain headwaters. On numerous occasions in the Pleistocene and Holocene past, Colorado River waters have through this process been diverted across the northern edge of the delta and the entire flow of the great stream has been captured by the Salton Trough.

When this happened, the basin quickly filled and formed Lake Cahuilla (or Lake LeConte or the Blake Sea, as it has variously been called), and, once the basin was filled, outflow back across the southwestern edge of the delta returned the excess Colorado River water to the sea. Calculations based on the historical annual flow volume of the Colorado River and on present-day evaporation rates for the Salton Sea indicate that the excess of inflow over evaporation would have maintained a great freshwater lake with a stable beach level along which an abundant shoreline biota would have quickly become established. For extended periods, apparently hundreds of years long, the lake remained in this condition. But eventually the Colorado silted in the area of its debouchment into the still waters of the lake, finally raising the gradient enough to cause a return of the river to its old distributary system across the southern slope of the delta and directly into the Gulf of California.

When it was no longer fed by the Colorado River, Lake Cahuilla dried up rapidly in a hyperarid setting, where the modern annual rate of evaporation has been calculated to be 1.8 m. The dramatic effect of this change on the human population, as inferred from biologic and cultural evidence, is lucidly summarized by Wilke (1978: 129).

> When the Colorado River was diverted once again to flow directly into the Gulf of California, which is believed to have occurred within 50 years of AD 1500, the entire complex of aquatic plants and animals disappeared within little more than 25 years, and the lake itself was gone by evaporation within about 55 years. The marsh vegetation and shellfish vanished within the first few years of the drying episode. Fishing continued at least on a seasonal basis for about 25 years, as documented by weirs constructed annually along the receding shorelines and by fishbones from house ruins at the same locality. But there is good reason to believe that within the first few years of the drying of the lake significant population movements into the nearby uplands occurred. Late prehistoric changes in settlement patterns appear to be documented in archaeological literature on inland southern California. These changes are seen as resulting in part from the decline of aquatic conditions in the Salton Basin. Available evidence suggests that the drying of Lake Cahuilla had grave consequences across much of southeastern California.

The abrupt, enforced shift from an adaptation based on the relative abundance of a great freshwater lake ecosystem to a way of life based on the much sparser and more scattered resources of a desert region in which the lake suddenly no longer existed would have required the rather large Lake Cahuilla population to adopt a more dispersed pattern of existence, and it may have brought about population shifts among neighboring groups as well. Archaeological evidence indicates a major increase in the abundance of small, ephemeral sites over much of southern California in late prehistoric times, and it is believed on both archaeological and linguistic grounds that a migration of the Luiseño and Diegueño peoples from the desert interior to the southern California coast took place within the last 1000 years or less (Meighan, 1954; True et al., 1974; Wallace, 1962). It has even been suggested that the pattern of endemic intertribal hostility for which the historical peoples of the lower Colorado River Delta are famous may have developed in resistance to heavy pressures put on the occupants of that area by populations displaced after the collapse of the Lake Cahuilla economic base (Aschmann, 1959).

The Lake Cahuilla phenomenon is perhaps the clearest and best-documented example to be found in the archaeology of the western United States of the effect of an environmental change on human affairs. The case brought together by Wilke (1978) invites further investigation and opens up the intriguing possibility that earlier lake cycles may be linked with other important prehistoric cultural events. For example, it might be speculated that the great northward expansion of Numic-speaking peoples, who are believed on the basis of linguistic evidence (Fowler, 1972; Lamb, 1958; Miller, 1966) to have dispersed from a homeland in the southern California desert within the last 1000 years or so, may have been given initial impetus by a similar population upheaval associated with the end of an earlier lacustral interval at Lake Cahuilla. (Any attempt to develop this notion into a coherent theory could not, of course, regard such an impetus as sufficient in itself to have propelled the Numic speakers all the way across the Great Basin like an arrow shot from a bow. But such an event might have catalyzed a reaction subsequently sustained by ecologic and social factors. Any exploration of these possibilities would also have to take into serious account a recent linguistic discussion by Goss [1977], in which he expresses grave doubts about the reality of the Numic expansion itself.)

Biogeography and Human Settlement

Several archaeological studies executed over the past decade in the Far West focus primarily on the geographic distribution of basic food resources within given human habitats and on the implications of these distributions for human occupation. In most cases, the modern environment has been used as a model for the past environments without significant recourse to paleoecologic reconstruction, so the proferred conclusions require further testing. Nevertheless, the approach has advanced our understanding of the prehistoric cultures concerned and of questions to be pursued in further investigations of such cases.

For a long time, the conventional wisdom in Great Basin studies has been that the hunting-gathering societies of the region were compelled, by the exigencies of adaptation to a land of sparse and scattered resources, to follow a lifeway of cyclical nomadism in small, highly mobile, unsettled groups (Aikens, 1970; Jennings, 1957; Steward, 1938). The most important single piece of research leading to a reassessment of that wisdom is a study by O'Connell (1975) of the archaeology of Surprise Valley, in northeastern California. Of note in the present context is the finding that between about 6500 and 4500 yr B.P. there existed in Surprise Valley small settlements of substantial semisubterranean earth lodges; this finding implies a greater degree of sedentariness than was believed by most archaeologists to be characteristic of the Great Basin Desert Culture. This discovery, however, is readily accounted for by the simple cartographic demonstration that in the particular geographic setting of Surprise

Valley, flanked by the high Warner Mountains on the west and the lower Hays Canyon Range on the east, the most important biotic resource zones were crowded together so closely that all could be exploited by a sedentary population ranging out over short distances from its settled villages on the valley floor. The mapping of lowland occupation sites, lowland temporary camps, and upland temporary camps in relation to modern biotic communities shows that all the regularly exploited resource zones of the area lay within 10 to 15 km (2 to 3 hours' walk) of the central settlements (O'Connell, 1975: Figures 2, 3, 5).

The archaeological record further shows that after about 4500 yr B.P. and continuing down to recent times the large earth-lodge structures passed out of use, to be replaced by smaller, more lightly made and apparently temporary brush structures. Despite this change, the basic general pattern of settlement and subsistence in Surprise Valley, as established by site-type distributions in relation to biotic zones, apparently remained the same as that of the earlier period. O'Connell also approaches this problem in biogeographic terms with a hypothesis suggesting that the shift to an apparently less sedentary residential pattern might have resulted from climatic change. Reduced effective moisture, he suggests, may have led to a reduction in the abundance and reliability of seed crops, with a concomitant effect on the numbers of ungulates and wild fowl. Under such circumstances, human groups may have shifted to a more mobile residential pattern, abandoning their large, substantial earth lodges in favor of more lightly and quickly built brush structures.

This hypothesis, plausible as it is, requires further testing on the basis of locally derived environmental data. It must be noted that the period of earth-lodge occupation, 6500 to 4500 yr B.P., was a time of relatively lower effective moisture throughout the West but the subsequent period, during which lighter structures were occupied, was one when effective moisture conditions seem generally to have improved somewhat or at least to have become no worse. This climatic sequence is the opposite of what O'Connell postulates for Surprise Valley. Perhaps it could be argued, as a partial reversal of his interpretation, that reduced effective moisture during the interval between 6500 and 4500 yr B.P. restricted the local availability of foods in economically harvestable quantities to the well-watered zone in the immediate vicinity of Surprise Valley. The observed sedentariness might then have been a response to the concentration of population through economic necessity. With the amelioration of the climate after 4500 yr B.P., preferred foods may have become available in exploitable quantities more broadly throughout the region, attracting people to a more mobile and wide-ranging lifeway and inducing them to give up their sedentary villages.

Further paleoecologic and experimental field work must be done to evaluate these alternatives. The problem is an intriguing one, with readily testable biogeographic and archaeological ramifications.

The degree to which Great Basin aborigines exploited the seeds of the Pinyon pine (*Pinus edulis*, *P. monophylla*) for food in prehistoric times has recently become a point of contention. Pinyon pine forests now cover vast areas in the intermediate elevations of Great Basin mountain ranges, extending from a northern limit in northern Nevada and Utah to well south of the international border with Mexico. Ethnographic evidence from the central Great Basin documents clearly the overriding importance of the autumn Pinyon harvest in providing an abundant, concentrated, and storable food resource to carry the historical people there through the winter and early spring (Steward, 1938). It has been widely assumed by archaeologists that this reliance on the Pinyon pine is ancient (Jennings, 1957; Thomas, 1974), but recent findings challenge this assumption.

Pollen evidence and evidence from other sources summarized by Madsen (1982), Madsen and Berry (1975), and Thomas (1982) now suggest that the modern abundance and widespread distribution of the Pinyon pine in the desert West may have been achieved only in later postglacial times, between 7000 and 3000 yr B.P., depending on the area. Thomas maintains, however, that Pinyon pine was sufficiently available throughout the period of archaeological record in the central Great Basin to have supported a pattern of exploitation there comparable to that known for the region in historical times.

A further complication of the problem is that, although it seems more than reasonable that such an important food resource would have been exploited by native peoples if it were available, a recent archaeological study of Owens Valley, in the southwestern corner of the Great Basin, shows that few harvesting camps discovered in the modern Pinyon forest zone are dated earlier than about 1400 B.P. and that the greatest number are dated within the last 700 or so years (Bettinger, 1977).

The resolution of this problem is obviously some distance away; until more data can be generated on the dated distribution of the Pinyon pine forest and until more archaeological contexts have been examined where Pinyon remains, if once present, might reasonably be expected to have been preserved, disagreement will be possible. The issue is mentioned here because, although proposed interpretations are inconclusive, the problem is one fundamental to an understanding of prehistory over a broad area of the West and undoubtedly more will be heard of it.

A comprehensive environmental perspective has been applied to archaeological research carried out over a number of years in the northern Rocky Mountains-Snake River Plain region of Idaho by Swanson (1962a, 1972). The broadest objective of the research program was to demonstrate that culture and environment are continuous or, in other words, that specific cultures can be seen as direct reflections of specific environments. If this were to be granted, it could be concluded that present environments permit the recognition of past cultures that were adapted to them and that the origin of cultural patterns is analogous to the origin of biogeographic patterns.

In Swanson's specific case, it was argued that just as the flora and fauna of the Great Basin province are simplified marginal derivatives of the richer and more complex Rocky Mountain ecologic system (Daubenmire, 1943) so are the simple cultures of the Great Basin marginal derivatives of the richer and more complex Northern Shoshoni culture of the Rocky Mountains. This was offered not simply as an analogy but as a statement of the historical, structural, and mechanical relationships between natural and cultural systems. Variations in the size and structure of human populations should correspond to variations in the natural setting. The effect of climatic change over time on biotic and cultural patterns was also part of the concept.

Swanson postulated a close relationship between patterning in the biota of the northern Rocky Mountains and patterning in the culture of the Northern Shoshoni who historically inhabited the region. Archaeological research then sought to show that this same culture-environment relationship existed throughout most of postglacial time. Proposed as archaeologically testable corollaries of this interpretive model were (1) that archaeological sequences displaying a continuum of intergrading traits over a long period of time are to be expected in areas where the environment has remained essentially

stable and (2) that archaeological sequences displaying a discontinuous series of discrete artifact assemblages are to be expected where intercultural relationships existed, as in cases where climatic changes have caused fluctuations in the boundaries of the biotic zones to which different human societies were adapted.

Excavations and surveys in Birch Creek, a long valley basin extending northward into the Rocky Mountains from the edge of the Snake River Plain, produced what Swanson saw as an essentially continuous cultural sequence that reached back some 8500 years into the past. At lower altitudes and toward the south, archaeological sequences were seen to be increasingly discontinuous. These continuities and discontinuities, and other variations in occupational intensity over time, seemed to correlate with environmental changes inferred from sedimentary sequences and to support Swanson's concept of culture-environment relationships and Northern Shoshoni prehistory.

Swanson's work has been controversial. His claim that the Northern Shoshoni's ancient homeland was in the northern Rocky Mountains was a minority opinion. Many feel that it is refuted by linguistic and archaeological evidence for a very recent south-to-north expansion of a series of Numic-speaking groups, of which Northern Shoshoni was one, out of the southern Great Basin deserts within the last 1000 or so years (Butler, 1978; Fowler, 1972; Lamb, 1958; Madsen, 1975; Miller, 1966), but that interpretation too is controversial (Goss, 1977; Heizer and Napton, 1970: 11). However the questions of ethnic and linguistic continuity in the northern Rocky Mountains may be finally resolved, Swanson did offer a theoretically interesting, cogent, and emphatic statement of the structural relationships between environment and culture as an adaptive system. Perhaps the theory cannot be applied as elegantly as Swanson was inclined to apply it, to the particularly difficult problem of tracing ethnic and linguistic identity, but it remains worthy of further examination as an attempt to understand prehistoric human ecology, and it continues to influence archaeological research in the region.

Another valuable perspective on the conditions of human existence in the desert West has been offered by an approach to the question of the carrying capacity of the sagebrush-grass steppe biome (Kuchler, 1964), which exists over much of southwestern Wyoming, southern Idaho, eastern Oregon and Washington, and northwestern Nevada (Butler, 1972, 1976). In historical times, much of this area was occupied by the Northern Shoshoni and their linguistic relatives, broad-spectrum hunters and gatherers who depended on wild seeds and roots and a wide variety of small and large game for subsistence. Farther back in time, there is evidence of an emphasis on bison hunting; and earlier than that, perhaps on elephant hunting. At Owl Cave, on the eastern Snake River Plain, a dense stratum of bison bone dated at about 8000 yr B.P. indicates a mass kill, while at a lower level elephant bones found with fragmentary fluted points and other artifacts are dated at about 11,000 yr B.P. This and other evidence is summarized by Butler (1978).

At Owl Cave, aptly named as a site where the winged predators have roosted throughout most of Holocene time, thousands upon thousands of bones of small mammals accumulated in an orderly stratigraphic sequence as the cave was gradually filled with nearly 4 m of windblown sediment (Guilday, 1969). Graphs based on the radiocarbon-dated stratigraphic distributions of over 7000 individual animals (the numbers of lower jaws were tabulated) show very high numbers of all species at levels dating from about 9000 and 8000 yr B.P., followed by a major decline around 7000 yr B.P. to a low level that was thereafter maintained until the record effectively ended at about 1500 yr B.P. This decline in the abundance of small-mammal remains, as well as species changes, has been related to middle-postglacial climatic change by Butler (1978: 38).

> Before 7200 years ago, the cool, moisture-loving grass-associated northern pocket gopher [*Thomomys talpoides*] was the dominant form; after 7200 years ago, the dry-tolerant, sagebrush-associated pygmy rabbit [*Sylvilagus idahoensis*] became the dominant form along with the dry-adapted ground squirrel [*Citellus townsendii*]. Evidently there was a quantitative change in the plant cover, with all of the plants becoming sparser, but with the grasses and forbs diminishing to a greater extent and more rapidly than the sagebrush. Such changes in the vegetation cover would have had an enormous impact on the large grazing and browsing animals inhabiting the upper Snake and Salmon River Country. They also imply a significant change in the climate of the region.

This shift in species frequency and general abundance of small mammals at Owl Cave is taken as a sign of a persistent lowering of the general carrying capacity of the sagebrush-steppe biome caused by climatic change after about 7000 years ago, a shift that should presumably have affected the economy and thereby the social organization and general density and dispersion of the human population throughout the area where this biotic association occurred (Butler, 1972). In the cultural history of southern Idaho, a transition from a big-game-hunting tradition to an Archaic tradition is placed at about this time, but data on dietary remains adequate to test the concept of a significant economic shift have not yet been recovered from archaeological contexts (Butler, 1978).

The concept of carrying capacity has also figured significantly, if implicitly, in discussions of Columbia Plateau prehistory. The Interior Plateau of the northwestern United States was in ethnographic times dominated by semisedentary riverine cultures that relied heavily for their subsistence on the anadromous salmon, which ran seasonally throughout the Columbia River system from early summer to late autumn. Excavations at The Dalles of the Columbia River, near the point where the river breaks through the Cascade Range on its way to the sea, have established the existence of native salmon fishing there as early as 10,000 years ago (Cressman et al., 1960). It has been suggested, however, that the well-developed salmon-fishing riverine way of life is relatively new to the Columbia system upstream of The Dalles. Swanson (1962b), for example, perceives a fully formed Plateau tradition as having emerged only about 700 years ago.

Attempts to account for the relative paucity of evidence for a similar lifeway along the river in early times postulate a dependence on primarily upland resources by the native people that was forced upon them by environmental factors limiting the availability of salmon. Sanger (1967) calls attention to the possible effects of the great Cascade landslide, which about 700 years ago poured such a volume of debris into the Columbia Gorge below The Dalles that the river level above the slide was raised by approximately 100 m. It has been speculated that, before this event, Celilo Falls at The Dalles, one of the most important aboriginal fisheries in the Northwest in historical times, had been too high to allow significant numbers of salmon to pass upriver. But, with the downstream damming and subsequent raising of the river level, the relative height of the falls was reduced enough to allow salmon to pass over it in numbers, providing the economic base necessary for the emergence of the Plateau culture in a mature form.

Subsequent archaeological research has established a date of at least 2500 B.P. for the widespread occurrence above The Dalles of riverside earth-lodge villages economically based on salmon exploitation, rendering moot Sanger's earlier speculations about the effect of the Cascade landslide (Leonhardy and Rice, 1970). The problem of the paucity of earlier evidence remains; it has simply been pushed farther back in time. Brauner (1976: 307-10) has proposed another environmental explanation, with appropriate new dates, suggesting that a surge in effective precipitation known to have affected the southern Columbia Plateau region between about 4000 and 3500 yr B.P. drastically reduced salmon populations for a long period by giving rise to rapid stream downcutting, which destroyed spawning beds either through erosion or siltation. Volcanic ash, eroded year after year out of side canyons to become suspended in the river waters, might have further limited the fish populations, killing many through the mechanical destruction of their gills by the volcanic glass shards in the ash.

It remains to be seen whether interpretations linking a lack of evidence for early salmon-fishing villages with postulated environmental constraints on early salmon populations can be successfully defended. Nelson (1973) has advocated, as a cultural-historical explanation, a late diffusionary spread of the pattern from a source area farther north. Schalk's (1977) analysis of salmon habitat requirements and spawning behavior implies that salmon fishing may simply not have become economically worthwhile in the upper reaches of the Columbia River system until the regional population had grown to the point that it put severe pressure on terrestrial resources. Yet another viewpoint is that the paucity of early winter village remains along the Columbia may simply be a result of removal or deep burial of the earlier record by the awesome floods to which the Columbia River is subject, with its huge drainage basin heading in the well-watered northern Rocky Mountains. Hammatt (1976) has summarized geologic evidence indicating active alluvial erosion and deposition along the Columbia and its tributary the Snake during the period 5000 to 2500 yr B.P. This is clearly a line of inquiry that will repay further research.

Volcanism

The fate of Pompeii and Herculaneum has long illustrated the power of volcanism to affect human affairs, and there is no dearth of additional examples (Sheets and Grayson, 1979). The cataclysmal eruption of Mount Mazama in Oregon, radiocarbon dated at shortly before 7000 yr B.P., and the eruption of Sunset Crater in Arizona, dendrochronologically dated to slightly more than 900 years ago (A.D. 1065-1066), are two volcanic events that have captured the imagination of prehistorians in the western United States.

The Mount Mazama eruption, which left behind the great Crater Lake caldera of the southern Oregon Cascades, cast some 30 km³ of pyroclastic material into the air. This settled to earth over a vast arc extending northward beyond the Canadian border, eastward at least as far as the Rocky and Wasatch Mountains in Montana and Utah, and southward into central Nevada (Lemke et al., 1975; Williams, 1942; Williams and Goles, 1968). Malde (1964a) has speculated that this eruption had disastrous effects on the human population, reducing the plants and animals on which people depended for food and forcing migration out of the areas most heavily affected.

Mazama ash occurs as a distinct stratum in geologic, archaeological, and paleoecologic sites throughout the Northwest, where it has served as a valuable time marker (Kittleman, 1973;

Wilcox, 1965). In the Fort Rock Valley, about 90 km northeast of Crater Lake, a stratum of Mazama ash 15 to 20 cm thick occurs at the Connley Caves, which provided a rich archaeological and faunal record for the period approximately 11,000 to 3000 yr B.P. The possibility of local catastrophic effects caused by the ash fall was endorsed by Bedwell, who followed the lead of Wilcox (1959) and Malde (1964a), but he was unable to document any concrete evidence of the imagined effects on the human population from the cultural record of the site (Bedwell, 1973: 61-68).

The problem was addressed again by Grayson (1979), whose detailed analysis of the small-mammal and bird remains from Connley Caves indicated a reduction in the amount of standing water near the caves after about 7000 yr B.P., when the ash fell, and a shift from a relatively mesic local flora dominated by herbaceous vegetation to a relatively xeric one dominated by shrubby vegetation. This shift could not, however, be directly linked to the ash fall. In fact, since it coincided with comparable shifts toward less effective moisture that are known from a number of distant localities, the change seems better explained as a result of widespread middle-postglacial climatic trends than as an effect of the Mount Mazama ash fall.

The eruption of Sunset Crater, near modern Flagstaff in central Arizona, has been seen as highly beneficial to the Puebloan occupants of that area during the 11th and 12th centuries A.D. It is theorized that the volcanic cinders thrown out by the eruption in A.D. 1065-1066 served as a moisture-retaining mulch over hundreds of square kilometers that allowed the native people to farm extensive new areas where horticulture had not been possible before. This effect, theoretically, attracted a substantial influx of people from the surrounding regions who remained until soil depletion and wind deflation of the volcanic mulch reduced the agricultural potential of the area and caused an abrupt decline in population near the end of the 12th century (Colton, 1932, 1960).

This interpretation, favored as a classic case by a generation of prehistorians, has been challenged by recent archaeological and paleoenvironmental research. Colton had estimated a 1100% increase in posteruptive habitation sites in the region, but new data from more extensive and systematic archaeological surveys suggest that the actual increase amounted only to about 19% (Pilles, 1979). Pollen studies, plant macrofossil remains, and tree-growth records show that the observed curve of human population growth during the 11th century and the decline during the 12th century are correlated with a trend first toward and then away from greater warmth and effective moisture (Hevly et al., 1979). Such a change might well have been sufficient in itself, particularly in the high, cool Flagstaff region, first to encourage and then to discourage agricultural populations. Indeed, it has been pointed out that a similar curve of rapid initial population growth followed by an equally rapid decline was common in regions of higher elevation throughout the northern part of the Southwest during this period (Euler et al., 1979).

There is a great deal of scientific excitement over the current eruptive phase of Mount St. Helens, in southwestern Washington, and many studies of diverse kinds are underway. An early study by Cook and others (1981) indicates that there have been significant effects from volcanic ash fall on insect populations and agricultural plants up to 400 km downwind of the vent, but it will be some time before the long-term effects on the human food chain can be fully understood. When the Mount St. Helens studies are completed, they may provide evidence for a fuller archaeological appreciation of such events as those at Mount Mazama and Sunset Crater. Until then, the human

importance of those fascinating ancient eruptions will remain speculative.

Conclusion

The foregoing accounts do not yet add up to a connected cultural-environmental prehistory of the western United States, but they do offer a series of vignettes that adumbrate the full picture to be developed by future research, and they illustrate the approaches through which this research may be forwarded. Comparison of this chapter with a review commissioned for the Seventh Congress of the International Association for Quaternary Research (Baumhoff and Heizer, 1965) reveals a substantial recent advance in the number and quality of examples of research available for correlating environmental and cultural phenomena.

One topic of potentially great importance to human ecology not yet addressed adequately by environmental archaeologists is fire. Students of Quaternary studies and of the modern biota have long been aware of the important environmental implications of fire, but archaeological attempts to take advantage of this understanding have lagged behind. One ethnologic account, Lewis's (1973) "Patterns of Indian Burning in California," offers a fascinating glimpse into the potential of such studies, which surely could contribute much to our understanding of the human occupation of the forested regions in the western United States, especially in the dense woods of the Pacific Northwest coastal zone.

References

Aikens, C. M. (1970). "Hogup Cave." University of Utah Anthropological Papers 93.

Aikens, C. M. (1976). Cultural hiatus in the eastern Great Basin? *American Antiquity* 41, 543-50.

Aikens, C. M. (1978). The Far West. *In* "Ancient Native Americans" (J. D. Jennings, ed.), pp. 130-81. Freeman, New York.

Aikens, C. M., Cole, D. L., and Stuckenrath, R. (1977) "Excavations at Dirty Shame Rockshelter, Southeastern Oregon." Tebiwa Miscellaneous Papers of the Idaho State Museum of Natural History 4.

Antevs, E. (1948). Climatic changes and pre-white man. *In* "The Great Basin, with Emphasis on Glacial and Post-glacial Times." Bulletin of the University of Utah 38 (20), Biological Series 10 (7), pp. 168-91.

Antevs, E. (1955). Geologic-climatic dating in the West. *American Antiquity* 20, 317-35.

Arnold, B. A. (1957). "Late Pleistocene and Recent Changes in Land Forms, Climate, and Archaeology in Central Baja California." University of California Publications in Geography 10 (4), pp. 201-318.

Aschmann, H. (1958). Great Basin climates in relation to human occupance. *University of California Archaeological Survey Reports* 47, 23-40.

Aschmann, H. (1959). The evolution of a wild landscape and its persistence in southern California. *Annals of the Association of American Geographers* 49 (3, part 2), 34-56.

Barrett, S. A. (1910). "The Material Culture of the Klamath Lake and Modoc Indians of Northeastern California and Southern Oregon." University of California Publications in American Archaeology and Ethnology 5 (4), pp. 239-92.

Baumhoff, M. A., and Heizer, R. (1965). Postglacial climate and archaeology in the desert West. *In* "The Quaternary of the United States" (H. E. Wright, Jr., and D. Frey, eds.), pp. 697-707. Princeton University Press, Princeton, N. J.

Bedwell, S. F. (1973). "Fort Rock Basin Prehistory and Environment." University of Oregon Books, Eugene.

Bedwell, S. F. and Cressman, L. S. (1971). Fort Rock report: Prehistory and environment of the pluvial Fort Rock Lake area of south-central Oregon. *In* "Great Basin

Anthropological Conference 1970: Selected Papers'' (C. M. Aikens, ed.), pp. 1-25. University of Oregon Anthropological Papers 1.

Bettinger, R. L. (1977). Aboriginal human ecology in Owens Valley: Prehistoric change in the Great Basin. *American Antiquity* 42, 3-17.

Brauner, D. R. (1976). Alpowai: The culture history of the Alpowa locality. Ph.D. dissertation, Washington State University, Pullman.

Broecker, W. S., and A. Kaufman, (1965). Radiocarbon chronology of Lake Lahontan and Lake Bonneville II, Great Basin. *Geological Society of America Bulletin* 76, 537-66.

Bryan, A. L., and Gruhn, R. A. (1964). Problems relating to the Neothermal climate sequence. *American Antiquity* 29, 307-15.

Butler, B. R. (1972). The Holocene in the desert West and its cultural significance. *In* ''Great Basin Cultural Ecology: A Symposium'' (D. D. Fowler, ed.), pp. 5-12. Desert Research Institute Publications in the Social Sciences 8.

Butler, B. R. (1976). The evolution of the modern sagebrush-grass steppe biome on the eastern Snake River Plain. *In* ''Holocene Environmental Change in the Great Basin'' (R. Elston, ed.), pp. 2-39. Nevada Archaeological Survey Research Paper 6.

Butler, B. R. (1978). ''A Guide to Understanding Idaho Archaeology: The Upper Snake and Salmon River Country.'' Special Publication of the Idaho Museum of Natural History.

Campbell, E. W. C., Campbell, W., Antevs, E., Amsden, C. A., Barbieri, J. A., and Bode, F. D. (1934). ''The Archaeology of Pleistocene Lake Mojave: A Symposium.'' Southwest Museum Papers 11.

Campbell, E. W. C., and Campbell, W. H. (1935). ''The Pinto Basin Site.'' Southwest Museum Papers 9.

Carter, G. F. (1958). Archaeology in the Reno area in relation to the age of man and the culture sequence in America. *Proceedings of the American Philosophical Society* 102, 174-92.

Clements, T., and Clements, L. (1953). Evidence of Pleistocene man in Death Valley, California. *Bulletin of the Geological Society of America* 64, 1189-204.

Colton, H. S. (1932). Sunset Crater: The effect of a volcanic eruption on an ancient Pueblo people. *Geographical Review* 22, 582-90.

Colton, H. S. (1960). ''Black Sand: Prehistory in Northern Arizona.'' University of New Mexico Press, Albuquerque.

Colton, H. S. (1965). Experiments in raising corn in the Sunset Crater ashfall area east of Flagstaff, Arizona. *Plateau* 37 (3), 77-79.

Cook, R. J., Barron, J. C., Papendick, R. I., and Williams, G. J., III (1981). Impact on agriculture of the Mount St. Helens eruptions. *Science* 211, 16-22.

Cressman, L. S., Baker, F. C., Conger, P. S., Hansen, H. P., and Heizer, R. E. (1942). ''Archaeological Researches in the Northern Great Basin.'' Carnegie Institution of Washington Publication 538.

Cressman, L. S., Cole, D. L., Davis, W. A., Newman, T. M., and Scheans, D. J. (1960). Cultural sequences at The Dalles, Oregon: A contribution to Pacific Northwest prehistory. *Transactions of the American Philosophical Society* 50, 1-108.

Cressman, L. S., Williams, H., and Krieger, A. D. (1940) ''Early Man in Oregon.'' University of Oregon Monographs, Studies in Anthropology, 3, pp. 1-78.

Daubenmire, R. F. (1943). Vegetation zonation in the Rocky Mountains. *Botanical Review* 9, 325-93.

Davall, J. D., and Venner, W. T. (1979). A statistical analysis of the lithics from the Calico site (SBCM 1500A), California. *Journal of Field Archaeology* 6, 455-62.

Euler, R. C., Gumerman, G. J., Karlstrom, T. N. V., Dean, J. S., and Hevly, R. H. (1979). The Colorado Plateaus: Cultural dynamics and paleoenvironment. *Science* 205, 1089-101.

Fowler, C. S. (1972). Some ecological clues to proto-numic homelands. *In* ''Great Basin Cultural Ecology: A Symposium'' (D. D. Fowler, ed.), pp. 105-21. Desert Research Institute Publications in the Social Sciences 8.

Fowler, D. D., Madsen, D. B., and Hattori, E. M. (1973). ''Prehistory of Southeastern Nevada.'' Desert Research Institute Publications in the Social Sciences 6.

Fry, G. F. (1976). ''Analysis of Prehistoric Coprolites from Utah.'' University of Utah Anthropological Papers 97.

Gayton, A. H. (1948). ''Yokuts and Western Mono Ethnography: I. Tulare Lake, Southern Valley, and Central Foothill Yokuts'' University of California at Berkeley Anthropological Records 10 (1).

Gilbert, G. K. (1890). ''Lake Bonneville.'' Monographs of the U.S. Geological Survey I.

Goss, J. A. (1977). Linguistic tools for the Great Basin prehistorian. *In* ''Models and Great Basin Prehistory: A Symposium,'' (D. D. Fowler, ed.), pp. 49-70. Desert Research Institute Publications in the Social Sciences 12.

Grayson, D. K. (1977). ''Paleoclimatic Implications of the Dirty Shame Rockshelter Mammalian Fauna.'' Tebiwa Miscellaneous Papers of the Idaho State University Museum of Natural History 9.

Grayson, D. K. (1979). Mount Mazama, climatic change, and Fort Rock Basin archaeofaunas. *In* ''Volcanic Activity and Human Ecology'' (P. D. Sheets and D. K. Grayson, eds.), pp. 427-57. Academic Press, New York.

Grayson, D. K. (1982). Toward a history of Great Basin mammals during the past 15,000 years. *In* ''Man and Environment in the Great Basin'' (D. B. Madsen and J. F. O'Connell, Jr., eds.), pp. 82-101. Society for American Archaeology Papers 2.

Green, J. P. (1972). Archaeology of the Rock Creek site 10CA33, Sawtooth National Forest, Cassia County, Idaho. M.S. thesis, Idaho State University, Pocatello.

Guilday, J. E. (1969). ''Small Mammal Remains from the Wasden Site (Owl Cave), Bonneville County, Idaho.'' Tebiwa, the Journal of the Idaho State University Museum 12 (1).

Gwynn, J. (1980). ''Great Salt Lake: A Scientific, Historical and Economic Overview.'' Utah Geological and Mineralogical Survey Bulletin 116.

Hammatt, H. H. (1976). Geological processes and apparent settlement densities along the lower Snake River: A Geo-centric view. Paper presented at the 29th Annual Meeting of the Northwest Anthropological Conference, Ellensburg, Wash.

Haynes, C. V., Jr. (1969). The earliest Americans. *Science* 166; 709-15.

Haynes. C. V., Jr. (1973). The Calico site: Artifacts or geofacts? *Science* 181, 305-10.

Heizer, R. F. (1951). Preliminary report on the Leonard Rockshelter site, Pershing County, Nevada. *American Antiquity* 17, 89-98.

Heizer, R. F., and Krieger, A. D. (1956). ''The Archaeology of Humboldt Cave, Churchill County, Nevada.'' University of California Publications in American Archaeology and Ethnology 47, pp. 1-190.

Heizer, R. F., and Napton, L. K. (1970). ''Archaeology and the Prehistoric Great Basin Lacustrine Subsistence Regime as Seen from Lovelock Cave, Nevada.'' Contributions of the University of California Archaeological Research Facility 10.

Hevly, R. H., Kelly, R. E., Anderson, G. A., and Olsen, S. J. (1979). Comparative effects of climatic change, cultural impact, and volcanism in the paleoecology of Flagstaff, Arizona, AD 900-1300. *In* ''Volcanic Activity and Human Ecology'' (P. D. Sheets and D. K. Grayson, eds.), pp. 487-523. Academic Press, New York.

Hunt, A. P. (1960). ''Archaeology of the Death Valley Salt Pan, California.'' University of Utah Anthropological Papers 47.

Irwin-Williams, C. (1967). Picosa: The Elementary Southwestern culture. *American Antiquity* 32, 441-57.

Irwin-Williams, C. (1973). ''The Oshara Tradition: Origins of Anasazi Culture.'' Eastern New Mexico University Contributions in Anthropology 5 (1).

Irwin-Williams, C. and Haynes, C. V., Jr. (1970). Climatic change and early population dynamics in the southwestern United States. *Quaternary Research* 1, 59-71.

Jennings, J. D. (1957). ''Danger Cave.'' University of Utah Anthropological Papers 27. (Also released as Society for American Archaeology Memoir 14.)

Jennings, J. D. (1964). The desert West. *In* ''Prehistoric Man in the New World'' (J. D. Jennings and E. Norbeck, eds.), pp. 149-74. University of Chicago Press, Chicago.

Jennings, J. D., Schroedl, A. R., and Holmer, R. (1980). ''Sudden Shelter.'' University of Utah Anthropological Papers 103.

Kittleman, L. R. (1973). Mineralogy, correlation, and grain-size distributions of Mazama Tephra and other postglacial pyroclastic layers, Pacific Northwest. *Geological Society of America Bulletin* 84, 2957-80.

Kuchler, A. E. (1964). ''Potential Natural Vegetation of the Coterminous United States.'' American Geographical Society Special Publication 36.

Lamb, S. M. (1958). Linguistic prehistory in the Great Basin. *International Journal of American Linguistics* 24, 95-100.

Leakey, L. S. B., Simpson, R. D., Clements, T., Berger, R., and Witthoft, J. (1972). ''Pleistocene Man at Calico.'' San Bernadino County Museum, San Bernadino, Calif.

Lemke, R. W., Mudge, M. R., Wilcox, R. E., and Powers, H. A. (1975). "Geologic Setting of the Glacier Peak and Mazama Ash-Bed Markers in West-Central Montana." U.S. Geological Survey Bulletin 1395-H.

Leonhardy, F. C., and Rice, D. G. (1970). A proposed culture typology for the lower Snake River region, southeastern Washington. *Northwest Anthropological Research Notes* 4, 1-29.

Lewis, H. T. (1973). "Patterns of Indian Burning in California." Ballena Press Anthropological Papers 1.

Loud, L. L., and Harrington, M. R. (1929) "Lovelock Cave." University of California Publications in American Archaeology and Ethnology 25 (1).

Lyneis, M. M. (1982). Prehistory in the southern Great Basin. *In* "Man and Environment in the Great Basin" (D. B. Madsen and J. F. O'Connell, Jr., eds.), pp. 172-85. Society for American Archaeology Papers 2.

Madsen, D. B. (1975). Dating Paiute-Shoshoni expansion in the Great Basin. *American Antiquity* 40, 82-86.

Madsen, D. B. (1982). Get it where the gettin's good: A variable model of Great Basin subsistence and settlement based on data from the eastern Great Basin. *In* "Man and Environment in the Great Basin" (D. B. Madsen and J. F. O'Connell, Jr., eds.), pp. 207-26. Society for American Archaeology Papers 2.

Madsen, D. B., and Berry, M. S. (1975). A reassessment of northeastern Great Basin prehistory. *American Antiquity* 49, 391-405.

Malde, H. E. (1964a). The ecologic significance of some unfamiliar geologic processes. *In* "The Reconstruction of Past Environments" (J. J. Hester and J. Schoenwetter, eds.), pp. 7-15. Fort Burgwin Research Center Publications 3.

Malde, H. E. (1964b). Environment and man in arid America. *Science* 145, 123-29.

Mehringer, P. J., Jr. (1977). Great Basin late Quaternary environments and chronology. *In* "Models and Great Basin Prehistory: A Symposium" (D. D. Fowler, ed.), pp. 113-67. Desert Research Institute Publications in the Social Sciences 12.

Meighan, C. W. (1954). A late complex in southern California prehistory. *Southwestern Journal of Anthropology* 10, 215-27.

Meighan, C. W., and Haynes, C. V., Jr. (1970). The Borax Lake site revisited. *Science* 167, 1213-21.

Miller, W. R. (1966). Anthropological linguistics in the Great Basin. *In* "The Current Status of Anthropological Research in the Great Basin: 1964" (W. L. d'Azevedo, W. A. Davis, D. D. Fowler, and W. Suttles, eds.), pp. 75-112. Desert Research Institute Publications in the Social Sciences and Humanities 1.

Morrison, R. B. (1965). Quaternary geology of the Great Basin. *In* "The Quaternary of the United States" (H. E. Wright, Jr., and D. G. Frey, eds.), pp. 265-85. Princeton University Press, Princeton, N.J.

Mulloy, W. (1958). "A Preliminary Historical Outline for the Northwestern Plains." University of Wyoming Publications 22 (1).

Nelson, C. M. (1973). Prehistoric culture change in the intermontane plateau of western North America. *In* "The Explanation of Culture Change: Models in Prehistory" (C. Renfrew, ed.), pp. 371-90. University of Pittsburgh Press, Pittsburgh.

O'Connell, J. F. (1975). "The Prehistory of Surprise Valley." Ballena Press Anthropological Papers 4.

Ore, H. T. and Warren C. N. (1971). Late Pleistocene-early Holocene geomorphic history of Lake Mojave, California. *Geological Society of America Bulletin* 82, 2553-62.

Orr, P. C. (1956). "Pleistocene Man in Fishbone Cave, Pershing County, Nevada." Nevada State Museum Department of Archaeology Bulletin 2.

Pilles, P. J., Jr. (1979). Sunset Crater and the Sinagua: A new interpretation. *In* "Volcanic Activity and Human Ecology" (P. D. Sheets and D. K. Grayson, eds.), pp. 459-85. Academic Press, New York.

Rogers, M. J. (1929). The stone art of the San Dieguito Plateau. *American Anthropologist* 31, 454-67.

Rogers, M. J. (1939). "Early Lithic Industries of the Lower Basin of the Colorado River and Adjacent Desert Areas." San Diego Museum Papers 3.

Rogers, M. J. (1958). San Dieguito implements from the terraces of the Rincon-Patano and Rillito drainage systems. *Kiva* 24, 1-23.

Russell, I. C. (1885). "Geological History of Lake Lahontan: A Quaternary Lake of Northwestern Nevada." Monographs of the U.S. Geological Survey XI.

Sanger, D. (1967). Prehistory of the Pacific Northwest plateau as seen from the interior of British Columbia. *American Antiquity* 32, 186-97.

Schalk, R. F. (1977). The structure of an anadromous fish resource. *In* "For Theory Building in Archaeology" (L. R. Binford, ed.), pp. 207-49.

Sheets, P. D., and Grayson, D. K. (eds.) (1979). "Volcanic Activity and Human Ecology." Academic Press, New York.

Simms, S. R. (1977). A mid-Archaic subsistence and settlement shift in the northeastern Great Basin. *In* "Models and Great Basin Prehistory: A Symposium" (D. D. Fowler, ed.), pp. 195-210. Desert Research Institute Publication in the Social Sciences 12.

Simpson, R. D. (1958). The Manix Lake archaeological survey. *Masterkey* 32, 4-10.

Simpson, R. D. (1960). Archaeological survey of the eastern Calico Mountains. *Masterkey* 34, 25-35.

Spier, L. (1930). "Klamath Ethnography." University of California Publications in American Archaeology and Ethnology 30.

Steward, J. H. (1938). "Basin-Plateau Aboriginal Sociopolitical Groups." Bureau of American Ethnology Bulletin 120.

Swanson, E. H., Jr. (1962a). Early cultures in northwestern America. *American Antiquity* 28, 151-58.

Swanson, E. H., Jr. (1962b). "The Emergence of Plateau Culture." Occasional Papers of the Idaho State College Museum 8.

Swanson, E. H., Jr. (1972). "Birch Creek: Human Ecology in the Cool Desert of the Northern Rocky Mountains 9000 BC-AD 1850." Idaho State University Press, Pocatello.

Thomas, D. H. (1974). An archaeological perspective on Shoshonean bands. *American Anthropologist* 76, 11-23.

Thomas, D. H. (1982). An overview of central Great Basin prehistory. *In* "Man and Environment in the Great Basin" (D. B. Madsen and J. F. O'Connell, Jr., eds.), pp. 156-71. Society for American Archaeology Papers 2.

True, D. L., Meighan, C. W., and Crew, H. (1974). "Archaeological Investigations at Molpa, San Diego County, California." University of California Publications in Anthropology 11.

Van Devender, T. R., and Spaulding, W. G. (1979). Development of vegetation and climate in the southwestern United States. *Science* 204, 701-10.

Wallace, W. J. (1962). Prehistoric cultural development in the southern California deserts. *American Antiquity* 28, 172-80.

Warner, T. J. (ed.) (1976). "The Dominguez-Escalante Journal." Brigham Young University Press, Provo, Utah.

Warren, C. N. (1967). The San Dieguito complex: A review and hypothesis. *American Antiquity* 32, 168-85.

Wedel, W. R. (1961). "Prehistoric Man on the Great Plains." University of Oklahoma Press, Norman.

Wheat, M. (1967). "Survival Arts of the Primitive Paiutes." University of Nevada Press, Reno.

Wilcox, R. E. (1959). "Some Effects of Recent Volcanic Ash Falls, with Especial Reference to Alaska." U.S. Geological Survey Bulletin 1028-N, pp. 409-76.

Wilcox, R. E. (1965). Volcanic ash chronology. *In* "The Quaternary of the United States" (H. E. Wright, Jr., and D. G. Frey, eds.), pp. 807-16. Princeton University Press, Princeton, N.J.

Wilke, P. J. (1978). "Late Prehistoric Human Ecology at Lake Cahuilla, Coachella Valley, California." Contributions of the University of California Archaeological Research Facility 38.

Wilke, P. J., and Lawton, H. W. (1976). "Early Observations on the Cultural Geography of Coachella Valley." Ballena Press Anthropological Papers 3, 9-43.

Williams, H. (1942). "The Geology of Crater Lake National Park." Carnegie Institution of Washington Publication 540.

Williams, H., and Goles, G. (1968). Volume of the Mazama ash-fall and the origin of Crater Lake caldera. *In* "Andesite Conference Guidebook" (H. M. Dole, ed.), pp. 37-41. Oregon Department of Geology and Mineral Industries Bulletin 62.

Williams, P. A., and Orlins, R. I. (1963). "The Corn Creek Dunes Site: A Dated Surface Site in Southern Nevada." Nevada State Museum Anthropological Papers 10.

CHAPTER 16

The Evolution of Human Ecosystems in the Eastern United States

James B. Stoltman and David A. Baerreis

Introduction

This chapter outlines out current understanding of the Holocene prehistoric human ecology of the United States east of the Rocky Mountains. The geographic scope of the chapter includes two major provinces, the Great Plains and the Eastern Woodlands, both widely recognized as natural and cultural areas (Kroeber, 1939; Stoltman, 1978; Wedel, 1961). The temporal limits of discussion will be 8000 B.C. to the time of contact with Europeans around A.D. 1600.

This chapter could be said to be oriented toward environmental archaeology, but only if that term were being used in its broadest sense. From our viewpoint, environmental archaeology involves much more than the reconstruction of past environments by archaeologists and paleoecologists; it involves the more important issue of the interaction between environment and culture.

This emphasis on the interaction of human societies with their environments is commonly considered the least common denominator of human ecology (Duncan, 1959: 680), and practitioners of this approach in archaeology are considered to have adopted the ecological approach. By adopting this approach, an archaeologist subscribes explicitly to a conceptual framework that acknowledges "the fundamental reciprocity of nature and culture," although it has been claimed that many subscribers either ignore or are unaware of the terms of their subscription (Butzer, 1975: 109). Even if archaeologists have been remiss in their use of the ecological approach, such carelessness cannot be taken as an indictment of the approach, for it still offers the most effective framework within which the anthropocentric bias that culture can be understood in isolation from natural phenomena can be avoided and within which a realisitic assault can be made on the goal of understanding "the natural history of the man" (Duncan, 1959: 680-81).

In reviewing the literature on the eastern United States relevant to the theme of environmental archaeology, or human paleoecology as we prefer to call it, we use as our unit of analysis the human ecosystem type composed of individual human ecosystems. Deriving inspiration from Duncan (1959), Odum (1969), and Smith (1976) and with modifications of our own, we define a human ecosystem as a discrete human population that has a shared cultural inventory of

technologic, social-organizational, and subsistence practices and is in interactive association with a specified environment. Figure 16-1 illustrates a simplified model of the human ecosystem as we define it. It is conceived as a system composed of five subsystems: subsistence, technology, social organization, population, and environment. In this context, environment is broadly defined as including not only such factors as flora, fauna, climate, and topography but also the other human populations sharing the same environment but differing in adaptive pattern. The five components incorporated into the model do not constitute an exhaustive list of subsystems that might be used to characterize a human ecosystem. They have been selected because of their relevance to archaeological reconstructions of human ecosystems; that is, each can be expected to leave interpretable physical traces in the archaeological record. At a higher level of abstraction, individual human ecosystems can be lumped together into ecosystem types. It is at this level that we organize the following discussion.

The basic focus of this chapter is cultural change as revealed in the Holocene archaeological record of the eastern United States. The human ecosystem model serves to underscore the complexity of cultural change while at the same time providing a framework that facilitates its analysis. A common and productive assumption con-

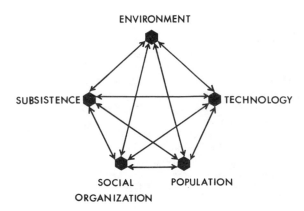

Figure 16-1. Simplified model of a human ecosystem.

cerning an ecosystem such as that illustrated in Figure 16-1 is that it is an equilibruim-seeking system that undergoes evolutionary changes as it is influenced by various dislocating forces (Duncan, 1959: 684). Whether the dislocating forces are of internal or external derivation, they can be conceptualized as arising or being introduced at any of the major points within the system. Because each of the five subsystems is "open" to influences from all other subsystems, cultural change must be envisioned as a complex process of interaction among multiple variables, and not simply as a cause-and-effect relationship between only the two variables of environment and culture. Viewed in this context, environment is but one of a number of interacting factors that must be considered in any effort to understand the trajectory of an evolving human ecosystem.

Pioneering Ecosystem Type

Although the Pioneering Ecosystem Type (PET) flowered during the late Pleistocene, it persisted into postglacial times on the Great Plains and in parts of the Eastern Woodlands in the United States. There has been considerable debate over whether or not the human participants in these ecosystems were the first settlers of the East, but we can safely sidestep that issue here as one irrelevant to the subject at hand. Any effort to understand the Holocene prehistory of the East must begin with Pioneering ecosystems, the cultural substratum into which the roots of all indigenous traditions ultimately extend.

The environmental component of most Pioneering ecosystems was the metamorphosing world of late-glacial and early-postglacial times in the East, when climatic conditions were becoming warmer (Webb

and Bryson, 1972) and plant distributions and associations were distinctly different from those of the present time (Davis, 1976). This was a period of time-transgressive change in floral composition, with boreal species such as spruce and jack pine spread well beyond their present ranges (King, 1981; Watts, 1970, 1980b; Wright, 1970, 1971). During the interval between 10,000 and 7000 B.C., the broad belt of the spruce-dominated boreal forest that extended across the northern half of the East from the Great Plains to the Atlantic retreated northward and was succeeded in the West by prairie and in the East first by pine and then by deciduous forest. In the northern Upper Great Lakes and the northern Northeast, coniferous forest persisted until as late as 4000 B.C. before giving way to northern hardwood forest (Bradstreet and Davis, 1975; Webb, 1974; Wright, 1974). In the Midcontinent region and the southern Applachians, cool-temperate, mixed mesic forest dominated by oak, ash, and hickory became widespread (Delcourt and Delcourt, 1979). Farther south, on the Coastal Plain, an oak-dominated forest prevailed until around 5500 or 5000 B.C., when the pine-dominated forest areas of the present were established (Delcourt and Delcourt, 1979; Watts, 1980b; Watts and Stuiver, 1980).

As it has been emphasized by a number of scholars, a salient quality of biotic conditions of late-glacial and early-postglacial times in the East was variability through both time and space (Brown and Cleland, 1968; Cushing, 1965; Fitting, 1968; Wright, 1968). Nonetheless, some generalizations can be made concerning the nature of the primary resources available for human exploitation. Big game (including caribou, musk-oxen, mastodons, mammoths, and longhorned bison) was relatively abundant, while edible plant resources

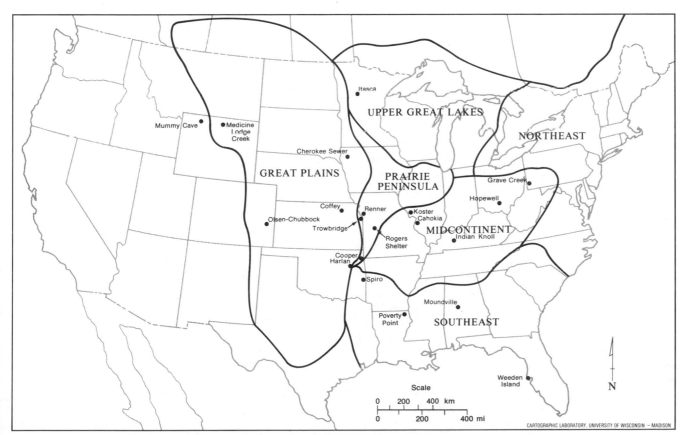

Figure 16-2. Map of the United States showing the areas and some of the locations discussed in the text.

were sparse (except for acorns in parts of the Midcontinent and the Southeast, which were not likely to have been an important early food source for humans because they are unpalatable without special processing). The relative importance of aquatic resources such as fish and shellfish was severely curtailed by a series of interrelated environmental conditions that adversely affected their availability: low sea levels that steepened the downstream gradients, unstable shoreline conditions, and in some rivers, cold temperatures and turbid water conditions related to glacial runoff. Even those aquatic species that were readily available probably were not particularly productive, especially compared to later times when warmer water temperatures afforded rich aquatic habitats (Matteson, 1960; Parmalee, 1968: 107-8).

Despite profound environmental variation, the technologic components of Pioneering ecosystems showed surprising uniformity over virtually all of the East. Indeed, at no other time in eastern prehistory can we observe such close cultural similarities over the Great Plains and Eastern Woodlands as are apparent from late-glacial to early-postglacial times. Shared over this vast area was the manufacture of lanceolate spear or dart points accompanied by such equally distinctive chipped stone tools as endscrapers, gravers, spokeshaves, and wedges and often by at least a rudimentary blade technology (Fitting, Devisscher, and Wahla, 1966; Gardner, 1974; Goodyear, 1974; Irwin and Wormington, 1970; Prufer, 1963; Wilmsen, 1970). On the Great Plains, where a combination of favorable soil conditions and rapid site burial enhances bone preservation, an inventory of finished bone tools emerges (Frison and Zeimens, 1980; Haynes and Hemmings, 1968; Lahren and Bonnichsen, 1974), as well as of crude bone choppers used in butchering bison (Frison, 1974: 51-57).

Efforts at making functional interpretations of the tool kits associated with the PET have mainly been subjective formal analyses; nonetheless, there can be no serious dispute with the opinion that the primary tasks reflected in the stone- and bone-tool inventories are hunting, butchering, hide processing, and bone/wood working (Frison, 1974; Gardner, 1974: 25; Goodyear, 1974: 14; Irwin and Wormington, 1970; Wilmsen, 1970). There is no convincing evidence in the tool asemblages to support the existence of fishing and plant-food processing as significant activities.

Attempts to reconstruct the subsistence component of the PET are hampered by the paucity of food remains recovered from suitable archaeological contexts. Outside the Great Plains, faunal remains are either absent or scarce and floral remains are virtually nonexistent throughout the East in this setting. There is evidence that suggests that some elements of the late-Pleistocene megafauna were hunted in the Woodlands (e.g., Bullen et al., 1970; Palmer and Stoltman, 1976; Wittry, 1965) and that, at least in the North, barren-ground caribou were hunted from late-glacial into early-postglacial times (Cleland, 1965; Funk et al., 1969). There is ample evidence that the principal game animal hunted south of the retreating range of caribou and moose from at least 8000 B.C. on in the Woodlands was the white-tailed deer (Logan, 1952: 63; Parmalee, 1962; Parmalee et al., 1976).

On the Great Plains, by contrast, there is abundant and unambiguous archaeological evidence of the importance of big-game hunting not just in the late Pleistocene but well into postglacial times as well. Accompanying the late-glacial expansion of the prairie and the consequent increase in the bison population in this area was the emergence of a distinctive human adaptation "based on mobility, portability of possessions, and prime reliance on bison hunting" (Wedel, 1978: 187). It is appropriate to regard this as a distinctive

variant of the PET, based as it was on an exploitation of the gregarious nature of the bison and their tendency toward mass stampedes when threatened, permitting periodic mass kills (e.g., Davis and Wilson, 1978; Dibble and Lorraine, 1968; Frison, 1974; Schultz, 1943: 244-48; Wheat, 1972). It is a variant that persisted in some portions of the Great Plains until the historical period.

Without minimizing the diversity that must have characterized PET subsistence patterns, especially the seasonal variations, the foregoing lines of evidence concerning environment, technology, and diet converge at the same conclusion: big-game hunting occupied an overall position of preeminence in PET subsistence. To such a subsistence pattern Cleland (1976: 68-69) has applied the term "Early Focal." Accepting this interpretation of PET subsistence also helps to explain the virtually pan-eastern sharing of styles of projectile points and tool kits, for big game hunting was part of an adaptive pattern that could transcend such environmental diversity.

Data on the population and social-organizational components of the PET are meager indeed, permitting little more than reasonable speculation. Small, widely scattered, mostly open-air sites with low artifact densities, no significant midden accumulation, no evidence of recurrent burial, and no evidence of structures suggest that local groups were small and highly mobile and that they had only ephemeral identifications with specific places. The archaeological evidence is entirely consistent with what could be expected of egalitarian societies characterized by a band level of social organization (Fried, 1967; Service, 1971). During much of the year, it can be presumed that family units or small multifamily units scoured the countryside in quest of game, with groups probably having an average of 25 individuals or fewer (Service, 1971: 58) and that during favorable seasons, especially in caribou and bison range, larger, multiband encampments were established. Wheat's carefully reasoned arguments, based on the quantity of meat actually processed at the Olsen-Chubbuck site, indicate that a group of 150 to 200 persons was involved in the exploitation of a mass bison kill. Undoubtedly, however, such large numbers only congregated seasonally.

High-quality siliceous rock seems to have constituted another basic resource within this ecosystem type, as evidenced by the intensity with which a number of localized, high-quality cherts and comparable materials were exploited and circulated widely (Gardner, 1974; Prufer, 1963; Wilmsen, 1970). In Wisconsin, and probably elsewhere in the East, we believe that participants in the PET possessed much higher proportions of exotic lithic materials (e.g., Knife River chalcedony from North Dakota, Upper Mercer cherts from Ohio, and Peoria and Burlington cherts from Illinois) than any subsequent occupants of the area except participants in the Hopewell Interaction Sphere. We attribute this pattern to a combination of factors, the most important of which were the continuing high demand for stone tools with everyday rather than esoteric uses and the low population densities and the high population mobility that facilitated intergroup interaction, often over great distances.

A Free-Wandering form of community patterning (Beardsley et al., 1956: 135-36) can be hypothesized for the initial stages of the Pioneering Ecosystem. The nearly universal stylistic and functional characteristics of the tool kits used by makers of fluted points throughout the East suggest that their expansion was rapid and unimpeded by significant resistance from preexisting populations (Fitting, 1975; Mason, 1962: 245). In the early stage of this ecosystem, there would have been minimal environmental and cultural constraints to population growth and areal expansion; when population pressure on local resources became too great, a segment of the population could simply branch off into new terrain. Eventually, of course, when the prime areas became filled to capacity, this solution was no longer tenable. With environmental constraints then fully operative, local populations no longer had the option of moving into new terrain; therefore, increasingly more stable regional adaptations can be hypothesized. A symptom of this trend would have been the emergence of regional style zones reflecting reduced regional interaction and increased exploitation of local resources. This trend had surely begun before the termination of the fluted-point tradition (Williams and Stoltman, 1965), but it was clearly in evidence by early-postglacial times, when the Dalton and Plano styles prevailed in largely mutually exclusive areas. With the progressive development of environmental constraints to expansion, community patterning must have shifted to Restricted Wandering and eventually to Central-Based Wandering (Beardsley et al., 1956: 136-40; Morse, 1971).

Foraging Ecosystem Type

As the successor of an ecosystem type whose longevity varied from region to region, the Foraging Ecosystem Type (FET) did not develop as a synchronous process across the East. The earliest cultural complex assignable to the new type of ecosystem, which was associated with Kirk Corner notched projectile points, dates to the last half of the eighth millenium B.C. in the Midcontinent and the Southeast (Broyles, 1966, 1971; Chapman, 1977, 1978; Coe, 1964). On the Great Plains, by contrast, the bison resources sustained a persisting Pioneering Ecosystem Type until the onset of the historical period, although trade with sedentary agriculturalists modified its characteristics. Likewise, in the pine-dominated forests of the Great Lakes and New England, it can be posited, the Pioneering ecosystems endured for as long as herds of barren-ground caribou penetrated the area, probably into the sixth millennium B.C.

The environmental component of the FET was characterized by the absence both of the last vestiges of the late-Pleistocene megafauna and of herding mammals such as bison and barren-ground caribou. In their place, the principal game animals were moose and woodland caribou in the coniferous forests and white-tailed deer in the deciduous forests; all were basically solitary animals. This pattern, combined with the presence of scattered human populations in most habitable regions of the East by early-postglacial times, mitigated against the persistence of Pioneering ecosystems when these species became the primary prey animals. A unique, last-gasp version of a PET based on the intensive exploitation of white-tailed deer may have characterized the initial stages of the Dalton culture of the Midcontinent region in early-postglacial times (DeJarnette et al., 1962; Ford, 1977: 167-71; Goodyear, 1974: 14). Such an adaptation would have been viable for small populations with large hunting territories; however, it seems unlikely that such conditions endured for very long. Before its end, the Dalton culture was probably undergoing change toward the FET (Chapman, 1975: 95-125).

We hypothesize that, since human populations had turned to deer, moose, and woodland caribou as primary prey animals and since population segments could no longer move off into uninhabited areas, more diversified and/or balanced subsistence systems would have evolved by necessity. These would have been marked by the more intensive exploitation of local resources, such as small game, nuts, fish, and shellfish, that formerly were either neglected or underexploited (Stoltman, 1978: 714-15). Important additions to the technologic in-

ventory that presumably were associated with the increasingly intensive exploitation of local resources included polished stone, milling/nutting stones, steatite and ceramic containers, and fishing gear such as harpoon heads, net weights, and hooks. As this pattern of regional intensification continued, the proliferation of localized technologies and style zones signified the progressive establishment of group territories and social boundaries and the reduction of interregional interaction. In addition, sites of increased size and complexity, often with houses and multiple burials testifying to their less ephemeral occupancy, became focal points in the community-patterning systems of the Restricted Wandering and Central-Based Wandering types that then prevailed (Beardsley et al., 1956: 136-40).

There is only minimal archaeological evidence concerning the population and social-organizational components of this ecosystem type. Generally, an increase in the number and size of sites across the East (Fitting, 1970: 64-67; Funk and Rippeteau, 1977: 29; Sanger, 1975: 61) argues for a notable increase in population. By contrast, there is no archaeological evidence to indicate that social-organizational complexity had progressed beyond the egalitarian band level.

Although some evidence of coexistence among Pioneering and Foraging ecosystems has been uncovered, the Foraging ecosystems were generally prevalent in the Eastern Woodlands from around 7500 B.C. to around 2500 B.C., with persistence occurring in northern and southern marginal areas up to until the time of contact with Europeans. On the Great Plains, however, the pattern of development of Foraging ecosystems differed from that prevalent in the Eastern Woodlands. Conditions in the Plains were markedly affected by the Altithermal or Hypsithermal climatic episode of 6500 to 3000 B.C. The grassland ecotone reached a maximum eastward expansion during this time, reflecting maximum Holocene aridity and/or elevated temperatures in the Prairie Peninsula (Baerreis and Bryson, 1965; Wendland and Bryson, 1974; Wright, 1968). Various scholars (e.g., Frison, 1975; Hurt, 1966; Reeves, 1973) have argued positions ranging from a nearly complete depopulation of the Plains to a somewhat reduced population maintaining the same pattern as in the earlier period, or they have contended that the population mainly existed on the fringes of the Plains with what we have called the FET. Although bison were a constant resource on the Great Plains, the herds fluctuated in size, depending on the impact of climate on the available grasses (Dillehay, 1974). The reconstruction of both the cultural patterns and the resource base is complicated by the possibility that alluviation and erosion on the floodplains during this interval has destroyed, covered, or moved evidence of occupation (Knox, 1976).

Data on both the character of the Hypsithermal episode and the human exploitation of local resources are provided by the Cherokee sewer site located on an alluvial fan at the margin of the Little Sioux River near Cherokee in northwest Iowa (Anderson and Semken, 1981; Shutler et al., 1974). The site contains three deeply buried cultural horizons associated with evidence of extensive bison processing dated 6400, 5300, and 4350 B.C. The general results of the varied approaches used in the study (micromammals, terrestrial gastropods, pollen, climatology) are remarkably consistent, and they suggest for 4400 B.C. an increase in aridity and an elevation in winter temperature of but 1 °C to 2 °C compared to the present. The climatic warmth during this time seems to be a consequence of the increased frequency of fronts drawing mild Pacific air, which also tended to reduce the seasonal range of temperature. Accompanying these climatic changes, some technologic innovations appeared at the

Cherokee sewer site, including the shift from lanceolate to notched projectile points, an elaboration in bone tools, and the appearance in horizon 11B of a metate fragment. Still, the overwhelming impression is one of cultural continuity.

Although the Cherokee sewer site documents a continuation of the PET with perhaps some acculturation in tool types from components of the FET, other more marginal sites indicate the presence of the FET's more diffuse exploitative pattern. Thus, the Coffey site in northeastern Kansas dated at 3100 B.C., in addition to limited bison use, showed a dependence on deer, fish, waterfowl, and a variety of wild seeds and berries (Schmits, 1976). The importance of grinding stones in the artifact assemblage clearly indicates that the charred seeds were not just fortuitously recovered but were an essential part of the economy. Similarly, Mummy Cave in northern Wyoming (Wedel et al., 1968), as evidenced by 38 distinct cultural levels beginning before 7300 B.C., had occupants whose subsistence was based largely on mountain sheep and deer. There is virtually no evidence of bison hunting. Although this site reflects an FET, it is only 64 km from the contemporary bison kill at the Horner site, where Scottsbluff and Eden points and other Pioneering attributes were found. The Medicine Lodge Creek site, 160 km east of Mummy Cave, also through its stratigraphic sequence, indicates the importance of the use of plants. and small animals during the same period, with some use of deer and mountain sheep but little use of bison (Frison and Wilson, 1975). Thus, a mixture of Pioneering and Foraging ecosystems is to be found on the periphery of the Great Plains.

The Itasca bison kill site on the eastern margin of the Plains (Shay, 1971) should be mentioned if only for its investigators' skillful use of multidisciplinary techniques. The site, dating from 5000 to 6000 B.C., includes a hunting camp and the associated remains of a bison kill incorporated in adjacent lake and swamp deposits. Pollen recovered in the swamp deposits indicates that an open pine forest prevailed in the area around 6300 B.C. but that it was replaced by 5500 B.C. by prairie and then oak savanna. The site appears to document the time of the maximum eastward expansion of prairie in this region. Although diverse faunal remains were present in the swamp sediment, the context made it difficult to ascertain whether the animals were present as a consequence of human hunting activities. Only the bison bones and turtle remains showed scratches or other butchering marks.

Rogers Shelter, an outstanding site with deep stratigraphy located near the ecotone between the oak-hickory forests of the Ozark Highland and the prairie areas of western Missouri, is of interest as a record of the expansion of bison to this area only in the horizon dating between 6600 and 5000 B.C. (Wood and McMillan, 1976: 224-25). The availability of limited quantities of bison, however, did not essentially modify the basic sequence of some 3000 years of FET adaptations on the margin of the Prairie Peninsula.

As on the Great Plains and in the Prairie Peninsula, the Woodland zone proper experienced environmental changes during the middle-postglacial times that expedited the maturation of Foraging ecosystems in many areas and facilitated their expansion in others. Among the more significant environmental changes to impinge upon the Woodland zone during this period were: (1) the final wastage of the Laurentide ice sheet, which not only released icy meltwaters into the ocean basins but also opened up vast areas of eastern Canada to caribou and human colonization (Bryson et al., 1969); (2) low but rising sea levels and warming water temperatures along the Atlantic seaboard, which permitted a steady northward expansion of improved fish and shellfish habitats (Braun, 1974; Emery and Edwards, 1966; Fairbridge, 1977; Snow, 1972; Turnbaugh, 1975); (3) low but rising water levels in the Great Lakes basin, which reduced tributary stream gradients and contributed to improved fishing conditions, especially as water temperatures also rose (Fitting, 1970: 68; Hough, 1963); (4) reduced stream gradients along the Atlantic coast, which must have lowered the rapids and other impediments to the upstream migration of anadromous fishes (Sanger et al., 1977: 469-70; (5) a shift in floodplain conditions along the Mississippi River and its major tributaries from rapidly aggrading, braided channels carrying sandy sediments in broad, flat bottomlands lacking levees and backwaters to meandering channels carrying finer sediments across a convex floodplain replete with levees and sluggish backwaters, which must have significantly improved human habitats (Butzer, 1977, 1978; Saucier, 1974); and (6) an expansion northward of deciduous tree species, especially nut-bearing trees such as oak, hickory, beech, and chestnut, into portions of the Upper Great Lakes and the Northeast that were formerly dominated by coniferous forest conditions, which increased the amount of plant foods available for human consumption in these areas (Bernabo and Webb, 1977; Bradstreet and Davis, 1975; Davis, 1976).

Considerable archaeological evidence suggests that the environmental changes enumerated above were key variables in the growth and expansion of Foraging ecosystems during the middle-postglacial times, especially in the Upper Great Lakes and the Northeast. Before around 5000 B.C., conifer-dominated forests prevailed over all but the southern portions of these regions (Bernabo and Webb, 1977: 85-86). A number of archaeologists have emphasized the low carrying capacity for human populations of these pine-dominated forests, and, by correlating this observation with the relative paucity of known archaeological sites of the same age, they have hypothesized that these regions were sparsely inhabited at the time (Cleland, 1966: 50-53; Fitting, 1968: 443, 1970: 67; Funk, 1977; Rippeteau, 1977; Ritchie, 1969: 213). Indeed, with minimal plant resources available for human consumption, it is inconceivable that substantial, stable human populations could have been sustained in these regions without the intensive exploitation of aquatic resources. Thus, the observed widespread increase in the size and density of sites in the northern regions after 4000 B.C., but even more so after 3000 B.C., correlates with the peoples' increased technologic sophistication in exploiting aquatic resources, combined with an increased availability of those resources that is presumably associated with rising sea and lake levels (Bourque, 1975; Braun, 1974; Fitting, 1970: 67-68; Funk, 1977: 328; Ritchie, 1969: 215-16; Ritchie and Funk, 1973: 41; Ritzenthaler, 1957).

Elsewhere in the Woodland zone, particularly in the Midcontinent and the Southeast, the environment is generally assumed by archaeologists to have been more or less stable throughout the middle-Holocene times. With the environmental component thus conceptually relegated to a passive role, the active points of change within the Foraging ecosystems of this time are widely assumed to have been the subsistence and technologic subsystems. The observed increase in the size and complexity of archaeological assemblages and sites is thus attributed to a progressively more efficient exploitation of the Woodlands, the establishment of a "primary forest efficiency" (Caldwell, 1958: 6-18; 1965).

There is mounting evidence, however, that indicates that the environment south of the Great Lakes and the Northeast was far from stable before 3000 B.C. For example, Butzer's geomorphologic studies in the lower Illinois Valley indicate that around 3000 B.C. there was a major transformation in floodplain conditions, with a braided stream pattern on a sandy, flat floodplain being replaced by a meandering stream pattern associated with levees and backwaters (Butzer, 1977, 1978). Since this environmental change coincided almost precisely with the termination of the Helton phase in this area, the possibility that the subsequent Titterington phase had a different subsistence system than its predecessor must be given serious consideration (Cook, 1976).

Similarly, recent pollen analyses in the contiguous portions of Georgia and Florida indicate that significant vegetational changes occurred during the interval between 5000 and 3000 B.C. (Watts, 1969, 1971, 1980b; Watts and Stuiver, 1980). Before this time, back to around 7500 B.C., a dry oak forest with significant amounts of hickory prevailed. Presumably, ideal conditions for browsing and grazing animals existed, and nut production in this forest must have been the greatest of any during the post-Pleistocene period (Watts, 1971). Beginning at about 5000 B.C., pine increased steadily at the expense of oak while hickory also declined until around 3000 B.C., when the pine-dominated forests of the area took on their modern configuration. Coincident with the establishment of pine-dominated forests in the Southeast were rising water tables that resulted in the establishment of swamps and marshes widely over the Coastal Plain from the Dismal Swamp south to the Okefenokee and beyond (Watts, 1980b). Upland conditions were deteriorating for Foraging ecosystems (more restricted deer habitat, less nut mast, and more wetlands), and more intensive coastal and riverine adaptations would have been encouraged as sea levels began to stabilize after 3000 B.C. Thus, it seems no coincidence that shell middens are found for the first time in the archaeological record of Georgia and Florida at around 3000 B.C. (Bullen, 1972; Stoltman, 1972). Of course, caution should be exercised in making evaluations of the negative evidence concerning coastal adaptations along the Atlantic and Gulf coasts because rising sea levels have inundated many areas that were once habitable (Goodyear and Warren, 1972; Ruppé, 1980). In any case, this and the foregoing examples should serve to underscore the risk involved in making the tacit simplified assumption that seems to pervade much of the archaeological literature: that, following the turn of the Pleistocene, the environment was a passive variable that could be largely ignored in analyses of the dynamics of Foraging ecosystems.

Cultivating Ecosystem Type

The Cultivating Ecosystem Type (CET), as it is described in this chapter, follows Bronson's (1977: 26) definition of cultivation as denoting "only that a useful species has been deliberately caused to reproduce by man," in contrast to agriculture, which "is reserved for contexts of substantial dependence on plants by humans." The basic distinction is not predicated upon the presence or absence of cultivated plants, but rather upon the relative importance of cultigens in the overall diet of a population. Such a determination usually is extremely difficult to make objectively on the basis of archaeological data alone. Yet, the theoretical value of such a construct coupled with a mounting body of evidence suggesting that cultivation in this sense was an important prelude to intensive agriculture in the East prompts us to include it as a diagnostic criterion of the ecosystem type that succeeded Foraging ecosystems in many eastern areas of the United States.

The earliest published evidence for plant cultivation in the East in-

volves the tropical cucurbits (squashes and gourds) that appear in the archaeological record of the Midcontinent in the third millennium B.C. (Chomko and Crawford, 1978; Kay et al., 1980; Marquardt and Watson, in press) but whose still earlier appearance is to be anticipated on the basis of current paleoethnobotanical research. This surprisingly early adoption of cultigens of ultimate Mesoamerican derivation by cultures of the Late Archaic stage has invalidated the previously widely accepted hypothesis that an indigenous seed plant cultivation developed in the East (Fowler, 1971; Struever, 1964; Struever and Vickery, 1973; see also Asch and Asch, 1978; Asch et al., 1979; Snook and Swartz, 1975). Although indigenous seeds were being harvested to some extent in the East before the arrival of the cucurbits, there is firm archaeological evidence for the intensive collecting of native seeds and for the domestication of local plants only after 2000 B.C. (Asch and Asch, 1978: 330; Asch et al., 1972; Ford, 1977; Yarnell, 1976, 1977).

Joining the cucurbits, in what Yarnell (1964: 103) once referred to as the "early eastern agricultural complex," were a number of native seed plants whose abundance, occurrence beyond their natural habitats, and/or increased seed sizes testify to their cultivation. Included in this complex were such small-seeded annual plants as goosefoot, knotweed, and maygrass and two species that definitely seem to have been domesticated, sunflower and sumpweed (Asch and Asch, 1978; Cowan, 1978; Yarnell, 1978). This mix of tropical (squash, gourd), native oily (sunflower, sumpweed), and native starchy (goosefoot, knotweed, maygrass) cultivated seed-producing plants constitutes a significant addition to the Foraging ecosystem and is the major feature distinguishing the CET.

It is probable that the progressively more intensive utilization of the various seed plants that can be observed in the Midcontinent from around 2500 B.C. to at least A.D. 400 reflects a much greater control over the propagation of these plants than is usually implied under such traditional appellations as "primary forest efficiency" and "intensive harvest collecting" (Caldwell, 1958, 1965; Struever, 1968). It now seems likely that, rather than having to depend primarily on wild stands of plants, humans prepared, planted, and cared for garden plots, which can be distinguished from fields by their smaller size and greater botanical diversity (Ford, 1979: 237).

This gardening activity took place mainly on the major alluvial valleys of the Midcontinent, where the critical resources and the majority of sites are found (Crites, 1978; Johnson, 1975; Kay, 1980; McCullough and Faulkner, 1973; Munson, Parmalee, and Yarnell, 1971; Roper, 1979; Seeman, 1979b; Struever, 1964, 1968). This, we feel, is a valid observation so long as it is applied to the full duration of the CET. In the past, the riverine adaptation of the Hopewellian cultures was portrayed as being qualitatively distinct from that of their predecessors (see, for example, Struever, 1968: 305); this no longer appears to be the case. Rather, Hopewell is better conceived as the culmination of an adaptive trend that extends back to the beginnings of the CET in the third millennium B.C. (See also Asch et al., 1979: 83).

Considering the limited nutritional value of the cucurbits and the relatively large energy expenditure required to harvest and to prepare the small-seeded native plants, it seems unlikely that yields from these gardens were substantial enough to be considered true staples (Asch and Asch, 1978: 312-15; Ford, 1979: 237). It seems more likely that the principal value of the gardens was as a source of easily storable commodities that could be called upon during times of special need or scarcity in a subsistence cycle still geared primarily to the seasonal exploitation of deer, nuts, fish, shellfish, and small game (Yarnell, 1965: 80; 1969: 49; 1974). Cleland (1976: 70-71) has referred to this subsistence type as "Late Diffuse."

It is also probable that the increasing dietary importance of seeds in general may have been tied to population trends. As population densities apparently increased during the Transitional II and Developmental periods (Stoltman, 1978: 714-21), group territories seem to have become more sharply defined, with group mobility correspondingly limited; in such circumstances, local populations may have become increasingly vulnerable to the unpredictability of local nut harvests (Ford, 1977: 173; 1979: 236) and seeds would have been an underexploited resource that could have made up the deficiency. A final bonus of a garden system, as it evolved into an increasingly more secure source of storable food reserves, was the increased measure of sedentariness afforded a segment of the population, especially childbearing women, children, and old people. This asset, in turn, might have been fed back into the population component of the ecosystem, contributing to further growth as a result of the expectable relaxation of fertility controls accompanying a more sedentary existence (Binford, 1968: 332; Sussman, 1972).

Excluding the controversial claim for the appearance of maize pollen at the Koster site before 3000 B.C. (Schoenwetter, 1979), corn seems to have been a late addition to the Cultivating ecosystem; corn appeared in the Midcontinent late in the first millennium B.C. (Adovasio and Johnson, 1981; Murphy, 1971; Shea, 1978; Whitehead, 1965). Yet, the spread of this new tropical cultigen was neither rapid nor universal, even in the Midcontinent, where some cultivators apparently resisted its use as late as the fifth century A.D. (Munson et al., 1971). For the early centuries of the Christian era, there is definite evidence that a number of Hopewellian communities in Ohio and Illinois had incorporated corn into their subsistence subsystems (Ford, 1979). Although the presence of corn in Hopewellian contexts is no longer in doubt, there is considerable evidence to suggest that it was still an unimportant dietary item. Ford (1979: 237) has outlined three lines of botanical evidence suggesting that corn was not a staple in Hopewellian societies, but much more persuasive direct evidence is now available in the form of carbon isotope analyses of human bone that document the paucity of corn in the diet of Hopewellian peoples in Ohio, Illinois, and Wisconsin (Bender et al., 1981). It is primarily on the basis of this newer evidence that Hopewellian societies can be assigned to the CET.

Elsewhere in the East (i.e., beyond the Midcontinent that encompasses the area from the Ozarks to the Appalachians and from southern Ohio to the Tennessee Valley), the assignment of specific regional complexes to the CET is severely hampered by the paucity of reliable subsistence data. Squash has been recovered in Early to Middle Woodland contexts as far afield as northern Ohio, central Michigan, eastern Kansas, and southern Louisiana (Byrd, 1976; Johnson, 1975; Yarnell, 1976); but otherwise few cultigens are known to have existed outside the Midcontinent until after A.D. 900. It is impossible to assign objectively many regional cultures of the Late Archaic, Early Woodland, Middle Woodland, and Late Woodland stages to specific ecosystem types. However, one thing seems certain: each of these four cultural stages included cultures of both the Foraging and Cultivating ecosystem types.

The notable increase in the size and complexity of sites within the Midcontinent region during the Transitional II and Developmental periods is consistent with the expectation that population size and stability and social-organizational complexity increased with the

establishment of the CET. For example, Poverty Point in Louisiana has embankment and mound construction involving nearly 750,000 m³ of earth (Webb, 1968: 318). The colossal Grave Creek Mound of probable Adena construction in West Virginia originally towered nearly 21 m above the surrounding terrain and had a basal diameter of 90 m (Norona, 1962: 15). The Hopewell site in Ohio has over 40 ha enclosed by earthen embankments and 38 associated mounds (Shetrone, 1926). These sites are unrivaled in size and complexity by any sites known to have been associated with the FET. Although the Central-Based Wandering community pattern persisted in some areas, it is clear that larger sites had become the focus of Semipermanent Sedentary and Simple Nuclear Centered settlement systems (Beardsley et al., 1956: 138-43).

Because such complex sites obviously grew by accretion, it is extremely difficult to estimate the size of the populations associated with them without more precise chronologic controls over the duration of occupancy than are currently available. The most promising technique for estimating population sizes is the careful analyses of burial populations (Asch, 1976; Buikstra, 1976; Howells, 1960; Jamison, 1971). A number of fragile suppositions concerning sample representativeness and duration of site occupancy underlie such analyses, however, and they must be considered in evaluating the results. Despite these limitations, the findings of three independent studies of burial populations assignable to the CET have been surprisingly uniform. Assuming a death rate of 5% per annum and a 500-year occupancy of Indian Knoll, a Late Archaic site dating to the third millennium B.C., Howells (1960) estimated an average population of 50. Buikstra (1976), using burial data from 2 Illinois Valley Hopewell mounds, and Jamison (1971), using comparable data from 13 mounds at a single Mississippi Valley Hopewell site, both estimated that the size of the populations contributing to these mounds was between 50 and 100. Needless to say, these estimates must be treated as tentative at best, but they are useful in giving quantitative expression to the population-size differential distinguishing sites of the FET (mean populations near 25) from sites of the CET (mean populations of 50 or more).

When consideration is shifted from population estimates of individual sites to those of whole regions, the data available are even more scanty. The most ambitious effort at estimating regional population size from archaeological data is that of Asch (1976), who was working in the lower Illinois Valley. Starting primarily from 17 extensively excavated mounds from two sites and extrapolating average numbers of burials per mound to the estimated total of comparable mounds within a 7469-km² research universe, he estimated maximum and minimum population densities for Hopewellian times and concluded that a maximum density range of 0.22 to 0.33 persons per square kilometer was reasonable (Asch, 1976: 59). At this time it is impossible to compare this estimate with other prehistoric data, but it is interesting to note that this estimate for Illinois Valley population density during the interval from 150 B.C. to A.D. 400 is comparable to levels reported in the Southeast during historical times (Asch, 1976: 68).

Efforts at reconstructing the social-organizational component of the CET have been based primarily on observations of the differential complexity of sites and burial programs. A number of scholars, impressed with the obviously greater size and complexity of sites and burial practices assignable to the CET, have hyposthesized that at least some of these social systems had attained the level of chiefdoms with ranked rather than egalitarian societies (Buikstra, 1976; Fried,

1967; Gibson, 1974; Seeman, 1979a, 1979b: 401-2; Struever and Houart, 1972). Other scholars, however, have felt that a lesser level of social complexity (i.e., the tribe with egalitarian status grading [Fried, 1967; Service, 1971]) characterized even the most complex societies of the CET (Ford, 1974: 402, 1979; Griffin, 1967: 184). Two recent analyses intensively and quantitatively reexamining the extant data support the latter position (Braun, 1979; Seeman, 1979b), although Seeman (1979b: 401-2) feels that in one culture of the CET, Ohio Hopewell, the chiefdom level was attained. Whether or not Ohio Hopewell had attained the chiefdom level, a legitimately debatable issue, it seems probable that most cultures of the CET were characterized by egalitarian social systems and that shamans, or "big men" (rather than chiefs or priests who occupied permanent offices), were the principal manipulators and beneficiaries of the active traffic in exotic materials so characteristic of these basically tribal societies.

Greater interregional interaction, as evidenced by the widespread sharing of specific artifact types and styles and the widespread circulation of such source-specific raw materials as Yellowstone obsidian, Great Lakes copper, and Gulf Coast shells, is one of the salient properties of the CET. Not since the time of early Pioneering ecosystems does the archaeological record of the East show such widespread cultural interaction. The motivating forces and mechanisms behind this interaction, however, were very different. High group mobility in a sparsely occupied environment facilitated the interaction of cultures within the Pioneering ecosystem, and the commodities circulated were high-quality siliceous rocks used to fabricate everyday tools and weapons. By contrast, the commodities circulated among cultures of the CET were much more varied, including metals, shells, and many nonsiliceous rocks, destined for use primarily as status symbols for the living and burial offerings for the dead. With population mobility then predictably constrained to limited geographic areas and since there is minimal evidence for population movements (Buikstra, 1979), the observed interregional interaction must be attributed primarily to trade. The extensiveness of trade among cultures of the CET has been amply documented by a number of authors (Dragoo, 1976; Seeman, 1979b; Struever and Houart, 1972; Webb, 1968; Winters, 1968).

What were the sources of this trade, and what were the functions that it served? There are still no clear-cut answers to these important questions. Viewed in the context of the CET, the causal forces of interregional interaction, it seems to us, must have arisen from within the subsistence, population, or social-organizational components. Brose (1979) suggests that interregional exchange arose because of inherent instabilities on the subsistence level and that trade functioned primarily to maintain open channels to external food resources needed in times of local scarcity. Considering the vast distances involved and the primitive transportation facilities available (mainly human carriers), it seems to us improbable that such an explanation could adequately account for trade contacts between the Midcontinent and such regions as the Lake Superior basin, the Gulf Coast, and the Rocky Mountains. (See also Ford, 1979: 238.) It is more likely that the primary impetus behind long-distance exchange of scarce resources emerged in the context of expanding populations, reduced mobility, and increasing social complexity that accompanied the transition from the Foraging to the Cultivating ecosystem type in the Midcontinent. Since local peoples already possessed a long-standing tradition of interring personal possessions with the dead, it is likely that, as social systems became larger and more complex, the demand among the living for status symbols would increase and more and

more of these symbols would accompany a growing number of individuals with special status to their graves, where archaeologists have been especially successful in recovering them (Struever, 1964: 88).

Efforts to understand the dynamics of the rise and fall of the CET have given little emphasis to technologic innovation as a contributing factor. There is good reason for this. It seems to be a fact that no major qualitative (as opposed to quantitative) technologic advances (except possibly the bow and arrow, which are discussed later) were associated with this ecosystem type in the East. By contrast, the role of the environment in the dynamics of Cultivating ecosystems has been afforded wider and more serious consideration. A pioneering formulation of Hopewell relationships to climatic factors was developed by Griffin (1960a, 1961 [see also Vickery, 1970]), who hypothesized that conditions favorable to maize agriculture were terminated by the onset of a cool, moist period around A.D. 200. Although our changing appreciation of Hopewell subsistence practices negates the link with agriculture, it does not rule out a climatic impact on other subsistence elements. Recent palynologic studies, especially those of varved sediments, provide enough fine detail to suggest that Hopewell might have been terminated with the onset of a drier episode rather than of a moister one (Baerreis et al., 1976). These data, however, are derived from geographic positions marginal to the major areas of Hopewell occupation, and, except for the potentialities inherent in synoptic climatological approaches, they do not give a clear picture of what was happening in the regions of major population concentration. Quite promising are the results of flotation and fine-screening techniques used on soil samples from sites on the western periphery. The detailed analysis of faunal and floral remains from the Trowbridge site (Johnson, 1975), a Kansas City Hopewell village, is of interest in that the major animal in the economy was deer (51.5% of the identified remains), whereas no bison are reported. The faunal remains also include three species no longer occurring in the vicinity of the site, the closest area of sympatry being southeastern Oklahoma, indicating somewhat warmer conditions during the time of the site's occupation. The absence of bison is not duplicated at the nearby Renner site (Roedl and Howard, 1957: 57; Wedel, 1943: 27), although deer again are the major element among the animal remains. At the Cooper site in northeastern Oklahoma (Bell and Baerreis, 1951: 27-33), teeth of both bison and deer were present, but poor bone preservation made it impossible to provide a quantitative estimate of their relative importance. It would appear, however, that at least during a portion of the existence of this marginal Hopewell manifestation a deer habitat was more critical than a bison habitat, despite the site's location. An indication of the importance of seeds and nuts in their economy (Johnson, 1979: 88) further supports the possibility that a wooded habitat had expanded westward during the time in question.

Agricultural Ecosystem Type

In contrast to Cultivating ecosystems, the Agricultural Ecosystem Type (AET) has a subsistence component characterized by a "substantial dependence on plants grown by humans" (Bronson, 1977: 26). Cleland (1976) has referred to this adaptive pattern as "Late Focal," although, this term encompasses considerable variation in subsistence patterns. The new ecosystem type was well established across most of the deciduous forest zone and even well out into the Great Plains by A.D. 900, but its background, when and where it first arose, it still obscure. Almost certainly the emergence of the AET should be sought in the temporal context of the Intermediate period, A.D. 400 to 900 (Stoltman, 1978), and in the geographic context of the Woodland zone. It is likely, however, that it arose over a broad geographic front rather than in a specific area (Clay, 1976).

Cleland (1966: 94-95) has made the interesting suggestion, endorsed by Hall (1973), that the demise of Hopewell and the emergence of the AET were causally related. The reasoning behind this hypothesis is that the Hopewellian economy was of the diffuse type and that the characteristic trade networks had arisen as a result of local scarcities in order to import outside foodstuffs, thus permitting the system to ramify beyond limits that the local area alone could have sustained. Following this reasoning further, when corn was introduced into Hopewellian economies, it permitted a shift to a focal economy in which the need for outside foodstuffs diminished along with the increasing importance of corn. If the Hopewellian Interaction Sphere dissolved as a result of the adoption of intensive corn agriculture, it could be anticipated that the archaeological record of the immediately post-Hopewellian period would be rich in corn-producing sites. Just the opposite is the case, however, and "there is a strange scarcity of corn from the time of Hopewell culture to about the time of Mississippian emergence . . ." (Cutler and Blake, 1969: 135). Moreover, there is direct evidence that, at least within the realm of the Illinois Hopewell, the CET endured into post-Hopewell times (Munson et al., 1971).

Of particular importance to the success of the AET in the East was the addition of beans to the culigens utilized. If the reported but unconfirmed occurrence of beans in Kansas City Hopewell contexts is disregarded (as many scholars are now prone to do), the earliest verified use of beans in the East was in New York in the 11th century A.D. (Ford, 1979; Munson, 1973; Yarnell, 1976). Since beans are undeniably of tropical origin, it is probable that earlier ones will eventually be uncovered farther to the south and west, most likely dating to the Intermediate period (Yarnell, 1976: 272). The nutritional importance of beans lies in their almost perfect complementarity to corn. Alone, corn, which is deficient in the critical amino acids lysine and trytophone, is an inadequate source of protein. But in combination with beans, which are rich in those two amino acids, corn provides protein of high biologic value for all but the most protein-sensitive members of a population, for example, lactating mothers (Kaplan, 1965).

It is almost certainly no coincidence that the widespread occurrence of stable Agricultural ecosystems corresponds closely with the introduction of beans into the East, but the development of improved strains of corn also facilitated the expansion of Agricultural ecosystems after A.D. 900. The earliest corn used in the East was characteristically of 12- and 14-row varieties resembling southwestern types (Yarnell, 1964: 111-13). Presumably, this type of corn required a long growing season in order to mature, probably nearly 200 frost-free days, for it never became popular north of the Ohio River (Hall, 1973; Yarnell, 1964). With the development of a hardier 8-row variety, the range of effective corn agriculture was expanded considerably, reaching as far north as the 120-day frost-free line (Cutler and Blake, 1969; Hall, 1973; Yarnell, 1964). South of the Ohio River, the length of the growing season was not likely to have been a constraint on the expansion of corn agriculture, so it can be assumed either that

Agricultural ecosystems began there earlier than in the Northeast or that it was the late arrival of beans that retarded agricultural development. The difficulty in evaluating these alternatives is that the paleo-ethnobotanical record of the Southeast is so deficient that any effort to estimate the first appearance of corn or beans or true Agricultural ecosystems can be nothing more than speculation. Although it seems improbable that Agricultural ecosystems existed in the Southeast during the Developmental period, 100 B.C. to A.D. 400 (Sears, 1971), the chances are much better that they existed during the following Intermediate period, A.D. 400 to 900, when the complex Weeden Island and Coles Creek cultures flourished in northern Florida and the lower Mississippi Valley, respectively (Milanich, 1980; Phillips, 1970). However, in the absence of paleoethnobotanical materials from such contexts, these can only be considered "possible" AET cultures. However, there can be no doubt that by A.D. 900 cultures of the new ecosystem type had become established all across the Southeast, including the South Appalachian, Gulf, Lower Mississippi Valley, and Caddoan provinces (Stoltman, 1978).

Two technologic innovation associated with Agricultural ecosystems are noteworthy because of the improved efficiency they afforded in certain food-extraction tasks: the bow and arrow and the hoe. On the basis of the reduced size of projectile points, it is widely believed that the bow and arrow was an innovation of the peoples of the Late Woodland stage during the immediate post-Hopewellian (i.e., Intermediate) period (Brain, 1976; Ford, 1977; Hall, 1973). Such evidence is far from compelling, but, if it is accepted provisionally, then the bow and arrow appeared first in the East in the context of the CET at its nadir in sociocultural complexity following the demise of the Hopewell Interaction Sphere. The bow and arrow did not immediately supplant the atlatl and dart, but the two apparently coexisted (e.g., Ford, 1951: 117; Munson et al., 1971). However, it was only in the context of the AET that the bow and arrow became prominant across the East, perhaps in part because of their value in warfare as well as in hunting. The Florescent period, A. D. 900 to around 1600, was certainly a time of increased social conflict across the East, as suggested by both the multiplication of fortified communities and the increased evidence of violent death. Bonafide hoes, whether of stone, bone, or shell, are confirmed only in the context of the Agricultural ecosystem and, indeed, may be taken as a sign of the more intensive form of food production that distinguishes the AET from the CET.

At the time of contact with Europeans Agricultural ecosystems existed along most of the major river valleys of the Plains and throughout the Woodland zone except in some parts of the Upper Great Lakes, Northeast, and Southeast (Gulf Coast), where Foraging and Cultivating ecosystems persisted in marginal habitats. In the western sectors of the Plains, both Pioneering and Foraging ecosystems also seem to have survived, but they were often modified through interaction with AET cultures. In general, however, the vast majority of the 16th-century natives of the East were participants in Agricultural ecosystems. To convey more accurately the variability within this ecosystem type, we conclude this section with a discussion of three major subtypes within the AET. Each of the subtypes represents a different pattern of adaptation to three broadly different environmental settings: major alluvial valleys of the deciduous forest zone, upland habitats of the deciduous forest zone, and alluvial valleys of the Plains. Although each of these subtypes possesses unique qualities in contrast to the others, it is important to note that a great deal of cultural interaction occurred among them and also between them and the surviving remnants of other ecosystem types.

Intensive Riverine Subtype

The largest population densities and highest levels of sociocultural complexity in eastern prehistory were achieved within the context of the Intensive Riverine Subtype (IRS) during the Florescent period. The geographic focus of this subtype was the great alluvial valleys of the Midcontinent and the Southeast, especially the Mississippi, Ohio, Tennessee, Arkansas, and Red Rivers and their major tributaries. In this setting, large temple centers burgeoned as the seats not only of religious influence but also of political and economic influence over the surrounding hinterland. Often stockaded, the core of these temple centers was a plaza bordered by from 1 or 2 up to 20 flat-topped platform mounds that served as substructures for religious and/or public buildings and elite residences. One of these centers, Cahokia, near East St. Louis, Illinois, is the largest prehistoric site north of Mexico, containing over 100 mounds within an area of roughly 13 km² and possessing a population in its heyday (around A.D. 1050 to 1250) of perhaps 10,000 people (Fowler, 1974).

Much cultural variation can be observed among different regional expressions of this ecosystem subtype, indicating both local autonomy and cultural continuity extending back to Intermediate times as well as interregional interaction during the Florescent period. Some of the better-known regional variants include such cultural complexes as Middle Mississippian, Plaquemine, Caddoan, South Appalachian, Fort Walton, and Fort Ancient (Griffin, 1967; Smith, 1978; Stoltman, 1978).

The subsistence component of this ecosystem subtype consisted of intensive hoe agriculture in alluvial valley settings complemented by hunting and gathering of a wide range of natural animal and plant resources of the floodplain and the adjacent upland deciduous forests (Ford, 1977: 179-82; Smith, 1975; Waselkov, 1978). As in early historical times, it is probable that the agricultural practices of these ecosystems involved both multiple cropping and intercropping (Hudson, 1976: 297). Multiple cropping consisted of planting two successive corn crops in the same field, each of which was harvested early and eaten green. Other fields were planted in one corn crop and harvested in the fall. In both cases, the fields were intercropped, that is, corn and beans were planted together, the bean vines being allowed to twine around the cornstalks while squashes, gourds, sunflowers, and chenopods were either planted or encouraged around the field margins. Such an agricultural system, which was technologically simple in that it required only axes, digging sticks, and hoes, was best suited to loamy soils in wooded or cane-covered areas; it generally was unsuited to heavy clayey soils, to light infertile sandy soils, and to areas blanketed with deep-rooted grasses. A number of archaeologists have noted that such conditions are most commonly met with within the major alluvial valleys of the East and that it is this association that helps account for the observed distribution of the major sites of this ecosystem subtype (Larson, 1971; Peebles, 1978; Ward, 1965).

Despite the association of this ecosystem subtype with rivers there is neither archaeological nor ethnographic evidence that any form of water management, whether for flood control or irrigation, existed.

The locations of dwellings and ceremonial sites were invariably in low-risk areas, either on levees or on bluffs adjacent to the floodplain. Presumably, the agricultural fields were located in similar settings nearby. On the basis of this observation, it has been argued that population pressure was still not sufficient enough to have forced the utilization of higher-risk terrain within the alluvial valleys (Lewis, 1974). Under the false supposition that most of the Eastern Woodlands had an essentially uniform distribution of agricultural resources, Kroeber (1939: 147) had concluded earlier that in the East, "there was a hundredfold surplus of potentially farmable land over farming population." This opinion notwithstanding, the widespread evidence for warfare, inlcuding the existence of ubiquitous fortified sites, suggests that population pressures did exist within this ecosystem subtype and that a major contributing factor was the scarcity of agricultural land suitable to the prevailing subsistence pattern (Larson, 1972).

The level of social complexity attained in ecosystems of this subtype surpassed all others in the East. It is generally recognized that these societies were chiefdoms characterized by rank rather than egalitarian status structures (Fried, 1967; Hudson, 1976; Service, 1971). In such societies, leaders occupied institutionalized, permanent offices that were filled not as a reward for personal achievement but by virtue of ascribed, usually hereditary, status. The enduring high status/low status distinction between chiefs and non-chiefs in such societies is given archaeological expression in predictable ways in differential burial practices and in differential site location, size, and complexity within the same social system (Peebles, 1972; Peebles and Kus, 1977; Steponaitis, 1978).

Findings from the two most intensively studied examples of prehistoric eastern chiefdoms, Cahokia and Moundville, indicate that an even more complex social system approaching that of a state with stratification was in the process of evolving by Late Florescent times (Fowler, 1974; Peebles, 1978; Steponaitis, 1978). Steponaitis (1978) draws the distinction between simple and complex chiefdoms (one level of superordinate political office versus multiple levels) to characterize this trend toward increasing complexity, but other scholars have argued that, at least in the case of Cahokia, social complexity had progressed beyond the chiefdom to the state (Gibbon, 1974; O'Brien, 1972; for an opposing view, see Ford, 1974: 406, and Hines, 1977). If true statehood had been achieved at Cahokia, it was both a unique and abortive development, for by the time of contact with the Europeans the chiefdom was the highest level of social complexity extant in the East.

Upland Subtype

The Upland Subtype (US) can be distinguished from the IRS mainly in terms of its less intensive pattern of agriculture and its primary adaptive niche, the upland deciduous forest biome rather than riverine floodplain. The same basic agricultural staples were grown, commonly following the same practice of intercropping maize, beans, and squashes characteristic of the IRS. One important difference, however, was that multiple cropping was not practiced by US cultures. Instead, the norm was shifting cultivation, so-called slash-and-burn, or swidden, agriculture, which involved planting a single corn crop per year in specially prepared forest clearings that were rotated every few years. Supplementing the yields of these agricultural fields was a wide range of wild foods. The relative importance of these wild foods varied considerably, depending upon the productivity of local resources as well as local agricultural potential. In most areas white-tailed deer were of paramount importance, but in many Midcontinent localities nuts were also of great importance seasonally. In the Upper Great Lakes and along the eastern seaboard, aquatic resources (fish, shellfish, waterfowl, and wild rice) rivaled deer in importance. Indeed, so important were wild foods among some northern, corn-growing peoples (e.g., Menomini, Abenaki) that some must be assigned to the CET.

US cultures held sway over most of the deciduous forest zone north of the Ohio River during the Florescent period. Virtually all cultures of this vast area from the Mississippi River to the Atlantic coast that have been assigned to the Late Woodland stage or to the Oneota cultural complex can be categorized as belonging to the US. An immense amount of cultural variability has been documented both within and among the material inventories of Late Woodland and Oneota cultures, and yet they can be seen to have shared a similar adaptive pattern in contrast to the FET or CET cultures that preceded them and the IRS cultures that paralleled them. Compared to the former, the US villages were larger and more stable (as evidenced by the stockades, houses, and storage pits that appear commonly in the archaeological record of the North for the first time). But, in contrast to the latter, US villages were structurally much less complex, lacking the plaza-platform mound complex or any of the other trappings of a chiefdom level of social organization. The available archaeological evidence is in accord with the ethnographic evidence from the North in suggesting that the tribal level of social organization with an egalitarian status structure characterized the US. Community patterning among US cultures was generally of the Semipermanent Sedentary type in which populations occupied stable villages but moved the villages periodically. By contrast, major communities of the IRS were of the Simple and Advanced Nuclear Centered types in which there was true permanence (Beardsley et al., 1956).

We suggest that the US subtype is best conceived as an extension of the AET into those portions of the Eastern Woodlands unsuited to the IRS pattern. The US subtype was an agriculturally based adaptation to the forest soils and/or short growing seasons of the temperate deciduous forest zone of the eastern United States. Viewed in this context, differences from the IRS, such as having smaller, autonomous villages (community sizes of a few hundred at most) and a simpler (tribal) form of social organization, can be seen as logical correlates of shifting cultivation combined with the intensive exploitation of a wide variety of locally and seasonally available wild plants and animals. The arrival or development of early-ripening, eight-row corn in conjunction with beans to provide a balanced protein intake obviously was a critical factor in opening up the Northeast to effective corn agriculture. Low population densities in the upland zones of the Northeast, along with increasingly severe environmental constraints to areal expansion within the major alluvial valleys, where suitable agricultural soils were finite, would have been factors favoring the growth and expansion of the US. If the Neo-Atlantic climatic episode had been somewhat moister or otherwise more favorable to corn agriculture, as some have suggested (Baerreis et al., 1976; Griffin, 1960b; Hall, 1973), there would have been an added inducement to the expansion of the US. As the US pattern expanded into northern areas where no established IRS competitors existed, alluvial valleys as well as upland habitats were naturally exploited, but there is

no indication of any significant departure from the general US pattern (e.g., McKern, 1945). When the geographic limits of effective agriculture were reached by cultures of the US, it is interesting to note, the cultures had the capacity to adopt specialized practices commonly associated with more complex ecosystems, for example, the construction of ridged fields. As suggested by Riley and Freimuth (1979), it is probable that the so-called garden beds of Wisconsin and Michigan were constructed for the purpose of frost drainage by agriculturalists from various cultures who were confronted with the common problem of early killing frosts along the northern limits of effective corn agriculture.

Plains Village Subtype

The Plains Village Subtype (PVS), with its dual dependence on agriculture and bison hunting, was securely established on the eastern margin of the Plains by A.D. 900. Its agriculture, based on the familiar combination of corn, beans, and squashes, was strictly a matter of bottomland farming, for more-adequate soil moisture was needed to compensate for the vagaries of summer precipitation, and upland grasses also provided great difficulties for hoe cultivation. This system, which in time came to rely on corn varieties adapted to climatic conditions on the Plains (Will and Hyde, 1917), seems to have emerged almost as early as the AET in the East and differed from its eastern contemporaries not only in the importance it placed on bison but also in a number of important technologic innovations.

The technologic innovations contributing to the emerging climax culture of the Plains accompanied a series of accretions from earlier traditions and formed a finely tuned adjustment to the distinctive Plains habitat. Villages of substantial houses, rectangular in the earlier traditions but generally shifting to circular around the 15th century A.D., when droughts led to extensive population movements from the central Plains into the Middle Missouri regions (Baerreis and Bryson, 1966; Lehmer, 1970; Wedel, 1953), provided protection from the contrasting Plains winter and summer weather. Particularly striking was the rich assemblage in bone, including a knowledge of bone-bending and bone-shaping techniques as well as effective combinations of stone inserted in bone, the stone later being replaced by metal in the composite tools. Of importance was the use of the scapula hoe, whose abundance at sites attests to the care and attention the people paid to their crops. The use of bows is indicated by the appearance of small, triangular projectile points. In some areas, the highly efficient sinew-backed bow greatly increased the killing power of the weapon. The abundance of planoconvex endscrapers in conjunction with a series of specialized bone tools comprised a portion of an extensive skin-dressing industry and indicates an elaboration in the production of skin containers, robes, and skin covers for portable dwellings, characteristic of the mobile, western hunters but known even to the sedentary groups who used them on extended hunting trips.

In the northern Plains, even the earlier sites of the Initial Middle Missouri tradition, including the Over, Anderson, Monroe, and Mill Creek groups, are characterized by a great abundance of bison bone within the village area, implying that the bison during some seasons were present in numbers not too distant from the village (this is documented by historical descriptions of villages in this area). In the central and southern Plains, bison were not initially present in comparable abundance. Indeed, Dillehay (1974) characterizes the period of A.D. 500 to A.D. 1200 to 1300 as one when bison populations were sparse in the southern Plains. Lehmer (1954) and Wood (1969) have suggested that the early Upper Republican sites in the central Plains, producing in their refuse a greater abundance of deer and smaller game than of bison, engaged seasonally in communal bison hunts in more western locales. Wedel (1970), however, feels that this suggestion still lacks acceptable supporting evidence. On the southeastern periphery the earlier Caddoan phases, Harlan and Spiro, covering a time span of about A.D. 900 to 1350, show little evidence of bison utilization or distinctive Plains traits. Yet, the later phases (such as Fort Coffee) occupying the same region during the following century had access to bison; this is indicated by the fact that their artifact inventory includes distinctive Plains tools manufactured from bison bone (Bell, 1972; Bell and Baerreis, 1951: 62-64; Lehmer, 1952).

The initial differentiation in adaptation between the northern and southern/central Plains confirms the fact that the Great Plains do not respond as a single unit to climatic variability but that stress conditions, especially in the adequacy of moisture, have a localized distribution within the Plains. The cultural consequences of this environmental factor were the extensive migrations of populations within the Plains that resulted in the rapid diffusion of innovations and stylistic changes within the region as well as local discontinuities in cultural traditions. In addition, the migrations may well have resulted in localized population pressures that accentuated the tendency toward militancy, as shown by the important role of warfare in the historical Plains societies. The early and frequent appearance of fortified villages in the archaeological record provides the line of evidence for tracing this development in time.

Although climatic factors have played an important role in interpretations of the events in the nearly-thousand-year duration of the PVS (Wedel, 1941, 1947, 1953, 1961), specific ecologically oriented analyses of climatic proxies have been few in number. Lehmer (1970, 1971) makes use of climatic models, particularly in reference to the Middle Missouri region, and has elaborated upon the relations between climate and culture. On the northeastern periphery in western Iowa, the Mill Creek culture, whose duration spans the earlier and moister Neo-Atlantic climatic episode of A.D. 900 to 1200 and the succeeding Pacific episode characterized by the appearance of increased zonal westerlies, provided an opportunity to investigate the extent to which field data supported the predicted climatic impact (Bryson and Baerreis, 1968). It is evident that continued attention directed toward such sensitive forms as rodents and terrestrial gastropods will produce a great deal of information on changing environmental conditions during the period of occupation.

Evidence concerning population changes is also difficult to secure. Within local regions, clear fluctuations in the number of sites can be traced over time. But, because this fluctuation in number is also associated with fluctuations in the size of the sites, it is not immediately evident whether or not these reflect substantial shifts in population size (Lehmer, 1971). Without the length of occupation of sites controlled, the basic datum of number is also suspect, for it now appears that in the Mill Creek culture of northwestern Iowa only a few simultaneously occupied villages were responsible for the entire series of sites (Tiffany, 1979). The need to shift villages frequently in order to secure supplies of wood for houses, fortifications, and fuel may have both created the appearance of a dense population and produced a population pressing on available resources. Although the

central Plains, and perhaps the southern Plains as well, have been viewed traditionally as the environment initiating the PVS, with settlement patterns involving isolated houses and clusters of two or three houses, Gradwohl (1969) has documented the presence of small nucleated villages and has indicated how the character of the early explorations in this region may have tended to underestimate the number of houses.

Unlike the record for the East, the archaeological record for the Great Plains provides no evidence for a level of sociopolitical complexity beyond that of the tribe, the pattern prevailing at the beginning of the historical record. Cemeteries show little evidence of status differentiation among individuals insofar as such differences are reflected in burial associations.

Conclusions

The primary objective of this chapter was to review our current understanding of the nature of the interaction between the prehistoric societies of the eastern United States and their environments. Insofar as it was possible, we relied upon research of an interdisciplinary nature, but, in order to make the narrative more nearly complete, it was necessary to delve into the purely archaeological and paleoecologic literature as well. In organizing our presentation, we elected to treat culture and environment as integral parts of a single domain: the human ecosystem or, at a higher level of abstraction, the human ecosystem type. The human ecosystem construct, herein defined as a totality composed of five subsystems (environment, subsistence, technology, population, and social organization) in mutual interaction, was introduced as a supplement to the traditional temporal and formal frameworks usually employed in archaeological syntheses (Stoltman, 1978).

We do not propose substituting the human ecosystem for temporal or formal frameworks because each of these has an important role to play in helping us to think about the past and to communicate our thoughts. But for certain problems it is suggested that human ecosystems are vastly superior to time units (e.g., eras, periods, and subperiods) and to form units (e.g., stages and traditions) as conceptual tools for organizing archaeological taxa such as components, phases, and cultures into higher-level analytical units that can enhance our understanding of the past.

We suggest that the main advantage of the human ecosystem concept is the more productive framework that it provides for the analysis of cultural change in prehistory. Formal frameworks (e.g., Archaic, Woodland, and Mississippi traditions) are best suited to addressing the question of *what* changes occurred, whereas temporal frameworks are most effective in dealing with the topic of *when* changes occurred. Neither framework, however, is equipped to attack the more important issue of *why* changes occurred, because critical variables such as population and environment are either excluded from the analysis altogether (see Butzer, 1978) or treated as external forces impinging on a culture. The drawback of the latter approach is that the focus of analysis is artificially structured in the format of a dyadic relationship between either population and culture or environment and culture, a framework that encourages the search for simplistic cause-and-effect solutions to complex problems. In contrast, the unit of analysis suggested here, the human ecosystem, explicitly incorporates population and environment along with culture (further subdivided into subsistence, technology, and social organization) into a single framework so that the search for the causes of change must be in the context of a matrix of multiple, mutually interacting variables. Such a context is more likely to produce comprehensive and satisfying explanations of the changes observed in the archaeological record.

Finally, we want to point out how the human ecosystem framework has provided a new perspective on the prehistory of the eastern United States in contrast to traditional approaches using culture, stage, tradition, or period as organizing concepts. The uniqueness and, we feel, the strength of the human ecosystem concept is its core quality, which is neither culture nor environment nor population but the *relationships* among them. We can refer to this series of relationships as an adaptive pattern that exists in dynamic equilibrium in an ideal situation, but in actuality it is constantly evolving as the subsystems adjust and readjust to both stochastic and directional changes in the other subsystems. Viewing eastern prehistory from the human-ecosystem perspective, we conclude: (1) that the traditional taxonomic distinction between Archaic and Woodland obscures rather than illuminates a major change in adaptive pattern, that from the Foraging to the Cultivating ecosystems; (2) that the Hopewellian achievement was not a new adaptation, as many have suggested, but the culmination of a long evolutionary development of the Cultivating Ecosystem Type; (3) that Late Woodland and eastern Oneota (popularly called Upper Mississippian) cultures share a fundamentally similar adaptive pattern that is generally overlooked in view of their ceramic and other material culture differences; and (4) that the Early, Middle, and Late Woodland stages all contained cultures of two or three different human ecosystem types, and thus we stress the need for a critical reappraisal of the proper role and utility of these concepts.

References

Adovasio, J. M., and Johnson, W. C. (1981). The appearance of cultigens in the upper Ohio Valley: A view from Meadowcroft Rockshelter. *Pennsylvania Archaeologist* 51, 63-80.

Anderson, D. C., and Semken, H. A. (eds.) (1981). "Mid-Holocene Ecology and Human Adaptation in Northwestern Iowa." Academic Press, New York.

Asch, D. (1976). "The Middle Woodland Population of the Lower Illinois Valley." Northwestern Archeological Program Scientific Papers 1.

Asch, D. L., and Asch, N. B. (1978). The economic potential of *Iva annua* and its prehistoric importance in the lower Illinois Valley. *In* "The Nature and Status of Ethnobotany" (R. I. Ford, ed.), pp. 300-41. Museum of Anthropology, University of Michigan, Anthropological Papers 67.

Asch, D., Farnsworth, K. B., and Asch, N. B. (1979). Woodland subsistence and settlement in west central Illinois. *In* "Hopewell Archaeology" (D. S. Brose and N. Greber, eds.), pp. 80-85. Kent State University Press, Kent, Ohio.

Asch, N. B., Ford, R. I., and Asch, D. L. (1972). "Paleoethnobotany of the Koster Site: The Archaic Horizons." Illinois State Museum Reports of Investigations 24.

Baerreis, D. A., and Bryson, R. A. (1965). Climatic episodes and the dating of the Mississippian cultures. *The Wisconsin Archeologist* 46, 203-20.

Baerreis, D. A., and Bryson, R. A. (1966). Dating the Panhandle Aspect cultures. *Bulletin of the Oklahoma Anthropological Society* 14, 105-16.

Baerreis, D. A., Bryson, R. A., and Kutzbach, J. E. (1976). Climate and culture in the western Great Lakes region. *Midcontinental Journal of Archaeology* 1, 39-57.

Beardsley, R. K., Holder, P., Krieger, A. D., Meggars, B. J., Rinaldo, J. B., and Kutsche, P. (1956). Functional and evolutionary implications of community patterning. *In* "Seminars in Archaeology" (R. Wauchope, ed.), pp. 129-57. Memoirs of the Society for American Archaeology 11.

Bell, R. E. (1972). "The Harlan Site, Ck-6: A Prehistoric Mound Center in Cherokee County, Eastern Oklahoma." Oklahoma Anthropological Society Memoir 2.

Bell, R. E., and Baerreis, D. A. (1951). A survey of Oklahoma archaeology. *Bulletin of the Texas Archaeological and Paleontological Society* 22, 7-100.

Bender, M. M., Baerreis, D. A., and Steventon, R. L. (1981). Further light on carbon isotopes and Hopewell agriculture. *American Antiquity* 46, 346-53.

Bernabo, J. C., and Webb, T., III (1977). Changing patterns in the Holocene pollen record of northeastern North America: A mapped summary. *Quaternary Research* 8, 64-96.

Binford, L. R. (1968). Post-Pleistocene adaptations. *In* "New Perspectives in Archeology" (S. R. Binford and L. R. Binford, eds.), pp. 313-41. Aldine, Chicago.

Bourque, B. J. (1975). Comments on the Late Archaic populations of central Maine: The view from the Turner farm. *Arctic Anthropology* 12, 35-45.

Bradstreet, T. E., and Davis, R. B. (1975). Mid-postglacial environments in New England with emphasis on Maine. *Arctic Anthropology* 12, 7-22.

Brain, J. P. (1976). The question of corn agriculture in the lower Mississippi Valley. *Southeastern Archaeological Conference Bulletin* 19, 57-60.

Braun, D. P. (1974). Explanatory models for the evolution of coastal adaptation in prehistoric eastern New England. *American Antiquity* 39, 582-96.

Braun, D. P. (1979). Illinois Hopewell burial practices and social organization: A re-examination of the Klunk-Gibson mound group. *In* "Hopewell Archaeology" (D. S. Brose and N. Greber, eds.), pp. 66-79. Kent State University Press, Kent, Ohio.

Brennan, L. A. (1974). The lower Hudson: A decade of shell middens. *Archaeology of North America* 2, 81-93.

Bronson, B. (1977). The earliest farming: Demography as cause and consequence. *In* "Origins of Agriculture" (C. A. Reed, ed.), pp. 23-48. Mouton, The Hague.

Brose, D. S. (1979). A speculative model of the role of exchange in the prehistory of the eastern woodlands. *In* "Hopewell Archaeology" (D. S. Brose and N. Greber, eds.), pp. 3-8. Kent State University Press, Kent, Ohio.

Brown, J., and Cleland, C. (1968). The late glacial and early postglacial in midwestern biomes newly opened to human adaptation. *In* "The Quaternary of Illinois" (R. E. Bergstrom, ed.) pp. 114-22. University of Illinois College of Agriculture Special Publication 14.

Broyles, B. J. (1966). Preliminary report: The St. Albans site (46 Ka 27), Kanawha County, West Virginia. *The West Virginia Archeologist* 19, 1-43.

Broyles, B. J. (1971). "Second Preliminary Report: The St. Albans Site, Kanawha County, West Virginia." West Virginia Geological and Economic Survey Report of Archeological Investigations 3.

Bryson, R. A., and Baerreis, D. A. (1968). Introduction and project summary. *In* "Climatic Change and the Mill Creek Culture of Iowa, Part I" (D. Henning, ed.), pp. 1-34. Journal of the Iowa Archeological Society 15.

Bryson, R. A., Wendland, W. M., Ives, J. D., and Andrews, J. T. (1969). Radiocarbon isochrones in the disintegration of the Laurentide ice sheet. *Arctic and Alpine Research* 1, 1-14.

Buikstra, J. (1976). "Hopewell in the Lower Illinois Valley." Northwestern Archaeological Program Scientific Papers 2.

Buikstra, J. (1979). Contributions of physical anthropologists to the concept of Hopewell: A historical perspective. *In* "Hopewell Archaeology" (D. S. Brose and N. Greber, eds.), pp. 220-33. Kent State University Press, Kent, Ohio.

Bullen, R. P. (1972). The Orange period of peninsular Florida. *In* "Fiber-Tempered Pottery in Southeastern United States and Northern Columbia: Its Origin, Context, and Significance" (R. P. Bullen and J. B. Stoltman, eds.), pp. 9-33. Florida Anthropological Society Publications 6.

Bullen, R. P., Webb, D. and Waller, B. I. (1970). A worked mammoth bone from Florida. *American Antiquity* 35, 203-5.

Butzer, K. W. (1975). The ecological approach to archaeology: Are we really trying? *American Antiquity* 40, 106-11.

Butzer, K. W. (1977). "Geomorphology of the Lower Illinois Valley as a Spatial-Temporal Context for the Koster Archaic Site." Illinois State Museum Report of Investigations 34.

Butzer, K. W. (1978). Changing Holocene environments at the Koster site: A geo-archaeological perspective. *American Antiquity* 43, 408-13.

Byrd, K. M. (1976). Tchefuncte subsistence: Information obtained from the excavation of the Morton shell mound, Iberia Parish, Louisiana. *Southeastern Archaeological Conference Bulletin* 18, 70-75.

Caldwell, J. R. (1958). "Trend and Tradition in the Prehistory of the Eastern United States." American Anthropological Association Memoir 88.

Caldwell, J. R. (1965). Primary forest efficiency. *Southeastern Archaeological Conference Bulletin* 3, 66-69.

Chapman, C. H. (1975). "The Archaeology of Missouri, I." University of Missouri Press, Columbia.

Chapman, J. (1977). "Archaic Period Research in the Lower Little Tennessee River Valley." University of Tennessee Department of Anthropology Report of Investigations 18.

Chapman, J. (1978). "The Bacon Farm Site and a Buried Site Reconnaissance." University of Tennessee Department of Anthropology Report of Investigations 21.

Chomko, S. A., and Crawford, G. W. (1978). Plant husbandry in prehistoric eastern North America: New evidence for its development. *American Antiquity* 43, 405-8.

Clay, R. B. (1976). Tactics, strategy, and operations: The Mississippian system responds to its environment. *Midcontinental Journal of Archaeology* 1, 137-62.

Cleland, C. E. (1965). Barren ground caribou (*Rangifer arcticus*) from an early man site in southeastern Michigan. *American Antiquity* 30, 350-51.

Cleland, C. E. (1966). "The Prehistoric Animal Ecology and Ethnozoology of the Upper Great Lakes Region." Museum of Anthropology, University of Michigan, Anthropological Papers 29.

Cleland, C. E. (1976). The focal-diffuse model: An evolutionary perspective on the prehistoric cultural adaptations of the eastern United States. *Midcontinental Journal of Archaeology* 1, 59-76.

Coe, J. L. (1964). The Formative Cultures of the Carolina Piedmont. *Transactions of the American Philosophical Society*, new series, 54 (5).

Cook, T. G. (1976). "Koster, an Artifact Analysis of Two Archaic Phases in West-Central Illinois." Northwestern Archaeological Program, Prehistoric Records 1.

Cowan, C. W. (1978). The prehistoric use and distribution of Maygrass in eastern North America: Cultural and phytogeographical implications. *In* "The Nature and Status of Ethnobotany" (R. I. Ford, ed.), pp. 263-88. Museum of Anthropology, University of Michigan, Anthropological Papers 67.

Crites, G. D. (1978). Plant food utilization patterns during the Middle Woodland Owl Hollow phase in Tennessee: A preliminary report. *Tennessee Anthropologist* 3, 79-92.

Cushing, E. J. (1965). Problems in the Quaternary phytogeography of the Great Lakes region. *In* "The Quaternary of the United States" (H. E. Wright, Jr., and D. G. Frey, eds.), pp. 403-16. Princeton University Press, Princeton, N.J.

Cutler, H. C., and Blake, L. W. (1969). Corn from Cahokia sites. *In* "Explorations into Cahokia Archaeology" (M. L. Fowler, ed.), pp. 122-36. Illinois Archaeological Survey Bulletin 7.

Davis, L. B., and Wilson, M. (eds.) (1978). "Bison Procurement and Utilization: A Symposium." Plains Anthropologist Memoir 14.

Davis, M. B. (1965). Phytogeography and palynology of northeastern United States. *In* "The Quaternary of the United States" (H. E. Wright, Jr., and D. G. Frey, eds.), pp. 377-401. Princeton University Press, Princeton, N.J.

Davis, M. B. (1967). Late-glacial climate in northern United States: A comparison of New England and the Great Lakes region. *In* "Quaternary Paleoecology" (E. J. Cushing and H. E. Wright, Jr., eds.), pp. 11-43. Yale University Press, New Haven, Conn.

Davis, M. B. (1976). Pleistocene biogeography of temperate deciduous forests. *Geoscience and Man* 18, 13-26.

DeJarnette, D. L., Kurjack, E. B., and Cambron, J. W. (1962). Stanfield-Worley Bluff Shelter Excavations. *Journal of Alabama Archaeology* 8, 1-111.

Delcourt, P. A., and Delcourt, H. R. (1979). Late Pleistocene and Holocene distributional history of the deciduous forest in the southeastern United States. *Veroffentlichungen des Geobotanischen Institutes der ETH, Stiftung Rübel, Zürich* 68, 79-107.

De Pratter, C. B. (1977). Environmental changes on the Georgia coast during the prehistoric period. *Early Georgia* 5, 1-14.

Dibble, D. S., and Lorraine, D. (1968). "Bonfire Shelter: A Stratified Bison Kill Site, Val Verde County, Texas." Texas Memorial Museum Miscellaneous Papers 1.

Dillehay, T. D. (1974). Late Quaternary bison population changes on the southern Plains. *Plains Anthropologist* 19, 180-96.

Dragoo, D. W. (1976). Adena and the eastern burial cult. *Archaeology of Eastern North America* 4, 1-9.

Duncan, O. D. (1959). Human ecology and population studies. *In* "The Study of Population" (P. M. Hauser and O. D. Duncan, eds.), pp. 678-716. University of Chicago Press, Chicago.

Emery, K. O., and Edwards, R. L. (1966). Archaeological potential of the Atlantic continental shelf. *American Antiquity* 31, 733-37.

Fairbridge, R. W. (1977). Discussion paper: Late Quaternary environments in northeastern coastal North America. *In* Amerinds and their paleoenvironments in northeastern North America (W. S. Newman and B. Salwen, eds.), pp. 90-92. *Annals of the New York Academy of Sciences* 288.

Fitting, J. E. (1968). Environmental potential and the postglacial readaptation in eastern North America. *American Antiquity* 33, 441-45.

Fitting, J. E. (1970). "The Archaeology of Michigan." Natural History Press, Garden City, N.Y.

Fitting, J. E. (1975). Climatic change and cultural frontiers in eastern North America. *The Michigan Archaeologist* 21, 25-39.

Fitting, J. E., Devisscher, J., and Wahla, E. J. (1966). "The Paleo-Indian Occupation of the Holcombe Beach." Museum of Anthropology, University of Michigan, Anthropological Papers 27.

Ford, J. A. (1951). "Greenhouse: A Troyville-Coles Creek Period Site in Avoyelles Parish, Louisiana." Anthropological Papers of the American Museum of Natural History 44 (1).

Ford, R. I. (1974). Northeastern archeology: Past and future directions. *Annual Review of Anthropology* 3, 385-413.

Ford, R. I. (1977). Evolutionary ecology and the evolution of human ecosystems: A case study from the midwestern U.S.A. *In* "Explanation of Prehistoric Change" (J. N. Hill, ed.), pp. 153-84. University of New Mexico Press, Albuquerque.

Ford, R. I. (1979). Gathering and gardening: Trends and consequences of Hopewell subsistence strategies. *In* "Hopewell Archaeology" (D. S. Brose and N. Greber, eds.), pp. 234-38. Kent State University Press, Kent, Ohio.

Fowler, M. L. (1971). The origin of plant cultivation in the central Mississippi Valley: A hypothesis. *In* "Prehistoric Agriculture" (S. Struever, ed.), pp. 122-28. Natural History Press, Garden City, N.Y.

Fowler, M. L. (1974). "Cahokia: Ancient Capital of the Midwest." Addison-Wesley Module in Anthropology 48, Menlo Park, Calif.

Fried, M. H. (1967). "The Evolution of Political Society." Random House, New York.

Frison, G. C. (ed.) (1974). "The Casper Site: A Hell Gap Bison Kill on the High Plains." Academic Press, New York.

Frison, G. C. (1975). Man's interaction with Holocene environments on the Plains. *Quaternary Research* 5, 289-300.

Frison, G. C. (1978). "Prehistoric Hunters of the High Plains." Academic Press, New York.

Frison, G. C., and Wilson, M. (1975). An introduction to Bighorn Basin archeology. *In* "Wyoming Geological Association Guidebook, 27th Annual Field Conference—1975," pp. 19-35. Laramie, Wyoming.

Frison, G. C., and Zeimens, G. M. (1980). Bone projectile points: An addition to the Folsom cultural complex. *American Antiquity* 45, 231-37.

Funk, R. E. (1977). Early cultures in the Hudson drainage basin. *In* Amerinds and their paleoenvironments in northeastern North America (W. S. Newman and B. Salwen, eds.), pp. 316-32. *Annals of the New York Academy of Sciences* 288.

Funk, R. E., and Rippeteau, B. E. (1977). "Adaptation, Continuity and Change in Upper Susquehanna Prehistory. Occasional Publications in Northeastern Anthropology 3.

Funk, R. E., Walters, G. R., and Ehlers, W. F., Jr. (1969). The archeology of Dutchess Quarry Cave, Orange County, New York. *Pennsylvania Archaeologist* 39, 7-22.

Gardner, W. M. (ed.) (1974). "The Flint Run Paleo-Indian Complex: A Preliminary Report 1971-73 Seasons." Archaeology Laboratory, Department of Anthropology, Catholic University of America, Occasional Publications 1.

Gibbon, G. E. (1974). A model of Mississippian development and its implications for the Red Wing area. *In* "Aspects of Upper Great Lakes Anthropology" (E. Johnson, ed.), pp. 129-37. Minnesota Prehistoric Archaeology Series 11, Minnesota Historical Society, St. Paul.

Gibson, J. L. (1974). Poverty Point, the first North American chiefdom. *Archaeology* 27, 96-105.

Goodyear, A. C. (1974). "The Brand Site: A Techno-functional Study of a Dalton Site in Northeast Arkansas." Arkansas Archeological Survey Research Series 7.

Goodyear, A. C., and Warren, L. O. (1972). Further observations on the submarine oyster shell deposits of Tampa Bay. *The Florida Anthropologist* 25, 52-66.

Gradwohl, D. M. (1969). "Prehistoric Villages in Eastern Nebraska." Nebraska State Historical Society Publications in Anthropology 4.

Griffin, J. B. (1960a). Climatic change: A contributory cause of the growth and decline of northern Hopewellian culture. *The Wisconsin Archeologist* 41, 21-33.

Griffin, J. B. (1960b). A hypothesis for the prehistory of the Winnebago. *In* "Culture in History: Essays in Honor of Paul Radin" (S. Diamond, ed.), pp. 809-65. Columbia University Press, New York.

Griffin, J. B. (1961). Some correlations of climatic and cultural change in eastern North American prehistory. *In* Solar variations, climatic change, and related geophysical problems (R. W. Fairbridge, ed.). *Annals of the New York Academy of Sciences* 95, 710-17.

Griffin, J. B. (1967). Eastern North American archaeology: A summary. *Science* 156, 175-91.

Guilday, J. E. (1967). The climatic significance of the Hosterman's Pit local fauna, Centre County, Pennsylvania. *American Antiquity* 32, 231-32.

Guilday, J. E., Martin, P. S., and McCrady, A. D. (1964). New Paris No. 4: A Pleistocene cave deposit in Bedford County, Pennsylvania. *Bulletin of the National Speleological Society* 26, 121-94.

Hall, R. L. (1973). An interpretation of the two-climax model of Illinois prehistory. Paper presented at the 9th International Congress of Anthropological and Ethnological Sciences, Chicago.

Hayes, C. V., Jr. (1964). Fluted projectile points: Their age and dispersion. *Science* 145, 1408-13.

Haynes, C. V., Jr., and Hemmings, E. T. (1968). Mammoth-bone shaft wrench from Murray Springs, Arizona. *Science* 159, 186-87.

Hines, P. (1977). On social organization in the Middle Mississippian: States or chiefdoms? *Current Anthropology* 18, 337-38.

Hough, J. L. (1963). The prehistoric Great Lakes of North America. *American Scientist* 51, 84-109.

Howells, W. W. (1960). Estimating population numbers through archaeological and skeletal remains. *In* "The Application of Quantitative Methods in Archaeology" (R. F. Heizer and S. F. Cook, eds.), pp. 158-85. Viking Fund Publications in Anthropology 28.

Hudson, C. (1976). "The Southeastern Indians." University of Tennessee Press, Knoxville.

Hurt, W. R. (1966). The Altithermal and the prehistory of the northern Plains. *Quaternaria* 8, 101-13.

Irwin, H. T. (1971). Developments in early man studies in western North America, 1960-1970. *Arctic Anthropologist* 8, 42-67.

Irwin, H. T., and Wormington, H. M. (1970). Paleo-Indian tool types in the Great Plains. *American Antiquity* 35, 24-34.

Jamison, P. L. (1971). A demographic and comparative analysis of the Albany Mounds (Illinois) Hopewell skeletons. *In* "The Indian Mounds at Albany, Illinois" (E. B. Herold, ed.), pp. 107-53.

Johnson, A. E. (1979). Kansas City Hopewell. *In* "Hopewell Archaeology: The Chillicothe Conference" (D. S. Brose and N. Greber, eds.), pp. 86-93. Kent State University Press, Kent, Ohio.

Kay, M., King, F. B., and Robinson, C. K. (1980). Cucurbits from Phillips Spring: New evidence and interpretations. *American Antiquity* 45, 806-822.

King, J. E. (1981). Late Quaternary vegetation history of Illinois. *Ecological Monographs* 51, 43-62.

Knox, J. C. (1976). Geomorphic and hydrological characteristics of Great Plains rivers during the Holocene. Paper presented at the Joint Plains-Midwest Archaeological Conference, Minneapolis.

Kroeber, A. L. (1939). "Cultural and Natural Areas of Native North America." University of California Publications in American Archaeology and Ethnology 30.

Lahren, L., and Bonnichsen, R. (1974). Bone foreshafts from a Clovis burial in southwestern Montana. *Science* 186, 147-50.

Larson, L. H. (1971). Settlement distribution during the Mississippi period. *Southeastern Archaeological Conference Bulletin* 13, 19-25.

Larson, L. H. (1972). Functional considerations of warfare in the Southeast during the Mississippi period. *American Antiquity* 37, 383-92.

Lehmer, D. J. (1952). The Turkey Bluff focus of the Fulton aspect. *American Antiquity* 17, 313-18.

Lehmer, D. J. (1954). The sedentary horizon of the northern Plains. *Southwestern Journal of Anthropology* 10, 139-59.

Lehmer, D. J. (1970). Climate and culture history in the middle Missouri Valley. *In* "Pleistocene and Recent Environments of the Central Great Plains" (W. Dort, Jr., and J. K. Jones, Jr., eds.), pp. 117-29. University Press of Kansas, Lawrence.

Lehmer, D. J. (1971). "Introduction to Middle Missouri Archaeology." National Park Service Anthropology Papers 1.

Lewis, R. B. (1974). "Mississippian Exploitative Strategies: A Southeast Missouri Example." Missouri Archaeological Society Research Series 11.

Lewis, T. M. N., and Lewis, M. K. (1961). "Eva: An Archaic Site." University of Tennessee Press, Knoxville.

Logan, W. D. (1952). "Graham Cave, an Archaic Site in Montgomery County, Missouri." Memoir of the Missouri Archaeological Society 2.

McCollough, M. C. R., and Faulkner, C. H. (1973). "Excavation of the Higgs and Doughty Sites, I-75 Salvage Archaeology." Tennessee Archaeological Society Miscellaneous Papers 12.

MacDonald, G. F. (1971). A review of research on paleo-Indian in eastern North America, 1960-1970. *Arctic Anthropology* 8, 32-41.

McKern, W. C. (1945). "Preliminary Report on the Upper Mississippi Phase in Wisconsin." Bulletin of the Public Museum of the City of Milwaukee 16, pp. 109-285.

Marquardt, W. H., and Watson, P. J. (in press). Excavation and recovery of biological remains from two Archaic shell middens in western Kentucky. *Southeastern Archaeological Conference Bulletin* 20.

Mason, R. J. (1962). The paleo-Indian tradition in eastern North America. *Current Anthropology* 3, 227-78.

Matteson, M. R. (1960). Reconstruction of prehistoric environments through the analysis of molluscan collections from shell middens. *American Antiquity* 26, 117-20.

Milanich, J. T. (1980). Weeden Island studies—Past, present, and future. *Southeastern Archaeological Conference Bulletin* 22, 11-18.

Morse, D. F. (1971). Recent indications of Dalton settlement pattern in northeast Arkansas. *Southeastern Archaeological Conference Bulletin* 13, 5-10.

Morse, D. F. (1973). Dalton culture in northeast Arkansas. *The Florida Anthropologist* 26, 23-38.

Munson, P. J. (1973). The origins and antiquity of maize-beans-squash agriculture in eastern North America: Some linguistic implications. *In* "Variation in Anthropology" (D. W. Lathrap and J. Douglas, eds.), pp. 107-35. Illinois Archaeological Survey, Urbana.

Munson, P. J., Parmalee, P. W., and Yarnell, R. A. (1971). Subsistence ecology of Scoville, a terminal Middle Woodland village. *American Antiquity* 36, 410-31.

Murphy, J. L. (1971). Maize from an Adena mound in Athens County, Ohio. *Science* 171, 897-98.

Norona, D. (1962). "Moundsville's Mammoth Mound." Privately published, Moundsville, Ala.

O'Brien, P. J. (1972). Urbanism, Cahokia and Middle Mississippian. *Archaeology* 25, 189-97.

Odum, E. P. (1969). The strategy of ecosystem development. *Science* 164, 262-70.

Palmer, H. A., and Stoltman, J. B. (1976). The Boaz mastodon: A possible association of man and mastodon in Wisconsin. *Midcontinental Journal of Archaeology* 1, 163-77.

Parmalee, P. W. (1962). Faunal remains from the Stanfield-Worley Bluff Shelter, Colbert County, Alabama. *Journal of Alabama Archaeology* 8, 112-14.

Parmalee, P. W. (1968). Cave and archaeological faunal deposits as indicators of post-Pleistocene animal populations and distribution in Illinois. *In* "The Quaternary of Illinois" (R. E. Bergstrom, ed.), pp. 104-13. University of Illinois College of Agriculture Special Publication 14.

Parmalee, P. W., McMillan, R. B., and King, F. B. (1976). Changing subsistence patterns at Rodgers Shelter. *In* "Prehistoric Man and His Environments: A Case Study in the Ozark Highlands" (W. R. Wood and R. B. McMillan, eds.), pp. 141-61. Academic Press, New York.

Peebles, C. S. (1972). Monothetic-Divisive analysis of the Moundville burials: An initial report. *Newsletter of Computer Archaeology* 8, 1-13.

Peebles, C. S. (1978). Determinants of settlement size and location in the Moundville phase. *In* ''Mississippian Settlement Patterns'' (B. D. Smith, ed.), pp. 369-416. Academic Press, New York.

Peebles, C. S., and Kus, S. M. (1977). Some archaeological correlates of ranked societies. *American Antiquity* 42, 421-48.

Phillips, P. (1970). ''Archaeological Survey in the Lower Yazoo Basin, Mississippi, 1949-1955.'' Papers of the Peabody Museum of Archaeology and Ethnology 60.

Prufer, O. H. (1963). ''The McConnell Site, a late Palaeo-Indian Workshop in Coshocton County, Ohio. Scientific Publications of the Cleveland Museum of Natural History (new series) 2 (1).

Quimby, G. I. (1960). ''Indian Life in the Upper Great Lakes, 11,000 B.C. to A.D. 1800.'' University of Chicago Press, Chicago.

Reeves, B. (1973). The concept of an Altithermal cultural hiatus in northern Plains prehistory. *American Anthropologist* 75, 1221-53.

Riley, T. J., and Freimuth, G. (1979). Field systems and frost drainage in the prehistoric agriculture of the Upper Great Lakes. *American Antiquity* 44, 271-85.

Rippeteau, B. E. (1977). New data and models for 8,000 years of Archaic societies and their environments in the Upper Susquehanna Valley. *In* Amerinds and their paleoenvironments in northeastern North America (W. S. Newman and B. Salwen, eds.). *Annals of the New York Academy of Sciences* 288, 392-99.

Ritchie, W. A. (1932). ''The Lamoka Lake Site.'' Researches and Transactions of the New York State Archaeological Association 7 (4).

Ritchie, W. A. (1969). ''The Archaeology of Martha's Vineyard.'' Natural History Press, Garden City, N.Y.

Ritchie, W. A., and Funk, R. E. (1973). ''Aboriginal Settlement Patterns in the Northeast.'' New York State Museum and Science Service Memoir 20.

Ritzenthaler, R. (ed.) (1957). The Old Copper culture of Wisconsin. *The Wisconsin Archeologist* 38, 185-329.

Robbins, M. (1960). ''Wapanucket No. 6, an Archaic Village in Middleboro, Massachusetts.'' Massachusetts Archaeological Society, Attleboro.

Roedl, L. J., and Howard, J. H. (1957). Archaeological investigations at the Renner site. *The Missouri Archaeologist* 18, 52-96.

Roper, D. C. (1979). ''Archaeological Survey and Settlement Pattern Models in Central Illinois.'' *Midcontinental Journal of Archaeology* Special Paper 2.

Ruppé, R. J. (1980). The archaeology of drowned terrestrial sites: A preliminary report. *Bureau of Historical Sites and Properties Bulletin* 4, 35-45.

Sanger, D. (1975). Culture change as an adaptive process in the Maine-Maritimes region. *Arctic Anthropology* 12, 60-75.

Sanger, D., Davis, R. B., MacKay, R. G., and Borns, H. W., Jr. (1977). The Hirundo archaeological project—An interdisciplinary approach to central Maine prehistory. *In* Amerinds and their paleoenvironments in northeastern North America (W. S. Newman and B. Salwen, eds.), *Annals the New York Academy of Sciences* 288, 457-71.

Saucier, R. T. (1974). ''Quaternary Geology of the Lower Mississippi Valley.'' Arkansas Archeological Survey Research Series 6.

Schmits, L. J. (1976). The Coffey site: Environment and cultural adaptation at a prairie plains Archaic site. M.A. thesis, University of Kansas, Lawrence.

Schoenwetter, J. (1979). Comment on ''Plant Husbandry in Prehistoric Eastern North America.'' *American Antiquity* 44, 600-601.

Schultz, C. B. (1943). Some artifact sites of early man in the Great Plains and adjacent areas. *American Antiquity* 8, 242-49.

Sears, W. H. (1971). Food production and village life in prehistoric southeastern United States. *Archaeology* 24, 322-29.

Seeman, M. F. (1979a). Feasting with the dead: Ohio Hopewell charnel house ritual as a context for redistribution. *In* ''Hopewell Archaeology.'' (D. S. Brose and N. Greber, eds.), pp. 39-46. Kent State University Press, Kent, Ohio.

Seeman, M. F. (1979b). ''The Hopewell Interaction Sphere: The Evidence for Interregional Trade and Structural Complexity.'' Indiana Historical Society Prehistory Research Series 5, pp. 235-438.

Service, E. R. (1971). ''Primitive Social Organization.'' 2nd ed. Random House, New York.

Shay, C. T. (1971). ''The Itasca Bison Kill Site: An Ecological Analysis.'' Minnesota Prehistoric Archaeology Series, Minnesota Historical Society, St. Paul.

Shea, A. B. (1978). An analysis of plant remains from the Middle Woodland and Mississippian components of the Banks V site and a paleoethnobotanical study of the native flora of the upper Duck Valley. *In* ''Fifth Report of the Normandy Archaeological Project'' (C. H. Faulkner and M. C. R. McCollough, eds.), pp. 596-699. Department of Anthropology, University of Tennessee, Report of Investigations 20.

Shetrone, H. C. (1926). Exploration of the Hopewell group of prehistoric earthworks. *Ohio State Archaeological and Historical Quarterly* 35, 1-227.

Shutler, R., Jr., Anderson, D. C., Hallberg, G. A., Hoyer, B. E., Miller, G. A., Butler, K. E., Semken, H. A., Jr., Baerreis, D. A., Koeppen, R. C., Conrad, L. A., and Tiffany, J. A. (1974). The Cherokee sewer site (13CK405): A preliminary report of a stratified paleo-Indian/Archaic site in northwestern Iowa. *Journal of the Iowa Archeological Society* 21, 1-175.

Smith, B. D. (1975). ''Middle Mississippi Exploitation of Animal Resources.'' Museum of Anthropology, University of Michigan, Anthropological Papers 57.

Smith, B. D. (ed.) (1978). ''Mississippian Settlement Patterns.'' Academic Press, New York.

Smith, R. L. (1976). Concept of the ecosystem. *In* ''The Ecology of Man: An Ecosystem Approach'' (R. L. Smith, ed.), 2nd ed., pp. 4-23. Harper and Row, New York.

Snook, J. C., and Swartz, B. K., Jr. (1975). An over extension of the mud-flats hypothesis? A comment on the Struever-Vickery hypothesis. *American Anthropologist* 77, 88.

Snow, D. R. (1969). ''A Summary of Excavations at the Hathaway Site in Passadumkeag, Maine, 1912, 1947, and 1968.'' Department of Anthropology, University of Maine, Orono.

Snow, D. R. (1972). Rising sea level and prehistoric cultural ecology in northern New England. *American Antiquity* 37, 211-21.

Steponaitis, V. P. (1978). Location theory and complex chiefdoms: A Mississippian example. *In* ''Mississippian Settlement Patterns'' (B. D. Smith, ed.), pp. 417-53. Academic Press, New York.

Stoltman, J. B. (1972). The Late Archaic in the Savannah River region. *In* ''Fiber-Tempered Pottery in Southeastern United States and Northern Columbia: Its Origins, Context, and Significance'' (R. P. Bullen and J. B. Stoltman, eds.), pp. 37-62. Florida Anthropological Society Publications 6.

Stoltman, J. B. (1978). Temporal models in prehistory: An example from eastern North America. *Current Anthropology* 19, 703-46.

Struever, S. (1968). Woodland subsistence-settlement systems in the lower Illinois Valley. *In* ''New Perspectives in Archaeology'' (S. R. Binford and L. R. Binford, eds.), pp. 285-312. Aldine, Chicago.

Struever, S. (1964). The Hopewell interaction sphere in riverine-western Great Lakes culture history. *In* ''Hopewellian Studies'' (J. R. Caldwell and R. L. Hall, eds.), pp. 86-106. Illinois State Museum Scientific Papers 12.

Struever, S., and Holton, F. A. (1979). ''Koster.'' Anchor Press/Doubleday, Garden City, N.Y.

Struever, S., and Houart, G. (1972). An analysis of the Hopewell interaction sphere. *In* ''Social Exchange and Interaction'' (E. Wilmsen, ed.), pp. 47-79. Museum of Anthropology, University of Michigan Anthropological Papers 46.

Struever, S., and Vickery, K. D. (1973). The beginnings of cultivation in the midwest-riverine area of the United States. *American Anthropologist* 75, 1197-220.

Sussman, R. W. (1972). Child transport, family size, and increase in human population during the Neolithic. *Current Anthropology* 13, 258-59.

Tiffany, J. A. (1979). An overview of Oneota sites in southeastern Iowa: A perspective from the ceramic analysis of the Schmeiser site, 13DM101, Des Moines County, Iowa. *Proceedings of the Iowa Academy of Sciences* 86, 89-101.

Turnbaugh, W. A. (1975). Toward an explanation of the broadpoint dispersal in eastern North American prehistory. *Journal of Anthropological Research* 31, 51-68.

Vickery, K. D. (1970). Evidence supporting the theory of climatic change and the decline of Hopewell. *The Wisconsin Archeologist* 51, 57-76.

Ward, T. (1965). Correlation of Mississippian sites and soil types. *Southeastern Archaeological Conference Bulletin* 3, 42-48.

Waselkov, G. A. (1978). Evolution of deer hunting in the Eastern Woodlands. *Midcontinental Journal of Archaeology* 3, 15-34.

Watts, W. A. (1969). A pollen diagram from Mud Lake, Marion County, north-central Florida. *Geological Society of America Bulletin* 80, 631-42.

Watts, W. A. (1970). The full-glacial vegetation of northwestern Georgia. *Ecology* 51, 17-33.

Watts, W. A. (1971). Postglacial and interglacial vegetation history of southern Georgia and central Florida. *Ecology* 52, 676-90.

Watts, W. A. (1980a). Late-Quaternary vegetation history at White Pond on the inner Coastal Plain of South Carolina. *Quaternary Research* 13, 187-99.

Watts, W. A. (1980b). The late-Quaternary vegetation history of the southeastern United States. *Annual Review of Ecology and Systematics* 11, 387-409.

Watts, W. A., and Stuiver, M. (1980). Late Wisconsin climate of northern Florida and the origin of species-rich deciduous forest. *Science* 210, 325-27.

Webb, C. H. (1968). The extent and content of Poverty Point culture. *American Antiquity* 33, 297-331.

Webb, T., III (1974). A vegetational history from northern Wisconsin: Evidence from modern and fossil pollen. *The American Midland Naturalist* 92, 12-34.

Webb, T., III, and Bryson, R. A. (1972). Late- and postglacial climatic change in the northern Midwest, USA: Quantitative estimates derived from fossil pollen spectra by multivariate statistical analysis. *Quaternary Research* 2, 70-115.

Wedel, W. R. (1941). "Environment and Native Subsistence Economies in the Central Great Plains." Smithsonian Miscellaneous Collections 101 (3).

Wedel, W. R. (1943). "Archaeological Investigations in Platte and Clay Counties, Missouri," U.S. National Museum Bulletin 183.

Wedel, W. R. (1947). Prehistory and Environment in the central Great Plains. *Transactions of the Kansas Academy of Science* 50, 1-18.

Wedel, W. R. (1953). Some aspects of human ecology in the central Plains. *American Anthropologist* 55, 499-514.

Wedel, W. R. (1961). "Prehistoric Man on the Great Plains." University of Oklahoma Press, Norman.

Wedel, W. R. (1970). Some observations on two house sites in the central Plains: An experiment in archaeology. *Nebraska History* 51, 225-52.

Wedel, W. R. (1978). The prehistoric Plains. *In* "Ancient North Americans" (J. D. Jennings, ed.), pp. 183-220. W. H. Freeman, San Francisco.

Wedel, W. R., Husted, W. M., and Moss, J. H. (1968). Mummy Cave: Prehistoric record from Rocky Mountains of Wyoming. *Science* 160, 184-85.

Wendland, W. M., and Bryson, R. A. (1974). Dating climatic episodes of the Holocene. *Quaternary Research* 4, 9-24.

Wheat, J. B. (1972). "The Olsen-Chubbuck Site: A Paleo-Indian Bison Kill." Memoirs of the Society for American Archaeology 26.

Whitehead, D. R. (1965). Prehistoric maize in southeastern Virginia. *Science* 150, 881-82.

Will, G. F., and Hyde, G. E. (1917). "Corn among the Indians of the Upper Missouri." W. H. Miner, St. Louis.

Williams, S., and Stoltman, J. B. (1965). An outline of southeastern United States prehistory with particular emphasis on the paleo-Indian era. *In* "The Quaternary of the United States" (H. E. Wright, Jr., and D. G. Frey, eds.), pp. 669-83. Princeton University Press, Princeton, N.J.

Wilmsen, E. N. (1970). "Lithic Analysis and Cultural Inference, a Paleo-Indian Case." Anthropological Papers of the University of Arizona 16.

Winters, H. D. (1968). Value systems and trade cycles of the Late Archaic in the Midwest. *In* "New Perspectives in Archeology" (S. R. Binford and L. R. Binford, eds.), pp. 175-221. Aldine, Chicago.

Wittry, W. L. (1965). The institute digs a mastodon. *Cranbrook Institute of Science News Letter* 35, 14-25.

Wood, W. R. (ed.) (1969). "Two House Sites in the Central Plains: An Experiment in Archaeology." Plains Anthropologist Memoir 6.

Wood, W. R., and McMillan, R. B. (eds.) (1976). "Prehistoric Man and His Environments: A Case Study in the Ozark Highland." Academic Press, New York.

Wormington, H. M. (1957). "Ancient Man in North America." 4th ed. Denver Museum of Natural History Popular Series 4.

Wright, H. E., Jr. (1968). History of the Prairie Peninsula. *In* "The Quaternary of Illinois" (R. E. Bergstrom, ed.), pp. 78-88. University of Illinois College of Agriculture Special Publication 14.

Wright, H. E., Jr. (1970). Vegetational history of the central Plains. *In* "Pleistocene and Recent Environments of the Central Great Plains" (W. Dort, Jr., and J. K. Jones, eds.), pp. 157-72. Department of Geology, University of Kansas, Special Publication 3.

Wright, H. E., Jr. (1971). Late Quaternary vegetational history of North America. *In* "The Late Cenozoic Glacial Ages" (K. K. Turekian, ed.), pp. 425-64. Yale University Press, New Haven, Conn.

Wright, H. E., Jr. (1974). The environment of early man in the Great Lakes region. *In* "Aspects of Upper Great Lakes Anthropology" (E. Johnson, ed.), pp. 8-14. Minnesota Prehistoric Archaeology Series 11, Minnesota Historical Society, St. Paul.

Yarnell, R. A. (1964). "Aboriginal Relationships between Culture and Plant Life in the Upper Great Lakes Region." Museum of Anthropology, University of Michigan, Anthropological Papers 23.

Yarnell, R. A. (1965). Early Woodland plant remains and the question of cultivation. *The Florida Anthropologist* 18, 77-82.

Yarnell, R. A. (1969). Contents of human paleofeces. *In* "The Prehistory of Salts Cave, Kentucky" (P. J. Watson, ed.), pp. 41-54. Illinois State Museum Reports of Investigations 16.

Yarnell, R. A. (1974). Plant food and cultivation of the Salts Cavers. *In* "Archaeology of the Mammoth Cave Area" (P. J. Watson, ed.), pp. 113-22. Academic Press, New York.

Yarnell, R. A. (1976). Early plant husbandry in eastern North America. *In* "Culture Change and Continuity" (C. E. Cleland, ed.), pp. 265-73. Academic Press, New York.

Yarnell, R. A. (1977). Native plant husbandry north of Mexico. *In* "Origins of Agriculture" (C. A. Reed, ed.), pp. 861-75. Mouton, The Hague.

Yarnell, R. A. (1978). Domestication of sunflower and sumpweed in eastern North America. *In* "The Nature and Status of Ethnobotany" (R. A. Ford, ed.), pp. 289-99. Museum of Anthropology, University of Michigan, Anthropological Papers 67.

Climatology

CHAPTER 17

Modeling of Holocene Climates

John E. Kutzbach

Introduction

The climatic events of the Holocene have been dramatic. To place the events in perspective, it is useful to consider the entire sequence of climates, starting with the glacial maximum conditions (around 18,000 yr B.P.) and proceeding through the deglaciation to the interglacial maximum (early to middle Holocene) and then to the climate of our time. The facts concerning the starting point of this climatic sequence have become increasingly well established, thanks in large part to the intensive efforts of the CLIMAP project (1976) in charting the sea-surface temperature, sea level, sea-ice extent, land albedo, and ice-sheet topography for the glacial world. A wide variety of paleoclimatic evidence from the land surface has been assembled and summarized for this period (Barry, 1983; Peterson et al., 1979).

These types of data are precisely those required for global models of the general atmospheric circulation, and several model experiments have been conducted (Williams et al., 1974; Gates, 1976a, 1976b; Manabe and Hahn, 1977). For example, Gates (1976b) compares a simulation of the ice-age climate with a simulation of present July conditions. He reports that the ice-age atmosphere was substantially cooler and drier (especially over the continents of the Northern Hemisphere) and corresponded to an enhanced anticyclonic circulation over the major ice sheets and a general weakening of the summer monsoonal circulation. In the vicinity of the major ice sheets, the middle-latitude westerlies were strengthened and systematically displaced southward.

These pioneering efforts in paleoclimatic analysis and modeling advanced our knowledge of past climates and indicated some problems that remain. The climatic conditions for 18,000 yr B.P. are still unknown in some regions (Peterson et al., 1979), and the period since the glacial maximum has large data gaps in both space and time. Climate models are in an early stage of development, and so far the simulated paleoclimates are not everywhere consistent with the paleoclimatic evidence. Different models may also produce different climates because of differences in the physics of the models, the conditions at the boundary, or the spatial resolution or temporal integration of the models.

The period since 18,000 yr B.P. has been punctuated with change:

the ice sheets retreated, the ocean fronts shifted poleward, areas of sea ice contracted, certain middle-latitude lakes became desiccated and, around 10,000 yr B.P., lake levels rose throughout much of Africa and the Middle East. By middle-Holocene time, the global ice volume had reached a minimum value, the North American ice sheet had disappeared, and certain types of vegetation had reached the limits of their poleward migration. Shortly thereafter, spruce forests moved southward, perhaps as northern temperate climates cooled, and tropical lakes dried. It is these diverse changes that present a challenge for the modeling of the climate of the Holocene.

This chapter emphasizes the possible effects of orbital variations and the consequent changes in solar radiation on the transition from maximum glacial to maximum interglacial climate, and it is concerned as well with subsequent changes. The starting point is the relationship between global ice-volume changes and solar-radiation anomalies resulting from orbital variations. Latitudinal-average energy-balance models add a level of detail to the treatment of the seasonal and latitudinal distribution of the solar-radiation anomalies. A variety of regional processes and feedback mechanisms that may eventually be included in Holocene climate models are described in the fourth section of the chapter. A global model of the general atmospheric circulation allows the most complete description of regional climates and is used to estimate the climate of the early Holocene for the solar-radiation distribution of 9000 yr B.P. The last section describes other possible explanations for the climatic changes of the Holocene. This chapter discusses selected recent articles and does not constitute a historical review.

Research grants to the University of Wisconsin at Madison from the National Science Foundation's Climate Dynamics Program (NSF grants numbers ATM79-16443 and ATM79-26039) supported this work. In connection with the results that are reported in the fifth section of the chapter, the author thanks B. Otto-Bliesner for collaborating in the use of the low-resolution general circulation model, P. Guetter for performing the computations, and E. Hopkins for providing the algorithm that was used to calculate the insolation for 9000 yr B.P. The model computations were carried out at the National Center for Atmospheric Research, Boulder, Colorado, with a computing grant from the NCAR Computing Facility. M. Woodworth and E. Seehawer are thanked for preparing the manuscript.

271

Global Ice-Volume Models: The Role of Orbital Variations

There is growing evidence that variations in the Earth's orbital parameters (obliquity, precession, and eccentricity) played a major role in glacial-interglacial climatic changes (Hays et al., 1976). This deterministic framework, if upheld by future findings, will provide an exciting opportunity for progress in climatic research. The known external changes in the Earth's orbital parameters represent the "input" to a complicated dynamical system; the observed climate is the "output." From a knowledge of the variations of the input and output through time, it may be possible to deduce important characteristics of the climatic system.

This approach has been used by Imbrie and Imbrie (1980) to develop a physically based and time-dependent model that simulated the glacial-interglacial variations of the past half-million years. The model relates the time-rate-of-change of the total volume of land ice (y) to orbital forcing (x) by an equation of the form $dy/dt = (x - y)/T_i$, where the rate of ice-volume change is inversely proportional to a time constant T_i. The time constant assumes one of two specified values, depending upon whether the climate is warming (T_w) or cooling (T_c). After these time constants are tuned for optimum fit, the model places the last glacial maximum (maximum volume of land ice) at around 18,000 yr B.P. and the maximum interglacial (minimum volume of land ice) at around 6000 yr B.P. The time constants (half-response times) for T_w and T_c are 7300 and 29,500 years, respectively. For the purpose of modeling Holocene climates, the important result is neither the exact date of the glacial or interglacial maximum nor the precise value of a time constant; rather, it is that the external forcing of the climate by the orbital variations produces a simulation of land-ice volume that closely matches the observed ice-volume record. The ice-volume record is inferred from oxygen-isotope curves (Figure 17-1).

A physically based model of the mass budget and flow of continental ice sheets (Birchfield et al., 1981) has been used to simulate the growth and decay of an ice sheet in response to solar-radiation changes that were associated with the orbital perturbations. The summer solar-radiation anomalies directly forced latitudinal displacement of the snow line.

Although these two models simulate certain general features of the global ice-volume record quite well, they underestimate the strength of the prominent 100,000-year period. This important deficiency has motivated the development of more-detailed physical models.

Latitudinal-Average Energy-Balance Models: Response to Orbital Variations

Budyko (1969), Suarez and Held (1976, 1979), Schneider and Thompson (1979), and Pollard and others (1980) have studied the latitudinal-average climatic response of energy-balance models to changes in solar radiation caused by variations in orbital parameters. Budyko's 1969 model is based on annual-average conditions; the more recent studies include the effects of seasonal variations.

Figure 17-2 shows the solar-radiation variations of the past 100,000 years for January and July as calculated by Berger (1978). The changes reflect the variations in obliquity of the Earth's axis (with a period of about 41,000 years), the precession of the equinox (with a period of about 21,000 years), and the eccentricity of the

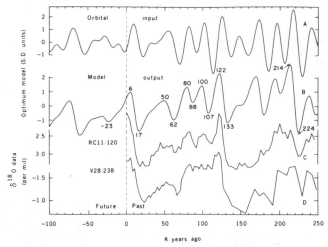

Figure 17-1. Input and output of Imbrie and Imbrie (1980) model compared with isotopic data on climate of the past 250,000 years. (A) Orbital input corresponding to an irradiation curve for July at latitude 65°N. (B) Output of a system function with a mean time constant of 17,000 years and a ratio of four to one between the time constants of glacial growth and melting. Ages of selected maxima and minima are given in thousands of years. According to this model, the influence of orbital variations over the next 23,000 years will enlarge the continental ice sheets. (C) Oxygen isotope curve for deep-sea core RC11-120 from the southern Indian Ocean. (D) Oxygen isotope curve for deep-sea core V-28-238 from the Pacific Ocean (Pee Dee belemnite [PDB] standard). (From Imbrie and Imbrie, 1980; *Science*, vol. 207 (February), 943-53; copyright 1980 by the American Association for the Advancement of Science.)

Earth's orbit (with a period of about 100,000 years). The precession effect tends to dominate in low latitudes. The orbital variations change the latitudinal and seasonal distribution of solar radiation; there is negligible change in the annual global-average amount. There is a strong tendency for compensation of seasonal extremes as with the increased radiation in July and the decreased radiation in January 10,000 yr B.P.

Because of the seasonal variations of climatic variables, such as albedo (snow cover) and ocean-heat storage (upper mixed layer), the response of a model to variations of the seasonal cycle of solar radiation may differ significantly from the response to variations of the annual-average radiation. The role of seasonal variations can be appreciated from an inspection of the energy-balance equation (Schneider and Thompson, 1979):

$$R \frac{\Delta T}{\Delta t} = Q (1 - a) - F_{IR} + H_{NET} ,$$

where all terms are functions of latitude and where T is surface temperature, $\Delta T/\Delta t$ is the change of temperature with time, R is the thermal inertia coefficient (often expressed in terms of the heat content of the ocean's mixed layer), Q is the solar radiation, a is the planetary albedo, F_{IR} is the infrared radiation going out to space (often expressed empirically as a linear function of surface temperature), and H_{NET} represents the net heat flux into a latitudinal zone from the combined processes of atmospheric and oceanic meridional heat transport. For example, if the planetary albedo (a) is allowed to vary as a function of temperature (being higher in winter and lower in summer because of the relation between snow cover and temperature) and if the seasonal cycle of Q is amplified because of increased axial tilt so that there is more insolation in summer and less in winter, then the in-phase changes of Q and $1-a$ cause an increase of annual-average absorbed solar radiation, $Q(1-a)$, even though the annual-average solar radiation is unchanged.

The energy-balance equation can also be used to illustrate the differential temperature response of land and ocean to a specified change of solar radiation (Q). Because the thermal inertia coefficient (R) of the ocean exceeds that of land, the temperature response ($\Delta T/\Delta t$) is larger for land than for ocean (Kutzbach, 1981).

Both the Suarez-Held and Schneider-Thompson models simulate glacial and interglacial maxima at times that were several thousand years earlier than the observations (for example, maximum interglacial at 12,000 to 10,000 yr B.P. instead of 6000 yr B.P.). This discrepancy has been attributed to the lack of an explicit ice sheet in the model that might require several thousand years to build or to decay.

Birchfield and others (1981) and Pollard and others (1980) include explicit ice sheets in their models, thereby improving the agreement between model and observation. However, the magnitude of the observed glacial-interglacial fluctuations is underestimated, and the fluctuations are not concentrated sufficiently at the 100,000-year period.

Regional Processes

Before global models of the general circulation of the atmosphere and ocean can be applied successfully to the detailed simulation of Holocene climates, a variety of regional processes must be studied. These processes may involve important feedback mechanisms that could influence the climatic response to orbital variations or other external forcing changes. Several examples from polar and tropical latitudes are mentioned.

POLAR LATITUDES

A better understanding of polar processes is important for modeling the changes in snow cover, ice sheets, ice shelves, sea ice, and ocean currents that have occurred since the time of the glacial maximum and throughout the Holocene.

Andrews (1973) and Hare (1976) have tried to explain the rapid wastage of the North American ice sheets between 10,000 and 7000 yr B.P. in climatological terms. Hare (1976) argues that even the combination of radiative and nonradiative processes (the latter including warm-air advection, condensation on the ice, and warm rain) would not be sufficient to account for the rapid deglaciation. Andrews (1973) suggests that wastage of the ice margin by calving into lakes and the sea was important. More recently, the concept of the disintegration of marine-based ice sheets (Hughes et al., 1977) has been used to explain the rapid deglaciation. Andrews and Barry (1978) argue that the final answer might involve several causes.

Johnson and McClure (1976) and Ruddiman and McIntyre (1981) propose rather detailed physical models for the growth and decay of the Laurentide ice sheet. Their conceptual models include a moisture-feedback loop that links the accumulation of snow on the ice sheet to the nearest moisture source—the North Atlantic Ocean. Ruddiman and McIntyre's model of the ice-decay sequence (Figure 17-3) is based upon their analysis of deep-ocean cores, and it involves the following sequence of events. Increased solar radiation in summer, from obliquity and precession changes, led to decreased continental ice volume and increased low-salinity glacial meltwater in the North Atlantic

Ocean. The combination of a stable, stratified low-salinity ocean surface layer and decreased solar radiation in winter (the orbital changes tend to pair increased summer radiation with decreased winter radiation) (Figure 17-2) led to the extension of winter sea ice in the North Atlantic. This southward shift of winter sea ice implied that the winter cyclonic storm track was shifted to the south. This resulted in decreased winter moisture flux onto the ice sheet and an acceleration of its decay. Ruddiman and McIntyre contend that this moisture feedback mechanism, coupled with the mechanical impact of rising sea levels on marine-based ice sheets (Denton and Hughes, 1981), would amplify the net rate of ice-volume decrease at terminations.

Another example of the possible importance of ice-sheet processes for climate models comes from the Antarctic. Thomas and Bentley (1978) apply a marine ice-sheet ice-shelf model to the case of a hypothetical "Ross ice sheet," which, 18,000 years ago, may have covered the present-day Ross Ice Shelf. By allowing the world sea-level to rise, they calculate retreat rates and find that most of the present-day Ross Ice Shelf was free of grounded ice by about 7000 yr B.P. According to their model, if ice shelves had not formed during the retreat, then most of the west Antarctic ice sheet would have collapsed. They concluded that the present day Ross Ice Shelf serves to stabilize the west Antarctic ice sheet.

The sensitivity of the climate of polar regions to changes in solar radiation has been demonstrated with general circulation-model experiments. Wetherald and Manabe (1975) report that a 2% increase in the solar constant produces a 4°C to 6°C increase of temperature poleward of latitude 65° compared to a 1° to 2°C temperature increase equatorward of latitude 30°. This increased high-latitude sensitivity resulted from the effects of the snow-cover feedback mechanism as well as from the suppression of vertical mixing by the stable lower troposphere. This sensitivity figured prominently in a model that showed that "instantaneous glacierization" could develop from an initial thin (and patchy) snow cover (Ives et al., 1975). An experiment with the British Meteorological Office's general circulation model compared May and June simulations for modern conditions with conditions for 10,000 yr B.P., when there was more solar radiation at high northern latitudes (Mason, 1976). The simulated climate of 10,000 yr B.P. was warmer than that of the present day,

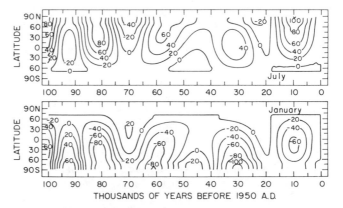

Figure 17-2. Time series of the zonal distribution of the solar radiation perturbation arising from variations in eccentricity, obliquity, and longitude of perihelion for January and July. Shown are the deviations of midmonth values of daily insolation (cal cm⁻² day⁻¹) from their A.D. 1950 values. (One cal cm⁻² day⁻¹ is equivalent to 0.484 W m⁻².) (From A. L. Berger, personal communication, 1977; as published in Schneider and Thompson, 1979.)

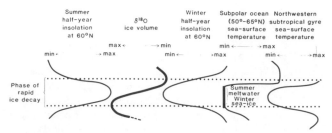

Figure 17-3. Sequence of changes associated with major ice-decay stages for the North Atlantic Ocean. Vertical axis is time. (From Ruddiman and McIntyre, 1981; *Science*, vol. 212 (March), pp. 617-27; copyright 1981 by the American Association for the Advancement of Science.)

with differences ranging from 6 °C over the Arctic to 4 °C at 30 °N, 2 °C at 30 °S, and 0 °C at the South Pole.

TROPICAL AND SUBTROPICAL LATITUDES

A better understanding of processes in the tropics and subtropics is important for modeling the changes in land surface hydrology (ground wetness, lake levels, runoff, evaporation), vegetation, and ocean currents that have occurred during the Holocene.

Global maps of lake-level fluctuations since the time before the glacial maximum (Street and Grove, 1979) illustrate that most tropical regions experienced low lake levels at or shortly after the glacial maximum and high lake levels during the maximum interglacial (10,000 to 5000 yr B.P.). The southwestern United States, on the other hand, experienced the highest lake levels near the glacial maximum (details of timing still need to be clarified) (Barry, 1983). Major shifts in lake-level regime occurred at approximately the same time in many parts of the world.

Regional climate models based on hydrologic and energy-balance equations have been used to estimate the changes in precipitation that would be required to account for the changes in lake level and lake size (Butzer et al., 1972; Street, 1979; Kutzbach, 1980). These models have been applied to lake basins in Ethiopia, West Africa, East Africa, and northwestern India. For the times of maximum lake extent, increases of precipitation of 200 to 400 mm per year compared to present were estimated; these absolute increases amounted to percentage increases of 20% to 80%, depending upon the modern value of annual precipitation. Similar estimates of precipitation have been obtained from pollen-climate calibration models (Bryson and Swain, 1981). Changes of precipitation and vegetation of this magnitude may also have been accompanied by changes in the albedo and ground wetness of the region surrounding the lakes. Experiments with general circulation models (Charney et al., 1975) have identified a so-called biosphere feedback mechanism, whereby an initial increase of rainfall could be amplified. Increased rainfall could be followed by a decrease of albedo (due to increased plant cover), an increase of heating on the land, and a strengthening of the monsoon. Increased rainfall could also cause an increase of ground wetness and additional recycling of water through multiple evaporation and precipitation cycles. (See also Cess, 1978.)

In view of the coincidence of increased precipitation over parts of the tropics and subtropics with increased solar radiation in Northern Hemisphere summer around 10,000 to 9000 yr B.P. (Figure 17-2), it is of interest that general circulation models show an extreme sensitivity in the intensity of the simulated hydrologic cycle to small changes of the solar constant. Wetherald and Manabe (1975) have found that a 6% increase of the solar constant (S) produced a 27% in-

crease of precipitation (P); that is, $\Delta P/P = 4.5\ \Delta S/S$. This large intensification of the hydrologic cycle in the model atmosphere resulted from an increase in the rate of evaporation, which, in turn, was caused by the following changes: (1) a reduction of the Bowen ratio due to the nonlinear dependence of the saturation vapor pressure on the temperature at the Earth's surface and (2) a decrease in the net upward long-wave radiation from the surface as a result of the increased moisture in the air (fixed *relative* humidity was assumed) and the reduction of the lapse rate. Of course, the orbital variations change the seasonal and latitudinal distribution of solar radiation, not the annual-average amount. Nevertheless, the results of the Wetherald-Manabe experiment suggest that increased solar radiation during the primary precipitation season would lead to an overall increase of precipitation.

A Model of the General Atmospheric Circulation for 9000 yr B.P.

Aspects of the preceding discussion provided the rationale for an experiment performed by Kutzbach (1981) and aimed at examining the possible role of orbital variations in providing at least a partial explanation for the increased tropical lake levels of the early Holocene. Previous work on orbital variations and monsoon climates has been summarized elsewhere (Kutzbach, 1983).

The general circulation model developed by Otto-Bliesner and others (1982) was used for the experiment. The model had five atmospheric levels and included atmospheric dynamics and thermodynamics, the orographic influences of mountains (and ice sheets), radiative and convective processes, condensation, and evaporation. A surface heat budget was computed over the land. Ocean-surface temperature, land albedo, and ice-sheet topography were specified. The model had lower spatial resolution (about 11 ° latitude by 11 ° longitude) and incorporated less-exact representations of certain physical processes than the models used by Gates (1976a, 1976b) and Manabe and Hahn (1977) for paleoclimatic simulation. Nevertheless, when tested with the CLIMAP project members' (1980) boundary conditions for August of 18,000 yr B.P., the simulated climate was similar to the results for 18,000 yr B.P. described by Gates (1976a, 1976b). (See Kutzbach and Wright, 1982.)

The orbital variations produced a maximum of solar radiation in Northern Hemisphere summer around 10,000 to 9000 yr B.P. (Figure 17-2). About 9000 yr B.P., obliquity was 24.23 ° (compared to 23.45 ° at present), perihelion was in Northern Hemisphere summer (July 30 compared to January 3 at present), and eccentricity was 0.0193 (compared to 0.0167 at present). These factors combined to produce increased solar radiation for July and decreased radiation for January. These changes exceeded 25 to 35 W per square meter, or about 7% of modern values, over a broad band of latitudes (Berger, 1978; Kutzbach, 1981).

Nine thousand years ago, the increased solar radiation for June to August warmed the model's land surface (relative to the ocean) and increased the land-ocean temperature constrast, compared to present conditions; in December to February, the decreased solar radiation cooled the model's land surface compared to present conditions. The model's ocean surface temperature distribution for 9000 yr B.P. was assumed to be the same as today's. This assumption was based on both observational evidence and the argument that the large heat capacity of the ocean would effectively damp the seasonal temperature response of the upper ocean to the altered seasonal radiation cycle.

The experiment with the solar radiation of 9000 yr B.P. included an entire year of model simulation. The method of simulation, the method for calculating the solar radiation, and the differences between the experiment for 9000 yr B.P. and the modern (control) experiment have been described in detail elsewhere (Kutzbach, 1981; Kutzbach and Otto-Bliesner, 1982). Only a few results are emphasized here.

Over much of Eurasia, the surface temperature for June to August was 2 to 4 °K higher for the 9000 yr B.P. simulation than for the modern simulation. The surface temperature was slightly lower than present in the tropics and subtropics primarily because cooler air was advected from over the ocean. In response to the increased solar radiation and the resultant higher temperature of the African-Eurasian land surface (relative to the surrounding ocean), the summer monsoon low intensified considerably compared to present (Figure 17-4). The low-level cyclonic inflow of air to the monsoon lands was strengthened, as were the Arabian Sea southwesterlies and the currents that flow from the Southern to the Northern Hemisphere over Africa and the western Indian Ocean. In the upper troposphere, the tropical easterly jet stream was strengthened. Precipitation and precipitation minus evaporation were increased over portions of North Africa, the Middle East, and Asia.

In January, the winter monsoon circulaion was slightly stronger than at present. The simulation results for the entire seasonal cycle were included in calculations of the surface temperature for Northern and Southern Hemisphere land (Table 17-1). The annual average temperature was changed less than the seasonal averages because the seasonal extremes tended to compensate each other.

A general-circulation-model experiment that incorporated both the North American ice sheet and the solar radiation regime for 9000 yr B.P. showed that the climatic impact of the ice sheet was restricted mainly to the North American continent (Kutzbach and Otto-Bliesner, 1982).

The climate that was simulated for the radiation conditions of 9000 yr B.P. agreed with many paleoclimatic observations (Kutzbach, 1981; Kutzbach and Otto-Bliesner, 1982). Most significantly, the *pattern* of increased precipitation and increased precipitation minus evaporation across North Africa/Asia agreed broadly with the maps of high lake level reported by Street and Grove (1979). The *magnitude* of the simulated precipitation change, or precipitation-minus-evaporation change, was consistent with the magnitude of the changes that have been inferred from paleolake studies. The simulated temperature change was consistent with observational evidence for a warmer growing-season climate during the period 10,000 to 5000 yr B.P. in parts of Eurasia. On the basis of pollen and macrofossil analyses that reflected primarily summer half-year conditions and that indicated northward shifts of 200 to 400 km of vegetational zones, Khotinski (1973) concludes that a pronounced warming occurred around 10,000 to 9000 yr B.P. in both western and eastern Siberia and was followed by cooling around 5000 yr B.P.

Further support for the hypothesis that orbital variations are related to variations in monsoon climates comes from comparison of time series of the orbitally induced radiation changes and the climatic changes. The peak in summertime insolation (compared to present) was around 10,000 yr B.P., and the largest radiation anomalies were confined to the period 15,000 to 5000 yr B.P. (See Figure 17-2.) The time and duration of this interval is in fair agreement with the corresponding interval of high lake levels. Prell (1981) infers from ocean-core studies that monsoonal upwelling in the Arabian Sea was more intense than at present from 10,000 to 4000 yr B.P. He attributes

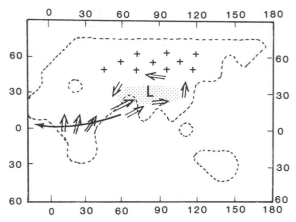

Figure 17-4. Schematic of simulated temperature, pressure, wind, and precipitation difference, 9000 yr B.P. minus present, for July. (Higher temperature is shown by plus signs, intensified monsoon low is shown by capital letter L, stronger low-level wind is marked by double arrows, stronger upper-level wind is marked by single arrow, and increased precipitation is indicated by stippling.) (From Kutzbach and Otto-Bliesner, 1982.)

this increase in upwelling to the increased summer insolation over southern Asia that resulted from the orbital parameter changes. Street and Grove (1979) report high lake levels prior to 25,000 yr B.P. This occurrence of high lake levels might possibly have coincided in time with the previous occurrence of perihelion in Northern Hemisphere summer around 35,000 to 30,000 yr B.P. (See Figure 17-2).

Other Possible Mechanisms for Holocene Climatic Change

This chapter has emphasized climate-modeling studies that were based on changes in external forcing resulting from long-period orbital parameter variations. Holocene climatic time series contain variance at periods much shorter than the 20,000-year orbital parameter period (Kutzbach and Bryson, 1974). For example, Denton and Karlén (1973) identify a neoglacial cycle with a characteristic period of roughly 2500 years. As Mitchell (1976) explains, the climate system may undergo internal oscillations (autooscillatory

Table 17-1.
Simulated Surface Temperature for Northern Hemisphere Land, Southern Hemisphere Land, and the Global Average of Land and Ocean for 9000 yr B.P. Compared to Modern Values

Space Average	9000 yr B.P. (°C)	Modern (°C)	Δ (K)	SL (%)
June to August Averages:				
Northern Hemisphere land	25.0	23.8	1.2	0.1
Southern Hemisphere land	2.5	1.7	0.8	—
Global land and ocean	17.8	17.5	0.3	1.0
Annual Averages:				
Northern Hemisphere land	13.3	13.6	−0.3	—
Southern Hemisphere land	7.5	7.2	0.3	—
Global land and ocean	15.9	15.9	0.0	—

Note: The difference between 9000 yr B.P. and the present is denoted by Δ. The significance level (SL) is determined from the ratio of Δ to the model standard deviation.

behavior) as well as responding to external forcing. Saltzman and Moritz (1980) have developed a time-dependent climatic feedback system involving sea-ice extent, ocean temperature, and atmospheric carbon dioxide content. For one set of parameters, damped oscillations of a period on the order of 1000 years were obtained.

Certain climatic variations of the Holocene may eventually be identified as autooscillatory behavior, or some external forcing mechanism may be identified. There are a number of candidates: short-period orbital parameter changes (i.e., shorter than the 20,000-year period), solar-lunar tidal forcing (Pettersson, 1914), solar variability (Eddy, 1976), and volcanic activity (Bryson and Goodman, 1980; Hansen et al., 1981; Lamb, 1970; Mitchell, 1961, 1975; Oliver, 1976). A simplified climate model, forced externally by an input time series representing volcanic activity and sunspot variations, simulated certain features of climatic time series of the past several centuries (Schneider and Mass, 1975). Citing 350 years of tree-ring data for the western United States, Mitchell and others (1979) report evidence of a 22-year rhythm of drought that was related to the 22-year sunspot cycle. They demonstrate "phase locking" between the drought and sunspot indexes and are able to confirm this sun-climate relationship at a high statistical confidence level.

In order to study the possible importance of external forcing mechanisms for explaining the climatic fluctuations of the Holocene, quantitative and complete indexes of insolation changes, tidal forces, volcanism, and solar activity are required. Significant progress has been made in developing some indexes. Bryson and Goodman (1980) have summarized the radiocarbon-dated volcanic eruptions in the Holocene. Acidity profiles from well-dated Greenland ice cores provide records of deposition of the acid volcanic gases from volcanic eruptions through the last 10,000 years (Hammer, 1977; Hammer et al., (1980). The volcanic index agrees chronologically with many known eruptions. Porter (1981) has shown a close relationship between glacier variations and volcanic eruptions over the past 100 years. Records of the ^{14}C content of tree rings provide estimates of atmospheric ^{14}C levels, which, in turn, are related to the flux of cosmic rays. By this means, a history of solar change can be inferred (Damon, 1977; Stuiver, 1980). The record of solar change obtained from these isotopic analyses agrees generally with the record obtained from historical sources. Pettersson (1914) identifies solar-lunar tidal-generating forces of periods around 2000 years and indicates a chronology of tidal maxima for the past 5000 years. These quantitative estimates of time series of potential external forcing mechanisms will be very useful for assessing the importance of external factors as causes of Holocene climatic change.

A recently identified feature of Holocene climates is the changing carbon dioxide concentration in the atmosphere. There was less carbon dioxide than now before 10,000 years ago (Berner et al., 1980; Delmas et al., 1980). Even if the carbon dioxide change did not initiate a climatic change, it could have amplified the climatic change. Further study may provide information about large-scale feedback processes that involve the carbon cycle and operate at the glacial-interglacial scale of climatic variations, a point made recently by Thompson and Schneider (1981).

A variety of climatic "scenarios" have been constructed from records of Holocene climate; see Barry (1983) for a review of empirical synoptic-dynamic climatological studies. These scenarios are based on modern instrumental records, paleoclimatic records, or combinations of the two. Some investigators have identified the middle Holocene as a possible scenario for a carbon dioxide-warmed Earth (Kellogg, 1978; Pittock and Salinger, 1982; Wigley et al. 1980;

Williams, 1980). Further study of this period of climatic history will be required to test this specific hypothesis. More generally, the analysis of the climatic record of the Holocene, along with accurate estimation of changing external conditions, should serve to develop explanations for Holocene climatic change.

Conclusions

The large climatic changes of the Late Wisconsin and the Holocene provide opportunities for a variety of climate-modeling studies. These studies deal with several topics: estimating the magnitude of the climatic change from evidence for environmental change, understanding internal climatic feedback processes, and testing for the influence of external factors (such as the changing of solar radiation) on Holocene climatic changes. This chapter has focused on the role of orbital parameter variations and the associated variations of the seasonal cycle of solar radiation as a partial explanation for the observed sequence of Holocene climates.

A goal of these modeling studies is improved understanding of the patterns and processes of climate. Advances in the development of a theory of climate may also have practical applications. The possible climatic effects of solar radiation changes produced by orbital variations are not necessarily analogous to the climatic effects of carbon dioxide changes, but our knowledge of the climatic response to known changes in solar radiation forcing may, nevertheless, be useful for studying the carbon dioxide problem. For this reason, studies of early- to middle-Holocene climates may improve our theory of climate and thereby help us to estimate the climate of the future.

References

Andrews, J. T. (1973). Late Pleistocene and Holocene climates: Some persistent problems. *Quaternary Research* 6, 185-99.

Andrews, J. T., and Barry, R. G. (1978). Glacial inception and disintegration during the last glaciation. *Annual Review of Earth and Planetary Sciences* 6, 205-28.

Barry, R. G. (1983). Late-Pleistocene climatology. *In* "Late Quaternary Environments of the United States," vol. 1, "The Late Pleistocene" (S. C. Porter, ed.), pp. 390-407. University of Minnesota Press, Minneapolis.

Berger, A. L. (1978). Long-term variations of caloric insolation resulting from the Earth's orbital elements. *Quaternary Research* 9, 139-67.

Berner, W., Oeschger, H., and Stauffer, B. (1980). Information on the CO_2 cycle from ice core studies. *Radiocarbon* 22, 227-35.

Birchfield, G. E., Weertman, J., and Lunde, A. T. (1981). A paleoclimate model of Northern Hemisphere ice sheets. *Quaternary Research* 15, 126-42.

Bryson, R. A., and Goodman, B. M. (1980). Volcanic activity and climatic changes. *Science* 207, 1041-44.

Bryson, R. A., and Swain, A. M. (1981). Holocene variations of monsoon rainfall in Rajasthan. *Quaternary Research* 16, 135-45.

Budyko, M. I. (1969). The effect of solar radiation variations on the climate of the Earth. *Tellus* 21, 611-19.

Butzer, K. W., Issac, G. L., Richardson, J. A., and Washbourn-Kamau, C. (1972). Radiocarbon dating of East African lake levels. *Science* 175, 1069-76.

Cess, R. D. (1978). Biosphere-albedo feedback and climate modeling. *Journal of the Atmospheric Sciences* 35, 1765-68.

Charney, J., Stone, P. M., and Quirk, W. J. (1975). Drought in the Sahara: A biogeophysical feedback mechanism. *Science* 187, 434-35.

CLIMAP Project Members (1976). The surface of the ice-age Earth. *Science* 191, 1131-36.

CLIMAP Project Members (1980). February and August boundary conditions for 18,000 yr B.P. Unpublished manuscript.

Damon, P. E. (1977). Solar induced variations of energetic particles at one AU. *In* "The Solar Output and its Variations" (O. R. White, ed.), pp. 429-48. Colorado Associated University Press, Boulder.

Delmas, R. J., Ascensio, J. M., and Legrand, M. (1980). Polar ice evidence that atmospheric CO_2 20,000 yr B.P. was 50% of present. *Nature (London)* 284, 155-57.

Denton, G. H., and Hughes, T. (eds.) (1981). "The Last Great Ice Sheets." Wiley Interscience Press, New York.

Denton, G. H., and Karlén, W. (1973). Holocene climatic variations: Their pattern and possible cause. *Quaternary Research* 3, 155-205.

Eddy, J. A. (1976). The Maunder minimum. *Science* 192, 1189-202.

Gates, W. L. (1976a). Modeling the ice-age climate. *Science* 191, 1138-44.

Gates, W. L. (1976b). The numerical simulation of ice-age climate with a global general circulation model. *Journal of the Atmospheric Sciences* 33, 1844-73.

Hammer, C. U. (1977). Past volcanism revealed by Greenland ice sheet impurities. *Nature (London)*, 270, 482-86.

Hammer, C. U., Clausen, H. B., and Dansgaard, W. (1981). Greenland ice sheet evidence of post-glacial volcanism and its climatic impact. *Nature (London)*, 288, 230-35.

Hansen, J., Johnson, D., Lacis, A., Lebedeff, S., Lee, P., Rind, D., and Russel, G. (1981). Climate impact of increasing atmospheric CO_2. *Science*, 213, 957-66.

Hare, F. K. (1976). Late Pleistocene and Holocene climates: Some persistent problems. *Quaternary Research* 6, 505-17.

Hays, J. D., Imbrie, J., and Shackleton, N. J. (1976). Variations in the Earth's orbit: Pacemaker of the ice ages. *Science* 194, 1121-32.

Hughes, T., Denton, G. H., and Grosswald, M. (1977). Was there a late-Würm Arctic ice sheet? *Nature (London)* 266, 596-602.

Imbrie, J., and Imbrie, J. Z. (1980). Modeling the climatic response to orbital variations. *Science* 207, 943-53.

Ives, J. D., Andrews, J. T., and Barry, R. G. (1975). Growth and decay of the Laurentide ice sheet and comparisons with Fenno-Scandinavia. *Naturwissenschaften* 62, 118-25.

Johnson, R. G., and McClure, B. T. (1976). A model for Northern Hemisphere continental ice sheet variation. *Quaternary Research* 6, 325-54.

Kellogg, W. W. (1978). Global influences of mankind on the climate. *In* "Climatic Change" (J. Gribbin, ed.), pp. 205-27. Cambridge University Press, Cambridge.

Khotinskii, N. A. (1973). Transkontinental'naia Korreliatsiia etapove istorii rastitel'nosti i Klimata severnogo; Evrazii v. Golotsene. "Problemy Palinologii Trudy III Mezhdunarodnoi Palinologicheskoi Konferentsii" (M. I. Neishtadt, ed.), pp. 116-23. Nauka, Moscow. (English translation by G. M. Peterson.)

Kutzbach, J. E. (1980). Estimates of past climate at paleolake Chad, North Africa, based on a hydrological and energy-balance model. *Quaternary Research* 14, 210-23.

Kutzbach, J. E. (1981). Monsoon climate of the early Holocene: Climatic experiment using the Earth's orbital parameters for 9000 years ago. *Science*, 214, 59-61.

Kutzbach, J. E. (1983). Monsoon rains of the late Pleistocene and early Holocene: Patterns, intensity, and possible cause. *In* "Variations in the Global Water Budget" (F. A. Street-Perrot, M. Beran, and R. Ratcliffe, eds.), pp. 371-89. D. Reidel, Dordrecht.

Kutzbach, J. E., and Bryson, R. A. (1974). Variance spectrum of Holocene climatic fluctuations in the North Atlantic sector. *Journal of the Atmospheric Sciences* 31, 1958-63.

Kutzbach, J. E., and Otto-Bliesner, B. L. (1982). The sensitivity of the African-Asian monsoonal climate to orbital parameter changes for 9000 yr B.P. in a low-resolution general circulation model. *Journal of the Atmospheric Sciences* 39, 1177-88.

Kutzbach, J. E., and Wright, H. E., Jr. (1982). The summer climate of North America 18,000 years ago: A comparison of observations and model simulations. Unpublished manuscript.

Lamb, H. H. (1970). Volcanic dust in the atmosphere: With a chronology and assessment of its meteorological significance. *Philosophical Transactions of the Royal Society of London* series A, 266, 425-533.

Manabe, S., and Hahn, D. G. (1977). Simulation of the tropical climate of an ice age. *Journal of Geophysical Research* 82, 3889-911.

Mason, B. J. (1976). Toward the understanding and prediction of climatic variations. *Quarterly Journal of the Royal Meteorological Society* 102, 473-98.

Mitchell, J. M., Jr. (1961). Recent secular changes of global temperature. *Annals of the New York Academy of Sciences* 95, 235-50.

Mitchell, J. M., Jr. (1975). Note on solar variability and volcanic activity as potential sources of climatic variability. *In* "The Physical Basis of Climate and Climate Modelling," pp. 127-310. Global Atmospheric Research Program Publications Series 16, World Meteorological Organization, Geneva.

Mitchell, J. M., Jr. (1976). An overview of climatic variability and its causal mechanisms. *Quaternary Research* 6, 481-94.

Mitchell, J. M., Jr., Stockton, C. W., and Meko, D. M. (1979). Evidence of a 22-year rhythm of drought in the western United States related to the Hale Solar Cycle since the 17th century. *In* "Solar Terrestrial Influences on Weather and Climate" (B. M. McCormac and T. A. Seliga, eds.), pp. 125-43. D. Reidel, Dordrecht.

Oliver, R. C. (1976). On the response of hemispheric mean temperature to stratospheric dust: An empirical approach. *Journal of Applied Meteorology*, 15, 933-50.

Otto-Bliesner, B. L., Branstator, G. W., and Houghton, D. D. (1982). A global low-order spectral general circulation model: Part I. Formulation and seasonal climatology. *Journal of the Atmospheric Sciences* 39, 929-48.

Peterson, G. M., Webb, T., III, Kutzbach, J. E., van der Hammen, T., Wijmstra, T. A., and Street, F. A. (1979). The continental record of environmental conditions at 18,000 yr B.P.: An initial evaluation. *Quaternary Research* 12, 47-82.

Pettersson, O. (1914). Climatic variations in historic and prehistoric time. *Svenska Hydrogr. Biol. Komm. Skr.* 5.

Pittock, A. B., and Salinger, M. J. (1981). Towards regional scenarios for a CO_2-warmed Earth. *Climatic Change* 4, 23-40.

Pollard, D. Ingersoll, A. P., and Lockwood, J. G. (1980). Response of a zoneal climate-ice sheet model to the orbital perturbations during the Quaternary ice ages. *Tellus* 32, 301-19.

Porter, S. C. (1981). Recent glacier variations and volcanic eruptions. *Nature* 291, 139-41.

Prell, W. L. (1981). Long-term variability (10^3-10^5 years) of monsoonal upwelling, western Arabian Sea: An oceanographic response to atmospheric forcing. Paper presented at the 61st annual meeting of the American Meteorological Society, January 19-22, 1981, San Diego, Calif.

Ruddiman, W. F., and McIntyre, A. (1981). Oceanic mechanisms for amplification of the 23,000-year ice-volume cycle. *Science* 212, 617-27.

Saltzman, B., and Moritz, R. E. (1980). A time-dependent climatic feedback system involving sea-ice extent, ocean temperature, and CO_2. *Tellus* 32, 93-118.

Schneider, S. H., and Mass, C. (1975). Volcanic dust, sunspots, and temperature trends. *Science* 190, 741-46.

Schneider, S. H., and Thompson, S. L. (1979). Ice ages and orbital variations: Some simple theory and modeling. *Quaternary Research* 12, 188-203.

Street, F. A. (1979). Late Quaternary precipitation estimates for the Ziway-Shala Basin, southern Ethiopia. *Palaeoecology of Africa* 11, 135-43.

Street, F. A., and Grove, A. T. (1979). Global maps of lake-level fluctuations since 30,000 yr B.P. *Quaternary Research* 12, 83-118.

Stuiver, M. (1980). Solar variability and climatic change during the current millennium. *Nature (London)*, 286, 868-71.

Suarez, M. J., and Held, I. M. (1976). Modeling climatic response to orbital parameter variations. *Nature (London)* 263, 46-47.

Suarez, M. J., and Held, I. M. (1979). The sensitivity of an energy balance climate model to variations in the orbital parameters. *Journal of Geophysical Research* 84, 4825-36.

Thomas, R. H., and Bentley, C. R. (1978). A model of Holocene retreat of the west Antarctic ice sheet. *Quaternary Research* 10, 150-70.

Thompson, S. L., and Schneider, S. H. (1981). Carbon dioxide and climate: Ice and ocean. *Nature (London)* 290, 9-10.

Wetherald, R. T., and Manabe, S. (1975). The effects of changing the solar constant on the climate of a general circulation model. *Journal of the Atmospheric Sciences* 32, 2044-59.

Wigley, T., Jones, P., and Kelly, P. (1980). Scenario for a high CO_2 world. *Nature* 283, 17-21.

Williams, J. (1980). Anomalies in temperature and rainfall during warm Arctic seasons as a guide to the formulation of climate scenarios. *Climatic Change* 2, 249-66.

Williams, J., Barry, R. G., and Washington, W. M. (1974). Simulation of the atmospheric circulation using the NCAR global circulation model with ice age boundary conditions. *Journal of Applied Meteorology* 13, 305-17.

H. E. Wright, Jr., is Regents' Professor of Geology, Ecology, and Botany at the University of Minnesota. Among the books he has co-edited are *Quaternary of the United States, Quaternary Paleoecology,* and *Pleistocene Extinctions: The Search for a Cause.*